A Handbook of Time-Series Analysis, Signal Processing and Dynamics

Signal Processing and its Applications

Books in the series

P. M. Clarkson and H. Stark, *Signal Processing Methods for Audio, Images and Telecommunications* (1995)

R. J. Clarke, *Digital Compression of Still Images and Video* (1995)

S-K. Chang and E. Jungert, *Symbolic Projection for Image Information Retrieval and Spatial Reasoning* (1996)

V. Cantoni, S. Levialdi and V. Roberto (eds.), *Artificial Vision* (1997)

R. de Mori, *Spoken Dialogue with Computers* (1998)

D. Bull, N. Canagarajah and A. Nix (eds.), *Insights into Mobile Multimedia Communications* (1999)

A Handbook of Time-Series Analysis, Signal Processing and Dynamics

D.S.G. POLLOCK

Queen Mary and Westfield College
The University of London
UK

ACADEMIC PRESS

San Diego • London • Boston • New York
Sydney • Tokyo • Toronto

This book is printed on acid-free paper.

Academic Press
24–28 Oval Road, London NW1 7DX, UK
http://www.hbuk.co.uk/ap/

Academic Press
A Harcourt Science and Technology Company
525 B Street, Suite 1900, San Diego, California 92101-4495, USA
http://www.apnet.com

ISBN 0-12-560990-6

A catalogue record for this book is available from the British Library

Typeset by Focal Image Ltd, London, in collaboration with the author Σπ

Printed in Great Britain by The University Press, Cambridge

99 00 01 02 03 04 CU 9 8 7 6 5 4 3 2 1

Series Preface

Signal processing applications are now widespread. Relatively cheap consumer products through to the more expensive military and industrial systems extensively exploit this technology. This spread was initiated in the 1960s by the introduction of cheap digital technology to implement signal processing algorithms in real-time for some applications. Since that time semiconductor technology has developed rapidly to support the spread. In parallel, an ever increasing body of mathematical theory is being used to develop signal processing algorithms. The basic mathematical foundations, however, have been known and well understood for some time.

Signal Processing and its Applications addresses the entire breadth and depth of the subject with texts that cover the theory, technology and applications of signal processing in its widest sense. This is reflected in the composition of the Editorial Board, who have interests in:

(i) Theory – The physics of the application and the mathematics to model the system;

(ii) Implementation – VLSI/ASIC design, computer architecture, numerical methods, systems design methodology, and CAE;

(iii) Applications – Speech, sonar, radar, seismic, medical, communications (both audio and video), guidance, navigation, remote sensing, imaging, survey, archiving, non-destructive and non-intrusive testing, and personal entertainment.

Signal Processing and its Applications will typically be of most interest to postgraduate students, academics, and practising engineers who work in the field and develop signal processing applications. Some texts may also be of interest to final year undergraduates.

Richard C. Green
The Engineering Practice,
Farnborough, UK

For Elizabeth Pollock

Contents

A complete bibliography can be found on the CD-Rom

PROGRAMS: Listed by Chapter

TEMPORAL SEQUENCES AND POLYNOMIAL ALGEBRA

(2.14) **procedure** *Convolution*(**var** *alpha, beta* : *vector*;
p, k : *integer*);

(2.19) **procedure** *Circonvolve*(*alpha, beta* : *vector*;
var *gamma* : *vector*;
n : *integer*);

(2.50) **procedure** *QuadraticRoots*(*a, b, c* : *real*);

(2.59) **function** *Cmod*(*a* : *complex*) : *real*;

(2.61) **function** *Cadd*(*a, b* : *complex*) : *complex*;

(2.65) **function** *Cmultiply*(*a, b* : *complex*) : *complex*;

(2.67) **function** *Cinverse*(*a* : *complex*) : *complex*;

(2.68) **function** *Csqrt*(*a* : *complex*) : *complex*;

(2.76) **procedure** *RootsToCoefficients*(*n* : *integer*;
var *alpha, lambda* : *complexVector*);

(2.79) **procedure** *InverseRootsToCoeffs*(*n* : *integer*;
var *alpha, mu* : *complexVector*);

RATIONAL FUNCTIONS AND COMPLEX ANALYSIS

(3.43) **procedure** *RationalExpansion*(*alpha* : *vector*;
p, k, n : *integer*;
var *beta* : *vector*);

(3.46) **procedure** *RationalInference*(*omega* : *vector*;
p, k : *integer*;
var *beta, alpha* : *vector*);

(3.49) **procedure** *BiConvolution*(**var** *omega, theta, mu* : *vector*;
p, q, g, h : *integer*);

POLYNOMIAL COMPUTATIONS

(4.11) **procedure** *Horner*(*alpha* : *vector*;

```
                 p : integer;
                 xi : real;
                 var gamma0 : real;
                 var beta : vector);

(4.16)   procedure ShiftedForm(var alpha : vector;
                 xi : real;
                 p : integer);

(4.17)   procedure ComplexPoly(alpha : complexVector;
                 p : integer;
                 z : complex;
                 var gamma0 : complex;
                 var beta : complexVector);

(4.46)   procedure DivisionAlgorithm(alpha, delta : vector;
                 p, q : integer;
                 var beta : jvector;
                 var rho : vector);

(4.52)   procedure RealRoot(p : integer;
                 alpha : vector;
                 var root : real;
                 var beta : vector);

(4.53)   procedure NRealRoots(p, nOfRoots : integer;
                 var alpha, beta, lambda : vector);

(4.68)   procedure QuadraticDeflation(alpha : vector;
                 delta0, delta1 : real;
                 p : integer;
                 var beta : vector;
                 var c0, c1, c2 : real);

(4.70)   procedure Bairstow(alpha : vector;
                 p : integer;
                 var delta0, delta1 : real;
                 var beta : vector);

(4.72)   procedure MultiBairstow(p : integer;
                 var alpha : vector;
                 var lambda : complexVector);

(4.73)   procedure RootsOfFactor(i : integer;
                 delta0, delta1 : real;
                 var lambda : complexVector);
```

DIFFERENTIAL AND DIFFERENCE EQUATIONS

MATRIX COMPUTATIONS

CLASSICAL REGRESSION ANALYSIS

(8.56) **procedure** *GaussianRegression*($k, Tcap$: *integer*;
var x, c : *matrix*);

(8.70) **procedure** *QRregression*($Tcap, k$: *integer*;
var $x, y, beta$: *matrix*;
var *varEpsilon* : *real*);

(8.71) **procedure** *Backsolve*(**var** r, x, b : *matrix*;
n, q : *integer*);

RECURSIVE LEAST-SQUARES METHODS

(9.26) **procedure** *RLSUpdate*(x : *vector*;
$k, sign$: *integer*;
$y, lambda$: *real*;
var h : *real*;
var $beta, kappa$: *vector*;
var p : *matrix*);

(9.34) **procedure** *SqrtUpdate*(x : *vector*;
k : *integer*;
$y, lambda$: *real*;
var h : *real*;
var $beta, kappa$: *vector*;
var s : *matrix*);

POLYNOMIAL TREND ESTIMATION

(10.26) **procedure** *GramSchmidt*(**var** phi, r : *matrix*;
n, q : *integer*);

(10.50) **procedure** *PolyRegress*(x, y : *vector*;
var $alpha, gamma, delta, poly$: *vector*;
q, n : *integer*);

(10.52) **procedure** *OrthoToPower*($alpha, gamma, delta$: *vector*;
var $beta$: *vector*;
q : *integer*);

(10.62) **function** *PolyOrdinate*(x : *real*;
$alpha, gamma, delta$: *vector*;
q : *integer*) : *real*;

(10.100) **procedure** *BSplineOrdinates*(*p* : *integer*;
x : *real*;
xi : *vector*;
var *b* : *vector*);

(10.101) **procedure** *BSplineCoefficients*(*p* : *integer*;
xi : *vector*;
mode : *string*;
var *c* : *matrix*);

SPLINE SMOOTHING

(11.18) **procedure** *CubicSplines*(**var** *S* : *SplineVec*;
n : *integer*);

(11.50) **procedure** *SplinetoBezier*(*S* : *SplineVec*;
var *B* : *BezierVec*;
n : *integer*);

(11.83) **procedure** *Quincunx*(*n* : *integer*;
var *u*, *v*, *w*, *q* : *vector*);

(11.84) **procedure** *SmoothingSpline*(**var** *S* : *SplineVec*;
sigma : *vector*;
lambda : *real*;
n : *integer*);

NONLINEAR OPTIMISATION

(12.13) **procedure** *GoldenSearch*(**function** *Funct*(*x* : *real*) : *real*;
var *a*, *b* : *real*;
limit : *integer*;
tolerance : *real*);

(12.20) **procedure** *Quadratic*(**var** *p*, *q* : *real*;
a, *b*, *c*, *fa*, *fb*, *fc* : *real*);

(12.22) **procedure** *QuadraticSearch*(**function** *Funct*(*lambda* : *real*;
theta, *pvec* : *vector*;
n : *integer*) : *real*;
var *a*, *b*, *c*, *fa*, *fb*, *fc* : *real*;
theta, *pvec* : *vector*;
n : *integer*);

(12.26) **function** *Check*(*mode* : *string*;
a, *b*, *c*, *fa*, *fb*, *fc*, *fw* : *real*) : *boolean*;

(12.27) **procedure** *LineSearch*(**function** *Funct*(*lambda* : *real*;
theta, *pvec* : *vector*;
n : *integer*) : *real*;
var *a* : *real*;
theta, *pvec* : *vector*;
n : *integer*);

(12.82) **procedure** *ConjugateGradients*(**function** *Funct*(*lambda* : *real*;
theta, *pvec* : *vector*;
n : *integer*) : *real*;
var *theta* : *vector*;
n : *integer*);

(12.87) **procedure** *fdGradient*(**function** *Funct*(*lambda* : *real*;
theta, *pvec* : *vector*;
n : *integer*) : *real*;
var *gamma* : *vector*;
theta : *vector*;
n : *integer*);

(12.119) **procedure** *BFGS*(**function** *Funct*(*lambda* : *real*;
theta, *pvec* : *vector*;
n : *integer*) : *real*;
var *theta* : *vector*;
n : *integer*);

THE FAST FOURIER TRANSFORM

(15.8) **procedure** *PrimeFactors*(*Tcap* : *integer*;
var *g* : *integer*;
var *N* : *ivector*;
var *palindrome* : *boolean*);

(15.45) **procedure** *MixedRadixCore*(**var** *yReal*, *yImag* : *vector*;
var *N*, *P*, *Q* : *ivector*;
Tcap, *g* : *integer*);

(15.49) **function** *tOfj*(*j*, *g* : *integer*;
P, *Q* : *ivector*) : *integer*;

(15.50) **procedure** *ReOrder*(*P*, *Q* : *ivector*;
Tcap, *g* : *integer*;
var *yImag*, *yReal* : *vector*);

(15.51) **procedure** *MixedRadixFFT*(**var** *yReal*, *yImag* : *vector*;
var *Tcap*, *g* : *integer*;
inverse : *boolean*);

(15.54) **procedure** *Base2FFT*(**var** *y* : *longVector*;
Tcap, *g* : *integer*);

(15.63) **procedure** *TwoRealFFTs*(**var** *f*, *d* : *longVector*
Tcap, *g* : *integer*);

(15.74) **procedure** *OddSort*(*Ncap* : *integer*;
var *y* : *longVector*);

(15.77) **procedure** *CompactRealFFT*(**var** *x* : *longVector*;
Ncap, *g* : *integer*);

LINEAR FILTERING

(16.20) **procedure** *GainAndPhase*(**var** *gain*, *phase* : *real*;
delta, *gamma* : *vector*;
omega : *real*;
d, *g* : *integer*);

(16.21) **function** *Arg*(*psi* : *complex*) : *real*;

LINEAR TIME-SERIES MODELS

(17.24) **procedure** *MACovariances*(**var** *mu*, *gamma* : *vector*;
var *varEpsilon* : *real*;
q : *integer*);

(17.35) **procedure** *MAParameters*(**var** *mu* : *vector*;
var *varEpsilon* : *real*;
gamma : *vector*;
q : *integer*);

(17.39) **procedure** *Minit*(**var** *mu* : *vector*;
var *varEpsilon* : *real*;
gamma : *vector*;
q : *integer*);

(17.40) **function** *CheckDelta*(*tolerance* : *real*;
q : *integer*;

```
                    var delta, mu : vector) : boolean;

(17.67)     procedure YuleWalker(p, q : integer;
                    gamma : vector;
                    var alpha : vector;
                    var varEpsilon : real);

(17.75)     procedure LevinsonDurbin(gamma : vector;
                    p : integer;
                    var alpha, pacv : vector);

(17.98)     procedure ARMACovariances(alpha, mu : vector;
                    var gamma : vector;
                    var varEpsilon : real;
                    lags, p, q : integer);

(17.106)    procedure ARMAParameters(p, q : integer;
                    gamma : vector;
                    var alpha, mu : vector;
                    var varEpsilon : real);
```

PREDICTION

```
(19.139)    procedure GSPrediction(gamma : vector;
                y : longVector;
                var mu : matrix;
                n, q : integer);
```

ESTIMATION OF THE MEAN AND THE AUTOCOVARIANCES

```
(20.55)     procedure Autocovariances(Tcap, lag : integer;
                    var y : longVector;
                    var acovar : vector);

(20.59)     procedure FourierACV(var y : longVector;
                    lag, Tcap : integer);
```

ARMA ESTIMATION: ASYMPTOTIC METHODS

```
(21.55)     procedure Covariances(x, y : longVector;
                    var covar : jvector;
                    n, p, q : integer);
```

(21.57) **procedure** *MomentMatrix*(*covarYY*, *covarXX*, *covarXY* : *jvector*;
p, *q* : *integer*;
var *moments* : *matrix*);

(21.58) **procedure** *RHSVector*(*moments* : *matrix*;
covarYY, *covarXY* : *jvector*;
alpha : *vector*;
p, *q* : *integer*;
var *rhVec* : *vector*);

(21.59) **procedure** *GaussNewtonARMA*(*p*, *q*, *n* : *integer*;
y : *longVector*;
var *alpha*, *mu* : *vector*);

(21.87) **procedure** *BurgEstimation*(**var** *alpha*, *pacv* : *vector*;
y : *longVector*;
p, *Tcap* : *integer*);

ARMA ESTIMATION: MAXIMUM-LIKELIHOOD METHODS

(22.40) **procedure** *ARLikelihood*(**var** *S*, *varEpsilon* : *real*;
var *y* : *longVector*;
alpha : *vector*;
Tcap, *p* : *integer*;
var *stable* : *boolean*);

(22.74) **procedure** *MALikelihood*(**var** *S*, *varEpsilon* : *real*;
var *y* : *longVector*;
mu : *vector*;
Tcap, *q* : *integer*);

(22.106) **procedure** *ARMALikelihood*(**var** *S*, *varEpsilon* : *real*;
alpha, *mu* : *vector*;
y : *longVector*;
Tcap, *p*, *q* : *integer*);

Preface

It is hoped that this book will serve both as a text in time-series analysis and signal processing and as a reference book for research workers and practitioners. Time-series analysis and signal processing are two subjects which ought to be treated as one; and they are the concern of a wide range of applied disciplines including statistics, electrical engineering, mechanical engineering, physics, medicine and economics.

The book is primarily a didactic text and, as such, it has three main aspects. The first aspect of the exposition is the mathematical theory which is the foundation of the two subjects. The book does not skimp this. The exposition begins in Chapters 2 and 3 with polynomial algebra and complex analysis, and it reaches into the middle of the book where a lengthy chapter on Fourier analysis is to be found.

The second aspect of the exposition is an extensive treatment of the numerical analysis which is specifically related to the subjects of time-series analysis and signal processing but which is, usually, of a much wider applicability. This begins in earnest with the account of polynomial computation, in Chapter 4, and of matrix computation, in Chapter 7, and it continues unabated throughout the text. The computer code, which is the product of the analysis, is distributed evenly throughout the book, but it is also hierarchically ordered in the sense that computer procedures which come later often invoke their predecessors.

The third and most important didactic aspect of the text is the exposition of the subjects of time-series analysis and signal processing themselves. This begins as soon as, in logic, it can. However, the fact that the treatment of the substantive aspects of the subject is delayed until the mathematical foundations are in place should not prevent the reader from embarking immediately upon such topics as the statistical analysis of time series or the theory of linear filtering. The book has been assembled in the expectation that it will be read backwards as well as forwards, as is usual with such texts. Therefore it contains extensive cross-referencing.

The book is also intended as an accessible work of reference. The computer code which implements the algorithms is woven into the text so that it binds closely with the mathematical exposition; and this should allow the detailed workings of the algorithms to be understood quickly. However, the function of each of the Pascal procedures and the means of invoking them are described in a reference section, and the code of the procedures is available in electronic form on a computer disc.

The associated disc contains the Pascal code precisely as it is printed in the text. An alternative code in the C language is also provided. Each procedure is coupled with a so-called driver, which is a small program which shows the procedure in action. The essential object of the driver is to demonstrate the workings of the procedure; but usually it fulfils the additional purpose of demonstrating some aspect the theory which has been set forth in the chapter in which the code of the procedure it to be found. It is hoped that, by using the algorithms provided in this book, scientists and engineers will be able to piece together reliable software tools tailored to their own specific needs.

The compact disc also contains a collection of reference material which includes the libraries of the computer routines and various versions of the bibliography of the book. The numbers in brackets which accompany the bibliographic citations refer to their order in the composite bibliography which is to be found on the disc. On the disc, there is also a bibliography which is classified by subject area.

A preface is the appropriate place to describe the philosophy and the motivation of the author in so far as they affect the book. A characteristic of this book, which may require some justification, is its heavy emphasis on the mathematical foundations of its subjects. There are some who regard mathematics as a burden which should be eased or lightened whenever possible. The opinion which is reflected in the book is that a firm mathematical framework is needed in order to bear the weight of the practical subjects which are its principal concern. For example, it seems that, unless the reader is adequately appraised of the notions underlying Fourier analysis, then the perplexities and confusions which will inevitably arise will limit their ability to commit much of the theory of linear filters to memory. Practical mathematical results which are well-supported by theory are far more accessible than those which are to be found beneath piles of technological detritus.

Another characteristic of the book which reflects a methodological opinion is the manner in which the computer code is presented. There are some who regard computer procedures merely as technological artefacts to be encapsulated in boxes whose contents are best left undisturbed for fear of disarranging them. An opposite opinion is reflected in this book. The computer code presented here should be read and picked to pieces before being reassembled in whichever way pleases the reader. In short, the computer procedures should be approached in a spirit of constructive play. An individual who takes such an approach in general will not be balked by the non-availability of a crucial procedure or by the incapacity of some large-scale computer program upon which they have come to rely. They will be prepared to make for themselves whatever tools they happen to need for their immediate purposes.

The advent of the microcomputer has enabled the approach of individualist self-help advocated above to become a practical one. At the same time, it has stimulated the production of a great variety of highly competent scientific software which is supplied commercially. It often seems like wasted effort to do for oneself what can sometimes be done better by purpose-built commercial programs. Clearly, there are opposing forces at work here—and the issue is the perennial one of whether we are to be the masters or the slaves of our technology. The conflict will never be resolved; but a balance can be struck. This book, which aims to help the reader to master one of the central technologies of the latter half of this century, places most of its weight on one side of the scales.

D.S.G. POLLOCK

Introduction

CHAPTER 1

The Methods of Time-Series Analysis

The methods to be presented in this book are designed for the purpose of analysing series of statistical observations taken at regular intervals in time. The methods have a wide range of applications. We can cite astronomy [539], meteorology [444], seismology [491], oceanography [232], [251], communications engineering and signal processing [425], the control of continuous process plants [479], neurology and electroencephalography [151], [540], and economics [233]; and this list is by no means complete.

The Frequency Domain and the Time Domain

The methods apply, in the main, to what are described as stationary or non-evolutionary time series. Such series manifest statistical properties which are invariant throughout time, so that the behaviour during one epoch is the same as it would be during any other.

When we speak of a weakly stationary or covariance-stationary process, we have in mind a sequence of random variables $y(t) = \{y_t; t = 0, \pm 1, \pm 2, \ldots\}$, representing the potential observations of the process, which have a common finite expected value $E(y_t) = \mu$ and a set of autocovariances $C(y_t, y_s) = E\{(y_t - \mu)(y_s - \mu)\} = \gamma_{|t-s|}$ which depend only on the temporal separation $\tau = |t - s|$ of the dates t and s and not on their absolute values. Usually, we require of such a process that $\lim(\tau \rightarrow \infty)\gamma_\tau = 0$, which is to say that the correlation between increasingly remote elements of the sequence tends to zero. This is a way of expressing the notion that the events of the past have a diminishing effect upon the present as they recede in time. In an appendix to the chapter, we review the definitions of mathematical expectations and covariances.

There are two distinct yet broadly equivalent modes of time-series analysis which may be pursued. On the one hand are the time-domain methods which have their origin in the classical theory of correlation. Such methods deal preponderantly with the autocovariance functions and the cross-covariance functions of the series, and they lead inevitably towards the construction of structural or parametric models of the autoregressive moving-average type for single series and of the transfer-function type for two or more causally related series. Many of the methods which are used to estimate the parameters of these models can be viewed as sophisticated variants of the method of linear regression.

On the other hand are the frequency-domain methods of spectral analysis. These are based on an extension of the methods of Fourier analysis which originate in the idea that, over a finite interval, any analytic function can be approximated, to whatever degree of accuracy is desired, by taking a weighted sum of sine and cosine functions of harmonically increasing frequencies.

Harmonic Analysis

The astronomers are usually given credit for being the first to apply the methods of Fourier analysis to time series. Their endeavours could be described as the search for hidden periodicities within astronomical data. Typical examples were the attempts to uncover periodicities within the activities recorded by the Wolfer sunspot index—see Izenman [266]—and in the indices of luminosity of variable stars.

The relevant methods were developed over a long period of time. Lagrange [306] suggested methods for detecting hidden periodicities in 1772 and 1778. The Dutchman Buijs-Ballot [86] propounded effective computational procedures for the statistical analysis of astronomical data in 1847. However, we should probably credit Sir Arthur Schuster [444], who in 1889 propounded the technique of periodogram analysis, with being the progenitor of the modern methods for analysing time series in the frequency domain.

In essence, these frequency-domain methods envisaged a model underlying the observations which takes the form of

$$\begin{aligned} y(t) &= \sum_j \rho_j \cos(\omega_j t - \theta_j) + \varepsilon(t) \\ &= \sum_j \{\alpha_j \cos(\omega_j t) + \beta_j \sin(\omega_j t)\} + \varepsilon(t), \end{aligned} \tag{1.1}$$

where $\alpha_j = \rho_j \cos\theta_j$ and $\beta_j = \rho_j \sin\theta_j$, and where $\varepsilon(t)$ is a sequence of independently and identically distributed random variables which we call a white-noise process. Thus the model depicts the series $y(t)$ as a weighted sum of perfectly regular periodic components upon which is superimposed a random component.

The factor $\rho_j = \sqrt{(\alpha_j^2 + \beta_j^2)}$ is called the amplitude of the jth periodic component, and it indicates the importance of that component within the sum. Since the variance of a cosine function, which is also called its mean-square deviation, is just one half, and since cosine functions at different frequencies are uncorrelated, it follows that the variance of $y(t)$ is expressible as $V\{y(t)\} = \frac{1}{2}\sum_j \rho_j^2 + \sigma_\varepsilon^2$ where $\sigma_\varepsilon^2 = V\{\varepsilon(t)\}$ is the variance of the noise.

The periodogram is simply a device for determining how much of the variance of $y(t)$ is attributable to any given harmonic component. Its value at $\omega_j = 2\pi j/T$, calculated from a sample $y_0, \ldots, y_{T-1}$ comprising T observations on $y(t)$, is given by

$$\begin{aligned} I(\omega_j) &= \frac{2}{T}\left[\left\{\sum_t y_t \cos(\omega_j t)\right\}^2 + \left\{\sum_t y_t \sin(\omega_j t)\right\}^2\right] \\ &= \frac{T}{2}\{a^2(\omega_j) + b^2(\omega_j)\}. \end{aligned} \tag{1.2}$$

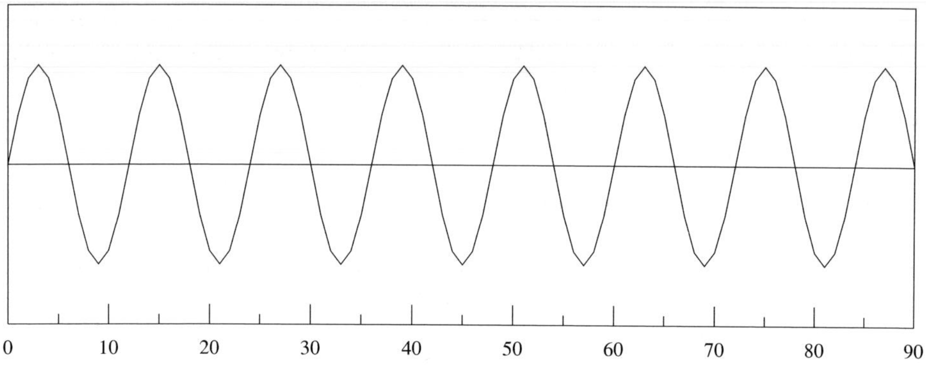

Figure 1.1. The graph of a sine function.

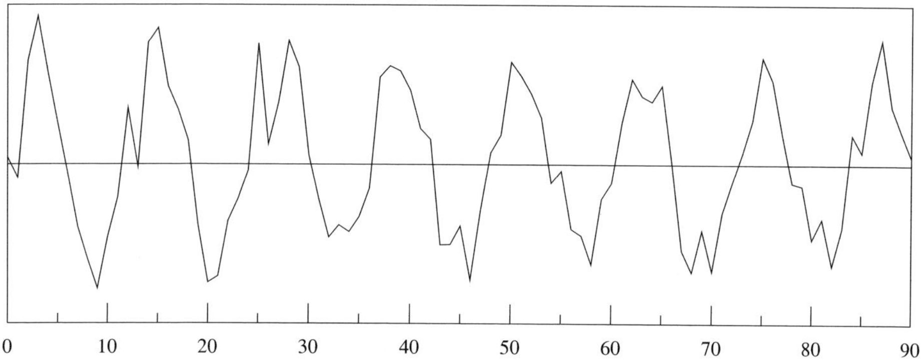

Figure 1.2. Graph of a sine function with small random fluctuations superimposed.

If $y(t)$ does indeed comprise only a finite number of well-defined harmonic components, then it can be shown that $2I(\omega_j)/T$ is a consistent estimator of ρ_j^2 in the sense that it converges to the latter in probability as the size T of the sample of the observations on $y(t)$ increases.

The process by which the ordinates of the periodogram converge upon the squared values of the harmonic amplitudes was well expressed by Yule [539] in a seminal article of 1927:

> If we take a curve representing a simple harmonic function of time, and superpose on the ordinates small random errors, the only effect is to make the graph somewhat irregular, leaving the suggestion of periodicity still clear to the eye (see Figures 1.1 and 1.2). If the errors are increased in magnitude, the graph becomes more irregular, the suggestion of periodicity more obscure, and we have only sufficiently to increase the errors to mask completely any appearance of periodicity. But, however large the errors, periodogram analysis is applicable to such a curve, and, given a sufficient number of periods, should yield a close approximation to the period and amplitude of the underlying harmonic function.

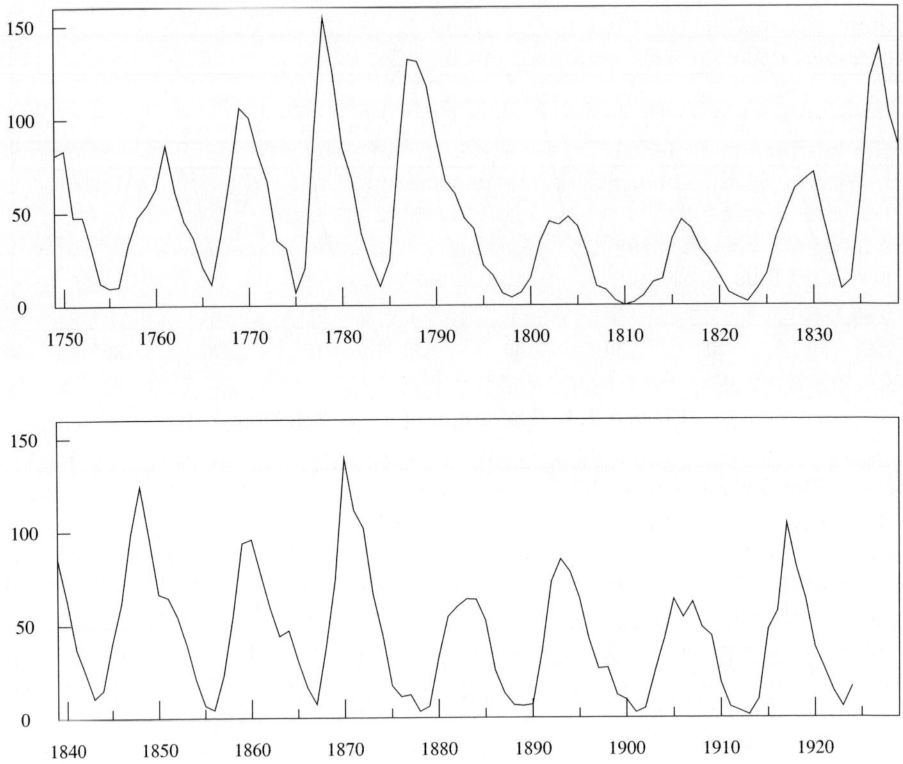

Figure 1.3. Wolfer's sunspot numbers 1749–1924.

We should not quote this passage without mentioning that Yule proceeded to question whether the hypothesis underlying periodogram analysis, which postulates the equation under (1.1), was an appropriate hypothesis for all cases.

A highly successful application of periodogram analysis was that of Whittaker and Robinson [515] who, in 1924, showed that the series recording the brightness or magnitude of the star T. Ursa Major over 600 days could be fitted almost exactly by the sum of two harmonic functions with periods of 24 and 29 days. This led to the suggestion that what was being observed was actually a two-star system wherein the larger star periodically masked the smaller, brighter star. Somewhat less successful were the attempts of Arthur Schuster himself [445] in 1906 to substantiate the claim that there is an 11-year cycle in the activity recorded by the Wolfer sunspot index (see Figure 1.3).

Other applications of the method of periodogram analysis were even less successful; and one application which was a significant failure was its use by William Beveridge [51], [52] in 1921 and 1922 to analyse a long series of European wheat prices. The periodogram of this data had so many peaks that at least twenty possible hidden periodicities could be picked out, and this seemed to be many more than could be accounted for by plausible explanations within the realm of economic history. Such experiences seemed to point to the inappropriateness to economic circumstances of a model containing perfectly regular cycles. A classic expression of disbelief was made by Slutsky [468] in another article of 1927:

> Suppose we are inclined to believe in the reality of the strict periodicity of the business cycle, such, for example, as the eight-year period postulated by Moore [352]. Then we should encounter another difficulty. Wherein lies the source of this regularity? What is the mechanism of causality which, decade after decade, reproduces the same sinusoidal wave which rises and falls on the surface of the social ocean with the regularity of day and night?

Autoregressive and Moving-Average Models

The next major episode in the history of the development of time-series analysis took place in the time domain, and it began with the two articles of 1927 by Yule [539] and Slutsky [468] from which we have already quoted. In both articles, we find a rejection of the model with deterministic harmonic components in favour of models more firmly rooted in the notion of random causes. In a wonderfully figurative exposition, Yule invited his readers to imagine a pendulum attached to a recording device and left to swing. Then any deviations from perfectly harmonic motion which might be recorded must be the result of errors of observation which could be all but eliminated if a long sequence of observations were subjected to a periodogram analysis. Next, Yule enjoined the reader to imagine that the regular swing of the pendulum is interrupted by small boys who get into the room and start pelting the pendulum with peas sometimes from one side and sometimes from the other. The motion is now affected not by superposed fluctuations but by true disturbances.

In this example, Yule contrives a perfect analogy for the autoregressive time-series model. To explain the analogy, let us begin by considering a homogeneous second-order difference equation of the form

$$y(t) = \phi_1 y(t-1) + \phi_2 y(t-2). \tag{1.3}$$

Given the initial values y_{-1} and y_{-2}, this equation can be used recursively to generate an ensuing sequence $\{y_0, y_1, \ldots\}$. This sequence will show a regular pattern of behaviour whose nature depends on the parameters ϕ_1 and ϕ_2. If these parameters are such that the roots of the quadratic equation $z^2 - \phi_1 z - \phi_2 = 0$ are complex and less than unity in modulus, then the sequence of values will show a damped sinusoidal behaviour just as a clock pendulum will which is left to swing without the assistance of the falling weights. In fact, in such a case, the general solution to the difference equation will take the form of

$$y(t) = \alpha\rho^t \cos(\omega t - \theta), \tag{1.4}$$

where the modulus ρ, which has a value between 0 and 1, is now the damping factor which is responsible for the attenuation of the swing as the time t elapses.

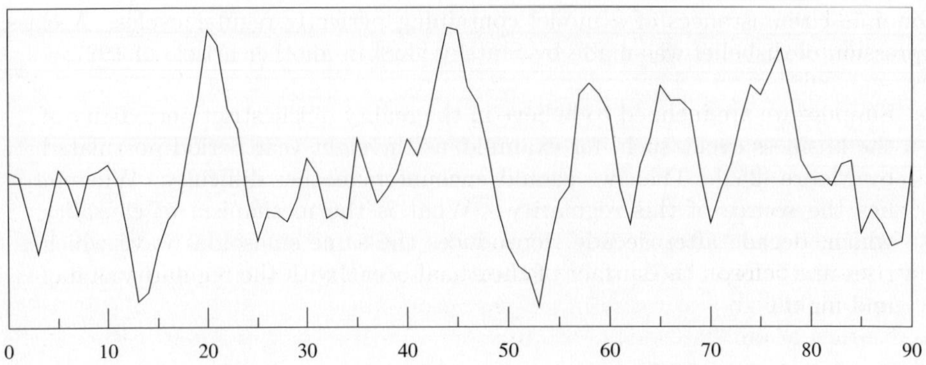

Figure 1.4. A series generated by Yule's equation
$y(t) = 1.343y(t-1) - 0.655y(t-2) + \varepsilon(t)$.

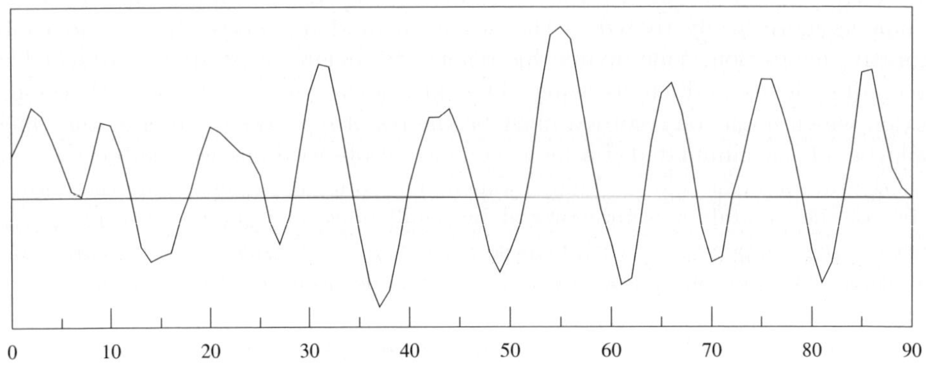

Figure 1.5. A series generated by the equation
$y(t) = 1.576y(t-1) - 0.903y(t-2) + \varepsilon(t)$.

The autoregressive model which Yule was proposing takes the form of

$$y(t) = \phi_1 y(t-1) + \phi_2 y(t-2) + \varepsilon(t), \tag{1.5}$$

where $\varepsilon(t)$ is, once more, a white-noise sequence. Now, instead of masking the regular periodicity of the pendulum, the white noise has actually become the engine which drives the pendulum by striking it randomly in one direction and another. Its haphazard influence has replaced the steady force of the falling weights. Nevertheless, the pendulum will still manifest a deceptively regular motion which is liable, if the sequence of observations is short and contains insufficient contrary evidence, to be misinterpreted as the effect of an underlying mechanism.

In his article of 1927, Yule attempted to explain the Wolfer index in terms of the second-order autoregressive model of equation (1.5). From the empirical autocovariances of the sample represented in Figure 1.3, he estimated the values

$\phi_1 = 1.343$ and $\phi_2 = -0.655$. The general solution of the corresponding homogeneous difference equation has a damping factor of $\rho = 0.809$ and an angular velocity of $\omega = 33.96$ degrees. The angular velocity indicates a period of 10.6 years which is a little shorter than the 11-year period obtained by Schuster in his periodogram analysis of the same data. In Figure 1.4, we show a series which has been generated artificially from Yule's equation, which may be compared with a series, in Figure 1.5, generated by the equation $y(t) = 1.576y(t-1) - 0.903y(t-2) + \varepsilon(t)$. The homogeneous difference equation which corresponds to the latter has the same value of ω as before. Its damping factor has the value $\rho = 0.95$, and this increase accounts for the greater regularity of the second series.

Neither of our two series accurately mimics the sunspot index; although the second series seems closer to it than the series generated by Yule's equation. An obvious feature of the sunspot index which is not shared by the artificial series is the fact that the numbers are constrained to be nonnegative. To relieve this constraint, we might apply to Wolf's numbers y_t a transformation of the form $\log(y_t + \lambda)$ or of the more general form $(y_t + \lambda)^{\kappa-1}$, such as has been advocated by Box and Cox [69]. A transformed series could be more closely mimicked.

The contributions to time-series analysis made by Yule [539] and Slutsky [468] in 1927 were complementary: in fact, the two authors grasped opposite ends of the same pole. For ten years, Slutsky's paper was available only in its original Russian version; but its contents became widely known within a much shorter period.

Slutsky posed the same question as did Yule, and in much the same manner. Was it possible, he asked, that a definite structure of a connection between chaotically random elements could form them into a system of more or less regular waves? Slutsky proceeded to demonstrate this possibility by methods which were partly analytic and partly inductive. He discriminated between coherent series whose elements were serially correlated and incoherent or purely random series of the sort which we have described as white noise. As to the coherent series, he declared that

> their origin may be extremely varied, but it seems probable that an especially prominent role is played in nature by the process of *moving summation* with weights of one kind or another; by this process coherent series are obtained from other coherent series or from incoherent series.

By taking, as his basis, a purely random series obtained by the People's Commissariat of Finance in drawing the numbers of a government lottery loan, and by repeatedly taking moving summations, Slutsky was able to generate a series which closely mimicked an index, of a distinctly undulatory nature, of the English business cycle from 1855 to 1877.

The general form of Slutsky's moving summation can be expressed by writing

$$y(t) = \mu_0\varepsilon(t) + \mu_1\varepsilon(t-1) + \cdots + \mu_q\varepsilon(t-q), \tag{1.6}$$

where $\varepsilon(t)$ is a white-noise process. This is nowadays called a qth-order moving-average model, and it is readily compared to an autoregressive model of the sort depicted under (1.5). The more general pth-order autoregressive model can be expressed by writing

$$\alpha_0 y(t) + \alpha_1 y(t-1) + \cdots + \alpha_p y(t-p) = \varepsilon(t). \tag{1.7}$$

Thus, whereas the autoregressive process depends upon a linear combination of the function $y(t)$ with its own lagged values, the moving-average process depends upon a similar combination of the function $\varepsilon(t)$ with its lagged values. The affinity of the two sorts of process is further confirmed when it is recognised that an autoregressive process of finite order is equivalent to a moving-average process of infinite order and that, conversely, a finite-order moving-average process is just an infinite-order autoregressive process.

Generalised Harmonic Analysis

The next step to be taken in the development of the theory of time series was to generalise the traditional method of periodogram analysis in such a way as to overcome the problems which arise when the model depicted under (1.1) is clearly inappropriate.

At first sight, it would not seem possible to describe a covariance-stationary process, whose only regularities are statistical ones, as a linear combination of perfectly regular periodic components. However, any difficulties which we might envisage can be overcome if we are prepared to accept a description which is in terms of a nondenumerable infinity of periodic components. Thus, on replacing the so-called Fourier sum within equation (1.1) by a Fourier integral, and by deleting the term $\varepsilon(t)$, whose effect is now absorbed by the integrand, we obtain an expression in the form of

$$y(t) = \int_0^{\pi} \big\{ \cos(\omega t) dA(\omega) + \sin(\omega t) dB(\omega) \big\}. \tag{1.8}$$

Here we write $dA(\omega)$ and $dB(\omega)$ rather than $\alpha(\omega)d\omega$ and $\beta(\omega)d\omega$ because there can be no presumption that the functions $A(\omega)$ and $B(\omega)$ are continuous. As it stands, this expression is devoid of any statistical interpretation. Moreover, if we are talking of only a single realisation of the process $y(t)$, then the generalised functions $A(\omega)$ and $B(\omega)$ will reflect the unique peculiarities of that realisation and will not be amenable to any systematic description.

However, a fruitful interpretation can be given to these functions if we consider the observable sequence $y(t) = \{y_t; t = 0, \pm 1, \pm 2, \ldots\}$ to be a particular realisation which has been drawn from an infinite population representing all possible realisations of the process. For, if this population is subject to statistical regularities, then it is reasonable to regard $dA(\omega)$ and $dB(\omega)$ as mutually uncorrelated random variables with well-defined distributions which depend upon the parameters of the population.

We may therefore assume that, for any value of ω,

$$\begin{aligned} &E\{dA(\omega)\} = E\{dB(\omega)\} = 0 \quad \text{and} \\ &E\{dA(\omega)dB(\omega)\} = 0. \end{aligned} \tag{1.9}$$

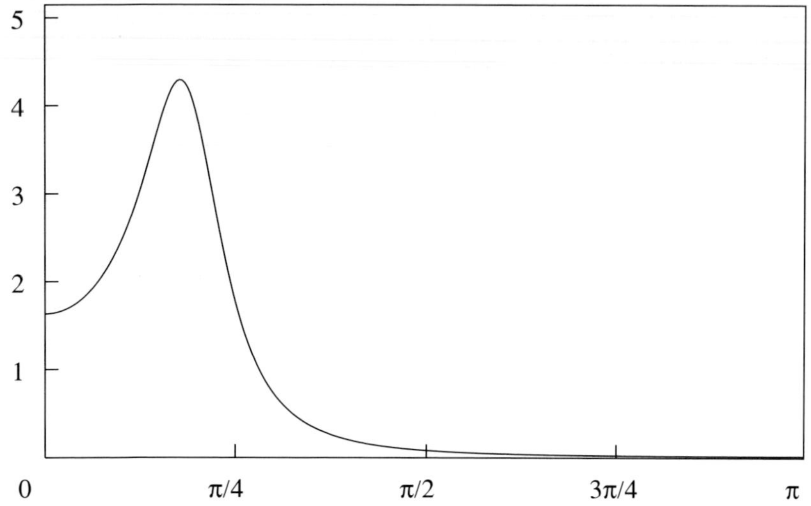

Figure 1.6. The spectrum of the process $y(t) = 1.343y(t-1) - 0.655y(t-2) + \varepsilon(t)$ which generated the series in Figure 1.4. A series of a more regular nature would be generated if the spectrum were more narrowly concentrated around its modal value.

Moreover, to express the discontinuous nature of the generalised functions, we assume that, for any two values ω and λ in their domain, we have

$$E\{dA(\omega)dA(\lambda)\} = E\{dB(\omega)dB(\lambda)\} = 0, \tag{1.10}$$

which means that $A(\omega)$ and $B(\omega)$ are stochastic processes—indexed on the frequency parameter ω rather than on time—which are uncorrelated in non-overlapping intervals. Finally, we assume that $dA(\omega)$ and $dB(\omega)$ have a common variance so that

$$V\{dA(\omega)\} = V\{dB(\omega)\} = dG(\omega). \tag{1.11}$$

Given the assumption of the mutual uncorrelatedness of $dA(\omega)$ and $dB(\omega)$, it therefore follows from (1.8) that the variance of $y(t)$ is expressible as

$$\begin{aligned} V\{y(t)\} &= \int_0^\pi \left[\cos^2(\omega t)V\{dA(\omega)\} + \sin^2(\omega t)V\{dB(\omega)\}\right] \\ &= \int_0^\pi dG(\omega). \end{aligned} \tag{1.12}$$

The function $G(\omega)$, which is called the spectral distribution, tells us how much of the variance is attributable to the periodic components whose frequencies range continuously from 0 to ω. If none of these components contributes more than an infinitesimal amount to the total variance, then the function $G(\omega)$ is absolutely continuous, and we can write $dG(\omega) = g(\omega)d\omega$ under the integral of equation (1.11). The new function $g(\omega)$, which is called the spectral density function or the spectrum, is directly analogous to the function expressing the squared amplitude which is associated with each component in the simple harmonic model discussed in our earlier sections. Figure 1.6 provides an an example of a spectral density function.

Smoothing the Periodogram

It might be imagined that there is little hope of obtaining worthwhile estimates of the parameters of the population from which the single available realisation $y(t)$ has been drawn. However, provided that $y(t)$ is a stationary process, and provided that the statistical dependencies between widely separated elements are weak, the single realisation contains all the information which is necessary for the estimation of the spectral density function. In fact, a modified version of the traditional periodogram analysis is sufficient for the purpose of estimating the spectral density.

In some respects, the problems posed by the estimation of the spectral density are similar to those posed by the estimation of a continuous probability density function of unknown functional form. It is fruitless to attempt directly to estimate the ordinates of such a function. Instead, we might set about our task by constructing a histogram or bar chart to show the relative frequencies with which the observations that have been drawn from the distribution fall within broad intervals. Then, by passing a curve through the mid points of the tops of the bars, we could construct an envelope that might approximate to the sought-after density function. A more sophisticated estimation procedure would not group the observations into the fixed intervals of a histogram; instead it would record the number of observations falling within a moving interval. Moreover, a consistent method of estimation, which aims at converging upon the true function as the number of observations increases, would vary the width of the moving interval with the size of the sample, diminishing it sufficiently slowly as the sample size increases for the number of sample points falling within any interval to increase without bound.

A common method for estimating the spectral density is very similar to the one which we have described for estimating a probability density function. Instead of being based on raw sample observations as is the method of density-function estimation, it is based upon the ordinates of a periodogram which has been fitted to the observations on $y(t)$. This procedure for spectral estimation is therefore called smoothing the periodogram.

A disadvantage of the procedure, which for many years inhibited its widespread use, lies in the fact that calculating the periodogram by what would seem to be the obvious methods be can be vastly time-consuming. Indeed, it was not until the mid 1960s that wholly practical computational methods were developed.

The Equivalence of the Two Domains

It is remarkable that such a simple technique as smoothing the periodogram should provide a theoretical resolution to the problems encountered by Beveridge and others in their attempts to detect the hidden periodicities in economic and astronomical data. Even more remarkable is the way in which the generalised harmonic analysis that gave rise to the concept of the spectral density of a time series should prove to be wholly conformable with the alternative methods of time-series analysis in the time domain which arose largely as a consequence of the failure of the traditional methods of periodogram analysis.

The synthesis of the two branches of time-series analysis was achieved independently and almost simultaneously in the early 1930s by Norbert Wiener [522] in America and A. Khintchine [289] in Russia. The Wiener–Khintchine theorem indicates that there is a one-to-one relationship between the autocovariance function of a stationary process and its spectral density function. The relationship is expressed, in one direction, by writing

$$(1.13)\qquad g(\omega)=\frac{1}{2\pi}\sum_{\tau=-\infty}^{\infty}\gamma_\tau\cos(\omega\tau);\qquad \gamma_\tau=\gamma_{-\tau},$$

where $g(\omega)$ is the spectral density function and $\{\gamma_\tau;\tau=0,1,2,\ldots\}$ is the sequence of the autocovariances of the series $y(t)$.

The relationship is invertible in the sense that it is equally possible to express each of the autocovariances as a function of the spectral density:

$$(1.14)\qquad \gamma_\tau=\int_0^\pi\cos(\omega\tau)g(\omega)d\omega.$$

If we set $\tau=0$, then $\cos(\omega\tau)=1$, and we obtain, once more, the equation (1.12) which neatly expresses the way in which the variance $\gamma_0=V\{y(t)\}$ of the series $y(t)$ is attributable to the constituent harmonic components; for $g(\omega)$ is simply the expected value of the squared amplitude of the component at frequency ω.

We have stated the relationships of the Wiener–Khintchine theorem in terms of the theoretical spectral density function $g(\omega)$ and the true autocovariance function $\{\gamma_\tau;\tau=0,1,2,\ldots\}$. An analogous relationship holds between the periodogram $I(\omega_j)$ defined in (1.2) and the sample autocovariance function $\{c_\tau;\tau=0,1,\ldots,T-1\}$ where $c_\tau=\sum(y_t-\bar{y})(y_{t-\tau}-\bar{y})/T$. Thus, in the appendix, we demonstrate the identity

$$(1.15)\qquad I(\omega_j)=2\sum_{t=1-T}^{T-1}c_\tau\cos(\omega_j\tau);\qquad c_\tau=c_{-\tau}.$$

The upshot of the Wiener–Khintchine theorem is that many of the techniques of time-series analysis can, in theory, be expressed in two mathematically equivalent ways which may differ markedly in their conceptual qualities.

Often, a problem which appears to be intractable from the point of view of one of the domains of time-series analysis becomes quite manageable when translated into the other domain. A good example is provided by the matter of spectral estimation. Given that there are difficulties in computing all T of the ordinates of the periodogram when the sample size is large, we are impelled to look for a method of spectral estimation which depends not upon smoothing the periodogram but upon performing some equivalent operation upon the sequence of autocovariances. The fact that there is a one-to-one correspondence between the spectrum and the

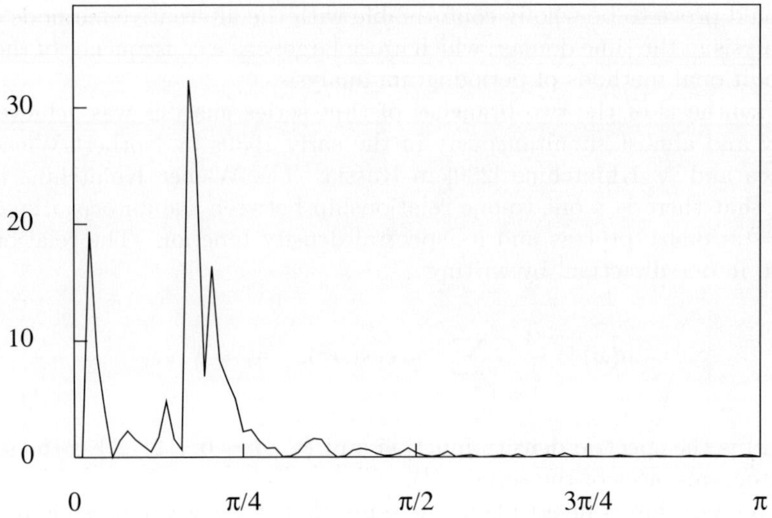

Figure 1.7. The periodogram of Wolfer's sunspot numbers 1749–1924.

sequence of autocovariances assures us that this equivalent operation must exist; though there is, of course, no guarantee that it will be easy to perform.

In fact, the operation which we perform upon the sample autocovariances is simple. For, if the sequence of autocovariances $\{c_\tau; \tau = 0, 1, \ldots, T-1\}$ in (1.15) is replaced by a modified sequence $\{w_\tau c_\tau; \tau = 0, 1, \ldots, T-1\}$ incorporating a specially devised set of declining weights $\{w_\tau; \tau = 0, 1, \ldots, T-1\}$, then an effect which is much the same as that of smoothing the periodogram can be achieved (compare Figures 1.7 and 1.8). Moreover, it may be relatively straightforward to calculate the weighted autocovariance function.

The task of devising appropriate sets of weights provided a major research topic in time-series analysis in the 1950s and early 1960s. Together with the task of devising equivalent procedures for smoothing the periodogram, it came to be known as spectral carpentry.

The Maturing of Time-Series Analysis

In retrospect, it seems that time-series analysis reached its maturity in the 1970s when significant developments occurred in both of its domains.

A major development in the frequency domain occurred when Cooley and Tukey [125] described an algorithm which greatly reduces the effort involved in computing the periodogram. The fast Fourier transform (FFT), as this algorithm has come to be known, allied with advances in computer technology, has enabled the routine analysis of extensive sets of data; and it has transformed the procedure of smoothing the periodogram into a practical method of spectral estimation.

The contemporaneous developments in the time domain were influenced by an important book by Box and Jenkins [70]. These authors developed the time-domain methodology by collating some of its major themes and by applying it

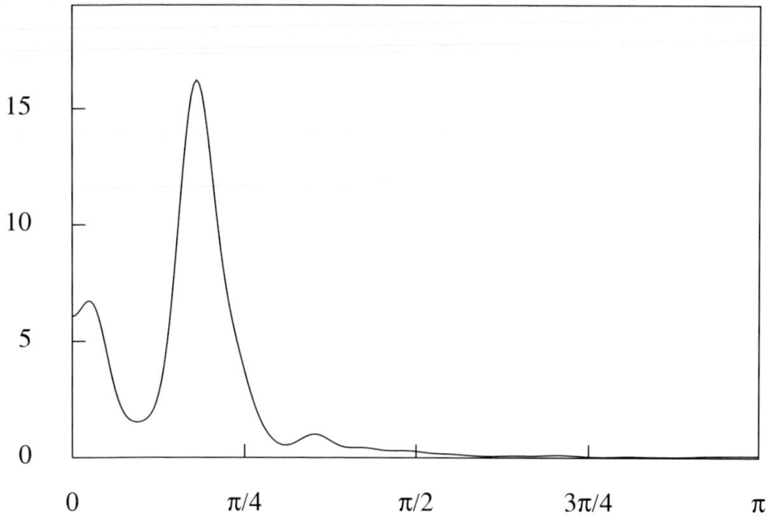

Figure 1.8. The spectrum of the sunspot numbers calculated from the autocovariances using Parzen's [383] system of weights.

to such important functions as forecasting and control. They demonstrated how wide had become the scope of time-series analysis by applying it to problems as diverse as the forecasting of airline passenger numbers and the analysis of combustion processes in a gas furnace. They also adapted the methodology to the computer.

Many of the current practitioners of time-series analysis have learnt their skills in recent years during a time when the subject has been expanding rapidly. Lacking a longer perspective, it is difficult for them to gauge the significance of the recent practical advances. One might be surprised to hear, for example, that, as late as 1971, Granger and Hughes [227] were capable of declaring that Beveridge's calculation of the periodogram of the wheat price index (see 14.4), comprising 300 ordinates, was the most extensive calculation of its type to date. Nowadays, computations of this order are performed on a routine basis using microcomputers containing specially designed chips which are dedicated to the purpose.

The rapidity of the recent developments also belies the fact that time-series analysis has had a long history. The frequency domain of time-series analysis, to which the idea of the harmonic decomposition of a function is central, is an inheritance from Euler (1707–1783), d'Alembert (1717–1783), Lagrange (1736–1813) and Fourier (1768–1830). The search for hidden periodicities was a dominant theme of nineteenth century science. It has been transmogrified through the refinements of Wiener's generalised harmonic analysis [522] which has enabled us to understand how cyclical phenomena can arise out of the aggregation of random causes. The parts of time-series analysis which bear a truly twentieth-century stamp are the time-domain models which originate with Slutsky and Yule and the computational technology which renders the methods of both domains practical.

The effect of the revolution in digital electronic computing upon the practicability of time-series analysis can be gauged by inspecting the purely mechanical devices (such as the Henrici–Conradi and Michelson–Stratton harmonic analysers invented in the 1890s) which were once used, with very limited success, to grapple with problems which are nowadays almost routine. These devices, some of which are displayed in London's Science Museum, also serve to remind us that many of the developments of applied mathematics which startle us with their modernity were foreshadowed many years ago.

Mathematical Appendix

Mathematical Expectations

The mathematical expectation or the expected value of a random variable x is defined by

$$(1.16)\qquad E(x) = \int_{-\infty}^{\infty} x dF(x),$$

where $F(x)$ is the probability distribution function of x. The probability distribution function is defined by the expression $F(x^*) = P\{x \leq x^*\}$ which denotes the probability that x assumes a value no greater than x^*. If $F(x)$ is a differentiable function, then we can write $dF(x) = f(x)dx$ in equation (1.16). The function $f(x) = dF(x)/dx$ is called the probability density function.

If $y(t) = \{y_t; t = 0, \pm 1, \pm 2, \ldots\}$ is a stationary stochastic process, then $E(y_t) = \mu$ is the same value for all t.

If $y_0, \ldots, y_{T-1}$ is a sample of T values generated by the process, then we may estimate μ from the sample mean

$$(1.17)\qquad \bar{y} = \frac{1}{T}\sum_{t=0}^{T-1} y_t.$$

Autocovariances

The autocovariance of lag τ of the stationary stochastic process $y(t)$ is defined by

$$(1.18)\qquad \gamma_\tau = E\{(y_t - \mu)(y_{t-\tau} - \mu)\}.$$

The autocovariance of lag τ provides a measure of the relatedness of the elements of the sequence $y(t)$ which are separated by τ time periods.

The variance, which is denoted by $V\{y(t)\} = \gamma_0$ and defined by

$$(1.19)\qquad \gamma_0 = E\left\{(y_t - \mu)^2\right\},$$

is a measure of the dispersion of the elements of $y(t)$. It is formally the autocovariance of lag zero.

If y_t and $y_{t-\tau}$ are statistically independent, then their joint probability density function is the product of their individual probability density functions so that $f(y_t, y_{t-\tau}) = f(y_t)f(y_{t-\tau})$. It follows that

$$\gamma_\tau = E(y_t - \mu)E(y_{t-\tau} - \mu) = 0 \quad \text{for all} \quad \tau \neq 0. \tag{1.20}$$

If $y_0, \ldots, y_{T-1}$ is a sample from the process, and if $\tau < T$, then we may estimate γ_τ from the sample autocovariance or empirical autocovariance of lag τ:

$$c_\tau = \frac{1}{T}\sum_{t=\tau}^{T-1}(y_t - \bar{y})(y_{t-\tau} - \bar{y}). \tag{1.21}$$

The Periodogram and the Autocovariance Function

The periodogram is defined by

$$I(\omega_j) = \frac{2}{T}\left[\left\{\sum_{t=0}^{T-1}\cos(\omega_j t)(y_t - \bar{y})\right\}^2 + \left\{\sum_{t=0}^{T-1}\sin(\omega_j t)(y_t - \bar{y})\right\}^2\right]. \tag{1.22}$$

The identity $\sum_t \cos(\omega_j t)(y_t - \bar{y}) = \sum_t \cos(\omega_j t)y_t$ follows from the fact that, by construction, $\sum_t \cos(\omega_j t) = 0$ for all j. Hence the above expression has the same value as the expression in (1.2). Expanding the expression in (1.22) gives

$$\begin{aligned} I(\omega_j) = &\frac{2}{T}\left\{\sum_t\sum_s \cos(\omega_j t)\cos(\omega_j s)(y_t - \bar{y})(y_s - \bar{y})\right\} \\ &+ \frac{2}{T}\left\{\sum_t\sum_s \sin(\omega_j t)\sin(\omega_j s)(y_t - \bar{y})(y_s - \bar{y})\right\}, \end{aligned} \tag{1.23}$$

and, by using the identity $\cos(A)\cos(B) + \sin(A)\sin(B) = \cos(A - B)$, we can rewrite this as

$$I(\omega_j) = \frac{2}{T}\left\{\sum_t\sum_s \cos(\omega_j[t - s])(y_t - \bar{y})(y_s - \bar{y})\right\}. \tag{1.24}$$

Next, on defining $\tau = t - s$ and writing $c_\tau = \sum_t (y_t - \bar{y})(y_{t-\tau} - \bar{y})/T$, we can reduce the latter expression to

$$I(\omega_j) = 2\sum_{\tau=1-T}^{T-1}\cos(\omega_j \tau)c_\tau, \tag{1.25}$$

which appears in the text as equation (1.15).

Bibliography

[10] Alberts, W.W., L.E. Wright and B. Feinstein, (1965), Physiological Mechanisms of Tremor and Rigidity in Parkinsonism, *Confinia Neurologica*, **26**, 318–327.

[51] Beveridge, W.H., (1921), Weather and Harvest Cycles, *Economic Journal*, **31**, 429–452.

[52] Beveridge, W.H., (1922), Wheat Prices and Rainfall in Western Europe, *Journal of the Royal Statistical Society*, **85**, 412–478.

[69] Box, G.E.P., and D.R. Cox, (1964), An Analysis of Transformations, *Journal of the Royal Statistical Society, Series B*, **26**, 211–243.

[70] Box, G.E.P., and G.M. Jenkins, (1970), *Time Series Analysis, Forecasting and Control*, Holden–Day, San Francisco.

[86] Buijs-Ballot, C.H.D., (1847), *Les Changements Périodiques de Température*, Kemink et Fils, Utrecht.

[125] Cooley, J.W., and J.W. Tukey, (1965), An Algorithm for the Machine Calculation of Complex Fourier Series, *Mathematics of Computation*, **19**, 297–301.

[151] Deistler, M., O. Prohaska, E. Reschenhofer and R. Volmer, (1986), Procedure for the Identification of Different Stages of EEG Background and its Application to the Detection of Drug Effects, *Electroencephalography and Clinical Neurophysiology*, **64**, 294–300.

[227] Granger, C.W.J., and A.O. Hughes, (1971), A New Look at Some Old Data: The Beveridge Wheat Price Series, *Journal of the Royal Statistical Society, Series A*, **134**, 413–428.

[232] Groves, G.W., and E.J. Hannan, (1968), Time-Series Regression of Sea Level on Weather, *Review of Geophysics*, **6**, 129–174.

[233] Gudmundsson, G., (1971), Time-series Analysis of Imports, Exports and other Economic Variables, *Journal of the Royal Statistical Society, Series A*, **134**, 383–412.

[251] Hassleman, K., W. Munk and G. MacDonald, (1963), Bispectrum of Ocean Waves, Chapter 8 in *Time Series Analysis*,, M. Rosenblatt (ed.), 125–139, John Wiley and Sons, New York.

[266] Izenman, A.J., (1983), J.R. Wolf and H.A. Wolfer: An Historical Note on the Zurich Sunspot Relative Numbers, *Journal of the Royal Statistical Society, Series A*, **146**, 311–318.

[289] Khintchine, A., (1934), Korrelationstheorie der Stationären Stochastischen Prozessen, *Mathematische Annalen*, **109**, 604–615.

[306] Lagrange, Joseph Louis, (1772, 1778), *Oeuvres*, 14 vols., Gauthier Villars, Paris, 1867–1892.

[352] Moore, H.L., (1914), *Economic Cycles: Their Laws and Cause*, Macmillan, New York.

[383] Parzen, E., (1957), On Consistent Estimates of the Spectrum of a Stationary Time Series, *Annals of Mathematical Statistics*, **28**, 329–348.

[425] Rice, S.O., (1963), Noise in FM Receivers, Chapter 25 in *Time Series Analysis*, M. Rosenblatt (ed.), John Wiley and Sons, New York.

[444] Schuster, A., (1898), On the Investigation of Hidden Periodicities with Application to a Supposed Twenty-Six Day Period of Meteorological Phenomena, *Terrestrial Magnetism*, **3**, 13–41.

[445] Schuster, A., (1906), On the Periodicities of Sunspots, *Philosophical Transactions of the Royal Society, Series A*, **206**, 69–100.

[468] Slutsky, E., (1937), The Summation of Random Causes as the Source of Cyclical Processes, *Econometrica*, **5**, 105–146.

[479] Tee, L.H., and S.U. Wu, (1972), An Application of Stochastic and Dynamic Models for the Control of a Papermaking Process, *Technometrics*, **14**, 481–496.

[491] Tukey, J.W., (1965), Data Analysis and the Frontiers of Geophysics, *Science*, **148**, 1283–1289.

[515] Whittaker, E.T., and G. Robinson, (1944), *The Calculus of Observations, A Treatise on Numerical Mathematics*, Fourth Edition, Blackie and Sons, London.

[522] Wiener, N., (1930), Generalised Harmonic Analysis, *Acta Mathematica*, **35,** 117–258.

[539] Yule, G.U., (1927), On a Method of Investigating Periodicities in Disturbed Series with Special Reference to Wolfer's Sunspot Numbers, *Philosophical Transactions of the Royal Society, Series A*, **226**, 267–298.

[540] Yuzuriha, T., (1960), The Autocorrelation Curves of Schizophrenic Brain Waves and the Power Spectrum, *Psychiatria et Neurologia Japonica*, **26**, 911–924.

Polynomial Methods

CHAPTER 2

Elements of Polynomial Algebra

In mathematical terminology, a time series is properly described as a temporal sequence; and the term series is reserved for power series. By transforming temporal sequences into power series, we can make use of the methods of polynomial algebra. In engineering terminology, the resulting power series is described as the z-transform of the sequence.

We shall begin this chapter by examining some of the properties of sequences and by defining some of the operations which may be performed upon them. Then we shall examine briefly the basic features of time-series models which consist of linear relationships amongst the elements of two or more sequences. We shall quickly reach the opinion that, to conduct the analysis effectively, some more mathematical tools are needed. Amongst such tools are a variety of linear operators defined on the set of infinite sequences; and it transpires that the algebra of the operators is synonymous with the algebra of polynomials of some indeterminate argument. Therefore, we shall turn to the task of setting forth the requisite results from the algebra of polynomials. In subsequent chapters, further aspects of this algebra will be considered, including methods of computation.

Sequences

An *indefinite sequence* $x(t) = \{x_t; t = 0, \pm 1, \pm 2, \ldots\}$ is any function mapping from the set of integers $\mathcal{Z} = \{t = 0, \pm 1, \pm 2, \ldots\}$ onto the real line $\mathcal{R}$ or onto the complex plane $\mathcal{C}$. The adjectives indefinite and infinite may be used interchangeably. Whenever the integers represents a sequence of dates separated by a unit time interval, the function $x(t)$ may be described as a time series. The value of the function at the point $\tau \in \mathcal{Z}$ will be denoted by $x_\tau = x(\tau)$. The functional notation will be used only when $\tau \in \mathcal{Z}$, which is to say when τ ranges over the entire set of positive and negative integers.

A *finite sequence* $\{\alpha_0, \alpha_1, \ldots, \alpha_p\}$ is one whose elements may be placed in a one-to-one correspondence with a finite set of consecutive integers. Such sequences may be specified by enumeration. Usually, the first (nonzero) element of a finite sequence will be given a zero subscript. A set of T observations on a time series $x(t)$ will be denoted by $x_0, x_1, \ldots, x_{T-1}$. Occasionally, t itself will denote a nonzero base index.

It is often convenient to extend a finite sequence so that it is defined over the entire set of integers $\mathcal{Z}$. An *ordinary extension* $\alpha(i)$ of a finite sequence

$\{\alpha_0, \alpha_1, \ldots, \alpha_p\}$ is obtained by extending the sequence on either side by an indefinite number of zeros. Thus

$$\alpha(i) = \begin{cases} \alpha_i, & \text{for } 0 \leq i \leq p; \\ 0, & \text{otherwise.} \end{cases} \tag{2.1}$$

A *periodic extension* $\tilde{\alpha}(i)$ of the sequence is obtained by replicating its elements indefinitely in successive segments. Thus

$$\tilde{\alpha}(i) = \begin{cases} \alpha_i, & \text{for } 0 \leq i \leq p; \\ \alpha_{(i \bmod [p+1])}, & \text{otherwise,} \end{cases} \tag{2.2}$$

where $(i \bmod [p+1])$ is the (positive) remainder after the division of i by $p+1$. The ordinary extension of the finite sequence $\{\alpha_0, \alpha_1, \ldots, \alpha_p\}$ and its periodic extension are connected by the following formula:

$$\tilde{\alpha}(i) = \sum_{j=-\infty}^{\infty} \alpha(i + [p+1]j). \tag{2.3}$$

It is helpful to name a few sequences which are especially important for analytic purposes. The sequence specified by the conditions

$$\delta(\tau) = \begin{cases} 1, & \text{if } \tau = 0; \\ 0, & \text{if } \tau \neq 0 \end{cases} \tag{2.4}$$

is called the unit impulse. The formulation which takes i as the index of this sequence and which sets $\delta(i-j) = 1$ when $i = j$ and $\delta(i-j) = 0$ when $i \neq j$ reminds us of Kronecker's delta. The continuous-time counterpart of the impulse sequence is known as Dirac's delta.

The unit-step sequence is specified by

$$u(\tau) = \begin{cases} 1, & \text{if } \tau \geq 0; \\ 0, & \text{if } \tau < 0. \end{cases} \tag{2.5}$$

This is related to the unit-impulse sequence by the equations

$$\delta(\tau) = u(\tau) - u(\tau - 1) \quad \text{and} \quad u(\tau) = \sum_{t=-\infty}^{\tau} \delta(t). \tag{2.6}$$

Also of fundamental importance are real and complex exponential sequences. A real exponential sequence is given by the function $x(t) = e^{rt} = a^t$ where $a = e^r$ is a real number. A complex exponential sequence is given by $x(t) = e^{i\omega t + \phi}$ with $i = \sqrt{-1}$. A sinusoidal sequence is a combination of conjugate complex exponential sequences:

$$x(t) = \rho\cos(\omega t - \theta) = \frac{1}{2}\rho\{e^{i(\omega t - \theta)} + e^{-i(\omega t - \theta)}\}. \tag{2.7}$$

Here ω, which is a number of radians, is a measure of the angular velocity or angular frequency of the sinusoid. The parameter ρ represents the amplitude or maximum value of the sinusoid, whilst θ is its phase displacement.

A sequence $x(t)$ is said to be periodic with a period of T if $x(t+T) = x(t)$ for all integer values t or, equivalently, if $x(t) = x(t \bmod T)$. The function $x(t) = \rho\cos(\omega t - \theta)$ is periodic in this sense only if $2\pi/\omega$ is a rational number. If $2\pi/\omega = T$ is an integer, then it is the period itself. In that case, its inverse $f = \omega/2\pi$ is the frequency of the function measured in cycles per unit of time.

In some cases, it is helpful to define the energy of a sequence $x(t)$ as the sum of squares of the moduli of its elements if the elements are complex valued, or simply as the sum of squares if the elements are real:

$$J = \sum |x_t|^2. \tag{2.8}$$

In many cases, the total energy will be unbounded, although we should expect it to be finite over a finite time interval.

The power of a sequence is the time-average of its energy. The concept is meaningful only if the sequence manifests some kind of stationarity. The power of a constant sequence $x(t) = a$ is just a^2. The power of the sequence $x(t) = \rho\cos(\omega t)$ is $\frac{1}{2}\rho^2$. This result can be obtained in view of the identity $\cos^2(\omega t) = \frac{1}{2}\{1+\cos(2\omega t)\}$; for the average of $\cos(2\omega t)$ over an integral number of cycles is zero.

In electrical engineering, the measure of the power of an alternating current is its so-called mean-square deviation. In statistical theory, the mean-square deviation of a finite sequence of values drawn from a statistical population is known as the sample variance. The notions of power and variance will be closely linked in this text.

When the condition that

$$\sum |x_t| < \infty \tag{2.9}$$

is fulfilled, the sequence $x(t) = \{x_t; t = 0, \pm 1, \pm 2, \ldots\}$ is said to be absolutely summable. A sequence which is absolutely summable has finite energy and vice versa.

There are numerous operations which may be performed upon sequences. Amongst the simplest of such operations are the scalar multiplication of the elements of a sequence, the pairwise addition of the elements of two sequences bearing the same index and the pairwise multiplication of the same elements. Thus, if λ is a scalar and $x(t) = \{x_t\}$ is a sequence, then $\lambda x(t) = \{\lambda x_t\}$ is the sequence obtained by scalar multiplication. If $x(t) = \{x_t\}$ and $y(t) = \{y_t\}$ are two sequences, then $x(t) + y(t) = \{x_t + y_t\}$ is the sequence obtained by their addition, and $x(t)y(t) = \{x_t y_t\}$ is the sequence obtained by their multiplication.

In signal processing, a multiplication of one continuous-time signal by another often corresponds to a process of amplitude modulation. This entails superimposing the characteristics of a (continuous-time) signal $y(t)$ onto a carrier $x(t)$ so that information in the signal can be transmitted by the carrier. Usually, the unmodulated carrier, which should contain no information of its own, has a periodic waveform.

Also of fundamental importance are the operations linear and circular convolution which are described in the next two sections.

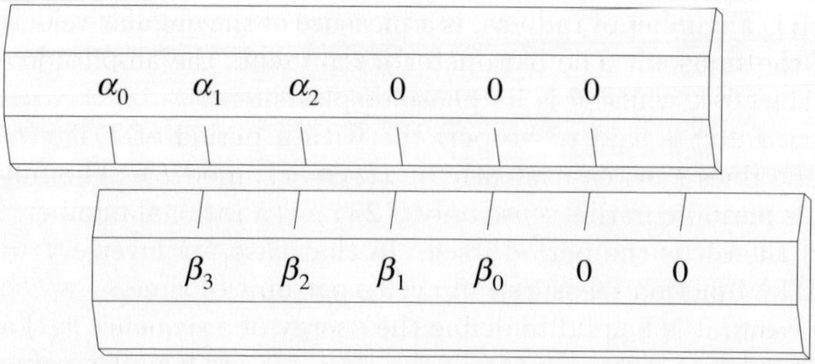

Figure 2.1. A method for finding the linear convolution of two sequences. The element $\gamma_4 = \alpha_1\beta_3 + \alpha_2\beta_2$ of the convolution may be formed by multiplying the adjacent elements on the two rulers and by summing their products.

Linear Convolution

Let $\{\alpha_0, \alpha_1, \ldots, \alpha_p\}$ and $\{\beta_0, \beta_1, \ldots, \beta_k\}$ be two finite sequences, and consider forming the pairwise products of all of their elements. The products can be arrayed as follows:

$$\begin{array}{ccccc} \alpha_0\beta_0 & \alpha_0\beta_1 & \alpha_0\beta_2 & \ldots & \alpha_0\beta_k \\ \alpha_1\beta_0 & \alpha_1\beta_1 & \alpha_1\beta_2 & \ldots & \alpha_1\beta_k \\ \alpha_2\beta_0 & \alpha_2\beta_1 & \alpha_2\beta_2 & \ldots & \alpha_2\beta_k \\ \vdots & \vdots & \vdots & & \vdots \\ \alpha_p\beta_0 & \alpha_p\beta_1 & \alpha_p\beta_2 & \ldots & \alpha_p\beta_k. \end{array} \tag{2.10}$$

Then a sequence $\gamma_0, \gamma_1, \ldots, \gamma_{p+q}$ can be defined whose elements are obtained by summing the elements of the array along each of the diagonals which run in the NE–SW direction:

$$\begin{aligned} \gamma_0 &= \alpha_0\beta_0, \\ \gamma_1 &= \alpha_0\beta_1 + \alpha_1\beta_0, \\ \gamma_2 &= \alpha_0\beta_2 + \alpha_1\beta_1 + \alpha_2\beta_0, \\ &\vdots \\ \gamma_{p+k} &= \alpha_p\beta_k. \end{aligned} \tag{2.11}$$

The sequence $\{\gamma_j\}$ is described as the convolution of the sequences $\{\alpha_j\}$ and $\{\beta_j\}$. It will be observed that

$$\begin{aligned} &\sum_{j=0}^{p+k} \gamma_j = \Big(\sum_{j=0}^{p} \alpha_j\Big)\Big(\sum_{j=0}^{k} \beta_j\Big) \qquad \text{and that} \\ &\sum_{j=0}^{p+k} |\gamma_j| \leq \Big(\sum_{j=0}^{p} |\alpha_j|\Big)\Big(\sum_{j=0}^{k} |\beta_j|\Big). \end{aligned} \tag{2.12}$$

Example 2.1. The process of linear convolution can be illustrated with a simple physical model. Imagine two rulers with adjacent edges (see Figure 2.1). The lower edge of one ruler is marked with the elements of the sequence $\{\alpha_0, \alpha_1, \ldots, \alpha_p\}$ at equally spaced intervals. The upper edge of the other ruler is marked with the elements of the reversed sequence $\{\beta_k, \ldots, \beta_1, \beta_0\}$ with the same spacing. At first, the rulers are placed so that α_0 is above β_0. The pair (α_0, β_0) is written down and the product $\gamma_0 = \alpha_0\beta_0$ is formed. Next the lower ruler is shifted one space to the right and the pairs (α_0, β_1) and (α_1, β_0) are recorded from which the sum of products $\gamma_1 = \alpha_0\beta_1 + \alpha_1\beta_0$ is formed. The lower ruler is shifted to the right again and γ_2 is formed. The process continues until the final product $\gamma_{p+k} = \alpha_p\beta_k$ is formed from the pair (α_p, β_k).

The need to form the linear convolution of two finite sequences arises very frequently, and a simple procedure is required which will perform the task. The generic element of the convolution of $\{\alpha_0, \alpha_1, \ldots, \alpha_p\}$ and $\{\beta_0, \beta_1, \ldots, \beta_k\}$, is given by

$$\gamma_j = \sum_{i=r}^{s} \alpha_i \beta_{j-i}, \quad \text{where} \qquad (2.13)$$
$$r = \max(0, j-k) \quad \text{and} \quad s = \min(p, j).$$

Here the restriction $r \leq i \leq s$ upon the index of the summation arises from the restrictions that $0 \leq i \leq p$ and that $0 \leq (j-i) \leq k$ which apply to the indices of α_i and β_{j-i}.

In the following procedure, which implements this formula, the elements γ_j are generated in the order of $j = p+k, \ldots, 0$ and they are written into the array *beta* which has hitherto contained the elements of $\{\beta_j\}$.

(2.14)
```
procedure Convolution(var alpha, beta : vector;
                         p, k : integer);

  var
    gamma : real;
    i, j, r, s : integer;

begin
  for j := p + k downto 0 do
    begin {j}
      s := Min(j, p);
      r := Max(0, j - k);
      gamma := 0.0;
      for i := r to s do
        gamma := gamma + alpha[i] * beta[j - i];
      beta[j] := gamma;
    end; {j}
end; {Convolution}
```

Some care must be exercised in extending the operation of linear convolution to the case of indefinite sequences, since certain conditions have to be imposed to ensure that the elements of the product sequence will be bounded. The simplest case concerns the convolution of two sequences which are absolutely summable:

(2.15) If $\alpha(i) = \{\alpha_i\}$ and $\beta(i) = \{\beta_i\}$ are absolutely summable sequences such that $\sum|\alpha_i| < \infty$ and $\sum|\beta_i| < \infty$, then their convolution product, which is defined by

$$\alpha(i) * \beta(i) = \sum_{i=-\infty}^{\infty} \alpha_i \beta(j-i) = \sum_{i=-\infty}^{\infty} \beta_i \alpha(j-i),$$

is also an absolutely summable sequence.

Here the absolute summability of the product sequence, which entails its boundedness, can be demonstrated by adapting the inequality under (2.12) to the case of infinite sequences.

Circular Convolution

Indefinite sequences which are obtained from the periodic extension of finite sequences cannot fulfil the condition of absolute summability; and the operation of linear convolution is undefined for them. However, it may be useful, in such cases, to define an alternative operation of circular convolution.

(2.16) Let $\tilde{\alpha}(i) = \{\tilde{\alpha}_i\}$ and $\tilde{\beta}(i) = \{\tilde{\beta}_i\}$ be the indefinite sequences which are formed by the periodic extension of the finite sequences $\{\alpha_0, \alpha_1, \ldots, \alpha_{n-1}\}$ and $\{\beta_0, \beta_1, \ldots, \beta_{n-1}\}$ respectively. Then the circular convolution of $\tilde{\alpha}(i)$ and $\tilde{\beta}(i)$ is a periodic sequence defined by

$$\tilde{\gamma}(j) = \sum_{i=0}^{n-1} \tilde{\alpha}_i \tilde{\beta}(j-i) = \sum_{i=0}^{n-1} \tilde{\beta}_i \tilde{\alpha}(j-i).$$

To reveal the implications of the definition, consider the linear convolution of the finite sequences $\{\alpha_j\}$ and $\{\beta_j\}$ which is a sequence $\{\gamma_0, \gamma_1, \ldots, \gamma_{2n-2}\}$. Also, let $\tilde{\alpha}_j = \alpha_{(j \bmod n)}$ and $\tilde{\beta}_j = \beta_{(j \bmod n)}$ denote elements of $\tilde{\alpha}(i)$ and $\tilde{\beta}(i)$. Then the generic element of the sequence $\tilde{\gamma}(j)$ is

$$\begin{aligned} \tilde{\gamma}_j &= \sum_{i=0}^{j} \tilde{\alpha}_i \tilde{\beta}_{j-i} + \sum_{i=j+1}^{n-1} \tilde{\alpha}_i \tilde{\beta}_{j-i} \\ &= \sum_{i=0}^{j} \alpha_i \beta_{j-i} + \sum_{i=j+1}^{n-1} \alpha_i \beta_{j+n-i} \\ &= \gamma_j + \gamma_{j+n}. \end{aligned} \tag{2.17}$$

The second equality depends upon the conditions that $\tilde{\alpha}_i = \alpha_i$ when $0 \leq i < n$, that $\tilde{\beta}_{j-i} = \beta_{j-i}$ when $0 \leq (j-i) < n$ and that $\tilde{\beta}_{j-i} = \beta_{(j-i) \bmod n} = \beta_{j+n-i}$ when

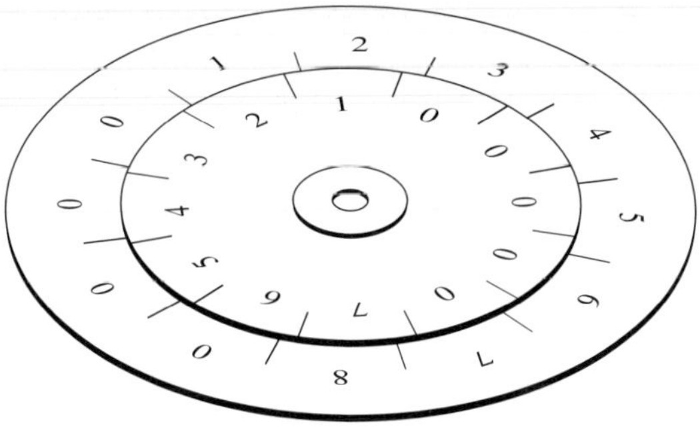

Figure 2.2. A device for finding the circular convolution of two sequences. The upper disc is rotated clockwise through successive angles of 30 degrees. Adjacent numbers on the two discs are multiplied and the products are summed to obtain the coefficients of the convolution.

$-n < (j - i) < 0$. Thus it can be seen that $\tilde{\gamma}(j)$ represents the periodic extension of the finite sequence $\{\tilde{\gamma}_0, \tilde{\gamma}_1, \ldots, \tilde{\gamma}_{n-1}\}$ wherein

$$(2.18) \qquad \tilde{\gamma}_j = \gamma_j + \gamma_{j+n} \quad \text{for} \quad j = 0, \ldots, n-2 \quad \text{and} \quad \tilde{\gamma}_{n-1} = \gamma_{n-1}.$$

Example 2.2. There is a simple analogy for the process of circular convolution which also serves to explain the terminology. One can imagine two discs placed one above the other on a common axis with the rim of the lower disc protruding (see Figure 2.2). On this rim, are written the elements of the sequence $\{\alpha_0, \alpha_1, \ldots, \alpha_{n-1}\}$ at equally spaced intervals in clockwise order. On the rim of the upper disc are written the elements of the sequence $\{\beta_0, \beta_1, \ldots, \beta_{n-1}\}$ equally spaced in an anticlockwise order. The circular disposition of the sequences corresponds to the periodic nature of the functions $\tilde{\alpha}(i)$ and $\tilde{\beta}(i)$ defined in (2.16).

At the start of the process of circular convolution, α_0 and β_0 are in alignment, and the pairs $(\alpha_0, \beta_0), (\alpha_1, \beta_{n-1}), \ldots, (\alpha_{n-1}, \beta_1)$ are read from the discs. Then, the upper disc is turned clockwise through an angle $2\pi/n$ radians and the pairs $(\alpha_0, \beta_1), (\alpha_1, \beta_0), \ldots, (\alpha_{n-1}, \beta_2)$ are read and recorded. The process continues until the $(n-1)$th turn when the pairs $(\alpha_0, \beta_{n-1}), (\alpha_1, \beta_{n-2}), \ldots, (\alpha_{n-1}, \beta_0)$ are read. One more turn would bring the disc back to the starting position. From what has been recorded, one can form the products $\tilde{\gamma}_0 = \alpha_0\beta_0 + \alpha_1\beta_{n-1} + \cdots + \alpha_{n-1}\beta_1$, $\tilde{\gamma}_1 = \alpha_0\beta_1 + \alpha_1\beta_0 + \cdots + \alpha_{n-1}\beta_2, \ldots, \tilde{\gamma}_{n-1} = \alpha_0\beta_{n-1} + \alpha_1\beta_{n-2} + \cdots + \alpha_{n-1}\beta_0$ which are the coefficients of the convolution.

The Pascal procedure which effects the circular convolution of two sequences is a straightforward one:

(2.19) **procedure** *Circonvolve*(*alpha*, *beta* : *vector*;
var *gamma* : *vector*;
n : *integer*);

```
    var
      i, j, k : integer;

  begin
    for j := 0 to n - 1 do
      begin {j}
        gamma[j] := 0.0;
          for i := 0 to n - 1 do
            begin
              k := j - i;
              if k < 0 then
                k := k + n;
              gamma[j] := gamma[j] + alpha[i] * beta[k];
            end;
          end; {j}
  end; {Circonvolve}
```

Time-Series Models

A time-series model is one which postulates a relationship amongst a number of temporal sequences or time series. Consider, for example, the regression model

$$y(t) = \beta x(t) + \varepsilon(t), \tag{2.20}$$

where $x(t)$ and $y(t)$ are observable sequences indexed by the time subscript t and $\varepsilon(t)$ is an unobservable sequence of independently and identically distributed random variables which are also uncorrelated with the elements of the explanatory sequence of $x(t)$. The purely random sequence $\varepsilon(t)$ is often described as white noise.

A more general model is one which postulates a relationship comprising any number of consecutive elements of $x(t)$, $y(t)$ and $\varepsilon(t)$. Such a relationship is expressed by the equation

$$\sum_{i=0}^{p} \alpha_i y(t-i) = \sum_{i=0}^{k} \beta_i x(t-i) + \sum_{i=0}^{q} \mu_i \varepsilon(t-i), \tag{2.21}$$

wherein the restriction $\alpha_0 = 1$ is imposed in order to identify $y(t)$ as the output of the model. The effect of the remaining terms on the LHS is described as feedback. Any of the sums in this equation can be infinite; but, if the model is to be viable, the sequences of coefficients $\{\alpha_i\}$, $\{\beta_i\}$ and $\{\mu_i\}$ must depend on a strictly limited number of underlying parameters. Notice that each of the terms of the equation represents a convolution product.

A model which includes an observable explanatory sequence or signal sequence $x(t)$ is described as a regression model. When $x(t)$ is deleted, the simpler unconditional linear stochastic models are obtained. Thus the equation

$$\sum_{i=0}^{p} \alpha_i y(t-i) = \sum_{i=0}^{q} \mu_i \varepsilon(t-i) \tag{2.22}$$

represents a so-called autoregressive moving-average (ARMA) process. When $\alpha_i = 0$ for all $i > 0$, this becomes a pure moving-average (MA) process. When $\mu_i = 0$ for all $i > 0$, it becomes a pure autoregressive (AR) process.

Transfer Functions

Temporal regression models are more easily intelligible if they can be represented by equations in the form of

$$y(t) = \sum_{i \geq 0} \omega_i x(t-i) + \sum_{i \geq 0} \psi_i \varepsilon(t-i), \tag{2.23}$$

where there is no lag scheme affecting the output sequence $y(t)$. This equation depicts $y(t)$ as a sum of a systematic component $h(t) = \sum \omega_i x(t-i)$ and a stochastic component $\eta(t) = \sum \psi_i \varepsilon(t-i)$. Both of these components comprise transfer-function relationships whereby the input sequences $x(t)$ and $\varepsilon(t)$ are translated, respectively, into output sequences $h(t)$ and $\eta(t)$.

In the case of the systematic component, the transfer function describes how the signal $x(t)$ is commuted into the sequence of systematic values which explain a major part of $y(t)$ and which may be used in forecasting it.

In the case of the stochastic component, the transfer function describes how a white-noise process $\varepsilon(t)$, comprising a sequence of independent random elements, is transformed into a sequence of serially correlated disturbances. In fact, the elements of $h(t)$ represent efficient predictors of the corresponding elements of $y(t)$ only when $\eta(t) = \psi_0 \varepsilon(t)$ is white noise.

A fruitful way of characterising a transfer function is to determine the response, in terms of its output, to a variety of standardised input signals. Examples of such signals, which have already been presented, are the unit-impulse $\delta(t)$, the unit-step $u(t)$ and the sinusoidal and complex exponential sequences defined over a range of frequencies.

The impulse response of the systematic transfer function is given by the sequence $h(t) = \sum_i \omega_i \delta(t-i)$. Since $i \in \{0, 1, 2, \ldots\}$, it follows that $h(t) = 0$ for all $t < 0$. By setting $t = \{0, 1, 2, \ldots\}$, a sequence is generated beginning with

$$\begin{aligned} h_0 &= \omega_0, \\ h_1 &= \omega_1, \\ h_2 &= \omega_2. \end{aligned} \tag{2.24}$$

The impulse-response function is nothing but the sequence of coefficients which define the transfer function.

The response of the transfer function to the unit-step sequence is given by $h(t) = \sum_i \omega_i u(t-i)$. By setting $t = \{0, 1, 2, \ldots\}$, a sequence is generated which begins with

$$\begin{aligned} h_0 &= \omega_0, \\ h_1 &= \omega_0 + \omega_1, \\ h_2 &= \omega_0 + \omega_1 + \omega_2. \end{aligned} \tag{2.25}$$

Thus the step response is obtained simply by cumulating the impulse response.

In most applications, the output sequence $h(t)$ of the transfer function should be bounded in absolute value whenever the input sequence $x(t)$ is bounded. This is described as the condition of bounded input–bounded output (BIBO) stability.

If the coefficients $\{\omega_0, \omega_1, \ldots, \omega_p\}$ of the transfer function form a finite sequence, then a necessary and sufficient condition for BIBO stability is that $|\omega_i| < \infty$ for all i, which is to say that the impulse-response function must be bounded. If $\{\omega_0, \omega_1, \ldots\}$ is an indefinite sequence, then it is necessary, in addition, that $|\sum \omega_i| < \infty$, which is the condition that the step-response function is bounded. Together, the two conditions are equivalent to the single condition that $\sum |\omega_i| < \infty$, which is to say that the impulse response is absolutely summable.

To confirm that the latter is a sufficient condition for stability, let us consider any input sequence $x(t)$ which is bounded such that $|x(t)| < M$ for some finite M. Then

$$(2.26)\qquad |h(t)| = \left|\sum \omega_i x(t-i)\right| \leq M\left|\sum \omega_i\right| < \infty,$$

and so the output sequence $h(t)$ is bounded. To show that the condition is necessary, imagine that $\sum |\omega_i|$ is unbounded. Then a bounded input sequence can be found which gives rise to an unbounded output sequence. One such input sequence is specified by

$$(2.27)\qquad x_{-i} = \begin{cases} \dfrac{\omega_i}{|\omega_i|}, & \text{if } \omega_i \neq 0; \\ 0, & \text{if } \omega_i = 0. \end{cases}$$

This gives

$$(2.28)\qquad h_0 = \sum \omega_i x_{-i} = \sum |\omega_i|,$$

and so $h(t)$ is unbounded.

A summary of this result may be given which makes no reference to the specific context in which it has arisen:

(2.29) The convolution product $h(t) = \sum \omega_i x(t-i)$, which comprises a bounded sequence $x(t) = \{x_t\}$, is itself bounded if and only if the sequence $\{\omega_i\}$ is absolutely summable such that $\sum_i |\omega_i| < \infty$.

In order to investigate the transfer-function characteristics of a relationship in the form of the general temporal model of equation (2.21), it is best to eliminate the lagged values of the output sequence $y(t)$ which represent feedback. This may be done in a number of ways, including a process of repeated substitution.

A simple example is provided by the equation

$$(2.30)\qquad y(t) = \phi y(t-1) + \beta x(t) + \varepsilon(t).$$

A process of repeated substitution gives

$$
\begin{aligned}
y(t) &= \phi y(t-1) + \beta x(t) + \varepsilon(t) \\
&= \phi^2 y(t-2) + \beta\{x(t) + \phi x(t-1)\} + \varepsilon(t) + \phi\varepsilon(t-1) \\
&\ \ \vdots \\
&= \phi^n y(t-n) + \beta\{x(t) + \phi x(t-1) + \cdots + \phi^{n-1}x(t-n+1)\} \\
&\qquad + \varepsilon(t) + \phi\varepsilon(t-1) + \cdots + \phi^{n-1}\varepsilon(t-n+1).
\end{aligned} \tag{2.31}
$$

If $|\phi| < 1$, then $\lim(n \to \infty)\phi^n = 0$; and it follows that, if $x(t)$ and $\varepsilon(t)$ are bounded sequences, then, as the number of repeated substitutions increases indefinitely, the equation will tend to the limiting form of

$$
y(t) = \beta\sum_{i=0}^{\infty}\phi^i x(t-i) + \sum_{i=0}^{\infty}\phi^i\varepsilon(t-i), \tag{2.32}
$$

which is an instance of the equation under (2.23).

For models more complicated than the present one, the method of repeated substitution, if pursued directly, becomes intractable. Thus we are motivated to use more powerful algebraic methods to effect the transformation of the equation.

The Lag Operator

The pursuit of more powerful methods of analysis begins with the recognition that the set of all time series $\{x(t); t \in \mathcal{Z}, x \in \mathcal{R}\}$ represents a vector space. Various linear transformations or operators may be defined over the space. The lag operator L, which is the primitive operator, is defined by

$$
Lx(t) = x(t-1). \tag{2.33}
$$

Now, $L\{Lx(t)\} = Lx(t-1) = x(t-2)$; so it makes sense to define L^2 by $L^2x(t) = x(t-2)$. More generally, $L^kx(t) = x(t-k)$ and, likewise, $L^{-k}x(t) = x(t+k)$. Other important operators are the identity operator $I = L^0$, the annihilator or zero operator $0 = I - I$, the forward-difference operator $\Delta = L^{-1} - I$, the backwards-difference operator $\nabla = L\Delta = I - L$ and the summation operator $S = (I + L + L^2 + \cdots)$.

The backwards-difference operator has the effect that

$$
\nabla x(t) = x(t) - x(t-1), \tag{2.34}
$$

whilst the summation operator has the effect that

$$
Sx(t) = \sum_{i=0}^{\infty} x(t-i). \tag{2.35}
$$

These two operators bear an inverse relationship to each other. On the one hand, there is the following subtraction:

$$
\begin{aligned}
S &= I + L + L^2 + \cdots \\
LS &= \phantom{I + {}} L + L^2 + \cdots \\
\hline
S - LS &= I.
\end{aligned} \tag{2.36}
$$

This gives $S(I-L) = S\nabla = I$, from which $S = \nabla^{-1}$. The result is familiar from the way in which the sum is obtained of a convergent geometric progression. On the other hand is expansion of $S = I/(I-L)$. A process of repeated substitution gives rise to

$$\begin{aligned} S &= I + LS \\ &= I + L + L^2 S \\ &= I + L + L^2 + L^3 S. \end{aligned} \tag{2.37}$$

If this process is continued indefinitely, then the original definition of the summation operator is derived. The process of repeated substitution is already familiar from the way in which the equation under (2.30), which stands for a simple temporal regression model, has been converted to its transfer-function form under (2.32).

Another way of expanding the operator $S = I/(I-L)$ is to use the algorithm of long division:

$$\begin{array}{r|l} & I + L + L^2 + \cdots \\ \hline I - L & I \\ & I - L \\ \hline & L \\ & L - L^2 \\ \hline & L^2 \\ & L^2 - L^3 \end{array} \tag{2.38}$$

If this process is stopped at any stage, then the results are the same as those from the corresponding stage of the process under (2.37). The binomial theorem can also be used in expanding $S = (I-L)^{-1}$.

To all appearances, the algebra of the lag operator is synonymous with ordinary polynomial algebra. In general, a polynomial of the lag operator of the form $p(L) = p_0 + p_1 L + \cdots + p_n L^n = \sum p_i L^i$ has the effect that

$$\begin{aligned} p(L)x(t) &= p_0 x(t) + p_1 x(t-1) + \cdots + p_n x(t-n) \\ &= \sum_{i=0}^{n} p_i x(t-i). \end{aligned} \tag{2.39}$$

The polynomial operator can be used to re-express the temporal regression model of (2.21) as

$$\alpha(L)y(t) = \beta(L)x(t) + \mu(L)\varepsilon(t). \tag{2.40}$$

In these terms, the conversion of the model to the transfer-function form of equation (2.23) is a matter of expanding the rational polynomial operators $\beta(L)/\alpha(L)$ and $\mu(L)/\alpha(L)$ in the expression

$$y(t) = \frac{\beta(L)}{\alpha(L)} x(t) + \frac{\mu(L)}{\alpha(L)} \varepsilon(t). \tag{2.41}$$

We shall be assisted in such matters by having an account of the relevant algebra and of the corresponding algorithms readily to hand.

Algebraic Polynomials

A polynomial of the pth degree in a variable z, which may stand for a real or a complex-valued variable, or which may be regarded as an indeterminate algebraic symbol, is an expression in the form of

$$\alpha(z) = \alpha_0 + \alpha_1 z + \cdots + \alpha_p z^p, \tag{2.42}$$

where it is understood that $\alpha_0, \alpha_p \neq 0$. When $z \in \mathcal{C}$ is a complex-valued variable, $\alpha(z)$ may be described as the z-transform of the sequence $\{\alpha_0, \alpha_1, \ldots, \alpha_1\}$. From another point of view, $\alpha(z)$ is regarded as the generating function of the sequence.

Let $\alpha(z) = \alpha_0 + \alpha_1 z + \cdots + \alpha_p z^p$ and $\beta(z) = \beta_0 + \beta_1 z + \cdots + \beta_k z^k$ be two polynomials of degrees p and k respectively. Then, if $k \geq p$, their sum is defined by

$$\begin{aligned}\alpha(z) + \beta(z) = (\alpha_0 + \beta_0) + (\alpha_1 + \beta_1)z + \cdots + (\alpha_p + \beta_p)z^p \\ + \beta_{p+1} z^{p+1} + \cdots + \beta_k z^k.\end{aligned} \tag{2.43}$$

A similar definition applies when $k < p$.

The product of the polynomials $\alpha(z)$ and $\beta(z)$ is defined by

$$\begin{aligned}\alpha(z)\beta(z) &= \alpha_0\beta_0 + (\alpha_0\beta_1 + \alpha_1\beta_0)z + \cdots + \alpha_p\beta_k z^{p+k} \\ &= \gamma_0 + \gamma_1 z + \gamma_2 z^2 + \cdots + \gamma_{p+k} z^{p+k} \\ &= \gamma(z).\end{aligned} \tag{2.44}$$

The sequence of coefficients $\{\gamma_i\}$ in the product is just the convolution of the sequences $\{\alpha_i\}$ and $\{\beta_i\}$ of the coefficients belonging to its factors.

These operations of polynomial addition and multiplication obey the simple rules of (i) associativity, (ii) commutativity and (iii) distributivity which are found in arithmetic. If $\alpha(z)$, $\beta(z)$ and $\gamma(z)$ are any three polynomials, then

$$\begin{aligned}
&\text{(i)} & \{\alpha(z)\beta(z)\}\gamma(z) &= \alpha(z)\{\beta(z)\gamma(z)\}, \\
& & \{\alpha(z) + \beta(z)\} + \gamma(z) &= \alpha(z) + \{\beta(z) + \gamma(z)\}; \\
&\text{(ii)} & \alpha(z) + \beta(z) &= \beta(z) + \alpha(z), \\
& & \alpha(z)\beta(z) &= \beta(z)\alpha(z); \\
&\text{(iii)} & \alpha(z)\{\beta(z) + \gamma(z)\} &= \alpha(z)\beta(z) + \alpha(z)\gamma(z).
\end{aligned} \tag{2.45}$$

Periodic Polynomials and Circular Convolution

If the polynomial argument z^j is a periodic function of the index j, then the set of polynomials is closed under the operation of polynomial multiplication. To be precise, let $\alpha(z) = \alpha_0 + \alpha_1 z + \cdots + \alpha_{n-1} z^{n-1}$ and $\beta(z) = \beta_0 + \beta_1 z + \cdots + \beta_{n-1} z^{n-1}$ be polynomials of degree $n-1$ at most in an argument which is n-periodic such that $z^{j+n} = z^j$ for all j, or equivalently $z \uparrow j = z \uparrow (j \bmod n)$. Then the product of $\gamma(z) = \alpha(z)\beta(z)$ is given by

$$(2.46) \qquad \begin{aligned} \gamma(z) &= \gamma_0 + \gamma_1 z + \cdots + \gamma_{2n-2} z^{2n-2} \\ &= \tilde{\gamma}_0 + \tilde{\gamma}_1 z + \cdots + \tilde{\gamma}_{n-1} z^{n-1}, \end{aligned}$$

where the elements of $\{\gamma_0, \gamma_1, \ldots, \gamma_{2n-2}\}$ are the products of the linear convolution of $\{\alpha_0, \alpha_1, \ldots, \alpha_{n-1}\}$ and $\{\beta_0, \beta_1, \ldots, \beta_{n-1}\}$, and where

$$(2.47) \qquad \tilde{\gamma}_j = \gamma_j + \gamma_{j+n} \quad \text{for} \quad j = 0, \ldots, n-2 \quad \text{and} \quad \tilde{\gamma}_{n-1} = \gamma_{n-1}.$$

These coefficients $\{\tilde{\gamma}_0, \tilde{\gamma}_1, \ldots, \tilde{\gamma}_{n-1}\}$ may be generated by applying a process of circular convolution to sequences which are the periodic extensions of $\{\alpha_0, \alpha_1, \ldots, \alpha_{n-1}\}$ and $\{\beta_0, \beta_1, \ldots, \beta_{n-1}\}$.

The circular convolution of (the periodic extensions of) two sequences $\{\alpha_0, \alpha_1, \ldots, \alpha_{n-1}\}$ and $\{\beta_0, \beta_1, \ldots, \beta_{n-1}\}$ may be effected indirectly via a method which involves finding their discrete Fourier transforms. The Fourier transforms of the sequences are obtained by evaluating their z-transforms $\alpha(z)$, $\beta(z)$ at n distinct points which are equally spaced around the circumference of a unit circle in the complex plane. These points $\{z_k; k = 0, \ldots, n-1\}$ come from setting $z_k = \exp(-i2\pi k/n)$. From the values $\{\alpha(z_k);\ k = 0, \ldots, n-1\}$ and $\{\beta(z_k);\ k = 0, \ldots, n-1\}$, the corresponding values $\{\gamma(z_k) = \alpha(z_k)\beta(z_k);\ k = 0, \ldots, n-1\}$ can be found by simple multiplication. The latter represent n ordinates of the polynomial product whose n coefficients are being sought. Therefore, the coefficients can be recovered by an application of an (inverse) Fourier transform.

It will be demonstrated in a later chapter that there are some highly efficient algorithms for computing the discrete Fourier transform of a finite sequence. Therefore, it is practical to consider effecting the circular convolution of two sequences first by computing their discrete Fourier transforms, then by multiplying the transforms and finally by applying the inverse Fourier transform to recover the coefficients of the convolution.

The Fourier method may also be used to affect the *linear* convolution of two sequences. Consider the sequences $\{\alpha_0, \alpha_1, \ldots, \alpha_p\}$ and $\{\beta_0, \beta_1, \ldots, \beta_k\}$ whose z-transforms may be denoted by $\alpha(z)$ and $\beta(z)$. If z^j is periodic in j with a period of $n > p + k$, then

$$(2.48) \qquad \alpha(z)\beta(z) = \gamma_0 + \gamma_1 z + \cdots + \gamma_{p+k} z^{p+k}$$

resembles the product of two polynomials of a non-periodic argument, and its coefficients are exactly those which would be generated by a linear convolution of the sequences. The reason is that the degree $p+k$ of the product is less than the period n of the argument.

In the context of the discrete Fourier transform, the period of the argument z corresponds to the length n of the sequence which is subject to the transformation. In order to increase the period, the usual expedient is to extend the length of the sequence by appending a number of zeros to the end of it. This is described as "padding" the sequence.

Consider the padded sequences $\{\alpha_0, \alpha_1, \ldots, \alpha_{n-1}\}$ and $\{\beta_0, \beta_1, \ldots, \beta_{n-1}\}$ in which $\alpha_{p+1} = \cdots = \alpha_{n-1} = 0$ and $\beta_{k+1} = \cdots = \beta_{n-1} = 0$. Let their z-transforms be denoted by $\tilde{\alpha}(z)$ and $\tilde{\beta}(z)$, and let the period of z be $n > p+k$. Then the product $\tilde{\gamma}(z) = \tilde{\alpha}(z)\tilde{\beta}(z)$, will entail the coefficients $\tilde{\gamma}_0 = \gamma_0$, $\tilde{\gamma}_1 = \gamma_1, \ldots, \tilde{\gamma}_{p+k} = \gamma_{p+k}$ of the linear convolution of $\{\alpha_0, \alpha_1, \ldots, \alpha_p\}$ and $\{\beta_0, \beta_1, \ldots, \beta_k\}$ together with some higher-order coefficients which are zeros. Nevertheless, these are the coefficients which would result from applying the process of circular convolution to the padded sequences; and, moreover, they can be obtained via the Fourier method.

The result can be made intuitively clear by thinking in terms of the physical model of circular convolution illustrated in Figure 2.2. If the sequences which are written on the rims of the two discs are padded with a sufficient number of zeros, then one sequence cannot engage both the head and the tail of the other sequence at the same time, and the result is a linear convolution.

Polynomial Factorisation

Consider the equation $\alpha_0 + \alpha_1 z + \alpha_2 z^2 = 0$. This can be factorised as $\alpha_2(z - \lambda_1)(z - \lambda_2)$ where λ_1, λ_2 are the roots of the equation which are given by the formula

$$\lambda = \frac{-\alpha_1 \pm \sqrt{\alpha_1^2 - 4\alpha_2\alpha_0}}{2\alpha_2}. \tag{2.49}$$

If $\alpha_1^2 \geq 4\alpha_2\alpha_0$, then the roots λ_1, λ_2 are real. If $\alpha_1^2 = 4\alpha_2\alpha_0$, then $\lambda_1 = \lambda_2$. If $\alpha_1^2 < 4\alpha_2\alpha_0$, then the roots are the conjugate complex numbers $\lambda = \alpha + i\beta$, $\lambda^* = \alpha - i\beta$ where $i = \sqrt{-1}$.

It is helpful to have at hand a Pascal procedure for finding the roots of a quadratic equation:

(2.50)
```
procedure QuadraticRoots(a, b, c : real);
  var
    discriminant, root1, root2, modulus : real;

  begin
    discriminant := Sqr(b) - 4 * a * c;
    if (discriminant > 0) and (a <> 0) then
      begin
        root1 := (-b + Sqrt(discriminant))/(2 * a);
        root2 := (-b - Sqrt(discriminant))/(2 * a);
        Writeln('Root(1) = ', root1 : 10 : 5);
        Writeln('Root(2) = ', root2 : 10 : 5);
      end;
    if (discriminant = 0) and (a <> 0) then
      begin
        root1 := -b/(2 * a);
        Writeln('The roots coincide at the value = ', root1 : 10 : 5);
      end;
    if (discriminant < 0) and (a <> 0) then
```

```
        begin
          root1 := -b/(2 * a);
          root2 := Sqrt(-discriminant)/(2 * a);
          modulus := Sqrt(Sqr(root1) + Sqr(root2));
          Writeln('We have conjugate complex roots');
          Writeln('The real part is ', root1 : 10 : 5);
          Writeln('The imaginary part is ', root2 : 10 : 5);
          Writeln('The modulus is ', modulus : 10 : 5);
        end;
    end; {QuadraticRoots}
```

Complex Roots

There are three ways of representing the conjugate complex numbers λ and λ^*:

$$\begin{aligned}\lambda &= \alpha + i\beta = \rho(\cos\theta + i\sin\theta) = \rho e^{i\theta},\\ \lambda^* &= \alpha - i\beta = \rho(\cos\theta - i\sin\theta) = \rho e^{-i\theta}.\end{aligned} \tag{2.51}$$

Here there are

$$\rho = \sqrt{\alpha^2 + \beta^2} \qquad \text{and} \qquad \tan\theta = \beta/\alpha. \tag{2.52}$$

The three representations are called, respectively, the Cartesian form, the trigonometrical form and the exponential form. The parameter $\rho = |\lambda|$ is the modulus of the roots and the parameter θ, which is sometimes denoted by $\theta = \arg(\lambda)$, is the argument of the exponential form. This is the angle, measured in radians, which λ makes with the positive real axis when it is interpreted as a vector bound to the origin. Observe that θ is not uniquely determined by this definition, since the value of the tangent is unaffected if $2n\pi$ is added to or subtracted from θ, where n is any integer. The principal value of $\arg(\lambda)$, denoted $\mathrm{Arg}(\lambda)$, is the unique value of $\theta \in (-\pi, \pi]$ which satisfies the definition.

The Cartesian and trigonometrical representations are understood by considering the Argand diagram (see Figure 2.3). The exponential form is understood by considering the series expansions of $\cos\theta$ and $i\sin\theta$ about the point $\theta = 0$:

$$\begin{aligned}\cos\theta &= \left\{1 - \frac{\theta^2}{2!} + \frac{\theta^4}{4!} - \frac{\theta^6}{6!} + \cdots\right\},\\ i\sin\theta &= \left\{i\theta - \frac{i\theta^3}{3!} + \frac{i\theta^5}{5!} - \frac{i\theta^7}{7!} + \cdots\right\}.\end{aligned} \tag{2.53}$$

Adding the series gives

$$\begin{aligned}\cos\theta + i\sin\theta &= \left\{1 + i\theta - \frac{\theta^2}{2!} - \frac{i\theta^3}{3!} + \frac{\theta^4}{4!} + \cdots\right\}\\ &= e^{i\theta}.\end{aligned} \tag{2.54}$$

Likewise, subtraction gives

$$\cos\theta - i\sin\theta = e^{-i\theta}. \tag{2.55}$$

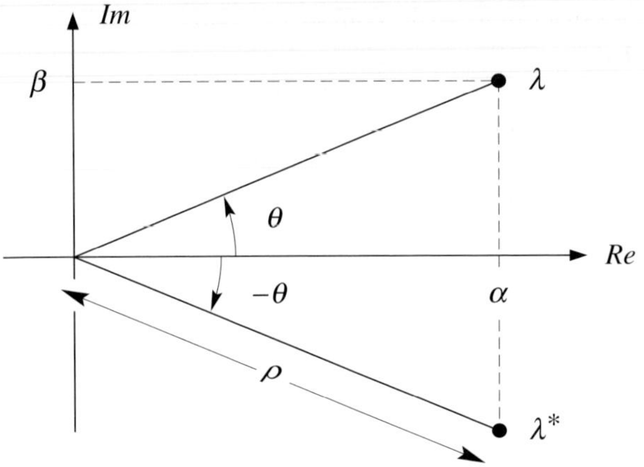

Figure 2.3. The Argand diagram showing a complex number $\lambda = \alpha + i\beta$ and its conjugate $\lambda^* = \alpha - i\beta$.

The equations (2.54) and (2.55) are known as Euler's formulae. The inverse formulae are

(2.56) $$\cos\theta = \frac{e^{i\theta} + e^{-i\theta}}{2}$$

and

(2.57) $$\sin\theta = -\frac{i}{2}(e^{i\theta} - e^{-i\theta}) = \frac{e^{i\theta} - e^{-i\theta}}{2i}.$$

In some computer languages—for example, in FORTRAN—complex numbers correspond to a predefined type; and the usual operations of complex arithmetic are also provided. This is not the case in Pascal; and it is helpful to define the complex type and to create a small library of complex operations. It is convenient to use a record type:

(2.58)
```
type
complex = record
    re, im : real;
  end;
```

The modulus of a complex number defined in (2.52) and its square are provided by the following two functions:

(2.59)
```
function Cmod(a : complex) : real;
begin
  Cmod := Sqrt(Sqr(a.re) + Sqr(a.im));
end; {Cmod}
```

```
function Cmodsqr(a : complex) : real;
begin
  Cmodsqr := Sqr(a.re) + Sqr(a.im);
end; {Cmodsqr}
```

The addition of a pair of complex numbers is a matter of adding their real and imaginary components:

$$(\alpha + i\beta) + (\gamma + i\delta) = (\alpha + \gamma) + i(\beta + \delta). \tag{2.60}$$

Functions are provided both for addition and for subtraction:

(2.61)

```
function Cadd(a, b : complex) : complex;
  var
    c : complex;
begin
  c.re := a.re + b.re;
  c.im := a.im + b.im;
  Cadd := c;
end; {Cadd}

function Csubtract(a, b : complex) : complex;
  var
    c : complex;
begin
  c.re := a.re - b.re;
  c.im := a.im - b.im;
  Csubtract := c;
end; {Csubtract}
```

The product of the numbers

$$\begin{aligned}\lambda &= \alpha + i\beta = \rho(\cos\theta + i\sin\theta) = \rho e^{i\theta},\\ \mu &= \gamma + i\delta = \kappa(\cos\omega + i\sin\omega) = \kappa e^{i\omega}\end{aligned} \tag{2.62}$$

is given by

$$\begin{aligned}\lambda\mu &= \alpha\gamma - \beta\delta + i(\alpha\delta + \beta\gamma)\\ &= \rho\kappa\big\{(\cos\theta\cos\omega - \sin\theta\sin\omega) + i(\cos\theta\sin\omega + \sin\theta\cos\omega)\big\}\\ &= \rho\kappa\big\{\cos(\theta+\omega) + i\sin(\theta+\omega)\big\}\\ &= \rho\kappa e^{i(\theta+\omega)},\end{aligned} \tag{2.63}$$

where two trigonometrical identities have been used to obtain the third equality.

In the exponential form, the product of the complex numbers comprises the product of their moduli and the sum of their arguments. The exponential representation clarifies a fundamental identity known as DeMoivre's theorem:

$$\big\{\rho(\cos\theta + i\sin\theta)\big\}^n = \rho^n\big\{\cos(n\theta) + i\sin(n\theta)\big\}. \tag{2.64}$$

In exponential form, this becomes $\{\rho e^{i\theta}\}^n = \rho^n e^{in\theta}$.

For the purposes of computing a complex multiplication, the Cartesian representation is adopted:

(2.65)

```
function Cmultiply(a, b : complex) : complex;
  var
    c : complex;
begin
  c.re := a.re * b.re - a.im * b.im;
  c.im := a.im * b.re + b.im * a.re;
  Cmultiply := c;
end; {Cmultiply}
```

The inverse of the number $\alpha + i\beta$ is

$$(\alpha + i\beta)^{-1} = \frac{\alpha - i\beta}{\alpha^2 + \beta^2}. \tag{2.66}$$

This is obtained from the identity $\lambda^{-1} = \lambda^*/(\lambda^*\lambda)$. A formula for the division of one complex number by another follows immediately; and the trigonometrical and polar forms of these identities are easily obtained. Separate code is provided for the operations of inversion and division:

(2.67)

```
function Cinverse(a : complex) : complex;
  var
    c : complex;
begin
  c.re := a.re/(Sqr(a.re) + Sqr(a.im));
  c.im := -a.im/(Sqr(a.re) + Sqr(a.im));
  Cinverse := c;
end; {Cinverse}

function Cdivide(a, b : complex) : complex;
  var
    c : complex;
begin
  c.re := (a.re * b.re + a.im * b.im)/(Sqr(b.re) + Sqr(b.im));
  c.im := (a.im * b.re - b.im * a.re)/(Sqr(b.re) + Sqr(b.im));
  Cdivide := c;
end; {Cdivide}
```

Finally, there is the code for computing the square root of a complex number. In this case, the polar representation is used:

(2.68)

```
function Csqrt(a : complex) : complex;

  const
```

```
    virtualZero = 1E - 12;
    pi = 3.1415926;

  var
    rho, theta : real;
    c : complex;

begin {complex square root}
  rho := Sqrt(Sqr(a.re) + Sqr(a.im));
  if Abs(a.re) < virtualZero then
    begin
      if a.im < 0 then
      theta := pi/2
      else
      theta := -pi/2
    end
  else if a.re < 0 then
    theta := ArcTan(a.im/a.re) + pi
  else
    theta := Arctan(a.im/a.re);
  c.re := Sqrt(rho) * Cos(theta/2);
  c.im := Sqrt(rho) * Sin(theta/2);
  Csqrt := c;
end; {Csqrt : complex square root}
```

The Roots of Unity

Consider the equation $z^n = 1$. This is always satisfied by $z = 1$; and, if n is even, then $z = -1$ is also a solution. There are no other solutions amongst the set of real numbers. The solution set is enlarged if z is allowed to be a complex number. Let $z = \rho e^{i\theta} = \rho\{\cos(\theta) + i\sin(\theta)\}$. Then $z^n = \rho^n e^{i\theta n} = 1$ implies that $\rho^n = 1$ and therefore $\rho = 1$, since the equality of two complex numbers implies the equality of their moduli. Now consider $z^n = e^{i\theta n} = \cos(n\theta) + i\sin(n\theta)$. Equating the real parts of the equation $z^n = 1$ shows that $\cos(n\theta) = 1$ which implies that $n\theta = 2\pi k$, where k is any integer. Equating the imaginary parts shows that $\sin(n\theta) = 0$ which, again, implies that $n\theta = 2\pi k$. Therefore, the solutions take the form of

$$z = \exp\left(\frac{i2\pi k}{n}\right) = \cos\frac{2\pi k}{n} + i\sin\frac{2\pi k}{n}. \tag{2.69}$$

Such solutions are called roots of unity.

Since $\cos\{2\pi k/n\}$ and $\sin\{2\pi k/n\}$ are periodic functions with a period of $k = n$, it makes sense to consider only solutions with values of k less than n. Also, if z is a root of unity, then so too is z^*. The nth roots of unity may be represented by n equally spaced points around the circumference of the unit circle in the complex plane (see Figure 2.4).

The roots of unity are entailed in the process of finding the discrete Fourier transform of a finite sequence. Later in the present chapter, and in Chapter 14,

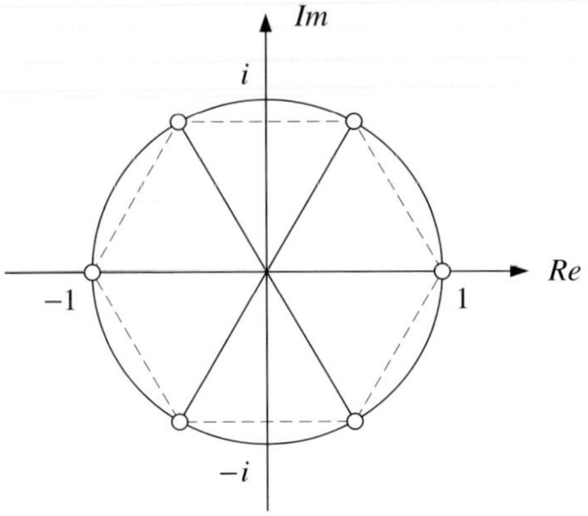

Figure 2.4. The 6th roots of unity inscribed in the unit circle.

where we consider the discrete Fourier transform of a sequence of data points $y_t; t = 0, \ldots, T-1$, we adopt the notation

$$W_T^{jt} = \exp\left(\frac{-i2\pi jt}{T}\right); \qquad t = 0, \ldots, T-1 \tag{2.70}$$

to describe the T points on the unit circle at which the argument z^j is evaluated.

The Polynomial of Degree n

Now consider the general equation of the nth degree:

$$\alpha_0 + \alpha_1 z + \cdots + \alpha_n z^n = 0. \tag{2.71}$$

On dividing by α_n, a monic polynomial is obtained which has a unit associated with the highest power of z. This can be factorised as

$$(z - \lambda_1)(z - \lambda_2) \cdots (z - \lambda_n) = 0, \tag{2.72}$$

where some of the roots may be real and others may be complex. If the coefficients of the polynomial are real-valued, then the complex roots must come in conjugate pairs. Thus, if $\lambda = \alpha + i\beta$ is a complex root, then there is a corresponding root $\lambda^* = \alpha - i\beta$ such that the product $(z - \lambda)(z - \lambda^*) = z^2 + 2\alpha z + (\alpha^2 + \beta^2)$ is real and quadratic. When the n factors are multiplied together, we obtain the expansion

$$0 = z^n - \sum_i \lambda_i z^{n-1} + \sum_{i \neq j} \lambda_i \lambda_j z^{n-2} - \cdots (-1)^n (\lambda_1 \lambda_2 \cdots \lambda_n). \tag{2.73}$$

This can be compared with the expression $(\alpha_0/\alpha_n) + (\alpha_1/\alpha_n)z + \cdots + z^n = 0$. By equating coefficients of the two expressions, it is found that $(\alpha_0/\alpha_n) = (-1)^n \prod \lambda_i$ or, equivalently,

$$\alpha_n = \alpha_0 \prod_{i=1}^{n} (-\lambda_i)^{-1}. \tag{2.74}$$

Thus the polynomial may be expressed in any of the following forms:

$$\begin{aligned}\sum \alpha_i z^i &= \alpha_n \prod (z - \lambda_i)\\ &= \alpha_0 \prod (-\lambda_i)^{-1} \prod (z - \lambda_i)\\ &= \alpha_0 \prod (1 - z/\lambda_i).\end{aligned} \tag{2.75}$$

The following procedure provides the means for compounding the coefficients of a monic polynomial from its roots. The n roots are contained in a complex array *lambda*. When all the complex roots are in conjugate pairs, the coefficients become the real elements of the complex array *alpha*.

(2.76)

```
procedure RootsToCoefficients(n : integer;
                     var alpha, lambda : complexVector);

  var
    j, k : integer;
    store : complex;

begin {RootsToCoefficients}
  alpha[0].re := 1.0;
  alpha[0].im := 0.0;

  for k := 1 to n do
    begin {k}
      alpha[k].im := 0.0;
      alpha[k].re := 0.0;
      for j := k downto 1 do
        begin {j}
          store := Cmultiply(lambda[k], alpha[j]);
          alpha[j] := Csubtract(alpha[j - 1], store);
        end; {j}
      alpha[0] := Cmultiply(lambda[k], alpha[0]);
      alpha[0].re := -alpha[0].re;
      alpha[0].im := -alpha[0].im
    end; {k}

end; {RootsToCoefficients}
```

Occasionally it is useful to consider, instead of the polynomial $\alpha(z)$ of (2.71), a polynomial in the form

$$\alpha'(z) = \alpha_0 z^n + \alpha_1 z^{n-1} + \cdots + \alpha_{n-1} z + \alpha_n. \tag{2.77}$$

This has the same coefficients as $\alpha(z)$, but is has declining powers of z instead of rising powers. Reference to (2.71) shows that $\alpha'(z) = z^n \alpha(z^{-1})$.

If λ is a root of the equation $\alpha(z) = \sum \alpha_i z^i = 0$, then $\mu = 1/\lambda$ is a root of the equation $\alpha'(z) = \sum \alpha_i z^{n-i} = 0$. This follows since $\sum \alpha_i \mu^{n-i} = \mu^n \sum \alpha_i \mu^{-i} = 0$ implies that $\sum \alpha_i \mu^{-i} = \sum \alpha_i \lambda^i = 0$. Confusion can arise from not knowing which of the two equations one is dealing with.

Another possibility, which may give rise to confusion, is to write the factorisation of $\alpha(z)$ in terms of the inverse values of its roots which are the roots of $\alpha(z^{-1})$. Thus, in place of the final expression under (2.75), one may write

$$\sum \alpha_i z^i = \alpha_0 \prod (1 - \mu_i z). \tag{2.78}$$

Since it is often convenient to specify $\alpha(z)$ in this manner, a procedure is provided for compounding the coefficients from the inverse roots. In the procedure, it is assumed that $\alpha_0 = 1$.

(2.79)
```
procedure InverseRootsToCoeffs(n : integer;
                        var alpha, mu : complexVector);

  var
    j, k : integer;
    store : complex;

begin
  alpha[0].re := 1.0;
  alpha[0].im := 0.0;

  for k := 1 to n do
    begin {k}
      alpha[k].im := 0.0;
      alpha[k].re := 0.0;
      for j := k downto 1 do
        begin {j}
          store := Cmultiply(mu[k], alpha[j - 1]);
          alpha[j] := Csubtract(alpha[j], store);
        end; {j}
    end; {k}

end; {InverseRootsToCoefficients}
```

To form the coefficients of a polynomial from its roots is a simple matter. To unravel the roots from the coefficients is generally far more difficult. The topic of polynomial factorisation is pursued is a subsequent chapter where practical methods are presented for extracting the roots of polynomials of high degrees.

Matrices and Polynomial Algebra

So far, in representing time-series models, we have used rational polynomial operators. The expansion of a rational operator gives rise to an indefinite power series. However, when it comes to the numerical representation of a model, one is constrained to work with finite sequences; and therefore it is necessary to truncate the power series. Moreover, whenever the concepts of multivariate statistical analysis are applied to the problem of estimating the parameters of time-series model, it becomes convenient to think in terms of the algebra of vectors and matrices. For these reasons, it is important to elucidate the relationship between the algebra of coordinate vector spaces and the algebra of polynomials.

Lower-Triangular Toeplitz Matrices

Some of the essential aspects of the algebra of polynomials are reflected in the algebra of lower-triangular Toeplitz matrices.

A Toeplitz matrix $A = [\alpha_{ij}]$ is defined by the condition that $\alpha_{ij} = \alpha_{i+k,j+k}$ for all i, j and k within the allowable range of the indices. The elements of a Toeplitz matrix vary only when the difference of the row and column indices varies; and, therefore, the generic element can be written as $\alpha_{ij} = \alpha_{i-j}$. The $n \times n$ matrix $A = [\alpha_{i-j}]$ takes the following form:

$$A = \begin{bmatrix} \alpha_0 & \alpha_{-1} & \alpha_{-2} & \dots & \alpha_{1-n} \\ \alpha_1 & \alpha_0 & \alpha_{-1} & \dots & \alpha_{2-n} \\ \alpha_2 & \alpha_1 & \alpha_0 & \dots & \alpha_{3-n} \\ \vdots & \vdots & \vdots & \ddots & \vdots \\ \alpha_{n-1} & \alpha_{n-2} & \alpha_{n-3} & \dots & \alpha_0 \end{bmatrix}. \tag{2.80}$$

A lower-triangular Toeplitz $A = [\alpha_{i-j}]$ has $\alpha_{i-j} = 0$ whenever $i < j$. Such a matrix is completely specified by its leading vector $\alpha = \{\alpha_0, \dots, \alpha_{n-1}\}$. This vector is provided by the equation $\alpha = Ae_0$ where e_0 is the leading vector of the identity matrix of order n which has a unit as its leading element and zeros elsewhere. Occasionally, when it is necessary to indicate that A is completely specified by α, we shall write $A = A(\alpha)$.

Any lower-triangular Toeplitz matrix A of order n can be expressed as a linear combination of a set of basis matrices $I, L, \dots, L^{n-1}$, where the matrix $L = [e_1, \dots, e_{n-2}, 0]$, which has units on the first subdiagonal and zeros elsewhere, is formed from the identity matrix $I = [e_0, e_1, \dots, e_{n-1}]$ by deleting the leading vector and appending a zero vector to the end of the array:

$$L = \begin{bmatrix} 0 & 0 & 0 & \dots & 0 & 0 \\ 1 & 0 & 0 & \dots & 0 & 0 \\ 0 & 1 & 0 & \dots & 0 & 0 \\ \vdots & \vdots & \vdots & \ddots & \vdots & \vdots \\ 0 & 0 & 0 & \dots & 0 & 0 \\ 0 & 0 & 0 & \dots & 1 & 0 \end{bmatrix}. \tag{2.81}$$

This is a matrix analogue of the lag operator. When $q < n$, the matrix L^q, which is the qth power of L, has units on the qth subdiagonal and zeros elsewhere. When $q \geq n$ the matrix L^q is null; and therefore L is said to be nilpotent of degree n. Thus the lower-triangular Toeplitz matrix A may be expressed as

$$\begin{aligned} A &= \alpha_0 I + \alpha_1 L + \dots + \alpha_{n-1} L^{n-1} \\ &= \alpha(L). \end{aligned} \tag{2.82}$$

This can be construed as polynomial whose argument is the matrix L. The notation is confusable with that of a polynomial in the lag operator L which operates on the set of infinite sequences. Distinctions can be made by indicating the order of the matrix via a subscript. The matrix L_∞ is synonymous with the ordinary lag operator.

According to the algebra of polynomials, the product of the pth degree polynomial $\alpha(z)$ and the kth degree polynomial $\beta(z)$ is a polynomial $\gamma(z) = \alpha(z)\beta(z) = \beta(z)\alpha(z)$ of degree $p+k$. However, in forming the matrix product $AB = \alpha(L)\beta(L)$ according the rules of polynomial algebra, it must be recognised that $L^q = 0$ for all $q \geq n$; which means that the product corresponds to a polynomial of degree $n-1$ at most. The matter is summarised as follows:

(2.83) If $A = \alpha(L)$ and $B = \beta(L)$ are lower-triangular Toeplitz matrices, then their product $\Gamma = AB = BA$ is also a lower-triangular Toeplitz matrix. If the order of Γ exceeds the degree of $\gamma(z) = \alpha(z)\beta(z) = \beta(z)\alpha(z)$, then the leading vector $\gamma = \Gamma e_1$ contains the complete sequence of the coefficients of $\gamma(z)$. Otherwise it contains a truncated version of the sequence.

If the matrices A, B and Γ were of infinite order, then the products of multiplying polynomials of any degree could accommodated.

The notable feature of this result is that lower-triangular Toeplitz matrices commute in multiplication; and this corresponds to the commutativity of polynomials in multiplication.

Example 2.3. Consider the polynomial product

$$\begin{aligned}\alpha(z)\beta(z) &= (\alpha_0 + \alpha_1 z + \alpha_2 z^2)(\beta_0 + \beta_1 z) \\ &= \alpha_0\beta_0 + (\alpha_0\beta_1 + \alpha_1\beta_0)z + (\alpha_1\beta_1 + \alpha_2\beta_0)z^2 + \alpha_2\beta_1 z^3.\end{aligned} \tag{2.84}$$

This may be compared with the following commutative matrix multiplication:

$$\begin{bmatrix} \alpha_0 & 0 & 0 & 0 \\ \alpha_1 & \alpha_0 & 0 & 0 \\ \alpha_2 & \alpha_1 & \alpha_0 & 0 \\ 0 & \alpha_2 & \alpha_1 & \alpha_0 \end{bmatrix} \begin{bmatrix} \beta_0 & 0 & 0 & 0 \\ \beta_1 & \beta_0 & 0 & 0 \\ 0 & \beta_1 & \beta_0 & 0 \\ 0 & 0 & \beta_1 & \beta_0 \end{bmatrix} = \begin{bmatrix} \gamma_0 & 0 & 0 & 0 \\ \gamma_1 & \gamma_0 & 0 & 0 \\ \gamma_2 & \gamma_1 & \gamma_0 & 0 \\ \gamma_3 & \gamma_2 & \gamma_1 & \gamma_0 \end{bmatrix}, \tag{2.85}$$

where

$$\begin{aligned}\gamma_0 &= \alpha_0\beta_0, \\ \gamma_1 &= \alpha_0\beta_1 + \alpha_1\beta_0, \\ \gamma_2 &= \alpha_1\beta_1 + \alpha_2\beta_0, \\ \gamma_3 &= \alpha_2\beta_1.\end{aligned} \tag{2.86}$$

The inverse of a lower-triangular Toeplitz matrix $A = \alpha(L)$ is defined by the identity

$$A^{-1}A = \alpha^{-1}(L)\alpha(L) = I. \tag{2.87}$$

Let $\alpha^{-1}(z) = \{\omega_0 + \omega_1 z + \cdots + \omega_{n-1}z^{n-1} + \cdots\}$ denote the expansion of the inverse polynomial. Then, when L in put in place of z and when it is recognised that $L^q = 0$ for $q \geq n$, it will be found that

(2.88) $$A^{-1} = \omega_0 + \omega_1 L + \cdots + \omega_{n-1}L^{n-1}.$$

The result may be summarised as follows:

(2.89) Let $\alpha(z) = \alpha_0 + \alpha_1 z + \cdots + \alpha_p z^p$ be a polynomial of degree p and let $A = \alpha(L)$ be a lower-triangular Toeplitz matrix of order n. Then the leading vector of A^{-1} contains the leading coefficients of the expansion of $\alpha^{-1}(z) = \{\omega_0 + \omega_1 z + \cdots + \omega_{n-1}z^{n-1} + \cdots\}$.

Notice that there is no requirement that $n \geq p$. When $n < p$, the elements of the inverse matrix A^{-1} are still provided by the leading coefficients of the expansion of $\alpha^{-1}(z)$, despite the fact that the original matrix $A = \alpha(L)$ contains only the leading coefficients of $\alpha(z)$.

Example 2.4. The matrix analogue of the product of $1 - \theta z$ and $(1 - \theta z)^{-1} = \{1 + \theta z + \theta^2 z^2 + \cdots\}$ is

(2.90) $$\begin{bmatrix} 1 & 0 & 0 & 0 \\ -\theta & 1 & 0 & 0 \\ 0 & -\theta & 1 & 0 \\ 0 & 0 & -\theta & 1 \end{bmatrix} \begin{bmatrix} 1 & 0 & 0 & 0 \\ \theta & 1 & 0 & 0 \\ \theta^2 & \theta & 1 & 0 \\ \theta^3 & \theta^2 & \theta & 1 \end{bmatrix} = \begin{bmatrix} 1 & 0 & 0 & 0 \\ 0 & 1 & 0 & 0 \\ 0 & 0 & 1 & 0 \\ 0 & 0 & 0 & 1 \end{bmatrix}.$$

This matrix equation is also the analogue of the product of $(1 - \theta z)/(1 - \theta^4 z^4)$ and $1 + \theta z + \theta^2 z^2 + \theta^3 z^3$.

Circulant Matrices

A circulant matrix is a Toeplitz matrix which has the general form of

(2.91) $$A = \begin{bmatrix} \alpha_0 & \alpha_{n-1} & \alpha_{n-2} & \cdots & \alpha_1 \\ \alpha_1 & \alpha_0 & \alpha_{n-1} & \cdots & \alpha_2 \\ \alpha_2 & \alpha_1 & \alpha_0 & \cdots & \alpha_3 \\ \vdots & \vdots & \vdots & \ddots & \vdots \\ \alpha_{n-1} & \alpha_{n-2} & \alpha_{n-3} & \cdots & \alpha_0 \end{bmatrix}.$$

The vectors of such a matrix are generated by applying a succession of cyclic permutations to the leading vector, which therefore serves to specify the matrix completely. The elements of the circulant matrix $A = [\alpha_{ij}]$ fulfil the condition that $\alpha_{ij} = \alpha\{(i-j) \bmod n\}$. Hence, the index for the supradiagonal elements, for which $1 - n < i - j < 0$, becomes $(i - j) \bmod n = n + (i - j)$.

Any circulant matrix of order n can be expressed as a linear combination of a set of basis matrices $I, K, \ldots, K^{n-1}$, where $K = [e_1, \ldots, e_{n-1}, e_0]$ is formed from the identity matrix $I = [e_0, e_1, \ldots, e_{n-1}]$ by moving the leading vector to the back of the array:

$$K = \begin{bmatrix} 0 & 0 & 0 & \ldots & 0 & 1 \\ 1 & 0 & 0 & \ldots & 0 & 0 \\ 0 & 1 & 0 & \ldots & 0 & 0 \\ \vdots & \vdots & \vdots & \ddots & \vdots & \vdots \\ 0 & 0 & 0 & \ldots & 0 & 0 \\ 0 & 0 & 0 & \ldots & 1 & 0 \end{bmatrix}. \tag{2.92}$$

This is a matrix operator which effects a cyclic permutation of the elements of any (column) vector which it premultiplies. Thus, an arbitrary circulant matrix A of order n can be expressed as

$$\begin{aligned} A &= \alpha_0 I + \alpha_1 K + \cdots + \alpha_{n-1} K^{n-1} \\ &= \alpha(K). \end{aligned} \tag{2.93}$$

The powers of K form an n-periodic sequence such that $K^{j+n} = K^j$ for all j or, equivalently, $K \uparrow j = K \uparrow (j \bmod n)$. The inverse powers of the operator K are defined by the condition that $K^{-q}K^q = K^0 = I$. It can be confirmed directly that $K^{-q} = K^{n-q}$. However, this also follows formally from the condition that $K^n = K^0 = I$. It may also be confirmed directly that the transpose of K is $K' = K^{n-1} = K^{-1}$.

It is easy to see that circulant matrices commute in multiplication, since this is a natural consequence of identifying them with polynomials. Thus

(2.94) If $A = \alpha(K)$ and $B = \beta(K)$ are circulant matrices, then their product $\Gamma = AB = BA$ is also a circulant matrix whose leading vector $\gamma = \Gamma e_0$ contains the coefficients of the circular convolution of the leading vectors $\alpha = Ae_0$ and $\beta = Be_0$.

Example 2.5. Consider the following product of circulant matrices:

$$\begin{bmatrix} \alpha_0 & 0 & \alpha_2 & \alpha_1 \\ \alpha_1 & \alpha_0 & 0 & \alpha_2 \\ \alpha_2 & \alpha_1 & \alpha_0 & 0 \\ 0 & \alpha_2 & \alpha_1 & \alpha_0 \end{bmatrix} \begin{bmatrix} \beta_0 & 0 & \beta_2 & \beta_1 \\ \beta_1 & \beta_0 & 0 & \beta_2 \\ \beta_2 & \beta_1 & \beta_0 & 0 \\ 0 & \beta_2 & \beta_1 & \beta_0 \end{bmatrix} = \begin{bmatrix} \gamma_0 & \gamma_3 & \gamma_2 & \gamma_1 \\ \gamma_1 & \gamma_0 & \gamma_3 & \gamma_2 \\ \gamma_2 & \gamma_1 & \gamma_0 & \gamma_3 \\ \gamma_3 & \gamma_2 & \gamma_1 & \gamma_0 \end{bmatrix}. \tag{2.95}$$

Here

$$\begin{aligned} \gamma_0 &= \alpha_0\beta_0 + \alpha_2\beta_2, \\ \gamma_1 &= \alpha_1\beta_0 + \alpha_0\beta_1, \\ \gamma_2 &= \alpha_2\beta_0 + \alpha_1\beta_1 + \alpha_0\beta_2, \\ \gamma_3 &= \alpha_2\beta_1 + \alpha_1\beta_2, \end{aligned} \tag{2.96}$$

represent the coefficients of the circular convolution of $\{\alpha_0, \alpha_1, \alpha_2, 0\}$ and $\{\beta_0, \beta_1, \beta_2, 0\}$. Notice that, with $\beta_2 = 0$, the coefficients $\{\gamma_0, \gamma_1, \gamma_2, \gamma_3\}$ would be the same as those from the linear convolution depicted under (2.85). Thus it is confirmed that the coefficients of the *linear* convolution of $\{\alpha_0, \alpha_1, \alpha_2\}$ and $\{\beta_0, \beta_1\}$ may be obtained by applying the process of *circular* convolution to the padded sequences $\{\alpha_0, \alpha_1, \alpha_2, 0\}$ and $\{\beta_0, \beta_1, 0, 0\}$.

If $A = \alpha(K)$ is a circulant matrix, then its inverse is also a circulant matrix which is defined by the condition

$$(2.97) \qquad A^{-1}A = \alpha^{-1}(K)\alpha(K) = I.$$

If the roots of $\alpha(z) = 0$ lie outside the unit circle, then coefficients of the expansion $\alpha(z)^{-1} = \{\omega_0 + \omega_1 z + \cdots + \omega_{n-1}z^{n-1} + \cdots\}$ form a convergent sequence. Therefore, by putting K in place of z and noting that $K \uparrow q = K \uparrow (q \bmod n)$, it is found that

$$(2.98) \qquad \begin{aligned} A^{-1} &= \sum_{j=0}^{\infty}\omega_{jn} + \left\{\sum_{j=0}^{\infty}\omega_{(jn+1)}\right\}K + \cdots + \left\{\sum_{j=0}^{\infty}\omega_{(jn+n-1)}\right\}K^{n-1} \\ &= \psi_0 + \psi_1 K + \cdots + \psi_{n-1}K^{n-1}. \end{aligned}$$

Given that $\omega_j \to 0$ as $j \to \infty$, it follows that the sequence $\{\psi_0, \psi_1, \ldots, \psi_{n-1}\}$ converges to the sequence $\{\omega_0, \omega_1, \ldots, \omega_{n-1}\}$ as n increases. If the roots of $\alpha(z) = 0$ lie inside the unit circle, then it becomes appropriate to express A as $A = K^{-1}(\alpha_{n-1} + \alpha_{n-2}K^{-1} + \cdots + \alpha_1 K^{2-n} + \alpha_0 K^{1-n}) = K^{-1}\alpha'(K^{-1})$ and to defined the inverse of A by the condition

$$(2.99) \qquad A^{-1}A = \alpha'^{-1}(K^{-1})\alpha'(K^{-1}) = I.$$

The expression under (2.98) must then be replaced by a similar expression in terms of a convergent sequence of coefficients from the expansion of $\alpha'(z^{-1})$.

The Factorisation of Circulant Matrices

The matrix operator K has a spectral factorisation which is particularly useful in analysing the properties of the discrete Fourier transform. To demonstrate this factorisation, we must first define the so-called Fourier matrix. This is a symmetric matrix $U = n^{-1/2}[W^{jt}; t, j = 0, \ldots, n-1]$ whose generic element in the jth row and tth column is $W^{jt} = \exp(-i2\pi tj/n)$. On taking account of the n-periodicity of $W^q = \exp(-i2\pi q/n)$, the matrix can be written explicitly as

$$(2.100) \qquad U = \frac{1}{\sqrt{n}}\begin{bmatrix} 1 & 1 & 1 & \ldots & 1 \\ 1 & W & W^2 & \ldots & W^{n-1} \\ 1 & W^2 & W^4 & \ldots & W^{n-2} \\ \vdots & \vdots & \vdots & & \vdots \\ 1 & W^{n-1} & W^{n-2} & \ldots & W \end{bmatrix}.$$

The second row and the second column of this matrix contain the nth roots of unity. The conjugate matrix is defined as $\bar{U} = n^{-1/2}[W^{-jt}; t, j = 0, \ldots, n-1]$;

and, by using $W^{-q} = W^{n-q}$, this can be written explicitly as

$$\bar{U} = \frac{1}{\sqrt{n}} \begin{bmatrix} 1 & 1 & 1 & \dots & 1 \\ 1 & W^{n-1} & W^{n-2} & \dots & W \\ 1 & W^{n-2} & W^{n-4} & \dots & W^2 \\ \vdots & \vdots & \vdots & & \vdots \\ 1 & W & W^2 & \dots & W^{n-1} \end{bmatrix}. \tag{2.101}$$

It is readily confirmed that U is a unitary matrix fulfilling the condition

$$\bar{U}U = U\bar{U} = I. \tag{2.102}$$

To demonstrate this result, consider the generic element in the rth row and the sth column of the matrix $U\bar{U} = [\delta_{rs}]$. This is given by

$$\begin{aligned} \delta_{rs} &= \frac{1}{n}\sum_{t=0}^{n-1} W^{rt}W^{-st} \\ &= \frac{1}{n}\sum_{t=0}^{n-1} W^{(r-s)t}. \end{aligned} \tag{2.103}$$

Here there is

$$\begin{aligned} W^{(r-s)t} = W^{qt} &= \exp(-i2\pi qt/n) \\ &= \cos(-i2\pi qt/n) - i\sin(-i2\pi qt/n). \end{aligned} \tag{2.104}$$

Unless $q = 0$, the sums of these trigonometrical functions over an integral number of cycles are zero, and therefore $\sum_t W^{qt} = 0$. If $q = 0$, then the sine and cosine functions assume the values of zero and unity respectively, and therefore $\sum_t W^{qt} = n$. It follows that

$$\delta_{rs} = \begin{cases} 1, & \text{if } r = s; \\ 0, & \text{if } r \neq s, \end{cases}$$

which proves the result.

Example 2.6. Consider the matrix

$$U = \frac{1}{2}\begin{bmatrix} 1 & 1 & 1 & 1 \\ 1 & W & W^2 & W^3 \\ 1 & W^2 & W^4 & W^6 \\ 1 & W^3 & W^6 & W^9 \end{bmatrix} = \frac{1}{2}\begin{bmatrix} 1 & 1 & 1 & 1 \\ 1 & W & W^2 & W^3 \\ 1 & W^2 & 1 & W^2 \\ 1 & W^3 & W^2 & W \end{bmatrix}. \tag{2.105}$$

The equality comes from the 4-period periodicity of $W^q = \exp(-\pi q/2)$. The conjugate matrix is

$$\bar{U} = \frac{1}{2}\begin{bmatrix} 1 & 1 & 1 & 1 \\ 1 & W^{-1} & W^{-2} & W^{-3} \\ 1 & W^{-2} & W^{-4} & W^{-6} \\ 1 & W^{-3} & W^{-6} & W^{-9} \end{bmatrix} = \frac{1}{2}\begin{bmatrix} 1 & 1 & 1 & 1 \\ 1 & W^3 & W^2 & W \\ 1 & W^2 & 1 & W^2 \\ 1 & W & W^2 & W^3 \end{bmatrix}. \tag{2.106}$$

With $W^q = \exp(-\pi q/2) = \cos(-\pi q/2) - i\sin(-\pi q/2)$, it is found that $W^0 = 1$, $W^1 = -i$, $W^2 = -1$ and $W^3 = i$. Therefore,

$$(2.107)\qquad U\bar{U} = \frac{1}{4}\begin{bmatrix} 1 & 1 & 1 & 1 \\ 1 & -i & -1 & i \\ 1 & -1 & 1 & -1 \\ 1 & i & -1 & -i \end{bmatrix}\begin{bmatrix} 1 & 1 & 1 & 1 \\ 1 & i & -1 & -i \\ 1 & -1 & 1 & -1 \\ 1 & -i & -1 & i \end{bmatrix} = \begin{bmatrix} 1 & 0 & 0 & 0 \\ 0 & 1 & 0 & 0 \\ 0 & 0 & 1 & 0 \\ 0 & 0 & 0 & 1 \end{bmatrix}.$$

Consider postmultiplying the unitary matrix U of (2.100) by a diagonal matrix

$$(2.108)\qquad D = \begin{bmatrix} 1 & 0 & 0 & \ldots & 0 \\ 0 & W^{n-1} & 0 & \ldots & 0 \\ 0 & 0 & W^{n-2} & \ldots & 0 \\ \vdots & \vdots & \vdots & \ddots & \vdots \\ 0 & 0 & 0 & \ldots & W \end{bmatrix}.$$

Then it is easy to see that

$$(2.109)\qquad UD = KU,$$

where K is the circulant operator from (2.92). From this it follows that $K = UD\bar{U}$ and, more generally, that

$$(2.110)\qquad K^q = UD^q\bar{U}.$$

By similar means, it can be shown that $K' = U\bar{D}\bar{U}$, where

$$(2.111)\qquad \bar{D} = \mathrm{diag}\{1, W, W^2, \ldots, W^{n-1}\}$$

is the conjugate of D. The following conclusions can be reached in a straightforward manner:

(2.112) If $A = \alpha(K)$ is a circulant matrix then

$$\begin{aligned} &\text{(i)} && A = \alpha(K) = U\alpha(D)\bar{U}, \\ &\text{(ii)} && A' = \alpha(K') = U\alpha(\bar{D})\bar{U}, \\ &\text{(iii)} && A^{-1} = \alpha(K) = U\alpha^{-1}(D)\bar{U}. \end{aligned}$$

(2.113) If the elements of the circulant matrix A are real numbers, then the matrix is its own conjugate and

$$A = \bar{A} = \bar{U}\alpha(\bar{D})U.$$

Notice that the set of matrices $D^k; k = 0, \ldots, n-1$ forms a basis for an n-dimensional complex vector space comprising all diagonal matrices of order n. Therefore, provided that its coefficients can be complex numbers, the polynomial

$\alpha(D)$ in the expressions above stands for an arbitrary diagonal matrix with real or complex elements. If the coefficients of $\alpha(D)$ are constrained to be real, then the jth element of the diagonal matrix takes the form of

$$\delta_j = \sum_j \alpha_j W^{jt} = \sum_j \alpha_j \{ \cos(\omega_j t) - i \sin(\omega_j t) \}, \tag{2.114}$$

where $\omega_j = 2\pi j/n$. In that case, the sequence of complex numbers $\{\delta_j; j = 0, 1, \ldots, n-1\}$ consists of a real part which is an even or symmetric function of t and an imaginary part which is an odd or anti-symmetric function.

Example 2.7. Consider the equation

$$X = Ux(D)\bar{U} = \bar{U}x(\bar{D})U, \tag{2.115}$$

which defines the real-valued circulant matrix X. The second equality follows from the fact that matrix is its own conjugate. Observe that, if $e_0 = [1, 0, \ldots, 0]'$ is the leading vector of the identity matrix of order n, then

$$\begin{aligned} Xe_0 &= x = [x_0, x_1, \ldots, x_{n-1}]', \\ Ue_0 &= \bar{U}e_0 = i = [1, 1, \ldots, 1]', \\ x(\bar{D})i &= \xi = [\xi_0, \xi_1, \ldots, \xi_{n-1}]' \quad \text{and} \\ x(D)i &= \xi^* = [\xi_{n-1}, \xi_{n-2}, \ldots, \xi_0]'. \end{aligned} \tag{2.116}$$

Here the vector ξ is the discrete Fourier transform of the vector x. Its elements are the values $\{\xi_k = x(z_k); k = 0, \ldots, n-1\}$ which come from setting $z = z_k = \exp(-2\pi k/n)$ in $x(z)$ which is the z-transform of the sequence. $\{x_0, x_1, \ldots, x_{n-1}\}$. Premultiplying the equation $X = \bar{U}x(\bar{D})U$ from (2.115) by U and postmultiplying it by e_0 gives

$$Ux = \xi, \tag{2.117}$$

which represents the direct Fourier transform of the vector x. Postmultiplying the equation by e_0 gives

$$x = \bar{U}\xi; \tag{2.118}$$

and this represents the inverse Fourier transform by which x is recovered from ξ.

Example 2.8. Consider the multiplication of two circulant matrices

$$\begin{aligned} A &= \alpha(K) = U\alpha(D)\bar{U} \quad \text{and} \\ B &= \alpha(K) = U\beta(D)\bar{U}. \end{aligned} \tag{2.119}$$

Their product is

$$\begin{aligned} AB &= U\alpha(D)\bar{U}U\beta(D)\bar{U} \\ &= U\alpha(D)\beta(D)\bar{U}. \end{aligned} \tag{2.120}$$

On the LHS, there is a matrix multiplication which has already been interpreted in Example 2.5 as the circular convolution of the sequences $\{\alpha_0,\ldots,\alpha_{n-1}\}$ and $\{\beta_0,\ldots,\beta_{n-1}\}$ which are the coefficients of the polynomials of $\alpha(z)$ and $\beta(z)$. On the RHS there is a matrix multiplication $\alpha(D)\beta(D)$ which represents the pairwise multiplication of the corresponding nonzero elements of the diagonal matrices in question. These diagonal elements are the values $\{\alpha(z_k); k = 0,\ldots,n-1\}$ and $\{\beta(z_k); k = 0,\ldots,n-1\}$ of the discrete Fourier transforms of the sequences; and they come from setting $z = z_k = \exp(-2\pi k/n)$ in $\alpha(z)$ and $\beta(z)$. Thus it can be demonstrated that a convolution product in the time domain is equivalent to a modulation product in the frequency domain.

Bibliography

[17] Anderson, T.W., (1977), Estimation for Autoregressive Moving Average Models in Time and Frequency Domains, *Annals of Statistics*, **5**, 842–865.

[29] Argand, Jean Robert, (1806), *Essai sur une manière de répresenter des quantités imaginaires dans les constructions géométriques*, Paris. Published 1874, G.J. Hoüel, Paris. Reprinted (Nouveau tirage de la 2e édition) 1971, Blanchard, Paris.

[123] Cooke, R.C., (1955), *Infinite Matrices and Sequence Spaces*, Dover Publications, New York.

[137] Davis, P.J., (1979), *Circulant Matrices*, John Wiley and Sons, New York.

[267] Jain, A.K., (1978), Fast Inversion of Banded Toeplitz Matrices by Circular Decompositions, *IEEE Transactions on Acoustics, Speech and Signal Processing*, **ASSP-26**, 121–126.

[273] Jury, E.I., (1964), *Theory and Applications of the z-Transform Method*, John Wiley and Sons, New York.

CHAPTER 3

Rational Functions and Complex Analysis

The results in the algebra of polynomials which were presented in the previous chapter are not, on their own, sufficient for the analysis of time-series models. Certain results regarding rational functions of a complex variable are also amongst the basic requirements.

Rational functions may be expanded as Taylor series or, more generally, as Laurent series; and the conditions under which such series converge are a matter for complex analysis.

The first part of this chapter provides a reasonably complete treatment of the basic algebra of rational functions. Most of the results which are needed in time-series analysis are accessible without reference to a more sophisticated theory of complex analysis of the sort which pursues general results applicable to unspecified functions of the complex variable. However, some of the classic works in time-series analysis and signal processing do make extensive use of complex analysis; and they are liable to prove inaccessible unless one has studied the rudiments of the subject.

Our recourse is to present a section of complex analysis which is largely self-contained and which might be regarded as surplus to the basic requirements of time-series analysis. Nevertheless, it may contribute towards a deeper understanding of the mathematical foundations of the subject.

Rational Functions

In many respects, the algebra of polynomials is merely an elaboration of the algebra of numbers, or of arithmetic in other words. In dividing one number by another lesser number, there can be two outcomes. If the quotient is restricted to be an integer, then there is liable to be a remainder which is also an integer. If no such restriction is imposed, then the quotient may take the form of a interminable decimal. Likewise, with polynomials, there is a process of synthetic division which generates a remainder, and there is a process of rational expansion which is liable to generate an infinite power-series.

It is often helpful in polynomial algebra, as much as in arithmetic, to be aware of the presence of factors which are common to the numerator and the denominator.

Euclid's Algorithm

Euclid's method, which is familiar from arithmetic, can be used to discover whether two polynomials $\alpha(z) = \alpha_0+\alpha_1 z+\cdots+\alpha_p z^p$ and $\beta(z) = \beta_0+\beta_1 z+\cdots+\beta_k z^k$ possess a polynomial factor in common. Assume that the degree of $\alpha(z)$ is no less than that of $\beta(z)$ so that $\beta(z)$ divides $\alpha(z)$ with a remainder of $r(z)$. Then $\beta(z)$ can be divided by $r(z)$ giving a remainder of $r_1(z)$, and then $r(z)$ can be divided by $r_1(z)$ giving a remainder of $r_2(z)$, and so on. Thus the following sequence is generated:

$$(3.1)\qquad \begin{aligned} \alpha(z) &= q(z)\beta(z) + r(z),\\ \beta(z) &= q_1(z)r(z) + r_1(z),\\ r(z) &= q_2(z)r_1(z) + r_2(z),\\ &\vdots\\ r_{p-2}(z) &= q_p(z)r_{p-1}(z) + r_p(z),\\ r_{p-1}(z) &= q_{p+1}(z)r_p(z) + c. \end{aligned}$$

Now, the polynomials in the sequence $\{\alpha(z), \beta(z), r(z), \ldots\}$ are of decreasing degree, so it follows that the process must terminate. The final element $r_{p+1}(z) = c$ in the sequence either vanishes or has a degree of zero, in which case c is a nonzero constant.

Take the first case where $r_{p+1}(z) = c = 0$ and all of its predecessors in the sequence are nonzero. Then $r_p(z)$ is a factor of $r_{p-1}(z)$. It is also a factor of $r_{p-2}(z) = q_p(z)r_{p-1}(z) + r_p(z)$ and of all other elements of the sequence including $\alpha(z)$ and $\beta(z)$.

Conversely, any factor which is common to $\alpha(z)$ and $\beta(z)$ is a factor of $r(z) = \alpha(z) - q(z)\beta(z)$ and similarly of $r_1(z) = \beta(z) - q_1(z)r(z)$, and so on down to $r_p(z)$. Hence $r_p(z)$ includes every common factor of $\alpha(z)$ and $\beta(z)$, and it must therefore be the highest common factor of $\alpha(z)$ and $\beta(z)$.

Now consider the case where $r_{p+1}(z) = c$ is a nonzero constant. As before, it must include every factor common to $\alpha(z)$ and $\beta(z)$. But, since c contains no nonzero power of z, it can be said that $\alpha(z)$ and $\beta(z)$ are relatively prime to each other.

Next we shall prove that

(3.2) If $r_p(z)$ is the highest common factor of $\alpha(z)$ and $\beta(z)$, then there exist polynomials $f(z)$ and $g(z)$ such that

$$r_p(z) = f(z)\alpha(z) + g(z)\beta(z).$$

Proof. The result is implicit in Euclid's algorithm. On setting $r_{p+1}(z) = c = 0$, the sequence of equations in (3.1) can be rewritten as

$$(3.3)\qquad \begin{aligned} r(z) &= \alpha(z) - q(z)\beta(z),\\ r_1(z) &= \beta(z) - q_1(z)r(z),\\ &\vdots\\ r_{p-1}(z) &= r_{p-3}(z) - q_{p-1}(z)r_{p-2}(z),\\ r_p(z) &= r_{p-2}(z) - q_p(z)r_{p-1}(z). \end{aligned}$$

Putting the expression for $r(z)$ into the expression for $r_1(z)$ gives $r_1(z) = f_1(z)\alpha(z) + g_1(z)\beta(z)$ with $f_1(z) = -q_1(z)$ and $g_1(z) = 1 + q(z)q_1(z)$. Putting this expression into $r_2(z)$ gives $r_2(z) = f_2(z)\alpha(z) + g_2(z)\beta(z)$. One can continue the substitutions until the expression $r_p(z) = f_p(z)\alpha(z) + g_p(z)\beta(z)$ is obtained. Then the equation under (3.2) comes from suppressing the subscripts on $f_p(z)$ and $g_p(z)$.

Other polynomials, which can play the roles of $f(z)$ and $g(z)$ in equation (3.2), can be derived from those generated by Euclid's algorithm. Thus, if we set

$$\begin{aligned} F(z) &= f(z) - p(z)\beta(z), \\ G(z) &= g(z) + p(z)\alpha(z), \end{aligned}$$

where $p(z)$ is an arbitrary polynomial, then we get

$$\begin{aligned} F(z)\alpha(z) + G(z)\beta(z) &= f(z)\alpha(z) + g(z)\beta(z) \\ &= r_p(z). \end{aligned}$$

The next result, which builds upon the previous one, is fundamental to the theory of partial fractions:

(3.4) If $\alpha(z)$ and $\beta(z)$ are relatively prime, then there exist unique polynomials $g(z)$ and $f(z)$ of degrees less than $\alpha(z)$ and $\beta(z)$ respectively such that

$$c = f(z)\alpha(z) + g(z)\beta(z).$$

Proof. If $\alpha(z)$ and $\beta(z)$ are relatively prime, then, by virtue of Euclid's algorithm, there exists a pair of polynomials $G(z)$ and $F(z)$ for which

$$c = F(z)\alpha(z) + G(z)\beta(z). \tag{3.5}$$

The object is to replace $F(z)$ and $G(z)$ by polynomials $f(z)$ and $g(z)$ whose degrees are less that those of $\beta(z)$ and $\alpha(z)$ respectively.

Dividing, by $\beta(z)$ and $\alpha(z)$ gives

$$\begin{aligned} F(z) &= q_\beta(z)\beta(z) + r_\beta(z), \\ G(z) &= q_\alpha(z)\alpha(z) + r_\alpha(z), \end{aligned} \tag{3.6}$$

where either or both of the quotients $q_\beta(z)$, $q_\alpha(z)$ may be zero. On multiplying these equations by $\alpha(z)$ and $\beta(z)$ respectively and rearranging them, we get

$$\begin{aligned} F(z)\alpha(z) - r_\beta(z)\alpha(z) &= q_\beta(z)\alpha(z)\beta(z), \\ G(z)\beta(z) - r_\alpha(z)\beta(z) &= q_\alpha(z)\alpha(z)\beta(z). \end{aligned} \tag{3.7}$$

Adding these and using (3.5) gives

$$c - r_\beta(z)\alpha(z) - r_\alpha(z)\beta(z) = \{q_\beta(z) + q_\alpha(z)\}\alpha(z)\beta(z). \tag{3.8}$$

But now, unless $q_\beta(z)+q_\alpha(z)=0$, the degree of the RHS cannot be less than that of $\alpha(z)\beta(z)$, whilst the degree of the LHS must be less than that of $\alpha(z)\beta(z)$—since the degree of $r_\beta(z)$ is less than that of $\beta(z)$ and the degree of $r_\alpha(z)$ is less than that of $\alpha(z)$. Therefore, the factor in question must be zero, and so we have

$$c = r_\beta(z)\alpha(z) + r_\alpha(z)\beta(z). \tag{3.9}$$

Now it has to be demonstrated that $r_\beta(z)=f(z)$ and $r_\alpha(z)=g(z)$ are unique. Imagine that $c=f(z)\alpha(z)+g(z)\beta(z)$, where $f(z)\neq r_\beta(z)$ and $g(z)\neq r_\alpha(z)$ have degrees less than and $\beta(z)$ and $\alpha(z)$ respectively. By subtracting, we should get $\{r_\beta(z)-f(z)\}\alpha(z)=\{g(z)-r_\alpha(z)\}\beta(z)$ or $\alpha(z)/\beta(z)=\{g(z)-r_\alpha(z)\}/\{r_\beta(z)-f(z)\}$. But then the RHS is expressed in polynomials of lower degree than those on the LHS; and, since $\alpha(z),\beta(z)$ are relatively prime, this is not possible. The problem is averted only if $r_\beta(z)=f(z)$ and $g(z)=r_\beta(z)$.

(3.10) If $\gamma_m(z)$ and $\gamma_n(z)$ are relatively prime polynomials of degrees m and n respectively, and if $\delta(z)$ is a polynomial of degree less than $m+n$, then there is a unique representation

$$\delta(z) = f(z)\gamma_m(z) + g(z)\gamma_n(z).$$

where $g(z)$ and $f(z)$ are polynomials of degrees less than m and n respectively.

Proof. We begin by asserting that there always exists a representation in the form of

$$\delta(z) = F(z)\gamma_m(z) + G(z)\gamma_n(z); \tag{3.11}$$

for, according to (3.4), we can always set $c=f(z)\gamma_m(z)+g(z)\gamma_n(z)$, whence, by multiplying throughout by $\delta(z)/c$ and defining $F(z)=f(z)\delta(z)/c$ and $G(z)=g(z)\delta(z)/c$, we get the desired result.

Next, we set about finding replacements for $F(z)$ and $G(z)$ which are of the requisite degrees. We proceed as in the proof of (3.4). Dividing, by $\gamma_n(z)$ and $\gamma_m(z)$ gives

$$\begin{aligned} F(z) &= q_F(z)\gamma_n(z) + r_F(z),\\ G(z) &= q_G(z)\gamma_m(z) + r_G(z). \end{aligned} \tag{3.12}$$

On multiplying these equations by $\gamma_m(z)$ and $\gamma_n(z)$ respectively and rearranging them, we get

$$\begin{aligned} F(z)\gamma_m(z) - r_F(z)\gamma_m(z) &= q_F(z)\gamma_m(z)\gamma_n(z),\\ G(z)\gamma_n(z) - r_G(z)\gamma_n(z) &= q_G(z)\gamma_m(z)\gamma_n(z). \end{aligned} \tag{3.13}$$

Adding these and using (3.11) gives

$$\delta(z) - r_F(z)\gamma_m(z) - r_G(z)\gamma_n(z) = \{q_F(z)+q_G(z)\}\gamma_m(z)\gamma_n(z). \tag{3.14}$$

But now, unless $q_F(z)+q_G(z)=0$, the degree of the RHS cannot be less than that of $\gamma_m(z)\gamma_n(z)$, whilst the degree of the LHS must be less than that of $\gamma_m(z)\gamma_n(z)$— since the degree of $r_F(z)$ is less than that of $\gamma_n(z)$ and the degree of $r_G(z)$ is less than that of $\gamma_m(z)$. Therefore, the factor in question must be zero, and so we have

$$\delta(z)=r_F(z)\gamma_m(z)+r_G(z)\gamma_n(z). \tag{3.15}$$

Finally, it can be demonstrated, as in the proof of (3.4), that $r_F(z)=f(z)$ and $r_G(z)=g(z)$ are unique.

Partial Fractions

The ratio $\rho(z)=\delta(z)/\gamma(z)$ of two polynomials $\delta(z)=\delta_0+\delta_1 z+\cdots+\delta_n z^n$ and $\gamma(z)=\gamma_0+\gamma_1 z+\cdots+\gamma_m z^m$ is described as a rational function. The degree of a rational function is defined as the degree of the numerator less the degree of the denominator. If the degree is negative, then $\rho(z)$ is a proper rational function, otherwise $\rho(z)$ is improper.

(3.16) If $\delta(z)/\gamma(z)=\delta(z)/\{\gamma_1(z)\gamma_2(z)\}$ is a proper rational function and if $\gamma_1(z)$ and $\gamma_2(z)$ have no common factors, then it is expressed uniquely as

$$\frac{\delta(z)}{\gamma(z)}=\frac{\delta_1(z)}{\gamma_1(z)}+\frac{\delta_2(z)}{\gamma_2(z)},$$

where $\delta_1(z)/\gamma_1(z)$ and $\delta_2(z)/\gamma_2(z)$ are proper rational functions.

Proof. According to (3.10), there is a unique expression in the form of $\delta(z)=\delta_1(z)\gamma_2(z)+\delta_2(z)\gamma_1(z)$ where the degrees of $\delta_1(z)$ and $\delta_2(z)$ are less than those of $\gamma_1(z)$ and $\gamma_2(z)$ respectively. Hence

$$\begin{aligned}\frac{\delta(z)}{\gamma(z)}&=\frac{\delta_1(z)\gamma_2(z)+\delta_2(z)\gamma_1(z)}{\gamma_1(z)\gamma_2(z)}\\&=\frac{\delta_1(z)}{\gamma_1(z)}+\frac{\delta_2(z)}{\gamma_2(z)},\end{aligned} \tag{3.17}$$

is a uniquely defined sum of proper fractions.

If $\gamma_1(z)$ or $\gamma_2(z)$ has polynomial factors, then the process which is represented by (3.16) can be repeated until all of the distinct factors of $\gamma(z)$ have been segregated. Thus the rational function $\delta(z)/\gamma(z)$ can be expressed as a sum of partial fractions in the form of

$$\frac{\delta(z)}{\gamma(z)}=\sum_j\frac{\delta_j(z)}{\gamma_j(z)}, \tag{3.18}$$

where no two denominators have any factors in common. These denominators will assume the generic forms of $(z^2+\rho z+\sigma)^r=\{(z-\lambda)(z-\lambda^*)\}^r$ and $(z-\lambda)^s$.

The simplest case is when $\gamma(z) = \prod(z - \lambda_i)$, where all the factors $(z - \lambda_i)$ are real and distinct. Then

$$\frac{\delta(z)}{\gamma(z)} = \frac{\kappa_1}{z - \lambda_1} + \frac{\kappa_2}{z - \lambda_2} + \cdots + \frac{\kappa_m}{z - \lambda_m}. \tag{3.19}$$

To evaluate the constant κ_1, for example, we may multiply by $z - \lambda_1$ throughout the equation to get

$$\frac{\delta(z)(z - \lambda_1)}{\gamma(z)} = \kappa_1 + \frac{\kappa_2(z - \lambda_1)}{z - \lambda_2} + \cdots + \frac{\kappa_m(z - \lambda_1)}{z - \lambda_m}. \tag{3.20}$$

On the left, there is $\delta(z)/\{(z - \lambda_2)(z - \lambda_3)\ldots(z - \lambda_m)\}$; and, from the right side, it appears that, when $z = \lambda_1$, this gives the value of κ_1.

Example 3.1. Consider the equation

$$\begin{aligned} \frac{3z}{1 + z - 2z^2} &= \frac{3z}{(1 - z)(1 + 2z)} \\ &= \frac{\kappa_1}{1 - z} + \frac{\kappa_2}{1 + 2z} \\ &= \frac{\kappa_1(1 + 2z) + \kappa_2(1 - z)}{(1 - z)(1 + 2z)}. \end{aligned} \tag{3.21}$$

Equating the terms of the numerator gives

$$3z = (2\kappa_1 - \kappa_2)z + (\kappa_1 + \kappa_2), \tag{3.22}$$

so that $\kappa_2 = -\kappa_1$, which gives $3 = (2\kappa_1 - \kappa_2) = 3\kappa_1$; and thus it is found that $\kappa_1 = 1$, $\kappa_2 = -1$. We have pursued a more laborious method of finding κ_1, κ_2 in this example that the method which has been suggested above. However, the method can be relied upon in all cases.

The case where $\gamma(z) = \gamma_1(z)\gamma_2(z)$ contains a factor $\gamma_2(z) = z^2 + \rho z + \sigma = (z - \lambda)(z - \lambda^*)$ is marginally more complicated than the case where the linear factors are real. Now there is a partial fraction in the form of $(az + b)/\{(z - \lambda)(z - \lambda^*)\}$. One should multiply throughout by $(z - \lambda)(z - \lambda^*)$ and proceed to set $z = \lambda$ to get

$$\frac{\delta(\lambda)}{\gamma_1(\lambda)} = a\lambda + b. \tag{3.23}$$

This is a complex equation which can be separated into its real and imaginary parts. The two parts constitute a pair of simultaneous equations which may be solved for a and b. Of course, the same pair of simultaneous equations will be obtained from (3.23) when λ is replaced by λ^*.

A partial fraction with a quadratic denominator can be decomposed into a pair of fractions with complex numbers in their denominators:

$$\frac{az + b}{(z - \lambda)(z - \lambda^*)} = \frac{\kappa}{z - \lambda} + \frac{\kappa^*}{z - \lambda^*}. \tag{3.24}$$

Then it is found that $\kappa = (a\lambda + b)/(\lambda - \lambda^*)$ and $\kappa^* = (a\lambda^* + b)/(\lambda^* - \lambda)$. These are conjugate complex numbers.

Example 3.2. Consider the equation

$$\frac{z^2}{(z+1)(z^2+4z+5)} = \frac{\kappa}{z+1} + \frac{az+b}{z^2+4z+5}, \tag{3.25}$$

wherein $z^2 + 4z + 5 = (z - \lambda)(z - \lambda^*)$ with $\lambda = -2 - i$. Multiplying throughout by $(z - \lambda)(z - \lambda^*)$ and setting $z = \lambda$ gives the complex equation $\lambda^2/(\lambda + 1) = a\lambda + b$ which can be written as $0 = (a - 1)\lambda^2 + (b + a)\lambda + b$. This amounts to two real equations. The first, which comes from the real terms, is $a - b - 3 = 0$. The second, which comes from the imaginary terms, is $3a - b - 4 = 0$. The solutions to the equations are $a = 1/2$ and $b = -5/2$. Also, $\kappa = 1/2$.

Now consider the case where one of the partial fractions is $\delta_1(z)/(z - \lambda)^s$. We can write $\delta_1(z) = \pi_0 + \pi_1(z - \lambda) + \cdots + \pi_{s-1}(z - \lambda)^{s-1}$. The coefficients of this polynomial correspond to the values generated in the process of synthetic division which is described in Chapter 4. Thus π_0 is the remainder term in $\delta_1(z) = \beta_1(z)(z - \lambda) + \pi_0$ and π_1 is the remainder term in $\beta_1(z) = \beta_2(z)(z - \lambda) + \pi_1$ and so on. Using this form of $\delta_1(z)$, the rational function can be written as

$$\frac{\delta(z)}{\gamma(z)} = \frac{\pi_0}{(z-\lambda)^s} + \frac{\pi_1}{(z-\lambda)^{s-1}} + \cdots + \frac{\pi_{s-1}}{(z-\lambda)} + \frac{\delta_2(z)}{\gamma_2(z)}. \tag{3.26}$$

Then, multiplying this by $\gamma_1(z) = (z - \lambda)^s$ gives

$$\frac{\delta(z)}{\gamma_2(z)} = \pi_0 + \pi_1(z-\lambda) + \cdots + \pi_{s-1}(z-\lambda)^{s-1} + \gamma_1(z)\frac{\delta_2(z)}{\gamma_2(z)}; \tag{3.27}$$

and setting $z = \lambda$ isolates the value of π_0. Next, if we differentiate with respect to z, we get

$$\begin{aligned} \frac{d}{dz}\left[\frac{\delta(z)}{\gamma_2(z)}\right] = \pi_1 &+ 2\pi_2(z-\lambda) + \cdots + (s-1)\pi_{s-1}(z-\lambda)^{s-2} \\ &+ \frac{d}{dz}\left[\gamma_1(z)\frac{\delta_2(z)}{\gamma_2(z)}\right]; \end{aligned} \tag{3.28}$$

and setting $z = \lambda$ isolates π_1. We can continue in this way until all of the coefficients have been found.

Finally, there might be a repeated quadratic factor in $\gamma(z)$. The corresponding partial fraction would be $\delta_1(z)/(z^2 + \rho z + \sigma)^r$. Since the denominator is of degree $2r$, the degree of $\delta_1(z)$ may be $2r - 1$ or less. By dividing $\delta_1(z)$ by $\theta(z) = z^2 + \rho z + \sigma$ in a process of synthetic division, we get remainders $\pi_0(z), \pi_1(z), \ldots, \pi_{r-1}(z)$ which are either linear functions of z or constants. With $\delta_1(z) = \pi_0(z) + \pi_1(z)\theta(z) + \cdots + \pi_{r-1}(z)\theta(z)^{r-1}$, the partial fraction can be written as

$$\frac{\delta_1(z)}{\theta(z)^r} = \frac{\pi_0(z)}{\theta(z)^r} + \frac{\pi_1(z)}{\theta(z)^{r-1}} + \cdots + \frac{\pi_{r-1}(z)}{\theta(z)}. \tag{3.29}$$

It is also possible to decompose the quadratic factor into two conjugate terms involving complex roots and complex numerators. Then the terms can be written in the form of (3.26).

The Expansion of a Rational Function

In time-series analysis, models are often encountered which contain transfer functions in the form of $y(t) = \{\delta(L)/\gamma(L)\}x(t)$. For this to have a meaningful interpretation, it is normally required that the rational operator $\delta(L)/\gamma(L)$ should obey the BIBO stability condition; which is to say that $y(t)$ should be a bounded sequence whenever $x(t)$ is bounded.

The necessary and sufficient condition for the boundedness of $y(t)$ is that the series expansion $\{\omega_0+\omega_1 z+\cdots\}$ of $\omega(z) = \delta(z)/\gamma(z)$ should be convergent whenever $|z| \leq 1$. We can determine whether or not the series will converge by expressing the ratio $\delta(z)/\gamma(z)$ as a sum of partial fractions.

Imagine that $\gamma(z) = \gamma_m \prod(z - \lambda_i) = \gamma_0 \prod(1 - z/\lambda_i)$ where the roots may be complex. Then, assuming that there are no repeated roots, and taking $\gamma_0 = 1$, the ratio can be written as

$$\text{(3.30)} \qquad \frac{\delta(z)}{\gamma(z)} = \frac{\kappa_1}{1 - z/\lambda_1} + \frac{\kappa_2}{1 - z/\lambda_2} + \cdots + \frac{\kappa_m}{1 - z/\lambda_m}.$$

Since any scalar factor of $\gamma(L)$ may be absorbed in the numerator $\delta(L)$, setting $\gamma_0 = 1$ entails no loss of generality.

If the roots of $\gamma(z) = 0$ are real and distinct, then the conditions for the convergence of the expansion of $\delta(z)/\gamma(z)$ are straightforward. For the rational function converges if and only if the expansion of each of its partial fractions in terms of ascending powers of z converges. For the expansion

$$\text{(3.31)} \qquad \frac{\kappa}{1 - z/\lambda} = \kappa\{1 + z/\lambda + (z/\lambda)^2 + \cdots\}$$

to converge for all $|z| \leq 1$, it is necessary and sufficient that $|\lambda| > 1$.

In the case where a real root occurs with a multiplicity of n, as in the expression under (3.26), a binomial expansion is available:

$$\text{(3.32)} \qquad \frac{1}{(1 - z/\lambda)^n} = 1 - n\frac{z}{\lambda} + \frac{n(n-1)}{2!}\left(\frac{z}{\lambda}\right)^2 - \frac{n(n-1)(n-2)}{3!}\left(\frac{z}{\lambda}\right)^3 + \cdots.$$

Once more, it is evident that $|\lambda| > 1$ is the necessary and sufficient condition for convergence when $|z| \leq 1$.

The expansion under (3.31) applies to complex roots as well as to real roots. To investigate the conditions of convergence in the case of complex roots, it is appropriate to combine the products of the expansion of a pair of conjugate factors. Therefore, consider following expansion:

$$\text{(3.33)} \qquad \begin{aligned} \frac{c}{1 - z/\lambda} + \frac{c^*}{1 - z/\lambda^*} &= c\{1 + z/\lambda + (z/\lambda)^2 + \cdots\} \\ &\quad + c^*\{1 + z/\lambda^* + (z/\lambda^*)^2 + \cdots\} \\ &= \sum_{t=0}^{\infty} z^t(c\lambda^{-t} + \lambda^{*-t}). \end{aligned}$$

The various complex quantities can be represented in terms of exponentials:

$$\begin{aligned} \lambda &= \kappa^{-1}e^{-i\omega}, & \lambda^* &= \kappa^{-1}e^{i\omega}, \\ c &= \rho e^{-i\theta}, & c^* &= \rho e^{i\theta}. \end{aligned} \tag{3.34}$$

Then the generic term in the expansion becomes

$$\begin{aligned} z^t(c\lambda^{-t} + c^*\lambda^{*-t}) &= z^t\{\rho e^{-i\theta}\kappa^t e^{i\omega t} + \rho e^{i\theta}\kappa^t e^{-i\omega t}\} \\ &= z^t\rho\kappa^t\{e^{i(\omega t-\theta)} + e^{-i(\omega t-\theta)}\} \\ &= z^t 2\rho\kappa^t \cos(\omega t - \theta). \end{aligned} \tag{3.35}$$

The expansion converges for all $|z| \leq 1$ if and only if $|\kappa| < 1$. But $|\kappa| = |\lambda^{-1}| = |\lambda|^{-1}$; so it is confirmed that the necessary and sufficient condition for convergence is that $|\lambda| > 1$.

The case of repeated complex roots can also be analysed to reach a similar conclusion. Thus a general assertion regarding the expansions of rational function can be made:

(3.36) The expansion $\omega(z) = \{\omega_0 + \omega_1 z + \omega_2 z^2 + \cdots\}$ of the rational function $\delta(z)/\gamma(z)$ converges for all $|z| \leq 1$ if and only if every root λ of $\gamma(z) = 0$ lies outside the unit circle such that $|\lambda| > 1$.

So far, the condition has been imposed that $|z| \leq 1$. The expansion of a rational function may converge under conditions which are either more or less stringent in the restrictions which they impose on $|z|$. If fact, for any series $\omega(z) = \{\omega_0 + \omega_1 z + \omega_2 z^2 + \cdots\}$, there exists a real number $r \geq 0$, called the radius of convergence, such that, if $|z| < r$, then the series converges absolutely with $\sum |\omega_i| < \infty$, whereas, if $|z| > r$, the series diverges.

In the case of the rational function $\delta(z)/\gamma(z)$, the condition for the convergence of the expansion is that $|z| < r = \min\{|\lambda_1|, \ldots, |\lambda_m|\}$, where the λ_i are the roots of $\gamma(z) = 0$.

The roots of the numerator polynomial $\delta(z)$ of a rational function are commonly described as the zeros of the function, whilst the roots of the denominator function polynomial $\gamma(z)$ are described as the poles.

In electrical engineering, the z-transform of a sequence defined on the positive integers is usually expressed in terms of negative powers of z. This leads to an inversion of the results given above. In particular, the condition for the convergence of the expansion of the function $\delta(z^{-1})/\gamma(z^{-1})$ is that $|z| > r = \max\{|\mu_1|, \ldots, |\mu_m|\}$, where $\mu_i = 1/\lambda_i$ is a root of $\gamma(z^{-1}) = 0$.

Example 3.3. It is often helpful to display a transfer function graphically by means of a pole–zero plot in the complex plane; and, for this purpose, there is an advantage in the form $\delta(z^{-1})/\gamma(z^{-1})$ which is in terms of negative powers of z (see Figure 3.1). Thus, if the function satisfies the BIBO stability condition, then the poles of $\delta(z^{-1})/\gamma(z^{-1})$ will be found within the unit circle. The numerator may also be subject to conditions which will place the zeros within the unit circle. On

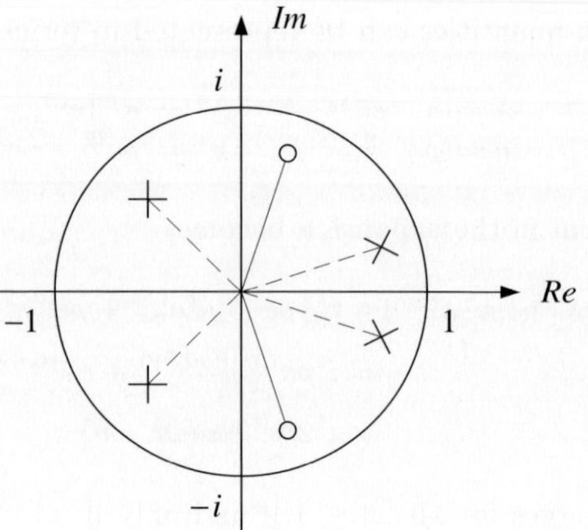

Figure 3.1. The pole–zero diagram of the stable transfer function.

$$\frac{\delta(z^{-1})}{\gamma(z^{-1})} = \frac{\{1-(0.25 \pm i0.75)z^{-1}\}}{\{1-(0.75 \pm i0.25)z^{-1}\}\{1+(0.5 \pm i0.5)z^{-1}\}}.$$

The poles are marked with crosses and the zeros with circles.

the other hand, the poles of $\delta(z)/\gamma(z)$ will fall outside the unit circle; and they may be located at a considerable distance from the origin, which would make a diagram inconvenient.

Because the pole–zero diagram can be of great assistance in analysing a transfer function, we shall adopt the negative-power z-transform whenever it is convenient to do so.

Recurrence Relationships

We have seen that, if z is not a root of $\alpha(z)$, then $\beta(z)/\alpha(z)$ may be expanded in powers of z:

$$\text{(3.37)} \qquad \frac{\beta(z)}{\alpha(z)} = \omega(z) = \{\omega_0 + \omega_1 z + \omega_2 z^2 + \cdots\}.$$

When $j \geq p$ and $j > k$, where p and k are the degrees of $\alpha(z)$ and $\beta(z)$ respectively, the sequences of coefficients $\{\omega_j, \omega_{j-1}, \ldots, \omega_{j-p},\}$ obey a recurrence relationship such that

$$\text{(3.38)} \qquad \alpha_0\omega_j + \alpha_1\omega_{j-1} + \cdots + \alpha_p\omega_{j-p} = 0.$$

Given the p consecutive values $\omega_{j-1}, \ldots, \omega_{j-p}$, the relationship can be used to generate the ensuing value ω_j. Thus

$$\text{(3.39)} \qquad \omega_j = -\frac{1}{\alpha_0}\{\alpha_1\omega_{j-1} + \cdots + \alpha_p\omega_{j-p}\}.$$

This feature can be used in deriving an effective algorithm for the expansion of the rational function.

A set of p instances of the relationship of (3.38) can also be used to infer the values of the denominator coefficients $\alpha_1, \ldots, \alpha_p$ on the assumption that $\alpha_0 = 1$; for then the relationships constitute a simple system of p linear equations in p unknowns. Once the denominator coefficients have been found, it is straightforward to find the values of the numerator coefficients $\beta_0, \ldots, \beta_k$ from $\omega_0, \ldots, \omega_k$.

In order to derive the algorithm for expanding the rational function, the equation $\beta(z)/\alpha(z) = \omega(z)$ may be written in the form of $\alpha(z)\omega(z) = \beta(z)$. Then the following expressions are obtained by equating the coefficients associated with the same powers of z on both sides:

(3.40)
$$\begin{aligned} \beta_j &= \sum_{i=0}^{r} \alpha_i \omega_{j-i}; \qquad 0 \leq j \leq k, \\ 0 &= \sum_{i=0}^{r} \alpha_i \omega_{j-i}; \qquad j > k, \end{aligned}$$

where $r = \min(p, j)$. The latter are rearranged to provide the equations for determining the coefficients of the expansion:

(3.41)
$$\begin{aligned} \omega_j &= (\beta_j - \sum_{i=1}^{r} \alpha_i \omega_{j-i})/\alpha_0; \qquad 0 \leq j \leq k, \\ \omega_j &= -\sum_{i=1}^{r} \alpha_i \omega_{j-i}/\alpha_0; \qquad j > k. \end{aligned}$$

Example 3.4. When $\alpha(z) = \alpha_0 + \alpha_1 z + \alpha_2 z^2 + \alpha_3 z^3$ and $\beta(z) = \beta_0 + \beta_1 z$, the following system arises:

(3.42)
$$\alpha_0 \begin{bmatrix} \omega_0 \\ \omega_1 \\ \omega_2 \\ \omega_3 \\ \omega_4 \end{bmatrix} = \begin{bmatrix} \beta_0 \\ \beta_1 \\ 0 \\ 0 \\ 0 \end{bmatrix} - \begin{bmatrix} 0 & 0 & 0 \\ \omega_0 & 0 & 0 \\ \omega_1 & \omega_0 & 0 \\ \omega_2 & \omega_1 & \omega_0 \\ \omega_3 & \omega_2 & \omega_1 \end{bmatrix} \begin{bmatrix} \alpha_1 \\ \alpha_2 \\ \alpha_3 \\ 0 \\ 0 \end{bmatrix}.$$

The algorithm for generating the coefficients of the expansion of $\beta(z)/\alpha(z)$ is implemented in the procedure which follows. The coefficients of the expansion are written in place of the coefficients of $\beta(z)$ in the array *beta*:

(3.43)
```
procedure RationalExpansion(alpha : vector;
                            p, k, n : integer;
                            var beta : vector);

var
  i, j, r : integer;
```

```
begin
  for j := 0 to n do
    begin {j}
      r := Min(p, j);
      if j > k then
        beta[j] := 0.0;
      for i := 1 to r do
        beta[j] := beta[j] - alpha[i] * beta[j - i];
      beta[j] := beta[j]/alpha[0]
    end; {j}
end; {RationalExpansion}
```

The algorithm which is used to recover the coefficients of $\alpha(z)$ and $\beta(z)$ from those of the expansion $\omega(z)$ is also derived from the equations under (3.40). Thus, by setting $\alpha_0 = 1$, which entails no loss of generality, and by letting $j = k+1, \ldots, k+p$ in the second equation of (3.40), a linear system is derived which can be solved for $\alpha_1, \ldots, \alpha_p$. Then, by substituting these values into the first equation of (3.40) and letting $j = 0, \ldots, k$, the values of $\beta_0, \ldots, \beta_k$ are obtained.

Example 3.5. When $\alpha(z) = \alpha_0 + \alpha_1 z + \alpha_2 z^2 + \alpha_3 z^3$ and $\beta(z) = \beta_0 + \beta_1 z + \beta_2 z^2$, we get the following system:

$$\begin{bmatrix} \omega_0 & 0 & 0 & 0 \\ \omega_1 & \omega_0 & 0 & 0 \\ \omega_2 & \omega_1 & \omega_0 & 0 \\ \omega_3 & \omega_2 & \omega_1 & \omega_0 \\ \omega_4 & \omega_3 & \omega_2 & \omega_1 \\ \omega_5 & \omega_4 & \omega_3 & \omega_2 \end{bmatrix} \begin{bmatrix} \alpha_0 \\ \alpha_1 \\ \alpha_2 \\ \alpha_3 \end{bmatrix} = \begin{bmatrix} \beta_0 \\ \beta_1 \\ \beta_2 \\ 0 \\ 0 \\ 0 \end{bmatrix}. \tag{3.44}$$

Given $\alpha_0 = 1$, we can take the last three equations in the form of

$$\begin{bmatrix} \omega_2 & \omega_1 & \omega_0 \\ \omega_3 & \omega_2 & \omega_1 \\ \omega_4 & \omega_3 & \omega_2 \end{bmatrix} \begin{bmatrix} \alpha_1 \\ \alpha_2 \\ \alpha_3 \end{bmatrix} = - \begin{bmatrix} \omega_3 \\ \omega_4 \\ \omega_5 \end{bmatrix} \tag{3.45}$$

and solve them for $\alpha_1, \alpha_2, \alpha_3$. Then we can use these values in the first three equations for finding the values of $\beta_0, \beta_1, \beta_2$.

The Pascal procedure which implements the method which we have just described invokes the procedure *LUSolve* of (7.28) for finding the solution of a set of linear equations.

(3.46)
```
procedure RationalInference(omega : vector;
                            p, k : integer;
                            var beta, alpha : vector);

var
  i, j, r : integer;
```

```
    v : vector;
    w : matrix;

  begin {RationalInference}

  {Solve for the alpha coefficients}
    for i := 1 to p do
      begin {i}
        v[i] := -omega[k + i];
        for j := 1 to p do
          begin {j}
            if k + i - j < 0 then
              w[i, j] := 0.0
            else
              w[i, j] := omega[k + i - j];
          end; {j}
      end; {i}
    LUsolve(1, p, w, alpha, v);
    alpha[0] := 1.0;

  {Find the beta coefficients}
    for j := 0 to k do
      begin {j}
        r := Min(p, j);
        beta[j] := 0.0;
        for i := 0 to r do
          beta[j] := beta[j] + alpha[i] * omega[j - i];
      end; {j}

  end; {RationalInference}
```

Laurent Series

For some purposes, it is necessary to consider a two-sided or bilateral z-transform of a sequence which extends over positive and negative integers. An example is provided by the generating function of the cross-covariances of the moving-average processes $y(t) = \mu(L)\varepsilon(t)$ and $q(t) = \theta(L)\varepsilon(t)$, where $\varepsilon(t)$ is a white-noise process with $V\{\varepsilon(t)\} = \sigma^2$. The cross-covariance of $y(t-\tau)$ and $q(t)$ is the coefficient associated with z^τ in the expansion of $\sigma^2\mu(z^{-1})\theta(z)$.

The product of $\mu(z^{-1})$ and $\theta(z)$ is given by

$$\begin{aligned}&\left(\mu_0 + \frac{\mu_1}{z} + \cdots + \frac{\mu_q}{z^q}\right)\left(\theta_0 + \theta_1 z + \cdots + \theta^h z^h\right)\\ &= \frac{\theta_0\mu_q}{z^q} + \frac{1}{z^{q-1}}(\theta_0\mu_{q-1} + \theta_1\mu_q) + \cdots + \frac{1}{z}(\theta_0\mu_1 + \theta_1\mu_2 + \cdots)\\ &\qquad + (\theta_0\mu_0 + \theta_1\mu_1 + \cdots)\\ &+ z(\mu_0\theta_1 + \mu_1\theta_2\mu_1 + \cdots) + \cdots + z^{h-1}(\mu_0\theta_{h-1} + \mu_1\theta_h) + z^h\mu_0\theta_h.\end{aligned} \tag{3.47}$$

For greater generality, the case should be considered where the initial index of $\mu(z^{-1})$ is p whilst the initial index of $\theta(z)$ is g. These initial indices may be positive or negative, individually. Then the product is given by

$$\left(\sum_{i=p}^{q}\mu_i z^{-i}\right)\left(\sum_{j=g}^{h}\theta_j z^{j}\right) = \sum_{k=g-q}^{h-p}\left(\sum_{i=m}^{n}\mu_i\theta_{i+k}\right)z^k = \sum_{k=g-q}^{h-p}\omega_k z^k, \tag{3.48}$$
$$\text{where}\quad m = \max(p, g-k) \quad\text{and}\quad n = \min(q, h-k).$$

The limits on the index i are obtained by combining the restrictions $p \leq i \leq q$ and $g \leq j = i + k \leq h$ or, equivalently, $g - k \leq i \leq h - k$; and the limits on $k = j - i$ are obtained by combining the restrictions $g \leq j \leq h$ and $-q \leq -i \leq -p$.

The Pascal procedure for forming the coefficients of the product is as follows:

(3.49)
```
procedure BiConvolution(var omega, theta, mu : vector;
                            p, q, g, h : integer);

  var
    i, k, m, n : integer;

begin
  for k := g - q to h do
    begin {k}
      m := Max(p, g - k);
      n := Min(q, h - k);
      omega[k] := 0.0;
      for i := m to n do
        omega[k] := omega[k] + mu[i] * theta[k + i];
    end; {k}
end; {BiConvolution}
```

A more complicated circumstance arises when it is required to form the cross-covariance generating function of the autoregressive moving-average processes $y(t)$ and $q(t)$ defined by $\alpha(L)y(t) = \mu(L)\varepsilon(t)$ and $\phi(L)q(t) = \theta(L)\varepsilon(t)$ respectively. Let

$$\begin{aligned}
\alpha(z^{-1}) &= 1 + \frac{\alpha_1}{z} + \cdots + \frac{\alpha_p}{z^p} = \prod_{i=1}^{p}(1 - \lambda_i z^{-1}),\\
\phi(z) &= 1 + \phi_1 z + \cdots + \phi_f z^f = \prod_{i=1}^{f}(1 - \kappa_i z),\\
\mu(z^{-1}) &= 1 + \frac{\mu_1}{z} + \cdots + \frac{\mu_q}{z^q} = \prod_{i=1}^{q}(1 - \rho_i z^{-1}),\\
\theta(z) &= 1 + \theta_1 z + \cdots + \theta_h z^h = \prod_{i=1}^{h}(1 - \nu_i z).
\end{aligned} \tag{3.50}$$

Then, on the assumption that $V\{\varepsilon(t)\} = 1$, the generating function for the cross-covariances of $y(t)$ and $q(t)$ is

$$\gamma(z) = \frac{\mu(z^{-1})\theta(z)}{\alpha(z^{-1})\phi(z)} = \frac{\omega(z)}{\psi(z)}. \tag{3.51}$$

The appropriate expansion is a Laurent series of the form

$$\sum_{i=-\infty}^{\infty} \gamma_i z^i = \left(\cdots + \frac{\gamma_{-2}}{z^2} + \frac{\gamma_{-1}}{z}\right) + \gamma_0 + \left(\gamma_1 z + \gamma_2 z^2 + \cdots\right). \tag{3.52}$$

For such a series to converge, it is necessary and sufficient that the component series in the parentheses should converge for a common value of z. A real number r_- can always be found such that, if $|z| > r_-$, then the series in negative powers converges absolutely. Likewise, a number r_+ can be found such that, if $|z| < r_+$, then the series in positive powers converges absolutely. Therefore, if $r_- < r_+$, then there exits an annulus bounded by concentric circles, in which the Laurent series as a whole converges.

The simplest case is where the transfer functions $\mu(z)/\alpha(z)$ and $\theta(z)/\phi(z)$ both fulfil the conditions of BIBO stability, which is to say that $|\lambda_i| < 1$ and $|\kappa_j| < 1$ for all i and j. Then the function of (3.51) has poles at $z = \lambda_i$ inside a circle within the unit circle, and poles at $z = \kappa_j^{-1}$ outside a circle containing the unit circle. Moreover, expansions at all points within the annulus containing the unit circle and none of the poles will have the same coefficients. In particular, the coefficient associated with z^τ in the Laurent expansion of $\gamma(z)$ is the cross-covariance of $y(t-\tau)$ and $q(t)$, which we shall call the cross-covariance of $y(t)$ and $q(t)$ at lag τ: the lag being associated with the first-named sequence.

Unless $\mu(z^{-1})/\alpha(z^{-1})$ and $\theta(z)/\phi(z)$ are both proper rational functions, it is easiest, in seeking to form the series $\gamma(z)$, to expand the numerator and denominator separately and then to form their product.

The partial-fraction expansion of $\alpha^{-1}(z^{-1})$ is given by

$$\frac{1}{\alpha(z^{-1})} = \frac{C_1}{1 - \lambda_1 z^{-1}} + \cdots + \frac{C_p}{1 - \lambda_p z^{-1}}, \tag{3.53}$$

where the generic coefficient is

$$C_i = \frac{\lambda_i^{p-1}}{\prod_{j \neq i}(\lambda_i - \lambda_j)}. \tag{3.54}$$

Likewise, for $\phi^{-1}(z)$, there is

$$\frac{1}{\phi(z)} = \frac{D_1}{1 - \kappa_1 z} + \cdots + \frac{D_f}{1 - \kappa_f z}. \tag{3.55}$$

It follows that the denominator of $\gamma(z)$ is

$$\frac{1}{\psi(z)} = \frac{1}{\alpha(z^{-1})\phi(z)} = \sum_i \sum_j \frac{C_i D_j}{(1 - \lambda_i z^{-1})(1 - \kappa_j z)}. \tag{3.56}$$

This expression may be evaluated using the result that

(3.57)
$$\frac{C_iD_j}{(1-\lambda_iz^{-1})(1-\kappa_jz)}=\frac{C_iD_j}{(1-\lambda_i\kappa_j)}\left\{\cdots+\frac{\lambda_i^2}{z^2}+\frac{\lambda_i}{z}+1+\kappa_jz+\kappa_j^2z^2+\cdots\right\}.$$

An expression for the numerator of $\gamma(z)$ is provided by the formula under (3.48).

Example 3.6. It will be useful for later reference to derive the variance of the ARMA(2, 1) process defined by

(3.58)
$$y(t)=\frac{\mu(L)}{\alpha(L)}\varepsilon(t)=\frac{1-\rho L}{(1-\lambda_1L)(1-\lambda_2L)}\varepsilon(t).$$

The partial-fraction decomposition of the rational function $\mu(z)/\alpha(z)$ is given by

(3.59)
$$\frac{1-\rho z}{(1-\lambda_1z)(1-\lambda_2z)}=\frac{C_1}{1-\lambda_1z}+\frac{C_2}{1-\lambda_2z},$$

wherein

(3.60)
$$C_1=\frac{\lambda_1-\rho}{\lambda_1-\lambda_2}\quad\text{and}\quad C_2=\frac{\lambda_2-\rho}{\lambda_2-\lambda_1}.$$

The variance of the process is given by

(3.61)
$$\begin{aligned}\gamma_0&=\sigma_\varepsilon^2\left\{\frac{C_1^2}{1-\lambda_1^2}+\frac{C_2^2}{1-\lambda_2^2}+\frac{2C_1C_2}{1-\lambda_1\lambda_2}\right\}\\&=\sigma_\varepsilon^2\frac{(1+\lambda_1\lambda_2)(1+\rho^2)-2\rho(\lambda_1+\lambda_2)}{(1-\lambda_1^2)(1-\lambda_2^2)(1-\lambda_1\lambda_2)},\end{aligned}$$

where the final equality is obtained by dint of some tedious manipulation. The same expression will be obtained later by a different method.

Analytic Functions

The complex-valued function $f(z)$ is said to have a limit ϕ as z approaches z_0, and we write $\lim(z\rightarrow z_0)f(z)=\phi$, if, for every real number ϵ, there exists a corresponding real δ such that $0<|z-z_0|<\delta$ implies $|f(z)-\phi|<\epsilon$. Notice that, according to the definition, the function need not be defined at z_0.

A function f is said to be continuous at the point z_0 if $\lim(z\rightarrow z_0)f(z)=\phi$ and if $f(z_0)=\phi$. Thus, if a function is to be continuous at a point, it must be defined at that point where it must also take a value equal to that of its limit.

Suppose that f and g are functions which are continuous at the point z_0. Then $f+g$ is also continuous at z_0. Moreover, if $g(z_0)\neq 0$, then the function f/g is continuous at z_0. Finally, if g is a function which is continuous at each point in a disc centred at $\phi_0=f(z_0)$, then the composition $g\{f(z)\}$ is continuous at z_0.

These facts indicate that any polynomial $\alpha(z) = \alpha_0 + \alpha_1 z + \cdots + \alpha_n z^n$ is continuous on the complex plane. Also, if $\beta(z)$ and $\alpha(z)$ are two polynomials, then their quotient $\beta(z)/\alpha(z) = \omega(z)$, which is described as a rational function, is continuous except at the points where $\alpha(z) = 0$, which are called the poles of $\omega(z)$.

We say that $f(z)$ is differentiable at z_0 if

$$\frac{\partial f(z_0)}{\partial z} = \lim_{h\to 0} \frac{f(z_0 + h) - f(z_0)}{h} \tag{3.62}$$

is uniquely defined, regardless of the manner in which h approaches zero. From the identity

$$\begin{aligned} f(z) &= f(z_0) + \frac{f(z) - f(z_0)}{z - z_0}(z - z_0) \\ &= f(z_0) + \frac{f(z) - f(z_0)}{h} h, \end{aligned} \tag{3.63}$$

it follows that, if f is differentiable at z_0, then $\lim(z \to z_0)f(z) = f(z_0)$, which is to say that f is also continuous at z_0. The converse is not true; for a function may be continuous but nowhere differentiable in the sense of the foregoing definition.

Example 3.7. Let $f(z) = z^*$ where $z = z^{re} + iz^{im}$ and $z^* = z^{re} - iz^{im}$. This is a continuous function; and it is found that

$$\begin{aligned} \frac{f(z_0 + h) - f(z_0)}{h} &= \frac{(z_0 + h)^* - z_0^*}{h} \\ &= \frac{h^{re} - ih^{im}}{h^{re} + ih^{im}}. \end{aligned} \tag{3.64}$$

Let $h^{im} = 0$, and let $h \to 0$ along the real line. Then the value of the expression is unity. On the other hand, let $h = it$ where t is real. Then $h^{re} = 0$ and $h \to 0$ along the imaginary line as $t \to 0$, and so the value is -1. Thus the limit of the expression does not exist for any z, and f is nowhere differentiable.

A function f is said to be analytic at the point z_0 if it is differentiable at every point in a neighbourhood of z_0. If it is analytic for every point in some domain $\mathcal{D}$, then it is said to be analytic in $\mathcal{D}$.

The essential feature of an analytic function is that the limit in (3.62) is uniquely defined no matter how z approaches z_0. This condition leads to a relationship between the partial derivatives of the real and imaginary parts of the function.

(3.65) Let $f = f^{re} + if^{im}$ be a complex function which is analytic in the neighbourhood of a point $z = x + iy$. The derivatives of f at that point satisfy the *Cauchy–Riemann equations:*

$$\frac{\partial f^{re}(x,y)}{\partial x} = \frac{\partial f^{im}(x,y)}{\partial y}, \qquad \frac{\partial f^{re}(x,y)}{\partial y} = -\frac{\partial f^{im}(x,y)}{\partial x}.$$

Proof. According to the definition of the derivative to be found under (3.62), h can approach zero along any path. Let $h \to 0$ along the real line. Then

$$\begin{aligned}\frac{\partial f(z)}{\partial z} &= \lim_{t\to 0} \frac{f^{re}(x+h,y) - f^{re}(x,y)}{h} + i \lim_{t\to 0} \frac{f^{im}(x+h,y) - f^{im}(x,y)}{h} \\ &= \frac{\partial f^{re}(x,y)}{\partial x} + i\frac{\partial f^{im}(x,y)}{\partial x}.\end{aligned}$$

Now let $h = it$, where t is real, so that $h \to 0$ along the imaginary line as $t \to 0$. Then

$$\begin{aligned}\frac{\partial f(z)}{\partial z} &= \lim_{h\to 0} \frac{f^{re}(x,y+t) - f^{re}(x,y)}{it} + i \lim_{h\to 0} \frac{f^{im}(x,y+t) - f^{im}(x,y)}{it} \\ &= -i\frac{\partial f^{re}(x,y)}{\partial y} + \frac{\partial f^{im}(x,y)}{\partial y}.\end{aligned}$$

Equating the real and imaginary parts of the two expressions for the derivative gives the Cauchy–Riemann equations.

A result which is the converse of (3.65) also holds. That is to say, if the four partial derivatives of the complex function f exist and are continuous in the neighbourhood of z_0, and if, in addition, they satisfy the Cauchy–Riemann equations, then f is differentiable at z_0. It follows that, if these conditions hold throughout a domain $\mathcal{D}$, then f is analytic in that domain.

Complex Line Integrals

The integral of a complex function along a path in the complex plane may be defined in much the same manner as a line integral in the real plane. We should begin by considering lines in the complex plane.

Let $a \leq b$ be points on the real line and let $\gamma(t) = x(t) + iy(t)$, with $t \in [a, b]$, be a continuous complex-valued function. Then the set of points $\{\gamma(t); a \leq t \leq b\}$, which is the range of γ, represents the trace of the curve γ in the complex plane. Notice that two curves may share the same trace. Thus the curve $\gamma(t)$ and its reverse $\bar{\gamma} = \gamma(a + b - t)$, where $a \leq t \leq b$, share the same trace. The function $z = \gamma(t)$ is described as the parametric equation of the curve.

A curve is γ is called simple if it does not cross itself; that is to say if $\gamma(t_1) \neq \gamma(t_2)$ when $a < t_1 < t_2 < b$. If $\gamma(a) = \gamma(b)$, then the curve is closed.

Let $\gamma_1(t)$ with $a_1 \leq t \leq b_1$ and $\gamma_2(t)$ with $a_2 \leq t \leq b_2$ be two curves with $\gamma_1(b_1) = \gamma_2(a_2)$. Then their sum is defined by

$$(3.66) \qquad \gamma_1 + \gamma_2 = \begin{cases} \gamma_1(t), & \text{if } a_1 \leq t \leq b_1; \\ \gamma_2(t + a_2 - b_1), & \text{if } b_1 \leq t \leq b_1 + b_2 - a_2. \end{cases}$$

If the function $\gamma(t)$ has a continuous derivative in every closed subset of $[a, b]$, then $\gamma(t)$ is described as a contour. If $\gamma(a) = \gamma(b)$, then it is a closed contour.

Now we are in a position to define a contour integral. Let $\gamma(t)$ be a contour in the complex plane and let $f(z)$ be a complex-valued function for which the composition $f\{\gamma(t)\}$ is a continuous function of t. Then the integral of f along the contour γ is defined by

$$\int_\gamma f(z)dz = \int_a^b f\{\gamma(t)\}\frac{d\gamma(t)}{dt}dt. \tag{3.67}$$

By decomposing $\gamma(t) = x(t) + iy(t)$ into its real and imaginary parts, we can write the contour integral as

$$\begin{aligned}\int_\gamma f(z)dz &= \int_a^b \left(f^{re} + if^{im}\right)\left(\frac{dx}{dt} + i\frac{dy}{dt}\right)dt \\ &= \int_\gamma \left(f^{re}dx - f^{im}dy\right) + i\int_\gamma \left(f^{im}dx + f^{re}dy\right).\end{aligned} \tag{3.68}$$

A closed contour has no evident beginning or end; and, when closure is assumed, it is appropriate to modify the notation by writing the integral as

$$\oint_\gamma f(z)dz. \tag{3.69}$$

It is usually understood that the integration follows a counterclockwise direction; and this is sometimes reflected in the symbol of the contour integral when a directional arrowhead is superimposed on the circle.

The following are some of the essential properties of contour integration:

(3.70) (i) For any pair of complex numbers κ and λ,

$$\int_\gamma \{\kappa f(z) + \lambda g(z)\}dz = \kappa\int_\gamma f(z)dz + \lambda\int_\gamma g(z)dz.$$

(ii) If γ_1, γ_2 and $\gamma = \gamma_1 + \gamma_2$ are contours, then

$$\int_\gamma f(z)dz = \int_{\gamma_1} f(z)dz + \int_{\gamma_2} f(z)dz.$$

(iii) Changing the direction of integration reverses the sign of the integral so that, if $\gamma(t)$ and $\bar{\gamma}(t)$ are the contour and its reverse, then

$$\int_\gamma f(z)dz = -\int_{\bar{\gamma}} f(z)dz.$$

The last of these follows from the fact that, if $\bar{\gamma}(t) = \gamma(a+b-t)$, then $d\bar{\gamma}(t)/dt = -d\gamma(a+b-t)/dt$. This indicates that

$$\begin{aligned}\int_{\bar{\gamma}} f(z)dz &= \int_a^b f\{\bar{\gamma}(t)\}\frac{d\bar{\gamma}(t)}{dt}dt \\ &= -\int_a^b f\{\gamma(a+b-t)\}\frac{d\gamma(a+b-t)}{dt}dt \\ &= -\int_a^b f\{\gamma(t)\}\frac{d\gamma(t)}{dt}dt = -\int_\gamma f(z)dz.\end{aligned} \tag{3.71}$$

Example 3.8. Let the contour of integration γ be a circle in the complex plane of radius ρ and centre z_0. The parametric equation $z = \gamma(t)$ for the circle takes the form of

$$
\begin{aligned}
z &= z_0 + \rho\{\cos(t) + i\sin(t)\} \\
&= z_0 + \rho e^{it},
\end{aligned}
\tag{3.72}
$$

with $0 \leq t \leq 2\pi$. Let $f(z) = (z - z_0)^m$. Along the contour, $(z - z_0)^m = \rho^m e^{imt}$ and $dz = i\rho e^{it}dt$. Therefore,

$$
\begin{aligned}
\oint_\gamma f(z)dz &= \int_0^{2\pi} \rho^m e^{imt} i\rho e^{it}dt = i\rho^{m+1}\int_0^{2\pi} e^{i(m+1)t}\iota t \\
&= i\rho^{m+1}\left\{\int_0^{2\pi} \cos\{(m+1)t\}dt + i\int_0^{2\pi} \sin\{(m+1)t\}dt\right\},
\end{aligned}
\tag{3.73}
$$

where the final equality follows from (2.54), which is Euler's equation. When $m \neq -1$, the integrals are zero, since each integrand is a whole number of cycles of a trigonometrical function. When $m = -1$, we have $e^{i(m+1)} = e^0 = 1$ and $\rho^{m+1} = \rho^0 = 1$, and the value of the integral is 2π. Thus

$$
\oint_\gamma (z - z_0)^m dz = \begin{cases} 2\pi i, & \text{if } m = -1; \\ 0, & \text{if } m \neq -1. \end{cases}
\tag{3.74}
$$

In general, if we integrate the function $f(z)$ from $z_a = \gamma(a)$ to $z_b = \gamma(b)$ along different paths, we get different values for the integral. Thus the value of a complex integral depends not only on the endpoints but also on the geometric shape of the path between them. However, with analytic functions it is different: it transpires that the value of the integral is the same whatever the shape of the contour.

The Cauchy Integral Theorem

Cauchy's integral theorem asserts that, under certain conditions, the integral of a complex function around a closed contour is zero. The consequences of the result are far reaching. One consequence is that integrals around certain paths may be replaced by integrals around quite different paths which may lend themselves more readily to analysis. This result, which is known as the invariance of complex integration under the deformation of its path, lends a topological flavour to complex analysis.

There are several versions of Cauchy's theorem, which are distinguished from each other by the degree of generality in the assumptions concerning the function to be integrated. The simplest version makes use of the fundamental result of calculus concerning definite integrals.

(3.75) Let $f(z) = dF(z)/dz$ be continuous in the domain $\mathcal{D}$ where $F(z)$ is analytic. Then, for any curve $\gamma(t)$ in $\mathcal{D}$ with $a \leq t \leq b$ and $z_a = \gamma(a)$, $z_b = \gamma(b)$, we have

$$
\int_{z_a}^{z_b} f(z)dz = F(z_b) - F(z_a).
$$

Moreover, if $\gamma(a) = \gamma(b)$ and if γ encloses only points in $\mathcal{D}$, then

$$\oint_\gamma f(z)dz = 0.$$

Proof. In terms of the parametric equation $z = \gamma(t)$, the integral may be written as

$$(3.76)\qquad \begin{aligned} \int_\gamma f(z)dz &= \int_a^b f\{\gamma(t)\}\frac{d\gamma(t)}{dt}dt \\ &= \int_a^b \frac{d}{dt}\{F(\gamma)\gamma(t)\}dt \\ &= F(z_b) - F(z_a). \end{aligned}$$

An instance of this result is provided by the previous example. For, when we integrate the function $f(z) = (z - z_0)^m$, wherein $m \geq 0$, around a circle centred on z_0, we find, according to (3.74), that the result is zero.

In fact, the condition that $f(z)$ is the derivative of an analytic function $F(z)$ implies that $f(z)$ is itself analytic. If we do not assume that there exists such a function $F(z)$ but assume, merely, that the derivative of $f(z)$ in $\mathcal{D}$ is continuous, then we have to prove the following theorem in place of the previous one:

(3.77) *Cauchy's Theorem.* Let f be an analytic function on the domain $\mathcal{D}$ which has a continuous derivative, and let γ be a closed curve in $\mathcal{D}$ whose inside also lies in $\mathcal{D}$. Then

$$\oint_\gamma f(z)dz = 0.$$

This may be proved by invoking Green's theorem which indicates that, if $f_1(x, y)$ and $f_2(x, y)$ are real-valued functions which, together with their partial derivatives, are continuous throughout a region $\mathcal{G}$ in the plane which is bounded by the closed contour γ, then

$$(3.78)\qquad \int_\gamma (f_1dx + f_2dy) = \iint_\mathcal{G} \left(\frac{df_2}{dx} - \frac{df_1}{dy}\right) dxdy.$$

This result enables us to evaluate a contour integral by integrating a volume and vice versa. Now, according to (3.68), the contour integral can be written as

$$(3.79)\qquad \int_\gamma f(z)dz = \int_\gamma (f^{re}dx - f^{im}dy) + i\int_\gamma (f^{im}dx + f^{re}dy).$$

In view of Green's theorem, this may be written as

$$(3.80)\qquad \int_\gamma f(z)dz = \iint_\mathcal{G} \left(-\frac{df^{im}}{dx} - \frac{df^{re}}{dy}\right) dxdy + i\iint_\mathcal{G} \left(\frac{df^{re}}{dx} - \frac{df^{im}}{dy}\right) dxdy.$$

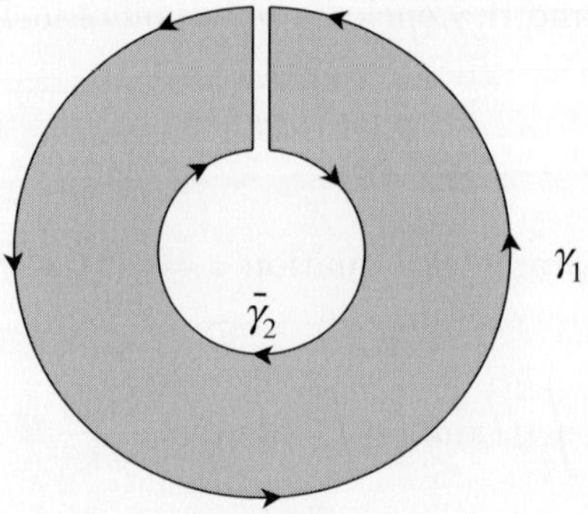

Figure 3.2. An annulus can be cut to form a simply connected domain.

But, according to the Cauchy–Riemann equations of (3.65), the integrands of these two double integrals are zero throughout $\mathcal{G}$, from which it follows that contour integral is zero.

A more general version of the integral theorem is the Cauchy–Goursat theorem. This proceeds without the assumption that f has a continuous first derivative and, as a consequence, the proof becomes more elaborate. Since, as we shall show, an analytic function possesses derivatives of all orders, it may seem that the extra generality of this theorem is not always worth the effort of achieving it.

Multiply Connected Domains

A domain $\mathcal{D}$ in the complex plane is said to be simply connected if every closed path γ encloses only points in $\mathcal{D}$. A domain which is not simply connected is said to be multiply connected or disconnected. An example of a doubly-connected domain is an annulus bounded by an inner and an outer circle with a common centre—see Figure 3.2.

In multiply connected domains, the versions of Cauchy's theorem stated above no longer apply. However, a multiply connected domain may be cut so that the resulting domain becomes simply connected. The annulus provides an example. The cut can be made along a radial line. The path surrounding the new simply connected domain traverses the circumference of the inner circle in a clockwise direction along the curve $\bar{\gamma}_2$ starting to the right of the cut. Then it follows the line of the radius to the left of the cut before traversing the circumference of the outer circle in a counterclockwise direction along the curve γ_1. It returns to the point of departure by following the line of the radius to the right of the cut. Since the line of the radius is traversed in both directions, the corresponding integrals cancel, and the value of the integral overall is given by

$$(3.81) \qquad \oint_{\gamma} f(z)dz = \int_{\gamma_1} f(z)dz + \int_{\bar{\gamma}_2} f(z)dz.$$

Reversing the direction of the integration around the inner circle from clockwise to counterclockwise gives

$$\oint_{\gamma} f(z)dz = \int_{\gamma_1} f(z)dz - \int_{\gamma_2} f(z)dz. \tag{3.82}$$

Integrals and Derivatives of Analytic Functions

The result, known as Cauchy's integral formula, indicates that, if a function f is analytic at every point enclosed by a contour γ and on the contour itself, then its values at the interior points are determined by its values on γ:

(3.83) Let $f(z)$ be analytic in a simply connected domain $\mathcal{D}$ and let γ be a closed path in $\mathcal{D}$. If z_0 is interior to γ, then

$$f(z_0) = \frac{1}{2\pi i}\oint_{\gamma} \frac{f(z)}{z - z_0} dz.$$

Proof. Consider an annulus within $\mathcal{D}$ which is centred on the point z_0. According to the arguments of the previous section, we can cut the annulus to form a simply connected domain; and, thereafter, we may apply Cauchy's integral theorem to show that

$$\oint_{\gamma} \frac{f(z)}{z - z_0} dz = \int_{\gamma_0} \frac{f(z)}{z - z_0} dz, \tag{3.84}$$

where γ is the circumference of the outer circle and γ_0 is the inner circle surrounding z_0 which is traversed in a counterclockwise direction. The inner circle can be represented parametrically by the equation $z = z_0 + \rho e^{it}$, and the radius ρ can be allowed to approach zero. Thus we have

$$\begin{aligned} \oint_{\gamma_0} \frac{f(z)}{z - z_0} dz &= \lim_{\rho \to 0} \int_0^{2\pi} \frac{f(z_0 + \rho e^{it})}{\rho e^{it}} \rho i e^{it} dt \\ &= i f(z_0) \int_0^{2\pi} dt = 2\pi i f(z_0). \end{aligned} \tag{3.85}$$

Putting the final expression in place of the RHS of (3.84) proves the theorem.

Cauchy's integral formula can be used to obtain expressions for the derivatives of an analytic function.

(3.86) If $f(z)$ is analytic in a domain $\mathcal{D}$, then its derivatives in $\mathcal{D}$ of all orders are also analytic functions; and the nth derivative at the point z_0, enclosed by γ in $\mathcal{D}$, is given by

$$f^{(n)}(z_0) = \frac{n!}{2\pi i}\oint_{\gamma} \frac{f(z)}{(z - z_0)^{n+1}} dz.$$

Proof. In view of the integral formula of (3.83), it can be seen that the derivative of $f(z)$ at z_0 is the limit, as $h \to 0$, of

$$\tag{3.87} \begin{aligned} \frac{f(z_0+h)-f(z_0)}{h} &= \frac{1}{2\pi ih}\left\{\oint \frac{f(z)}{z-z_0-h}dz - \oint \frac{f(z)}{z-z_0}dz\right\} \\ &= \frac{1}{2\pi ih}\oint \frac{hf(z)}{(z-z_0-h)(z-z_0)}dz. \end{aligned}$$

Thus

$$\tag{3.88} f'(z_0) = \frac{1}{2\pi i}\oint_\gamma \frac{f(z)}{(z-z_0)^2}dz,$$

which is simply the result of differentiating the equation of (3.83) under the integral sign with respect to z_0. This technique for obtaining the derivative may be repeated with $f'(z_0+h)$ and $f'(z_0)$ replacing $f(z_0+h)$ and $f(z_0)$ in equation (3.87); and it will be found that

$$\tag{3.89} f^{(2)}(z_0) = \frac{2}{2\pi i}\oint_\gamma \frac{f(z)}{(z-z_0)^3}dz.$$

The general formula or the nth derivative may be obtained by induction.

It is significant that the requirement that $f(z)$ be analytic is sufficient for the existence of the derivatives of all orders.

Series Expansions

A power series with a nonzero radius of convergence represents an analytic function at every point interior to the circle defined by that radius. Moreover, the derivative of the analytic function can be obtained by differentiating the power series term by term. We can also show that, conversely, every analytic function can be represented by a power series; and for this purpose we use the integral formula.

(3.90) *The Taylor Series.* Let $f(z)$ be analytic for every point in the open set $\mathcal{D} = \{z; |z-z_0| < r\}$ which is a disc of radius r centred on z_0. Then f may be represented uniquely as a power series

$$f(z) = \sum_{n=0}^{\infty} a_n(z-z_0)^n \quad \text{with} \quad a_n = \frac{1}{n!}f^{(n)}(z_0).$$

Proof. Let s be a point on the circle γ of radius r and centre z_0 which bounds the set $\mathcal{D}$ containing z. Then, according to the integral formula of (3.83),

$$\tag{3.91} \begin{aligned} f(z) &= \frac{1}{2\pi i}\oint_\gamma \frac{f(s)}{s-z}ds \\ &= \frac{1}{2\pi i}\oint_\gamma \frac{f(s)}{(s-z_0)-(z-z_0)}ds \\ &= \frac{1}{2\pi i}\oint_\gamma \frac{f(s)}{(s-z_0)\{1-(z-z_0)/(s-z_0)\}}ds. \end{aligned}$$

The denominator of the final expression can be expanded using

$$\begin{aligned}\frac{1}{1-(z-z_0)/(s-z_0)} &= 1+\frac{z-z_0}{s-z_0}+\frac{(z-z_0)^2}{(s-z_0)^2}+\cdots\\ &= \sum_{n=0}^{\infty}\frac{(z-z_0)^n}{(s-z_0)^n},\end{aligned} \tag{3.92}$$

where convergence is guaranteed by the condition that $|z-z_0|<|s-z_0|$. Therefore, equation (3.91) can be written as

$$\begin{aligned}f(z) &= \frac{1}{2\pi i}\oint_\gamma \sum_{n=0}^{\infty}\frac{(z-z_0)^n f(s)}{(s-z_0)^{n+1}}ds\\ &= \frac{1}{2\pi i}\sum_{n=0}^{\infty}(z-z_0)^n\oint_\gamma \frac{f(s)}{(s-z_0)^{n+1}}ds\\ &= \sum_{n=0}^{\infty}(z-z_0)^n\left\{\frac{f^{(n)}(z_0)}{n!}\right\},\end{aligned} \tag{3.93}$$

where the final equality comes from (3.86).

It is appropriate, at this stage, to prove a result which, hitherto, we have taken for granted:

(3.94) If the power series $\sum_{n=0}^{\infty}a_n(z-z_0)^n$ converges at some point z_1 and diverges at some point z_2, then it converges absolutely for all z such that $|z-z_0|<|z_1-z_0|$ and it diverges for all $|z-z_0|>|z_2-z_0|$.

Proof. Since $\sum_{n=0}^{\infty}a_n(z_1-z_0)^n$ converges, there exists a positive number δ such that $|a_n(z_1-z_0)^n|<\delta$ for all n. Now let z be any point closer to z_0 than is z_1. Then $|z-z_0|<|z_1-z_0|$ and

$$\sum_{n=0}^{\infty}|a_n(z-z_0)^n| = \sum_{n=0}^{\infty}|a_n(z_1-z_0)^n|\left|\frac{z-z_0}{z_1-z_0}\right|^n < \delta\sum_{n=0}^{\infty}\left|\frac{z-z_0}{z_1-z_0}\right|^n.$$

But $|(z-z_0)/(z_1-z_0)|<1$, so it follows that the series converges absolutely at z. The second part of this proposition is proved likewise.

Let $f(z)$ be a function which is analytic at z_0 for which the first $m-1$ derivatives $f', f^{(2)},\ldots,f^{m-1}$ are all zero at z_0. Then f is said to have a zero of the mth order at z_0.

This condition has an immediate consequence for the Taylor series expansion of f about z_0, for it implies that

$$\begin{aligned}f(z) &= \{a_m(z-z_0)^m+a_{m+1}(z-z_0)^{m+1}+\cdots\}\\ &= (z-z_0)^m\{a_m+a_{m+1}(z-z_0)+\cdots\}\\ &= (z-z_0)^m g(z),\end{aligned} \tag{3.95}$$

where $g(z)$ is analytic and $a_m\neq 0$.

If $f(z) = (z - z_0)^m g(z)$, is analytic at a point z_0, which is a zero of f, then there is a neighbourhood of z_0 throughout which f has no other zeros unless it is identically zero. That is to say, the zeros of f are isolated. This is a consequence of the condition that f is continuous at z_0, whereby there exists a positive real number δ such that $|z - z_0| < \delta$ implies $|g(z) - a_m| < |a_m|$. From this, it follows that $g(z) \neq 0$ for all points z such that $|z - z_0| < \delta$; for otherwise $|a_m| < |a_m|$, which is a contradiction.

Example 3.9. It is easy to confirm that, if $\alpha(z)$ is a polynomial of degree p which has m roots or zeros equal to λ, then the first $m-1$ derivatives are zero at λ whilst the remaining derivatives are nonzero at λ. First we write $\alpha(z) = (z - \lambda)^m \gamma(z)$, where $\gamma(\lambda) \neq 0$. Differentiating once gives $\alpha'(z) = m(z-\lambda)^{m-1}\gamma(z) + (z-\lambda)^m \gamma'(z)$; so $\alpha'(z)$ contains the factor $(z - \lambda)^{m-1}$ in association with a cofactor $m\gamma(z) + (z - \lambda)\gamma'(z)$ which becomes $m\gamma(\lambda) \neq 0$ when $z = \lambda$. Differentiating a second time shows, likewise, that $\alpha^{(2)}(z)$ contains the factor $z - \lambda$ with a multiplicity of $m - 2$. We can proceed to show that the $(m - 1)$th derivative contains the factor once and that the mth derivative is free of it.

It is sometimes necessary to expand a function $f(z)$ around points at which it may be singular. In that case, a Laurent expansion is called for. This is a representation which is valid in an annulus bounded by an inner circle γ_2 and an outer circle γ_1 which are concentric; and $f(z)$ may be singular both at points inside γ_2 and at points outside γ_1.

(3.96) *The Laurent Series.* Let $f(z)$ be analytic for every point in the open set $\{z; r_2 < |z - z_0| < r_1\}$, which is an annulus enclosing a circle γ of radius r centred on z_0 with $r_2 < r < r_1$. Then f may be represented by a Laurent series

$$f(z) = \sum_{n=0}^{\infty} a_i (z - z_0)^n + \sum_{n=1}^{\infty} \frac{b_n}{(z - z_0)^n} \quad \text{with}$$

$$a_n = \frac{1}{2\pi i} \oint_{\gamma} \frac{f(z)}{(z - z_0)^{n+1} dz}, \quad b_n = \frac{1}{2\pi i} \oint_{\gamma} f(z)(z - z_0)^{n-1} dz.$$

Proof. The annulus may be converted to a simply connected domain $\mathcal{D}$ by cutting it along a radial line in the manner described in a previous section and illustrated by Figure 3.2. The contour, which bounds $\mathcal{D}$, comprises the inner circle $\bar{\gamma}_2$, which is traversed in a clockwise direction, the outer circle γ_1, which is traversed in a counterclockwise direction, and the lines running back and forth along either side of the cut. The integrals over the latter will cancel each other.

When z is within the annulus, it follows from Cauchy's integral formula (3.83) that

$$f(z) = \frac{1}{2\pi i} \oint_{\gamma_1} \frac{f(s)}{s - z} ds - \frac{1}{2\pi i} \oint_{\gamma_2} \frac{f(s)}{s - z} ds, \tag{3.97}$$

where the negative sign on the second integral comes from reversing the direction of integration on the inner circle from clockwise to counterclockwise. The denominators may be written as $(s - z_0) - (z - z_0)$.

The points s on γ_1 satisfy $|s - z_0| > |z - z_0|$. Therefore, the denominator of the integral on γ_1 may be expanded as

$$\begin{aligned} \frac{1}{(s - z_0) - (z - z_0)} &= \frac{1/(s - z_0)}{1 - (z - z_0)/(s - z_0)} \\ &= \frac{1}{(s - z_0)} \sum_{n=0}^{\infty} \frac{(z - z_0)^n}{(s - z_0)^n}. \end{aligned} \tag{3.98}$$

The points on γ_2, on the other hand, satisfy $|s - z_0| < |z - z_0|$. Therefore, the denominator of the integral on γ_2 may be expanded as

$$\begin{aligned} \frac{1}{(s - z_0) - (z - z_0)} &= \frac{-1/(z - z_0)}{1 - (s - z_0)/(z - z_0)} \\ &= -\sum_{n=1}^{\infty} \frac{(s - z_0)^{n-1}}{(z - z_0)^n}, \end{aligned} \tag{3.99}$$

where it should be noted that summation begins at $n = 1$ and not at $n = 0$ as in (3.98). It follows that

$$\begin{aligned} f(z) = &\frac{1}{2\pi i} \sum_{n=0}^{\infty} (z - z_0)^n \oint_{\gamma_1} \frac{f(s)}{(s - z_0)^{n+1}} ds \\ &+ \frac{1}{2\pi i} \sum_{n=1}^{\infty} (z - z_0)^{-n} \oint_{\gamma_2} (s - z_0)^{n-1} f(s) ds. \end{aligned} \tag{3.100}$$

The portion of the Laurent series involving negative powers of $z - z_0$ is called the principal part of f at z_0. The portion involving positive powers is called the analytic part of f.

If the Laurent expansion takes the form of

$$f(z) = \sum_{n=0}^{\infty} a_n (z - z_0)^n + \frac{b_1}{z - z_0} + \frac{b_2}{(z - z_0)^2} + \cdots + \frac{b_m}{(z - z_0)^m}, \tag{3.101}$$

where $b_m \neq 0$, then the singular point z_0 is described as a pole of order m. In particular, if $g(z)$ is analytic at z_0 and has a zero of order m at that point, then $1/g(z)$ has a pole of order m at z_0. The same is true of $p(z)/g(z)$ if $p(z)$ is analytic at z_0 and $p(z_0) \neq 0$.

To see how the expression under (3.101) may arise, consider the function $f(z) = (z - z_0)^m g(z)$ of (3.95) which has a zero of the mth order at z_0. Here $g(z_0) \neq 0$ by assumption, and g and f are analytic in some domain $\mathcal{D} = \{z; 0 < |z - z_0| < r\}$. Also $h(z) = 1/g(z)$ is analytic on the disc $|z - z_0| < r$. Thus

$$\frac{1}{f(z)} = \frac{1}{(z - z_0)^m g(z)} = \frac{h(z)}{(z - z_0)^m}; \tag{3.102}$$

and, by taking the Taylor-series expansion of $h(z)$ about the point z_0, we obtain an expression in the form of (3.101).

Residues

The Cauchy integral theorem indicates that, if the function f is analytic at all points enclosed by the contour γ, then its integral around the contour is zero-valued. However, if the contour encloses a finite number of isolated singular points where the function is not analytic, then each of these points will contribute to the integral a specific value, called a residue.

Let z_0 be an isolated singular point of f. Then there will be a set of points $\{z; 0 < |z - z_0| < r\}$ not including z_0, described as a punctured disc, where f is analytic and where, consequently, it may be represented by the Laurent series

$$f(z) = \sum_{n=0}^{\infty} a_n(z - z_0)^n + \left\{ \frac{b_1}{z - z_0} + \frac{b_2}{(z - z_0)^2} + \cdots \right\}. \tag{3.103}$$

Here, according to (3.96), we have

$$b_n = \frac{1}{2\pi i} \oint_{\gamma} f(z)(z - z_0)^{n-1} dz, \tag{3.104}$$

where γ encloses z_0. Setting $n = 1$, we see that

$$b_1 = \frac{1}{2\pi i} \oint_{\gamma} f(z) dz. \tag{3.105}$$

The number b_1, which is the coefficient of $1/(z - z_0)$ in the expansion, is described as the residue of f at z_0. It is commonly denoted as $b_1 = \text{Res}(f, z_0)$.

The equation above provides a useful way of evaluating the contour integral; for it is often easier to find the terms of the Laurent expansion than it is to perform the integration directly.

If the singular point z_0 is a pole of the function f, then there are other ways of obtaining the residue which do not require us to form the expansion explicitly. Imagine that z_0 is a pole of order m. Then the expansion would take the form given under (3.101). Multiplying by $(z - z_0)^m$ gives

$$\begin{aligned}(z - z_0)^m f(z) = {} & b_m + b_{m-1}(z - z_0) + \cdots + b_1(z - z_0)^{m-1} \\ & + \sum_{n=0}^{\infty} a_n(z - z_0)^{m+n}.\end{aligned} \tag{3.106}$$

This is just the Taylor series expansion of $g(z) = (z - z_0)^m f(z)$; and the residue $b_1 = \text{Res}(f, z_0)$ has become the series coefficient associated with $(z - z_0)^{m-1}$. It follows from (3.90), where the basic results concerning Taylor series are to be found, that

$$b_1 = \frac{1}{(m-1)!} g^{(m-1)}(z_0); \tag{3.107}$$

The effective way of isolating the coefficient b_1 is to differentiate $g(z) = (z - z_0)^m f(z)$ with respect to z $m - 1$ times and then to set $z = z_0$ to eliminate the other terms which would remain in a series expansion. Thus

(3.108) $$\text{Res}(f, z_0) = \frac{1}{(m-1)!} \frac{d^{m-1}}{dz^{m-1}} \left[(z - z_0)^m f(z) \right]_{z=z_0}.$$

In the case of a pole of order one, the result simplifies to

(3.109) $$b_1 = \text{Res}(f, z_0) = \lim_{z \to z_0} (z - z_0) f(z).$$

This is evident in the light of the Laurent expansion (3.101) which, for a pole of order one, has just one term associated with a negative power of $z - z_0$: namely the term $b_1/(z - z_0)$.

Example 3.10. Consider the function $f(z) = 1/(z^3 - z^4)$. Writing $z^3 - z^4 = z^3(1 - z)$ shows that f has singular points at $z = 1$ and $z = 0$. To integrate the function around a circle γ defined by $|z| = 1/2$, we can find the Laurent expansion about the origin which converges for $0 < |z| < 1$:

$$\frac{1}{z^3 - z^4} = \frac{1}{z^3} + \frac{1}{z^2} + \frac{1}{z} + \{1 + z + z^2 + \cdots\}.$$

This comes from multiplying the expansion $(1 - z)^{-1} = \{1 + z + z^2 + \cdots\}$ by z^{-3}. The residue, which is the coefficient associated with $1/z$ in the Laurent expansion, is 1. Therefore, the value of the integral is 2π.

So far, we have considered the evaluation of integrals around contours which enclose only one isolated singular point of the function. There is a simple extension to the case where the contour encloses several isolated singularities of the integrand:

(3.110) *The Residue Theorem.* Let $f(z)$ be a function which is analytic on a simple closed contour γ and at the points inside γ with the exception of a finite number of points $z_1, \ldots, z_k$. Then

$$\oint_\gamma f(z) dz = 2\pi i \sum_{i=1}^{k} \text{Res}(f, z_i).$$

To prove this, we may use the technique of cutting which was used in finding the integral of a contour within an annulus surrounding one singular point. Now each of the k singular points is surrounded by a clockwise circle $\bar{\gamma}_i$ small enough not to intersect with any of the circles surrounding neighbouring singular points. By excluding the points interior to the circles, a multiply connected domain is created. Each of these circles can be reached from the boundary contour γ by a straight path which is described twice in opposite directions. The paths are the lines along which the multiply connected domain is cut to form a simply connected domain—see Figure 3.3. Since the integrals in opposite directions along the path cancel, it

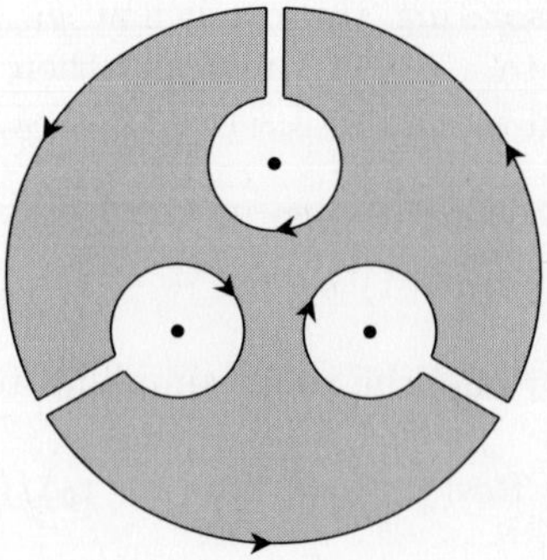

Figure 3.3. Isolated singularities enclosed by a contour may be encircled and excluded from the domain. Then paths can be cut to so as to form a simply connected domain.

follows from Cauchy's integral theorem that the integral along the boundary of the simply connected domain which winds around the circles is just

$$\oint_{\gamma} f(z)dz + \sum_{i=1}^{k} \oint_{\bar{\gamma}_i} f(z)dz = 0. \tag{3.111}$$

By reversing the direction of the integration around the circles, we get

$$\begin{aligned} \oint_{\gamma} f(z)dz &= \sum_{i=1}^{k} \oint_{\gamma_i} f(z)dz \\ &= \sum_{i=1}^{k} \mathrm{Res}(f, z_i). \end{aligned} \tag{3.112}$$

The Autocovariance Generating Function

The autocovariance generating function for the ARMA process described by the equation $\alpha(L)y(t) = \mu(L)\varepsilon(t)$ is given by

$$\begin{aligned} \gamma(z) &= \frac{\mu(z)\mu(z^{-1})}{\alpha(z)\alpha(z^{-1})} = \sigma_\varepsilon^2 \frac{\prod_{j=1}^{q}(1-\rho_j z)(1-\rho_j z^{-1})}{\prod_{j=1}^{p}(1-\lambda_j z)(1-\lambda_j z^{-1})} \\ &= \sigma_\varepsilon^2 z^{p-q} \frac{\prod_{j=1}^{q}(1-\rho_j z)(z-\rho_j)}{\prod_{j=1}^{p}(1-\lambda_j z)(z-\lambda_j)}. \end{aligned} \tag{3.113}$$

The condition that the process be stationary imposes the restriction that the roots of $\alpha(z) = \prod_j (1 - \lambda z_j) = 0$ must lie outside the unit circle, which is to say that $|1/\lambda_j| > 1$ for all j, whereas the restriction that the process be invertible imposes the condition that the roots of $\mu(z) = \prod_j (1 - \rho z_j) = 0$ must lie outside the unit circle, which is to say that $|1/\rho_j| > 1$ for all j. The same conditions, when stated in terms of the roots of $\alpha(z^{-1}) = 0$ and $\mu(z^{-1}) = 0$, require that $|\lambda_j| < 1$ and $|\rho_j| < 1$.

It follows that the function $\gamma(z)$ is analytic in some annulus around the unit circle which includes neither the values λ_j which are encircled by the annulus, nor their reciprocals, which are outside the annulus. If $q > p$, which is to say that the moving-average order is greater that the autoregressive order, then $\gamma(z)$ has a pole of multiplicity $q - p$ at zero.

The autocovariance generating function may be expanded as a Laurent series

$$\gamma(z) = \cdots + \frac{\gamma_2}{z^2} + \frac{\gamma_1}{z} + \gamma_0 + \gamma_1 z + \gamma_2 z^2 + \cdots \tag{3.114}$$

wherein the generic coefficient $\gamma_{-\tau} = \gamma_\tau$ represents the autocovariance at lag τ. If the coefficient is taken in association with $z^{-\tau}$, then, according to the result under (3.96) concerning the coefficients of a Laurent series, we have

$$\begin{aligned} \gamma_{-\tau} &= \frac{1}{2\pi i}\oint z^{\tau-1}\gamma(z)dz \\ &= \frac{\sigma_\varepsilon^2}{2\pi i}\oint z^{p-q+\tau-1}\left\{\frac{\prod_{j=1}^q(1-\rho_j z)(z-\rho_j)}{\prod_{j=1}^p(1-\lambda_j z)(z-\lambda_j)}\right\}dz, \end{aligned} \tag{3.115}$$

where the contour of integration is taken to be the unit circle; and this is also the sum of the residues of $f_\tau(z) = z^{\tau-1}\gamma(z)$ at the points of singularity which lie within the unit circle. These residues correspond to the points of singularity of f_τ which are associated with the autoregressive roots $\lambda_1, \ldots, \lambda_p$ and with the element $z^{p-q+\tau-1}$ whenever its exponent is negative. According to the result under (3.109), the residue associated with a simple pole at λ_j inside the circle is

$$\begin{aligned} \text{Res}(f_\tau, \lambda_k) &= \lim_{z\to\lambda_k}(z-\lambda_k)f(z) \\ &= \lambda_k^{p-q+\tau-1}\left\{\frac{\prod_{j=1}^q(1-\rho_j\lambda_k)(\lambda_k-\rho_j)}{\prod_{j=1}^p(1-\lambda_j\lambda_k)\prod_{\substack{j=1\\ j\neq k}}^p(\lambda_k-\lambda_j)}\right\}. \end{aligned} \tag{3.116}$$

According to the result under (3.108), the residue associated with the pole at zero of multiplicity of $q - p - \tau + 1$ in the case where $p < q + 1$ is given by

$$\text{Res}(f, 0) = \lim_{z\to 0}\frac{1}{(q-p-\tau)!}\frac{d^{q-p-\tau}}{dz^{p-q-\tau}}\left\{\frac{\prod_{j=1}^q(1-\rho_j z)(z-\rho_j)}{\prod_{j=1}^p(1-\lambda_j z)(z-\lambda_j)}\right\}. \tag{3.117}$$

However, if $p \geq q + 1$, then there are no poles at zero in the function $\gamma(z)$ and the problem is simplified. In that case, there are p poles within the unit circle which are due to the autoregressive operator, and we get

$$\gamma_\tau = \sum_{k=1}^p \text{Res}(f_\tau, \lambda_k), \tag{3.118}$$

where $\text{Res}(f_\tau, \lambda_k)$ is defined in (3.116). Note that, if $p = q$, then there remains a single pole at zero in the function $f_\tau(z) = z^{\tau-1}\gamma(z)$ when $\tau = 0$.

Example 3.11. Let us consider the case where the autoregressive order p exceeds the moving-average order q. The variance of the process $y(t)$ is given by combining the equations (3.116) and (3.118):

$$\gamma_0 = \sum_{k=1}^{p} \operatorname{Res}(f_0, \lambda_k) = \sum_{k=1}^{p} \left\{ \frac{\lambda_k^{p-q-1} \prod_{j=1}^{q} (1 - \rho_j \lambda_k)(\lambda_k - \rho_j)}{\prod_{j=1}^{p} (1 - \lambda_j \lambda_k) \prod_{\substack{j=1 \\ j \neq k}}^{p} (\lambda_k - \lambda_j)} \right\}. \tag{3.119}$$

In the case of the ARMA(2, 1) process

$$(1 - \lambda_1 L)(1 - \lambda_2 L) y(t) = (1 - \rho L)\varepsilon(t), \tag{3.120}$$

we have

$$\begin{aligned} \gamma_0 &= \frac{\sigma_\varepsilon^2 (1 - \rho\lambda_1)(\lambda_1 - \rho)}{(1 - \lambda_1^2)(1 - \lambda_1\lambda_2)(\lambda_1 - \lambda_2)} + \frac{\sigma_\varepsilon^2 (1 - \rho\lambda_2)(\lambda_2 - \rho)}{(1 - \lambda_2^2)(1 - \lambda_1\lambda_2)(\lambda_2 - \lambda_1)} \\ &= \sigma_\varepsilon^2 \frac{(1 + \lambda_1\lambda_2)(1 + \rho^2) - 2\rho(\lambda_1 + \lambda_2)}{(1 - \lambda_1^2)(1 - \lambda_2^2)(1 - \lambda_1\lambda_2)}. \end{aligned} \tag{3.121}$$

This result has been derived already by another method, and it has been displayed under (3.61).

The Argument Principle

Consider a function $f(z)$ which is analytic in the domain $\mathcal{D}$ except at a finite number of poles. Let γ be a contour in $\mathcal{D}$ which encloses P poles and N zeros of $f(z)$ and which passes through none of the poles or zeros.

If z_0 is a zero of f of multiplicity or order r, then

$$f(z) = (z - z_0)^r g(z), \tag{3.122}$$

where $g(z)$ in a function which is analytic in a neighbourhood of z_0 with $g(z_0) \neq 0$. Taking logarithms of this function and differentiating gives

$$\frac{d}{dz} \log f(z) = \frac{f'(z)}{f(z)} = \frac{r}{z - z_0} + \frac{g'(z)}{g(z)}. \tag{3.123}$$

Since $g'(z)/g(z)$ is analytic at z_0, it follows that $f'(z)/f(z)$ has a simple pole at z_0 which corresponds to a residue of $\operatorname{Res}\{f'(z)/f(z), z_0\} = r$.

On the other hand, if w_0 is a pole of f of multiplicity m, then

$$f(z) = \frac{h(z)}{(z - w_0)^m}, \tag{3.124}$$

where $h(z)$ in a function which is analytic in a neighbourhood of w_0 with $h(w_0) \neq 0$. Then the derivative of the logarithm of f may be expressed as

$$\frac{d}{dz} \log f(z) = \frac{f'(z)}{f(z)} = \frac{h'(z)}{h(z)} - \frac{m}{z - w_0}. \tag{3.125}$$

Since $h'(z)/h(z)$ is analytic at w_0, it follows that $f'(z)/f(z)$ has a simple pole at w_0 which corresponds to a residue of $\mathrm{Res}\{f'(z)/f(z), w_0\} = -m$.

Thus it follows from the residue theorem that

(3.126)
$$\begin{aligned}\frac{1}{2\pi i}\oint_\gamma \frac{f'(z)}{f(z)}dz &= \sum_j \mathrm{Res}\left(\frac{f'}{f}, z_j\right) - \sum_k \mathrm{Res}\left(\frac{f'}{f}, w_k\right)\\ &= N - P,\end{aligned}$$

which is the number of zeros $N = \sum_j r_j$, including multiplicities, enclosed by the contour γ less the the number of poles $P = \sum_k m_k$, including multiplicities, enclosed by γ.

The result which we have just obtained may be used in proving a theorem which is important in investigating the stability of linear systems—as we shall do in Chapter 5—and in demonstrating the phase effect of linear filters—which is a major concern of Chapter 16:

(3.127) *The Argument Principle.* Let $f(z)$ be a function which is analytic on a domain $\mathcal{D}$ except at a finite number of poles. Let γ be a contour in $\mathcal{D}$ which encloses P poles and N zeros of $f(z)$—where P and N include multiplicities—and which passes through none of the poles or zeros. Then, as z travels once around the contour γ in an anticlockwise direction, the argument of $f(z)$ will change by

$$\Delta_\gamma \mathrm{arg} f(z) = 2\pi(N - P).$$

Proof. Consider the polar exponential representation of $f(z)$ together with its logarithm. These are

(3.128)
$$f(z) = |f(z)| \exp\{i \mathrm{arg} f(z)\} \qquad \text{and}$$
$$\ln f(z) = \ln|f(z)| + i\mathrm{arg} f(z).$$

Let z_a be the point on γ where the path of integration begins and let z_b be the point where it ends. Then $\ln|f(z_a)| = \ln|f(z_b)|$, and so

(3.129)
$$\begin{aligned}\frac{1}{2\pi i}\oint_\gamma \frac{f'(z)}{f(z)}dz &= \frac{1}{2\pi i}\oint_\gamma \left\{\frac{d}{dz}\ln f(z)\right\}dz\\ &= \frac{1}{2\pi i}\Big\{i\mathrm{arg} f(z_b) - i\mathrm{arg} f(z_a)\Big\}\\ &= \frac{1}{2\pi}\Delta_\gamma \mathrm{arg} f(z).\end{aligned}$$

But, according to equation (3.126), the expression on the LHS has the value of $N - P$, and so the theorem is proved.

Bibliography

[27] Archbold, J.W., (1964), *Algebra, Third Edition*, Pitman Press, London.

[109] Churchill, R.V., and J.W. Brown, (1984), *Complex Variables and Applications, Fourth Edition*, McGraw-Hill Book Company, New York.

[185] Fisher, S.D., (1990), *Complex Variables, Second Edition*, Wadsworth and Brooks Cole, Pacific Grove California.

[231] Grove, E.A., and G. Ladas, (1974), *Introduction to Complex Variables*, Houghton Mifflin Company, Boston.

[459] Silverman, R.A., (1972), *Introductory Complex Analysis*, Dover Publications, New York.

[492] Turnbull, H.W., (1953), *The Theory of Equations, Fifth Edition*, Oliver and Boyd, Edinburgh.

CHAPTER 4

Polynomial Computations

Amongst the topics of numerical analysis which are of enduring interest are the problems of polynomial root finding and of polynomial interpolation. These are the principal concerns of the present chapter. Both topics have a history which extends over hundreds of years, and both have received considerable attention in recent years.

The need to find the roots of a polynomial arises in many areas of scientific work. Much of the interest is due to the fact that the dynamic properties of linear systems can be accounted for in terms of the roots of characteristic polynomial equations.

The principle of superposition, which is the cornerstone of linear system theory, indicates that the output of a higher-order dynamic system can be represented by taking a linear combination of the outputs of a set of independent or "decoupled" second-order systems. The behaviour of a second-order system can be accounted for in terms of the roots of a characteristic quadratic equation.

This simple fact belies the practical difficulties which are involved in resolving a higher-order system into its linear or quadratic components. These are precisely the difficulties of root finding. Whereas closed-form solutions are available for polynomial equations of degrees three and four, there are no generally applicable formulae for finding the roots when the degree is greater than four. Therefore, we are bound to consider iterative methods for finding the roots of polynomials in general.

The intractability of polynomial equations, before the advent of electronic computers, encouraged mathematicians and engineers to look for other means of characterising the behaviour of linear dynamic systems. The well-known criterion of Routh [431] was the result of a prolonged search for a means of determining whether or not a linear differential equation is stable without solving its characteristic equation.

In recent years, a wide variety of root-finding methods have been implemented on computers. Roughly speaking, the methods fall into two categories. In the first category are the so-called Newton methods. So long as they are provided with adequate starting values, these methods are generally fast and efficient. However, to find such starting values is not always an easy task; and the occasional annoyance at having to resort to trial and error can be avoided only at the cost of a considerable effort in programming. We shall make no attempt to find accurate starting values.

Much attention has been focused recently on the pathologies of the Newton methods; for it has been discovered that the boundaries, within the complex plane, which separate the regions of convergence for the various roots, are liable to have a complicated fractal nature—see, for example, Curry, Garnett and Sullivan [134] and Shub and Smale[453] and, in particular, the colour plate in Gleick [213, p. 114].

In the second category are the so-called fail-safe methods which are guaranteed to converge to a root from any starting point. Their disadvantage is that they can be time consuming. We shall implement an algorithm which is due to Müller [356]; and we shall regard this method as our workhorse.

Müller's method, which depends upon a quadratic approximation to the polynomial function, provides an example of the technique of polynomial interpolation which is the subject of the final section of this chapter. The topic is developed more fully in the later chapters which deal with polynomial regression and cubic-spline interpolation.

Before embarking on the major topics, we must lay some groundwork.

Polynomials and their Derivatives

In this section, we shall develop the means for evaluating a polynomial and its derivatives at an arbitrary point, as well as the means for dividing one polynomial by another. First, we consider shifted polynomials.

When the polynomial

$$\alpha(x) = \alpha_0 + \alpha_1 x + \alpha_2 x^2 + \cdots + \alpha_p x^p \tag{4.1}$$

is written as

$$\alpha(x) = \gamma_0 + \gamma_1(x-\xi) + \gamma_2(x-\xi)^2 + \cdots + \gamma_p(x-\xi)^p, \tag{4.2}$$

it is said to be in shifted power form with its centre at ξ.

The usefulness of the shifted form can be recognised when it is compared with the Taylor-series expansion of $\alpha(x)$ about the point ξ:

$$\begin{aligned}\alpha(x) = \alpha(\xi) + \frac{\partial\alpha(\xi)}{\partial x}(x-\xi) + \frac{1}{2!}\frac{\partial^2\alpha(\xi)}{\partial x^2}(x-\xi)^2 + \cdots \\ \cdots + \frac{1}{p!}\frac{\partial^p\alpha(\xi)}{\partial x^p}(x-\xi)^p.\end{aligned} \tag{4.3}$$

By comparing (4.2) and (4.3), it can be see that

$$\gamma_0 = \alpha(\xi) \quad \text{and} \quad \gamma_r = \frac{1}{r!}\frac{\partial^r\alpha(\xi)}{\partial x^r}; \quad r = 1, \ldots, p. \tag{4.4}$$

The coefficients $\gamma_0, \gamma_1, \ldots, \gamma_r$ of the shifted form are obtained by the process of synthetic division. First $\alpha(x)$ is divided by $x - \xi$ to give a quotient $\beta_1(x) = \beta_{10} + \beta_{11}x + \cdots + \beta_{1,p-1}x^{p-1}$ and a remainder γ_0. Then the quotient $\beta_1(x)$ is divided by $x - \xi$ to give a quotient $\beta_2(x)$ and a remainder γ_1. By continuing in this way, the following scheme is generated:

$$\begin{aligned}\alpha(x) &= \gamma_0 + \beta_1(x)(x-\xi), \\ &= \gamma_0 + \gamma_1(x-\xi) + \beta_2(x)(x-\xi)^2, \\ &= \gamma_0 + \gamma_1(x-\xi) + \gamma_2(x-\xi)^2 + \beta_3(x)(x-\xi)^3, \\ &\;\;\vdots \\ &= \gamma_0 + \gamma_1(x-\xi) + \gamma_2(x-\xi)^2 + \cdots + \gamma_p(x-\xi)^p.\end{aligned} \tag{4.5}$$

Here, $\beta_r(x)$ stands for a polynomial in x of degree $p-r$. We proceed from the rth line to the $(r+1)$th line via the equation

$$\beta_r(x) = \gamma_r + \beta_{r+1}(x)(x - \xi). \tag{4.6}$$

Setting $x = \xi$ in the first equation of (4.5) gives $\alpha(\xi) = \gamma_0$. This result, which has been seen already under (4.4), is the subject of a well-known theorem:

(4.7) *The Remainder Theorem* states that the remainder obtained by dividing the polynomial $\alpha(x)$ by $x - \xi$ is $\alpha(\xi)$.

To derive an algorithm for synthetic division, consider writing the first of the equations of (4.5) explicitly to give

$$\alpha_0 + \alpha_1 x + \cdots + \alpha_p x^p = \gamma_0 + \bigl\{\beta_{10} + \beta_{11}x + \cdots + \beta_{1,p-1}x^{p-1}\bigr\}(x - \xi). \tag{4.8}$$

By equating coefficients associated with the same powers of x on either side of the equation, we obtain the following identities:

$$\begin{aligned} \alpha_p &= \beta_{1,p-1}, \\ \alpha_{p-1} &= \beta_{1,p-2} - \beta_{1,p-1}\xi, \\ \alpha_{p-2} &= \beta_{1,p-3} - \beta_{1,p-2}\xi, \\ &\ \vdots \\ \alpha_1 &= \beta_{10} - \beta_{11}\xi, \\ \alpha_0 &= \gamma_0 - \beta_{10}\xi. \end{aligned} \tag{4.9}$$

These can be rearranged to give

$$\begin{aligned} \beta_{1,p-1} &= \alpha_p, \\ \beta_{1,p-2} &= \beta_{1,p-1}\xi + \alpha_{p-1}, \\ \beta_{1,p-3} &= \beta_{1,p-2}\xi + \alpha_{p-2}, \\ &\ \vdots \\ \beta_{10} &= \beta_{11}\xi + \alpha_1, \\ \gamma_0 &= \beta_{10}\xi + \alpha_0. \end{aligned} \tag{4.10}$$

Here is a simple recursion which can be used to generate the coefficients β_{10}, $\beta_{11}, \ldots, \beta_{1,p-1}$ of the quotient polynomial $\beta_1(\xi)$ as well as the value $\gamma_0 = \alpha(\xi)$. The recursion is known as Horner's method of nested multiplication; and the code for implementing it is as follows:

(4.11)
```
procedure Horner(alpha : vector;
                 p : integer;
                 xi : real;
                 var gamma0 : real;
```

```
                    var beta : vector);

    var
      i : integer;

  begin
    beta[p - 1] := alpha[p];
    for i := 1 to p - 1 do
      beta[p - i - 1] := beta[p - i] * xi + alpha[p - i];
    gamma0 := beta[0] * xi + alpha[0];
  end; {Horner}
```

If the only concern is to find $\gamma_0 = \alpha(\xi)$, then the intermediate products $\beta_{1,p-1}, \ldots, \beta_{11}, \beta_{10}$, may be eliminated via a process of repeated substitution running from top to bottom of (4.10). The result will be a nested expression in the form of

$$\alpha(\xi) = \left[\cdots\left\{(\alpha_p\xi + \alpha_{p-1})\xi + \alpha_{p-2}\right\}\xi + \cdots + \alpha_1\right]\xi + \alpha_0, \tag{4.12}$$

which may be evaluated by performing successive multiplications, beginning with the innermost parentheses. The code of the procedure may be modified accordingly to eliminate the array *beta*.

Once the coefficients of $\beta_1(\xi)$ have been found via the recursion in (4.10), the value of the first derivative $\gamma_1 = \partial\alpha(\xi)/\partial x$ can be generated by a further recursion of the same nature. From (4.6), it follows that

$$\beta_1(x) = \gamma_1 + \beta_2(x)(x - \xi). \tag{4.13}$$

Setting $x = \xi$ shows that $\gamma_1 = \beta_1(\xi)$; and, to find this value, Horner's algorithm can be used in the form of

$$\begin{aligned}\beta_{2,p-2} &= \beta_{1,p-1},\\ \beta_{2,p-3} &= \beta_{2,p-2}\xi + \beta_{1,p-2},\\ \beta_{2,p-4} &= \beta_{2,p-3}\xi + \beta_{1,p-3},\\ &\vdots\\ \beta_{20} &= \beta_{21}\xi + \beta_{11},\\ \gamma_1 &= \beta_{20}\xi + \beta_{10}.\end{aligned} \tag{4.14}$$

It should be easy to see how the succeeding values of the sequence $\gamma_2, \ldots, \gamma_p$ could be generated.

It is interesting to recognise that the recursion of (4.14) can also be obtained by differentiating directly each term of the recursion under (4.10) to give

$$
\begin{aligned}
\frac{\partial(\beta_{1,p-1})}{\partial\xi} &= 0,\\
\frac{\partial(\beta_{1,p-2})}{\partial\xi} &= \beta_{1,p-1},\\
\frac{\partial(\beta_{1,p-3})}{\partial\xi} &= \frac{\partial(\beta_{1,p-2})}{\partial\xi}\xi + \beta_{1,p-2},\\
&\;\vdots\\
\frac{\partial(\beta_{10})}{\partial\xi} &= \frac{\partial(\beta_{11})}{\partial\xi} + \beta_{11},\\
\frac{\partial(\gamma_{0})}{\partial\xi} &= \frac{\partial(\beta_{10})}{\partial\xi}\xi + \beta_{10}.
\end{aligned}
\tag{4.15}
$$

A version of the algorithm of nested multiplication can be presented which generates all of the coefficients $\gamma_0, \ldots, \gamma_p$ of the shifted power form under (4.2). In this case, the coefficients $\alpha_0, \ldots, \alpha_p$ of the original form may be overwritten by the new coefficients. Thus, in comparison with the code under (4.11), $beta[p-i-1]$ is replaced by $alpha[p-i]$ and $beta[p-i]$ is replaced by $alpha[p-i+1]$. Also, the code is surrounded by an additional loop wherein j runs from 0 to $p-1$:

(4.16)

```
procedure ShiftedForm(var alpha : vector;
                      xi : real;
                      p : integer);

  var
    i, j : integer;
begin
  for j := 0 to p - 1 do
    for i := 1 to p - j do
      alpha[p - i] := alpha[p - i] + alpha[p - i + 1] * xi;
end; {ShiftedForm}
```

There are occasions when a polynomial must be evaluated whose argument is a complex number. This can be achieved by modifying the procedure *Horner* so that the operations of addition and multiplication are defined for complex numbers. In some computer languages, such as FORTRAN, the necessary complex operations are predefined. In Pascal, they must be defined by the user; and a collection of functions which perform complex operations has been provided already under (2.58)–(2.68).

(4.17)

```
procedure ComplexPoly(alpha : complexVector;
                      p : integer;
                      z : complex;
                      var gamma0 : complex;
```

```
                    var beta : complexVector);

  var
    i : integer;

begin
  beta[p - 1] := alpha[p];
  for i := 1 to p - 1 do
    begin
      beta[p - i - 1] := Cmultiply(beta[p - i], z);
      beta[p - i - 1] := Cadd(beta[p - i - 1], alpha[p - i]);
    end;
  gamma0 := Cmultiply(beta[0], z);
  gamma0 := Cadd(gamma0, alpha[0]);
end; {ComplexPoly}
```

One can avoid using complex operations by computing the real and the imaginary parts of a complex number separately. Consider the generic equation of the recursion under (4.10) which can be written as $\beta_{j-1} = \beta_j z + \alpha_j$. If β_j, α_j and z are complex numbers, then the equation can be expanded to give

$$\begin{aligned}\beta_{j-1}^{re} + i\beta_{j-1}^{im} &= (\beta_j^{re} + i\beta_j^{im})(z^{re} + iz^{im}) + \alpha_j^{re} + i\alpha_j^{im} \\ &= (\beta_j^{re} z^{re} - \beta_j^{im} z^{im}) + \alpha_j^{re} \\ &\quad + i(\beta_j^{im} z^{re} + \beta_j^{re} z^{im}) + i\alpha_j^{im}.\end{aligned} \tag{4.18}$$

By equating the real and the imaginary terms on both sides, we find that

$$\begin{aligned}\beta_{j-1}^{re} &= \beta_j^{re} z^{re} - \beta_j^{im} z^{im} + \alpha_j^{re}, \\ \beta_{j-1}^{im} &= \beta_j^{im} z^{re} + \beta_j^{re} z^{im} + \alpha_j^{im}.\end{aligned} \tag{4.19}$$

The Division Algorithm

So far, we have considered only the problem of dividing a polynomial $\alpha(x)$ by the term $x - \xi$. It is instructive to consider the general problem of dividing a polynomial of degree p by another polynomial of degree $q \leq p$. Let

$$\alpha(x) = \alpha_p x^p + \alpha_{p-1} x^{p-1} + \cdots + \alpha_1 x + \alpha_0 \tag{4.20}$$

be a polynomial of degree p and let

$$\delta(x) = \delta_q x^q + \delta_{q-1} x^{q-1} + \cdots + \delta_1 x + \delta_0 \tag{4.21}$$

be a polynomial of degree $q \leq p$. The object is to divide $\alpha(x)$ by $\delta(x)$ so as to obtain an equation in the form of

$$\alpha(x) = \delta(x)\beta(x) + \rho(x), \tag{4.22}$$

where

$$\beta(x) = \beta_{p-q}x^{p-q} + \beta_{p-q-1}x^{p-q-1} + \cdots + \beta_1 x + \beta_0 \tag{4.23}$$

is the quotient polynomial of degree $q - p$, and

$$\rho(x) = \rho_{q-1}x^{q-1} + \rho_{q-2}x^{q-2} + \cdots + \rho_1 x + \rho_0 \tag{4.24}$$

is the remainder polynomial of degree $q - 1$ at most.

The operation of dividing $\alpha(x)$ by $\delta(x)$ may be performed by the process known at school as long division. First $\delta(x)$ is multiplied by $\{\alpha_p/\delta_q\}x^{p-q}$. The result is a new polynomial with the same leading term as $\alpha(x)$. Then the new polynomial is subtracted from $\alpha(x)$ to yield

$$\gamma_1(x) = \alpha(x) - \beta_{p-q}x^{p-q}\delta(x), \quad \text{where} \quad \beta_{p-q} = \alpha_p/\delta_q. \tag{4.25}$$

The resulting polynomial $\gamma_1(x)$ will have a degree of $g_1 \leq p - 1$ and a leading term of $(\alpha_{p-1} - \beta_{p-q}\delta_{q-1})x^{p-q-1}$. If $g_1 < q$, then the process terminates here and $\gamma_1(x) = \rho(x)$ is designated the remainder. Otherwise, with $g_1 \geq q$, we can proceed to the next step which is to form a new polynomial by multiplying $\delta(x)$ by $\{(\alpha_{p-1} - \beta_{p-q}\delta_{q-1})/\delta_q\}x^{p-q-1}$. The new polynomial is subtracted from $\gamma_1(x)$ to yield

$$\begin{aligned} \gamma_2(x) &= \gamma_1(x) - \beta_{p-q-1}x^{p-q-1}\delta(x), \\ \text{where} \quad \beta_{p-q-1} &= (\alpha_{p-1} - \beta_{p-q}\delta_{q-1})/\delta_q. \end{aligned} \tag{4.26}$$

The process may continue through further steps based on the generic equation

$$\gamma_{n+1}(x) = \gamma_n(x) - \beta_{p-q-n}x^{p-q-n}\delta(x); \tag{4.27}$$

but, ultimately, it must terminate when $n = p - q$.

When the process terminates, the results from each stage are substituted into the next stage; and thus an expression is derived which corresponds to equation (4.22):

$$\begin{aligned} \gamma_{p-q-1}(x) &= \alpha(x) - (\beta_{p-q}x^{p-q} + \beta_{p-q-1}x^{p-q-1} + \cdots + \beta_1 x + \beta_0)\delta(x) \\ &= \rho(x). \end{aligned} \tag{4.28}$$

Example 4.1. The familiar form in which long division is presented may be illustrated as follows:

$$\begin{array}{r|l} & 4x^2 + 2x + 1 \\ \hline 4x^2 - 2x + 1 & 16x^4 \qquad\qquad + 4x^2 + x \\ & 16x^4 - 8x^3 + 4x^2 \\ \hline & \qquad 8x^3 \qquad\qquad + x \\ & \qquad 8x^3 - 4x^2 + 2x \\ \hline & \qquad\qquad 4x^2 - x \\ & \qquad\qquad 4x^2 - 2x + 1 \\ \hline & \qquad\qquad\qquad x - 1. \end{array} \tag{4.29}$$

The result can be expressed in the manner of (4.22) by writing

$$16x^4 + 4x^2 + x = (4x^2 - 2x + 1)(4x^2 + 2x + 1) + (x - 1). \tag{4.30}$$

An alternative way of developing an algorithm to represent the process of long division is to adopt the method of detached coefficients which has been employed in the previous section in connection with the process of synthetic division. Consider writing the equation (4.22) as

$$\sum_{j=0}^{p} \alpha_j x^j = \sum_{j=0}^{p} \Big(\sum_{i=h}^{s} \beta_i \delta_{j-i} \Big) x^j + \sum_{j=0}^{q-1} \rho_j x^j, \tag{4.31}$$

where

$$h = \max(0, j - q) \quad \text{and} \quad s = \min(j, p - q). \tag{4.32}$$

These limits on the index i are implied by the restriction $0 \leq i \leq p - q$, which corresponds to the range of the index i in the sequence $\beta_0, \ldots, \beta_i, \ldots, \beta_{p-q}$, together with the restriction $0 \leq j - i \leq q$, which corresponds to the range of the index $j - i$ in the sequence $\delta_0, \ldots, \delta_{j-i}, \ldots, \delta_q$. By equating coefficients associated with the same powers of x on both sides of the equation (4.31), it is found that

$$\alpha_j = \sum_{i=h}^{s} \beta_i \delta_{j-i} \quad \text{if} \quad j \geq q, \tag{4.33}$$

and that

$$\alpha_j = \sum_{i=h}^{s} \beta_i \delta_{j-i} + \rho_j \quad \text{if} \quad j < q. \tag{4.34}$$

By rearranging these results, expressions for the coefficients of the quotient and remainder polynomials are found in the form of

$$\beta_{j-q} = \frac{1}{\delta_q} \Big(\alpha_j - \sum_{i=j-q+1}^{s} \beta_i \delta_{j-i} \Big) \tag{4.35}$$

and

$$\rho_j = \alpha_j - \sum_{i=h}^{s} \beta_i \delta_{j-i}. \tag{4.36}$$

An alternative and perhaps a neater expression for the division algorithm may be derived from an equation which sets

$$\alpha_j = \sum_{i=j-q}^{s} \beta_i \delta_{j-i} \quad \text{for} \quad j = 0, \ldots, p. \tag{4.37}$$

Here, in comparison with the expression under (4.33), the lower bound on the index i has been relaxed so as to create the additional coefficients $\beta_{-1}, \beta_{-2}, \ldots, \beta_{-q}$. In effect, these new coefficients replace the coefficients $\rho_0, \rho_1, \ldots, \rho_{q-1}$ of the remainder polynomial $\rho(x)$ which is reparametrised as follows:

$$\begin{aligned}\rho(x) &= \rho_{q-1}x^{q-1} + \rho_{q-2}x^{q-2} + \cdots + \rho_1 x + \rho_0\\ &= \beta_{-1}(\delta_q x^{q-1} + \delta_{q-1}x^{q-2} + \cdots + \delta_2 x + \delta_1)\\ &\quad + \beta_{-2}(\delta_q x^{q-2} + \delta_{q-1}x^{q-3} + \cdots + \delta_2)\\ &\quad + \cdots\\ &\quad + \beta_{1-q}(\delta_q x + \delta_{q-1})\\ &\quad + \beta_{-q}\delta_q.\end{aligned} \tag{4.38}$$

The relationship between the two sets of parameters is expressed in an identity

$$\begin{bmatrix}\rho_{q-1}\\ \rho_{q-2}\\ \vdots\\ \rho_1\\ \rho_0\end{bmatrix} = \begin{bmatrix}\delta_q & 0 & \ldots & 0 & 0\\ \delta_{q-1} & \delta_q & \ldots & 0 & 0\\ \vdots & \vdots & \ddots & \vdots & \vdots\\ \delta_2 & \delta_3 & \ldots & \delta_q & 0\\ \delta_1 & \delta_2 & \ldots & \delta_{q-1} & \delta_q\end{bmatrix}\begin{bmatrix}\beta_{-1}\\ \beta_{-2}\\ \vdots\\ \beta_{1-q}\\ \beta_{-q}\end{bmatrix} \tag{4.39}$$

which entails a one-to-one mapping.

The new remainder parameters are found using equation (4.35). As j runs from p down to q, the coefficients of the quotient polynomial are generated; and, as j runs from $q-1$ to 0, the coefficients $\beta_{-1}, \ldots, \beta_{-q}$, of the reparametrised form of the remainder polynomial materialise. Once the full set of parameters $\beta_{p-q}, \ldots, \beta_{-q}$, has been generated, which includes the set of alternative parameters for the remainder polynomial, we can obtain the ordinary parameters of the remainder from the expression

$$\rho_j = \sum_{i=j-q}^{-1} \beta_i \delta_{j-i}. \tag{4.40}$$

Example 4.2. Consider

$$\begin{aligned}&\alpha_4 x^4 + \alpha_3 x^3 + \alpha_2 x^2 + \alpha_1 x + \alpha_0\\ &= (\beta_2 x^2 + \beta_1 x + \beta_0)(\delta_2 x^2 + \delta_1 x + \delta_0) + (\rho_1 x + \rho_0).\end{aligned} \tag{4.41}$$

The coefficients of the quotient polynomial are found by the recursion

$$\begin{aligned}\beta_2 &= \alpha_4/\delta_2,\\ \beta_1 &= (\alpha_3 - \beta_2\delta_1)/\delta_2,\\ \beta_0 &= (\alpha_2 - \beta_2\delta_0 - \beta_1\delta_1)/\delta_2.\end{aligned} \tag{4.42}$$

Then the coefficients of the remainder are found via

$$\begin{aligned}\rho_1 &= \alpha_1 - \beta_1\delta_0 - \beta_0\delta_1,\\ \rho_0 &= \alpha_0 - \beta_0\delta_0.\end{aligned} \tag{4.43}$$

Alternatively, if we set

$$\rho_1 x + \rho_0 = \beta_{-1}(\delta_2 x + \delta_1) + \beta_{-2}\delta_2, \tag{4.44}$$

then we can extend the recursion to generate the alternative coefficients of the remainder:

$$\begin{aligned}\beta_{-1} &= (\alpha_1 - \beta_1\delta_0 - \beta_0\delta_1)/\delta_2,\\ \beta_{-2} &= (\alpha_0 - \beta_0\delta_0 - \beta_{-1}\delta_1)/\delta_2.\end{aligned} \tag{4.45}$$

The following procedure for the division algorithm generates the alternative parameters of the remainder and then it transforms them into the ordinary parameters:

(4.46)

```
procedure DivisionAlgorithm(alpha, delta : vector;
                            p, q : integer;
                            var beta : jvector;
                            var rho : vector);

  var
    store : real;
    h, s, j, i : integer;

begin

  for j := p downto 0 do
    begin {j}
      s := Min(j, p - q);
      store := 0.0;
      for i := j - q + 1 to s do
        store := store + beta[i] * delta[j - i];
      beta[j - q] := (alpha[j] - store)/delta[q]
    end; {j}

  for j := 0 to q - 1 do
    begin {j}
      rho[j] := 0.0;
      for i := j - q to - 1 do
        rho[j] := rho[j] + beta[i] * delta[j - i];
    end; {j}

end; {DivisionAlgorithm}
```

Roots of Polynomials

A root of a polynomial $\alpha(x)$ is any real or complex value λ such that $\alpha(\lambda) = 0$.

(4.47) If $\lambda_1, \lambda_2, \ldots, \lambda_r$ are distinct roots of the polynomial $\alpha(x)$, then $\alpha(x) = (x - \lambda_1)(x - \lambda_2)\cdots(x - \lambda_r)\beta_r(x)$ for some polynomial $\beta_r(x)$.

Proof. According to the remainder theorem, we can write $\alpha(x) = \beta_1(x)(x-\xi) + \alpha(\xi)$. Setting $\xi = \lambda_1$, and using $\alpha(\lambda_1) = 0$, gives $\alpha(x) = \beta_1(x)(x-\lambda_1)$; and this proves the theorem for a single root.

Now assume that the theorem is true for $r-1$ roots so that $\alpha(x) = (x-\lambda_1)(x-\lambda_2)\cdots(x-\lambda_{r-1})\beta_{r-1}(x)$ for some polynomial $\beta_{r-1}(x)$. Since $\alpha(\lambda_r) = 0$ by assumption, and since $(\lambda_r - \lambda_1), \ldots, (\lambda_r - \lambda_{r-1})$ are all nonzero, it follows that $\beta_{r-1}(\lambda_r) = 0$. Therefore, setting $\xi = \lambda_r$ in $\beta_{r-1}(x) = \beta_r(x)(x-\xi) + \beta_{r-1}(\xi)$ shows that $\beta_{r-1}(x) = \beta_r(x)(x-\lambda_r)$; and thus the theorem is proved.

The fundamental theorem of algebra asserts that every polynomial which is defined over the complex plane has at least one root in that domain. From this, it can be inferred that any polynomial of degree $p > 1$ is equal to the product of p linear factors with complex coefficients.

The case of degree 1 requires no proof. The general case follows easily by induction. Assume that the theorem holds for all polynomials of degrees less than p, and let $\alpha(x)$ have a degree of p. Then, since $\alpha(x)$ has a root λ, it follows that $\alpha(x) = (x-\lambda)\beta(x)$, where $\beta(x)$ has a degree of $p-1$. But, by assumption, $\beta(x)$ has $p-1$ linear factors so $\alpha(x)$ must have p such factors.

On gathering together the factors of the polynomial $\alpha(x)$, we obtain

$$\begin{aligned} \alpha(x) &= \alpha_p x^p + \cdots + \alpha_1 x + \alpha_0 \\ &= \alpha_p (x-\lambda_1)^{r_1} \cdots (x-\lambda_s)^{r_s}. \end{aligned} \tag{4.48}$$

Here $\lambda_1, \ldots, \lambda_s$ are the distinct roots of $\alpha(x)$ whilst $r_1 + \cdots + r_s = p$ is the sum of their multiplicities.

We shall take the view that, in practical applications, multiple roots rarely arise, unless they represent a feature of the design of a model. In that case, their values may well be known in advance. Therefore, in the algorithms for root finding which we shall present in the following sections, we shall make no special provision for multiple roots. We shall begin by presenting a time-honoured algorithm for finding the real roots of a polynomial. Then we shall describe the more general algorithms which are capable of finding both real and complex roots.

Real Roots

A common procedure for finding a real-valued solution or root of the polynomial equation $\alpha(x) = 0$ is the Newton–Raphson procedure which depends upon approximating the curve $y = \alpha(x)$ by its tangent at a point near the root. Let this point be $[x_0, \alpha(x_0)]$. Then the equation of the tangent is

$$y = \alpha(x_0) + \frac{\partial \alpha(x_0)}{\partial x}(x - x_0) \tag{4.49}$$

and, on setting $y = 0$, we find that this line intersects the x-axis at

$$x_1 = x_0 - \left[\frac{\partial \alpha(x_0)}{\partial x}\right]^{-1} \alpha(x_0). \tag{4.50}$$

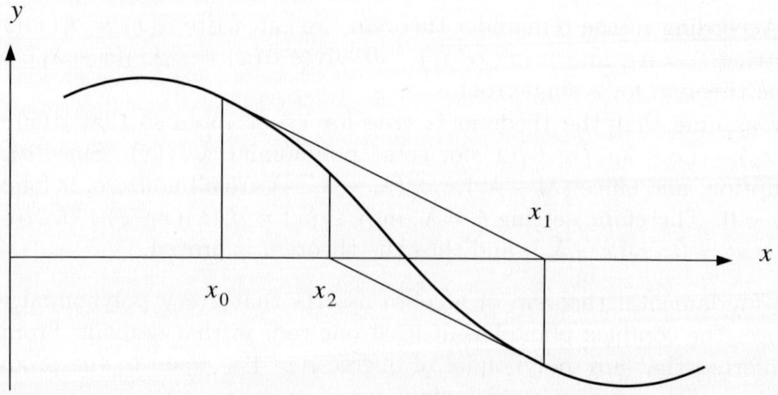

Figure 4.1. If x_0 is close to the root of the equation $\alpha(x) = 0$, then we can expect x_1 to be closer still.

If x_0 is close to the root λ of the equation $\alpha(x) = 0$, then we can expect x_1 to be closer still (see Figure 4.1). To find an accurate approximation to λ, we generate a sequence of approximations $\{x_0, x_1, \ldots, x_r, x_{r+1}, \ldots\}$ according to the algorithm

$$x_{r+1} = x_r - \left[\frac{\partial\alpha(x_r)}{\partial x}\right]^{-1}\alpha(x_r). \tag{4.51}$$

One should beware that the Newton–Raphson algorithm will not work well when two roots coincide; because, then, both the function $\alpha(x)$ and its derivative will vanish in the neighbourhood of the roots.

In order to implement the algorithm, an efficient method is required for evaluating the polynomial $\alpha(x)$ and is derivative $\partial\alpha(x)/\partial x$ at an arbitrary point ξ. This is provided by Horner's method of nested multiplication which has been described in a previous section. The following procedure takes in the values of the coefficients of $\alpha(x)$ together with an initial estimate of a root λ. It returns an accurate value of the root together with the coefficients of the quotient polynomial $\beta(x)$ which is defined by the equation $\alpha(x) = (x - \lambda)\beta(x)$.

(4.52)
```
procedure RealRoot(p : integer;
                   alpha : vector;
                   var root : real;
                   var beta : vector);

var
  x, oldx, f, fprime : real;
  iterations : integer;
  quotient : vector;
  convergence : boolean;
```

```
begin {Real Root}
  x := root;
  iterations := 0;
  convergence := false;

  repeat
    Horner(alpha, p, x, f, beta);
    Horner(beta, p - 1, x, fprime, quotient);
    oldx := x;
    x := x - f/fprime;
    iterations := iterations + 1;
    if Abs(x - oldx) < Abs(x) * 1E - 5 then
      convergence := true
  until (convergence) or (iterations > 20);

  root := x;
  if not convergence then
    Writeln('The program failed to converge');
end; {RealRoot}
```

If λ_1 is a root of the polynomial $\alpha(x)$, then the remainder on division by $(x-\lambda_1)$ is zero, and so $\alpha(x) = (x-\lambda_1)\beta_1(x)$. Therefore, further roots of $\alpha(x)$ can be sought by seeking the roots of the so-called deflated polynomial $\beta_1(x)$. The coefficients of $\beta_1(x)$ are already available from the procedure above as the elements of the array *beta*. To find a second root, the procedure can be applied a second time with *beta* in place of *alpha*; and successive roots could be found in like manner.

The disadvantage of this method is that it is liable to lead to an accumulation of errors. Therefore, once an initial estimate of a root λ_{i+1} has been found from a deflated polynomial $\beta_i(x)$, a more accurate estimate should be sought by applying the procedure again to the original polynomial $\alpha(x)$ taking the initial estimate as a starting value. Once the accurate estimate has been found, it can be used as a starting value in a further application of the procedure aimed at producing a refined estimate of the deflated polynomial $\beta_{i+1}(x)$ which is to be used in the next round of root extraction. This strategy is represented by the following procedure which invokes the preceding procedure *RealRoot*:

(4.53)

```
procedure NRealRoots(p, NofRoots : integer;
                     var alpha, beta, lambda : vector);

  var
    i : integer;
    root : real;
    q : vector;

begin {NRealRoots}
  beta := alpha;
  for i := 0 to NofRoots - 1 do
```

```
    begin
      root := 0;
      RealPolyRoots(p - i, beta, root, q); {Initial value of root}
      RealPolyRoots(p, alpha, root, q); {Refined value of root}
      lambda[i + 1] := root;
      RealPolyRoots(p - i, beta, root, q); {Refined value of beta}
      beta := q;
    end;
end; {NRealRoots}
```

A disadvantage of the Newton–Raphson procedure is that it cannot be relied upon, in general, to converge to a root of a polynomial unless the starting value is sufficiently close. However, if all of the roots are real, then the procedure will find them. In particular, it can be proved that

(4.54) If $\alpha(x)$ is a polynomial of degree $n \geq 2$ with real coefficients, and if all the roots $\lambda_n \geq \lambda_{n-1} \geq \cdots \geq \lambda_1$ of $\alpha(x) = 0$ are real, then the Newton–Raphson method yields a convergent strictly decreasing sequence for any initial value $x_0 > \lambda_n$.

A proof is given by Stoer and Bulirsh [474, p. 272].

The result implies that, if we start by finding the largest root, then we can proceed, via a process of deflation, to find the next largest root with certainty, and so on, down to the smallest root. However, in the procedure *NRealRoots*, the starting value is invariably set to zero, with the effect that we tend to find the roots in an ascending order of magnitude. In practice, this strategy will also work reliably when all of the roots are real.

According to the version of Horner's algorithm presented above, the coefficients of the deflated polynomial $\beta_0 + \beta_1 x + \cdots + \beta_{n-1}x^{n-1}$ are computed in the order $\beta_{n-1}, \beta_{n-2}, \ldots, \beta_0$. This procedure, which is know as forward deflation, is numerically stable when the deflating factor contains the smallest absolute root. If we were to deflate the polynomial by the factor containing the largest root, then it would be appropriate to calculate the coefficients in reverse order. The largest root of $\beta(x) = 0$ is, of course, the reciprocal of the smallest root of $x^{n-1}\beta(x^{-1}) = \beta_0 x^{n-1} + \beta_1 x^{n-2} + \cdots + \beta^{n-1} = 0$; and this helps to explain why the procedure of backwards deflation is appropriate to the latter case.

There are numerous results which can help in locating approximate values of the real roots in less favourable cases. The problem of determining the number of real roots of a polynomial equation engaged the attention of mathematicians for almost two hundred years. The period begins in 1637 with Descartes' rule of signs which establishes upper limits to the numbers of positive and negative real roots. It ends with the solution of Sturm [476] which was announced in 1829 and published in 1835. According to Sturm's theorem,

(4.55) There exists a sequence of real polynomials $f(x), f_1(x), \ldots, f_p(x)$, whose degrees are in descending order, such that, if $b > a$, then the number of distinct real roots of $\alpha(z) = 0$ between a and b is equal to the excess of the number of changes of signs in the sequence $f, f_1, \ldots, f_p$ when $x = a$ over the number of changes of sign when $x = b$.

There can be many Sturm sequences which fulfil these prescriptions. To show how to construct one such sequence, let $f(x) = 0$ be a polynomial equation with distinct roots. Then $f(x)$ and its derivative $f_1(x)$ have only a nonzero constant f_p as their highest common factor. The Sturm sequence can be generated by the usual process for finding the common factor which relies upon an adaptation of Euclid's algorithm which has been presented under (3.1):

$$\begin{aligned} f(x) &= q_1(x)f_1(x) - f_2(x), \\ f_1(x) &= q_2(x)f_2(x) - f_3(x), \\ &\vdots \\ f_{p-2}(x) &= q_{p-1}(x)f_{p-1}(x) - f_p. \end{aligned} \tag{4.56}$$

Observe that, for a given values of x, no two consecutive polynomials of this Sturm sequence can vanish since, otherwise, there would be $f_p = 0$. Moreover, if λ is a root of $f_i(x) = 0$, then (4.56) shows that

$$f_{i-1}(\lambda) = -f_{i+1}(\lambda); \tag{4.57}$$

and, from the continuity of the functions, it follows that $f_{i-1}(x)$ and $f_{i+1}(x)$ have opposite signs in the neighbourhood of λ. Therefore, as x passes through the value of λ, which is to say, as the sign of $f_i(x)$ changes, the sequence

$$f_{i-1}(x), f_i(x), f_{i+1}(x) \tag{4.58}$$

continues to display a single change of sign. Hence, when x increases through the value of a root of any $f_i(x)$, the number of sign changes in the sequence remains unaltered.

On the other hand, as x increases through a root of $f(x) = 0$, the signs of $f(x)$ and of its derivative $f_1(x)$ change from being opposite to being the same; and hence the number of sign changes in the Sturm sequence decreases by one as x passes a root of $f(x) = 0$. This establishes that the Sturm sequence of (4.56) fulfils the prescriptions of (4.55).

For a fuller exposition of the theory of Sturm sequences, one may consult the texts on the theory of equations of Todhunter [486], Dickson [157] and Uspensky [495], or the more recent text in numerical analysis of Ralston and Rabinowitz [419].

In time-series analysis, root-finding methods are commonly applied to polynomial equations of the form $\alpha(L) = 0$ wherein L is the lag operator. Usually a stability condition prevails which restricts the roots of the equation $\alpha(z^{-1}) = 0$ to lie within the unit circle. The Newton–Raphson procedure can be expected to perform well under such circumstances, even when there is no attempt at finding appropriate staring values.

In the next chapter, we shall present a means of assessing whether or not the roots of $\alpha(z^{-1}) = 0$ to lie within the unit circle which relieves of the need to find these roots directly. This can help us, sometimes, to avoid some heavy labour.

Complex Roots

A more sophisticated version of the Newton–Raphson algorithm, which uses complex arithmetic, can be employed for finding complex roots.

However, in the case of a polynomial with real coefficients, the complex roots must occur in conjugate pairs which are to be found within quadratic factors of the polynomial. Therefore, instead of looking for one complex root at a time, we may look for the roots by isolating the quadratic factors. This idea leads to the method of Bairstow.

Consider dividing $\alpha(x)$ by the quadratic $\delta(x) = x^2 + \delta_1 x + \delta_0$. If the alternative representation of the remainder polynomial is used which has been developed in the context of the division algorithm, then this gives

$$(4.59)\qquad \begin{aligned}\alpha_p x^p + \alpha_{p-1}x^{p-1} + \cdots + \alpha_1 x + \alpha_0 &= \beta_{-1}(x+\delta_1) + \beta_{-2}\\ &+ (x^2+\delta_1 x+\delta_0)(\beta_{p-2}x^{p-2} + \beta_{p-3}x^{p-3} + \cdots + \beta_1 x + \beta_0).\end{aligned}$$

Here the terms $\beta_{-1}(x+\delta_1)+\beta_{-2}$ constitute the linear remainder; and, if the divisor $\delta(x)$ is indeed a quadratic factor of $\alpha(x)$, then the remainder must be zero. In effect, values of δ_1 and δ_0 must be sought which will satisfy the equations

$$(4.60)\qquad \begin{aligned}\beta_{-1}(\delta_1,\delta_0) &= 0,\\ \beta_{-2}(\delta_1,\delta_0) &= 0.\end{aligned}$$

Let d_1 and d_0 be values which come close to satisfying both equations. Then the equations $y_1 = \beta_{-1}(\delta_1,\delta_0)$ and $y_2 = \beta_{-2}(\delta_1,\delta_0)$, can be approximated in the neighbourhood of the point (d_1, d_0) by the following linear functions:

$$(4.61)\qquad \begin{aligned}y_1 &= \beta_{-1}(d_1,d_0) + \frac{\partial\beta_{-1}}{\partial\delta_1}(\delta_1 - d_1) + \frac{\partial\beta_{-1}}{\partial\delta_0}(\delta_0 - d_0),\\ y_2 &= \beta_{-2}(d_1,d_0) + \frac{\partial\beta_{-2}}{\partial\delta_1}(\delta_1 - d_1) + \frac{\partial\beta_{-2}}{\partial\delta_0}(\delta_0 - d_0).\end{aligned}$$

Here it is understood that the derivatives are also evaluated at the point (d_1, d_0). Setting $y_1,\ y_2 = 0$ and putting the equations in a matrix format gives

$$(4.62)\qquad \begin{bmatrix}\beta_{-1}\\ \beta_{-2}\end{bmatrix} = \begin{bmatrix}\dfrac{\partial\beta_{-1}}{\partial\delta_1} & \dfrac{\partial\beta_{-1}}{\partial\delta_0}\\ \dfrac{\partial\beta_{-2}}{\partial\delta_1} & \dfrac{\partial\beta_{-2}}{\partial\delta_0}\end{bmatrix}\begin{bmatrix}d_1-\delta_1\\ d_0-\delta_0\end{bmatrix};$$

and the solution is

$$(4.63)\qquad \begin{bmatrix}\delta_1\\ \delta_0\end{bmatrix} = \begin{bmatrix}d_1\\ d_0\end{bmatrix} - \begin{bmatrix}\dfrac{\partial\beta_{-1}}{\partial\delta_1} & \dfrac{\partial\beta_{-1}}{\partial\delta_0}\\ \dfrac{\partial\beta_{-2}}{\partial\delta_1} & \dfrac{\partial\beta_{-2}}{\partial\delta_0}\end{bmatrix}^{-1}\begin{bmatrix}\beta_{-1}\\ \beta_{-2}\end{bmatrix}.$$

Of course, the values of δ_1 and δ_0 which are determined by these equations are still only approximations to the ones which satisfy the equations (4.60), yet they are expected to be better approximations than d_1 and d_0 respectively. Equation (4.63) is in the form of a two-dimensional version of the Newton–Raphson algorithm, and it may be used to generate a succession of improving approximations to the solution of the equations (4.60).

To implement the algorithm for finding the parameters of the quadratic factor, we must be able to generate the values of the functions $\beta_{-1}(\delta_1, \delta_0)$ and $\beta_{-2}(\delta_1, \delta_0)$ and their derivatives corresponding to arbitrary values of δ_1 and δ_0. The division algorithm may be used to generate β_{-1} and β_{-2}. Setting $q = 2$ in equation (4.37) an letting j run from p down to 0 generates the following sequence which ends with the requisite values:

$$
\begin{aligned}
\beta_{p-2} &= \alpha_p, \\
\beta_{p-3} &= \alpha_{p-1} - \delta_1 \beta_{p-2}, \\
\beta_{p-4} &= \alpha_{p-2} - \delta_1 \beta_{p-3} - \delta_0 \beta_{p-2}, \\
&\vdots \\
\beta_0 &= \alpha_2 - \delta_1 \beta_1 - \delta_0 \beta_2, \\
\beta_{-1} &= \alpha_1 - \delta_1 \beta_0 - \delta_0 \beta_1, \\
\beta_{-2} &= \alpha_0 - \delta_1 \beta_{-1} - \delta_0 \beta_0.
\end{aligned}
\tag{4.64}
$$

The derivatives, may be found by differentiating the recurrence relationship. To simplify the notation, let $(\beta_j)_1 = \partial\beta_j / \partial\delta_1$. Then

$$
\begin{aligned}
(\beta_{p-2})_1 &= 0, \\
(\beta_{p-3})_1 &= -\beta_{p-2}, \\
(\beta_{p-4})_1 &= -\beta_{p-3} - \delta_1 (\beta_{p-3})_1, \\
(\beta_{p-5})_1 &= -\beta_{p-4} - \delta_1 (\beta_{p-4})_1 - \delta_0 (\beta_{p-3})_1, \\
&\vdots \\
(\beta_0)_1 &= -\beta_1 - \delta_1 (\beta_1)_1 - \delta_0 (\beta_2)_1 = c_0, \\
(\beta_{-1})_1 &= -\beta_0 - \delta_1 (\beta_0)_1 - \delta_0 (\beta_1)_1 = c_1, \\
(\beta_{-2})_1 &= -\beta_{-1} - \delta_1 (\beta_{-1})_1 - \delta_0 (\beta_0)_1 = c_2;
\end{aligned}
\tag{4.65}
$$

and the last two terms generated by this recurrence are two of the sought-after derivatives: $(\beta_{-1})_1 = \partial\beta_{-1}/\partial\delta_1$ and $(\beta_{-2})_1 = \partial\beta_{-2}/\partial\delta_1$.

Next, by differentiating the recurrence relationship of (4.64) with respect to

δ_0, we get

$$(4.66)\qquad \begin{aligned}(\beta_{p-2})_0 &= 0,\\ (\beta_{p-3})_0 &= 0,\\ (\beta_{p-4})_0 &= -\beta_{p-2},\\ (\beta_{p-5})_0 &= -\delta_1(\beta_{p-4})_0 - \beta_{p-3},\\ &\vdots\\ (\beta_0)_0 &= -\delta_1(\beta_1)_0 - \delta_0(\beta_2)_0 - \beta_2,\\ (\beta_{-1})_0 &= -\delta_1(\beta_0)_0 - \delta_0(\beta_1)_0 - \beta_1,\\ (\beta_{-2})_0 &= -\delta_1(\beta_{-1})_0 - \delta_0(\beta_0)_0 - \beta_0.\end{aligned}$$

This provides us with the remaining two derivatives $(\beta_{-1})_0 = \partial\beta_{-1}/\partial\delta_0$, and $(\beta_{-2})_0 = \partial\beta_{-2}/\partial\delta_0$. Notice, however, that the schemes under (4.65) and (4.66) generate the same values, with $(\beta_j)_0 = (\beta_{j+1})_1$. Therefore, the scheme under (4.66) is actually redundant; and a single recurrence serves to generate the three distinct values which are found within the matrix

$$(4.67)\qquad \begin{bmatrix}\dfrac{\partial\beta_{-1}}{\partial\delta_1} & \dfrac{\partial\beta_{-1}}{\partial\delta_0}\\[2ex] \dfrac{\partial\beta_{-2}}{\partial\delta_1} & \dfrac{\partial\beta_{-2}}{\partial\delta_0}\end{bmatrix} = \begin{bmatrix}(\beta_{-1})_1 & (\beta_0)_1\\[2ex] (\beta_{-2})_1 & (\beta_{-1})_1\end{bmatrix} = \begin{bmatrix}c_1 & c_0\\[2ex] c_2 & c_1\end{bmatrix}.$$

The values in (4.67), together with the values of β_{-1} and β_{-2}, are generated by the following procedure which implements the recursions of (4.64) and (4.66):

(4.68)

```
procedure QuadraticDeflation(alpha : vector;
                             delta0, delta1 : real;
                             p : integer;
                             var beta : vector;
                             var c0, c1, c2 : real);

var
  i : integer;

begin {QuadraticDeflation}
  beta[p - 2] := alpha[p];
  beta[p - 3] := alpha[p - 1] - delta1 * beta[p - 2];
  c1 := 0;
  c2 := -beta[p - 2];

  for i := 4 to p + 2 do
    begin
      beta[p - i] := alpha[p - i + 2] - delta1 * beta[p - i + 1];
      beta[p - i] := beta[p - i] - delta0 * beta[p - i + 2];
```

```
        c0 := c1;
        c1 := c2;
        c2 := -beta[p - i + 1] - delta1 * c1 - delta0 * c0;
      end;
    end; {QuadraticDeflation}
```

The procedure *QuadraticDeflation* provides the various elements which are needed in implementing the Newton–Raphson procedure depicted in equation (4.63). Now this equation can be rewritten as

(4.69)
$$\delta_1 = d_1 - \frac{c_1\beta_{-1} - c_0\beta_{-2}}{c_1^2 - c_0c_2} \quad \text{and}$$
$$\delta_1 = d_1 - \frac{c_1\beta_{-2} - c_2\beta_{-1}}{c_1^2 - c_0c_2}.$$

The following procedure implements the algorithm in a simple way:

(4.70)
```
procedure Bairstow(alpha : vector;
                   p : integer;
                   var delta0, delta1 : real;
                   var beta : vector);

  var
    iterations : integer;
    c0, c1, c2, det : real;
    convergence : boolean;

begin {Bairstow}
  iterations := 0;
  convergence := false;

  repeat
    QuadraticDeflation(alpha, delta0, delta1, p, beta, c0, c1, c2);
    det := Sqr(c1) - c0 * c2;
    delta1 := delta1 - (c1 * beta[-1] - c0 * beta[-2])/det;
    delta0 := delta0 - (c1 * beta[-2] - c2 * beta[-1])/det;
    iterations := iterations + 1;
    if (Abs(beta[-1]) < 1E - 5) and (Abs(beta[-2]) < 1E - 5) then
      convergence := true;
  until (convergence) or (iterations > 30);

  if not convergence then
    Writeln('The program failed to converge');

end; {Bairstow}
```

The procedure delivers the coefficients of the quadratic factor $x^2 + \delta_1 x + \delta_0$; and, from these, the roots are calculated readily as

$$\lambda_1, \lambda_2 = \frac{-\delta_1 \pm \sqrt{\delta_1^2 - 4\delta_0}}{2}. \tag{4.71}$$

The procedure also returns the coefficients of the deflated polynomial $\beta(x)$.

We can proceed to extract further quadratic factors from the deflated polynomial; but, in doing so, we should use the same processes of refinement as we have applied to the calculation of real roots. That is to say, once an initial estimate of a quadratic factor has been found from the deflated polynomial, it should be recalculated from the original polynomial using the initial estimate as the starting value. Then the deflated polynomial should be recalculated. These steps should be taken in order to avoid an accumulation of rounding errors.

(4.72)
```
procedure MultiBairstow(p : integer;
                 var alpha : vector;
                 var lambda : complexVector);

  var
    i, j, r : integer;
    c0, c1, c2, delta0, delta1 : real;
    beta, quotient : vector;

begin {MultiBairstow}

  if Odd(p) then
    r := (p - 1) div 2
  else
    r := p div 2;
  beta := alpha;

  for i := 0 to r - 1 do
    begin
      j := p - 2 * i;
      delta1 := 0;
      delta0 := 0;
      Bairstow(beta, j, delta0, delta1, quotient); {Initial value}
      Bairstow(alpha, p, delta0, delta1, quotient); {Refined value}
      RootsOfFactor(j, delta0, delta1, lambda);
      QuadraticDeflation(beta, delta0, delta1, p, quotient, c0, c1, c2);
      beta := quotient;
    end;

  if Odd(p) then
    begin
      lambda[1].re := -quotient[0];
```

```
        lambda[1].im := 0.0
      end;

  end; {MultiBairstow}
```

In order to extract the roots from the quadratic factors and to place them within the complex vector *lambda*, the following procedure is called by *MultiBairstow*:

(4.73)
```
procedure RootsOfFactor(i : integer;
                   delta0, delta1 : real;
                   var lambda : complexVector);

begin

  if Sqr(delta1) <= 4 * delta0 then
    begin {complex roots}
      lambda[i].re := -delta1/2;
      lambda[i - 1].re := -delta1/2;
      lambda[i].im := Sqrt(4 * delta0 - Sqr(delta1))/2;
      lambda[i - 1].im := -Sqrt(4 * delta0 - Sqr(delta1))/2;
    end

  else if Sqr(delta1) > 4 * delta0 then
    begin {real roots}
      lambda[i].re := (-delta1 + Sqrt(Sqr(delta1) - 4 * delta0))/2;
      lambda[i - 1].re := (-delta1 - Sqrt(Sqr(delta1) - 4 * delta0))/2;
      lambda[i].im := 0.0;
      lambda[i - 1].im := 0.0;
    end;

end; {RootsOfFactor}
```

Müller's Method

Now we must face the problems which arise when Bairstow's procedure fails to converge from the rough starting values which we have provided.

There is an extensive and difficult literature relating to the problem of locating the complex roots of a polynomial (see Marden [331], [332], for example); and it should be possible to incorporate the lessons of these studies within a computer program with the object of finding accurate starting values. However, this would entail very extensive coding.

The approach which we shall follow here is to resort, in the case of an initial failure, to a robust and slow method which is guaranteed to find a root no matter which starting values are taken. We shall implement a method of Müller [356] which is sometimes used in "hands-off" applications where there can be no intercession from the user to guide the process of root finding.

Müller's method discovers the roots one at a time by ranging over the complex plane. It uses quadratic approximations to the polynomial which are valid in the vicinity of a root. If the root of the polynomial is real, then it may be approximated by one of the roots of the quadratic. If the root of the polynomial is complex, then the quadratic is also likely to have a complex root which can serve as an approximation. In either case, if the approximation to the root is inadequate, then a new quadratic function is calculated.

There is no need for an accurate initial approximation. All that is required, in the beginning, is a set of three points z_0, z_1 and z_2 in the complex plane together with the corresponding values $f_0 = \alpha(z_0)$, $f_1 = \alpha(z_1)$ and $f_2 = \alpha(z_2)$ of the polynomial.

It is straightforward to find the quadratic function which interpolates the coordinates (z_0, f_0), (z_1, f_1) and (z_2, f_2). The matter is simplified by taking the quadratic in the shifted form of $q(z) = a(z - z_2)^2 + b(z - z_2) + c$ with z_2 as the centre. The parameters a, b and c are found by solving the following equations:

$$\begin{aligned} f_0 &= a(z_0 - z_2)^2 + b(z_0 - z_2) + c, \\ f_1 &= a(z_1 - z_2)^2 + b(z_1 - z_2) + c, \\ f_2 &= a(z_2 - z_2)^2 + b(z_2 - z_2) + c \\ &= c. \end{aligned} \tag{4.74}$$

On setting $c = f_2$, these may be reduced to a pair of equations which, in matrix form, are

$$\begin{bmatrix} (z_0 - z_2)^2 & (z_0 - z_2) \\ (z_1 - z_2)^2 & (z_1 - z_2) \end{bmatrix} \begin{bmatrix} a \\ b \end{bmatrix} = \begin{bmatrix} f_0 - f_2 \\ f_1 - f_2 \end{bmatrix}. \tag{4.75}$$

It is easy to verify that

$$\begin{aligned} a &= \frac{(z_1 - z_2)(f_0 - f_2) - (z_0 - z_2)(f_1 - f_2)}{(z_0 - z_2)(z_1 - z_2)(z_0 - z_1)}, \\ b &= \frac{(z_0 - z_2)^2(f_1 - f_2) - (z_1 - z_2)^2(f_0 - f_2)}{(z_0 - z_2)(z_1 - z_2)(z_0 - z_1)}. \end{aligned} \tag{4.76}$$

The root of the interpolating quadratic, which we denote by z_3, may be determined from the formula

$$z_3 - z_2 = \frac{-2c}{b \pm \sqrt{b^2 - 4ac}}. \tag{4.77}$$

This formula relates to the problem of finding the roots of the auxiliary equation $a + bw + cw^2 = 0$ as opposed to the roots of the primary equation $aw^2 + bw + c = 0$. In fact, we are seeking the roots of the primary equation. These are the reciprocals of the roots of the auxiliary equation; and this accounts for the fact that the formula appears to have been written upside down.

The purpose of using the inverse formula is to allow us more easily to isolate the root of the quadratic which has the smaller absolute value. This is a matter of choosing the sign in (4.77) so that the absolute value of the denominator will be as large as possible. When a and b are real numbers, the absolute value of $a \pm b$ is maximised by taking $a + b$, when $ab > 0$, and $a - b$, when $ab < 0$. More generally, when a and b are complex numbers, the modulus of $a \pm b$ is maximised by taking $a + b$, when $a^{re}b^{re} + a^{im}b^{im} > 0$, and $a - b$, when $a^{re}b^{re} + a^{im}b^{im} < 0$.

If z_3 is not an adequate approximation to the root of the polynomial, then z_0 is discarded and new quadratic is found which interpolates the points (z_1, f_1), (z_2, f_2) and (z_3, f_3), where $f_3 = \alpha(z_3)$. The smallest root of the new quadratic is liable to be a better approximation to the root of the polynomial.

The following procedure implements Müller's method. The need to invoke various functions to perform operations in complex arithmetic detracts from the appearance of the code which, otherwise, would be more compact.

(4.78)
```
procedure Mueller(p : integer;
                  poly : complexVector;
                  var root : complex;
                  var quotient : complexVector);

  const
    almostZero = 1E - 15;

  var
    iterations, exit : integer;
    convergence : boolean;
    a, b, c, z0, z1, z2, z3, h, h1, h2, h3, f0, f1, f2, f3 : complex;
    delta1, delta2, discrim, denom, store1, store2 : complex;

begin {Mueller}

  {Set the initial values}
  z2.re := -0.75 * root.re - 1;
  z2.im := root.im;
  z1.re := 0.75 * root.re;
  z1.im := 1.2 * root.im + 1;
  z0 := root;
  ComplexPoly(poly, p, z0, f0, quotient);
  ComplexPoly(poly, p, z1, f1, quotient);
  ComplexPoly(poly, p, z2, f2, quotient);

  iterations := 0;
  exit := 0;
  convergence := false;

  while (not convergence) and (exit = 0)
   and (iterations < 60) do
```

```
begin
  delta1 := Csubtract(f1, f2);
  delta2 := Csubtract(f0, f2);
  h1 := Csubtract(z1, z0);
  h2 := Csubtract(z2, z0);
  h3 := Csubtract(z1, z2);
  h := Cmultiply(h1, h2);
  h := Cmultiply(h, h3);
  if Cmod(h) < almostZero then
    exit := 1; {Cannot fit a quadratic to these points}

  if exit = 0 then
    begin {Calculate coefficients of the quadratic}
      store1 := Cmultiply(h1, delta1);
      store2 := Cmultiply(h2, delta2);
      a := Csubtract(store2, store1);
      a := Cdivide(a, h);    {a = (h2δ2 − h1δ1)/h}
      store1 := Cmultiply(store1, h1);
      store2 := Cmultiply(store2, h2);
      b := Csubtract(store1, store2);
      b := Cdivide(b, h);    {b = (h1²δ1 − h2²δ2)/h}
      c := f0;
    end;

  if (Cmod(a) <= almostZero)
   and (Cmod(b) <= almostZero) then
    exit := 2; {Test if parabola is really a constant}

  discrim := Cmultiply(b, b);    {b²}
  h := Cmultiply(a, c);
  discrim.re := discrim.re − 4 * h.re;    {b² − 4ac}
  discrim.im := discrim.im − 4 * h.im;
  discrim := Csqrt(discrim);    {√(b² − 4ac)}

  if (discrim.re * b.re + discrim.im * b.im > 0) then
    denom := Cadd(b, discrim)
  else
    denom := Csubtract(b, discrim);

  if Cmod(denom) < almostZero then
    begin    {if b ± √(b² − 4ac) ≃ 0}
      z3.re := 0;
      z3.im := 0;
    end
  else
    begin
```

```
        h := Cadd(c, c);
        h := Cdivide(h, denom);
        z3 := Csubtract(z0, h);
      end;

    ComplexPoly(poly, p, z3, f3, quotient);
    if (Cmod(h) < 1E - 10) or (Cmod(f3) < 1E - 10) then
      convergence := true;

    z2 := z1;
    f2 := f1;
    z1 := z0;
    f1 := f0;
    z0 := z3;
    f0 := f3;

    iterations := iterations + 1;
  end; {while}
root := z3;

if not convergence then
  Writeln('The program failed to converge');

end; {Mueller}
```

In implementing the procedure, we have had to guard against the possibility that $h = (z_0 - z_2)(z_1 - z_2)(z_0 - z_1)$ will come close to zero as a result of the virtual coincidence of two or three successive values of z. Since this term represents the denominator of the formulae for the quadratic coefficients a and b, the effect would be a numerical overflow. However, if h is virtually zero, then the values of x are likely to be close to the root of the polynomial, and the iterations may be terminated.

An alternative way of calculating the quadratic coefficients which is less sensitive to these problems is provided in an example in a subsequent section which treats the topic of divided differences.

To find the full set of roots for a polynomial, the procedure above can be driven by a further procedure similar to the one which drives the procedure *Bairstow* of (4.70). This is provided below:

(4.79)
```
procedure ComplexRoots(p : integer;
                       var alpha, lambda : complexVector);

  var
    i : integer;
    root : complex;
    quotient, beta : complexVector;
```

```
begin {ComplexRoots}

  beta := alpha;
  for i := 0 to p - 2 do
    begin
      root.re := 0;
      root.im := 0;
      Mueller(p - i, beta, root, quotient);
      lambda[i + 1] := root;
      beta := quotient;
    end;
  lambda[p].re := -beta[0].re;
  lambda[p].im := -beta[0].im;

end; {ComplexRoots}
```

In this instance, we are ignoring the injunction to recalculate the roots from the original polynomial after their values have been found from the deflated polynomial. To compensate for this, the tolerances within the procedure *Mueller* of (4.78) have been given stringent values. The procedure can be amended easily.

Polynomial Interpolation

The theory of interpolation has been important traditionally in numerical analysis since it shows how to find values for functions at points which lie in the interstices of tables. Nowadays, with the widespread availability of computers, this role has all but vanished since it is easy to evaluate the function anew at an arbitrary point whenever the demand arises. However, the theory of polynomial interpolation has acquired a new importance in engineering design and in statistics where it provides a basis for a wide variety of curve-fitting algorithms.

Imagine that we are given a set of $n+1$ points $(x_0, y_0), \ldots, (x_n, y_n)$, with strictly increasing values of x, which relate to a function $y = y(x)$ which is continuously differentiable n times. The object is to develop the facility for finding a value for y to correspond to an arbitrary value of x in the interval $[x_0, x_n]$. For this purpose, we may replace the unknown function $y = y(x)$ by a polynomial $P(x)$ of degree n which passes through the $n+1$ points. Thus, at the point (x_i, y_i), we have

$$\begin{aligned} y_i &= P(x_i) \\ &= \alpha_0 + \alpha_1 x_i + \cdots + \alpha_n x_i^n. \end{aligned} \tag{4.80}$$

By letting i run from 0 to n, we generate a set of $n+1$ equations which can be solved for the polynomial coefficients $\alpha_0, \alpha_1, \ldots, \alpha_n$; and, therefore, the problem of finding the interpolating polynomial appears to be amenable to a straightforward solution.

The system of $n+1$ equations may be written in matrix form as

$$\begin{bmatrix} 1 & x_0 & x_0^2 & \dots & x_0^n \\ 1 & x_1 & x_1^2 & \dots & x_1^n \\ 1 & x_2 & x_2^2 & \dots & x_2^n \\ \vdots & \vdots & \vdots & & \vdots \\ 1 & x_n & x_n^2 & \dots & x_n^n \end{bmatrix} \begin{bmatrix} \alpha_0 \\ \alpha_1 \\ \alpha_2 \\ \vdots \\ \alpha_n \end{bmatrix} = \begin{bmatrix} y_0 \\ y_1 \\ y_2 \\ \vdots \\ y_n \end{bmatrix}. \tag{4.81}$$

The matrix of this system is known as the Vandermonde matrix. Its determinant is given by

$$\begin{aligned} \prod_{i>k}(x_i - x_k) = (x_1 - x_0)(x_2 - x_0) & \cdots (x_n - x_0) \\ \times (x_2 - x_1) & \cdots (x_n - x_1) \\ & \cdots\cdots\cdots \\ & \times (x_n - x_{n-1}). \end{aligned} \tag{4.82}$$

The formula is easily verified for $n = 1, 2$. It is established in general by Uspensky [495, p. 214], amongst others. The value of the determinant is manifestly nonzero, which proves that the matrix is nonsingular and that the solution of the equations (4.81) is unique. Thus it follows that there is only one polynomial of degree less than or equal to n which interpolates $n+1$ distinct points.

The matrix of (4.81) is liable to be ill-conditioned, which discourages us from attempting to solve the equations directly in pursuit of the coefficients of the interpolating polynomial.

Lagrangean Interpolation

The principal object of interpolation is not to isolate the coefficients $\alpha_0, \dots, \alpha_n$ of the interpolating polynomial $P(x)$. Rather, it is to calculate values in the range of $P(x)$; and most methods avoid finding the values of the coefficients explicitly. A classical method of constructing an interpolating polynomial is Lagrange's method.

A function which interpolates the points $(x_0, y_0), \dots, (x_n, y_n)$ must be capable of being written as

$$P(x) = y_0 l_0(x) + y_1 l_1(x) + \cdots + y_n l_n(x), \tag{4.83}$$

where

$$l_j(x_i) = \begin{cases} 0, & \text{if } j \neq i; \\ 1, & \text{if } j = i. \end{cases}$$

The Lagrangean polynomials, which satisfy these conditions, can be written as

$$\begin{aligned} l_j(x) &= \frac{(x - x_0) \cdots (x - x_{j-1})(x - x_{j+1}) \cdots (x - x_n)}{(x_j - x_0) \cdots (x_j - x_{j-1})(x_j - x_{j+1}) \cdots (x_j - x_n)} \\ &= \prod_{i \neq j} \frac{(x - x_i)}{(x_j - x_i)}. \end{aligned} \tag{4.84}$$

Putting the latter into (4.83) gives

$$P(x) = \sum_{j=0}^{n} y_j \prod_{i \neq j} \frac{(x - x_i)}{(x_j - x_i)}. \tag{4.85}$$

From this, it can be seen that the leading coefficient associated with x^n is

$$\beta_n = \sum_{j=0}^{n} \frac{y_j}{\prod_{i \neq j}(x_j - x_i)}. \tag{4.86}$$

To reveal an interesting feature of the Lagrangean polynomials, we may consider expanding the generic polynomial to give

$$l_j(x_i) = \sum_{k=0}^{n} \beta_{jk} x_i^k = \delta_{ij}, \tag{4.87}$$

where δ_{ij} is Kronecker's delta. Letting $i, j = 0, \ldots, n$ generates the following system:

$$\begin{bmatrix} 1 & x_0 & \ldots & x_0^n \\ 1 & x_1 & \ldots & x_1^n \\ \vdots & \vdots & & \vdots \\ 1 & x_n & \ldots & x_n^n \end{bmatrix} \begin{bmatrix} \beta_{00} & \beta_{10} & \ldots & \beta_{n0} \\ \beta_{01} & \beta_{11} & \ldots & \beta_{n1} \\ \vdots & \vdots & & \vdots \\ \beta_{0n} & \beta_{1n} & \ldots & \beta_{nn} \end{bmatrix} = \begin{bmatrix} 1 & 0 & \ldots & 0 \\ 0 & 1 & \ldots & 0 \\ \vdots & \vdots & \ddots & \vdots \\ 0 & 0 & \ldots & 1 \end{bmatrix}; \tag{4.88}$$

which shows that the coefficients of the jth Lagrangean polynomial are the elements in the jth column of the inverse of the Vandermonde matrix of (4.81).

Some useful algebraic identities can be derived by setting $y_i = x_i^q$ in the definition of the interpolating polynomial $P(x)$ under (4.83). Then $P(x)$ interpolates the points $(x_0, x_0^q), \ldots, (x_n, x_n^q)$. But there is only one polynomial of degree less that or equal to n which interpolates $n+1$ points; so it follows that, if $q \leq n$, then x^q and $P(x)$ must coincide. Thus

$$x^q = P(x) = x_0^q l_0(x) + x_1^q l_1(x) + \cdots + x_n^q l_n(x). \tag{4.89}$$

Moreover, setting $q = 0$ shows that

$$\sum_{j=0}^{n} l_j(x) = 1. \tag{4.90}$$

Next consider expanding the numerator of $l_j(x)$ in (4.84) to give

$$l_j(x) = \frac{x^n - x^{n-1}\sum_{i \neq j} x_i + \cdots (-1)^n \prod_{i \neq j} x_i}{\prod_{i \neq j}(x_j - x_i)}. \tag{4.91}$$

Putting this into equation (4.89) shows that the coefficient of x^n is

$$\sum_{j=0}^{n} \frac{x_j^q}{\prod_{i \neq j}(x_j - x_i)} = \delta_{qn}, \tag{4.92}$$

where δ_{qn} is Kronecker's delta. This accords with the result under (4.86). It follows, by setting $q = 0$, that

$$\sum_j \frac{1}{\prod_{i \neq j}(x_i - x_j)} = 0. \tag{4.93}$$

Divided Differences

The method of Lagrangean interpolation is rarely the most efficient way of evaluating an interpolating polynomial $P(x)$. Often, a more effective way is to use Newton's form of the polynomial. Consider a sequence of polynomials of increasing degree defined recursively as follows:

$$\begin{aligned} P_0(x) &= \gamma_0, \\ P_1(x) &= P_0(x) + \gamma_1(x - x_0), \\ P_2(x) &= P_1(x) + \gamma_2(x - x_0)(x - x_1), \\ &\vdots \\ P_n(x) &= P_{n-1}(x) + \gamma_n(x - x_0)\cdots(x - x_n). \end{aligned} \tag{4.94}$$

A process of substitution leads to the expression

$$\begin{aligned} P_n(x) &= \gamma_0 + \gamma_1(x - x_0) + \gamma_2(x - x_0)(x - x_1) + \ldots \\ &\qquad + \gamma_n(x - x_0)\cdots(x - x_{n-1}) \\ &= \sum_{j=0}^{n} \gamma_j \prod_{i=0}^{j-1}(x - x_i), \end{aligned} \tag{4.95}$$

which is Newton's form for the polynomial of degree n. The polynomial can also be written is a nested form:

$$\begin{aligned} P_n(x) = \gamma_0 + (x - x_0)\Big[\gamma_1 + \cdots \\ + (x - x_{n-2})\big\{\gamma_{n-1} + (x - x_{n-1})\gamma_n\big\}\cdots\Big]; \end{aligned} \tag{4.96}$$

and, given the parameters $\gamma_0, \gamma_1, \ldots, \gamma_n$, the value of $P_n(x)$ may be generated recursively as follows:

$$\begin{aligned} q_n &= \gamma_n, \\ q_{n-1} &= q_n(x - x_{n-1}) + \gamma_{n-1}, \\ &\vdots \\ P_n(x) = q_0 &= q_1(x - x_0) + \gamma_0. \end{aligned} \tag{4.97}$$

The coefficients $\gamma_0, \ldots, \gamma_n$ of an interpolating polynomial in Newton's form may themselves be generated recursively. If $P_n(x)$ is to interpolate the points $(x_0, y_0), \ldots, (x_n, y_n)$, then it must satisfy a sequence of $n + 1$ equations

$$\begin{aligned} y_0 &= P_n(x_0) = \gamma_0, \\ y_1 &= P_n(x_1) = \gamma_0 + \gamma_1(x_1 - x_0), \\ y_2 &= P_n(x_2) = \gamma_0 + \gamma_1(x_2 - x_0) + \gamma_2(x_2 - x_0)(x_2 - x_1), \\ &\vdots \\ y_n &= P_n(x) = \gamma_0 + \gamma_1(x_n - x_0) + \gamma_2(x_n - x_0)(x_n - x_1) + \cdots \\ &\qquad + \gamma_n(x_n - x_0)\cdots(x_n - x_{n-1}). \end{aligned} \tag{4.98}$$

Reference to (4.94) shows that the generic equation of this sequence may be written as

$$y_{k+1} = P_k(x_{k+1}) + \gamma_{k+1} \prod_{i=0}^{k} (x_{k+1} - x_i),$$

from which

$$\gamma_{k+1} = \frac{y_{k+1} - P_k(x_{k+1})}{\prod_{i=0}^{k}(x_{k+1} - x_i)}. \tag{4.99}$$

Thus one of the advantages of Newton's form is that it enables us to construct the polynomial $P_{k+1}(x)$ which interpolates the $k+1$ points $(x_i, y_i), i = 0, \ldots, k$ by adding a term to the kth polynomial $P_k(x)$ which interpolates the first k points.

The coefficients of Newton's form are often described as divided differences:

(4.100) The coefficient γ_k of x^k in the kth degree polynomial $P_k(x)$ which interpolates the points $(x_0, y_0), \ldots, (x_k, y_k)$ is said to be a divided difference of order k, and it is denoted by $\gamma_k = f[x_0, \ldots, x_k]$.

The following theorem justifies this terminology; and it indicates an alternative recursive procedure for calculating these coefficients:

(4.101) Let $f[x_0, \ldots, x_k]$ and $f[x_1, \ldots, x_{k+1}]$ be divided differences of order k and let $f[x_0, \ldots, x_{k+1}]$ be a divided difference of order $k+1$. Then

$$f[x_0, \ldots, x_{k+1}] = \frac{f[x_1, \ldots, x_{k+1}] - f[x_0, \ldots, x_k]}{x_{k+1} - x_0}.$$

Proof. Let P_k be the polynomial of degree k which interpolates the points $(x_0, y_0), \ldots, (x_k, y_k)$ and let Q_k be the polynomial of degree k which interpolates the points $(x_1, y_1), \ldots, (x_{k+1}, y_{k+1})$. Then the function

$$P_{k+1}(x) = \frac{(x - x_0)Q_k(x) + (x_{k+1} - x)P_k(x)}{x_{k+1} - x_0}$$

clearly interpolates all $k+1$ points $(x_0, y_0), \ldots, (x_{k+1}, y_{k+1})$. Moreover, since $f[x_0, \ldots, x_k]$ is the coefficient of x^k in P_k and $f[x_1, \ldots, x_{k+1}]$ is the coefficient of x^k in Q_k, it follows that $f[x_0, \ldots, x_{k+1}]$, as it is defined above, is indeed the coefficient of x^{k+1} in the interpolating polynomial P_{k+1}.

A scheme for computing the divided differences is given in the following table wherein the first column contains values of the function $f(x)$ on a set of strictly increasing values of x:

$$\begin{array}{ccccccc}
f(x_1) & & & & & & \\
 & \searrow & f[x_1, x_2] & & & & \\
 & \nearrow & & \searrow & & & \\
f(x_2) & & & & f[x_1, x_2, x_3] & & \\
 & \searrow & f[x_2, x_3] & \nearrow & & \searrow & f[x_1, x_2, x_3, x_4] \\
 & \nearrow & & \searrow & & \nearrow & \\
f(x_3) & & & & f[x_2, x_3, x_4] & & \\
 & \searrow & f[x_3, x_4] & \nearrow & & & \\
 & \nearrow & & & & & \\
f(x_4) & & & & & &
\end{array} \tag{4.102}$$

An elementary property of the divided difference $f[x_i, \ldots, x_{i+n}]$ of the nth order is that it vanishes if $f(x)$ is a polynomial of degree less than n.

Example 4.3. Consider the problem, which we have already faced in connection with Müller's method, of interpolating a quadratic $q(x)$ through the points (x_0, f_0), (x_1, f_1) and (x_2, f_2). Using divided differences and taking the points in reversed order, we have

$$q(x) = f_2 + f[x_2, x_1](x - x_2) + f[x_2, x_1, x_0](x - x_2)(x - x_1); \tag{4.103}$$

where

$$\begin{aligned} f[x_2, x_1] &= \frac{f_1 - f_2}{x_1 - x_2} \qquad \text{and} \\ f[x_2, x_1, x_0] &= \frac{1}{x_0 - x_2}\left\{\frac{f_0 - f_1}{x_0 - x_1} - \frac{f_1 - f_2}{x_1 - x_2}\right\}. \end{aligned} \tag{4.104}$$

But

$$(x - x_2)(x - x_1) = (x - x_2)^2 + (x - x_2)(x_2 - x_1). \tag{4.105}$$

Therefore,we can write (4.103) as

$$q(x) = c + b(x - x_2) + a(x - x_2)^2, \tag{4.106}$$

where

$$\begin{aligned} c &= f_2, \\ b &= f[x_2, x_1] + f[x_2, x_1, x_0](x_2 - x_1), \\ a &= f[x_2, x_1, x_0]. \end{aligned} \tag{4.107}$$

It may be confirmed that these are the same as the coefficients specified under (4.74) and (4.75). A reason for preferring the latter formulae in the context of Müller's algorithm is that they enable us more easily to ensure that the coefficients are well-determined by checking that the numerator $(x_0 - x_2)(x_1 - x_2)(x_0 - x_1)$ is not too close to zero.

Bibliography

[134] Curry, J.H., L. Garnett and D. Sullivan, (1983), On the Iteration of Rational Functions: Computer Experiments with Newton's Method, *Communications in Mathematical Physics*, **91**, 267–277.

[157] Dickson, E.L., (1914), *Elementary Theory of Equations*, John Wiley, New York.

[213] Gleick, J., (1987), *Chaos: Making a New Science*, Viking Penguin Inc., New York.

[331] Marden, M., (1949), *The Geometry of the Zeros of a Polynomial in a Complex Variable, Mathematical Surveys, Number III*, The American Mathematical Society.

[332] Marden, M., (1966), *Geometry of Polynomials, No. 3 of Mathematical Surveys of the AMS*, American Mathematical Society, Providence, Rhode Island.

[356] Müller, W.E., (1956), A Method of Solving Algebraic Equations Using an Automatic Computer, *Mathematical Tables and Other Aids to Computation (MTAC)*, **10**, 208–215.

[419] Ralston, A., and P. Rabinowitz, (1978), *A First Course in Numerical Analysis, Second Edition*, McGraw-Hill Kogakusha, Tokyo.

[431] Routh, E.J., 1831–1907, (1877), *A Treatise on the Stability of a Given State of Motion*, (being the essay which was awarded the Adams prize in 1877, in the University of Cambridge), Macmillan and Co., London.

[432] Routh, E.J., 1831–1907, (1905), *The Elementary Part of a Treatise on the Dynamics of a System of Rigid Bodies*, (being part 1 of a treatise on the whole subject), 7th edition, revised and enlarged, Macmillan and Co., London. Reprinted 1960, Dover Publications, New York.

[433] Routh, E.J., 1831–1907, (1905), *The Advanced Part of a Treatise on the Dynamics of a System of Rigid Bodies*, (being part 2 of a treatise on the whole subject), 6th edition, revised and enlarged, Macmillan and Co., London. Reprinted, Dover Publications, New York.

[453] Shub, M., and S. Smale, (1985), Computational Complexity. On the Geometry of Polynomials and the Theory of Cost, *Annales Scientifiques de L'École Normal Supérieure (ASENAH)*, 107–142.

[474] Stoer, J., and R. Bulirsh, (1980), *Introduction to Numerical Analysis*, Springer-Verlag, New York.

[476] Sturm, J.C.F., (1829), *Mémoire sur la Resolution des Équations Numeriques*, published in 1835 in Mémoires Présenté par des Savants Étrangers, Paris.

[486] Todhunter, I., (1882), *An Elementary Treatise on the Theory of Equations*, Macmillan and Co., London.

[492] Turnbull, H.W., (1953), *The Theory of Equations, Fifth Edition*, Oliver and Boyd, Edinburgh.

[495] Uspensky, J.V., (1948), *Theory of Equations, Fifth Edition*, McGraw-Hill Book Co., New York.

CHAPTER 5

Difference Equations and Differential Equations

This chapter is concerned with the analysis of linear dynamic systems which are driven by nonstochastic inputs. When time is treated as a continuum, the analysis entails differential equations; when it is treated as a succession of discrete instants separated by unit intervals, the analysis entails difference equations.

Until recently, it was common to find in textbooks of signal processing, and even in research papers, passages declaring that difference equations are best understood in reference to differential equations to which they are closely related. This was a natural point of view to take when signal processing depended mainly on analogue equipment. As the use of digital computers in signal processing increased, the emphasis began to shift; and the point has now been reached where difference equations deserve priority.

The concept of discrete time comes naturally to statisticians and time-series analysts whose data are sampled at regular intervals; and, although it is well-grounded in theory, a stochastic process in continuous time is hard to imagine. It is also true that economists have shown a marked tendency over many years to express their models in discrete time.

The question arises of whether difference and differential equations can be used interchangeably in modelling continuous processes which are observed at regular intervals. Given that the general solutions to both varieties of equations are represented, in the homogeneous case, by a sum of real and complex exponential terms, the answer would seem to be in the affirmative. However, there is still the question of whether periodic observations can reveal all that is happening in continuous time. It transpires that, if none of the cyclical, or complex-exponential, components which are present in the continuous process complete their cycles in less time that it takes to make two observations, then no information is lost in the process of sampling.

Our exposition of difference equations is mainly in terms of the lag operator L. In other branches of mathematics, the difference operator is used instead. The forward-difference operator Δ bears a familiar relationship to the operator D which produces derivatives; and, for the purpose of comparing differential and difference equations, we show how to make the conversion from L to Δ.

The final sections of the chapter are concerned with the conditions which are necessary and sufficient for the stability of linear systems. The stability of a system depends upon the values taken by the roots of a characteristic polynomial equation. Since it is time-consuming to evaluate the roots, we look for equivalent conditions which can be expressed in terms of the ordinary parameters of the systems.

The classical stability conditions for differential equations are the well-known Routh–Hurwitz conditions (see [431] and [263]) which were discovered at the end of the nineteenth century. From these, we may deduce the corresponding conditions for the stability of difference equations, which are commonly know as the Samuelson [436] conditions by economists and as the Schur–Cohn conditions by others (see [443] and [118]). Whereas we shall state the Routh–Hurwitz conditions without proof, we shall take care to establish the stability conditions for difference equations from first principles. These conditions will also emerge elsewhere in the text as the side products of other endeavours.

Linear Difference Equations

A pth-order linear difference equation with constant coefficients is a relationship amongst $p+1$ consecutive elements of a sequence $y(t)$ of the form

$$\alpha_0 y(t) + \alpha_1 y(t-1) + \cdots + \alpha_p y(t-p) = u(t). \tag{5.1}$$

Here $u(t)$ is a specified sequence of inputs which is known as the forcing function. The equation can also be written as

$$\alpha(L)y(t) = u(t), \tag{5.2}$$

where

$$\alpha(L) = \alpha_0 + \alpha_1 L + \cdots + \alpha_p L^p. \tag{5.3}$$

If p consecutive values of $y(t)$, say $y_0, y_1, \ldots, y_{p-1}$, are given, then equation (5.1) may be used to find the next value y_p. So long as $u(t)$ is fully specified, successive elements of the sequence can be found in this way, one after another. Likewise, values of the sequence prior to $t = 0$ can be generated; and thus, in effect, any number of elements of $y(t)$ can be deduced from the difference equation. However, instead of a recursive solution, we often seek an analytic expression for $y(t)$.

The analytic solution of the difference equation is a function $y(t; c)$ comprising a set of p coefficients in $c = [c_1, c_2, \ldots, c_p]'$ which can be determined once p consecutive values of $y(t)$ are given which are called initial conditions. The same values would serve to initiate a recursive solution. The analytic solution can be written as the sum $y(t; c) = x(t; c) + w(t)$, where $x(t)$ is the general solution of the homogeneous equation $\alpha(L)x(t) = 0$, and $w(t) = \alpha^{-1}(L)u(t)$ is a particular solution of the inhomogeneous equation (5.2).

The difference equation may be solved in three steps. The first step is to find the general solution of the homogeneous equation; and this embodies the unknown coefficients. The next step is to find the particular solution $w(t)$ which contains no unknown quantities. Finally, the p initial values of y may be used to determine the coefficients $c_1, c_2, \ldots, c_p$. We shall begin by discussing the solution of the homogeneous equation.

Solution of the Homogeneous Difference Equation

If λ_j is a root of the equation $\alpha(z) = \alpha_0 + \alpha_1 z + \cdots + \alpha_p z^p = 0$ such that $\alpha(\lambda_j) = 0$, then $y_j(t) = (1/\lambda_j)^t$ is a solution of the equation $\alpha(L)y(t) = 0$. This can be seen by considering the expression

$$\begin{aligned}\alpha(L)y_j(t) &= (\alpha_0 + \alpha_1 L + \cdots + \alpha_p L^p)(1/\lambda_j)^t \\ &= \alpha_0(1/\lambda_j)^t + \alpha_1(1/\lambda_j)^{t-1} + \cdots + \alpha_p(1/\lambda_j)^{t-p} \\ &= (\alpha_0 + \alpha_1\lambda_j + \cdots + \alpha_p\lambda_j^p)(1/\lambda_j)^t \\ &= \alpha(\lambda_j)(1/\lambda_j)^t.\end{aligned} \tag{5.4}$$

Alternatively, one may consider the factorisation $\alpha(L) = \alpha_0\prod_i(1 - L/\lambda_i)$. Within this product, there is the term $1 - L/\lambda_j$; and, since

$$(1 - L/\lambda_j)(1/\lambda_j)^t = (1/\lambda_j)^t - (1/\lambda_j)^t = 0, \tag{5.5}$$

it follows that $\alpha(L)(1/\lambda_j)^t = 0$.

Imagine that $\alpha(z) = 0$ has p distinct roots $\lambda_1, \lambda_2, \ldots, \lambda_p$, some of which may be conjugate complex numbers. Then the general solution of the homogeneous equation is given by

$$x(t;c) = c_1(1/\lambda_1)^t + c_2(1/\lambda_2)^t + \cdots + c_p(1/\lambda_p)^t, \tag{5.6}$$

where $c_1, c_2, \ldots, c_p$ are the coefficients which are determined by the initial conditions.

In the case where two roots coincide at a value of λ, the equation $\alpha(L)y(t) = 0$ has the solutions $y_1(t) = (1/\lambda)^t$ and $y_2(t) = t(1/\lambda)^t$. To show this, let us extract the term $(1 - L/\lambda)^2$ from the factorisation $\alpha(L) = \alpha_0\prod_j(1 - L/\lambda_j)$. Then, according to the previous argument, we have $(1 - L/\lambda)^2(1/\lambda)^t = 0$; but it is also found that

$$\begin{aligned}(1 - L/\lambda)^2 t(1/\lambda)^t &= (1 - 2L/\lambda + L^2/\lambda^2)t(1/\lambda)^t \\ &= \big\{t - 2(t-1) + (t-2)\big\}(1/\lambda)^t \\ &= 0.\end{aligned} \tag{5.7}$$

More generally, it can be asserted that

(5.8) If $\alpha(z) = \gamma(z)(1 - z/\lambda)^r$, which is to say that $\alpha(z) = 0$ has a repeated root of multiplicity r, then each of the r functions $(1/\lambda)^t$, $t(1/\lambda)^t, \ldots,$ $t^{r-1}(1/\lambda)^t$ is a solution of the equation $\alpha(L)y(t) = 0$.

Proof. This is proved by showing that, if $r > n$, then $(1 - L/\lambda)^r t^n(1/\lambda)^t = 0$. Let $f_n(t) = t^n$, and consider

$$\begin{aligned}(1 - L/\lambda)t^n(1/\lambda)^t &= (1 - L/\lambda)f_n(t)(1/\lambda)^t \\ &= \big\{f_n(t) - f_n(t-1)\big\}(1/\lambda)^t \\ &= f_{n-1}(t)(1/\lambda)^t.\end{aligned} \tag{5.9}$$

Here $f_{n-1}(t) = f_n(t) - f_n(t-1)$ is a polynomial in t of degree $n-1$. Next, consider

$$\begin{aligned}(1 - L/\lambda)^2 t^n (1/\lambda)^t &= (1 - L/\lambda) f_{n-1}(t)(1/\lambda)^t \\ &= \{f_{n-1}(t) - f_{n-1}(t-1)\}(1/\lambda)^t \\ &= f_{n-2}(t)(1/\lambda)^t.\end{aligned} \tag{5.10}$$

Here $f_{n-2}(t)$ is a polynomial in t of degree $n-2$. After n steps, it will be found that

$$(1 - L/\lambda)^n t^n (1/\lambda)^t = g(1/\lambda)^t, \tag{5.11}$$

where g is a polynomial of degree zero, which is a constant in other words. It follows that $(1 - L/\lambda)^r t^n (1/\lambda)^t$ vanishes when $r > n$; and this is what had to be proved.

A root of multiplicity r in the polynomial equation $\alpha(z) = 0$ gives rise to r different solutions of the homogeneous difference equation. If each root is counted as many times as its multiplicity, then it can be said that the number of solutions of a difference equation is equal to the number of roots of $\alpha(z) = 0$, with is p. If these solutions are denoted by $y_1(t), y_2(t), \ldots, y_p(t)$, then the general solution of the homogeneous equation may be expressed as

$$x(t; c) = c_1 y_1(t) + c_2 y_2(t) + \cdots + c_p y_p(t), \tag{5.12}$$

where $c_1, c_2, \ldots, c_p$ are the coefficients which are determined by the initial conditions.

Example 5.1. For some purposes, it is more convenient to describe the solution of a difference equation in terms of the roots of the *auxiliary* equation $\alpha'(z) = \alpha_0 z^p + \alpha_1 z^{p-1} + \cdots + \alpha_p = 0$ than in terms of the roots of the *primary* equation $\alpha(z) = \alpha_0 + \alpha_1 z + \cdots + \alpha_p z^p = 0$. Since $\alpha'(z) = z^p \alpha(z^{-1})$, it follows that, if λ is a root of the equation $\alpha(z) = 0$ such that $\alpha(\lambda) = 0$, then $\mu = 1/\lambda$ is a root of the auxiliary equation such that $\alpha'(\mu) = 0$. The auxiliary equation of a second-order difference equation takes the form of

$$\begin{aligned}\alpha_0 z^2 + \alpha_1 z + \alpha_2 &= \alpha_0 (z - \mu_1)(z - \mu_2) \\ &= \alpha_0 \{z^2 - (\mu_1 + \mu_2) z + \mu_1 \mu_2\} = 0,\end{aligned} \tag{5.13}$$

where the roots are

$$\mu_1, \mu_2 = \frac{-\alpha_1 \pm \sqrt{\alpha_1^2 - 4\alpha_0 \alpha_2}}{2\alpha_0}. \tag{5.14}$$

Complex Roots

The treatment of complex roots may be developed further to take account of the fact that, if the coefficients of $\alpha(z)$ are real-valued, then such roots must occur in conjugate pairs.

Imagine that the equation $\alpha(z) = 0$ has the conjugate complex roots $\lambda = 1/\mu$ and $\lambda^* = 1/\mu^*$. The complex numbers may be expressed in various ways:

$$
\begin{aligned}
\mu &= \gamma + i\delta = \kappa(\cos\omega + i\sin\omega) = \kappa e^{i\omega}, \\
\mu^* &= \gamma - i\delta = \kappa(\cos\omega - i\sin\omega) = \kappa e^{-i\omega}.
\end{aligned} \tag{5.15}
$$

The roots will contribute the following expression to the general solution of the difference equation:

$$
\begin{aligned}
q(t) &= c\mu^t + c^*(\mu^*)^t \\
&= c(\kappa e^{i\omega})^t + c^*(\kappa e^{-i\omega})^t.
\end{aligned} \tag{5.16}
$$

This stands for a real-valued sequence; and, since a real variable must equal its own conjugate, it follows that c and c^* are conjugate numbers in the forms of

$$
\begin{aligned}
c^* &= \rho(\cos\theta + i\sin\theta) = \rho e^{i\theta}, \\
c &= \rho(\cos\theta - i\sin\theta) = \rho e^{-i\theta}.
\end{aligned} \tag{5.17}
$$

Thus we have

$$
\begin{aligned}
q(t) &= \rho\kappa^t \left\{ e^{i(\omega t-\theta)} + e^{-i(\omega t-\theta)} \right\} \\
&= 2\rho\kappa^t \cos(\omega t - \theta).
\end{aligned} \tag{5.18}
$$

To analyse the final expression, consider first the factor $\cos(\omega t - \theta)$. This is a displaced cosine wave. The value ω, which is a number of radians per unit period, is called the angular velocity or the angular frequency of the wave. The value $f = \omega/2\pi$ is its frequency in cycles per unit period. If time is measured in seconds, then f becomes a number of hertz. The duration of one cycle, also called the period, is $r = 2\pi/\omega$. The term θ is called the phase displacement of the cosine wave, and it serves to shift the cosine function along the axis of t. The function $\cos(\omega t)$ attains it maximum value when $t = 0$, whereas a function $\cos(\omega t - \theta)$ defined over the real line has a peak when $t = \theta/\omega$.

Next consider the term κ^t wherein $\kappa = \sqrt{(\gamma^2 + \delta^2)}$ is the modulus of the complex roots. When κ has a value of less than unity, it becomes a damping factor which serves to attenuate the cosine wave as t increases.

Finally, the factor 2ρ determines the initial amplitude of the cosine wave. Since ρ is the modulus of the complex numbers c and c^*, and since θ is their argument, amplitude and the phase are determined by the initial conditions.

The solution of a second-order homogeneous difference equation in the case of complex roots is illustrated in Figure 5.1.

The condition which is necessary and sufficient for $q(t)$ of (5.18) to tend to zero as t increases is that $\kappa = |\mu| < 0$. When it is expressed in terms of $\lambda = 1/\mu$, the condition is that $|\lambda| > 0$. The same result applies to the contribution of the real roots on the understanding that the notation $|\lambda| > 0$ refers to an absolute value. The cases of repeated roots, whether they be real or complex, are no different from the cases of distinct roots since, within the products $t^j(1/\lambda)^t; j = 1, \ldots, r-1$, the term $(1/\lambda)^t$ is dominant. Thus it can be asserted that

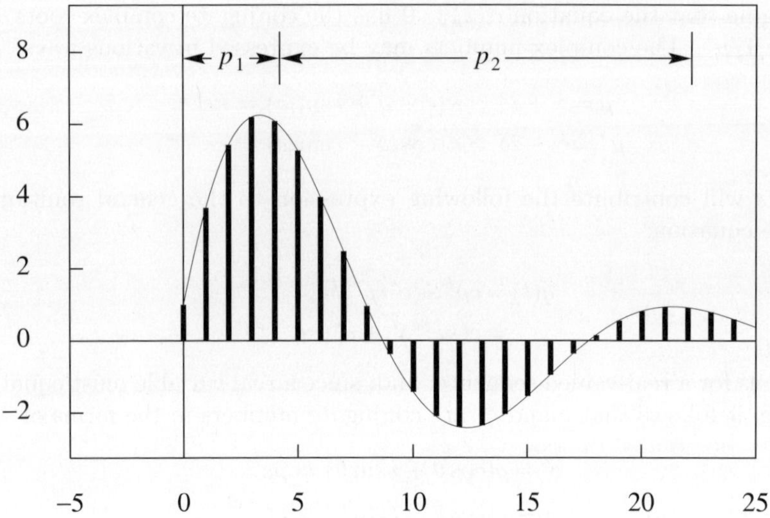

Figure 5.1. The solution of the homogeneous difference equation $(1 - 1.69L + 0.81L^2)y(t) = 0$ for the initial conditions $y_0 = 1$ and $y_1 = 3.69$. The time lag of the phase displacement p_1 and the duration of the cycle p_2 are also indicated.

(5.19) The general solution of the homogeneous equation $\alpha(L)y(t) = 0$ tends to zero as t increases if and only if all of the roots of $\alpha(z) = 0$ lie outside the unit circle. Equivalently, it tends to zero if and only if all of the roots of $\alpha'(z) = z^p\alpha(z^{-1}) = 0$ lie inside the unit circle.

Particular Solutions

The general solution of the difference equation $\alpha(L)y(t) = u(t)$ is obtained by adding the general solution $x(t; c)$ of the homogeneous equation $\alpha(L)x(t) = 0$ to a particular solution $w(t) = u(t)/\alpha(L)$ of the complete equation.

If the homogeneous equation is stable, then its contribution to the general solution will be a transient component which will vanish with time. The component which will persist in the long run is the particular solution which is liable to be described as the steady-state solution or as the equilibrium time path of the system.

The business of evaluating the particular solution, so as to obtain an analytic form, may be problematic if the forcing function $u(t)$ is not of a tractable nature. Matters are simplified when $u(t)$ is itself the solution of a homogeneous difference equation such that $\theta(L)u(t) = 0$ for some $\theta(L)$. This is so whenever $u(t)$ is a real or complex exponential function or a polynomial in t, or some linear combination of such functions.

In such cases, one approach to finding the particular solution is via the augmented homogeneous equation $\theta(L)\alpha(L)y(t) = \theta(L)u(t) = 0$. The particular solution $w_p(t, c)$ of the original equation is that part of the general solution of the augmented equation which corresponds to the roots of $\theta(L)$. The coefficients in $w_p(t, c)$ are determined so as to satisfy the equation $\alpha(L)w_p(t, c) = u(t)$.

The approach which we shall follow in finding particular solutions deals directly with the product $w(t) = u(t)/\alpha(L)$. We may begin by considering some basic results concerning a polynomial operator in L:

$$\text{(5.20)}\qquad \begin{aligned} &\text{(i)} && \delta(L)\lambda^{-t} - \delta(\lambda)\lambda^{-t}, \\ &\text{(ii)} && \delta(L)\{\lambda^{-t}u(t)\} = \lambda^{-t}\delta(\lambda L)u(t), \\ &\text{(iii)} && \delta(L)\{v(t) + w(t)\} = \delta(L)v(t) + \delta(L)w(t). \end{aligned}$$

The first of these, which we have used already, comes from the identity $L^n\lambda^{-t} = \lambda^{n-t} = \lambda^n\lambda^{-t}$, which can be applied to the terms of the polynomial operator. The second result comes from the fact that $L^n\{\lambda^{-t}u(t)\} = \lambda^{n-t}u(t-n) = \lambda^{-t}\{\lambda L\}^n u(t)$. The third result indicates that $\delta(L)$ is a linear operator.

The same results are also available when the operator $\delta(L)$ is an infinite series such as would result from the expansion of a rational function of L. It follows that

$$\text{(5.21)}\qquad \begin{aligned} &\text{(i)} && \frac{1}{\gamma(L)}\lambda^{-t} = \frac{1}{\gamma(\lambda)}\lambda^{-t} \quad \text{if} \quad \gamma(\lambda) \neq 0, \\ &\text{(ii)} && \frac{1}{\gamma(L)}\{\lambda^{-t}u(t)\} = \lambda^{-t}\frac{1}{\gamma(\lambda L)}u(t), \\ &\text{(iii)} && \frac{1}{\gamma(L)}\{v(t) + w(t)\} = \frac{1}{\gamma(L)}v(t) + \frac{1}{\gamma(L)}w(t). \end{aligned}$$

The case of $\gamma(\lambda) = 0$, which affects the result under (i), arises when $\gamma(L) = (1 - L/\lambda)^r\gamma_1(L)$, where $\gamma_1(\lambda) \neq 0$ and r is the multiplicity of the root λ. Then another result may be used:

$$\text{(5.22)}\qquad \begin{aligned} \frac{1}{\gamma(L)}\lambda^{-t} &= \frac{1}{(1 - L/\lambda)^r}\left\{\frac{\lambda^{-t}}{\gamma_1(L)}\right\} = \frac{1}{\gamma_1(\lambda)}\left\{\frac{\lambda^{-t}}{(1 - L/\lambda)^r}\right\} \\ &= \frac{t^r\lambda^{-t}}{\gamma_1(\lambda)}. \end{aligned}$$

Here the penultimate equality comes from (i). The final equality depends upon the result that

$$\text{(5.23)}\qquad \frac{\lambda^{-t}}{(I - L/\lambda)^r} = t^r\lambda^{-t}.$$

This comes from applying the inverse operator $(I - L/\lambda)^{-r}$ to both sides of the identity $(I - L/\lambda)^r t^r\lambda^{-t} = \lambda^{-t}$ which is verified by using (5.20)(ii).

The result under (5.22) is used in finding a particular solution to the inhomogeneous equation $\alpha(L)y(t) = \lambda^{-t}$ in the case where λ is a root of $\alpha(z) = 0$ of multiplicity r. This can also be understood by considering the augmented homogeneous equation $(1 - L/\lambda)\alpha(L)y(t) = 0$. The general solution of the latter contains the terms $(1/\lambda)^t, t(1/\lambda)^t, \ldots, t^r(1/\lambda)^t$. Of these, the first r are attributed to the general solution of the homogeneous equation $\alpha(L)x(t) = 0$, whilst the final term is attributed to the particular solution of the inhomogeneous equation.

The following examples illustrate the method of finding particular solutions.

Example 5.2. Let the difference equation be

$$(5.24)\qquad (6 - 5L + L^2)\, y(t) = \left(\frac{1}{4}\right)^t.$$

Then the result under (5.21)(i) can be invoked to show that the particular solution is

$$(5.25)\qquad \begin{aligned} w(t) &= \frac{4^{-t}}{6 - 5L + L^2} \\ &= \frac{4^{-t}}{6 - 5 \times 4 + 4^2} \\ &= \frac{1}{2}\left(\frac{1}{4}\right)^t. \end{aligned}$$

Example 5.3. Consider the difference equation $(1 - \phi L)y(t) = \cos(\omega t)$. We can write

$$\cos(\omega t) - i\sin(\omega t) = \bigl(\cos\omega + i\sin\omega\bigr)^{-t} = \lambda^{-t},$$

where λ is a point on the unit circle in the complex plane. Since $\cos(\omega t) = \mathrm{Re}(\lambda^{-t})$, the particular solution of the difference equation can be expressed as

$$(5.26)\qquad y(t) = \frac{\cos(\omega t)}{1 - \phi L} = \mathrm{Re}\left\{\frac{\lambda^{-t}}{1 - \phi L}\right\} = \mathrm{Re}\left\{\frac{\lambda^{-t}}{1 - \phi\lambda}\right\},$$

where the final equality is by virtue of (5.21)(i). The inverse of the complex number $1 - \phi\lambda = (1 - \phi\cos\omega) - i\phi\sin\omega$ is

$$(5.27)\qquad \frac{1}{1 - \phi\lambda} = \frac{(1 - \phi\cos\omega) + i\phi\sin\omega}{1 - 2\phi\cos\omega + \phi^2}.$$

Therefore, the particular solution is

$$(5.28)\qquad \begin{aligned} y(t) &= \frac{\mathrm{Re}\Bigl[\bigl\{(1 - \phi\cos\omega) + i\phi\sin\omega\bigr\}\bigl\{\cos(\omega t) - i\sin(\omega t)\bigr\}\Bigr]}{1 - 2\phi\cos\omega + \phi^2} \\ &= \frac{(1 - \phi\cos\omega)\cos(\omega t) + \phi\sin\omega\sin(\omega t)}{1 - 2\phi\cos\omega + \phi^2}. \end{aligned}$$

The weighted sum of a sine and a cosine function of the same argument may be expressed as a cosine function with a phase displacement. Therefore, the particular or steady-state solution of the difference equation also takes the form of

$$(5.29)\qquad y(t) = \frac{1}{\sqrt{1 - 2\phi\cos\omega + \phi^2}}\cos(\omega t - \theta),$$

where

$$\theta = \tan^{-1}\left(\frac{\phi\sin\omega}{1-\phi\cos\omega}\right). \tag{5.30}$$

This result shows the two effects of applying the linear filter $1/(1-\phi L)$ to a cosine signal. The first of these, which is described as the gain effect, is to alter the amplitude of the signal. The second, which is described as the phase effect, is to displace the cosine by θ radians.

Example 5.4. Let $(1+\frac{1}{2}L)y(t)=t^2$. Then

$$\begin{aligned} w(t) &= \frac{t^2}{1+\frac{1}{2}(I-\nabla)} \\ &= \frac{2}{3}\frac{t^2}{(I-\frac{1}{3}\nabla)} \\ &= \frac{2}{3}\Big\{1+\frac{1}{3}\nabla+\frac{1}{9}\nabla^2+\frac{1}{27}\nabla^3+\cdots\Big\}t^2. \end{aligned} \tag{5.31}$$

But $\nabla t^2 = 2t-1$, $\nabla^2 t^2 = 2$ and $\nabla^n t^2 = 0$ for $n>2$, so this gives

$$w(t) = \frac{2}{27}\{9t^2+6t-1\}. \tag{5.32}$$

Solutions of Difference Equations with Initial Conditions

The general solution of a pth-order difference equation contains p arbitrary coefficients. In order to obtain a fully specified analytic solution, which is described as a complete solution, a set of p additional conditions is required. These conditions can take various forms, but, usually, they are provided by a set of consecutive observations on $y(t)$ described as initial conditions. We shall assume that a set of values are given such as $y_{-1},\ldots,y_{-p}$ or $y_0,\ldots,y_{p-1}$, and we shall describe alternative ways of using them to obtain the arbitrary constants which are to be found within the analytic expression for the general solution of the difference equation. The classical method for incorporating the initial conditions can be described adequately through an example:

Example 5.5. Consider the inhomogeneous second-order difference equation $(\alpha_0+\alpha_1L+\alpha_2L^2)y(t)=\gamma^{-t}$, and let λ_1,λ_2 be the roots of the equation $\alpha_0+\alpha_1z+\alpha_2z^2=0$. Then the general solution of the difference equation is $y(t;c)=x(t;c)+w(t)$, where

$$x(t;c) = c_1\left(\frac{1}{\lambda_1}\right)^t + c_2\left(\frac{1}{\lambda_2}\right)^t \tag{5.33}$$

is the general solution of the corresponding homogeneous equation and

$$w(t) = \frac{\gamma^{-t}}{\alpha_0+\alpha_1\gamma+\alpha_2\gamma^2} = \frac{\delta}{\gamma^t} \tag{5.34}$$

is the particular solution of the inhomogeneous equation. The method for obtaining the particular solution has been illustrated in Example 5.2.

Imagine that the values y_0 and y_1 have been given. Then, by setting $t = 0, 1$ in the analytic expression $y(t; c)$, the following system is derived:

$$\begin{bmatrix} 1 & 1 \\ \frac{1}{\lambda_1} & \frac{1}{\lambda_2} \end{bmatrix} \begin{bmatrix} c_1 \\ c_2 \end{bmatrix} = \begin{bmatrix} y_0 - \delta \\ y_1 - \frac{\delta}{\gamma} \end{bmatrix}; \tag{5.35}$$

and this is readily solved to provide the values of c_1 and c_2. The matrix on the LHS of the equation is known as the Wronskian.

In the case of Example 5.2, where the roots of the associated homogeneous equation are $\lambda_1 = 2$ and $\lambda_2 = 3$ and where $\gamma = 4$ and $\delta = 1/2$, the general solution is found to be

$$y(t) = c_1 \left(\frac{1}{2}\right)^t + c_2 \left(\frac{1}{3}\right)^t + \frac{1}{2}\left(\frac{1}{4}\right)^t. \tag{5.36}$$

The initial conditions $y_0 = 1$ and $y_1 = 1/3$ imply that $c_1 = c_2 = 1/4$.

An alternative method for finding a complete analytic solution satisfying given initial conditions makes use of the z-transform of the one-sided sequence which is formed from $y(t)$ by discarding all of the elements prior to time $t = 0$. Let $y_+(t) = \{y_0, y_1, y_2, \ldots\}$ denote the resulting sequence. Then, in advancing the sequence by one period to form $y_+(t+1) = \{y_1, y_2, y_3, \ldots\}$, we must delete the element y_0; and, in lagging the sequence to form $y_+(t-1) = \{y_{-1}, y_0, y_1, \ldots\}$, we must add the element y_{-1}.

In forming the z-transforms of the lagged and the advanced sequences, we must likewise take account of these end conditions. The correspondence between the sequences and their z-transforms can be illustrated by a few instances. Let $y_+(z) = \{y_0 + y_1 z + y_2 z^2 + \cdots\}$. Then

$$\begin{aligned} y_+(t-2) &\longleftrightarrow z^2 y_+(z) + z y_{-1} + y_{-2}, \\ y_+(t-1) &\longleftrightarrow z y_+(z) + y_{-1}, \\ y_+(t) &\longleftrightarrow y_+(z), \\ y_+(t+1) &\longleftrightarrow z^{-1} y_+(z) - z^{-1} y_0, \\ y_+(t+2) &\longleftrightarrow z^{-2} y_+(z) - z^{-1} y_1 - z^{-2} y_0. \end{aligned} \tag{5.37}$$

More generally, we have

$$\begin{aligned} y_+(t-j) &\longleftrightarrow z^j y_+(z) + \sum_{i=0}^{j-1} z^i y_{i-j} \quad \text{and} \\ y_+(t+j) &\longleftrightarrow z^{-j} y_+(z) - \sum_{i=1}^{j} z^{-i} y_{j-i}. \end{aligned} \tag{5.38}$$

Now consider the case of a pth-order homogeneous difference equation. The one-sided version is

$$\alpha_0 y_+(t) + \alpha_1 y_+(t-1) + \alpha_2 y_+(t-2) + \cdots + \alpha_p y_+(t-p) = 0. \tag{5.39}$$

The elements of the sum may be transformed separately and their transforms combined in the z-domain. The elements and their transforms are

$$\begin{aligned} \alpha_0 y_+(t) &\longleftrightarrow \alpha_0 y_+(z), \\ \alpha_1 y_+(t-1) &\longleftrightarrow \alpha_1 z y_+(z) + \alpha_1 y_{-1}, \\ \alpha_2 y_+(t-2) &\longleftrightarrow \alpha_2 z^2 y_+(z) + \alpha_2 z y_{-1} + \alpha_2 y_{-2}, \\ &\vdots \\ \alpha_p y_+(t-p) &\longleftrightarrow \alpha_p z^p y_+(z) + \alpha_p z^{p-1} y_{-1} + \cdots + \alpha_p y_{-p}. \end{aligned} \tag{5.40}$$

Adding the transforms on the RHS gives

$$\alpha(z) y_+(z) + Q(z) = 0, \tag{5.41}$$

where

$$Q(z) = \sum_{i=0}^{p-1}\left(\sum_{j=i+1}^{p} \alpha_j y_{i-j}\right) z^i = Q_0 + Q_1 z + \cdots + Q_{p-1} z^{p-1}. \tag{5.42}$$

This polynomial $Q(z)$, which owes its existence to the end-effects in forming the delayed sequences $y_+(t-1), y_+(t-2), \ldots, y_+(t-p)$, embodies the p initial conditions $y_{-1}, y_{-2}, \ldots, y_{-p}$.

The solution of the difference equation in the z-domain is provided by

$$y_+(z) = -\frac{Q(z)}{\alpha(z)}; \tag{5.43}$$

and, since the degree of $Q(z)$ is $p-1$, it follows that $Q(z)/\alpha(z)$ is a proper rational function. Given the factorisation $\alpha(z) = \alpha_0(1 - z/\lambda_1)\cdots(1 - z/\lambda_p)$, and assuming that $\alpha_0 = 1$, we can write the following partial-fraction expansion:

$$-\frac{Q(z)}{\alpha(z)} = \frac{c_1}{1 - z/\lambda_1} + \cdots + \frac{c_p}{1 - z/\lambda_p}. \tag{5.44}$$

To find the solution in the time domain, we have to apply the inverse of the z-transform. In view of the relationship

$$c_i\left(\frac{1}{\lambda_i}\right)^t \longleftrightarrow \frac{c_i}{1 - z/\lambda_i}, \tag{5.45}$$

it can be seen that equation (5.43) corresponds exactly to the general solution of the homogeneous difference equation given under (5.6). However, the coefficients $c_1, \ldots, c_p$ of equation (5.44), which embody the initial conditions, are fully determined, whereas the same coefficients in equation (5.6) were regarded as values which remained to be determined.

Example 5.6. Consider the homogeneous difference equation $y(t) - \phi^2 y(t-2) = 0$ together with the initial conditions y_{-1} and y_{-2}. With reference to (5.37), it can be seen that the appropriate z-transform is $y_+(z) - \phi^2\{z^2 y_+(z) + zy_{-1} + y_{-2}\} = 0$. This gives

$$(5.46)\qquad \begin{aligned} y_+(z) &= \frac{z\phi^2 y_{-1} + \phi^2 y_{-2}}{1 - z^2\phi^2} \\ &= \frac{c_1}{1 - \phi z} + \frac{c_2}{1 + \phi z}, \end{aligned}$$

where

$$(5.47)\qquad c_1 = \frac{\phi y_{-1} + \phi^2 y_{-2}}{2} \quad \text{and} \quad c_2 = \frac{\phi^2 y_{-2} - \phi y_{-1}}{2}.$$

The latter are the constants in the general solution of the difference equation which is $y(t) = c_1\phi^t + c_2(-\phi)^t$. The same values for c_1 and c_2 could also be obtained by solving the equation

$$(5.48)\qquad \begin{bmatrix} y_{-1} \\ y_{-2} \end{bmatrix} = \begin{bmatrix} \dfrac{1}{\phi} & \dfrac{1}{-\phi} \\ \dfrac{1}{\phi^2} & \dfrac{1}{\phi^2} \end{bmatrix} \begin{bmatrix} c_1 \\ c_2 \end{bmatrix},$$

which comes from setting $t = -1, -2$ in the general solution.

Now consider the inhomogeneous difference equation

$$(5.49)\qquad \alpha(L)y(t) = (1 + \alpha_1 L + \cdots + \alpha_p L^p)y(t) = u(t).$$

The corresponding one-sided z-transform is

$$(5.50)\qquad \alpha(z)y_+(z) = u_+(z) - Q(z),$$

and the solution is provided by

$$(5.51)\qquad y_+(z) = \frac{u_+(z) - Q(z)}{\alpha(z)}.$$

Often we are assisted in finding the solution by discovering an expression for $u_+(z)$ in a table.

Example 5.7. The previous example may be elaborated by adding a forcing function which is a geometric sequence. This gives an equation in the form of $\alpha(L)y(t) = u(t)$ where $\alpha(L) = 1 - \phi^2 L^2$ and $u(t) = \gamma^t$. The z-transform of $u_+(t) = \{1, \gamma, \gamma^2, \ldots\}$ is $u_+(z) = 1/(1 - \gamma z)$. The term, which is due to the forcing function, has the following partial-fraction expansion:

$$(5.52)\qquad \frac{u_+(z)}{\alpha(z)} = \frac{d_1}{1 - \phi z} + \frac{d_2}{1 + \phi z} + \frac{d_3}{1 - \gamma z},$$

where

$$d_1 = \frac{\phi}{2(\phi - \gamma)}, \quad d_2 = \frac{\phi}{2(\phi + \gamma)} \quad \text{and} \quad d_3 = \frac{\gamma^2}{\gamma^2 - \phi^2}. \tag{5.53}$$

The expression for $Q(z)/\alpha(z)$ may be taken from the previous example. By combining some of the terms of $Q(z)/\alpha(z)$ and $u_+(z)/\alpha(z)$, we derive the expression

$$y_+(z) = \frac{g_1}{1 - \phi z} + \frac{g_2}{1 + \phi z} + \frac{d_3}{1 - \gamma z}, \tag{5.54}$$

where $g_1 = c_1 + d_1$ and $g_2 = c_2 + d_2$ incorporate the constants defined in (5.47) and (5.53). On translating this expression from the z-domain to the time domain, we find that the complete solution to the difference equation is

$$y(t) = g_1 \phi^t + g_2(-\phi)^t + \frac{\gamma^{t+2}}{\gamma^2 - \phi^2}. \tag{5.55}$$

In practice, it is easier to obtain the partial-fraction decomposition of the expression on the RHS of (5.51) without breaking it into its elements $u_+(z)/\alpha(z)$ and $Q(z)/\alpha(z)$ as we have done in the foregoing example. One of the advantages of using the z-transform method in solving difference equations is that the problem of incorporating the initial conditions may be solved without first determining a general solution for the difference equation. Thus an inhomogeneous difference equation may be solved without first solving the corresponding homogeneous equation.

Alternative Forms for the Difference Equation

Difference equations derive their name from the fact that, in classical applications, they are commonly expressed in terms of the forward-difference operator $\Delta = L^{-1} - I = F - I$ or the backward-difference operator $\nabla = I - L$.

It is always possible to derive expressions for a difference equation in terms of any of the operators L, F, Δ and ∇. It seems natural to use the operators L and ∇ in representing a recurrence relation when this is used to generate successive values of a sequence. For then the current value of the sequence will be expressed as a function of the preceding values. However, if we wish to emphasise the affinities between difference equations and differential equations, then we should employ the forward-difference operator Δ, since this has a familiar relationship with the differential operator D.

To convert a difference equation expressed in terms of powers of L to one which incorporates powers of ∇, we may use the following binomial expansion:

$$\begin{aligned} L^n &= (I - \nabla)^n \\ &= I - n\nabla + \frac{n(n-1)}{2!}\nabla^2 - \cdots (-1)^n \nabla^n. \end{aligned} \tag{5.56}$$

To derive an expression in terms of Δ is more problematic, since the expansion of $L^n = (\Delta + I)^{-n}$ gives rise to an infinite series. The appropriate recourse is to express the original difference equation in terms of the operator $F = L^{-1}$ and then to use the expansion

$$(5.57)\qquad \begin{aligned} F^n &= (\Delta + I)^n \\ &= \Delta^n + n\Delta^{n-1} + \frac{n(n-1)}{2!}\Delta^{n-2} + \cdots + I. \end{aligned}$$

Consider the pth-order difference equation $\alpha(L)y(t) = u(t)$. Multiplying both sides by the operator L^{-p}, gives $L^{-p}\alpha(L)y(t) = u(t+p)$. Now

$$(5.58)\qquad \begin{aligned} &L^{-p}\alpha(L) = F^p\alpha(F^{-1}) = \alpha'(F), \\ \text{where}\quad &\alpha'(F) = \alpha_0 F^p + \alpha_1 F^{p-1} + \cdots + \alpha_p; \end{aligned}$$

so the equation can be written as $\alpha'(F)y(t) = u(t+p)$. Then the expansion of $F^n = (\Delta + I)^n$ can be used to recast it into the form of $\phi(\Delta)y(t) = u(t+p)$, where $\phi(\Delta) = \phi_p + \phi_{p-1}\Delta + \phi_{p-2}\Delta^2 + \cdots + \phi_0\Delta^p$. We should take note of the fact that $\phi_0 = \alpha_0$, which also transpires in the following example.

Example 5.8. Consider the equation

$$(5.59)\qquad \begin{aligned} u(t+2) &= \alpha_0 y(t+2) + \alpha_1 y(t+1) + \alpha_2 y(t) \\ &= (\alpha_0 + \alpha_1 L + \alpha_2 L^2)y(t+2). \end{aligned}$$

An alternative form is

$$(5.60)\qquad \begin{aligned} u(t+2) &= \alpha_2 y(t) + \alpha_1 y(t+1) + \alpha_0 y(t+2) \\ &= (\alpha_2 + \alpha_1 F + \alpha_0 F^2)y(t). \end{aligned}$$

Using $F = I + \Delta$ and $F^2 = I + 2\Delta + \Delta^2$, we can rewrite this as

$$(5.61)\qquad u(t+2) = (\phi_2 + \phi_1\Delta + \phi_0\Delta^2)y(t),$$

with

$$(5.62)\qquad \begin{aligned} \phi_2 &= \alpha_2 + \alpha_1 + \alpha_0, \\ \phi_1 &= \alpha_1 + 2\alpha_0, \\ \phi_0 &= \alpha_0. \end{aligned}$$

It is clear that nothing essential is changed by recasting the equations in this way. Thus, if $y(t) = (1/\lambda)^t$ is a solution of the equation $\alpha(L)y(t) = 0$ when λ is a root of the equation $\alpha(z) = 0$, then it must also be a solution of the equation $\phi(\Delta)y(t) = 0$. However, if we were to adopt the latter form of the difference equation, then it would be convenient to express the solution in terms of the roots of the equation $\phi(z) = 0$.

If κ is a root of the equation $\phi(z) = \phi_p + \phi_{p-1}z + \phi_{p-2}z^2 + \cdots + \phi_0 z^p = 0$ such that $\phi(\kappa) = 0$, then $y(t) = (1+\kappa)^t$ is a solution of the equation $\phi(\Delta)y(t) = 0$. This follows in consequence of the fact that $\Delta(1+\kappa)^t = \kappa(1+\kappa)^t$, which implies that $\Delta^p(1+\kappa)^t = \kappa^p(1+\kappa)^t$; for we have

$$\begin{aligned}\phi(\Delta)(1+\kappa)^t &= (\phi_p + \phi_{p-1}\kappa + \phi_{p-2}\kappa^2 + \cdots + \phi_0\kappa^p)(1+\kappa)^t \\ &= \phi(\kappa)(1+\kappa)^t = 0.\end{aligned} \tag{5.63}$$

The connection between the roots of the equations $\phi(z) = 0$ and $\alpha(z) = 0$, which is already indicated by the fact that both $y(t) = (1+\kappa)^t$ and $y(t) = (1/\lambda)^t$ are solutions for the difference equation, can be established using the identity $L^{-p}\alpha(L) = \alpha'(F) = \phi(\Delta)$. On the one hand, using $\phi_0 = \alpha_0$, we find that

$$\begin{aligned}\phi(\Delta) &= \phi_0 \prod_{i=1}^{p}(\Delta - \kappa_i) \\ &= \alpha_0 \prod \{L^{-1} - (1+\kappa_i)\} \\ &= \alpha_0 L^{-p} \prod \{1 - (1+\kappa_i)L\}.\end{aligned} \tag{5.64}$$

On the other hand, there is

$$L^{-p}\alpha(L) = \alpha_0 L^{-p} \prod_{i=1}^{p}(1 - \mu_i L), \tag{5.65}$$

where $\mu_i = 1/\lambda_i$. The comparison shows that $\mu_i = 1 + \kappa_i$.

Linear Differential Equations

An pth-order linear differential equation with constant coefficients is a linear relationship amongst a continuous function $y(t)$ and its derivatives $dy(t)/dt = Dy(t), d^2y(t)/dt^2 = D^2y(t), \ldots, d^py(t)/dt^p = D^py(t)$. The differential equation may be presented in the form of

$$\phi_0 \frac{d^py(t)}{dt^p} + \phi_1 \frac{d^{p-1}y(t)}{dt^{p-1}} + \cdots + \phi_p y(t) = u(t), \tag{5.66}$$

where $u(t)$, which is known as the forcing function, is a specified function of t. The equation can also be written as

$$\phi(D)y(t) = u(t), \tag{5.67}$$

where

$$\phi(D) = \phi_0 D^p + \phi_1 D^{p-1} + \cdots + \phi_p. \tag{5.68}$$

The variable $y(t)$ and its p derivatives provide a complete description of the state of a physical system at a point in time. Knowing these values and the relationship which prevails amongst them at one point in time should enable us to predict the value of $y(t)$ and hence the state of the system at any other point. Equivalent information, equally appropriate for the purpose of predicting $y(t)$, would be provided by observations on this variable at p separate instants. Such observations, whether they relate to values of $y(t)$ at different times or to its derivatives at a point in time, are called initial conditions.

The function $y(t;c)$, expressing the analytic solution of the differential equation, will comprise a set of p constants in $c = [c_1, c_2, \ldots, c_p]'$ which can be determined once a set of p initial conditions has been specified. The general analytic solution of the equation $\phi(D)y(t) = u(t)$ may be expressed as $y(t;c) = x(t;c) + w(t)$, where $x(t;c)$ is the general solution of the homogeneous equation $\phi(D)x(t) = 0$, and $w(t) = \phi^{-1}(D)u(t)$ is a particular solution of the inhomogeneous equation.

The differential equation may be solved in the same manner as a difference equation. There are three steps. First the general solution of the homogeneous equation is found. Next, a particular solution $w(t)$ is obtained which contains no unknown quantities. Finally, the constants $c_1, c_2, \ldots, c_p$ are determined in view of the p initial values of y and its derivatives.

Solution of the Homogeneous Differential Equation

To assist in finding the solution of a differential equation, some results concerning a polynomial operator in D may be used which are analogous to those which have been given under (5.20) in connection with the lag operator L:

$$\text{(5.69)}\qquad \begin{aligned} &\text{(i)} & \phi(D)e^{\kappa t} &= \phi(\kappa)e^{\kappa t}, \\ &\text{(ii)} & \phi(D)\big\{e^{\kappa t}u(t)\big\} &= e^{\kappa t}\phi(D+\kappa)u(t), \\ &\text{(iii)} & \phi(D)\big\{v(t)+w(t)\big\} &= \phi(D)v(t) + \phi(D)w(t). \end{aligned}$$

The first of these results is proved by observing that $De^{\kappa t} = \kappa e^{\kappa t}$ and that, more generally, $D^n e^{\kappa t} = \kappa^n e^{\kappa t}$. The second result comes from observing that, according to the product rule, $De^{\kappa t}u(t) = \kappa e^{\kappa t}u(t) + e^{\kappa t}Du(t) = e^{\kappa t}(D+\kappa)u(t)$. Applying the result recursively gives $D^2 e^{\kappa t}u(t) = D\big\{e^{\kappa t}(D+\kappa)u(t)\big\} = e^{\kappa t}(D+\kappa)^2 u(t)$, and so on. The result under (ii) is an immediate generalisation. The property (iii) comes from the fact that $D\big\{v(t)+w(t)\big\} = Dv(t) + Dw(t)$.

The result under (i) indicates that, if κ_j is a root of the auxiliary equation $\phi(z) = \phi_0 z^p + \phi_1 z^{p-1} + \cdots + \phi_p = 0$ such that $\phi(\kappa_j) = 0$, then $y_j(t) = e^{\kappa_j t}$ is a solution of the equation $\phi(D)y(t) = 0$. Thus

$$\text{(5.70)}\qquad \phi(D)e^{\kappa_j t} = \phi(\kappa_j)e^{\kappa_j t} = 0.$$

Alternatively, we can consider the factorisation $\phi(D) = \phi_0 \prod_i (D - \kappa_i)$. Within this product, there is the term $D - \kappa_j$; and, since

$$\text{(5.71)}\qquad (D - \kappa_j)e^{\kappa_j t} = \kappa_j e^{\kappa_j t} - \kappa_j e^{\kappa_j t} = 0,$$

it follows that $\phi(D)e^{\kappa_j t} = 0$.

The general solution in the case where $\phi(z) = 0$ has distinct roots is given by

$$\text{(5.72)}\qquad y(t;c) = c_1 e^{\kappa_1 t} + c_2 e^{\kappa_2 t} + \cdots + c_p e^{\kappa_p t},$$

where $c_1, c_2, \ldots, c_p$ are the constants which are determined in view of the initial conditions.

In the case where two roots coincide at a value of κ, the equation $\phi(D)y(t) = 0$ has the solutions $y_1(t) = e^{\kappa t}$ and $y_2(t) = te^{\kappa t}$. We know already that $y_1(t)$ is a solution. To show that $y_2(t)$ is also a solution, let us consider the factorisation $\phi(D) = \phi_0 \prod_i (D - \kappa_j)$. If κ is a repeated root, then, from the expression $\phi(D)y_2(t) = \phi(D)te^{\kappa t}$, we can extract the factor

$$(D - \kappa)^2 te^{\kappa t} = (D^2 - 2\kappa D + \kappa^2)te^{\kappa t}. \tag{5.73}$$

But now the result under (5.69)(ii) serves to show that this is

$$\begin{aligned}(D^2 - 2\kappa D + \kappa^2)te^{\kappa t} &= e^{\kappa t}\{(D+\kappa)^2 - 2\kappa(D+\kappa) + \kappa^2\}t \\ &= e^{\kappa t} D^2 t \\ &= 0.\end{aligned} \tag{5.74}$$

The result, in summary, is that $(D - \kappa)^2 te^{\kappa t} = e^{\kappa t}D^2 t = 0$; and this could be inferred directly from (5.69)(ii). A more general result is that

$$(D - \kappa)^n t^{n-1} e^{\kappa t} = e^{\kappa t} D^n t^{n-1} = 0; \tag{5.75}$$

and this can be used to show that, if there are r repeated roots, then $e^{\kappa t}$, $te^{\kappa t}$, $t^2 e^{\kappa t}, \ldots, t^{r-1}e^{\kappa t}$ are all solutions to the equation $\phi(D)y(t) = 0$.

Differential Equation with Complex Roots

Imagine that the equation $\phi(z) = 0$ has conjugate complex roots $\kappa = \gamma + i\omega$ and $\kappa^* = \gamma - i\omega$. These will contribute to the general solution of the differential equation a term in the form of

$$\begin{aligned}q(t) &= ce^{(\gamma+i\omega)t} + c^* e^{(\gamma-i\omega)t} \\ &= e^{\gamma t}\{ce^{i\omega t} + c^* e^{-i\omega t}\}.\end{aligned} \tag{5.76}$$

This is a real-valued function; and, since a real term must equal its own conjugate, c and c^* must be conjugate numbers of the form

$$\begin{aligned}c^* &= \rho(\cos\theta + i\sin\theta) = \rho e^{i\theta}, \\ c &= \rho(\cos\theta - i\sin\theta) = \rho e^{-i\theta}.\end{aligned} \tag{5.77}$$

It follows that

$$\begin{aligned}q(t) &= \rho e^{\gamma t}\{e^{i(\omega t-\theta)} + e^{-i(\omega t-\theta)}\} \\ &= 2\rho e^{\gamma t}\cos(\omega t - \theta).\end{aligned} \tag{5.78}$$

The condition which is necessary and sufficient for $q(t)$ to tend to zero as t increases is that $\text{Re}\{\kappa\} = \gamma < 0$, which is to say that the root κ must lie in the left half of the complex plane. The condition applies both to real and to complex roots. Thus

(5.79) The general solution of the homogeneous equation $\phi(D)y(t) = 0$ tends to zero as t increases if and only if all of the roots of $\phi(z) = 0$ lie in the left half-plane.

Example 5.9. An idealised physical model of an oscillatory system consists of a weight of mass m suspended from a helical spring of negligible mass which exerts a force proportional to its extension. Let y be the displacement of the weight from its position of rest and let h be Young's modulus, which, according to Hooke's law, is the force exerted by the spring per unit of extension. Then Newton's second law of motion gives the equation

$$m\frac{d^2y}{dt^2} + hy = 0. \tag{5.80}$$

This is an instance of a second-order differential equation. The solution is

$$y(t) = 2\rho\cos(\omega_n t - \theta), \tag{5.81}$$

where $\omega_n = \sqrt{h/m}$ is the so-called natural frequency and ρ and θ are constants determined by the initial conditions. There is no damping or frictional force in the system and its motion is perpetual.

In a system which is subject to viscous damping, the resistance to the motion is proportional to its velocity. Then the differential equation becomes

$$m\frac{d^2y}{dt^2} + c\frac{dy}{dt} + hy = 0, \tag{5.82}$$

where c is the damping coefficient. The auxiliary equation of the system is

$$\begin{aligned} mz^2 + cz + h &= m(z - \kappa_1)(z - \kappa_2) \\ &= 0, \end{aligned} \tag{5.83}$$

and the roots κ_1, κ_2 are given by

$$\kappa_1, \kappa_2 = \frac{-c \pm \sqrt{c^2 - 4mh}}{2m}. \tag{5.84}$$

The character of the system's motion depends upon the discriminant $c^2 - 4mh$. If $c^2 < 4mh$, then the motion will be oscillatory, whereas, if $c^2 \geq 4mh$, the displaced weight will return to its position of rest without overshooting. If $c^2 = 4mh$, then the system is said to be critically damped. The critical damping coefficient is defined by

$$c_c = 2\sqrt{mh} = 2m\omega_n, \tag{5.85}$$

where ω_n is the natural frequency of the undamped system. On defining the so-called damping ratio $\zeta = c/c_c$, we may write equation (5.84) as

$$\kappa_1, \kappa_2 = -\zeta\omega_n \pm \omega_n\sqrt{\zeta^2 - 1}. \tag{5.86}$$

In the case of light damping, where $\zeta < 1$, the equation of the roots becomes

$$\begin{aligned} \kappa, \kappa^* &= -\zeta\omega_n \pm i\omega_n\sqrt{1-\zeta^2} \\ &= \gamma \pm i\omega; \end{aligned} \tag{5.87}$$

and the motion of the system is given by

$$\begin{aligned} y(t) &= 2\rho e^{\gamma t}\cos(\omega t - \theta) \\ &= 2\rho e^{-\zeta\omega_n t}\cos\left\{(1-\zeta^2)^{1/2}\omega_n t - \theta\right\}. \end{aligned} \tag{5.88}$$

Particular Solutions for Differential Equations

If $u(t)$ is a polynomial in t or an exponential function or a combination of sines and cosines, then it is a relatively simple matter to find the particular solution of the equation $\phi(D)y(t) = u(t)$ which takes the form of $y(t) = u(t)/\phi(D)$. With other types of function, the particular solution has to be expressed as a definite integral.

To show how the more tractable problems may be solved, let us state the inverse results corresponding to those given under (5.69):

$$\begin{aligned} &\text{(i)} && \frac{1}{\phi(D)}e^{\kappa t} = \frac{1}{\phi(\kappa)}e^{\kappa t} \quad \text{if} \quad \phi(\kappa) \neq 0, \\ &\text{(ii)} && \frac{1}{\phi(D)}\left\{e^{\kappa t}u(t)\right\} = e^{\kappa t}\frac{1}{\phi(D+\kappa)}u(t), \\ &\text{(iii)} && \frac{1}{\phi(D)}\left\{u(t)+v(t)\right\} = \frac{1}{\phi(D)}u(t) + \frac{1}{\phi(D)}u(t). \end{aligned} \tag{5.89}$$

These are closely analogous to the results under (5.21) which concern the lag operator. The case of $\phi(\kappa) = 0$, which affects (i), arises when $\phi(D) = (D-\kappa)^r\phi_1(D)$, where $\phi_1(D) \neq 0$ and r is the multiplicity of the root κ. Then (i) may be replaced by the last of the following expressions:

$$\begin{aligned} \frac{1}{\phi(D)}e^{\kappa t} &= \left\{\frac{1}{(D-\kappa)^r}\right\}\left\{\frac{e^{\kappa t}}{\phi_1(D)}\right\} = \left\{\frac{e^{\kappa t}}{(D-\kappa)^r}\right\}\left\{\frac{1}{\phi_1(\kappa)}\right\} \\ &= \left\{\frac{e^{\kappa t}}{D^r}\right\}\left\{\frac{1}{\phi_1(\kappa)}\right\} = \frac{t^r\ e^{\kappa t}}{r!\ \phi_1(\kappa)}. \end{aligned} \tag{5.90}$$

Here the penultimate equality comes from applying (ii) to the expression $e^{\kappa t}/(D-\kappa)^r$ to give $e^{\kappa t}/D^r$. The final equality depends upon $1/D^r = t^r/r!$ which is the result of integrating unity r times in respect of t.

The results above are used in the following examples which run parallel to those which have illustrated the solution of difference equations.

Example 5.10. Consider the equation

$$\frac{d^2y}{dt^2}+5\frac{dy}{dt}+6y=e^{3t}. \tag{5.91}$$

According (5.89)(i), the particular solution is

$$\begin{aligned} w(t) &= \frac{1}{D^2+5D+6}e^{3t} \\ &= \frac{1}{3^2+5.3+6}e^{3t}=\frac{1}{30}e^{3t}. \end{aligned} \tag{5.92}$$

Example 5.11. Let $(D+3)y(t)=t^3$. Then the particular solution is

$$\begin{aligned} w(t) &= \frac{1}{3+D}t^3=\frac{1}{3}.\frac{1}{1+\frac{1}{3}D}t^3 \\ &= \frac{1}{3}\left(1-\frac{1}{3}D+\frac{1}{9}D^2-\frac{1}{27}D^3+\cdots\right)t^3 \\ &= \frac{1}{3}t^3-\frac{1}{3}t^2+\frac{2}{9}t-\frac{2}{27}. \end{aligned} \tag{5.93}$$

Here the expansion of $1/(1-\frac{1}{3}D)$ has been carried no further than the term in D^3, since all the higher-order derivatives of t^3 vanish.

Example 5.12. Consider the differential equation

$$(D^2+\phi_1D+\phi_2)y(t)=\delta\cos(\omega t). \tag{5.94}$$

Using the identity $e^{i\omega t}=\cos(\omega t)+i\sin(\omega t)$, we can take $\cos(\omega t)=\text{Re}(e^{i\omega t})$, which is the real part of the complex function. This gives

$$\begin{aligned} y(t) &= \text{Re}\left\{\frac{\delta}{D^2+\phi_1D+\phi_2}e^{i\omega t}\right\} \\ &= \text{Re}\left\{\frac{\delta}{(\phi_2-\omega^2)+i\phi_1\omega}e^{i\omega t}\right\}, \end{aligned} \tag{5.95}$$

where (5.89)(i) has been used to obtain the second equality. By using the result that $(\alpha+i\beta)^{-1}=(\alpha-i\beta)/(\alpha^2+\beta^2)$ and by writing the complex exponential in terms of a sine and a cosine, we get

$$\begin{aligned} y(t) &= \delta\text{Re}\left\{\frac{(\phi_2-\omega^2)-i\phi_1\omega}{(\phi_2-\omega^2)^2+\phi_1^2\omega^2}\big[\cos(\omega t)+i\sin(\omega t)\big]\right\} \\ &= \delta\frac{(\phi_2-\omega^2)\cos(\omega t)+\phi_1\omega\sin(\omega t)}{(\phi_2-\omega^2)^2+\phi_1^2\omega^2}. \end{aligned} \tag{5.96}$$

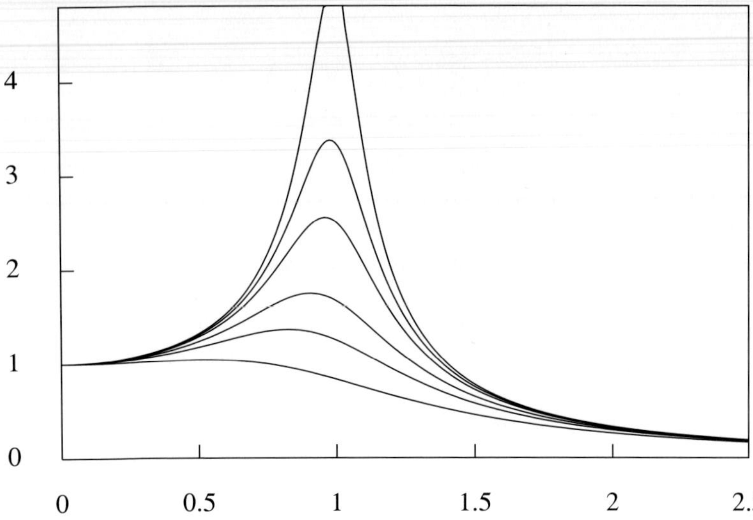

Figure 5.2. The frequency response of a second-order system with various damping ratios. On the horizontal axis is the relative frequency ω/ω_n. The six curves, from the highest to the lowest, correspond to the damping ratios $\zeta = 0.1, 0.15, 0.2, 0.3, 0.4, 0.6$.

This can also be written as

$$y(t) = \frac{\delta \cos(\omega t - \theta)}{\sqrt{(\phi_2 - \omega^2)^2 + \phi_1^2\omega^2}}, \tag{5.97}$$

where

$$\theta = \tan^{-1}\left(\frac{\phi_1\omega}{\phi_2 - \omega^2}\right). \tag{5.98}$$

This result may be applied to the problem of finding the steady-state solution for a simple damped system which is driven by a sinusoidal forcing function. The differential equation is an elaboration of equation (5.82):

$$m\frac{d^2y}{dt^2} + c\frac{dy}{dt} + hy = \beta\cos(\omega t). \tag{5.99}$$

Setting $\phi_1 = c/m$, $\phi_2 = h/m$ and $\delta = \beta/m$ in equations (5.97) and (5.98) shows that the steady-state solution is given by

$$y(t) = \gamma\cos(\omega t - \theta), \tag{5.100}$$

where

$$\gamma = \frac{\beta}{\sqrt{(h - m\omega^2)^2 + (c\omega)^2}} \tag{5.101}$$

and

$$\theta = \tan^{-1}\left(\frac{c\omega}{h - m\omega^2}\right). \tag{5.102}$$

The essential result, which is confirmed by common experience, is that an oscillatory input at a given frequency gives rise to an oscillatory output at the same frequency. Indeed, this result—under the guise of a trial solution—is the premise upon which many texts of mechanical engineering base their derivation of the formulae of (5.101) and (5.102).

The formulae may be expressed in terms of the following engineering quantities:

$$\begin{aligned} &\text{(i)} && \omega_n = \sqrt{\frac{h}{m}} && \text{the natural frequency,} \\ &\text{(ii)} && c_c = 2m\omega_n && \text{the critical damping coefficient,} \\ &\text{(iii)} && \zeta = \frac{c}{c_c} && \text{the damping ratio.} \end{aligned} \tag{5.103}$$

Then the steady-state amplitude becomes

$$\gamma = \frac{\beta/h}{\left[\left\{1 - (\omega/\omega_n)^2\right\}^2 + 4\zeta^2 (\omega/\omega_n)^2\right]^{1/2}}, \tag{5.104}$$

whilst the phase displacement is given by

$$\tan\theta = \frac{2\zeta(\omega/\omega_n)}{1 - (\omega/\omega_n)^2}. \tag{5.105}$$

In a lightly-damped system, the amplitude γ of the forced motion is greatest when the frequency ω of the forcing function is in the vicinity of the natural frequency ω_n of the undamped system which is depicted in equation (5.80). The large-amplitude oscillations of a system, which can result from a low-powered driving at such a frequency, is described as resonance. The phenomenon is illustrated in Figure 5.2, which shows the gain $\gamma h/\beta$ in the amplitude of the output as a function of the frequency ratio ω/ω_n.

Example 5.13. Simple electric circuits containing elements of resistance, capacitance and inductance are governed by second-order differential equations. There are many electrical components which combine these characteristic in varying degrees; but, in describing how such circuits function, some stylised components spring to mind.

A resistor R is thought of as a long piece of thin wire, often tightly wound in the shape of a cylinder, which impedes the flow of current. A capacitor or condenser C, which stores electric charge, is thought of as a sandwich consisting of two large conducting plates separated by an insulator. Equal electric charges of opposite sign will accumulate on these plates if the capacitor is placed in a

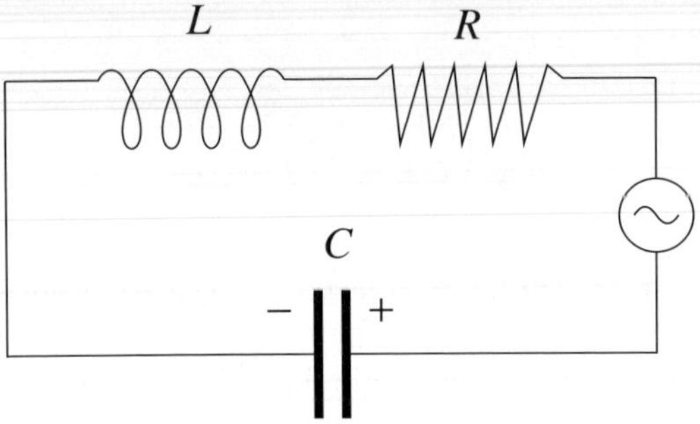

Figure 5.3. An LCR circuit incorporating an induction coil L, a resistor R and a capacitor C. The electrical input is a fluctuating voltage.

circuit and if a difference in voltage between the two plates is established. An inductance L is represented by the effect of a coil of wire of no intrinsic resistance which, nevertheless, serves to impede the flow of current on account of an induced electromotive force which acts in the opposite direction. This impedance is due to the electromagnetic flux generated by the coil when the current varies.

The formulae relating the flow of current i to the voltage drop across these components are

$$\text{(5.106)} \qquad \begin{aligned} &\text{(i)} && V_R = iR && \text{for resistance,} \\ &\text{(ii)} && V_L = L\frac{di}{dt} && \text{for induction,} \\ &\text{(iii)} && C\frac{dV_C}{dt} = i && \text{for a capacitor.} \end{aligned}$$

We shall use the results of Example 5.12 to analyse the case of a so-called LCR circuit where the electrical input is a voltage fluctuating at a frequency of ω (see Figure 5.3). This input may described as a signal. Since the components of the circuit are wired in series, the sum of the potential differences or voltage drops across each of them is equal to the signal voltage. Thus

$$\text{(5.107)} \qquad V_C(t) + V_R(t) + V_L(t) = V\cos(\omega t).$$

To simplify the analysis, we shall consider the differential equation for the charge q on the capacitor as a function of time instead of the equation for the current $i = dq/dt$. The equation is

$$\text{(5.108)} \qquad L\frac{d^2q}{dt^2} + R\frac{dq}{dt} + \frac{1}{C}q = V\cos(\omega t).$$

This may be assimilated to the equation (5.94) by setting $\phi_1 = R/L$, $\phi_2 = 1/(CL)$ and $\delta = V/L$. Then the steady-state solution for the system is found to be

$$\text{(5.109)} \qquad q(t) = \frac{V}{LQ}\cos(\omega t - \theta),$$

where

$$Q = \sqrt{\left(\frac{1}{CL} - \omega^2\right)^2 + \frac{R^2}{L^2}\omega^2} \tag{5.110}$$

and $\theta = RC\omega/(1 - C\omega)$.

The importance of the LCR circuit lies in the ease with which its natural resonance frequency $\omega_n = \sqrt{\{1/(CL)\}}$ may be adjusted by varying the capacitance C. We may observe that the voltage gain $V_C/V = 1/(LQ)$ is greatest when the resonance frequency is close to the signal frequency. If the signal is a radio transmission, then the circuit can be tuned to discriminate in favour of this frequency by amplifying it markedly in comparison with neighbouring frequencies. In this way, a radio receiver can be tuned to one transmitter at a time.

Solutions of Differential Equations with Initial Conditions

A complete solution of a pth-order differential equation is achieved when the arbitrary constants in the analytic expression $y(t; c)$ of the general solution are replaced by values which have been determined in the light of the initial conditions. To find these values, we can proceed in a variety of ways which run parallel to those which we have already described in connection with difference equations. The initial conditions usually take the form of the value of $y(t)$ and of its first $p-1$ derivatives at the time $t = 0$. A common way of incorporating this information is to solve a set of linear equations

Example 5.14. Let the difference equation be $(D^2 + 5D + 6)y(t) = 12e^t$, and assume that the initial conditions at time $t = 0$ are given by $y(0) = 2$ and $Dy(0) = 1$. In the manner of Example 5.10, the particular solution is found to be $w(t) = e^t$. The general solution of the equation is

$$y(t) = c_1e^{-2t} + c_2e^{-3t} + e^t. \tag{5.111}$$

Differentiating gives

$$Dy(t) = e^t - 2c_1e^{-2t} - 3c_2e^{-3t}. \tag{5.112}$$

Substituting the values at $t = 0$ into the two equations gives $2 = 1 + c_1 + c_2$ and $1 = 1 - 2c_1 - 3c_2$, from which $c_1 = 3$ and $c_2 = -2$.

Notice that this method of finding the coefficients can be adapted easily to accommodate cases where the initial conditions are a sequence of observations on $y(t)$.

An alternative method for finding a complete solution, which can be applied when the initial conditions are in the usual form of a sequence of derivatives, uses the Laplace transformation.

(5.113) If $y(t)$ is a function defined for $t \geq 0$, then its Laplace transform is $\mathcal{L}y(t) = y_+(s) = \int_0^\infty e^{-st}y(t)dt$ where $s = \sigma + i\omega$ is a complex number wherein $\sigma > 0$ is chosen so as to ensure that the integral will converge.

The Laplace transform is the analogue of the one-sided z-transform; and it will be used in seeking the solutions of differential equations in much the same way as the z-transform has been used in connection with difference equations.

The properties of the transform which are used for incorporating the initial conditions in the general solution are analogous to those for the z-transform which are listed under (5.37). They concern the Laplace transform of the first $p-1$ derivatives of the function $y(t)$:

(5.114)
$$\begin{aligned} y(t) &\longleftrightarrow y_+(s), \\ Dy(t) &\longleftrightarrow sy_+(s) - y(0), \\ D^2y(t) &\longleftrightarrow s^2y_+(s) - sy(0) - Dy(0), \\ D^3y(t) &\longleftrightarrow s^3y_+(s) - s^2y(0) - sDy(0) - D^2y(0). \end{aligned}$$

In general, we have

(5.115)
$$D^jy(t) \longleftrightarrow s^jy_+(s) - \sum_{i=1}^{j} s^{j-i}D^{i-1}y(0).$$

To demonstrate the formula for the transform of the first derivative of $y(t)$, we integrate $e^{-st}Dy(t)$ by parts to give

(5.116)
$$\begin{aligned} \int_0^\infty e^{-st}Dy(t)dt &= s\int_0^\infty e^{-st}y(t)dt + \left[e^{-st}y(t)\right]_0^\infty \\ &= sy_+(s) - y(0). \end{aligned}$$

The result can also be expressed by writing

(5.117)
$$\mathcal{L}Dy(t) = s\mathcal{L}y(t) - y(0).$$

To establish the transform of the second derivative, we may write

(5.118)
$$\begin{aligned} \mathcal{L}D^2y(t) &= \mathcal{L}D\{Dy(t)\} \\ &= s\mathcal{L}\{Dy(t)\} - \{Dy(0)\} \\ &= s^2\mathcal{L}y(t) - sy(0) - Dy(0). \end{aligned}$$

The transforms of the higher-order derivatives can be found by proceeding recursively.

Since the Laplace transform involves a linear operator, it is straightforward to define the transform of the function $\phi(D)y(t)$ wherein $\phi(D) = \phi_0D^p + \phi_1D^{p-1} + \cdots + \phi_0$ is a polynomial of degree p in the differentiating operator. Thus we have

(5.119)
$$\phi(D)y(t) \longleftrightarrow \phi(s)y_+(s) - Q(s),$$

where

(5.120)
$$\begin{aligned} Q(s) &= \sum_{j=1}^{p} \phi_{p-j}\left\{\sum_{i=1}^{j} s^{j-i}D^{i-1}y(0)\right\} \\ &= q_0y(0) + q_1Dy(0) + \cdots + q_{p-1}D^{p-1}y(0) \end{aligned}$$

is a polynomial of degree $p-1$ in s which incorporates the values of $y(t)$ and of its first $p-1$ derivatives at $t=0$.

Before applying the Laplace transform to the problem in hand, we should also show how the transforms of certain elementary functions may be obtained which are liable to arise in the search for particular solutions. Consider the function $e^{\kappa t}$ where κ is a real-valued constant. Then

$$\mathcal{L}e^{\kappa t} = \int_0^\infty e^{-st}e^{\kappa t}dt = \int_0^\infty e^{-(s-\kappa)t}dt = \left[\frac{-e^{-(s-\kappa)t}}{s-\kappa}\right]_0^\infty; \tag{5.121}$$

and, therefore,

$$e^{\kappa t} \longleftrightarrow \frac{1}{s-\kappa}. \tag{5.122}$$

Differentiating the functions on both sides of the relationship n times with respect to κ gives

$$t^n e^{\kappa t} \longleftrightarrow \frac{n!}{(s-\kappa)^{n-1}}. \tag{5.123}$$

If $\kappa = \gamma - i\omega$ is a complex number, then we get

$$\begin{aligned}\mathcal{L}e^{(\gamma-i\omega)t} &= \mathcal{L}e^{\gamma t}(\cos\omega t - i\sin\omega t)\\ &= \frac{1}{s-\gamma+i\omega} = \frac{s-\gamma-i\omega}{(s-\gamma)^2+\omega^2}.\end{aligned} \tag{5.124}$$

Taking the real and imaginary parts separately, we find that

$$\begin{aligned} e^{\gamma t}\cos\omega t &\longleftrightarrow \frac{s-\gamma}{(s-\gamma)^2+\omega^2} \quad \text{and}\\ e^{\gamma t}\sin\omega t &\longleftrightarrow \frac{\omega}{(s-\gamma)^2+\omega^2}.\end{aligned} \tag{5.125}$$

We can strip away the exponential factor by setting $\gamma = 0$.

Now let us consider using the Laplace transform in solving a differential equation in the form of

$$\phi(D)y(t) = u(t). \tag{5.126}$$

The Laplace transform of the equation is given by $\phi(s)y_+(s) - Q(s) = u_+(s)$, where $u_+(s) = \mathcal{L}u(t)$ stands for the Laplace transform of the forcing function and where $Q(s)$ is the function embodying the initial conditions. It follows that the Laplace transform of the complete solution is given by

$$y_+(s) = \frac{u_+(s) + Q(s)}{\phi(s)}. \tag{5.127}$$

To express the complete solution in the time domain, we must apply the inverse of the Laplace transform. This may be done by first expressing the RHS of (5.127) in partial fractions in order to use the appropriate standard forms of the transform.

Example 5.15. Consider again the equation $(D^2+5D+6)y(t) = 12e^t$ of Example 5.14 for which the initial conditions are $y(0) = 2$ and $Dy(0) = 1$. By applying the results under under (5.114) to the LHS of the equation and the result under (5.122) to the RHS, we find that its transform is

$$\{s^2y_+(s) - 2s - 1\} + 5\{sy_+(s) - 2\} + 6y_+(s) = \frac{12}{s-1}. \tag{5.128}$$

By solving this, and using a partial-fraction expansion, we find that

$$\begin{aligned} y_+(s) &= \frac{2s^2+9s+1}{(s-1)(s+2)(s+3)} \\ &= \frac{1}{s-1} + \frac{3}{s+2} - \frac{2}{s+3}. \end{aligned} \tag{5.129}$$

The inverse of the Laplace transformation, which depends solely on the result under (5.122), yields the following time-domain solution:

$$y(t) = e^t + 3e^{-2t} - 2e^{-3t}. \tag{5.130}$$

This agrees with result which was derived in Example 5.14 by the classical method.

Difference and Differential Equations Compared

It is interesting to compare the solution $y(t) = (1+\gamma)^t$ of the first-order difference equation $(\Delta - \gamma)y(t) = 0$ with the solution of the corresponding differential equation. In the case of the difference equation, the parameter γ may be construed as the proportional change in $y(t)$ from one period to the next. It might represent, for example, the rate of return on a financial investment which is compounded annually. An investment which is compounded twice a year has a growth factor of $(1+\frac{1}{2}\gamma)^2$, and one which is compounded each quarter has an annual growth factor of $(1+\frac{1}{4}\gamma)^4$. If an investment were compounded continuously, then its growth factor would be $\lim(n \to \infty)(1+\frac{1}{n}\gamma)^n = e^\gamma$. This is exactly the factor which is entailed in the solution of the first-order differential equation $(D-\gamma)y(t) = 0$ which is $y(t) = \rho e^{\gamma t}$.

The issue arises of whether difference and differential equations may be used interchangeably in representing continuous-time dynamic processes. Let us compare the equations of (5.18) and (5.78) which represent the sinusoidal motions generated respectively by difference and differential equations of the second order. By setting $\kappa = e^\gamma$, the two equations are rendered identical. However, in the differential equation, the argument t is a continuous variable whereas, in the difference equation, it is integer-valued.

When t is continuous, there is a one-to-one correspondence between the set of positive frequency values and the set of cosine functions $y(t) = \cos(\omega t)$. When the values of t are discrete, there are infinitely many frequency values which generate the same ordinates of $y(t)$. That is to say, when $t \in \{0, \pm1, \pm2, \ldots\}$, the identity

$$y(t) = \cos(\omega t) = \cos(2\pi jt + \omega t) = \cos(2\pi jt - \omega t) \tag{5.131}$$

holds for any positive or negative integer j. Thus the set

(5.132) $$\Omega(\omega) = \{2\pi j \pm \omega; j = 0, \pm 1, \pm 2, \ldots\}$$

defines a class of equivalent frequencies. Moreover, since the set of equivalence classes $\{\Omega(\omega); \omega \in [-\pi, \pi)\}$ defines a partition of the real line $\mathcal{R} = \{\omega; -\infty < \omega < \infty\}$, the equivalence class of any frequency is completely specified by a value of ω in the interval $[-\pi, \pi)$.

When we take account of the symmetry of the cosine function which implies that $\cos(\omega t) = \cos(-\omega t)$, and of the fact that $\cos(\omega_1 t) \neq \cos(\omega_2 t)$ when $\omega_1, \omega_2 \in [0, \pi)$ are distinct values, it follows that, to each class of equivalent frequencies, there corresponds a unique value of ω in the lesser interval $[0, \pi)$.

The upshot of these results is that the cyclical components of a process in continuous time can be identified uniquely from observations taken at unit intervals only if their frequencies are known to be confined to the band $[0, \pi)$. This means that a component must take at least two periods to complete its cycle if it is not to be confused with another component of lesser frequency which is in the same equivalence class. In the event of such a confusion, the only recourse is to increase the rate of sampling to an extent which succeeds in reducing the highest of the frequencies amongst the components of the process to somewhat less than π radians per sample period.

This result is the essence of the famous Nyquist–Shannon sampling theorem (see [368] and [450]), to which we shall return in the chapters devoted to Fourier analysis.

Conditions for the Stability of Differential Equations

In this section, we shall present, without proof, the procedure of Routh (see [431] and [433]) for establishing whether or not a homogeneous differential equation is stable. Let us continue to write the differential equation as $\phi(D)y(t) = 0$. Then, as we have already established, the necessary and sufficient condition for stability is that the polynomial equation $\phi(z) = \phi_0 z^p + \phi_1 z^{p-1} + \cdots + \phi_p = 0$ has all of its roots $\kappa_j = \gamma_j + i\omega_j$ in the left half of the complex plane, which is to say that $\gamma_j < 0$ for all $j = 1, \ldots, p$.

The roots of the polynomial equation $\phi'(z) = \phi_0 + \phi_1 z + \cdots + \phi_p z^p = 0$ are just the inverse values $\kappa_j^{-1} = (\gamma_j - i\omega_j)/(\gamma_j^2 + \omega_j^2)$. It follows that it is equivalent to require that $\phi'(z) = 0$ has all of its roots in the left half-plane. What this implies for the Routh test is that it makes no odds if the coefficients of the polynomial are taken in reverse order.

There is a preliminary test which should always be applied before embarking on the Routh test. The test makes use of the fact that

(5.133) If $\phi(z) = \phi_0 z^p + \phi_1 z^{p-1} + \cdots + \phi_p = \phi_0 \prod_j (z - \kappa_j) = 0$ is to have all of its roots in the left half-plane when $\phi_0 > 0$, then it is necessary that $\phi_j > 0$ for all j.

Proof. The roots of $\phi(z) = 0$ are either real or else they occur in conjugate pairs $\kappa, \kappa^* = \gamma \pm i\omega$. If $\kappa = \gamma$ is a real root, then it contributes to the polynomial a linear factor $z - \gamma$ which has a positive coefficient if $\gamma < 0$. A conjugate pair of complex roots contributes a quadratic factor $(z - \gamma - i\omega)(z - \gamma + i\omega) = z^2 - 2\gamma z + \gamma^2 + \omega^2$ which also has positive coefficients if $\gamma < 0$. Thus the condition that $\gamma_j < 0$ for every root $\kappa_j = \gamma_j + i\omega_j$ implies that the coefficients of $\phi(z)$ must all be positive.

Example 5.16. When $p = 2$, the conditions $\phi_2, \phi_1, \phi_0 > 0$ are sufficient as well as necessary for stability. In that case, the roots of $\phi_0 z^2 + \phi_1 z + \phi_2 = 0$ are given by

$$\kappa, \kappa^* = \frac{-\phi_1 \pm \sqrt{\phi_1^2 - 4\phi_0\phi_2}}{2\phi_0}. \tag{5.134}$$

If the roots are real, then they must both be negative since $\sqrt{}(\phi_1^2 - 4\phi_0\phi_2) < \phi_1$. If they are complex, then their real part is $-\phi_1/(2\phi_0)$ which is also negative. When $p = 3$, the conditions on the coefficients no longer guarantee that the real parts of the roots are negative. For a counterexample, we may take the polynomial

$$\begin{aligned} z^3 + 2z^2 + 2z + 40 &= (z + 4)(z^2 - 2z + 10) \\ &= (z + 4)\big(z - \{1 - i3\}\big)\big(z - \{1 + i3\}\big). \end{aligned} \tag{5.135}$$

The roots of the quadratic factor are the conjugate complex numbers $1 \pm i3$, which have a positive real part.

The test of Routh depends upon constructing an array whose rows are formed from the coefficients of the polynomial by the repeated application of a simple rule until a final row is generated which has a single element. The first and second rows are formed by taking respectively the coefficients with even and odd indices. The third row is formed from the first two in a way which should be evident:

$$\begin{array}{cccc} \phi_0 & \phi_2 & \phi_4 & \phi_6 \ \cdots \\ \phi_1 & \phi_3 & \phi_5 & \phi_7 \ \cdots \\ \phi_2 - \dfrac{\phi_0}{\phi_1}\phi_3 & \phi_4 - \dfrac{\phi_0}{\phi_1}\phi_5 & \phi_6 - \dfrac{\phi_0}{\phi_1}\phi_7 & \cdots \end{array} \tag{5.136}$$

The fourth row of the array is formed from the second and the third rows in the same manner as the third row is formed from the first and second. The process is continued as far as the $(p+1)$th row which has a single nonzero element. The system is stable according to Routh's criterion if and only if all of the elements in the first column of the array have the same sign.

One should note that, if any of the coefficients of the polynomial are zeros, then the preliminary test indicates that the conditions of stability are violated. The test breaks down if a zero is encountered in the first column of the array before the pth row has been reached. In that case, the array cannot be completed since, in attempting to form the next row, there would be a division by zero. A method of dealing with such cases is given by Gantmacher [201].

We guard against such eventualities in the following algorithm by aborting the process if a zero is encountered. Whereas such a condition may well arise when, by design, the coefficients of the polynomial are small integers, it is unlikely to arise when they are the products of an empirical estimation.

The structure of our algorithm for implementing Routh's test is complicated by the fact that successive rows of the array are stored in a single vector by overwriting previous elements which are no longer needed.

(5.137)

```
procedure RouthCriterion(phi : vector;
                         p : integer;
                         var stable : boolean);

  var
    i, j : integer;
    fact : real;

begin
  stable := true;

  {Preliminary Testing}
  for i := 1 to p do
    if phi[i] * phi[i - 1] <= 0 then
      stable := false;

  if stable = true then
    begin {Further Testing}
      i := 2;
      phi[p + 1] := 0;
      repeat
        j := i;
        fact := phi[i - 2]/phi[i - 1];
        repeat
          phi[j] := phi[j] - fact * phi[j + 1];
          j := j + 2;
        until j > p;
        if phi[i] * phi[i - 1] <= 0 then
          stable := false;
        i := i + 1
      until (i = p + 1) or (stable = false)
    end; {Further Testing}

end; {RouthCriterion}
```

The conditions for the stability of a differential equation were given in terms of determinants by Hurwitz [263] some years after Routh [431] had published his results.

The rth-order determinant of Hurwitz is defined by

$$\delta_r = \det \begin{bmatrix} \phi_1 & \phi_3 & \phi_5 & \ldots & \phi_{2r-1} \\ \phi_0 & \phi_2 & \phi_4 & \ldots & \phi_{2r-2} \\ 0 & \phi_1 & \phi_3 & \ldots & \phi_{2r-3} \\ 0 & \phi_0 & \phi_2 & \ldots & \phi_{2r-4} \\ \vdots & \vdots & \vdots & & \vdots \\ 0 & 0 & 0 & \ldots & \phi_2 \end{bmatrix}. \tag{5.138}$$

Having placed ϕ_1 in the leading position, the rule for forming the remainder of the array is to increase the indices on successive elements in each row by two and to decrease the indices of successive elements in each column by one. If the index exceeds the value of p, which is the degree of the polynomial, or if it becomes negative, then a zero element is put in place. Assuming that $\phi_0 > 0$, the rule of Hurwitz is that the system is stable if and only if each element in the sequence of the determinants is positive:

$$\delta_1 > 0, \delta_2 > 0, \ldots, \delta_p > 0. \tag{5.139}$$

It can be shown that the sequence of coefficients within the first column of the array generated in Routh's test is equivalent to the following sequence of ratios of determinants:

$$\delta_1, \quad \frac{\delta_2}{\delta_1}, \quad \frac{\delta_3}{\delta_2}, \ldots, \frac{\delta_p}{\delta_{p-1}}. \tag{5.140}$$

From this, the equivalence of the two criteria can be established. Indeed the two criteria are often referred to jointly as the Routh–Hurwitz criterion.

Conditions for the Stability of Difference Equations

Now consider a difference equation of the form $\alpha(L)y(t) = 0$. It has already been established that the difference equation is stable if and only if all of the roots λ_i of the primary polynomial equation $\alpha(z) = \alpha_0 + \alpha_1 z + \cdots + \alpha_p z^p$ lie outside the unit circle. Equivalently, all of the roots $\mu_i = 1/\lambda_i$ of the auxiliary equation $\alpha'(z) = z^p \alpha(z^{-1}) = 0$ must fall inside the unit circle.

One way of assessing the stability of the difference equation without evaluating its roots is to convert the polynomial $\alpha'(z)$ to a form to which the Routh–Hurwitz test may be applied. This requires converting the complex variable z into another variable $s = \gamma + i\delta$ by a transformation which maps the unit circle into the left half of the complex plane. This is achieved by the bilinear Möbius transformation which is given, together with its inverse, by

$$s = \frac{z+1}{z-1} \quad \text{and} \quad z = \frac{s+1}{s-1}. \tag{5.141}$$

The restriction that z lies inside the unit circle may be expressed in terms of the components of s:

$$|z| = \left|\frac{s+1}{s-1}\right| = \left|\frac{\gamma + i\delta + 1}{\gamma + i\delta - 1}\right| < 1. \tag{5.142}$$

Squaring the moduli gives

$$(5.143)\qquad \begin{gathered}(\gamma+1)^2+\delta^2<(\gamma-1)^2+\delta^2,\\ \text{whence}\quad \gamma<0,\end{gathered}$$

which is the restriction that s lies in the left half-plane.

Substituting the expression for $z = z(s)$ into the equation $\alpha'(z) = \alpha_0 z^p + \alpha_1 z^{p-1} + \cdots + \alpha_{p-1}z + \alpha_p = 0$ gives

$$(5.144)\qquad \alpha_0\left(\frac{s+1}{s-1}\right)^p + \alpha_1\left(\frac{s+1}{s-1}\right)^{p-1} + \cdots + \alpha_{p-1}\frac{s+1}{s-1} + \alpha_p = 0.$$

To clear the fractions, this may be multiplied throughout by $(s-1)^p$. The result is a polynomial

$$(5.145)\qquad \phi(s) = \phi_0 s^p + \phi_1 s^{p-1} + \cdots + \phi_{p-1}s + \phi_p = 0,$$

to which Routh's test may be applied.

The difficulty with this approach is the amount of algebraic manipulation which is necessary to obtain the $\phi(s)$ from $\alpha(z)$. The approach was followed by Samuelson [436], [437] in connection with economic dynamics and by Wise [529] in a statistical context. However, it appears that the solution of the problem by means of the Möbius transformation was reached originally by Herglotz [254] some twenty years earlier.

Example 5.17. When $p = 2$, the transformed polynomial becomes

$$(5.146)\qquad \begin{aligned}&\alpha_0(s+1)^2 + \alpha_1(s+1)(s-1) + \alpha_2(s-1)^2\\ &= (\alpha_0+\alpha_1+\alpha_2)s^2 + 2(\alpha_0-\alpha_2)s + (\alpha_0-\alpha_1+\alpha_2)\\ &= \phi_0 s^2 + \phi_1 s + \phi_2.\end{aligned}$$

From the previous example, we know that, on the assumption that $\phi_0 > 0$, the necessary and sufficient condition for the roots to lie in the left half-plane when $p = 2$ is that the other coefficients are also positive. Therefore, for the stability of the difference equation $\alpha_0 y(t) + \alpha_1 y(t-1) + \alpha_2 y(t-2) = 0$, it is necessary and sufficient that

$$(5.147)\qquad \begin{aligned}&\text{(i)} && \alpha_0+\alpha_1+\alpha_2>0,\\ &\text{(ii)} && \alpha_0-\alpha_1+\alpha_2>0,\\ &\text{(iii)} && \alpha_0-\alpha_2>0.\end{aligned}$$

Here (iii) may be replaced by $\alpha_0 > \alpha_2 > -\alpha_0$ to obtain the set of conditions which are quoted, for example, by Box and Jenkins [70, p. 58]. The additional result that $\alpha_2 > -\alpha_0$ is obtained by adding (i) and (ii).

There is indeed a wide variety of alternative ways of expressing the stability conditions. The following versions are often quoted:

$$(5.148)\qquad \begin{aligned}&\text{(a)} && \alpha_0>0,\\ &\text{(b)} && \alpha_0^2>\alpha_2^2,\\ &\text{(c)} && (\alpha_0+\alpha_2)^2>\alpha_1^2.\end{aligned}$$

To see that these imply the conditions under (5.147), first note that (b), which can be written as $\alpha_0^2 - \alpha_2^2 = (\alpha_0 - \alpha_2)(\alpha_0 + \alpha_2) > 0$, implies $(\alpha_0 - \alpha_2), (\alpha_0 + \alpha_2) > 0$, on the condition that $\alpha_0 > 0$, and hence $\alpha_0 > \alpha_2 > -\alpha_0$, which entails (iii). But now (c) implies that $\alpha_0 + \alpha_2 > \pm\alpha_1$ which gives (i) and (ii). It is easy to prove that, conversely, the conditions under (5.147) imply those under (5.148); and thus an equivalence may be established.

The condition that $\alpha_0^2 > \alpha_2^2$, which is necessary to ensure that a quadratic equation $\alpha_0 + \alpha_1 z + \alpha_2 z^2 = 0$ has its roots outside the unit circle, is readily generalised to higher-order cases.

(5.149) If $\alpha(z) = \alpha_0 + \alpha_1 z + \cdots + \alpha_p z^p = 0$ is to have all of its roots outside the unit circle, then it is necessary that $\alpha_0^2 > \alpha_p^2$.

To see this, we may consider the following factorisation of the pth-degree polynomial:

$$\begin{aligned}\alpha_0 + \alpha_1 z + \cdots + \alpha_p z^p &= \alpha_p \prod_{i=1}^{p} (z - \lambda_i)\\ &= \alpha_0 \prod_{i=1}^{p} (1 - z/\lambda_i).\end{aligned} \tag{5.150}$$

Here we have $\alpha_0 = \alpha_p \prod_i (-\lambda_i)$; and, if $|\lambda_i| > 1$ for all i, then we must have $|\alpha_0| > |\alpha_p|$ or, equivalently,

$$\delta_p = \alpha_0^2 - \alpha_p^2 > 0. \tag{5.151}$$

A criterion for the stability of a pth-order difference equation may be derived which applies analogous conditions to a sequence of polynomials of decreasing degrees which are derived from the pth-degree polynomial via the repeated application of a simple rule. Let the $f_p(z) = \alpha(z)$ and $f'_p(z) = z^p \alpha(z^{-1})$ stand for the primary polynomial and for the corresponding auxiliary polynomial. Then the first of the derived polynomials is

$$\begin{aligned}f_{p-1}(z) &= \alpha_0 f_p(z) - \alpha_p f'_p(z)\\ &= \delta_p + (\alpha_0\alpha_1 - \alpha_p\alpha_{p-1})z + \cdots + (\alpha_0\alpha_{p-1} - \alpha_p\alpha_1)z^{p-1},\end{aligned} \tag{5.152}$$

and the corresponding auxiliary polynomial is $f'_{p-1}(z) = z^{p-1} f_{p-1}(z^{-1})$. The rule is used to derive a sequence of polynomials $f_{p-1}(z), f_{p-2}(z), \ldots, f_0(z)$ together with the corresponding auxiliary polynomials. If the constant terms of these polynomials are $\delta_p, \delta_{p-1}, \ldots, \delta_1$, then the necessary and sufficient condition for all of the roots of $f_p(z) = 0$ to lie outside the unit circle is that $\delta_p, \delta_{p-1}, \ldots, \delta_1 > 0$.

This result may be established with the help of the following lemma:

(5.153) Let $f_n(z) = \alpha_0 + \alpha_1 z + \cdots + \alpha_n z^n$ be a polynomial with q zeros within the unit circle and $n - q$ zeros outside the circle, and let $f'(z) = z^n f(z^{-1})$. Let δ_n be the coefficient associated with z^0 in the derived polynomial $f_{n-1} = \alpha_0 f_n(z) - \alpha_n f'_n(z)$ which is of degree $n-1$. If $\delta_n > 0$, then $f_{n-1}(z)$ has q zeros inside the unit circle and $n - q - 1$ zeros outside, whereas, if $\delta_n < 0$, then $f_{n-1}(z)$ has $n - q$ zeros inside and $q - 1$ outside.

Proof. First assume that $\alpha_0^2 - \alpha_n^2 = \delta_n > 0$ and consider the rational function

$$\phi(z) = \frac{f_{n-1}(z)}{\alpha_0 f_n(z)} = 1 - \frac{\alpha_n f'_n(z)}{\alpha_0 f_n(z)}. \tag{5.154}$$

On the unit circle, we have

$$|f'_n(z)| = |z^n||f_n(z^{-1})| = |f_n(z^{-1})| = |f_n(z)|, \tag{5.155}$$

and, given that $|\alpha_n/\alpha_0| < 1$, it follows that

$$\left|\frac{a_n f'_n(z)}{a_0 f_n(z)}\right| < 1 \quad \text{and, therefore,} \quad \text{Re}\{\phi(z)\} > 0. \tag{5.156}$$

As z travels around the unit circle, the map of $\phi(z)$ defines a contour which lies in the right half-plane and which does not enclose the origin. It follows from the argument theorem of (3.127) that, if N and P are respectively the number of zeros and poles of $\phi(z)$ which lie within the unit circle, then $N - P = 0$. The P poles of $\phi(z)$, which are the q zeros of $f_n(z)$, are equal in number to the N zeros of $\phi(z)$, which are the zeros of $f_{n-1}(z)$. So $f_{n-1}(z)$ has q zeros inside the unit circle, and its remaining $n - q - 1$ zeros fall outside.

When $\alpha_0^2 - \alpha_n^2 = \delta_n < 0$, we may consider the rational function

$$\theta(z) = \frac{f_{n-1}(z)}{\alpha_n f'_n(z)} = \frac{\alpha_0 f_n(z)}{\alpha_n f'_n(z)} - 1. \tag{5.157}$$

Given that $|\alpha_0/\alpha_n| < 1$ and that $|f_n(z)| = |f'_n(z)|$ on the unit circle, it follows that

$$\left|\frac{\alpha_0 f_n(z)}{\alpha_n f'_n(z)}\right| < 1 \quad \text{and, therefore,} \quad \text{Re}\{\theta(z)\} < 0. \tag{5.158}$$

An argument can now be applied to show that, within the unit circle, the poles of $\theta(z)$, which are the zeros of $f'_n(z)$, are equal in number to the zeros of $\theta(z)$, which are the zeros of $f_{n-1}(z)$. Thus $f_{n-1}(z)$ has $n - q$ zeros inside the unit circle and $q - 1$ outside.

Armed with this lemma, it is easy to establish the conditions for the stability of the pth-order difference equation. First, the necessity of the conditions $\delta_p, \delta_{p-1}, \ldots, \delta_1 > 0$ is established by a recursion. We know that, if all the roots lie outside the unit circle, then $\delta_p > 0$. In that case, the lemma serves to demonstrate that the first of the derived polynomials $f_{p-1}(z)$ has all of its roots outside the unit circle; from which it follows that $\delta_{p-1} > 0$. Further deductions follow likewise.

To prove the sufficiency of the conditions, it must be shown that, if some of the roots of $\alpha(z) = 0$ lie inside the unit circle, then some of the inequalities will be reversed. Imagine that $f_p(z)$ has q zeros inside the unit circle and $p-q$ outside, and let $f_{p-1}(z), \ldots, f_{q+1}(z)$ be a sequence of $p-q-1$ derived polynomials. Imagine that the corresponding delta values are all positive: $\delta_p, \delta_{p-1}, \ldots, \delta_{q+1} > 0$. Then, according to the lemma, the next derived polynomial $f_q(z)$ has q zeros inside the unit circle and none outside; and it follows that $\delta_q < 0$. Thus there is a reversal in the sequence of signs; and it is clear that, if a reversal does not occur before δ_q is generated, then it must occur at that stage.

In fact, the lemma gives rise to a more sophisticated result. Let $P_j = \delta_p\delta_{p-1}\cdots\delta_{p-j+1}$ be the product of the first j delta values and let $j = 1, \ldots, p$. It may be deduced from the lemma that, if q of the products P_j are negative and if the remaining $p-q$ are positive, then $\alpha(z) = 0$ has q of its roots inside the unit circle and $p-q$ outside the circle. Identical proofs of this proposition are given by Marden [332, p. 196] and by Jury [273, p. 125]. The conditions of stability are commonly known, amongst electrical engineers, as the Jury–Blanchard [275] conditions.

To assist in evaluating the conditions of stability, a table may be constructed which is analogous to that of the Routh test. The leading rows of the table are as follows:

$$\begin{array}{cccccc} \alpha_0 & \alpha_1 & \alpha_2 & \cdots & \alpha_{p-1} & \alpha_p \\ \alpha_p & \alpha_{p-1} & \alpha_{p-2} & \cdots & \alpha_1 & \alpha_0 \\ \beta_0 & \beta_1 & \beta_2 & \cdots & \beta_{p-1} & \\ \beta_{p-1} & \beta_{p-2} & \beta_{p-3} & \cdots & \beta_0 & \end{array} \tag{5.159}$$

The third row is formed from the first and second by the rule $\beta_i = \alpha_0\alpha_i - \alpha_p\alpha_{p-i}$. This product may be regarded as the determinant of a 2 by 2 matrix formed from the first and the $(p-i)$th columns within the first two rows. The fourth row of the table is the third row reversed. Subsequent rows are generated in like manner from the rows immediately above them until the $(2p-1)$th row is reached or a negative element is found in the first column in an even-numbered row. In the former case, the difference equation is stable since the delta values, which are the leading elements of the even-numbered rows, are all positive. In the latter case, the difference equation is unstable.

Before embarking on the calculation of the delta values, a pair of preliminary tests may be performed which depend upon calculating an ordinary sum of coefficients and a sum of signed coefficients:

$$\begin{aligned} \alpha(1) &= \alpha_0 + \alpha_1 + \alpha_2 + \cdots + \alpha_n = \alpha_0 \prod_{i=1}^{p}\left(1 - \frac{1}{\lambda_i}\right), \\ \alpha(-1) &= \alpha_0 - \alpha_1 + \alpha_2 - \cdots + (-1)^p\alpha_n = \alpha_0 \prod_{i=1}^{p}\left(1 + \frac{1}{\lambda_i}\right). \end{aligned} \tag{5.160}$$

If $|\lambda_i| > 1$ for all i, then both of these sums must be positive since the factors in the products on the RHS will all be positive. Instances of these two conditions are to be found under (5.147).

The following procedure conducts the preliminary tests before calculating as many delta values as are necessary for assessing stability. The procedure avoids using unnecessary storage by overwriting the elements of the original array *alpha* with the coefficients of successive derived polynomials.

(5.161)

```
procedure JuryCriterion(alpha : vector;
                        p : integer;
                        var stable : boolean);

  var
    i, j, n, fact : integer;
    Q, R, a0, an, ai, anmi, temp : real;

begin
  stable := true;

  {Preliminary Testing}
  if Abs(alpha[0]) <= Abs(alpha[p]) then
    stable := false;
  Q := alpha[0];
  R := alpha[0];
  fact := 1;
  for i := 1 to p do
    begin
      fact := -1 * fact;
      Q := Q + alpha[i];
      R := R + fact * alpha[i];
    end;
  if (Q <= 0) or (R <= 0) then
    stable := false;

  if (stable = true) and (p > 2) then
    begin {Further Testing}
      n := p;
      repeat
        begin {repeat}
          a0 := alpha[0];
          an := alpha[n];
          alpha[0] := a0 * a0 - an * an;
          for i := 1 to n div 2 do
            begin {i}
              anmi := alpha[n - i];
              ai := alpha[i];
              alpha[i] := a0 * ai - an * anmi;
              alpha[n - i] := a0 * anmi - an * ai;
            end; {i}
```

```
            if Abs(alpha[0]) <= Abs(alpha[n - 1]) then
              stable := false;
            n := n - 1;
          end; {repeat}
        until (n = 2);
      end; {Further Testing}

  end; {JuryCriterion}
```

The essential results concerning the number of zeros of a polynomial within the unit circle are due to Schur [443] and Cohn [118], who expressed them in a determinant form. According to Schur, the necessary and sufficient conditions for the polynomial $\alpha(z) = \alpha_0 + \alpha_1 z + \cdots + \alpha_p z^p$ to have all of its roots lying outside the unit circle is that the determinants of the matrices

$$(5.162)\quad \Delta_j = \left[\begin{array}{cccc|cccc} \alpha_0 & 0 & \ldots & 0 & \alpha_p & \alpha_{p-1} & \ldots & \alpha_{p-j+1} \\ \alpha_1 & \alpha_0 & \ldots & 0 & 0 & \alpha_p & \ldots & \alpha_{p-j+2} \\ \vdots & \vdots & \ddots & \vdots & \vdots & \vdots & \ddots & \vdots \\ \alpha_{j-1} & \alpha_{j-2} & \cdots & \alpha_0 & 0 & 0 & \ldots & \alpha_p \\ \hline \alpha_p & 0 & \ldots & 0 & \alpha_0 & \alpha_1 & \ldots & \alpha_{j-1} \\ \alpha_{p-1} & \alpha_p & \ldots & 0 & 0 & \alpha_0 & \ldots & \alpha_{j-2} \\ \vdots & \vdots & \ddots & \vdots & \vdots & \vdots & \ddots & \vdots \\ \alpha_{p-j+1} & \alpha_{p-j+2} & \ldots & \alpha_p & 0 & 0 & \ldots & \alpha_0 \end{array}\right]$$

where $j = 1, \ldots, p$, should all be positive. The contribution of Cohn was to generalise these conditions by proving that $\alpha(z) = 0$ has q zeros within the circle and $p - q$ outside if the sequence of matrices $I, \Delta_1, \Delta_2, \ldots, \Delta_p$ gives rise to a sequence of determinants with q variations of sign.

As in the case of the analogous criterion of Hurwitz for the stability of differential equations, it is inconvenient to have to evaluate a sequence of determinants, since the computation is burdensome whenever the maximum order is greater than thee or four. However, it is possible to ease the burden by using the following determinant identity which is proved, for example, by Rao [421, p. 32]:

$$(5.163)\qquad \det\begin{bmatrix} E & F \\ G & H \end{bmatrix} = |EH - EGE^{-1}F| = |HE - HFH^{-1}G|.$$

This relates to a partitioned matrix wherein the submatrices E and H are square and nonsingular.

In terms of a summary notation, the determinant of the matrix Δ_p defined in (5.162) above becomes

$$(5.164)\qquad \det\begin{bmatrix} A & A_* \\ A_*' & A' \end{bmatrix} = |AA' - AA_*'A^{-1}A_*| = |AA' - A_*'A_*|.$$

Here the first equality follows from the first of the identities of (5.163). The second equality follows from the identity $AA'_* = A'_*A$ which is due to the commutativity in multiplication of lower-triangular Toeplitz matrices. There is also a further identity

$$AA' - A'_*A_* = A'A - A_*A'_* \tag{5.165}$$

affecting the matrix on the RHS of (5.164) which is due to its bisymmetric nature and which is indicated by the second identity under (5.163).

The matrix $AA' - A'_*A_*$ is positive definite if and only if the determinants of its principal minors are all positive. Since these determinants are identical to those entailed by the conditions of Schur, it follows that the latter are equivalent to the condition that the matrix be positive definite. The positive definiteness of the matrix is easily evaluated by finding its Cholesky decomposition in the manner described in Chapter 7.

This result is of particular interest in time-series analysis since the matrix in question has the form of the inverse of the dispersion matrix of an autoregressive process of order p, which is remarkable.

Bibliography

[11] Anderson, B.D.O., (1967), Application of the Second Method of Lyapanov to the Proof of the Markov Stability Criterion, *International Journal of Control*, **5**, 473–482.

[70] Box, G.E.P., and G.M. Jenkins, (1976), *Time Series Analysis: Forecasting and Control, Revised Edition*, Holden Day, San Francisco.

[80] Brown, B.M., (1965), *The Mathematical Theory of Linear Systems*, Chapman and Hall, London.

[118] Cohn, A., (1922), Über die Anzahl der Wurzeln einer algebraischen Gleichung in einem Kreise, *Mathematische Zeitschrift*, **14**, 110–148.

[201] Gantmacher, F.R., (1959), *Applications of the Theory of Matrices*, Interscience Publishers, New York.

[254] Herglotz, G., (1934), Über die Worzelanzahl algebraischer Gleichungen innerhalb und auf dem Einheitkreis, *Mathematische Zeitschrift*, **19**, 26–34.

[263] Hurwitz, A., (1895), Über die Bedingungen, unter welchen eine Gleichung nur Wurzeln mit negativen reellen Theilen besitz, *Mathematische Annalen*, **46**, 273–284.

[273] Jury, E.I., (1964), *Theory and Applications of the z-Transform Method*, John Wiley and Sons, New York.

[274] Jury, E.I., (1964), *On the Roots of a Real Polynomial inside the Real Circle and a Stability Criterion for Linear Discrete Systems*, Proceedings of the Second IFAC Congress (Theory), 142–153.

[275] Jury, E.I., and J. Blanchard, (1961), Stability Test for Linear Discrete Systems in Table Form, *Proceedings of the Institute of Radio Engineers*, **49**, 1947–1948.

[327] Main, I., (1984), *Vibrations and Waves in Physics, Second Edition*, Cambridge University Press, Cambridge.

[331] Marden, M., (1949), *The Geometry of the Zeros of a Polynomial in a Complex Variable, Mathematical Surveys, Number III*, The American Mathematical Society.

[332] Marden, M., (1966), *Geometry of Polynomials, No. 3 of Mathematical Surveys of the AMS*, American Mathematical Society, Providence, Rhode Island.

[368] Nyquist, H., (1928), Certain Topics in Telegraph Transmission Theory, *AIEE Journal*, **47**, 214–216.

[370] Okuguchi, K., and K. Irie, (1990), The Schur and Samuelson Conditions for a Cubic Equation, *The Manchester School*, **58**, 414–418.

[378] Parks, P.C., (1964), Liapanov and Schur–Cohn Stability Criteria, *IEEE Transactions on Automatic Control*, **AC-9**, 121.

[379] Parks, P.C., (1966), Analytic Methods for Investigating Stability—Linear and Nonlinear Systems. A Survey, *Proceedings of the Institute of Mechanical Engineers*, **178**, 7–17.

[421] Rao, C.R., (1973), *Linear Statistical Inference and its Applications, Second Edition*, John Wiley and Sons, New York.

[424] Reuter, G.E.H., (1958), *Elementary Differential Equations and Operators*, Routledge Kegan and Paul, London.

[431] Routh, E.J., 1831–1907, (1877), *A Treatise on the Stability of a Given State of Motion*, (being the essay which was awarded the Adams prize in 1877, in the University of Cambridge), Macmillan and Co., London.

[432] Routh, E.J., 1831–1907, (1905), *The Elementary Part of a Treatise on the Dynamics of a System of Rigid Bodies*, (being part 1 of a treatise on the whole subject), 7th edition, revised and enlarged, Macmillan and Co., London. Reprinted 1960, Dover Publications, New York.

[433] Routh, E.J., 1831–1907, (1905), *The Advanced Part of a Treatise on the Dynamics of a System of Rigid Bodies*, (being part 2 of a treatise on the whole subject), 6th edition, revised and enlarged, Macmillan and Co., London. Reprinted, Dover Publications, New York.

[436] Samuelson, P.A., (1941), Conditions that the Roots of a Polynomial be less than Unity in Absolute Value, *Annals of Mathematical Statistics*, **12**, 360–364.

[437] Samuelson, P.A., (1947), *Foundations of Economic Analysis*, Harvard University Press, Cambridge Mass.

[443] Schur, I., (1917), Über Potensreihen in Innern des Einhheitskreises Beschränkt Sind, *Journal für die Reine und Angewante Mathematik*, **147**, 205–232. English translation reprinted in *Operator Theory: Advances and Applications*, **18** 31–60, 1986.

[450] Shannon, C.E., (1949), Communication in the Presence of Noise, *Proceedings of the IRE*, **37**, 10–21.

[478] Talbot, A., (1960), The Number of Zeros of a Polynomial in the Half-Plane, *Proceedings of the Cambridge Philosophical Society*, **56**, 132–147.

[480] Thomson, W.T., (1981), *Theory of Vibration, Second Edition*, George Allen and Unwin, London.

[529] Wise, J., (1956), Stationarity Conditions for Stochastic Processes of the Autoregressive and Moving Average Type, *Biometrika*, **43**, 215–219.

CHAPTER 6

Vector Difference Equations and State-Space Models

Modern control theory deals with systems which may have many inputs and outputs which may be interrelated in a complicated time-varying manner. To analyse such systems and to devise methods for controlling them, it is essential to reduce the complexity of their mathematical expression. Therefore, control theory has resorted increasingly to the so-called state-space methods which depict such systems in terms of first-order vector differential or difference equations.

Since most electronic control systems are nowadays based on digital processors, attention has been focused mainly on discrete-time systems involving difference equations.

An nth-order difference equation can be represented as a first-order vector equation with a state vector of n elements. Therefore, state-space analysis can be applied to problems which might otherwise be treated by the classical methods presented in the previous chapter. Much of the present chapter is devoted to the task of finding appropriate state-space representations for scalar difference equations with the object of facilitating the application of state-space analysis.

This book shows a preference for treating single-equation problems by single-equation methods. It appears that, whenever a state-space method is available for treating a single-equation problem, a corresponding method can be discovered which draws upon the classical analysis. Moreover, such single-equation methods are often simpler from a conceptual point of view. Nevertheless, it is undeniable that the development of single-equation methods has sometimes been motivated by discoveries in the realms of state-space analysis.

We shall begin the analysis of this chapter by considering a simple first-order vector equation.

The State-Space Equations

Consider the first-order difference equation

$$\xi(t) = \Phi\xi(t-1) + \nu(t), \tag{6.1}$$

wherein $\xi(t) = [\xi_1(t), \xi_2(t), \ldots, \xi_n(t)]'$ and $\nu(t) = [\nu_1(t), \nu_2(t), \ldots, \nu_n(t)]'$ are vectors of time series and Φ is a matrix of order $n \times n$. This is called a transition equation or a process equation. The vector $\xi_\tau = \xi(\tau)$ is described as a state vector because it provides a complete description of the state of the system at a single instant τ. Moreover, if one knows the value of the transition matrix Φ and the values of the elements of the sequence $\nu(t)$ for all t, then knowing the state vector ξ_τ at an instant τ should enable one to infer the state of the system at any other time.

The vector sequence $\nu(t)$ is described as the input sequence or the forcing function. In some applications, $\nu(t)$ is a function of a set of control variables within a vector $u(t)$ of order m whose values may be manipulated so as to achieve a desired outcome for $\xi(t)$. In that case, it is appropriate to represent the system by the equation

$$\xi(t) = \Phi\xi(t-1) + Bu(t), \tag{6.2}$$

where B is a matrix of order $n \times m$ and $Bu(t) = \nu(t)$. This matrix serves to distribute the effects of the control variables amongst the equations which determine the variables of $\xi(t)$. In a more elaborate model, one might find an additional set of input variables which are not subject to control.

The information conveyed by the state vector $\xi(t)$ is not affected in any fundamental way when $\xi(t)$ is premultiplied by a nonsingular matrix T. The system which determines the transformed vector $\zeta(t) = T\xi(t)$ is described by the equation

$$\begin{aligned} \zeta(t) &= T\{\Phi\xi(t-1) + \nu(t)\} \\ &= \{T\Phi T^{-1}\}\{T\xi(t-1)\} + T\nu(t) \\ &= \Psi\zeta(t-1) + \upsilon(t); \end{aligned} \tag{6.3}$$

and it is said to be equivalent to the original system of (6.1). The matrix $\Psi = T\Phi T^{-1}$ is said to be similar to the matrix Φ, and T is described as a similarity transformation. The two matrices Ψ and Φ have the same characteristic roots and they have a common characteristic equation. Thus, if $\Lambda = \text{diag}(\lambda_1, \ldots, \lambda_n)$ is the matrix of the characteristic roots of Φ and if $\Phi = Q\Lambda Q^{-1}$, then $\Psi = R\Lambda R^{-1}$, where $R = TQ$.

A system without a forcing function is said to be free or homogeneous. Given an initial value of ξ_0, the homogeneous system $\xi(t) = \Phi\xi(t-1)$ can be solved recursively to generate the ensuing values:

$$\begin{aligned} \xi_1 &= \Phi\xi_0, \\ \xi_2 &= \Phi^2\xi_0, \\ &\vdots \\ \xi_\tau &= \Phi^\tau\xi_0. \end{aligned} \tag{6.4}$$

These form a convergent sequence if and only if the sequence of matrices $\{\Phi, \Phi^2, \ldots, \Phi^\tau, \ldots\}$ converges.

The matter of convergence may be investigated in terms of the factorisation $\Phi = Q\Lambda Q^{-1}$. It can be seen that $\Phi^2 = Q\Lambda^2 Q^{-1}$ and, more generally, that $\Phi^\tau = Q\Lambda^\tau Q^{-1}$. It follows that the sequence $\{\Phi, \Phi^2, \ldots, \Phi^\tau, \ldots\}$ converges if and only if all elements of the diagonal matrix Λ, which are the roots of Φ, lie inside the unit circle.

The sequence of values generated by the forced or driven system in (6.1) can be derived in the same way as the sequence generated by the free system. Thus, given the initial value ξ_0 together with the values of the sequence $\nu(t)$, the ensuing values can be generated:

$$\begin{aligned}\xi_1 &= \Phi\xi_0 + \nu_1,\\ \xi_2 &= \Phi^2\xi_0 + \{\nu_2 + \Phi\nu_1\},\\ &\vdots\\ \xi_\tau &= \Phi^\tau\xi_0 + \{\nu_\tau + \Phi\nu_{\tau-1} + \cdots + \Phi^{\tau-1}\nu_1\}.\end{aligned} \tag{6.5}$$

A fully-fledged state-space system of the sort studied by control engineers usually comprises a measurement or output equation which shows how the observations on the system are related to the state variables. Also, the parameters of the system are allowed to vary with time. For a linear time-invariant system, the transition equation and the measurement equation may be written as

$$\begin{aligned}\xi(t) &= \Phi\xi(t-1) + Bu(t),\\ y(t) &= \Gamma\xi(t) + \Delta u(t).\end{aligned} \tag{6.6}$$

In this case, $\xi(t)$ is the vector of state variables, $u(t)$ is the vector of inputs and $y(t)$ is the vector of measured outputs. Ostensibly, the transition equation and the measurement equation receive the same inputs. However, the matrices B and Δ may be structured so that the two equations have no inputs in common. In the sequel, we shall deal only with cases where $y(t)$ is a scalar sequence.

Conversions of Difference Equations to State-Space Form

Ordinary scalar difference equations can be converted easily to equivalent systems of first-order vector equations. Therefore, much of the theory concerning state-space models, such as the theory of Kalman filtering, can be applied to autoregressive moving-average models which entail stochastic difference equations.

Before demonstrating how a difference equation is converted into a first-order vector equation, it is useful to consider alternative ways of generating a sequence of values which satisfy the difference equation.

Let us consider an equation in the form of

$$\begin{aligned}&y(t) + \alpha_1 y(t-1) + \cdots + \alpha_r y(t-r)\\ &\qquad = \mu_0\varepsilon(t) + \mu_1\varepsilon(t-1) + \cdots + \mu_{r-1}\varepsilon(t-r+1),\end{aligned} \tag{6.7}$$

which can also be written in a rational transfer-function form:

$$y(t) = \frac{\mu(L)}{\alpha(L)}\varepsilon(t) = \frac{\mu_0 + \mu_1 L + \cdots + \mu_r L^{r-1}}{1 + \alpha_1 L + \cdots + \alpha_r L^r}\varepsilon(t). \tag{6.8}$$

Here the autoregressive order r and the moving-average order $r-1$ are maximum orders. The equation is designed to accommodate autoregressive moving-average

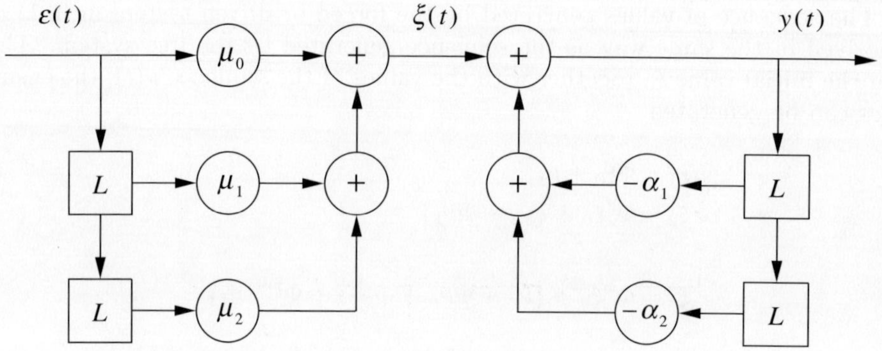

Figure 6.1. The direct realisation of the ARMA(2, 2) equation.

models of arbitrary orders; and this may be achieved by setting some of the higher-order parameters in either part of the equation to zeros.

By separating the numerator and denominator of (6.8), the mapping from $\varepsilon(t)$ to $y(t)$ can be depicted as the product of two successive operations. Thus

$$\begin{aligned} y(t) &= \frac{1}{\alpha(L)}\{\mu(L)\varepsilon(t)\} \\ &= \frac{1}{\alpha(L)}\xi(t); \qquad \xi(t) = \mu(L)\varepsilon(t). \end{aligned} \tag{6.9}$$

This form suggests that, in generating $y(t)$, one should begin by calculating an element of $\xi(t) = \mu(L)\varepsilon(t)$ and proceed to calculate the corresponding element of $y(t) = \alpha^{-1}(L)\xi(t)$. Therefore, a two-stage algorithm might be devised which would realise the following equations:

$$\begin{aligned} &\text{(i)} \qquad \xi(t) = \mu_0\varepsilon(t) + \mu_1\varepsilon(t-1) + \cdots + \mu_{r-1}\varepsilon(t-r+1), \\ &\text{(ii)} \qquad y(t) = \xi(t) - \{\alpha_1 y(t-1) + \cdots + \alpha_r y(t-r)\}. \end{aligned} \tag{6.10}$$

Since the operations involved in forming $y(t)$ are linear and time-invariant, the order in which they are applied may be reversed. That is to say, an element of $\xi(t) = \alpha^{-1}(L)\varepsilon(t)$ might be generated followed by an element of $y(t) = \mu(L)\xi(t)$. The net result of the two operations is $y(t) = \alpha^{-1}(L)\mu(L)\varepsilon(t) = \mu(L)\alpha^{-1}(L)\varepsilon(t)$, regardless of their ordering. It follows that equation (6.9) can be rewritten as

$$\begin{aligned} y(t) &= \mu(L)\left\{\frac{1}{\alpha(L)}\varepsilon(t)\right\} \\ &= \mu(L)\xi(t); \qquad \xi(t) = \alpha^{-1}(L)\varepsilon(t). \end{aligned} \tag{6.11}$$

Then, in place of the equations under (6.10), there would be

$$\begin{aligned} &\text{(i)} \qquad \xi(t) = \varepsilon(t) - \big\{\alpha_1\xi(t-1) + \cdots + \alpha_r\xi(t-r)\big\}, \\ &\text{(ii)} \qquad y(t) = \mu_0\xi(t) + \mu_1\xi(t-1) + \cdots + \mu_{r-1}\xi(t-r+1). \end{aligned} \tag{6.12}$$

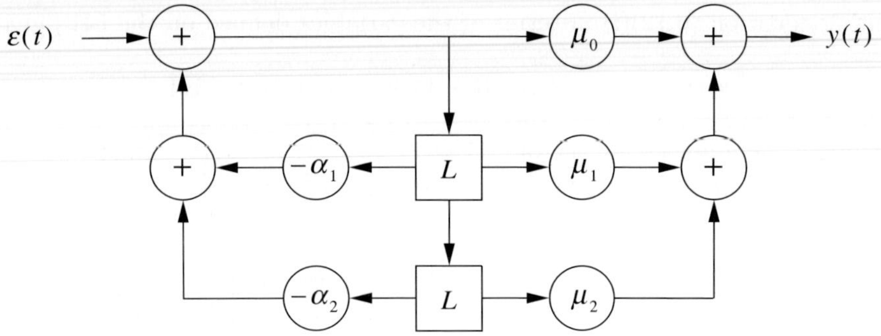

Figure 6.2. The direct realisation of the ARMA(2, 2) equation using common delay registers.

The advantage of basing the computations upon the equations of (6.12) is that their realisation, in terms of computer memory or other hardware, is bound to be more economical. For, whereas a recursion based on the equations of (6.10) would require values of $\varepsilon(t),\ldots,\varepsilon(t-r)$ and of $y(t-1),\ldots,y(t-r)$ to be stored, the recursions based on (6.12) would require only the storage of values from $\varepsilon(t),\xi(t-1),\ldots,\xi(t-r)$.

Example 6.1. Consider the ARMA(2, 2) equation

$$y(t)+\alpha_1 y(t-1)+\alpha_2 y(t-2)=\mu_0\varepsilon(t)+\mu_1\varepsilon(t-1)+\mu_2\varepsilon(t-2). \tag{6.13}$$

The algorithm corresponding to the equations of (6.10) in this case can be represented in a block diagram (Figure 6.1) which portrays, in succession, the filtering operations which give rise to $\xi(t)=\mu(L)\varepsilon(t)$ and $y(t)=\alpha^{-1}(L)\xi(t)$. In the diagram, the blocks labelled L are delay registers which give effect to the lag operator. The circles correspond to operations of scalar multiplication and addition.

The alternative algorithm, which corresponds to the equations of (6.12), can be depicted in a block diagram (Figure 6.2) which reverses the order of the ladders in Figure 6.1 and which merges the two adjacent rails so as to halve the number of delay registers.

Controllable Canonical State-Space Representations

Given that the values of $y(t)$ are to be generated according to the equations under (6.12), there remains the question of how to implement the recursion under (6.12)(i). There is a choice of two schemes. The first scheme, which is described as the direct method, requires that a set of r state variables should be defined as follows:

$$\begin{aligned}\xi_1(t)&=\xi(t),\\ \xi_2(t)&=\xi_1(t-1)=\xi(t-1),\\ &\;\;\vdots\\ \xi_r(t)&=\xi_{r-1}(t-1)=\xi(t-r+1).\end{aligned} \tag{6.14}$$

Rewriting equation (6.12)(i) in terms of the variables defined on the LHS gives

$$\xi_1(t) = \varepsilon(t) - \{\alpha_1\xi_1(t-1) + \cdots + \alpha_r\xi_r(t-1)\}. \tag{6.15}$$

Therefore, by defining a state vector $\xi(t) = [\xi_1(t), \xi_2(t), \ldots, \xi_r(t)]'$ and by combining (6.14) and (6.15), a linear system can be constructed in the form of

$$\begin{bmatrix} \xi_1(t) \\ \xi_2(t) \\ \vdots \\ \xi_r(t) \end{bmatrix} = \begin{bmatrix} -\alpha_1 & \ldots & -\alpha_{r-1} & -\alpha_r \\ 1 & \ldots & 0 & 0 \\ \vdots & \ddots & \vdots & \vdots \\ 0 & \ldots & 1 & 0 \end{bmatrix} \begin{bmatrix} \xi_1(t-1) \\ \xi_2(t-1) \\ \vdots \\ \xi_r(t-1) \end{bmatrix} + \begin{bmatrix} 1 \\ 0 \\ \vdots \\ 0 \end{bmatrix} \varepsilon(t). \tag{6.16}$$

The sparse matrix on the RHS of this equation is an example of a so-called companion matrix. The accompanying measurement equation which corresponds to equation (6.1)(ii) is given by

$$y(t) = \mu_0\xi_1(t) + \cdots + \mu_{r-1}\xi_r(t). \tag{6.17}$$

In summary notation, equations (6.16) and (6.17) are represented respectively by

$$\xi(t) = \Phi\xi(t-1) + \beta\varepsilon(t), \tag{6.18}$$

and

$$y(t) = \gamma'\xi(t), \tag{6.19}$$

where $\gamma' = [\mu_0, \ldots, \mu_{r-1}]$.

Equation (6.16) is often presented in the alternative form of

$$\begin{bmatrix} \xi_r(t) \\ \vdots \\ \xi_2(t) \\ \xi_1(t) \end{bmatrix} = \begin{bmatrix} 0 & 1 & \ldots & 0 \\ \vdots & \vdots & \ddots & \vdots \\ 0 & 0 & \ldots & 1 \\ -\alpha_r & -\alpha_{r-1} & \ldots & -\alpha_1 \end{bmatrix} \begin{bmatrix} \xi_r(t-1) \\ \vdots \\ \xi_2(t-1) \\ \xi_1(t-1) \end{bmatrix} + \begin{bmatrix} 0 \\ \vdots \\ 0 \\ 1 \end{bmatrix} \varepsilon(t), \tag{6.20}$$

for which the accompanying measurement equation is

$$y(t) = \mu_{r-1}\xi_r(t) + \cdots + \mu_0\xi_1(t). \tag{6.21}$$

Equations (6.20) and (6.21) may be represented, respectively, by

$$\tilde{\xi}(t) = \tilde{\Phi}\tilde{\xi}(t-1) + \tilde{\beta}\varepsilon(t) \tag{6.22}$$

and by

$$y(t) = \tilde{\gamma}'\tilde{\xi}(t), \tag{6.23}$$

where $\tilde{\xi}(t)$ contains the elements of $\xi(t)$ in reverse order.

To formalise the relationship between equations (6.18) and (6.22), a matrix J of order r may be introduced which has units running along the NE–SW diagonal and zeros elsewhere. When J premultiplies another matrix, the effect is to reverse the order of the rows. When it postmultiplies a matrix, the effect is to reverse the order of the columns. Multiplying J by itself gives $JJ = I$. Moreover $J\xi(t) = \tilde{\xi}(t)$, $J\beta = \tilde{\beta}$ and $J\Phi J = \tilde{\Phi}$. Therefore, when equation (6.18) is premultiplied by J, the result is

$$(6.24)\qquad \begin{aligned} J\xi(t) &= J\Phi\xi(t-1) + J\beta\varepsilon(t) \\ &= \{J\Phi J\}\{J\xi(t-1)\} + J\beta\varepsilon(t) \\ &= \tilde{\Phi}\tilde{\xi}(t-1) + \tilde{\beta}\varepsilon(t), \end{aligned}$$

which is equation (6.22). To establish the relationship between (6.19) and (6.23) is also straightforward.

An alternative way of generating $y(t)$ in accordance with equation (6.12)(i) depends upon a nested procedure. A simple recursion may be constructed using the following definitions:

$$(6.25)\qquad \begin{aligned} \xi_1(t) &= -\alpha_1\xi_1(t-1) + \xi_2(t-1) + \varepsilon(t), \\ \xi_2(t) &= -\alpha_2\xi_1(t-1) + \xi_3(t-1), \\ &\;\;\vdots \\ \xi_{r-1}(t) &= -\alpha_{r-1}\xi_1(t-1) + \xi_r(t-1), \\ \xi_r(t) &= -\alpha_r\xi_1(t-1). \end{aligned}$$

By a process of substitutions running from the bottom to the top of the list, equation (6.12)(i) may be recovered in the form of

$$(6.26)\qquad \begin{aligned} \xi_1(t) &= \varepsilon(t) - \{\alpha_1\xi_1(t-1) + \cdots + \alpha_r\xi_1(t-r)\} \\ &= \varepsilon(t) - \{\alpha_1\xi_1(t-1) + \cdots + \alpha_r\xi_r(t-1)\}. \end{aligned}$$

The state-space representation for this system of equations (6.25) and (6.26) is as follows:

$$(6.27)\qquad \begin{bmatrix} \xi_1(t) \\ \vdots \\ \xi_{r-1}(t) \\ \xi_r(t) \end{bmatrix} = \begin{bmatrix} -\alpha_1 & 1 & \dots & 0 \\ \vdots & \vdots & \ddots & \vdots \\ -\alpha_{r-1} & 0 & \dots & 1 \\ -\alpha_r & 0 & \dots & 0 \end{bmatrix} \begin{bmatrix} \xi_1(t-1) \\ \vdots \\ \xi_{r-1}(t-1) \\ \xi_r(t-1) \end{bmatrix} + \begin{bmatrix} 1 \\ 0 \\ \vdots \\ 0 \end{bmatrix} \varepsilon(t).$$

The measurement equation continues to be written as

$$(6.28)\qquad y(t) = \mu_0\xi_1(t) + \mu_1\xi_2(t) + \cdots + \mu_{r-1}\xi_r(t).$$

The system under (6.20) and (6.21) and its variant under (6.27) and (6.28) are examples of the so-called controllable canonical forms of the state-space equations. We shall be able to explain this terminology later.

Observable Canonical Forms

Alternative canonical forms for the state-space representation of the difference equations can be derived which are as parsimonious in their use of delay registers as the previous representations. These are the so-called observable forms.

Consider writing the equation of an ARMA$(r, r-1)$ model as

$$y(t) = \{\mu_0\varepsilon(t) - \alpha_1 y(t-1)\} + \cdots + \{\mu_{r-1}\varepsilon(t-r+1) - \alpha_r y(t-r)\}. \tag{6.29}$$

This gives rise to a recursion in the form of

$$\begin{aligned} y(t) &= -\alpha_1 y(t-1) + \xi_2(t-1) + \mu_0\varepsilon(t), \\ \xi_2(t) &= -\alpha_2 y(t-1) + \xi_3(t-1) + \mu_1\varepsilon(t), \\ &\vdots \\ \xi_{r-1}(t) &= -\alpha_{r-1} y(t-1) + \xi_r(t-1) + \mu_{r-2}\varepsilon(t), \\ \xi_r(t) &= -\alpha_r y(t-1) + \mu_{r-1}\varepsilon(t). \end{aligned} \tag{6.30}$$

Equation (6.29) can be recovered by a process of substitutions running from the bottom to the top of the list.

On defining $\xi_1(t) = y(t)$ and $\xi_1(t-1) = y(t-1)$, the corresponding state-space transition equation can be constructed:

$$\begin{bmatrix} \xi_1(t) \\ \vdots \\ \xi_{r-1}(t) \\ \xi_r(t) \end{bmatrix} = \begin{bmatrix} -\alpha_1 & 1 & \ldots & 0 \\ \vdots & \vdots & \ddots & \vdots \\ -\alpha_{r-1} & 0 & \ldots & 1 \\ -\alpha_r & 0 & \ldots & 0 \end{bmatrix} \begin{bmatrix} \xi_1(t-1) \\ \vdots \\ \xi_{r-1}(t-1) \\ \xi_r(t-1) \end{bmatrix} + \begin{bmatrix} \mu_0 \\ \vdots \\ \mu_{r-2} \\ \mu_{r-1} \end{bmatrix} \varepsilon(t). \tag{6.31}$$

The measurement equation is provided by

$$y(t) = \xi_1(t). \tag{6.32}$$

Another canonical form which should be considered is the observable-form counterpart of the equation under (6.20). Its derivation begins with the transfer function $\omega(L) = \mu(L)/\alpha(L)$. Multiplying both sides by $\alpha(L)$ gives $\alpha(L)\omega(L) = \mu(L)$. Equating coefficients from both sides of the latter equation which are associated with the same powers of L gives the following identities:

$$\begin{aligned} \mu_0 &= \alpha_0\omega_0, \\ \mu_1 &= \alpha_0\omega_1 + \alpha_1\omega_0, \\ &\vdots \\ \mu_{r-2} &= \alpha_0\omega_{r-2} + \alpha_1\omega_{r-3} + \cdots + \alpha_{r-2}\omega_0, \\ \mu_{r-1} &= \alpha_0\omega_{r-1} + \alpha_1\omega_{r-2} + \cdots + \alpha_{r-1}\omega_0. \end{aligned} \tag{6.33}$$

When $\alpha_0 = 1$, this becomes compatible with equation (6.7).

Now consider using these expressions in the equation for an ARMA$(r, r-1)$ model written as

$$\begin{aligned}\alpha_0 y(t+r-1) + \alpha_1 y(t+r-2) + \cdots + \alpha_{r-1}y(t) + \alpha_r y(t-1)\\ = \mu_0\varepsilon(t+r-1) + \mu_1\varepsilon(t+r-2) + \cdots + \mu_{r-2}\varepsilon(t+1) + \mu_{r-1}\varepsilon(t).\end{aligned} \tag{6.34}$$

A straightforward substitution of the expressions from (6.33) into the RHS of (6.34) gives

$$\begin{aligned}&\mu_0\varepsilon(t+r-1) + \mu_1\varepsilon(t+r-2) + \cdots + \mu_{r-2}\varepsilon(t+1) + \mu_{r-1}\varepsilon(t)\\ &= \alpha_0\big[\omega_0\varepsilon(t+r-1) + \omega_1\varepsilon(t+r-2) + \cdots + \omega_{r-1}\varepsilon(t)\big]\\ &\quad + \alpha_1\big[\omega_0\varepsilon(t+r-2) + \omega_1\varepsilon(t+r-3) + \cdots + \omega_{r-2}\varepsilon(t)\big]\\ &\quad\ \vdots\\ &\quad + \alpha_{r-2}\big[\omega_0\varepsilon(t+1) + \omega_1\varepsilon(t)\big]\\ &\quad + \alpha_{r-1}\omega_0\varepsilon(t).\end{aligned} \tag{6.35}$$

Therefore, on carrying the RHS across the equals sign, the ARMA$(r, r-1)$ equation can be expressed as

$$\begin{aligned}&\alpha_0\big\{y(t+r-1) - \big[\omega_0\varepsilon(t+r-1) + \cdots + \omega_{r-1}\varepsilon(t)\big]\big\}\\ &+ \alpha_1\big\{y(t+r-2) - \big[\omega_0\varepsilon(t+r-2) + \cdots + \omega_{r-2}\varepsilon(t)\big]\big\}\\ &\ \vdots\\ &+ \alpha_{r-2}\big\{y(t+1) - \big[\omega_0\varepsilon(t+1) + \omega_1\varepsilon(t)\big]\big\}\\ &+ \alpha_{r-1}\big\{y(t) - \omega_0\varepsilon(t)\big\}\\ &+ \alpha_r y(t-1) = 0.\end{aligned} \tag{6.36}$$

Within this equation, the following variables can be defined:

$$\begin{aligned}\xi_1(t-1) &= y(t-1),\\ \xi_2(t-1) &= y(t) - \omega_0\varepsilon(t),\\ \xi_3(t-1) &= y(t+1) - \big[\omega_0\varepsilon(t+1) + \omega_1\varepsilon(t)\big],\\ &\ \vdots\\ \xi_r(t-1) &= y(t+r-2) - \big[\omega_0\varepsilon(t+r-2) + \cdots + \omega_{r-2}\varepsilon(t)\big].\end{aligned} \tag{6.37}$$

These variables obey a simple recursion in the form of

$$\begin{aligned}\xi_1(t) &= \xi_2(t-1) + \omega_0\varepsilon(t),\\ \xi_2(t) &= \xi_3(t-1) + \omega_1\varepsilon(t),\\ &\ \vdots\\ \xi_{r-1}(t) &= \xi_r(t-1) + \omega_{r-2}\varepsilon(t).\end{aligned} \tag{6.38}$$

Also, by substituting from (6.37) into (6.36), the following expression is obtained for that equation:

$$\alpha_0\big\{\xi_r(t) - \omega_{r-1}\varepsilon(t)\big\} + \alpha_1\xi_r(t-1) + \cdots + \alpha_{r-1}\xi_2(t-1) + \alpha_r\xi_1(t-1) = 0. \tag{6.39}$$

The equations under (6.38) and (6.39), with $\alpha_0 = 1$, may be assembled into a state-space transition equation:

$$\begin{bmatrix} \xi_1(t) \\ \vdots \\ \xi_{r-1}(t) \\ \xi_r(t) \end{bmatrix} = \begin{bmatrix} 0 & 1 & \ldots & 0 \\ \vdots & \vdots & \ddots & \vdots \\ 0 & 0 & \ldots & 1 \\ -\alpha_r & -\alpha_{r-1} & \ldots & -\alpha_1 \end{bmatrix} \begin{bmatrix} \xi_1(t-1) \\ \vdots \\ \xi_{r-1}(t-1) \\ \xi_r(t-1) \end{bmatrix} + \begin{bmatrix} \omega_0 \\ \vdots \\ \omega_{r-2} \\ \omega_{r-1} \end{bmatrix} \varepsilon(t). \tag{6.40}$$

The accompanying measurement equation is

$$y(t) = \xi_1(t). \tag{6.41}$$

Reduction of State-Space Equations to a Transfer Function

It should always be possible to find the transfer function which summarises the mapping from the input $\varepsilon(t)$ to the output $y(t)$ which is effected by the state-space equations. Consider writing the transition equation and the corresponding measurement equation as

$$\begin{aligned} \xi(t) &= \Phi\xi(t-1) + \beta\varepsilon(t) \qquad \text{and} \\ y(t) &= \gamma'\xi(t) + \delta\varepsilon(t). \end{aligned} \tag{6.42}$$

The lag operator can be used to rewrite the first of these as $(I - \Phi L)\xi(t) = \beta\varepsilon(t)$, which gives $\xi(t) = (I - \Phi L)^{-1}\beta\varepsilon(t)$. Putting this into the second equation, gives rise to an expression for the transfer function:

$$\begin{aligned} y(t) &= \{\gamma'(I - \Phi L)^{-1}\beta + \delta\}\varepsilon(t) \\ &= \omega(L)\varepsilon(t). \end{aligned} \tag{6.43}$$

When the state-space equations assume one or other of the canonical forms, this reduction is usually accomplished with ease.

Example 6.2. Consider the case of an ARMA(3, 2) model for which the ordinary difference equation is written as

$$\begin{aligned} y(t) + \alpha_1 y(t-1) + \alpha_2 y(t-2) + \alpha_3 y(t-3) \\ = \mu_0\varepsilon(t) + \mu_1\varepsilon(t-1) + \mu_2\varepsilon(t-2). \end{aligned} \tag{6.44}$$

The corresponding transition equation, written in the form of (6.16), is

$$\begin{aligned} \begin{bmatrix} \xi_1(t) \\ \xi_2(t) \\ \xi_3(t) \end{bmatrix} &= \begin{bmatrix} -\alpha_1 & -\alpha_2 & -\alpha_3 \\ 1 & 0 & 0 \\ 0 & 1 & 0 \end{bmatrix} \begin{bmatrix} \xi_1(t-1) \\ \xi_2(t-1) \\ \xi_3(t-1) \end{bmatrix} + \begin{bmatrix} 1 \\ 0 \\ 0 \end{bmatrix} \varepsilon(t) \\ &= \Phi\xi(t) + \beta\varepsilon(t). \end{aligned} \tag{6.45}$$

The measurement equation, which corresponds to (6.17), takes the form of

$$\begin{aligned} y(t) &= \begin{bmatrix} \mu_0 & \mu_1 & \mu_2 \end{bmatrix} \begin{bmatrix} \xi_1(t) \\ \xi_2(t) \\ \xi_3(t) \end{bmatrix} \\ &= \gamma'\xi(t). \end{aligned} \tag{6.46}$$

Here it is found that, in comparison with the measurement equation under (6.42), the term $\delta\varepsilon(t)$ is absent.

An expression for the transfer function

$$\begin{aligned}\omega(L) &= \gamma'(I - \Phi L)^{-1}\beta \\ &= \gamma' v(L),\end{aligned} \tag{6.47}$$

is to be found by developing an expression for $v(L) = (I - \Phi L)^{-1}\beta$. Therefore, consider recasting the equation $(I - \Phi L)v(L) = \beta$ in the form $v(L) = \Phi L v(L) + \beta$ which, in the present instance, gives rise to following expression:

$$\begin{bmatrix} v_1(L) \\ v_2(L) \\ v_3(L) \end{bmatrix} = \begin{bmatrix} -\alpha_1 & -\alpha_2 & -\alpha_3 \\ 1 & 0 & 0 \\ 0 & 1 & 0 \end{bmatrix} \begin{bmatrix} Lv_1(L) \\ Lv_2(L) \\ Lv_3(L) \end{bmatrix} + \begin{bmatrix} 1 \\ 0 \\ 0 \end{bmatrix}. \tag{6.48}$$

These equations embody a recursion $v_2(L) = Lv_1(L)$, $v_3(L) = Lv_2(L) = L^2 v_1(L)$, which is represented separately in the equation

$$\begin{bmatrix} v_1(L) \\ v_2(L) \\ v_3(L) \end{bmatrix} = v_1(L) \begin{bmatrix} 1 \\ L \\ L^2 \end{bmatrix}. \tag{6.49}$$

The recursion also enables the leading equation within (6.48) to be written as

$$\begin{aligned}1 &= v_1(L) + \alpha_1 L v_1(L) + \alpha_2 L v_2(L) + \alpha_3 L v_3(L) \\ &= v_1(L) + \alpha_1 L v_1(L) + \alpha_2 L^2 v_1(L) + \alpha_3 L^3 v_1(L),\end{aligned} \tag{6.50}$$

which gives

$$v_1(L) = \frac{1}{1 + \alpha_1 L + \alpha_2 L^2 + \alpha_3 L^3}. \tag{6.51}$$

The latter can be used in equation (6.49), which can be carried, in turn, to equation (6.47). Then it can be seen that

$$\begin{aligned}\omega(L) &= \begin{bmatrix} \mu_0 & \mu_1 & \mu_2 \end{bmatrix} \begin{bmatrix} v_1(L) \\ v_2(L) \\ v_3(L) \end{bmatrix} \\ &= \frac{\mu_0 + \mu_1 L + \mu_2 L^2}{1 + \alpha_1 L + \alpha_2 L^2 + \alpha_3 L^3}.\end{aligned} \tag{6.52}$$

Controllability

Consider, once more, the system of state equations

$$\begin{aligned}\xi(t) &= \Phi\xi(t-1) + \beta u(t) \qquad \text{and} \\ y(t) &= \gamma'\xi(t) + \delta u(t),\end{aligned} \tag{6.53}$$

where Φ is a matrix of order $r \times r$ and $y(t)$ and $u(t)$ are scalar sequences. We shall imagine that the elements of $u(t)$ are within our control.

(6.54) The system (6.53) is controllable if, by an appropriate choice of a number of successive input values $u_1, \ldots, u_\tau$, the value of the state vector can be changed from $\xi_0 = 0$ at time $t = 0$ to $\bar{\xi}$ at time τ, where $\bar{\xi}$ has an arbitrary finite value.

It can be proved that

(6.55) The system (6.53) is controllable if and only if the matrix $Q = [\beta, \Phi\beta, \ldots, \Phi^{r-1}\beta]$ has $\text{Rank}(Q) = r$.

To see that the rank condition is sufficient for controllability, consider the recursion

$$\begin{aligned} \xi_1 &= \Phi\xi_0 + \beta u_1, \\ \xi_2 &= \Phi^2\xi_0 + \{\beta u_2 + \Phi\beta u_1\}, \\ &\vdots \\ \xi_r &= \Phi^r\xi_0 + \{\beta u_r + \Phi\beta u_{r-1} + \cdots + \Phi^{r-1}\beta u_1\}. \end{aligned} \tag{6.56}$$

When $\xi_0 = 0$, the final equation can be written as

$$\begin{aligned} \xi_r &= [\beta, \Phi\beta, \ldots, \Phi^{r-1}\beta] \begin{bmatrix} u_r \\ u_{r-1} \\ \vdots \\ u_1 \end{bmatrix} \\ &= Qu. \end{aligned} \tag{6.57}$$

Given an arbitrary target value of $\xi_r = \bar{\xi}$, the fulfilment of the rank condition guarantees that equation (6.57) can always be solved for the vector u which contains the controllable input values. Therefore, the condition guarantees that the target value can be reached in r periods.

To show that the rank condition is necessary for controllability, it must be shown that, if the target cannot be reached in r periods, then there is no guarantee that it can ever be reached. In this connection, it should be recognised that, if $\text{rank}[\beta, \Phi\beta, \ldots, \Phi^{r-1}\beta] \leq r$, then $\text{rank}[\beta, \Phi\beta, \ldots, \Phi^{\tau-1}\beta] \leq r$ for all τ. The latter implies that, whatever the value of τ, there will always exist an unattainable value $\bar{\xi}$ for which the equation $\bar{\xi} = [\beta, \Phi\beta, \ldots, \Phi^{\tau-1}\beta]u_{(\tau)}$ has no solution in terms of $u_{(\tau)} = [u_\tau, u_{\tau-1}, \ldots, u_1]'$. It follows that the proposition can be established if it can be shown that

(6.58) If Φ is a matrix of order $r \times r$, then $\text{rank}\,[\beta, \Phi\beta, \ldots, \Phi^{k-1}\beta] = \text{rank}\,[\beta, \Phi\beta, \ldots, \Phi^{r-1}\beta]$ for all $k \geq r$.

In proving this, consider, at first, the case where $Q_{(r)} = [\beta, \Phi\beta, \ldots, \Phi^{r-1}\beta]$ has the full rank of r. It is clear that $Q_{(r+1)} = [\beta, \Phi\beta, \ldots, \Phi^r\beta] = [Q_{(r)}, \Phi^r\beta]$, which comprises $Q_{(r)}$ as a submatrix, must also have a rank of r. The same is true of the succeeding matrices $Q_{(r+2)},\ldots,Q_{(k)}$. Next, consider the case where $\text{rank}\{Q_{(r)}\} = p < r$. Then, since the columns of $Q_{(r)}$ are linearly dependent, there exists a set of scalars $\mu_1,\mu_2,\ldots,\mu_r$ such that $\mu_1\beta + \mu_2\Phi\beta + \cdots + \mu_r\Phi^{r-1}\beta = 0$. Multiplying this equation by Φ gives $\mu_1\Phi\beta + \mu_2\Phi^2\beta + \cdots + \mu_r\Phi^r\beta = 0$. This shows that $\Phi^r\beta$ is linearly dependent on the columns of $Q_{(r)}$, so $\text{Rank}\{Q_{(r+1)}\} = \text{Rank}\{Q_{(r)}\}$. The argument may be extended to show that all succeeding matrices $Q_{(r+1)},\ldots,Q_{(k)}$ have the same rank as $Q_{(r)}$.

In the definition of controllability, the assumption has been made that the initial state of the system is $\xi_0 = 0$. This assumption is less restrictive than it might appear to be; for it is clear from equation (6.56) that, if $\xi_0 \neq 0$, then the sequence of inputs which will drive the system from ξ_0 to ξ_r is the same as the sequence which would drive the system from 0 to $\xi_r - \Phi^r\xi_0$, which is certainly available if the condition of controllability if fulfilled.

The next objective is to show that, if the transition equation of (6.53) can be cast in the form of equation (6.16) or in the form of equation (6.27), which have been described as controllable canonical forms, then the system does indeed satisfy the condition of controllability under (6.58).

Let Φ stand for the transition matrix in equation (6.16). Then it is readily confirmed that the matrix

(6.59)
$$[\beta,\Phi\beta,\ldots,\Phi^{r-1}\beta] = \begin{bmatrix} 1 & -\alpha_1 & \ldots & q & r \\ 0 & 1 & \ldots & p & q \\ \vdots & \vdots & \ddots & \vdots & \vdots \\ 0 & 0 & \ldots & 1 & -\alpha_1 \\ 0 & 0 & \ldots & 0 & 1 \end{bmatrix}$$

is nonsingular, and therefore the condition of (6.58) is satisfied. A similar demonstration can be given when Φ stands for the transition matrix in equation (6.27).

It can also be shown that any state-space system which satisfies the condition of controllability of (6.58) can be reduced to a controllable canonical form by means of a similarity transformation R. Suppose that the equation

(6.60)
$$\xi(t) = \Phi\xi(t-1) + \beta u(t)$$

belongs to a controllable system. Then the matrix

(6.61)
$$R = [\Phi^{r-1}\beta,\ldots,\Phi\beta,\beta]$$

is nonsingular, and it can be used in forming the equivalent system

(6.62)
$$R^{-1}\xi(t) = \{R^{-1}\Phi R\}\{R^{-1}\xi(t-1)\} + R^{-1}\beta u(t).$$

It transpires immediately that

(6.63)
$$R^{-1}\beta = \begin{bmatrix} 0 \\ \vdots \\ 0 \\ 1 \end{bmatrix};$$

since this is just the trailing vector of the identity matrix $R^{-1}[\Phi^{r-1}\beta, \ldots, \Phi\beta, \beta] = I$. Also, it follows that

$$R^{-1}\Phi R = [\Phi^r\beta, \Phi^{r-1}\beta, \ldots, \Phi\beta]$$

$$= \begin{bmatrix} -\alpha_1 & 1 & \ldots & 0 \\ \vdots & \vdots & \ddots & \vdots \\ -\alpha_{r-1} & 0 & \ldots & 1 \\ -\alpha_r & 0 & \ldots & 0 \end{bmatrix}, \tag{6.64}$$

where the leading vector of the matrix is the only one which does not come from the above-mentioned identity matrix.

On putting these results together, it is found that the transformed transition equation of (6.62) takes the form of

$$\begin{bmatrix} \xi_1(t) \\ \vdots \\ \xi_{r-1}(t) \\ \xi_r(t) \end{bmatrix} = \begin{bmatrix} -\alpha_1 & 1 & \ldots & 0 \\ \vdots & \vdots & \ddots & \vdots \\ -\alpha_{r-1} & 0 & \ldots & 1 \\ -\alpha_r & 0 & \ldots & 0 \end{bmatrix} \begin{bmatrix} \xi_1(t-1) \\ \vdots \\ \xi_{r-1}(t-1) \\ \xi_r(t-1) \end{bmatrix} + \begin{bmatrix} 0 \\ \vdots \\ 0 \\ 1 \end{bmatrix} \varepsilon(t). \tag{6.65}$$

Although equation (6.65) does represent a controllable canonical form, this is not one of the forms which we have attributed to the ARMA$(r, r-1)$ model. In the controllable-form representation under (6.27), the vector associated with the disturbance $\varepsilon(t)$ is $e_1 = [1, 0, \ldots, 0]'$, whereas, in the present equation, it is $e_r = [0, \ldots, 0, 1]'$. The difference is minor, but it cannot be overlooked.

Some further effort is required in order to discover the similarity transformation, represented by some matrix $P = [p_{r-1}, p_{r-2}, \ldots, p_0]$, which will transform the transition equation of (6.53) to the canonical form under (6.20).

Consider, therefore, the characteristic equation of the matrix Φ. This is

$$\det(\lambda I - \Phi) = \lambda^n + \alpha_1\lambda^{r-1} + \cdots + \alpha_{r-1}\lambda + \alpha_r = 0. \tag{6.66}$$

At this stage, we need not foresee the connection between the coefficients in this equation and the parameters of the ARMA model which are denoted by the same symbols. The coefficients of the characteristic equation are used to define the following recursion:

$$\begin{aligned} p_0 &= \beta, \\ p_1 &= \Phi p_0 + \alpha_1 p_0, \\ p_2 &= \Phi p_1 + \alpha_2 p_0, \\ &\vdots \\ p_r &= \Phi p_{r-1} + \alpha_r p_0. \end{aligned} \tag{6.67}$$

By a succession of substitutions running from top to bottom, the following sequence of vectors is obtained:

$$
\begin{aligned}
p_0 &= \beta, \\
p_1 &= \Phi\beta + \alpha_1\beta, \\
p_2 &= \Phi^2\beta + \alpha_1\Phi\beta + \alpha_2\beta, \\
&\vdots \\
p_r &= \Phi^r\beta + \alpha_1\Phi^{r-1}\beta + \cdots + \alpha_{r-1}\Phi\beta + \alpha_r\beta \\
&= 0.
\end{aligned} \tag{6.68}
$$

The final equality $p_r = 0$ follows from the Cayley–Hamilton theorem which indicates that

$$\Phi^r + \alpha_1\Phi^{r-1} + \cdots + \alpha_{r-1}\Phi + \alpha_r = 0; \tag{6.69}$$

which is to say that the matrix Φ satisfies its own characteristic equation.

The set of vectors from (6.68) may be gathered into the matrix

$$P = [p_{r-1}, p_{r-2}, \ldots, p_0]. \tag{6.70}$$

For this matrix to be nonsingular, it is necessary and sufficient that the condition of controllability under (6.55) is satisfied. The final column of the matrix is $Pe_r = [p_{r-1}, p_{r-2}, \ldots, p_0]e_r = \beta$; and, given that P is nonsingular, it follows that

$$P^{-1}\beta = e_r = \begin{bmatrix} 0 \\ \vdots \\ 0 \\ 1 \end{bmatrix}. \tag{6.71}$$

Now consider

$$
\begin{aligned}
P\begin{bmatrix} 0 & 1 & \ldots & 0 \\ \vdots & \vdots & \ddots & \vdots \\ 0 & 0 & \ldots & 1 \\ -\alpha_r & -\alpha_{r-1} & \ldots & -\alpha_1 \end{bmatrix} &= \bigl[-p_0\alpha_r, p_{r-1} - \alpha_{r-1}p_0, \ldots, p_1 - \alpha_1 p_0\bigr] \\
&= \bigl[\Phi p_{r-1}, \Phi p_{r-2}, \ldots, \Phi p_0\bigr] \\
&= \Phi P,
\end{aligned} \tag{6.72}
$$

where the second equality depends upon the definitions in (6.67) and the condition that $p_r = 0$ from (6.68). This shows that $P^{-1}\Phi P$ is in the form of a companion matrix. On putting these results together, it is found that the transformed transition equation

$$P^{-1}\xi(t) = \{P^{-1}\Phi P\}\{P^{-1}\xi(t-1)\} + P^{-1}\beta u(t) \tag{6.73}$$

has exactly the canonical form which was given under (6.20).

Finally, by using the usual Laplace expansion for determinants, it can be shown that the coefficients of characteristic equation $\det(\lambda I - P^{-1}\Phi P) = 0$ are exactly the parameters $\alpha_1, \ldots, \alpha_r$. Given that $\det(\lambda I - P^{-1}\Phi P) = 0$ and $\det(\lambda I - \Phi) = 0$ are the same equation, we now have the justification for the notation which has been used in equation (6.66).

Observability

A system is said to be observable if it is possible to infer its initial state by observing its input $u(t)$ and its output $y(t)$ over a finite period of time. If the initial state can be discovered and if all the relevant values of $u(t)$ are known, then it should be possible to obtain the values of $y(t)$ for any time. A formal definition is as follows:

(6.74) The system (53) is observable if, by setting $u_0 = \cdots = u_\tau = 0$ and by observing the output values $y_0, \ldots, y_\tau$, where τ denotes a finite number of periods, it is possible to infer the initial state ξ_0.

It can be proved that

(6.75) The system (53) is observable if and only if the $r \times r$ matrix

$$S = \begin{bmatrix} \gamma' \\ \gamma'\Phi \\ \vdots \\ \gamma'\Phi^{r-1} \end{bmatrix} \quad \text{has} \quad \text{rank}(S) = r.$$

To see that the rank condition is sufficient for observability, consider the sequence of observations generated by the equations of (6.53):

$$\begin{aligned} y_0 &= \gamma'\xi_0 + \delta u_0, \\ y_1 &= \gamma'\Phi\xi_0 + \gamma'\beta u_1 + \delta u_1, \\ &\vdots \\ y_{r-1} &= \gamma'\Phi^{r-1}\xi_0 + \gamma'\{\beta u_{r-1} + \Phi\beta u_{r-2} + \cdots + \Phi^{r-2}\beta u_1\} + \delta u_{r-1}. \end{aligned} \tag{6.76}$$

Setting $u_0 = \cdots = u_{r-1} = 0$ gives the equation

$$\begin{bmatrix} y_0 \\ y_1 \\ \vdots \\ y_{r-1} \end{bmatrix} = \begin{bmatrix} \gamma' \\ \gamma'\Phi \\ \vdots \\ \gamma'\Phi^{r-1} \end{bmatrix} \xi_0, \tag{6.77}$$

from which it follows that ξ_0 can be inferred if the rank condition is fulfilled. To see that the rank condition is necessary for observability, it needs to be recognised that, if $\text{Rank}(S) = q < r$, then $\text{Rank}[\gamma, \Phi'\gamma, \ldots, (\Phi^{\tau-1})'\gamma] = q$ for all $\tau \geq q$. This is demonstrated in the same way as the result under (6.58). The implication is that, if the initial state cannot be inferred from r observations on $y(t)$, then it can never be determined from such observations alone.

By checking the rank condition, it is straightforward to show that any system which can be represented by the canonical forms under (6.31) and (6.32) or (6.40) and (6.41) is observable. It can also be shown quite easily that any system which is observable can be transformed so as to conform with one of the observable canonical representations.

Consider the system

$$\begin{aligned} S\xi(t) &= \{S\Phi S^{-1}\}\{S\xi(t-1)\} + S\beta u(t), \\ y(t) &= \{\gamma' S^{-1}\}\{S\xi(t)\} + \delta u(t). \end{aligned} \tag{6.78}$$

Here there is

$$\gamma' S^{-1} = [1, 0, \ldots, 0], \tag{6.79}$$

since this is the leading row of the identity matrix $SS^{-1} = I$. Also, there is

$$S\Phi S^{-1} = \begin{bmatrix} \gamma'\Phi \\ \vdots \\ \gamma'\Phi^{r-1} \\ \gamma'\Phi^{r} \end{bmatrix} = \begin{bmatrix} 0 & 1 & \ldots & 0 \\ \vdots & \vdots & \ddots & \vdots \\ 0 & 0 & \ldots & 1 \\ -\alpha_r & -\alpha_{r-1} & \ldots & -\alpha_1 \end{bmatrix}. \tag{6.80}$$

Thus it transpires that the transformed equation has exactly the canonical form which has been given under (6.41) and (6.40).

Bibliography

[9] Akaike, H., (1974), Markovian Representation of Stochastic Processes and its Application to the Analysis of Autoregressive Moving Average Processes, *Annals of the Institute of Statistical Mathematics*, **26**, 363–387.

[26] Aoki, M., (1987), *State Space Modelling of Time Series*, Springer-Verlag, Berlin.

[161] DiStefano, J.J., A.R. Stubberud and I.J. Williams, (1967), *Feedback and Control Systems, Schaum's Outline Series*, McGraw-Hill Book Company, New York.

[193] Franklin, G.F., and D.J. Powell, (1980), *Digital Control of Dynamic Systems*, Addison Wesley, Reading, Massachusetts.

[369] Ogata, K., (1987), *Discrete-Time Control Systems*, John Wiley and Sons, New York.

[434] Rugh, W.J., (1975), *Mathematical Description of Linear Systems*, Marcel Dekker, New York.

[521] Wiberg, D.M., (1971), *State Space and Linear Systems, Schaum's Outline Series in Engineering*, McGraw-Hill Book Company, New York.

Least-Squares Methods

CHAPTER 7

Matrix Computations

The purpose of this chapter is to provide some of the staple routines of matrix computation which will be used in implementing many of the algorithms which appear subsequently in the text.

One of the most common problems in matrix computation is that of solving a set of consistent and independent linear equations where the number of equations is equal to the number of unknowns. If nothing more is known about the structure of the equations, then the efficient method of solution is that of Gaussian elimination which formalises and extends the method used in school to solve a pair of simultaneous equations.

The method of Gaussian elimination can be subjected to a number of elaborations and sophistications which are aimed, primarily, at improving its numerical stability. Since it is important to have a robust all-purpose method for solving linear equations, we shall devote considerable effort to this particular procedure.

In ordinary least-squares regression, the so-called normal equations, which are solved to obtain the regression parameters, embody a symmetric positive-definite matrix which is the cross-product of the data matrix and its own transpose. The properties of this matrix may be exploited to achieve a procedure for solving the equations which is simpler and more efficient than one which depends upon Gaussian elimination. This is the Cholesky procedure which involves finding the factors of the symmetric matrix in the form of a lower-triangular matrix and its transpose.

The simple Cholesky factorisation is available only when the symmetric matrix is positive definite. If this property cannot be guaranteed, then another factorisation must be used which interpolates a diagonal matrix as an additional factor between the lower-triangular matrix and its transpose. The importance of this factorisation is that it provides a test of the positive-definiteness of the original matrix; for the matrix is positive definite if and only if all the elements of the diagonal factor are positive. This factorisation also provides a means of calculating the determinant of the matrix by forming the products of the elements of the diagonal factor matrix. Such a facility is useful in testing the stability of linear dynamic systems.

The final algorithm to be presented in this chapter is a so-called Q–R decomposition of a matrix of full column rank which is due to Householder [261]. The Q–R decomposition can be used to advantage in the calculation of least-squares regression estimates when it is feared that solution of the normal equations is ill-determined. This alternative method of computation applies the Q–R decomposition directly to the data matrix rather than to the matrix of cross-products which is comprised by the normal equations.

The majority of the algorithms which are presented in this chapter are exploited in the following chapter on linear regression analysis. Several other algorithms of linear computation, which might have been placed in the present chapter, are dispersed throughout the text where they arise in specialised contexts.

Solving Linear Equations by Gaussian Elimination

Our aim is to solve a system equations in the form of $Ax = b$ for an unknown vector x of order n when $A = [a_{ij}]$ is a known matrix of order $n \times n$ and b is a known vector of order n.

The method of Gaussian elimination applies a sequence of transformations to both sides of the equation so that A is reduced to an upper-triangular matrix U whilst b is transformed into some vector q. Then the transformed system $Ux = q$ is solved quite easily for x by a process of back-substitution.

The matrix A is reduced to U by subjecting it to a series of elementary row operations involving the permutation of pairs of rows and the subtraction of a multiple of one row from another. These operations can be effected by premultiplying A by a succession of elementary matrices whose generic form is

(7.1) $$E(\lambda, u, v) = I - \lambda uv'.$$

The elementary permutation matrix which interchanges the ith and the jth rows is defined by

(7.2) $$P = I - (e_i - e_j)(e_i - e_j)',$$

where e_i and e_j denote vectors with units in the ith and the jth positions respectively and with zeros elsewhere. This matrix P may be formed from the identity matrix by interchanging the ith and the jth rows. By performing the operation on P itself, the identity matrix is recovered. Thus $P^2 = I$, and P is its own inverse.

The elementary matrix which can be used to subtract λ times the elements of the jth row from those of the ith row takes the form of

(7.3) $$\Lambda = I - \lambda e_i e_j'.$$

This is obtained from the identity matrix by replacing the zero element in the ijth position by $-\lambda$. The inverse of this matrix is just

(7.4) $$\Lambda^{-1} = I + \lambda e_i e_j'.$$

Notice that, when $j \neq k$, the product of two such matrices is

(7.5) $$(I - \lambda_{ij} e_i e_j')(I - \lambda_{kl} e_k e_l') = I - \lambda_{ij} e_i e_j' - \lambda_{kl} e_k e_l';$$

and there is no difficulty in representing the product of several elementary operations.

Example 7.1. To illustrate the use of elementary row operations in solving linear equations, let us consider the system $Ax = b$ for which

$$(7.6)\qquad \begin{bmatrix} A & b \end{bmatrix} = \begin{bmatrix} a_{11} & a_{12} & a_{13} & b_1 \\ a_{21} & a_{22} & a_{23} & b_2 \\ a_{31} & a_{32} & a_{33} & b_3 \end{bmatrix} = \begin{bmatrix} 6 & 3 & 9 & 9 \\ 4 & 2 & 10 & 10 \\ 2 & 4 & 9 & 3 \end{bmatrix}.$$

On premultiplying $[A, b]$ by the elimination matrix

$$(7.7)\qquad \Lambda = \begin{bmatrix} 1 & 0 & 0 \\ -a_{21}/a_{11} & 1 & 0 \\ -a_{31}/a_{11} & 0 & 1 \end{bmatrix} = \begin{bmatrix} 1 & 0 & 0 \\ -\frac{2}{3} & 1 & 0 \\ -\frac{1}{3} & 0 & 1 \end{bmatrix},$$

we get

$$(7.8)\qquad \Lambda \begin{bmatrix} A & b \end{bmatrix} = \begin{bmatrix} 6 & 3 & 9 & 9 \\ 0 & 0 & 4 & 4 \\ 0 & 3 & 6 & 0 \end{bmatrix}.$$

The elimination matrix Λ is just an instance of the product under (7.5) with $\lambda_{21} = a_{21}/a_{11}$ and $\lambda_{31} = a_{31}/a_{11}$ in place of λ_{ij} and λ_{kl}. When the matrix $\Lambda[A, b]$ is premultiplied by the permutation matrix

$$(7.9)\qquad P = \begin{bmatrix} 1 & 0 & 0 \\ 0 & 0 & 1 \\ 0 & 1 & 0 \end{bmatrix},$$

it becomes

$$(7.10)\qquad \begin{bmatrix} U & q \end{bmatrix} = \begin{bmatrix} u_{11} & u_{12} & u_{13} & q_1 \\ 0 & u_{22} & u_{23} & q_2 \\ 0 & 0 & u_{33} & q_3 \end{bmatrix} = \begin{bmatrix} 6 & 3 & 9 & 9 \\ 0 & 3 & 6 & 0 \\ 0 & 0 & 4 & 4 \end{bmatrix}.$$

Notice that $[U, q]$ may also be obtained from A first by applying the permutation P and then by applying the elimination matrix

$$(7.11)\qquad \Delta = P\Lambda P = \begin{bmatrix} 1 & 0 & 0 \\ -a_{31}/a_{11} & 1 & 0 \\ -a_{21}/a_{11} & 0 & 1 \end{bmatrix} = \begin{bmatrix} 1 & 0 & 0 \\ -\frac{1}{3} & 1 & 0 \\ -\frac{2}{3} & 0 & 1 \end{bmatrix}.$$

The identity $\Delta PA = (P\Lambda P)PA = P\Lambda A$ follows, of course, from the fact that $P^2 = I$.

The solution of the system $Ux = q$ is given by the following process of back-substitution:

$$(7.12)\qquad \begin{aligned} x_3 &= q_3/u_{33} = 1, \\ x_2 &= (q_2 - u_{23}x_3)/u_{22} = -2, \\ x_1 &= (q_1 - u_{13}x_3 - u_12x_2)/u_{11} = 1. \end{aligned}$$

On carrying these results back to equation (7.6), it can be confirmed that $Ax = b$.

To describe the general procedure for reducing a nonsingular $n \times n$ matrix A to an upper-triangular matrix, let us imagine that a matrix has already been created in which the subdiagonal elements of the first $i-1$ columns are zeros:

$$(7.13) \qquad A = \begin{bmatrix} a_{11} & \dots & a_{1,i-1} & a_{1i} & \dots & a_{1n} \\ \vdots & \ddots & \vdots & \vdots & & \vdots \\ 0 & \dots & a_{i-1,i-1} & a_{i-1,i} & \dots & a_{i-1,n} \\ 0 & \dots & 0 & a_{ii} & \dots & a_{in} \\ 0 & \dots & 0 & a_{i+1,i} & \dots & a_{i+1,n} \\ \vdots & & \vdots & \vdots & & \vdots \\ 0 & \dots & 0 & a_{ni} & \dots & a_{nn} \end{bmatrix}.$$

If the element a_{ii} is nonzero, then the subdiagonal elements $a_{i+1,i}, \dots, a_{ni}$ of the ith column can be set to zero by subtracting the ith row multiplied by $\lambda_{ki} = a_{ki}/a_{ii}$ from each of the rows indexed by $k = i+1, \dots, n$. These operations on the subdiagonal elements of the ith column can be performed by premultiplying the matrix of (7.13) by a lower-triangular matrix which is defined by

$$(7.14) \qquad \Lambda_i = \prod_{k=i+1}^{n} (I - \lambda_{ki} e_k e_i') = I - \sum_{k=i+1}^{n} \lambda_{ki} e_k e_i',$$

and which can be represented explicitly by

$$(7.15) \qquad \Lambda_i = \begin{bmatrix} 1 & \dots & 0 & 0 & \dots & 0 \\ \vdots & \ddots & \vdots & \vdots & & \vdots \\ 0 & \dots & 1 & 0 & \dots & 0 \\ 0 & \dots & 0 & 1 & \dots & 0 \\ 0 & \dots & 0 & -\lambda_{i+1,i} & \dots & 0 \\ \vdots & & \vdots & \vdots & & \vdots \\ 0 & \dots & 0 & -\lambda_{ni} & \dots & 1 \end{bmatrix}.$$

The inverse matrix Λ_i^{-1} is obtained simply by changing the signs of the subdiagonal elements of Λ_i.

The process by which A is reduced to an upper-triangular matrix may be summarised by writing

$$(7.16) \qquad (\Lambda_{n-1} \cdots \Lambda_2 \Lambda_1) A = \Lambda A = U,$$

where $\Lambda = \Lambda_{n-1} \cdots \Lambda_2 \Lambda_1$. A fragment of a Pascal procedure which implements the process on the assumption that none of the diagonal elements of A are zeros is given below.

(7.17)
```
for i := 1 to n - 1 do
  begin {i}
    for k := i + 1 to n do
```

```
        begin {k}
          lambda := a[k, i]/a[i, i];
          for j := i + 1 to n do
            a[k, j] := a[k, j] − lambda * a[i, j]
        end; {k}
    end; {i}
```

Applying the same sequence of operations to the vector b gives

(7.18) $$(\Lambda_{n-1}\cdots\Lambda_2\Lambda_1)b = \Lambda b = q.$$

Once q is available, the equations $Ux = q$ can be solved for x by the process of back-substitution. The following fragment provides the algorithm:

(7.19)
```
for i := n downto 1 do
  begin {i}
    x[i] := q[i];
    for j := i + 1 to n do
      x[i] := x[i] − u[i, j] * x[j];
    x[i] := x[i]/u[i, i]
  end; {i}
```

If the matrix $L = \Lambda^{-1}$ is available, then q can be obtained without operating upon b. For, premultiplying $q = \Lambda b$ by L gives

(7.20) $$Lq = b;$$

and, since L is a lower-triangular matrix, this can be solved for q easily by a process of forward-substitution which is the mirror image of the process described in (7.19).

To show how the matrix L is generated, let us write

(7.21) $$\Lambda_i^{-1} = I + \sum_{k=i+1}^{n} \lambda_{ki} e_k e_i'.$$

Then $L = \Lambda_1^{-1}\Lambda_2^{-1}\cdots\Lambda_{n-1}^{-1}$ can be written as

(7.22) $$\begin{aligned} L &= \prod_{i=1}^{n-1}\Big(I + \sum_{k=i+1}^{n-1} \lambda_{ki} e_k e_i'\Big) \\ &= I + \sum_{i=1}^{n-1}\sum_{k=i+1}^{n-1} \lambda_{ki} e_k e_i'; \end{aligned}$$

and this is nothing but a lower-triangular matrix containing the multipliers used in the process of elimination:

(7.23) $$L = \begin{bmatrix} 1 & 0 & \ldots & 0 \\ \lambda_{21} & 1 & \ldots & 0 \\ \vdots & \vdots & \ddots & \vdots \\ \lambda_{n1} & \lambda_{n2} & \ldots & 1 \end{bmatrix}.$$

These multipliers may be stored in place of the elements of A which are reduced to zeros.

If the ith diagonal element a_{ii} of A is zero, then a preliminary operation is required which interchanges the ith row with some row indexed by $l > i$. The row which is to be moved to the ith position may be chosen by selecting the element amongst $a_{i+1,i}, \ldots, a_{ni}$ which has the largest absolute value. The selected element is described as the pivot.

In order to enhance the numerical accuracy of the elimination procedure, it is worthwhile searching for an appropriate pivotal element even when the original diagonal element is nonzero. Also, to allow for those cases where the scale of the rows of A varies widely, it is best to choose the pivot by finding the element amongst $a_{i+1,i}, \ldots, a_{ni}$ which has the largest absolute value relative to the scale of the other elements in its own row. If the scale of the kth row is measured by the value of its largest element, then the element will be chosen which maximises the function

$$|a_{ki}|/d_k; \qquad d_k = \max_j |a_{kj}| \tag{7.24}$$

over the set of indices $k = i + 1, \ldots, n$.

If a strategy of pivotal selection is adopted, then the reduction of A is liable to be accomplished by an alternating sequence of permutations and eliminations in a manner which can be represented by writing

$$\Lambda_{n-1}P_{n-1} \cdots \Lambda_2 P_2 \Lambda_1 P_1 A = U. \tag{7.25}$$

This equation can also be written as

$$\begin{aligned}(\Delta_{n-1} \cdots \Delta_2 \Delta_1)(P_{n-1} \cdots P_2 P_1)A &= \Delta P A \\ &= U,\end{aligned} \tag{7.26}$$

where $\Delta_i = (P_{n-1} \cdots P_{i+1})\Lambda_i(P_{i+1} \cdots P_{n-1})$. The matrix Δ_i differs from Λ_i of (7.15) only by virtue of a permutation of the elements $\lambda_{i+1,i}, \ldots, \lambda_{ni}$ which are to be found below the diagonal in the ith column. Thus the matrix Δ_i can be generated by making successive amendments to the order of the subdiagonal elements of Λ_i, and the matrix $L = \Delta^{-1} = \Delta_1^{-1}\Delta_2^{-1} \cdots \Delta_{n-1}^{-1}$ can be accumulated in the places vacated by the subdiagonal elements A. The actual order of the elements of Λ_1^{-1} within the computer's memory is not altered in practice. Only their notional ordering is altered; and this is recorded in a vector of order n representing the product of the permutations entailed so far by the strategy of pivotal selection. Thus, in the Pascal program, an array $p[i]$, whose elements are just the integers $1, 2, \ldots, n$ in a permuted order, records the actual location of the row which is currently designated as the ith row.

When a strategy of pivotal selection is employed, the equation $Ux = q$ will have $U = \Delta P A$ and $q = \Delta P b$. Premultiplying both sides of the latter by $L = \Delta^{-1}$ gives

$$Lq = Pb. \tag{7.27}$$

Since Pb is obtained simply by re-indexing the elements of B, solving equation (7.27) for q is no more difficult than solving equation (7.20).

The following is the code of a Pascal procedure for solving the system $Ax = b$ which incorporates a strategy of pivotal selection and which obtains the matrix q by solving the equation $Lq = Pb$ by forward-substitution:

(7.28)

```
procedure LUSolve(start, n : integer;
                  var a : matrix;
                  var x, b : vector);

  var
    v, w, pivot, lambda : real;
    i, j, k, pivotRow, g, h, finish : integer;
    p : ivector;
    d : vector;

begin {LUSolve}

  finish := start + n − 1;
  for i := start to finish do
    begin {i; determine the scale factors}
      p[i] := i;
      d[i] := 0.0;
      for j := start to finish do
        if d[i] < Abs(a[i, j]) then
          d[i] := Abs(a[i, j]);
    end; {i}

  for i := start to finish − 1 do
    begin {i; begin the process of reduction}

      pivot := a[p[i], i];
      for k := i + 1 to finish do
        begin {k; search for a better pivot}
          v := Abs(pivot)/d[p[i]];
          w := Abs(a[p[k], i])/d[p[k]];
          if v < w then
            begin {interchange rows if better pivot is found}
              pivot := a[p[k], i];
              pivotRow := p[k];
              p[k] := p[i];
              p[i] := pivotRow;
            end; {end interchange}
        end; {k; end the search for a pivot}

      for k := i + 1 to finish do
        begin {k; eliminate a[k, i]}
```

```
            lambda := a[p[k], i]/pivot;
            for j := i + 1 to finish do
              a[p[k], j] := a[p[k], j] - lambda * a[p[i], j];
            a[p[k], i] := lambda; {save the multiplier}
          end; {k}

      end; {i; reduction completed}

    for i := start to finish do
      begin {i; forward-substitution}
        x[i] := b[p[i]];
        for j := i - 1 downto start do
          x[i] := x[i] - a[p[i], j] * x[j];
      end; {i; forward-substitution}

    for i := finish downto start do
      begin {i; back-substitution}
        for j := i + 1 to finish do
          x[i] := x[i] - a[p[i], j] * x[j];
        x[i] := x[i]/a[p[i], i];
      end; {i; back-substitution}

  end; {LUSolve}
```

It should be noted that the initial and terminal values of the index $j = 1, \ldots, n$ have been replaced in this procedure by *start* and *finish* = *start* + $(n-1)$ respectively. This adds a measure of flexibility which will allow the initial and terminal indices to be set 0 and $n-1$, respectively, when the occasion demands.

Inverting Matrices by Gaussian Elimination

The need to find the explicit inverse of a numerical matrix A is rare. One way of finding the inverse would be to use an extended version of the procedure *LUsolve* to solve the equation $AX = I$. For this purpose, one would have to accommodate the identity matrix I in place of the vector b of the system $Ax = b$. A procedure of this nature would begin by transforming the matrix A into an upper-triangular matrix U. In effect, the equations $AX = I$ would be transformed into the equivalent equations $UA^{-1} = Q$. In the second stage of the procedure, the solution A^{-1} would be found by a process of back-substitution.

An alternative way of finding the inverse is to use the method of Gaussian elimination to reduce the matrix A completely to the identity matrix instead of reducing it partially to an upper triangle U. In effect, the equations $AX = I$ are transformed into the equations $IX = A^{-1}$. Thus, when the operations of Gaussian elimination are applied to the combined matrix $[A, I]$, what emerges, at length, is the matrix $[I, A^{-1}]$. At the ith stage of the transformation, the ith column e_i of the identity matrix would appear in the place originally occupied by the ith column of matrix A. At the same time, the column e_i would disappear from its original position in the identity matrix on the RHS. In practice, there is no need to store any of the columns of the identity matrix; and so the inverse matrix A^{-1} may be formed in the space originally occupied by A.

The following procedure inverts a matrix directly by the methods of Gaussian elimination. Since no attention is paid to the problem of selecting appropriate pivotal elements, it should be used only to invert well-conditioned matrices. In fact, the procedure will be used only for inverting a symmetric positive-definite matrix wherein the diagonal elements are all units and the off-diagonal elements are constrained to lie in the open interval $(-1, 1)$.

For inverting a matrix in its entirely, the parameter *stop* should be set to value of n in calling the procedure. If *stop* is set to $p < n$ then only p steps of the process of inversion will be accomplished. The effect will be to invert the leading minor of order p within the matrix A.

(7.29)

```
procedure GaussianInversion(n, stop : integer;
                            var a : matrix);

  var
    lambda, pivot : real;
    i, j, k : integer;

begin {GaussianInversion}

  for i := 1 to stop do
    begin {i}
      pivot := a[i, i];
      for k := 1 to n do
        begin {k}
          if k <> i then
            begin
              lambda := a[k, i]/pivot;
              for j := 1 to n do
                a[k, j] := a[k, j] − lambda * a[i, j];
              a[k, i] := −lambda
            end;
        end; {k}
      for j := 1 to n do
        a[i, j] := a[i, j]/pivot;
      a[i, i] := 1/pivot;
    end; {i; reduction completed}

end; {GaussianInversion}
```

The Direct Factorisation of a Nonsingular Matrix

Our method of solving the equation $Ax = b$ by Gaussian elimination entails the factorisation $A = LU$ where $L = [l_{ij}]$ is a lower-triangular matrix with units on the diagonal and $U = [u_{ij}]$ is an upper-triangular matrix.

It is important to recognise that the L–U factorisation of a nonsingular matrix A is unique if the factor L is constrained to have only units on the diagonal. To show this, let us imagine that there are two such factorisations: $LU = A$ and $L_1U_1 = A$. Then there are $L^{-1} = UA^{-1}$ and $L_1 = AU_1^{-1}$; and so

$$\begin{aligned} L^{-1}L_1 &= UA^{-1}AU_1^{-1} \\ &= UU_1^{-1}. \end{aligned} \tag{7.30}$$

Now, if L and L_1 are lower-triangular matrices with unit diagonals, then so are L^{-1} and $L^{-1}L_1$. On the other hand, UU_1^{-1} is an upper-triangular matrix. It follows that the only way that the equality in (7.30) can be maintained is when $L^{-1}L_1 = I$ and $UU_1^{-1} = I$. But this implies that $L = L_1$ and $U = U_1$.

Gaussian elimination is only one of several ways of calculating the L–U factorisation of A. Another way is to obtain $L = [l_{ij}]$ and $U = [u_{ij}]$ directly by solving the equations $LU = A$ element by element. Let us consider the equations in more detail:

$$\begin{aligned} &\begin{bmatrix} l_{11} & 0 & 0 & \ldots & 0 \\ l_{21} & l_{22} & 0 & \ldots & 0 \\ l_{31} & l_{32} & l_{33} & \ldots & 0 \\ \vdots & & \vdots & \ddots & \vdots \\ l_{n1} & l_{n2} & l_{n3} & \ldots & l_{nn} \end{bmatrix} \begin{bmatrix} u_{11} & u_{12} & u_{13} & \ldots & u_{1n} \\ 0 & u_{22} & u_{23} & \ldots & u_{2n} \\ 0 & 0 & u_{33} & \ldots & u_{3n} \\ \vdots & & \vdots & \ddots & \vdots \\ 0 & 0 & 0 & \ldots & u_{nn} \end{bmatrix} \\ &\qquad = \begin{bmatrix} a_{11} & a_{12} & a_{13} & \ldots & a_{1n} \\ a_{21} & a_{22} & a_{23} & \ldots & a_{2n} \\ a_{31} & a_{32} & a_{33} & \ldots & a_{3n} \\ \vdots & & \vdots & & \vdots \\ a_{n1} & a_{n2} & a_{n3} & \ldots & a_{nn} \end{bmatrix}. \end{aligned} \tag{7.31}$$

If a_{ij} is on or above the diagonal with $i \leq j$, then the generic equation takes the form of

$$a_{ij} = l_{i1}u_{1j} + \cdots + l_{i,i-1}u_{i-1,j} + l_{ii}u_{ij}, \tag{7.32}$$

whilst, if a_{ij} is below the diagonal with $i > j$, it takes the form of

$$a_{ij} = l_{i1}u_{1j} + \cdots + l_{i,j-1}u_{j-1,j} + l_{ij}u_{jj}. \tag{7.33}$$

Since $l_{ii} = 1$ for all i, equation (7.32) gives

$$u_{ij} = a_{ij} - l_{i1}u_{1j} - \cdots - l_{i,i-1}u_{i-1,j}, \tag{7.34}$$

whereas, equation (7.33) gives

$$l_{ij} = (a_{ij} - l_{i1}u_{1j} - \cdots - l_{i,j-1}u_{j-1,j})/u_{jj}. \tag{7.35}$$

Equations (7.34) and (7.35) can be solved in any order which ensures that the values required on the RHS are available when needed. One way is to calculate alternately the qth row of U and the qth column of L in the sequence $q = 1, \ldots, n$. This could be accomplished by a Pascal procedure incorporating the following fragment:

(7.36)
```
for q := 1 to n do
  begin {q}

    for j := q to n do
      begin {j; find the qth row of U}
        u[q, j] := a[q, j];
        for k := 1 to q - 1 do
          u[q, j] := u[q, j] - l[q, k] * u[k, j];
      end; {j}

    for i := q + 1 to n do
      begin {i; find the qth column of L}
        l[i, q] := a[i, q];
        for k := 1 to q - 1 do
          l[i, q] := l[i, q] - l[i, k] * u[k, q];
        l[i, q] := l[i, q]/u[q, q];
      end; {i}

  end; {q}
```

In practice, computer memory should be conserved by writing the elements of U and the subdiagonal elements of L in place of the corresponding elements of A. There is no need to record the diagonal elements of L. To make this procedure wholly practical, we would need to support it by a method of pivotal selection. Instead of elaborating the procedure further, we shall consider only a special case in which pivotal selection is unnecessary.

The Cholesky Decomposition

As Golub and Van Loan [220, p. 81] have asserted, one of the basic principles of matrix computation is that one should seek to exploit any available knowledge of the structure of a matrix with a view to making an algorithm quicker in execution and more economical in its storage requirements.

There is scope for improving the efficiency of the algorithm for the L–U decomposition of a nonsingular matrix whenever the matrix is symmetric. When the matrix is also positive definite, the algorithm can be further simplified.

Recall that, if A is a nonsingular matrix, then there exits a unique decomposition $A = LU$ wherein L is a lower-triangular matrix with units on the diagonal and U is an unrestricted upper-triangular matrix. By interpolating a diagonal matrix $D = \mathrm{diag}\{d_1, \ldots, d_n\}$, it is possible to obtain a factorisation $A = LDM'$ where M' has units on the diagonal and $DM' = U$.

Now imagine that A is also a symmetric matrix with $A = A'$. Then $LDM' = MDL'$, where L and M are lower-triangular and DM' and DL' are upper-triangular. It follows from the uniqueness of the L–U factorisation that $M = L$; and so the symmetric matrix can be written as $A = A' = LDL'$.

The matrix $A = A'$ is positive definite if $x'Ax > 0$ for any nonzero value of x. If $A = LDL'$ is positive definite, then so too is $D = L^{-1}AL^{-1\prime}$; which implies that all of the elements of $D = \text{diag}\{d_1, \ldots, d_n\}$ must be positive. This result can be deduced from the fact that every principal submatrix of a positive-definite matrix is also positive definite.

Conversely, it is easy to see that that, if $d_i > 0$ for all i, then the condition $x'Ax = x'LDL'x = q'Dq > 0$ prevails for all nonzero values of q or x. If $d_i > 0$, then $\sqrt{d_i}$ is real-valued and, therefore, a matrix $D^{1/2} = \text{diag}\{\sqrt{d_1}, \ldots, \sqrt{d_n}\}$ can be defined such that $A = (LD^{1/2})(LD^{1/2})'$. Thus a symmetric positive-definite matrix can be factorised uniquely into the product of a lower-triangular matrix and its transpose. This factorisation, which can be written more simply as $A = A' = LL'$ by redefining L, is called the Cholesky decomposition.

The elements of the Cholesky factor L can be obtained from equations similar to those of (7.35). A difference arises from the fact that, in general, the diagonal elements of L are not units. Moreover, because of the symmetry of A, the amount of arithmetic in computing the factorisation is halved.

Consider an element a_{ij} of the symmetric matrix A which is on or below the diagonal such that $i \geq j$:

$$a_{ij} = l_{i1}l_{j1} + \cdots + l_{i,j-1}l_{j,j-1} + l_{ij}l_{jj}. \tag{7.37}$$

When $i > j$, this gives

$$l_{ij} = (a_{ij} - l_{i1}l_{j1} - \cdots - l_{i,j-1}l_{j,j-1})/l_{jj}, \tag{7.38}$$

which is just a specialisation of equation (7.35) which comes from setting $u_{ij} = l_{ji}$. When $i = j$, equation (7.37) becomes

$$a_{ii} = l_{i1}^2 + \cdots + l_{i,i-1}^2 + l_{ii}^2; \tag{7.39}$$

and this gives

$$l_{ii} = \sqrt{(a_{ii} - l_{i1}^2 - \cdots - l_{i,i-1}^2)}. \tag{7.40}$$

The matrix L can generated by progressing through its *columns* in a manner similar to that which is indicated in second half of the fragment under (7.36):

(7.41)
```
for j := 1 to n do
  for i := j to n do
    begin {i, j ; find the jth column of L}
      l[i, j] := a[i, j];
      for k := 1 to j - 1 do
        l[i, j] := l[i, j] - l[i, k] * l[j, k];
      if i = j then
        l[i, j] := Sqrt(l[i, j])
      else
        l[i, j] := l[i, j]/l[j, j]
    end; {i, j}
```

We shall embed the Cholesky decomposition within a procedure for solving an equation system $Ax = b$, where A is positive definite. The procedure uses the method of forward-substitution and back-substitution which is used within the more general procedure *LUSolve* of (7.28) which solves an equation system by Gaussian elimination. To explain this method, let us substitute the factorisation $A = LL'$ into the equation $Ax = b$ to obtain $LL'x = b$. Setting

$$L'x = q \tag{7.42}$$

within the equation, gives

$$Lq = b. \tag{7.43}$$

Once the matrix L has been obtained, the equation (7.43) can be solved for q by forward-substitution. When q has been drafted into equation (7.42), the latter can be solved for x by back-substitution. The Pascal procedure for accomplishing these operations is listed below:

(7.44)
```
procedure Cholesky(n : integer;
                   var a : matrix;
                   var x, b : vector);

  var
    l : real;
    i, j, k : integer;

begin {Cholesky}

  for j := 1 to n do
    for i := j to n do
      begin {i; find the jth column of L}
        l := a[i, j];
        for k := 1 to j - 1 do
          l := l - a[i, k] * a[j, k];
        if i = j then
          a[i, j] := Sqrt(l)
        else
          a[i, j] := l/a[j, j];
      end; {i}

  for i := 1 to n do
    begin {i; forward-substitution}
      x[i] := b[i];
      for j := i - 1 downto 1 do
        x[i] := x[i] - a[i, j] * x[j];
      x[i] := x[i]/a[i, i];
    end; {i; forward-substitution}
```

```
for i := n downto 1 do
  begin {i; back-substitution}
    for j := i + 1 to n do
      x[i] := x[i] - a[j, i] * x[j];
    x[i] := x[i]/a[i, i];
  end; {i; back-substitution}

end; {Cholesky}
```

The Cholesky decomposition of a symmetric matrix A depends crucially upon the condition that A is positive definite. If A is not positive definite, then the procedure will fail at the point where, in calculating one of the diagonal elements of a Cholesky triangle, an attempt is made to find the square root of a nonpositive number.

If the matrix A is indefinite, then the more general factorisation $A = LDL'$ is called for, where L is a lower-triangular matrix with units on the diagonal and D is a diagonal matrix.

Consider a generic element of $A = [a_{ij}]$ which is on or below the diagonal of the matrix such that $i \geq j$. It is readily confirmed that

$$a_{ij} = \sum_{k=1}^{j} d_k l_{ik} l_{jk}, \tag{7.45}$$

where d_k is the kth element of D and l_{ik} is an element from L. This equation gives rise to a generic expression for the subdiagonal elements of the jth column of L, and to an expression for the jth element of the diagonal matrix D:

$$\begin{aligned} l_{ij} &= \frac{1}{d_j}\left\{a_{ij} - \sum_{k=1}^{j-1} d_k l_{ik} l_{jk}\right\}, \\ d_j &= a_{jj} - \sum_{k=1}^{j-1} d_k l_{jk}^2. \end{aligned} \tag{7.46}$$

The following procedure uses the above equations in the process of factorising $A = LDL'$. The complete matrix A is passed to the procedure. It is returned with the subdiagonal elements of L replacing its own subdiagonal elements, and with the elements of D along its principal diagonal. From the returned matrix, it is easy to calculate the determinant of the original matrix A by forming the product of the elements of D. Notice that this procedure generates successive *rows* of the lower-triangular matrix L, which is in contrast to previous versions which generate successive *columns*.

(7.47)
```
procedure LDLprimeDecomposition(n : integer;
                               var a : matrix);

  var
    i, j, k : integer;

begin

  for i := 1 to n do
    for j := 1 to i do
      begin {i, j}
        for k := 1 to j - 1 do
          a[i, j] := a[i, j] - a[k, k] * a[i, k] * a[j, k];
        if i > j then
          begin
            a[i, j] := a[i, j]/a[j, j];
            a[j, i] := 0.0;
          end;
      end; {i, j}

end; {LDLprimeDecomposition}
```

Householder Transformations

An elementary reflector, or Householder transformation, is an orthonormal matrix H defined by

$$H = I - 2uu' \quad \text{with} \quad u'u = 1. \tag{7.48}$$

For any vector $a \in \mathcal{R}^n$, the effect of the transformation is to reverse the direction of the component which lies along the axis of the vector u.

Let $a = \lambda u + v$, where λ is some scalar and v is a vector which lies in the subspace $\mathcal{V}$ which represents the orthogonal complement of the axis of u. Then $Hv = v$ and $Hu = -u$; and, therefore,

$$\begin{aligned} z &= Ha \\ &= H(\lambda u + v) \\ &= -\lambda u + v. \end{aligned} \tag{7.49}$$

The mapping of a into z is depicted in Figure 7.1 which shows that $z = Ha$ is the reflection of a about the subspace $\mathcal{V}$ orthogonal to the vector u.

The Householder transformation H is completely determined by the pair of vectors a and $z = Ha$; and the vector u, which is found in the definition under (7.48), may be expressed in terms of a and z. When $a = \lambda u + v$ and $z = -\lambda u + v$, there is $a - z = 2\lambda u$. Now $u'u = 1$, which is to say that u has unit length; and, to obtain u, it is sufficient to normalise the vector $w = a - z$, or, in other words, to

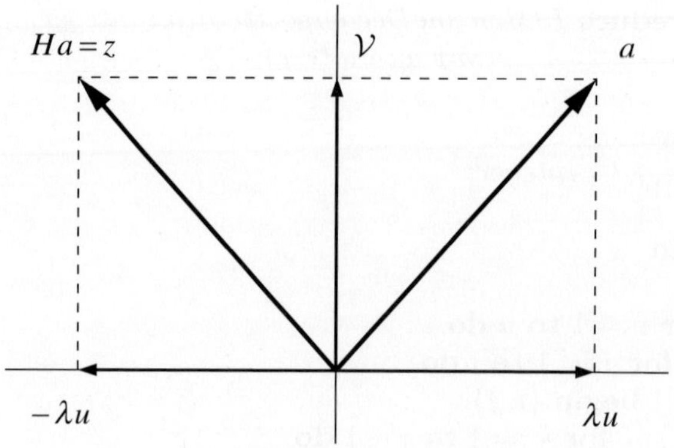

Figure 7.1. The vector $z = Ha$ is the reflection of a about the subspace $\mathcal{V}$ orthogonal to the vector u.

rescale it so that its length becomes unity. Thus

(7.50)
$$\begin{aligned} u &= \frac{(a-z)}{\sqrt{(a-z)'(a-z)}} \\ &= \frac{w}{\sqrt{w'w}}. \end{aligned}$$

Substituting from (7.50) into equation (7.48) gives

(7.51)
$$\begin{aligned} H &= I - 2\frac{(a-z)(a-z)'}{(a-z)'(a-z)} \\ &= I - 2\frac{ww'}{w'w}. \end{aligned}$$

This expression can be used in devising a transformation which will map the vector a into the axis of some specified vector y so as to obtain $Ha = z = \kappa y$. To find the value of z to be used in constructing the desired transformation in the form of (7.51), it is only necessary to find the scalar value κ. From the fact that H is an orthonormal matrix, it follows that $a'H'Ha = a'a = \kappa^2 y'y$. Thus it is found that

(7.52)
$$\kappa = \pm\sqrt{\left(\frac{a'a}{y'y}\right)}.$$

The Q–R Decomposition of a Matrix of Full Column Rank

Let A be an $m \times n$ matrix with $m \geq n$ and $\text{Rank}(A) = n$. Then there exists an orthonormal transformation Q of order $m \times m$ such that

(7.53)
$$Q'A = \begin{bmatrix} R \\ 0 \end{bmatrix},$$

where R is an upper-triangular matrix of order n.

This reduction, which is described as a Q–R decomposition can be accomplished by premultiplying A by a succession of orthonormal Householder transformations $P_1, \ldots, P_n$ which are effective in eliminating the subdiagonal elements from successive columns of the matrix and which together form the product $Q' = P_n \cdots P_2 P_1$.

Consider the first of these transformations:

$$P_1 = I_m - 2u_1 u_1'. \tag{7.54}$$

Its effect is to reduce the leading vector $a_{.1}$ of $A = [a_{.1}, \ldots, a_{.n}]$ to a vector $\kappa_1 e_1$ which has a scalar κ_1 in the leading position and zeros elsewhere. Setting $z = \kappa_1 e_1$ and $a = a_{.1}$ in (7.50) gives

$$u_1 = \frac{(a_{.1} - \kappa_1 e_1)}{\sqrt{(a_{.1} - \kappa_1 e_1)'(a_{.1} - \kappa_1 e_1)}}; \tag{7.55}$$

and, according to (7.52), there is $\kappa_1 = \pm\sqrt{(a_{.1}' a_{.1})}$ since $y'y = e_1' e_1 = 1$. Therefore, P_1 is now specified apart from the choice of sign for κ_1. The sign of κ_1 is chosen so that the leading term $a_{11} - \kappa_1$ within the vector $a_{.1} - \kappa_1 e_1$ is a sum rather than a difference. This is to avoid any undesirable cancellation which might bring the term close to zero, thereby prejudicing the numerical accuracy of the procedure. Thus

$$\kappa_1 = -\operatorname{sgn}(a_{11})\sqrt{a_{.1}' a_{.1}}, \tag{7.56}$$

whence it follows that

$$\begin{aligned}(a_{.1} - \kappa_1 e_1)'(a_{.1} - \kappa_1 e_1) &= a_{.1}' a_{.1} - 2\kappa_1 e_1' a_{.1} + \kappa_1^2 e_1' e_1 \\ &= 2\left(\kappa_1^2 + |\kappa_1 a_{11}|\right).\end{aligned} \tag{7.57}$$

Now that the subdiagonal elements have been eliminated from the first column, the first row and column of $P_1 A$ may be ignored and attention may be turned to the remaining submatrix. A transformation in the form of $H_2 = I_{m-1} - 2u_2 u_2'$, can be applied to this submatrix so as to reduce its leading vector to one with a leading nonzero element and with zeros elsewhere. Equivalently, a Householder transformation

$$P_2 = \begin{bmatrix} 1 & 0 \\ 0 & H_2 \end{bmatrix} \tag{7.58}$$

can be applied to the full matrix $P_1 A$ to obtain a matrix $P_2 P_1 A$ in which the first and second columns contain only zero elements below the principal diagonal. Proceeding in this way through n steps, we obtain

$$P_n \cdots P_2 P_1 A = Q' A = \begin{bmatrix} R \\ 0 \end{bmatrix}, \tag{7.59}$$

where $P_1, \ldots, P_n$ and Q' are all orthonormal matrices.

To illustrate the jth step of the procedure in detail, let us define $A_{j-1} = P_{j-1} \cdots P_2 P_1 A$, and let us consider $P_j A_{j-1} = A_j$. The latter can be partitioned to give

$$\begin{bmatrix} I_{j-1} & 0 \\ 0 & H_j \end{bmatrix} \begin{bmatrix} U & D \\ 0 & C \end{bmatrix} = \begin{bmatrix} U & D \\ 0 & H_j C \end{bmatrix}. \tag{7.60}$$

This shows that, at the jth step, we are operating upon a submatrix C of order $(m-j+1) \times (m-j+1)$.

In the Pascal procedure which follows, the Householder transformation $H_j = I_{m-j+1} - 2u_j u_j'$ is written in the alternative form of

$$H_j = I_{m-j+1} - \beta_j w_j w_j', \tag{7.61}$$

where

$$\begin{aligned} w_j &= \begin{bmatrix} a_{jj} - \kappa_j \\ a_{j+1,j} \\ \vdots \\ a_{mj} \end{bmatrix}, \\ \kappa_j &= -\operatorname{sgn}(a_{jj})\sqrt{\left(\sum_{i=j}^{m} a_{i,j}^2\right)} \quad \text{and} \\ \beta_j &= \left(\kappa_j^2 + |\kappa_j a_{jj}|\right)^{-1} = 2(w_j' w_j)^{-1}. \end{aligned} \tag{7.62}$$

The transformation of the matrix A_{j-1} by P_j entails the transformation of C in (7.60) by H_j. This is carried out by calculating

$$y_j = \beta_j w_j' C$$

and then modifying C to give

$$H_j C = C - w_j y_j.$$

The jth diagonal element a_{jj} of A_j is just κ_j, as can be seen from the appropriate substitutions. The subdiagonal elements of $a_{.j}$ are mapped into zeros. Of course, these results follow from the fact that H_j has been constructed precisely for the purpose of mapping the vector $[a_{jj}, \ldots, a_{mj}]'$ into the axis of the leading vector of the identity matrix I_{m-j+1}.

The procedure has a provision for subjecting an auxiliary matrix B of order $m \times q$ to the same series of transformations as A. For example, one might specify $B = I_m$. Then the procedure would return the matrix Q' in place of B. Finally, it should noted that, in implementing the procedure, nothing is gained by setting the subdiagonal elements of A to zero. The discarded by-products of the calculations can be allowed to accumulate in the positions where, in theory, there should be zeros.

(7.63)

```
procedure Householder(var a, b : matrix;
                      m, n, q : integer);

  var
    i, j, k : integer;
    S, sigma, kappa, beta, yPrime : real;

begin
  for j := 1 to n do
    begin {major loop}

      sigma := 0.0; {find the value of kappa}
      for i := j to m do
        sigma := sigma + Sqr(a[i, j]);
      S := Sqrt(sigma);
      beta := 1/(sigma + Abs(S * a[j, j]));
      if a[j, j] < 0 then
        kappa := S
      else
        kappa := -S;
      a[j, j] := a[j, j] - kappa;

      for k := j + 1 to n do
        begin {k}
          yPrime := 0.0;
          for i := j to m do
            yPrime := yPrime + a[i, j] * a[i, k];
          yPrime := beta * yPrime;
          for i := j to m do
            a[i, k] := a[i, k] - a[i, j] * yPrime;
        end; {k}

      for k := 1 to q do
        begin {k}
          yPrime := 0.0;
          for i := j to m do
            yPrime := yPrime + a[i, j] * b[i, k];
          yPrime := yPrime * beta;
          for i := j to m do
            b[i, k] := b[i, k] - a[i, j] * yPrime;
        end; {k}

      a[j, j] := kappa;
    end; {major loop}

end; {Householder}
```

Bibliography

[220] Golub, G., and C.F. Van Loan, (1983), *Matrix Computations*, John Hopkins University Press, Baltimore, Maryland.

[235] Hagger, J.M., (1988), *Applied Numerical Linear Algebra*, Prentice-Hall, Englewood Cliffs, New Jersey.

[261] Householder, A.S., (1958), Unitary Triangularization of a Nonsymmetric Matrix, *Communications of the ACM*, **5**, 339–342.

[262] Householder, A.S., (1964), *The Theory of Matrices in Numerical Analysis*, Blaidswell Publishing Co., New York.

[307] Lancaster, P., and M. Tismenetsky, (1985), *The Theory of Matrices*, Academic Press, New York.

[312] Lawson, C. L., and R.J. Hanson, (1974), *Solving Least Squares Problems*, Prentice-Hall, Englewood Cliffs, New Jersey.

[524] Wilkinson, J.H., and C. Reinsch, (1971), *Handbook for Automatic Computation, Vol. 2, Linear Algebra*, Springer-Verlag, Berlin.

CHAPTER 8

Classical Regression Analysis

In this chapter, we shall present the basic theory of the classical statistical method of regression analysis. The method can be applied in its own right to numerous problems within time-series analysis; and this alone should justify our giving it an extensive treatment. However, its importance is greatly enhanced by the fact that it contains many of the elements from which sophisticated methods are constructed for analysing time series in the time domain.

The routines of regression analysis which are presented in this chapter make use of the staple procedures for orthogonalising, triangularising and inverting matrices which have been presented in the previous chapter.

The Linear Regression Model

A regression equation of the form

$$\begin{aligned} y_t &= x_{t1}\beta_1 + x_{t2}\beta_2 + \cdots + x_{tk}\beta_k + \varepsilon_t \\ &= x_{t.}\beta + \varepsilon_t \end{aligned} \tag{8.1}$$

explains the value of a dependent variable y_t in terms of a set of k observable variables in $x_{t.} = [x_{t1}, x_{t2}, \ldots, x_{tk}]$ and an unobservable random variable ε_t. The vector $\beta = [\beta_1, \beta_2, \ldots, \beta_k]'$ contains the parameters of a linear combination of the variables in $x_{t.}$. A set of T successive realisations of the regression relationship, indexed by $t = 1, 2, \ldots, T$, can be compiled into a system

$$y = X\beta + \varepsilon, \tag{8.2}$$

wherein $y = [y_1, y_2, \ldots, y_T]'$ and $\varepsilon = [\varepsilon_1, \varepsilon_2, \ldots, \varepsilon_T]'$ are vectors of order T and $X = [x_{tk}]$ is a matrix of order $T \times k$. We shall assume that X is a nonstochastic matrix with $\text{rank}(X) = k$ which requires that $T \geq k$.

According to the classical assumptions, the elements of the disturbance vector ε are distributed independently and identically with expected values of zero and a common variance of σ^2. Thus

$$E(\varepsilon) = 0 \qquad \text{and} \qquad D(\varepsilon) = E(\varepsilon\varepsilon') = \sigma^2 I_T. \tag{8.3}$$

The matrix $D(\varepsilon)$, which is described as the variance–covariance matrix or the dispersion matrix of ε, contains the common variance $\sigma^2 = E[\{\varepsilon_t - E(\varepsilon_t)\}^2]$ in each of its diagonal locations. Its other locations contain zero-valued elements, each of which corresponds to the covariance $E[\{\varepsilon_t - E(\varepsilon_t)\}\{\varepsilon_s - E(\varepsilon_s)\}']$ of two distinct elements of ε.

The value of β may be estimated according to the principle of ordinary least-squares regression by minimising the quadratic function

$$S = \varepsilon'\varepsilon = (y - X\beta)'(y - X\beta). \tag{8.4}$$

The problem can be envisaged as one of finding a value for $\mu = X\beta$ residing, at a minimum distance from the vector y, in the subspace or the manifold spanned by the columns of X. This interpretation comes from recognising that the function $S = (y - X\beta)'(y - X\beta)$ represents the square of the Euclidean distance between the two vectors.

The minimising value of β is found by differentiating the function $S(\beta)$ with respect to β and setting the result to zero. This gives the condition

$$\frac{\partial S}{\partial \beta} = 2\beta'X'X - 2y'X = 0. \tag{8.5}$$

By rearranging the condition, the so-called normal equations are obtained

$$X'X\beta = X'y, \tag{8.6}$$

whose solution is the ordinary least-squares estimate of the regression parameters:

$$\hat{\beta} = (X'X)^{-1}X'y. \tag{8.7}$$

The estimate of the systematic component of the regression equations is

$$\begin{aligned} X\hat{\beta} &= X(X'X)^{-1}X'y \\ &= Py. \end{aligned} \tag{8.8}$$

Here $P = X(X'X)^{-1}X'$, which is called the orthogonal or perpendicular projector on the manifold of X, is a symmetric idempotent matrix with the properties that $P = P' = P^2$.

The Decomposition of the Sum of Squares

Ordinary least-squares regression entails the decomposition of the vector y into two mutually orthogonal components. These are the vector $Py = X\hat{\beta}$, which estimates the systematic component of the regression equation, and the residual vector $e = y - X\hat{\beta}$, which estimates the disturbance vector ε. The condition that e should be orthogonal to the manifold of X in which the systematic component resides, such that $X'e = X'(y - X\hat{\beta}) = 0$, is precisely the condition which is expressed by the normal equations (8.6).

Corresponding to the decomposition of y, there is a decomposition of the sum of squares $S = y'y$. To express the latter, let us write $X\hat{\beta} = Py$ and $e = y - X\hat{\beta} =$

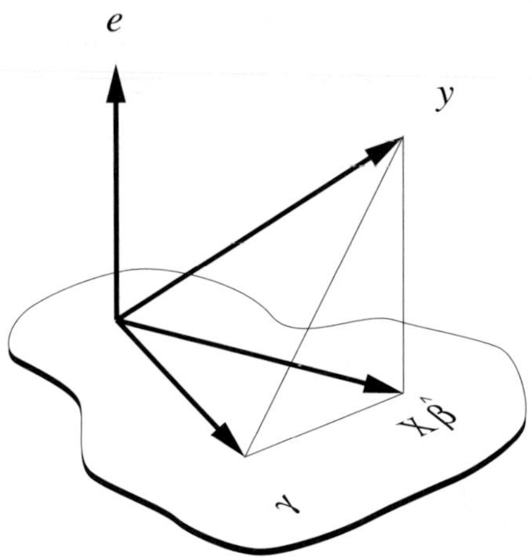

Figure 8.1. The vector $Py = X\hat{\beta}$ is formed by the orthogonal projection of the vector y onto the subspace spanned by the columns of the matrix X.

$(I - P)y$. Then, in consequence of the condition $P = P' = P^2$ and the equivalent condition $P'(I - P) = 0$, it follows that

$$(8.9)\qquad \begin{aligned} y'y &= \{Py + (I - P)y\}'\{Py + (I - P)y\} \\ &= y'Py + y'(I - P)y \\ &= \hat{\beta}'X'X\hat{\beta} + e'e. \end{aligned}$$

This is simply an instance of Pythagoras theorem; and the identity is expressed by saying that the total sum of squares $y'y$ is equal to the regression sum of squares $\hat{\beta}'X'X\hat{\beta}$ plus the residual or error sum of squares $e'e$. A geometric interpretation of the orthogonal decomposition of y and of the resulting Pythagorean relationship is given in Figure 8.1.

It is clear from intuition that, by projecting y perpendicularly onto the manifold of X, the distance between y and $Py = X\hat{\beta}$ is minimised (see Figure 8.1). In order to establish this point formally, imagine that $\gamma = Pg$ is an arbitrary vector in the manifold of X. Then the Euclidean distance from y to γ cannot be less than the distance from y to $X\hat{\beta}$. The square of the former distance is

$$(8.10)\qquad \begin{aligned} (y-\gamma)'(y-\gamma) &= \{(y - X\hat{\beta}) + (X\hat{\beta} - \gamma)\}'\{(y - X\hat{\beta}) + (X\hat{\beta} - \gamma)\} \\ &= \{(I - P)y + P(y - g)\}'\{(I - P)y + P(y - g)\}. \end{aligned}$$

The properties of the projector P which have been used in simplifying equation (8.9), indicate that

$$(8.11)\qquad \begin{aligned} (y-\gamma)'(y-\gamma) &= y'(I - P)y + (y - g)'P(y - g) \\ &= e'e + (X\hat{\beta} - \gamma)'(X\hat{\beta} - \gamma). \end{aligned}$$

Since the squared distance $(X\hat{\beta}-\gamma)'(X\hat{\beta}-\gamma)$ is nonnegative, it follows that $(y-\gamma)'(y-\gamma)\geq e'e$, where $e=y-X\hat{\beta}$; and this proves the assertion.

A summary measure of the extent to which the ordinary least-squares regression accounts for the observed vector y is provided by the coefficient of determination. This is defined by

$$(8.12)\qquad \begin{aligned} R^2 &= \frac{\hat{\beta}'X'X\hat{\beta}}{y'y} \\ &= \frac{y'Py}{y'y}. \end{aligned}$$

The measure is just the square of the cosine of the angle between the vectors y and $Py=X\hat{\beta}$; and the inequality $0\leq R^2\leq 1$ follows from the fact that the cosine of any angle must lie between -1 and $+1$.

Some Statistical Properties of the Estimator

The expectation, or mean, of the vector $\hat{\beta}$, and its dispersion matrix as well, may be found from the expression

$$(8.13)\qquad \begin{aligned} \hat{\beta} &= (X'X)^{-1}X'(X\beta+\varepsilon) \\ &= \beta+(X'X)^{-1}X'\varepsilon. \end{aligned}$$

On the assumption that the elements of X are nonstochastic, the expectation is given by

$$(8.14)\qquad \begin{aligned} E(\hat{\beta}) &= \beta+(X'X)^{-1}X'E(\varepsilon) \\ &= \beta. \end{aligned}$$

Thus $\hat{\beta}$ is an unbiased estimator. The deviation of $\hat{\beta}$ from its expected value is $\hat{\beta}-E(\hat{\beta})=(X'X)^{-1}X'\varepsilon$. Therefore the dispersion matrix, which contains the variances and covariances of the elements of $\hat{\beta}$, is

$$(8.15)\qquad \begin{aligned} D(\hat{\beta}) &= E\left[\{\hat{\beta}-E(\hat{\beta})\}\{\hat{\beta}-E(\hat{\beta})\}'\right] \\ &= (X'X)^{-1}X'E(\varepsilon\varepsilon')X(X'X)^{-1} \\ &= \sigma^2(X'X)^{-1}. \end{aligned}$$

The Gauss–Markov theorem asserts that $\hat{\beta}$ is the unbiased linear estimator of least dispersion. This dispersion is usually characterised in terms of the variance of an arbitrary linear combination of the elements of $\hat{\beta}$, although it may also be characterised in terms of the determinant of the dispersion matrix $D(\hat{\beta})$. Thus

(8.16) If $\hat{\beta}$ is the ordinary least-squares estimator of β in the classical linear regression model, and if β^* is any other linear unbiased estimator of β, then $V(q'\beta^*)\geq V(q'\hat{\beta})$ where q is any constant vector of the appropriate order.

Proof. Since $\beta^* = Ay$ is an unbiased estimator, it follows that $E(\beta^*) = AE(y) = AX\beta = \beta$, which implies that $AX = I$. Now set $A = (X'X)^{-1}X' + G$. Then $AX = I$ implies that $GX = 0$. Given that $D(y) = D(\varepsilon) = \sigma^2 I$, it follows that

$$\begin{aligned} D(\beta^*) &= AD(y)A' \\ &= \sigma^2\{(X'X)^{-1}X' + G\}\{X(X'X)^{-1} + G'\} \\ &= \sigma^2(X'X)^{-1} + \sigma^2 GG' \\ &= D(\hat{\beta}) + \sigma^2 GG'. \end{aligned} \tag{8.17}$$

Therefore, for any constant vector q of order k, there is the identity

$$\begin{aligned} V(q'\beta^*) &= q'D(\hat{\beta})q + \sigma^2 q'GG'q \\ &= V(q'\hat{\beta}) + \sigma^2 q'GG'q; \end{aligned} \tag{8.18}$$

and this implies the inequality $V(q'\beta^*) \geq V(q'\hat{\beta})$.

Estimating the Variance of the Disturbance

The principle of least squares does not, of itself, suggest a means of estimating the disturbance variance $\sigma^2 = V(\varepsilon_t)$. However, it is natural to estimate the moments of a probability distribution by their empirical counterparts. Given that $e_t = y - x_{t.}\hat{\beta}$ is an estimate of ε_t, it follows that $T^{-1}\sum_t e_t^2$ may be used to estimate σ^2. However, it transpires that this is biased. An unbiased estimate is provided by

$$\begin{aligned} \hat{\sigma}^2 &= \frac{1}{T-k}\sum_{t=1}^{T} e_t^2 \\ &= \frac{1}{T-k}(y - X\hat{\beta})'(y - X\hat{\beta}). \end{aligned} \tag{8.19}$$

The unbiasedness of this estimate may be demonstrated by finding the expected value of $(y - X\hat{\beta})'(y - X\hat{\beta}) = y'(I - P)y$. Given that $(I - P)y = (I - P)(X\beta + \varepsilon) = (I - P)\varepsilon$ in consequence of the condition $(I - P)X = 0$, it follows that

$$E\{(y - X\hat{\beta})'(y - X\hat{\beta})\} = E(\varepsilon'\varepsilon) - E(\varepsilon' P\varepsilon). \tag{8.20}$$

The value of the first term on the RHS is given by

$$E(\varepsilon'\varepsilon) = \sum_{t=1}^{T} E(e_t^2) = T\sigma^2. \tag{8.21}$$

The value of the second term on the RHS is given by

$$\begin{aligned} E(\varepsilon' P\varepsilon) &= \text{Trace}\{E(\varepsilon' P\varepsilon)\} = E\{\text{Trace}(\varepsilon' P\varepsilon)\} = E\{\text{Trace}(\varepsilon\varepsilon' P)\} \\ &= \text{Trace}\{E(\varepsilon\varepsilon')P\} = \text{Trace}\{\sigma^2 P\} = \sigma^2\text{Trace}(P) \\ &= \sigma^2 k. \end{aligned} \tag{8.22}$$

The final equality follows from the fact that $\text{Trace}(P) = \text{Trace}(I_k) = k$. Putting the results of (8.21) and (8.22) into (8.20), gives

$$E\{(y - X\hat{\beta})'(y - X\hat{\beta})\} = \sigma^2(T - k); \tag{8.23}$$

and, from this, the unbiasedness of the estimator in (8.19) follows directly.

The Partitioned Regression Model

In testing hypotheses, it is helpful to have explicit expressions for the subvectors within $\hat{\beta} = [\hat{\beta}_1', \hat{\beta}_2']$. To this end, the equations of (8.2) may be written as $y = X_1\beta_1 + X_2\beta_2 + \varepsilon$, and the normal equations of (8.6) may be partitioned conformably to give

$$\begin{aligned} X_1'X_1\beta_1 + X_1'X_2\beta_2 &= X_1'y \qquad \text{and} \\ X_2'X_1\beta_1 + X_2'X_2\beta_2 &= X_2'y. \end{aligned} \tag{8.24}$$

Premultiplying the first of these by $X_2'X_1(X_1'X_1)^{-1}$ and subtracting it from the second gives

$$\{X_2'X_2 - X_2'X_1(X_1'X_1)^{-1}X_1'X_2\}\beta_2 = X_2'y - X_2'X_1(X_1'X_1)^{-1}X_1'y. \tag{8.25}$$

When the projector $P_1 = X_1(X_1'X_1)^{-1}X_1'$ is defined, the equation may be written more intelligibly as $X_2'(I - P_1)X_2\beta_2 = X_2'(I - P_1)y$. The estimate of β_2 is given by

$$\hat{\beta}_2 = \{X_2'(I - P_1)X_2\}^{-1}X_2'(I - P_1)y. \tag{8.26}$$

An analogous expression is available for $\hat{\beta}_1$. However, knowing the value of $\hat{\beta}_2$ enables us to obtain $\hat{\beta}_1$ alternatively from the expression

$$\hat{\beta}_1 = (X_1'X_1)^{-1}X_1'(y - X_2\hat{\beta}_2) \tag{8.27}$$

which comes directly from the first equation of (8.24).

Some Matrix Identities

The estimators of β_1 and β_2 may also be derived by using the partitioned form of the matrix $(X'X)^{-1}$. This is given by

$$\begin{aligned} &\begin{bmatrix} X_1'X_1 & X_1'X_2 \\ X_2'X_1 & X_2'X_2 \end{bmatrix}^{-1} \\ &= \begin{bmatrix} \{X_1'(I - P_2)X_1\}^{-1} & -\{X_1'(I - P_2)X_1\}^{-1}X_1'X_2(X_2'X_2)^{-1} \\ -\{X_2'(I - P_1)X_2\}^{-1}X_2'X_1(X_1'X_1)^{-1} & \{X_2'(I - P_1)X_2\}^{-1} \end{bmatrix} \end{aligned} \tag{8.28}$$

The result is easily verified by postmultiplying the matrix on the RHS by the partitioned form of $X'X$ to give a partitioned form of the identity matrix.

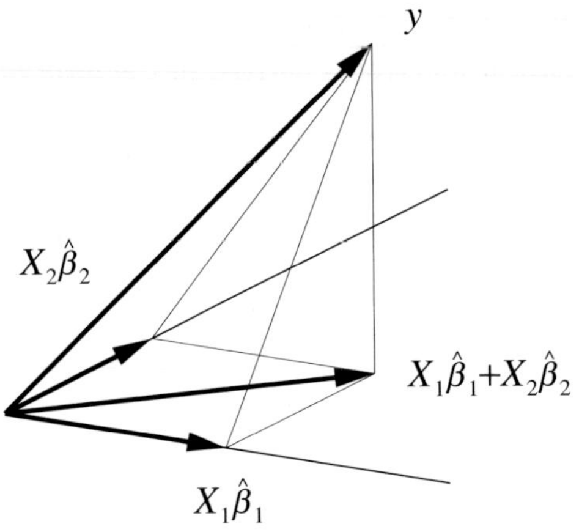

Figure 8.2. The decomposition of the vector $Py = X\hat{\beta} = X_1\hat{\beta}_1 + X_1\hat{\beta}_1$.

By forming the projector $P = X(X'X)^{-1}X'$ from $X = [X_1, X_2]$ and from the partitioned form of $(X'X)^{-1}$, it may be shown that

$$
\begin{aligned}
&P = P_{1/2} + P_{2/1}, \qquad \text{where} \\
&P_{1/2} = X_1\{X_1'(I - P_2)X_1\}^{-1}X_1'(I - P_2) \qquad \text{and} \\
&P_{2/1} = X_2\{X_2'(I - P_1)X_2\}^{-1}X_2'(I - P_1).
\end{aligned} \tag{8.29}
$$

In the notation of the regression model, the identity $Py = P_{1/2}y + P_{2/1}y$ is expressed as $X\hat{\beta} = X_1\hat{\beta}_1 + X_2\hat{\beta}_2$ (see Figure 8.2).

The restriction of the transformation $P_{1/2}$ to the manifold of X may be described as the oblique projection onto the manifold of X_1 along the manifold of X_2. This means that the manifold of X_2 falls within the null space of the projector. The corresponding conditions $P_{1/2}X_1 = X_1$ and $P_{1/2}X_2 = 0$ are readily confirmed. Thus

$$
\begin{aligned}
P_{1/2}P_1 &= P_1, \\
P_{1/2}P_2 &= 0.
\end{aligned} \tag{8.30}
$$

Likewise, $P_{2/1}X_2 = X_2$ and $P_{2/1}X_1 = 0$. These conditions indicate that

$$
\begin{aligned}
PP_1 &= (P_{1/2} + P_{2/1})P_1 \\
&= P_1 \\
&= P_1P.
\end{aligned} \tag{8.31}
$$

The final equality follows from the symmetry of P_1 and P.

Now consider premultiplying and postmultiplying the partitioned form of $(X'X)^{-1}$ by $(I - P_2)X = [(I - P_2)X_1, 0]$ and its transpose respectively. Reference to (8.28) shows that this gives

$$
\begin{aligned}
(I - P_2)X(X'X)^{-1}X'(I - P_2) &= (I - P_2)P(I - P_2) \\
&= (I - P_2)X_1\{X_1'(I - P_2)X_1\}^{-1}X_1'(I - P_2).
\end{aligned} \tag{8.32}
$$

But the conditions $PP_2 = P_2P = P_2$ can be used to show that $(I - P_2)P(I - P_2) = P - P_2$. Thus an important identity is derived in the form of

$$(I - P_2)X_1\{X_1'(I - P_2)X_1\}^{-1}X_1'(I - P_2) = P - P_2. \tag{8.33}$$

This result, which can also be found in another book by Pollock [397] which treats regression analysis, will be used in the sequel.

Computing a Regression via Gaussian Elimination

The traditional way of computing the ordinary least-squares regression estimates, which is still the prevalent way, is to solve the normal equations $X'X\beta = X'y$ as they stand. This can be done using the Cholesky algorithm, presented under (7.44) in the previous chapter, which finds the L–U decomposition of the symmetric matrix $X'X = LL'$. Having created the lower-triangular matrix L, the procedure solves the equation $Lq = X'y$ by a recursive process of forward-substitution to obtain $q = L'\hat{\beta}$. Then the equation $q = L'\hat{\beta}$ is solved for $\hat{\beta}$ by a converse process of back-substitution.

As an alternative to the Cholesky method, the Gaussian algorithm for matrix inversion, which appears under (7.29) in the previous chapter, can be used to find the inverse matrix $(X'X)^{-1}$. Then the product $(X'X)^{-1}X'y = \hat{\beta}$ may be formed.

In a notional sense, the Gaussian algorithm for inverting $X'X$ uses elementary row operations to transform the pair $[X'X, I]$ into the pair $[I, (X'X)^{-1}]$. At each stage of the procedure, a column belonging to the identity matrix of order k disappears from the right-hand matrix and is recreated in the left-hand matrix where the contents of the nondiagonal cells of one of the columns are swept out and a unit is placed in the diagonal cell. However, in practice, there is no need to store the columns of the identity; and so the inverse matrix $(X'X)^{-1}$ can be formed in the space originally occupied by the matrix $X'X$.

By a minor elaboration of the Gaussian inversion procedure, some of the other regression quantities can be generated as by-products. Consider replacing the matrix $[X'X, I]$, by the augmented matrix

$$\left[\begin{array}{cc|cc} X'X & X'y & I & 0 \\ y'X & y'y & 0 & 1 \end{array}\right]. \tag{8.34}$$

The matrices on either side of the major partition have $k + 1$ rows and columns. The first k columns of the matrix on the left may be swept out by applying k steps of the inversion procedure. The effect of this operation is summarised in the following equation:

$$\begin{aligned} &\begin{bmatrix} (X'X)^{-1} & 0 \\ -y'X(X'X)^{-1} & 1 \end{bmatrix} \left[\begin{array}{cc|cc} X'X & X'y & I & 0 \\ y'X & y'y & 0 & 1 \end{array}\right] \\ &= \left[\begin{array}{cc|cc} I & (X'X)^{-1}X'y & (X'X)^{-1} & 0 \\ 0 & y'y - y'X(X'X)^{-1}X'y & -y'X(X'X)^{-1} & 1 \end{array}\right]. \end{aligned} \tag{8.35}$$

Here the aggregate effect of the elementary row operations is achieved by premultiplying the matrix of (8.34) by the appropriate partitioned matrix. If the new columns which appear on the left of the major partition are accumulated in place of columns of the identity which appear on the right, then, after k steps, the matrix will take the form of

$$\text{(8.36)} \quad \begin{bmatrix} (X'X)^{-1} & (X'X)^{-1}X'y \\ -y'X(X'X)^{-1} & y'y - y'X(X'X)^{-1}X'y \end{bmatrix} = \begin{bmatrix} (X'X)^{-1} & \hat{\beta} \\ -\hat{\beta}' & \hat{\sigma}^2(T-k) \end{bmatrix}.$$

Apart from the ordinary least-squares estimate of β, the matrix contains the elements from which the estimates of $V(\varepsilon_t) = \sigma^2$ and $D(\hat{\beta}) = \sigma^2(X'X)^{-1}$ can be constructed immediately.

In many instances of the regression model, the leading column of the matrix X is a vector of units $i = [1, \ldots, 1]'$ which is associated with a so-called intercept parameter β_1. In such cases, it is appropriate to set $X = [i, Z]$ and $\beta = [\beta_1, \beta_z']'$ so as to express the regression equations as

$$\text{(8.37)} \qquad y = i\beta_1 + Z\beta_z + \varepsilon.$$

From the formulae given under (8.26) and (8.27), it follows that the estimates of the parameters can be expressed as

$$\text{(8.38)} \qquad \begin{aligned} \hat{\beta}_z &= \{Z'(I - P_i)Z\}^{-1}Z'(I - P_i)y \quad \text{and} \\ \hat{\beta}_1 &= (i'i)^{-1}i'(y - Z\hat{\beta}_z), \end{aligned}$$

where $P_i = i(i'i)^{-1}i'$.

If $x = [x_1, \ldots, x_T]'$ is any vector of sample values, then $(i'i)^{-1}i'x = \bar{x}$ is the sample mean and $(I - P_i)x = [x_1 - \bar{x}, \ldots, x_T - \bar{x}]'$ is the vector of their deviations about the mean. Moreover, as a result of the symmetry and idempotency of the matrix $(I - P_i)$, it follows that, if w is any other matrix of T elements, then $\{(I - P_i)w\}'\{(I - P_i)x\} = w'(I - P_i)x$. In view of these results, it can be seen that $\hat{\beta}_z$ may be obtained by applying the regression procedure to variables which are in deviation form. Also, the intercept term, which can be written as

$$\text{(8.39)} \qquad \begin{aligned} \hat{\beta}_1 &= \bar{y} - T^{-1}i'Z\hat{\beta}_z \\ &= \bar{y} - (\bar{x}_2\hat{\beta}_2 + \cdots + \bar{x}_k\hat{\beta}_k), \end{aligned}$$

is a function only of the sample means and of the estimated parameters in $\hat{\beta}_z = [\hat{\beta}_2, \ldots, \hat{\beta}_k]'$. Therefore it is readily calculated once $\hat{\beta}_z$ becomes available.

The dispersion of $\hat{\beta}_z$ is given by the matrix

$$\text{(8.40)} \qquad D(\hat{\beta}_z) = \sigma^2\{Z'(I - P_i)Z\}^{-1}.$$

This follows from the partitioned form of $D(\hat{\beta}) = \sigma^2(X'X)^{-1}$ when $X = [i, Z]$. The variance of $\hat{\beta}_1$ is given by

$$\text{(8.41)} \qquad \begin{aligned} D(\hat{\beta}_1) &= \sigma^2\{i'(I - P_z)i\}^{-1} \\ &= \sigma^2\frac{1}{T^2}[T + i'Z\{Z'(I - P_i)Z\}^{-1}Z'i]. \end{aligned}$$

The first form follows in the same way as the form under (8.40). The second form follows from setting $X_1 = i$ and $X_2 = Z$ in the identity

$$\begin{aligned}(X_1'X_1)^{-1} + (X_1'X_1)^{-1}X_1'X_2\{X_2'(I-P_1)X_2\}^{-1}X_2'X_1(X_1'X_1)^{-1} \\ = \{X_1'(I-P_2)X_1\}^{-1}.\end{aligned} \tag{8.42}$$

The identity itself is deduced from

$$I = X_1'X_1\{X_1'(I-P_2)X_1\}^{-1} - X_1'X_2\{X_2'(I-P_1)X_2\}^{-1}X_2'X_1(X_1'X_1)^{-1}, \tag{8.43}$$

which comes from the partitioned form of the identity $I = X'X(X'X)^{-1}$ which may be constructed from equation (8.28).

The covariance of $\hat{\beta}_z$ and $\hat{\beta}_1$ is given by

$$C(\hat{\beta}_z, \hat{\beta}_1) = -\sigma^2 T^{-1}\{Z'(I-P_i)Z\}^{-1}Z'i; \tag{8.44}$$

and this follows from the partitioned form of $D(\hat{\beta}) = \sigma^2(X'X)^{-1}$.

One is not bound to pay any special attention to the fact that a regression model contains an intercept term; for $\beta = [\beta_1, \beta_z']'$ can be estimated by applying the Gaussian inversion procedure to the matrix

$$\begin{bmatrix} X'X & X'y \\ y'X & y'y \end{bmatrix}, \tag{8.45}$$

wherein $X = [i, Z]$. However, β_z can be estimated on its own by applying the procedure to the matrix

$$\begin{bmatrix} Z'(I-P_i)Z & Z'(I-P_i)y \\ y'(I-P_i)Z & y'(I-P_i)y \end{bmatrix}, \tag{8.46}$$

which is simply T times the empirical variance–covariance matrix of the variables. The estimate of β_1 can then be obtained from the formula in (8.39).

When the Gaussian procedure is applied to the matrix of (8.45), it generates, as a by-product, the quantity

$$\begin{aligned} y'y - y'X(X'X)^{-1}X'y &= (y - X\hat{\beta})'(y - X\hat{\beta}) \\ &= (T-k)\hat{\sigma}^2. \end{aligned} \tag{8.47}$$

When the procedure is applied to the matrix of (8.46), it generates

$$\begin{aligned} y'(I-P_i)y - y'(I-P_i)Z\{Z'(I-P_i)Z\}^{-1}Z'(I-P_i)y \\ = \{(I-P_i)(y - Z\hat{\beta}_z)\}'\{(I-P_i)(y - Z\hat{\beta}_z)\}. \end{aligned} \tag{8.48}$$

To demonstrate that this is the same quantity, we need only confirm that

$$\begin{aligned} (I-P_i)(y - Z\hat{\beta}_z) &= y - Z\hat{\beta}_z - i(i'i)^{-1}i(y - Z\hat{\beta}_z) \\ &= y - i\hat{\beta}_1 - Z\hat{\beta}_z \\ &= y - X\hat{\beta}. \end{aligned} \tag{8.49}$$

The second of these equalities follows from the second equation under (8.38).

The advantage of taking deviations of the variables before estimating the regression parameters is that this reduces the disparities in the scale of the cross-products of the variables. This should diminish the rounding errors which beset the subsequent calculations. The accuracy of these calculations can be further enhanced by using the correlation matrix in place of the matrix of corrected sums of squares and cross-products of (8.46). The values of all the elements of the correlation matrix lie in the interval $[-1, 1]$.

Using the correlation matrix in calculating the regression parameters is tantamount to dividing each of the regressors by the appropriate standard deviation calculated from the T sample observations. The associated regression parameters are multiplied by the standard deviations. If the dependent variable is also scaled by its standard deviation, then the regression equation of (8.1) is transformed into

$$(8.50) \qquad \frac{y_t}{s_y} = \frac{x_{t1}}{s_1}\Big\{\frac{s_1}{s_y}\beta_1\Big\} + \cdots + \frac{x_{tk}}{s_k}\Big\{\frac{s_k}{s_y}\beta_k\Big\} + \frac{\varepsilon_t}{s_y}.$$

On the completion of the calculations, the the original scales must be restored to the estimated regression parameters.

Calculating the Corrected Sum of Squares

The use of the method of Gaussian elimination in calculating the ordinary least-squares regression estimates has been criticised on the grounds of numerical inaccuracy. The method works best when the elements of the cross-product matrix have a limited range of values; and, to reduce this range, it is best, whenever possible, to take the variables in deviation form. However, some care must be exercised in calculating the corrected sums of squares and cross-products if they are not to become an additional source of inaccuracy.

To show the nature of the problem, we may consider the matter of calculating the variance s^2 of a sample $[x_1, \ldots, x_T]$. This can be calculated from the sum of squares of the deviations of the sample values about their mean m:

$$(8.51) \qquad s^2 = \frac{1}{T}\sum_{t=1}^{T}(x_t - m)^2; \qquad m = \frac{1}{T}\sum_{t=1}^{T}x_t.$$

Alternatively, it may be calculated by adjusting the raw sum of squares of the sample values:

$$(8.52) \qquad s^2 = \frac{1}{T}\Big\{\sum_{t=1}^{T}x_t^2 - Tm^2\Big\}.$$

The latter formula is commonly recommended for hand calculations since it entails less labour. However, it can lead to significant errors if it is implemented on a computer whose arithmetic operations are of limited precision.

To understand this, consider a random sample of independently and identically distributed elements, $x_t = \mu + \varepsilon_t$; $t = 1, \ldots, T$. Let $E(\varepsilon_t) = 0$ and $V(\varepsilon_t) = \sigma^2$. Then the expected values of the two components in the formula of (8.52) are $E(\sum x_t^2) = T(\mu^2+\sigma^2)$ and $E(Tm^2) \simeq T\mu^2$. If the coefficient of variation σ/μ is small, then there may be insufficient digits in the ordinary floating-point representations to reflect accurately the small but crucial difference between the values of $\sum x_t^2$ and Tm^2. Hence the subtraction of one from the other may give an inaccurate value for the corrected sum of squares. This problem is likely to be a serious one only in extreme cases.

The problem of rounding error manifests itself more acutely in the business of cumulating the raw sum of squares $\sum x_t^2$. (See, amongst others, Gregory [228], Ling [319] and Malcolm [329].) As the number of the elements which have been assimilated increases, the size of the running total grows relative to that of its increments; and there may be a point in the sample beyond which the terms in ε_t within the increment $x_t^2 = (\mu+\varepsilon_t)^2$ cease to have any numerical significance.

One way of dealing with these problems is to use extra digits in the registers to which the sum $\sum x_t = Tm$ and the sum of squares $\sum x_t^2$ are accumulated. Another way is to resort to the method of calculating the corrected sum of squares which is indicated by the formula in (8.51). If there is no concern over the time taken in the calculations, then one might consider calculating the following sequence of values:

(8.53)
$$\begin{aligned} m_1 &= \frac{1}{T}\sum_{t=1}^{T} x_t, \\ m_2 &= m_1 + \frac{1}{T}\sum_{t=1}^{T}(x_t - m_1) \quad \text{and} \\ s^2 &= \frac{1}{T}\sum_{t=1}^{T}(x_t - m_2)^2. \end{aligned}$$

Recalculating the mean as m_2 after finding a trial value m_1 is an effective way of overcoming the problems of cumulation which can arise when T is large and when the number of digits in the register is limited.

The formulae under (8.53) are used in the following Pascal procedure which is for calculating the correlation matrix corresponding to n data series contained in a matrix of order $T \times n$.

(8.54)
```
procedure Correlation(n, Tcap : integer;
                      var x, c : matrix;
                      var scale, mean : vector);

  var
    i, j, t, d : integer;
    proMean : real;

begin
  for j := 1 to n do
    begin {j; form the jth sample mean}
      proMean := 0.0;
```

```
      for t := 1 to Tcap do
        proMean := proMean + x[t, j];
      proMean := proMean/Tcap;
      mean[j] := 0.0;
      for t := 1 to Tcap do
        mean[j] := mean[j] + (x[t, j] - proMean);
      mean[j] := mean[j]/Tcap + proMean;
    end; {j}

  d := 0;
  while d < n do
    begin {while}
      for i := d + 1 to n do
        begin {i}
          j := i - d;
          c[i, j] := 0.0;
          for t := 1 to Tcap do
            c[i, j] := c[i, j] + (x[t, i] - mean[i]) * (x[t, j] - mean[j]);
          if i <> j then
            c[i, j] := c[i, j]/Sqrt(c[i, i] * c[j, j]);
        end; {i}
      d := d + 1;
    end; {while}
  for i := 1 to n do
    begin
      scale[i] := Sqrt(c[i, i]/Tcap);
      c[i, i] := 1.0;
    end;

  for i := 1 to n do
    for j := i + 1 to n do
      c[i, j] := c[j, i];

end; {Correlation}
```

If the correlation matrix is used in place of the matrix of corrected cross-products, then what emerges from the Gaussian inversion procedure is the matrix

$$(8.55)\qquad \begin{bmatrix} S'_z\{Z'(I-P_i)Z\}^{-1}S_z & \hat{\beta}_z\dfrac{S'_z}{s_y} \\ -\hat{\beta}'_z\dfrac{S_z}{s_y} & \dfrac{(T-K)}{T}\dfrac{\hat{\sigma}^2}{s_y^2} \end{bmatrix},$$

where S_z is a diagonal matrix of scale factors which are the standard deviations of the variables in Z and where s_y is the standard deviation of the dependent variable.

The following procedure calculates the elements of the matrix of (8.55) which are then rescaled to obtain the estimates of β and σ^2.

(8.56)

```
procedure GaussianRegression(k, Tcap : integer;
                       var x, c : matrix);

  var
    i, j : integer;
    intercept : real;
    scale, mean : vector;

begin

  Correlation(k, Tcap, x, c, scale, mean);
  GaussianInversion(k, k - 1, c);
  for i := 1 to k do
    for j := 1 to k do
      begin
        if (i < k) and (j < k) then
          c[i, j] := c[i, j]/(scale[i] * scale[j])
        else if (i < k) and (j = k) then
          c[i, j] := (c[i, j] * scale[j])/scale[i]
        else if (j < k) and (i = k) then
          c[i, j] := (c[i, j] * scale[i])/scale[j]
        else
          c[i, j] := c[i, j] * Tcap * Sqr(scale[i])/(Tcap - k);
      end;
  intercept := mean[k];
  for i := 1 to k - 1 do
    intercept := intercept - mean[i] * c[i, k];
  c[k, 1] := intercept;

end; {GaussianRegression}
```

The procedure takes as one of its parameters the array x which contains the variables of $X = [Z, y]$. On completion, it delivers the moment matrix $Z'(I - P_i)Z$ in the locations $c[i, j]$; $i, j := 1$ **to** $k - 1$ and the elements of $\hat{\beta}$ in $c[i, k]$; $i := 1$ **to** $k - 1$. The intercept term $\hat{\alpha}$ is contained in $c[k, 1]$ whilst the value of $\hat{\sigma}^2$ is found in $c[k, k]$.

In some applications, it may be desirable to calculate the correlation matrix in a single pass, even at the cost of sacrificing numerical accuracy. This can be achieved by calculating the sample moments recursively according to an updating method which revises the estimates as each new data point is added.

To derive a recursive formula for the sample mean, consider the expressions

$$\begin{aligned}(t-1)m_{t-1} &= x_1 + \cdots + x_{t-1} \quad \text{and} \\ tm_t &= x_1 + \cdots + x_{t-1} + x_t \\ &= (t-1)m_{t-1} + x_t.\end{aligned} \tag{8.57}$$

Dividing the second equation by t and rearranging gives

$$m_t = m_{t-1} + \frac{1}{t}(x_t - m_{t-1}), \tag{8.58}$$

which indicates the appropriate algorithm. Here the term $x_t - m_{t-1}$ may be construed as a prediction error; and the recursive scheme which is based upon the equation is the simplest example of a prediction-error algorithm.

The sample variance s_t^2, calculated from t observations, is defined by

$$\begin{aligned} ts_t^2 &= \sum_{i=0}^{t}(x_i - m_t)^2 \\ &= (x_t - m_t)^2 + \sum_{i=0}^{t-1}\Big\{(x_i - m_{t-1}) + (m_{t-1} - m_t)\Big\}^2 \\ &= (x_t - m_t)^2 + \sum_{i=0}^{t-1}(x_i - m_{t-1})^2 + (t-1)(m_{t-1} - m_t)^2. \end{aligned} \tag{8.59}$$

Here, the third equality is by virtue of a vanishing cross-product. In the final expression, there are

$$\begin{aligned} (x_t - m_t)^2 &= \frac{(t-1)^2}{t^2}(m_{t-1} - x_t)^2, \\ \sum_{i=0}^{t-1}(x_i - m_{t-1})^2 &= (t-1)s_{t-1}^2, \\ (t-1)(m_{t-1} - m_t)^2 &= \frac{t-1}{t^2}(x_t - m_{t-1})^2. \end{aligned} \tag{8.60}$$

The first of these comes directly from (8.58) as does the third. The second provides a definition of s_{t-1}^2, which is the sample variance calculated from $t-1$ observations. The three terms on the RHS of the equations combine to give the following recursive formula which expresses s_t^2 in terms of s_{t-1}^2:

$$ts_t^2 = (t-1)s_{t-1}^2 - \frac{t-1}{t}(m_{t-1} - x_t)^2. \tag{8.61}$$

Some algebra of a more complicated nature will serve to show that the covariance c_t of the vectors $[x_1, \ldots, x_t]$ and $[y_1, \ldots, y_t]$ may be calculated recursively via the formula

$$tc_t = (t-1)c_{t-1} + \frac{t-1}{t}(x_t - m_{t-1})(y_t - n_{t-1}), \tag{8.62}$$

where n_{t-1} stands for the mean of $[y_1, \ldots, y_{t-1}]$.

Recursive methods for calculating the ordinary least-squares estimates are treated at some length in the following chapter where the calculation of the vector $\hat{\beta}$ itself is the subject of a recursion.

Computing the Regression Parameters via the *Q–R* Decomposition

Experience of the numerical inaccuracies of poorly coded procedures which use Gaussian elimination in calculating ordinary least-squares regression estimates has led some authorities to declare that the normal equations ought never to be formed. (See, for example, Chambers [99, p. 106].)

The methods which avoid the formation of the normal equations depend upon finding the so-called Q–R decomposition of the matrix X. This is a matter of premultiplying X by a series of orthonormal matrices $P_1, \ldots, P_k$ so as to eliminate the subdiagonal elements from successive columns. By these means, a matrix is generated in the form of

$$(8.63) \qquad Q'X = \begin{bmatrix} Q'_rX \\ Q'_eX \end{bmatrix} = \begin{bmatrix} R \\ 0 \end{bmatrix},$$

where $Q' = P_k \cdots P_2P_1$ is also an orthonormal matrix such that $Q'Q = QQ' = I$, and where R is a nonsingular upper-triangular matrix of order $k \times k$.

An orthonormal matrix represents an isometric transformation. If the vectors of the matrix $X = [x_{.1}, x_{.2}, \ldots, x_{.k}]$ are subjected to such a transformation, then their lengths and the angles between them are unaffected, and only their orientation relative to the axes of a fixed coordinate system is altered.

It is clear that an orthonormal transformation P_1 can be found which brings the leading vector $x_{.1}$ into alignment with the first axis $e_1 = [1, 0, \ldots, 0]'$ of the natural coordinate system. Once this has been accomplished, a second transformation P_2 can be found which leaves the first vector unaffected and which brings the second vector $x_{.2}$ into alignment with the plane spanned jointly by e_1 and the vector $e_2 = [0, 1, \ldots, 0]'$ which lies along the second axis of the coordinate system.

In general, a transformation P_j can be found, to be used at the jth stage of the procedure, which leaves the leading $j-1$ vectors unaffected and brings the jth vector into alignment with e_j. By taking k steps of this procedure, one can obtain the Q–R decomposition of X which is represented by (8.63) above.

The Pascal code for the Householder method of Q–R decomposition was presented in the previous chapter under (7.63). Other methods which are available are the Givens procedure and the Gram–Schmidt procedure. The code for the Gram–Schmidt procedure is presented in Chapter 10 under (10.26).

To understand the use of the Q–R decomposition in calculating the regression estimates, consider premultiplying the equation (8.63) by $Q = [Q_r, Q_e]$. In consequence of the condition $QQ' = I$, this gives

$$(8.64) \qquad X = Q_rR.$$

Substituting the expression on the right into the normal equations $X'X\beta = X'y$ gives

$$(8.65) \qquad R'Q'_rQ_rR\hat{\beta} = R'Q'_ry.$$

By premultiplying this equation by R^{-1} and by using the condition $Q'_rQ_r = I$, an equivalent system is derived in the form of

$$(8.66) \qquad R\hat{\beta} = Q'_ry.$$

Since R is an upper-triangular matrix, the latter equations can be solved easily for $\hat{\beta} = [\hat{\beta}_1, \ldots, \hat{\beta}_k]'$ by a process of back-substitution beginning with $\hat{\beta}_k$.

The elements of equation (8.66) are obtained by subjecting the augmented matrix $[X, y]$ to the series of Householder transformations comprised in $Q' = P_k \cdots P_2 P_1$. The result is the matrix

(8.67) $$Q'[X, y] = \begin{bmatrix} R & Q'_r y \\ 0 & Q'_e y \end{bmatrix}.$$

In calculating $Q'[X, y]$, a quantity $Q'_e y$ is obtained from which the estimate of the variance $\sigma^2 = V(\varepsilon_t)$ of the disturbances is readily formed. Consider the expression

(8.68) $$\begin{aligned}(T-k)\hat{\sigma}^2 &= y'(I-P)y \\ &= y'\{I - X(X'X)^{-1}X'\}y \\ &= y'(I - Q_r Q'_r)y.\end{aligned}$$

Given that $Q_r Q'_r + Q_e Q'_e = QQ' = I$, it follows that $I - Q_r Q'_r = Q_e Q'_e$. Hence

(8.69) $$\hat{\sigma}^2 = \frac{y' Q_e Q'_e y}{T-k}.$$

In addition to the estimates of β and σ^2, the value of $\hat{\sigma}^2(X'X)^{-1} = \hat{\sigma}^2 R^{-1} R'^{-1}$ is sought which is the estimate of the dispersion matrix $D(\hat{\beta})$. For this purpose, the matrix $B = R^{-1}$ may be found by solving the equation $RB = I$ by back-substitution. The matrix $X'X$, which might also be needed, is available in the form of $R'R$.

It is interesting to note that the expression $X'X = R'R$ is just an instance of the Cholesky decomposition of the moment matrix. In the previous chapter, the decomposition has been written as $X'X = LL'$ where L is a lower-triangular matrix; and it has been asserted that L is uniquely determined. The uniqueness is maintained by specifying that the diagonal elements of L are all positive. If this condition is not imposed on the matrix R, then $R' = LJ$, where J is a diagonal matrix whose nonzero elements have values of 1 and -1.

Presented below is an unadorned procedure for calculating the regression estimates $\hat{\beta}$ and $\hat{\sigma}^2$ using the Householder method for performing the Q–R decomposition of X. To accommodate a model with an intercept term, it is necessary to fill the leading column of X with units.

(8.70)
```
procedure QRregression(Tcap.k : integer;
                       var x, y, beta : matrix;
                       var varEpsilon : real);

  var
    i, j, t : integer;

  begin

    Householder(x, y, Tcap, k, 1);
```

```
    Backsolve(x, beta, y, k, 1);

    varEpsilon := 0.0;
    for t := k + 1 to Tcap do
      varEpsilon := varEpsilon + y[t, 1] * y[t, 1];
    varEpsilon := varEpsilon/(Tcap - k);

  end; {QRregression}
```

The procedure calls upon a subsidiary procedure for solving the equations $R\beta = Q_1'y$ by back-substitution:

(8.71)
```
  procedure Backsolve(var r, x, b : matrix;
                       n, q : integer);

    var
      i, j, k : integer;

  begin {Backsolve}

    for j := 1 to q do
      begin {j}
        for k := n downto 1 do
          begin {k}
            x[k, j] := b[k, j];
            for i := k + 1 to n do
              x[k, j] := x[k, j] - r[k, i] * x[i, j];
            x[k, j] := x[k, j]/r[k, k];
          end; {k}
      end; {j}

  end; {Backsolve}
```

The Normal Distribution and the Sampling Distributions

It is often appropriate to assume that the elements of the disturbance vector ε within the regression equations $y = X\beta + \varepsilon$ are distributed independently and identically according to a normal law. Under this assumption, the sampling distributions of the estimates may be derived and various hypotheses relating to the underlying parameters may be tested.

To denote that x is a normally distributed random variable with a mean of $E(x) = \mu$ and a dispersion matrix of $D(x) = \Sigma$, we shall write $x \sim N(\mu, \Sigma)$. A vector $z \sim N(0, I)$ with a mean of zero and a dispersion matrix of $D(z) = I$ is described as a standard normal vector. Any normal vector $x \sim N(\mu, \Sigma)$ can be standardised:

(8.72) If T is a transformation such that $T\Sigma T' = I$ and $T'T = \Sigma^{-1}$ and if $x \sim N(\mu, \Sigma)$, then $T(x - \mu) \sim N(0, I)$.

Associated with the normal distribution are a variety of so-called sampling distributions which occur frequently in problems of statistical inference. Amongst these are the chi-square distribution, the F distribution and the t distribution.

If $z \sim N(0, I)$ is a standard normal vector of n elements, then the sum of squares of its elements has a chi-square distribution of n degrees of freedom; and this is denoted this by $z'z \sim \chi^2(n)$. With the help of the standardising transformation of (8.72), it can be shown that,

(8.73) If $x \sim N(\mu, \Sigma)$ is a vector of order n, then $(x - \mu)'\Sigma^{-1}(x - \mu) \sim \chi^2(n)$.

The sum of any two independent chi-square variates is itself a chi-square variate whose degrees of freedom equal the sum of the degrees of freedom of its constituents. Thus,

(8.74) If $u \sim \chi^2(m)$ and $v \sim \chi^2(n)$ are independent chi-square variates of m and n degrees of freedom respectively, then $(u + v) \sim \chi^2(m + n)$ is a chi-square variate of $m + n$ degrees of freedom.

The ratio of two independent chi-square variates divided by their respective degrees of freedom has a F distribution which is completely characterised by these degrees of freedom. Thus

(8.75) If $u \sim \chi^2(m)$ and $v \sim \chi^2(n)$ are independent chi-square variates, then the variate $F = (u/m)/(v/n)$ has an F distribution of m and n degrees of freedom; and this is denoted by writing $F \sim F(m, n)$.

The sampling distribution which is most frequently used is the t distribution. A t variate is a ratio of a standard normal variate and the root of an independent chi-square variate divided by its degrees of freedom. Thus

(8.76) If $z \sim N(0, 1)$ and $v \sim \chi^2(n)$ are independent variates, then $t = z/\sqrt{(v/n)}$ has a t distribution of n degrees of freedom; and this is denoted by writing $t \sim t(n)$.

It is clear that $t^2 \sim F(1, n)$.

Hypothesis Concerning the Complete Set of Coefficients

We shall develop the common hypothesis tests of the classical model in two versions. The first version reflects the algebra of the Q–R decomposition and the second reflects that of the regression estimates which are obtained via the method of Gaussian elimination.

A start is made by considering the orthogonal decomposition of the disturbance vector $(y - X\beta) = \varepsilon \sim N(0, \sigma^2 I_T)$. Let $Q = [Q_r, Q_e]$ be the orthonormal matrix of equation (8.63) which is effective in reducing X to an upper-triangular matrix. Since $Q'Q = QQ' = I$, it follows that $Q'\varepsilon \sim N(0, \sigma^2 I_T)$; and, on partitioning, this becomes

$$\begin{bmatrix} Q_r' \\ Q_e' \end{bmatrix} \varepsilon \sim N\left(\begin{bmatrix} 0 \\ 0 \end{bmatrix}, \sigma^2 \begin{bmatrix} I_k & 0 \\ 0 & I_{T-k} \end{bmatrix} \right). \tag{8.77}$$

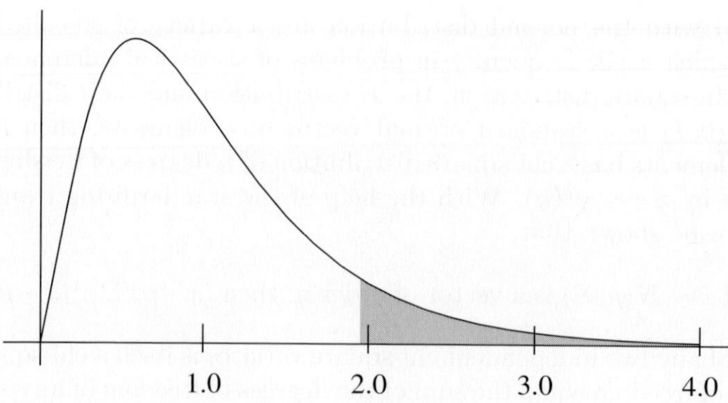

Figure 8.3. The critical region, at the 10% significance level, of an $F(5, 60)$ statistic.

Thus it can be seen that $Q_r'\varepsilon \sim N(0, \sigma^2 I_k)$ and $Q_e'\varepsilon \sim N(0, \sigma^2 I_{T-k})$ are uncorrelated normal vectors which are therefore statistically independent. From this result, it follows that the equation

$$\frac{\varepsilon'\varepsilon}{\sigma^2} = \frac{\varepsilon' Q_r Q_r'\varepsilon}{\sigma^2} + \frac{\varepsilon' Q_e Q_e'\varepsilon}{\sigma^2} \tag{8.78}$$

represents the decomposition of the chi-square variate $\varepsilon'\varepsilon/\sigma^2 \sim \chi^2(T)$ into two independent chi-square variates which are $\varepsilon' Q_r Q_r'\varepsilon/\sigma^2 \sim \chi^2(k)$ and $\varepsilon' Q_e Q_e'\varepsilon/\sigma^2 \sim \chi^2(T-k)$.

As an immediate corollary to this result, it can be deduced that the ratio

$$F = \left\{ \frac{\varepsilon' Q_r Q_r'\varepsilon}{k} \Bigg/ \frac{\varepsilon' Q_e Q_e'\varepsilon}{T-k} \right\} \tag{8.79}$$

has an $F(T-k, k)$ distribution.

To see how this ratio can be used in testing an hypothesis relating to the parameter vector β, consider applying the identities $Q_r'X = R$ and $Q_e'X = 0$ to the equations $Q_r'y = Q_r'X\beta + Q_r'\varepsilon$ and $Q_e'y = Q_e X\beta + Q_e'\varepsilon$. It will be found that

$$\begin{aligned} Q_r'y &= R\beta + Q_r'\varepsilon = R\hat{\beta}, \\ Q_e'y &= \quad Q_e'\varepsilon. \end{aligned} \tag{8.80}$$

The first of these indicates that $R(\hat{\beta} - \beta) = Q_r'\varepsilon$, from which it follows that $(\hat{\beta} - \beta)'R'R(\hat{\beta} - \beta) = \varepsilon' Q_r Q_r'\varepsilon$. The second indicates that $y' Q_e Q_e' y = \varepsilon' Q_e Q_e'\varepsilon$. It follows that a test the hypothesis that $\beta = \beta_\diamond$, where $\beta_\diamond$ is some specified value, can be performed by assessing the value of the statistic

$$F = \left\{ \frac{(\hat{\beta} - \beta_\diamond)'R'R(\hat{\beta} - \beta_\diamond)}{k} \Bigg/ \frac{y' Q_e Q_e' y}{T-k} \right\}, \tag{8.81}$$

which will be distributed as an $F(k, T-k)$ variate if the hypothesis is true. If the value of this F statistic falls in the critical region in the upper tail of the $F(k, T-k)$ distribution, then the hypothesis is liable to be rejected (see Figure 8.3).

An alternative expression for the statistic, which is compatible with the notation used in describing the method of Gaussian elimination, is

$$
\begin{aligned}
F &= \left\{ \frac{(\hat{\beta}-\beta_\diamond)'X'X(\hat{\beta}-\beta_\diamond)}{k} \bigg/ \frac{(y-X\hat{\beta})'(y-X\hat{\beta})}{T-k} \right\} \\
&= \frac{1}{\hat{\sigma}^2 k}(\hat{\beta}-\beta_\diamond)'X'X(\hat{\beta}-\beta_\diamond).
\end{aligned} \tag{8.82}
$$

This form of the statistic, which may be understood in reference to equations (8.47), (8.68) and (8.69), is more intelligible than the form under (8.81), since it indicates that the test is based on a measure of the distance between the hypothesised value $X\beta_\diamond$ of the systematic component of the regression and the value $X\hat{\beta}$ which is suggested by the data. If the two values are remote from each other, then we may suspect that the hypothesis is at fault.

In the case of the hypothesis $\beta_\diamond = 0$, the test statistic assumes a particularly simple form; for then the numerator of (8.81) becomes $y'Q_rQ_r'y = \hat{\beta}'R'R\hat{\beta} = \hat{\beta}'X'X\hat{\beta}$. However, it is unusual to postulate that all the elements of β are zeros. It is more usual to allow one nonzero element in association with a vector of units, which is tantamount to maintaining the hypothesis that the elements of the vector y have a common nonzero expected value.

Hypotheses Concerning a Subset of the Coefficients

It is usual to suppose that a subset of the elements of the parameter vector β are zeros. This represents an instance of a class of hypotheses which specify values for a subvector β_2 within the partitioned model $y = X_1\beta_1 + X\beta_2 + \varepsilon$ without asserting anything about the values of the remaining elements in the subvector β_1. An appropriate test statistic can be derived by refining the equations (8.80) so as to take account of the partitioning of $X = [X_1, X_2]$ wherein X_1 has k_1 columns and X_2 has $k_2 = k - k_1$ columns:

$$
\begin{aligned}
Q_{r1}'y &= R_{11}\beta_1 + R_{12}\beta_2 + Q_{r1}'\varepsilon = R_{11}\hat{\beta}_1 + R_{12}\hat{\beta}_2, \\
Q_{r2}'y &= \qquad\quad R_{22}\beta_2 + Q_{r2}'\varepsilon = \qquad\quad R_{22}\hat{\beta}_2, \\
Q_e'y &= \qquad\qquad\quad Q_e'\varepsilon.
\end{aligned} \tag{8.83}
$$

The second equation indicates that $R_{22}(\hat{\beta}_2 - \beta_2) = Q_{r2}'\varepsilon$. Since Q_{r2} is a matrix of k_2 orthonormal columns, it follows that $Q_{r2}'\varepsilon \sim N(0, \sigma^2 I)$, where I now stands for the identity matrix of order k_2. Therefore, $\varepsilon'Q_{r2}Q_{r2}'\varepsilon/\sigma^2 \sim \chi^2(k_2)$ is a chi-square variate which is independent of $\varepsilon'Q_eQ_e'\varepsilon/\sigma^2 = y'Q_eQ_e'y/\sigma^2 \sim \chi^2(T-k)$. It follows that the hypothesis that $\beta_2 = \beta_{2\diamond}$ can be tested by assessing the value of the statistic

$$
F = \left\{ \frac{(\hat{\beta}_2-\beta_{2\diamond})'R_{22}'R_{22}(\hat{\beta}_2-\beta_{2\diamond})}{k_2} \bigg/ \frac{y'Q_eQ_e'y}{T-k} \right\}, \tag{8.84}
$$

which will be distributed as an $F(k_2, T-k)$ variate if the hypothesis is true. In the case of the hypothesis that $\beta_2 = 0$, the numerator of this statistic can be rendered as $\hat{\beta}_2' R_{22}' R_{22} \hat{\beta}_2 / k_2 = y' Q_{r2} Q_{r2}' y / k_2$.

These results may be expressed in alternative forms which are more appropriate when the regression is computed via the method of Gaussian elimination. Consider the identity

$$\begin{aligned}\varepsilon'\varepsilon &= \varepsilon' Q_{r1} Q_{r1}' \varepsilon + \varepsilon' Q_{r2} Q_{r2}' \varepsilon + \varepsilon' Q_e Q_e' \varepsilon \\ &= \varepsilon' P_1 \varepsilon + \varepsilon'(P - P_1)\varepsilon + \varepsilon'(I - P)\varepsilon.\end{aligned} \tag{8.85}$$

For testing an hypothesis relating to β_2, the relevant term of this decomposition is

$$\begin{aligned}\varepsilon' Q_{r2} Q_{r2}' \varepsilon &= \varepsilon'(P - P_1)\varepsilon \\ &= (P\varepsilon)'(I - P_1)P\varepsilon.\end{aligned} \tag{8.86}$$

Given that $P\varepsilon = P(y - X\beta) = X\hat{\beta} - X\beta$ and that $(I - P_1)(X\hat{\beta} - X\beta) = (I - P_1)(X_2\hat{\beta}_2 - X_2\beta_2)$ since $(I - P_1)X_1 = 0$, it follows that

$$\varepsilon' Q_{r2} Q_{r2}' \varepsilon = (\hat{\beta}_2 - \beta_2)' X_2' (I - P_1) X_2 (\hat{\beta}_2 - \beta_2). \tag{8.87}$$

Therefore an alternative expression for the statistic for testing the hypothesis that $\beta_2 = \beta_{2\diamond}$ is

$$F = \frac{1}{\hat{\sigma}^2 k_2} (\hat{\beta}_2 - \beta_{2\diamond})' X_2' (I - P_1) X_2 (\hat{\beta}_2 - \beta_{2\diamond}). \tag{8.88}$$

Reference to (8.28) shows that the matrix $X_2'(I - P_1)X_2$ may be obtained by inverting a principal minor of the matrix $(X'X)^{-1}$. This a laborious operation compared with the ease with which $R_{22}' R_{22} = X_2'(I - P_1)X_2$ can be formed from the products of the Q–R decomposition of X.

A limiting case of the F statistic concerns the test of an hypothesis affecting a single element β_i within the vector β. By specialising the expression under (8.88), a statistic may be derived in the form of

$$F = \frac{(\hat{\beta}_i - \beta_{i\diamond})^2}{\hat{\sigma}^2 w_{ii}}, \tag{8.89}$$

wherein w_{ii} stands for the ith diagonal element of $(X'X)^{-1} = (R'R)^{-1}$. If the hypothesis is true, then this will be distributed according to the $F(1, T-k)$ law. However, the usual way of assessing such an hypothesis is to relate the value of the statistic

$$t = \frac{(\hat{\beta}_i - \beta_{i\diamond})}{\sqrt{(\hat{\sigma}^2 w_{ii})}} \tag{8.90}$$

to the tables of the $t(T-k)$ distribution. The advantage of the t statistic is that it shows the direction in which the estimate of β_i deviates from the hypothesised value as well as the size of the deviation.

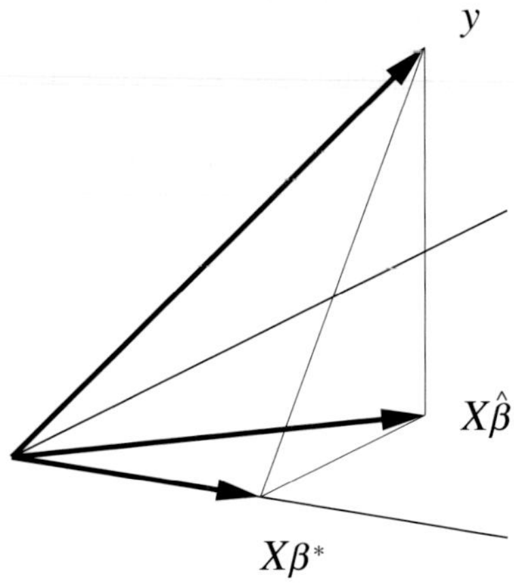

Figure 8.4. The test of the hypothesis $\beta_2 = \beta_{2\diamond}$ is based on a measure of the proximity of the restricted estimate $X\beta^*$, and the unrestricted estimate $X\hat{\beta}$. The USS is the squared distance $\|y - X\hat{\beta}\|^2$. The RSS is the squared distance $\|y - X\beta^*\|^2$.

An Alternative Formulation of the F statistic

An alternative way of forming the F statistic uses the products of two separate regressions. Consider the identity

$$\varepsilon'(P - P_1)\varepsilon = \varepsilon'(I - P_1)\varepsilon - \varepsilon'(I - P)\varepsilon. \tag{8.91}$$

The term of the LHS is the quadratic product which appears in the numerator of the F statistic of (8.84) and (8.88). The first term on the RHS can be written as

$$\begin{aligned}\varepsilon'(I - P_1)\varepsilon &= (y - X\beta)'(I - P_1)(y - X\beta)\\ &= (y - X_2\beta_2)'(I - P_1)(y - X_2\beta_2).\end{aligned} \tag{8.92}$$

Under the hypothesis that $\beta_2 = \beta_{2\diamond}$, the term amounts to the residual sum of squares from the regression of $y - X_2\beta_{2\diamond}$ on X_1. It may be described as the restricted residual sum of squares and denoted by RSS. The second term on the RHS of (8.91) is just the ordinary residual sum of squares

$$\begin{aligned}\varepsilon'(I - P)\varepsilon &= (y - X\beta)'(I - P)(y - X\beta)\\ &= y'(I - P)y.\end{aligned} \tag{8.93}$$

This may be obtained, equally, from the regression of y on X or from the regression of $y - X_2\beta_{2\diamond}$ on X; and it may be described as the unrestricted residual sum of squares and denoted by USS. From these considerations, it follows that the statistic for testing the hypothesis that $\beta_2 = \beta_{2\diamond}$ can also be expressed as

$$F = \left\{ \frac{RSS - USS}{k_2} \bigg/ \frac{USS}{T - k} \right\}. \tag{8.94}$$

As a matter of interpretation, it is interesting to note that the numerator of the F statistic is also the square of the distance between $X\beta^*$, which is the estimate of the systematic component from the restricted regression, and $X\hat{\beta}$, which is its estimate from the unrestricted regression (see Figure 8.4). The restricted estimate is

$$\begin{aligned} X\beta^* &= P_1(y - X_2\beta_{2\diamond}) + X_2\beta_{2\diamond} \\ &= P_1 y + (I - P_1)X_2\beta_{2\diamond}, \end{aligned} \tag{8.95}$$

and the unrestricted estimate is

$$\begin{aligned} X\hat{\beta} &= X_1\hat{\beta}_1 + X_2\hat{\beta}_2 \\ &= Py. \end{aligned} \tag{8.96}$$

The difference between the two estimates is

$$\begin{aligned} X\hat{\beta} - X\beta^* &= (P - P_1)y - (I - P_1)X_2\beta_{2\diamond} \\ &= (I - P_1)(Py - X_2\beta_{2\diamond}) \\ &= (I - P_1)(X_2\hat{\beta}_2 - X_2\beta_{2\diamond}). \end{aligned} \tag{8.97}$$

Here the final identity comes from the fact that $(I - P_1)X_1\hat{\beta}_1 = 0$. It then follows from the idempotency of $(I - P_1)$ that the square of the distance between $X\beta^*$ and $X\hat{\beta}$ is

$$(X\hat{\beta} - X\beta^*)'(X\hat{\beta} - X\beta^*) = (\hat{\beta}_2 - \beta_{2\diamond})'X_2'(I - P_1)X_2(\hat{\beta}_2 - \beta_{2\diamond}). \tag{8.98}$$

The expression on the RHS repeats the expression found in (8.88).

Bibliography

[99] Chambers, J.M., (1977), *Computational Methods for Data Analysis*, John Wiley and Sons, New York.

[177] Farebrother, R.W., (1988), *Linear Least Squares Computations*, Marcel Dekker, New York and Basel.

[228] Gregory, J., (1972), A Comparison of Floating Point Summation Methods, *Communications of the ACM*, **15**, 838.

[312] Lawson, C.L., and R.J. Hanson, (1974), *Solving Least Squares Problems*, Prentice-Hall, Englewood Cliffs, New Jersey.

[319] Ling, R.F., (1974), Comparison of Several Algorithms for Computing Sample Means and Variances, *Journal of the American Statistical Association*, **69**, 859–866.

[320] Linz, P., (1970), Accurate Floating-Point Summation, *Communications of the ACM*, **13**, 361–362.

[329] Malcolm, M.A., (1971), On Accurate Floating-Point Summation, *Communications of the ACM*, **14**, 731–736.

[358] Neeley, P.M., (1966), Comparison of Several Algorithms for Computation of Means, Standard Deviations and Correlation Coefficients, *Communications of the ACM*, **7**, 496–499.

[397] Pollock, D.S.G., (1979), *The Algebra of Econometrics*, John Wiley and Sons, Chichester.

[448] Seber, G.A.F., (1977), *Linear Regression Analysis*, John Wiley and Sons, New York.

[538] Youngs, E.A., and E.M. Cramer, (1971), Some Results Relevant to Choice of Sum and Sum-of-Product Algorithms, *Technometrics*, **13**, 657–665.

CHAPTER 9

Recursive Least-Squares Estimation

In this chapter, we shall develop recursive algorithms which facilitate the revision of least-squares estimates when new observations become available.

As Young [537] has observed, the theory of recursive least-squares estimation was first expounded by Gauss [204] in his original tract on the method of least squares. However, little attention was paid to this aspect of least-squares theory, which lay dormant for almost a century and a half before it was rediscovered on two separate occasions. The first rediscovery was by Plackett [395] in 1950, which was before the advent of efficient on-line electronic computing; and this also passed almost unnoticed. It was the second rediscovery of the recursive algorithms in 1960 in the context of control theory which was the cue to a rapid growth of interest. Stemming from the papers of Kalman [281] and Kalman and Bucy [282], a vast literature on Kalman filtering has since accumulated.

We shall begin the chapter by developing the recursive formulae for estimating the parameters of an ordinary regression model in the manner of Plackett. Then we shall develop more general versions of the formulae within the wider context of the model of Kalman.

Recursive Least-Squares Regression

Imagine that we have already calculated the ordinary least-squares estimate $\hat{\beta}_{t-1}$ of β in the model $(Y_{t-1}; X_{t-1}\beta, \sigma^2 I)$, where $Y'_{t-1} = [y_1, \ldots, y_{t-1}]$ is a vector of $t-1$ scalar observations and $X'_{t-1} = [x'_{1.}, \ldots, x'_{t-1.}]$ is a matrix of order $k \times (t-1)$ comprising $t-1$ successive observations of a vector of k explanatory variables. In this notation, $x_{t.} = [x_{t1}, \ldots, x_{tk}]$ stands for a *row* vector of k observations taken at the time t. Given the new information which is provided by the observations y_t, $x_{t.}$, we wish to form a revised or updated estimate of β in the manner which makes best use of the previous calculations.

The existing ordinary least-squares estimator $\hat{\beta}_{t-1}$ may be defined as the solution of the equation

$$X'_{t-1}X_{t-1}\hat{\beta}_{t-1} = X'_{t-1}Y_{t-1}, \tag{9.1}$$

which may be written as

$$M_{t-1}\hat{\beta}_{t-1} = q_{t-1}, \tag{9.2}$$

where $M_{t-1} = X'_{t-1}X_{t-1}$ and $q_{t-1} = X'_{t-1}Y_{t-1}$. If we define

$$M_t = M_{t-1} + x'_{t.}x_{t.} \quad \text{and} \quad q_t = q_{t-1} + x'_{t.}y_t, \tag{9.3}$$

then the equations from which the new estimate $\hat{\beta}_t$ is derived may be expressed as

$$M_t\hat{\beta}_t = q_t. \tag{9.4}$$

On the RHS of this expression, there is

$$\begin{aligned} q_t &= q_{t-1} + x'_{t.}y_t \\ &= (M_t - x'_{t.}x_{t.})\hat{\beta}_{t-1} + x'_{t.}y_t \\ &= M_t\hat{\beta}_{t-1} + x'_{t.}(y_t - x_{t.}\hat{\beta}_{t-1}). \end{aligned} \tag{9.5}$$

On putting the final expression into (9.4) and rearranging the result, we find that

$$\begin{aligned} \hat{\beta}_t &= \hat{\beta}_{t-1} + M_t^{-1}x'_{t.}(y_t - x_{t.}\hat{\beta}_{t-1}) \\ &= \hat{\beta}_{t-1} + \kappa_t(y_t - x_{t.}\hat{\beta}_{t-1}); \end{aligned} \tag{9.6}$$

and it can be seen that the updated estimate $\hat{\beta}_t$ differs from the previous estimate $\hat{\beta}_{t-1}$ by a function of the error $h_t = y_t - x_{t.}\hat{\beta}_{t-1}$ which comes from predicting y_t by $x_{t.}\hat{\beta}_{t-1}$.

The method by which the revised estimate of β is obtained may be described as a filtering process which maps the sequence of prediction errors into a sequence of revisions; and the vector $\kappa_t = M_t^{-1}x'_{t.}$ may be described as the gain of the filter. It is notable that, as the value of t increases, the values of the elements in M_t^{-1}, and therefore the values of those in κ_t, will decrease. Thus, the impact of successive prediction errors upon the values of the estimate of β will diminish as the amount of information already incorporated in the estimate increases.

The Matrix Inversion Lemma

The burden of computation can be eased by employing a scheme for calculating the inverse matrix M_t^{-1} by modifying the value of M_{t-1}^{-1}. The scheme depends upon the so-called matrix inversion lemma which provides an expression for the inverse of the matrix sum

$$A = C'DC + B, \tag{9.7}$$

wherein B and D are nonsingular matrices. To find the inverse, we may begin by premultiplying the sum by A^{-1} and postmultiplying it by B^{-1} which gives

$$B^{-1} = A^{-1}C'DCB^{-1} + A^{-1}. \tag{9.8}$$

Then, if we postmultiply by C', we get

$$\begin{aligned} B^{-1}C' &= A^{-1}C'DCB^{-1}C' + A^{-1}C' \\ &= A^{-1}C'D(CB^{-1}C' + D^{-1}), \end{aligned} \tag{9.9}$$

which leads to the identity

$$\begin{aligned}B^{-1}C'(CB^{-1}C'+D^{-1})^{-1}&=A^{-1}C'D\\&=(C'DC+B)^{-1}C'D.\end{aligned}\tag{9.10}$$

Postmultiplying again by CB^{-1} gives

$$B^{-1}C'(CB^{-1}C'+D^{-1})^{-1}CB^{-1}=A^{-1}C'DCB^{-1}.\tag{9.11}$$

When the expression on the LHS is used in equation (9.8), we can derive the identity

$$\begin{aligned}A^{-1}&=(C'DC+B)^{-1}\\&=B^{-1}-B^{-1}C'(CB^{-1}C'+D^{-1})^{-1}CB^{-1}.\end{aligned}\tag{9.12}$$

This formula can be applied directly to show that

$$\begin{aligned}M_t^{-1}&=(M_{t-1}+x'_{t.}x_{t.})^{-1}\\&=M_{t-1}^{-1}-M_{t-1}^{-1}x'_{t.}(x_{t.}M_{t-1}^{-1}x'_{t.}+1)^{-1}x_{t.}M_{t-1}^{-1}.\end{aligned}\tag{9.13}$$

Given that M_{t-1}^{-1} has a known value, the only inversion which is entailed in finding M_t^{-1} concerns the scalar quantity $1+x_{t.}M_{t-1}^{-1}x'_{t.}$.

The expression under (9.13) may be substituted into the formula for the recursive least-squares estimator $\hat{\beta}_t$ which is to be found under (9.6). In fact, the formula contains the factor $\kappa_t=M_t^{-1}x'_{t.}$. The identity under (9.10) serves to show that

$$\begin{aligned}\kappa_t&=M_t^{-1}x'_{t.}\\&=(M_{t-1}+x'_{t.}x_{t.})^{-1}x'_{t.}\\&=M_{t-1}^{-1}x'_{t.}(x_{t.}M_{t-1}^{-1}x'_{t.}+1)^{-1}.\end{aligned}\tag{9.14}$$

Using this expression in (9.6), we get

$$\hat{\beta}_t=\hat{\beta}_{t-1}+M_{t-1}^{-1}x'_{t.}(x_{t.}M_{t-1}^{-1}x'_{t.}+1)^{-1}(y_t-x_{t.}\hat{\beta}_{t-1}).\tag{9.15}$$

Prediction Errors and Recursive Residuals

Consider more closely the error of predicting y_t as $x_{t.}\hat{\beta}_{t-1}$. Let the vector of the disturbances, which are independently and identically distributed, be written as $\mathcal{E}_{t-1}=[\varepsilon_1,\ldots,\varepsilon_{t-1}]'$. Then the prediction error is

$$\begin{aligned}y_t-x_{t.}\hat{\beta}_{t-1}&=y_t-x_{t.}(X'_{t-1}X_{t-1})^{-1}X'_{t-1}Y_{t-1}\\&=(x_{t.}\beta+\varepsilon_t)-x_{t.}(X'_{t-1}X_{t-1})^{-1}X'_{t-1}(X_{t-1}\beta+\mathcal{E}_{t-1})\\&=\varepsilon_t-x_{t.}M_{t-1}^{-1}X'_{t-1}\mathcal{E}_{t-1}\\&=h_t.\end{aligned}\tag{9.16}$$

We shall assume that there is no prior information about the parameter vector β. Then, if the $t \times k$ matrix M_t has $\text{rank}(M_t) = \min\{t, k\}$, the first k observations can used in forming an initial estimate of β. Given that $E(\varepsilon_t) = 0$ for all t, and assuming that the recursion starts at $t = k + 1$, it is clear that $E(h_t) = 0$ for all $t \geq k + 1$. Also, it follows from (9.16) that

$$\text{(9.17)} \qquad \begin{aligned} V(h_t) &= V(\varepsilon_t) + x_{t.}M_{t-1}^{-1}X_{t-1}'D(\mathcal{E}_{t-1})X_{t-1}M_{t-1}^{-1}x_{t.}' \\ &= \sigma^2(1 + x_{t.}M_{t-1}^{-1}x_{t.}'), \end{aligned}$$

since $D(\mathcal{E}_{t-1}) = \sigma^2 I_{t-1}$ and $C(\varepsilon_t, \mathcal{E}_{t-1}) = 0$.

The prediction errors are uncorrelated. The covariance of the errors h_t and h_s is given by

$$\text{(9.18)} \quad C(h_t, h_s) = E\left[\left\{\varepsilon_t - x_{t.}M_{t-1}^{-1}\sum_{j=1}^{t-1}x_{j.}'\varepsilon_j\right\}\left\{\varepsilon_s - x_{s.}M_{s-1}^{-1}\sum_{l=1}^{s-1}x_{l.}'\varepsilon_l\right\}'\right].$$

When $t < s$, the terms of the product, disregarding their signs, are

$$\text{(9.19)} \qquad \begin{aligned} &E(\varepsilon_t\varepsilon_s) = 0, \\ &E\left[\left\{x_{t.}M_{t-1}^{-1}\textstyle\sum_{j=1}^{t-1}x_{j.}'\varepsilon_t\right\}\varepsilon_s\right] = 0, \\ &E\left[\varepsilon_t\left\{x_{s.}M_{s-1}^{-1}\textstyle\sum_{l=1}^{s-1}x_{l.}'\varepsilon_l\right\}'\right] = \sigma^2 x_{s.}M_{s-1}^{-1}x_{t.}', \\ &E\left[\left\{x_{t.}M_{t-1}^{-1}\textstyle\sum_{j=1}^{t-1}x_{j.}'\varepsilon_j\right\}\left\{x_{s.}M_{s-1}^{-1}\textstyle\sum_{l=1}^{s-1}x_{l.}'\varepsilon_l\right\}'\right] = \sigma^2 x_{t.}M_{s-1}^{-1}x_{s.}'. \end{aligned}$$

Taking account of the signs of the terms, we find that

$$\text{(9.20)} \qquad C(h_t, h_s) = 0.$$

Imagine now that the disturbances ε_t are distributed independently, identically and normally for all t. Then it follows that the standardised prediction errors $\omega_t = \sigma^2 h_t/V(h_t)$ are also distributed normally, independently and identically with a variance of σ^2.

We can also write a decomposition of the residual sum of squares in terms of the prediction errors. Consider the following identities:

$$\text{(9.21)} \qquad \begin{aligned} S_t &= (Y_t - X_t\hat{\beta}_t)'(Y_t - X_t\hat{\beta}_t) \\ &= (Y_t - X_t\hat{\beta}_{t-1})'(Y_t - X_t\hat{\beta}_{t-1}) - (\hat{\beta}_t - \hat{\beta}_{t-1})'X_t'X_t(\hat{\beta}_t - \hat{\beta}_{t-1}) \\ &= \left\{S_{t-1} + (y_t - x_{t.}\hat{\beta}_{t-1})^2\right\} - x_{t.}(X_t'X_t)^{-1}x_{t.}'(y_t - x_{t.}\hat{\beta}_{t-1})^2. \end{aligned}$$

Here the second equality depends upon the identity $X_t'Y_t = X_t'X_t\hat{\beta}_t$, whilst the final equality uses the identity from (9.6). On rearranging the terms, we get

$$\text{(9.22)} \qquad S_t = S_{t-1} + \left(1 - x_{t.}M_t^{-1}x_{t.}'\right)(y_t - x_{t.}\hat{\beta}_{t-1})^2.$$

Here the recursion starts when $t = k + 1$ with $S_k = 0$ as the initial condition.

Now observe that (9.14) implies that

$$1 - x_{t.}M_t^{-1}x'_{t.} = \frac{1}{x_{t.}M_{t-1}^{-1}x'_{t.} + 1}. \tag{9.23}$$

It follows that (9.22) may be rewritten as

$$\begin{aligned} S_t &= S_{t-1} + \frac{(y_t - x_{t.}\hat{\beta}_{t-1})^2}{x_{t.}M_{t-1}^{-1}x'_{t.} + 1} \\ &= S_{t-1} + w_t^2 = \sum_{j=k+1}^{t} w_j^2. \end{aligned} \tag{9.24}$$

In interpreting this formula, we should note that $\sigma^2 M_{t-1}^{-1} = D(\hat{\beta}_{t-1})$ is the dispersion matrix of the pre-existing estimate of β, whilst $\sigma^2(1 + x_{t.}M_{t-1}^{-1}x'_{t.}) = V(h_t)$ is the variance of the prediction error $h_t = y_t - x_{t.}\hat{\beta}_{t-1}$. It follows that $w_{k+1}, w_{k+2}, \ldots, w_t$ is a sequence of uncorrelated errors with $E(w_j) = 0$ and $E(w_j) = \sigma^2$ for all j. These are commonly described as recursive residuals—see Brown *et al.* [82].

The Updating Algorithm for Recursive Least Squares

At this stage, it is appropriate to provide the code which will serve to generate the updated values M_t^{-1} and $\hat{\beta}_t$ from the previous values M_{t-1}^{-1} and $\hat{\beta}_{t-1}$ and from the observations y_t and $x_{t.}$. The algorithm may be generalised slightly by the inclusion of an additional parameter λ_t, together with a choice of sign, whose purpose will become apparent in subsequent sections. The updated values may be obtained via the following sequence of computations:

$$\begin{aligned}
&\text{(i)} & h_t &= y_t - x_{t.}\hat{\beta}_{t-1}, \\
&\text{(ii)} & g_t &= M_{t-1}^{-1}x'_{t.}, \\
&\text{(iii)} & f_t &= x_{t.}g_t \pm \lambda_t \\
& & &= x_{t.}M_{t-1}^{-1}x'_{t.} \pm \lambda_t, \\
&\text{(iv)} & \kappa_t &= g_t f_t^{-1} \\
& & &= M_{t-1}^{-1}x'_{t.}(x_{t.}M_{t-1}^{-1}x'_{t.} \pm \lambda_t)^{-1}, \\
&\text{(v)} & \hat{\beta}_t &= \hat{\beta}_{t-1} + \kappa_t h_t \\
& & &= \hat{\beta}_{t-1} + \kappa_t(y_t - x_{t.}\hat{\beta}_{t-1}), \\
&\text{(vi)} & M_t^{-1} &= \frac{1}{\lambda}\Big\{M_{t-1}^{-1} - \kappa_t g'_t\Big\} \\
& & &= \frac{1}{\lambda}\Big\{M_{t-1}^{-1} - M_{t-1}^{-1}x'_{t.}(x_{t.}M_{t-1}^{-1}x'_{t.} \pm \lambda_t)^{-1}x_{t.}M_{t-1}^{-1}\Big\}.
\end{aligned} \tag{9.25}$$

In the following code, which implements these computations, the elements of the matrix M^{-1} are contained in the array P.

(9.26)
```
procedure RLSUpdate(x : vector;
                    k, sign : integer;
                    y, lambda : real;
                    var h : real;
                    var beta, kappa : vector;
                    var p : matrix);

  var
    f : real;
    g : vector;
    i, j : integer;

begin {RLSUpdate}
  h := y;
  f := sign * lambda;

  for i := 1 to k do
    begin {i}
      g[i] := 0.0;
      for j := 1 to k do
        g[i] := g[i] + p[i, j] * x[j];
      f := f + g[i] * x[i];
      h := h - x[i] * beta[i];
    end; {i}

  for i := 1 to k do
    begin {i}
      kappa[i] := g[i]/f;
      beta[i] := beta[i] + kappa[i] * h;
      for j := i to k do
        begin
          p[i, j] := (p[i, j] - kappa[i] * g[j])/lambda;
          p[j, i] := p[i, j];
        end;
    end; {i}

end; {RLSUpdate}
```

Experience with this algorithm indicates that it is sensitive to the effects of rounding error which occur when two quantities of the same sign are subtracted. It is also possible that the computed values of the matrix M^{-1}, or P as it is represented in the code, might lose the property of positive-definiteness. This may occur if some of the values of $\hat{\beta}$ become virtually constant in consequence of an abundance of data.

To avert such problems, one may use the so-called square-root filter which is commonly regarded as being, numerically, the most stable updating algorithm. The square-root algorithm depends upon a factorisation of the form $M^{-1} = SS'$ which enables one to write the updated moment matrix

$$M_t^{-1} = \frac{1}{\lambda}\Big\{M_{t-1}^{-1} - M_{t-1}^{-1}x_{t.}'(x_{t.}M_{t-1}^{-1}x_{t.}' + \lambda_t)^{-1}x_{t.}M_{t-1}^{-1}\Big\} \tag{9.27}$$

as

$$\begin{aligned} S_tS_t' &= \frac{1}{\lambda}S_{t-1}\Big\{I - S_{t-1}'x_{t.}'(x_{t.}S_{t-1}S_{t-1}'x_{t.}' + \lambda_t)^{-1}x_{t.}S_{t-1}\Big\}S_{t-1}' \\ &= \frac{1}{\lambda}S_{t-1}\Big\{I - \frac{g_tg_t'}{g_t'g_t + \lambda_t}\Big\}S_{t-1}', \end{aligned} \tag{9.28}$$

where $g_t = S_{t-1}'x_{t.}'$. Using the factorisation

$$I - \frac{gg'}{g'g + \lambda} = (I - \alpha gg')^2, \tag{9.29}$$

one may form the updated value of S according to

$$S_t = \frac{1}{\sqrt{\lambda}}S_{t-1}(I - \alpha_t g_t g_t'). \tag{9.30}$$

To find the value for the scalar α, one must solve a quadratic equation in the form of

$$\begin{aligned} &\alpha^2 g'g - 2\alpha + (\lambda + g'g)^{-1} = 0 \quad \text{or, equivalently,} \\ &\alpha^2(f - \lambda) - 2\alpha + f^{-1} = 0, \end{aligned} \tag{9.31}$$

where $f = \lambda + g'g$. The solution is

$$\alpha = \frac{1 \pm \sqrt{\lambda f^{-1}}}{f - \lambda} = \frac{1}{f \pm \sqrt{f\lambda}}; \tag{9.32}$$

and, to avoid cancellation, one should take the positive square root in the final expression.

The updated values S_t, β_t may be obtained from the previous values S_{t-1}, β_{t-1} and from the observations y_t, $x_{t.}$ by pursuing the following sequence of computations:

(9.33)

$$\begin{aligned}
&\text{(i)} & h_t &= y_t - x_{t.}\hat{\beta}_{t-1},\\
&\text{(ii)} & g_t &= S'_{t-1}x'_{t.},\\
&\text{(iii)} & f_t &= \lambda_t + g'_t g_t\\
& & &= \lambda_t + x_{t.}M^{-1}_{t-1}x'_{t.},\\
&\text{(iv)} & \rho_t &= S_{t-1}g_t\\
& & &= M^{-1}_{t-1}x'_{t.},\\
&\text{(v)} & \hat{\beta}_t &= \hat{\beta}_{t-1} + \rho_t f_t^{-1}h_t\\
& & &= \hat{\beta}_{t-1} + \kappa_t(y_t - x_{t.}\hat{\beta}_{t-1}),\\
&\text{(vi)} & \alpha_t &= (f + \sqrt{f\lambda})^{-1},\\
&\text{(vii)} & \sqrt{\lambda}S_t &= S_{t-1} - \alpha_t\rho_t g'_t\\
& & &= S_{t-1}(I - \alpha_t g_t g'_t).
\end{aligned}$$

The computations are implemented in the following code:

(9.34)

```
procedure SqrtUpdate(x : vector;
                     k : integer;
                     y, lambda : real;
                     var h : real;
                     var beta, kappa : vector;
                     var s : matrix);

  var
    f, alpha, sqrtlambda : real;
    g, rho : vector;
    i, j : integer;

begin {RLSUpdate}
  h := y;
  f := lambda;
  for i := 1 to k do
    begin {i}
      g[i] := 0.0;
      for j := 1 to k do
        g[i] := g[i] + s[j, i] * x[j];
      f := f + g[i] * g[i];
      h := h - x[i] * beta[i];
    end; {i}

  alpha := 1/(f + Sqrt(f * lambda));
  sqrtlambda := Sqrt(lambda);
```

```
for i := 1 to k do
  begin {i}
    rho[i] := 0;
    for j := 1 to k do
      rho[i] := rho[i] + s[i, j] * g[j];
    kappa[i] := rho[i]/f;
    beta[i] := beta[i] + rho[i] * h/f;
    for j := 1 to k do
      S[i, j] := (S[i, j] - alpha * rho[i] * g[j])/sqrtlambda;
  end; {i}

end; {SqrtUpdate}
```

Initiating the Recursion

It is necessary to specify some starting values for the recursive least-squares algorithm. Here we have wide discretion. If the object is simply to replicate the values of the ordinary least-squares estimates of $\beta = [\beta_1, \ldots, \beta_k]'$ for each value of the sample size in excess of $t = k$, then we must begin the recursion at the point $t = k + 1$ using the initial values

(9.35)
$$\begin{aligned}\hat{\beta}_k &= (X_k'X_k)^{-1}X_k'Y_k \\ &= X_k^{-1}Y_k \quad \text{and} \\ M_k &= X_k'X_k.\end{aligned}$$

Here it is assumed that $\text{rank}(X_k) = k$, which is to say that the $k \times k$ matrix $X_k' = [x_{1.}', \ldots, x_{k.}']$ is nonsingular and is therefore capable of generating an estimate.

On the other hand, we may be prepared to attribute to β a prior value $\hat{\beta}_0$, even before making any observations. This can be done by attributing to the parameter vector a complete prior probability density function with $\hat{\beta}_0$ as the expected value. In that case, a dispersion matrix $\sigma^2 M_0^{-1} = D(\hat{\beta}_0 - \beta)$ must also be specified. If there is doubt about the accuracy of $\hat{\beta}_0$, then large values should be given to the elements of $D(\hat{\beta}_0 - \beta)$. In this way, the prior assumptions are prevented from having too great an effect upon the subsequent estimates of β.

The business of incorporating the prior assumptions into the initial recursive estimates is straightforward in the case of a normal prior probability density function; for it is simply a matter of estimating the parameter β in the system

(9.36)
$$\begin{bmatrix}\hat{\beta}_0 \\ y_1\end{bmatrix} = \begin{bmatrix}I_k \\ x_{1.}\end{bmatrix}\beta + \begin{bmatrix}\eta_0 \\ \varepsilon_1\end{bmatrix},$$

where $\eta_0 = \hat{\beta}_0 - \beta$. The dispersion matrix for the combined disturbance term is

(9.37)
$$D\begin{bmatrix}\eta_0 \\ \varepsilon_1\end{bmatrix} = \sigma^2\begin{bmatrix}M_0^{-1} & 0 \\ 0 & 1\end{bmatrix}.$$

For a regression equation in the form of $g = W\beta + \varepsilon$, where ε has $D(\varepsilon) = \sigma^2 Q$, the efficient generalised least-squares estimator of β is given by

$$\hat{\beta} = (W'Q^{-1}W)^{-1}W'Q^{-1}g, \tag{9.38}$$

whilst the dispersion of the estimator is given by

$$D(\hat{\beta}) = \sigma^2(W'Q^{-1}W)^{-1}. \tag{9.39}$$

It follows that the efficient estimator of β in equation (9.36) is given by

$$\begin{aligned}\hat{\beta}_1 &= (M_0 + x'_{1.}x_{1.})^{-1}(M_0\hat{\beta}_0 + x'_{1.}y_1)\\ &= (M_0 + x'_{1.}x_{1.})^{-1}(q_0 + x'_{1.}y_1)\\ &= M_1^{-1}q_1.\end{aligned} \tag{9.40}$$

This is the estimator which one might expect in view of equations (9.3) and (9.4).

Estimators with Limited Memories

The form of the ordinary least-squares estimator indicates that the data comprised in the estimate $\hat{\beta}$ are equally weighted, with the effect that recent data and ancient data are valued equally. This is appropriate if the process generating the data is invariant. However, if there is doubt about the constancy of the regression parameters, then it may be desirable to give greater weight to the more recent data; and it might even be appropriate to discard data which has reached a certain age and has passed its retirement date.

The simplest way of accommodating parametric variability is to base the estimate on only the most recent portion of the data. As each new observation is acquired, another observation may be removed; so that, at any instant, the estimator comprises only n data points.

The recursive updating of the estimate can be accomplished in two stages. Imagine that an estimate calculated at time $t-1$ is at hand which is given by

$$\hat{\beta}_{t-1} = M_{t-1}^{-1}q_{t-1}, \tag{9.41}$$

where

$$M_{t-1} = \sum_{j=1}^{n} x'_{t-j.}x_{t-j.} \tag{9.42}$$

and

$$q_{t-1} = \sum_{j=1}^{n} x'_{t-j.}y_{t-j}. \tag{9.43}$$

The first step is to remove the data which was acquired at time $t-n$. Let

$$M_t^* = M_{t-1} - x'_{t-n.}x_{t-n.} \tag{9.44}$$

and

$$q_t^* = q_{t-1} - x'_{t-n.}y_{t-n}. \tag{9.45}$$

Then an intermediate estimate is defined by

$$\hat{\beta}_t^* = M_t^{*-1}q_t^*. \tag{9.46}$$

Here the term q_t^* may be expressed as follows:

$$\begin{aligned} q_t^* &= q_{t-1} - x'_{t-n.}y_{t-n} \\ &= M_{t-1}\hat{\beta}_{t-1} - x'_{t-n.}y_{t-n} \\ &= (M_t^* + x'_{t-n.}x_{t-n.})\hat{\beta}_{t-1} - x'_{t-n.}y_{t-n} \\ &= M_t^*\hat{\beta}_{t-1} - x'_{t-n.}(y_{t-n} - x_{t-n.}\hat{\beta}_{t-1}). \end{aligned} \tag{9.47}$$

Therefore, the intermediate estimator is given by

$$\begin{aligned} \hat{\beta}_t^* &= \hat{\beta}_{t-1} - M_t^{*-1}x'_{t-n.}(y_{t-n} - x_{t-n.}\hat{\beta}_{t-1}) \\ &= \hat{\beta}_{t-1} + M_{t-1}^{-1}x'_{t-n.}(x_{t-n.}M_{t-1}^{-1}x'_{t-n.} - 1)^{-1}(y_{t-n} - x_{t-n.}\hat{\beta}_{t-1}), \end{aligned} \tag{9.48}$$

where the second equality is by virtue of the identity

$$\begin{aligned} M_t^{*-1}x'_{t-n.} &= (M_{t-1} - x'_{t-n.}x_{t-n.})x'_{t-n.} \\ &= -M_{t-1}^{-1}x'_{t-n.}(x_{t-n.}M_{t-1}^{-1}x'_{t-n.} - 1)^{-1}, \end{aligned} \tag{9.49}$$

which can be demonstrated using (9.10).

The term M_t^{*-1} in isolation is found by applying the matrix inversion formula of (9.12) to (9.44):

$$M_t^{*-1} = M_{t-1}^{-1} - M_{t-1}^{-1}x'_{t-n.}(x_{t-n.}M_{t-1}^{-1}x'_{t-n.} - 1)^{-1}x_{t-n.}M_{t-1}^{-1}. \tag{9.50}$$

In the second stage, the new information is included. Let

$$M_t = M_t^* + x'_{t.}x_{t.} \tag{9.51}$$

and

$$q_t = q_{t-1}^* + x'_{t.}y_t. \tag{9.52}$$

Then the updated estimator is defined by

$$\hat{\beta}_t = M_t^{-1}q_t. \tag{9.53}$$

The latter is calculated as

$$\hat{\beta}_t = \hat{\beta}_t^* + M_t^{*-1}x'_{t.}(x_{t.}M_t^{*-1}x'_{t.} + 1)^{-1}(y_t - x_{t.}\hat{\beta}_t^*), \tag{9.54}$$

whilst the inverse of the moment matrix is calculated according to

$$M_t^{-1} = M_t^{*-1} - M_t^{*-1}x_{t.}'(x_{t.}M_t^{*-1}x_{t.}' + 1)^{-1}x_{t.}M_t^{*-1}. \tag{9.55}$$

The two updating steps of rolling regression, depicted respectively by the equations (9.48) and (9.54), can both be realised via the *RLSUpdate* procedure of (9.26).

Discarding observations which have passed a date of expiry is an appropriate procedure when the processes generating the data are liable, from time to time, to undergo sudden structural changes. For it ensures that any misinformation which is conveyed by data which predate a structural change will not be kept on record permanently. However, if the processes are expected to change gradually in a more or less systematic fashion, then a gradual discounting of old data may be more appropriate. An exponential weighting scheme might serve this purpose.

Let the cross-product matrices of the discounted data be represented by

$$M_t = \sum_{j=0}^{t-1} \lambda^j x_{t-j.}'x_{t-j.} + \lambda^t M_0 \tag{9.56}$$

and

$$q_t = \sum_{j=0}^{t-1} \lambda^j x_{t-j.}'y_{t-j} + \lambda^t q_0. \tag{9.57}$$

Then the corresponding estimate of β at time t may be defined, once more, by an equation in the form of (9.53). However, if $M_0 = 0$ and $q_0 = 0$, then the estimator, which cannot be calculated before $t = k$, can also be written in the form of a generalised least-squares estimator:

$$\hat{\beta}_t = (X_t'\Lambda X_t)^{-1}X_t'\Lambda Y_t, \tag{9.58}$$

with $\Lambda = \mathrm{diag}(\lambda^{t-1}, \ldots, \lambda, 1)$.

The moment matrices can be expressed in an incremental form. Consider subtracting

$$\lambda M_{t-1} = \sum_{j=1}^{t-1} \lambda^j x_{t-j.}'x_{t-j.} + \lambda^t M_0 \tag{9.59}$$

from M_t defined above in (9.56). This gives $x_{t.}'x_{t.} = M_t - \lambda M_{t-1}$ or

$$M_t = x_{t.}'x_{t.} + \lambda M_{t-1}. \tag{9.60}$$

Likewise, by subtracting

$$\lambda q_{t-1} = \sum_{j=1}^{t-1} \lambda^j x_{t-j.}'q_{t-j} + \lambda^t q_0 \tag{9.61}$$

from q_t, one finds that

$$q_t = x'_{t.}y_t + \lambda q_{t-1}. \tag{9.62}$$

To derive a recursive formula, we may write the estimator in the form of

$$\hat{\beta}_t = \hat{\beta}_{t-1} + M_t^{-1}x'_{t.}(y_t - x_{t.}\hat{\beta}_{t-1}), \tag{9.63}$$

which is familiar from equation (9.6). Then we may use the identities of (9.10) and (9.12) to show that

$$\begin{aligned} M_t^{-1} &= (\lambda M_{t-1} + x'_{t.}x_{t.})^{-1} \\ &= \frac{1}{\lambda}\Big\{M_{t-1}^{-1} - M_{t-1}^{-1}x'_{t.}(x_{t.}M_{t-1}^{-1}x'_{t.} + \lambda)^{-1}x_{t.}M_{t-1}^{-1}\Big\}, \end{aligned} \tag{9.64}$$

and that

$$M_t^{-1}x'_{t.} = M_{t-1}^{-1}x'_{t.}(x_{t.}M_{t-1}^{-1}x'_{t.} + \lambda)^{-1}. \tag{9.65}$$

The latter may be used in equation (9.63) to give

$$\hat{\beta}_t = \hat{\beta}_{t-1} + M_{t-1}^{-1}x'_{t.}(x_{t.}M_{t-1}^{-1}x'_{t.} + \lambda)^{-1}(y_t - x_{t.}\hat{\beta}_{t-1}). \tag{9.66}$$

We should end this section by noting that it is possible to combine the two memory processes which we have described by applying an exponential weighting scheme to the recent data and by discarding data which has become too old.

The Kalman Filter

We turn now to the topic of the Kalman filter. This may be regarded as a natural extension of the preceding topic which is the recursive estimation of the classical regression model; and it should be possible to build upon the results which are already established. However, in the ensuing sections, we shall provide a self-contained account of the Kalman filter; and a new notation will be adopted.

The Kalman filter is a system which accommodates a very wide range of models; and by adopting a new notation one can avoid making references automatically to the regression model. As will be seen in an example, only a few elaborations are needed to develop the regression model and the associated system of recursive estimation into a fully-fledged Kalman system. The consequence is that some of the recursive formulae which have been derived in the context of the regression model—such as those under (9.63)–(9.66)—will reappear in the more general context of the Kalman filter.

The technique of Kalman filtering depends upon a model consisting of two vector equations. The first equation describes the evolution of a vector ξ_t whose elements record the state of a system at a point in time. This so-called state-transition equation has the form of

$$\xi_t = \Phi_t\xi_{t-1} + \nu_t, \tag{9.67}$$

wherein Φ_t is the transition matrix and ν_t is a vector of stochastic disturbances which is independent of ξ_{t-1} and which has $E(\nu_t) = 0$ and $D(\nu_t) = \Psi_t$. It may be assumed that the values of Φ_t and Ψ_t are known.

In general, the state variables will not be directly observable. Instead, the information on the state of the system is conveyed by a vector of observations y_t which is related to the state vector ξ_t via the measurement equation

$$y_t = H_t\xi_t + \eta_t. \tag{9.68}$$

This is the second of the two equations. Here H_t, which has a known value, is the so-called measurement matrix and η_t is a vector of measurement errors. It is assumed that η_t is independent of ξ_t and that it has $E(\eta_t) = 0$ and $D(\eta_t) = \Omega_t$, where Ω_t is known.

In many applications, the quantities Φ_t, Ψ_t, H_t and Ω_t will be constant, and their temporal subscripts may be deleted.

The aim of the Kalman filter is to estimate the state vector ξ_t. A process of estimation which keeps pace with the data by generating an estimate of the current state vector ξ_t with each new observation y_t is described as *filtering*. The retrospective enhancement of a state estimate using data which has arisen subsequently is described as *smoothing*. The estimation of a future state vector is described as *prediction*. We shall treat each of these matters in turn.

Example 9.1. An example of the equations (9.67) and (9.68) is provided by a regression model with time-varying coefficients. The equations of this model at time t are

$$\beta_t = \Phi\beta_{t-1} + \nu_t, \tag{9.69}$$

$$y_t = x_{t.}\beta_t + \varepsilon_t, \tag{9.70}$$

where ν_t is a random variable with $V(\nu_t) = \lambda$ for all t. The first of these equations, which indicates that the regression coefficients follow a first-order vector autoregressive process, corresponds to the state-transition equation (9.67). The second equation, which is in the form of the ordinary regression equation, corresponds to the measurement equation (9.68).

The flexibility of the state-space formulation is demonstrated when equation (9.69) is replaced by the more general equation

$$\beta_t - \mu = \Phi(\beta_{t-1} - \mu) + \nu_t. \tag{9.71}$$

This specification is to be preferred whenever it is reasonable to assume that the distribution of β_t is centred on a nonzero value of μ. By defining $\delta_t = \beta_t - \mu$ and $\mu_t = \mu$ for all t, we can write the system comprising equations (9.70) and (9.71) as

$$\begin{bmatrix} \mu_t \\ \delta_t \end{bmatrix} = \begin{bmatrix} I & 0 \\ 0 & \Phi \end{bmatrix} \begin{bmatrix} \mu_{t-1} \\ \delta_{t-1} \end{bmatrix} + \begin{bmatrix} 0 \\ \nu_t \end{bmatrix}, \tag{9.72}$$

$$y_t = \begin{bmatrix} x_{t.} & x_{t.} \end{bmatrix} \begin{bmatrix} \mu_t \\ \delta_t \end{bmatrix} + \varepsilon_t. \tag{9.73}$$

Filtering

The object of Kalman filtering is to find unbiased estimates of the sequence of the state vectors ξ_t via a recursive process of estimation. The process starts at time $t = 1$; and it is assumed that prior information on the previous state vector ξ_0 is available in the form of an unbiased estimate x_0 which has been drawn from a distribution with a mean of ξ_0 and a dispersion matrix of P_0.

If the process described by equation (9.68) is stationary, then we should set $x_0 = E(\xi_0) = 0$, which is the unconditional expectation; and it should be possible to infer the corresponding dispersion matrix P_0 from the other parameters of the model.

If the process is nonstationary, then it may be necessary to estimate the initial state vector from data which is set aside for the purpose. At the same time, large values should be attributed to the elements of P_0 to reflect the low precision of the initial estimate. In the terminology of Bayesian statistics, this is a matter of attributing a diffuse prior distribution to ξ_0; and, recently, the procedure has been formalised in a model described as the diffuse Kalman filter—see De Jong [148] and Ansley and Kohn [25].

In each time period, new information on the system is provided by the vector y_t; and estimates of ξ_t may be formed both before and after the receipt of this information. The estimate of the state at time t formed without a knowledge of y_t will be denoted by $x_{t|t-1}$; whilst the estimate which incorporates the information of y_t will be denoted by x_t.

In the absence of the information of y_t, the estimate $x_{t|t-1}$ of ξ_t comes directly from equation (9.67) when ξ_{t-1} is replaced by x_{t-1} and ν_t is replaced by $E(\nu_t) = 0$. Thus

$$x_{t|t-1} = \Phi_t x_{t-1}. \tag{9.74}$$

The mean-square-error dispersion matrix of this estimator will be denoted by

$$P_{t|t-1} = E\{(\xi_t - x_{t|t-1})(\xi_t - x_{t|t-1})'\}, \tag{9.75}$$

whilst that of the updated estimator x_t will be denoted by

$$P_t = E\{(\xi_t - x_t)(\xi_t - x_t)'\}. \tag{9.76}$$

These dispersion matrices may be given a classical interpretation by considering ξ_t to have a fixed unknown value and by imagining its estimates $x_{t|t-1}$ and x_t to be subject to sampling variability—see Duncan and Horn [163], for example. In a subsequent Bayesian reinterpretation, ξ_t becomes a random variable with $x_{t|t-1}$ and x_t as its conditional expectations.

To derive the expression for $P_{t|t-1}$ in terms of P_{t-1}, we subtract equation (9.74) from equation (9.67) to give

$$\xi_t - x_{t|t-1} = \Phi_t(\xi_{t-1} - x_{t-1}) + \nu_t. \tag{9.77}$$

Then, since $\xi_{t-1} - x_{t-1}$ and ν_t are statistically independent, and since $E(\nu_t \nu_t') = D(\nu_t) = \Psi_t$, it follows that

$$P_{t|t-1} = \Phi_t P_{t-1} \Phi_t' + \Psi_t. \tag{9.78}$$

Before learning its value, we may predict y_t from equation (9.68) by replacing ξ_t by its estimate $x_{t|t-1}$ and replacing η_t by $E(\eta_t) = 0$. This gives

$$\hat{y}_{t|t-1} = H_t x_{t|t-1}. \tag{9.79}$$

The mean-square-error dispersion matrix of this prediction is

$$F_t = E\{(y_t - \hat{y}_{t|t-1})(y_t - \hat{y}_{t|t-1})'\}. \tag{9.80}$$

To express F_t in terms of $P_{t|t-1}$, we subtract equation (9.79) from equation (9.68) to give

$$\begin{aligned} e_t &= y_t - \hat{y}_{t|t-1} \\ &= H_t(\xi_t - x_{t|t-1}) + \eta_t. \end{aligned} \tag{9.81}$$

Then, since $\xi_t - x_{t|t-1}$ and η_t are statistically independent, and since $E(\eta \eta_t') = D(\eta_t) = \Omega_t$, it follows that

$$F_t = H_t P_{t|t-1} H_t' + \Omega_t. \tag{9.82}$$

The business of incorporating the new information provided by y_t into the estimate of the state vector may be regarded as a matter of estimating the parameter ξ_t in the system

$$\begin{bmatrix} x_{t|t-1} \\ y_t \end{bmatrix} = \begin{bmatrix} I_k \\ H_t \end{bmatrix} \xi_t + \begin{bmatrix} \zeta_t \\ \eta_t \end{bmatrix}, \tag{9.83}$$

where $\zeta_t = x_{t|t-1} - \xi_t$. The system is similar to the regression equation under (9.36), but it is distinguished from the latter by the fact that ξ_t is not a constant parameter but is, instead, a value realised by a random variable. The dispersion matrix for the combined disturbance term is

$$D\begin{bmatrix} \zeta_t \\ \eta_t \end{bmatrix} = \begin{bmatrix} P_{t|t-1} & 0 \\ 0 & \Omega_t \end{bmatrix}. \tag{9.84}$$

By applying the method of generalised least squares, we may obtain an estimating equation for ξ_t in the form of

$$\begin{aligned} x_t &= (P_{t|t-1}^{-1} + H_t'\Omega_t^{-1}H_t)^{-1}(P_{t|t-1}^{-1}x_{t|t-1} + H_t'\Omega_t^{-1}y_t) \\ &= P_t(P_{t|t-1}^{-1}x_{t|t-1} + H_t'\Omega_t^{-1}y_t), \end{aligned} \tag{9.85}$$

where

$$P_t = (P_{t|t-1}^{-1} + H_t'\Omega_t^{-1}H_t)^{-1} \tag{9.86}$$

is the dispersion matrix of the estimator. Using the matrix inversion lemma, we can rewrite this as

$$P_t = P_{t|t-1} - P_{t|t-1}H_t'(H_tP_{t|t-1}H_t' + \Omega_t)^{-1}H_tP_{t|t-1}, \tag{9.87}$$

which is a generalisation of the equation to be found under (9.13).

Combining equation (9.87) with the expression for $P_{t+1|t}$ which is indicated by equation (9.78) gives the so-called Riccati equation:

$$P_{t+1|t} = \Phi_{t+1}\Big\{P_{t|t-1} - P_{t|t-1}H_t'(H_tP_{t|t-1}H_t' + \Omega_t)^{-1}H_tP_{t|t-1}\Big\}\Phi_{t+1}' + \Psi_{t+1}. \tag{9.88}$$

This is a difference equation which provides a means for generating recursively the dispersion of the state prediction.

To give equation (9.85) a form which is amenable to a recursive procedure, we may consider the identity

$$\begin{aligned} P_{t|t-1}^{-1}x_{t|t-1} + H_t'\Omega^{-1}y_t &= (P_t^{-1} - H_t'\Omega_t^{-1}H_t)x_{t|t-1} + H_t'\Omega_t^{-1}y_t \\ &= P_t^{-1}x_{t|t-1} + H_t'\Omega_t^{-1}(y_t - H_tx_{t|t-1}). \end{aligned} \tag{9.89}$$

Using this on the RHS of equation (9.85) gives

$$\begin{aligned} x_t &= x_{t|t-1} + P_tH_t'\Omega_t^{-1}(y_t - H_tx_{t|t-1}) \\ &= x_{t|t-1} + K_t(y_t - H_tx_{t|t-1}) \\ &= (I - K_tH_t)x_{t|t-1} + K_ty_t, \end{aligned} \tag{9.90}$$

wherein $K_t = P_tH_t'\Omega_t^{-1}$ is commonly described as the Kalman gain. Using (9.86) and the identity of (9.10), we can show that

$$\begin{aligned} K_t &= P_tH_t'\Omega_t^{-1} \\ &= (P_{t|t-1}^{-1} + H_t'\Omega_t^{-1}H_t)^{-1}H_t'\Omega_t^{-1} \\ &= P_{t|t-1}H_t'(H_tP_{t|t-1}H_t' + \Omega_t)^{-1}. \end{aligned} \tag{9.91}$$

Therefore the estimating equation can be written as

$$x_t = x_{t|t-1} + P_{t|t-1}H_t'(H_tP_{t|t-1}H_t' + \Omega_t)^{-1}(y_t - H_tx_{t|t-1}). \tag{9.92}$$

Example 9.2. A specialised version of equation (9.92) has already appeared under (9.15) in the context of the recursive estimation of the ordinary regression model. Another, more general, version the equation is to be found under (9.66) where it relates to the discounted least-squares estimator which applies an exponentially decaying memory to the data. The performance of discounted least squares is similar to that of the estimator which arises from applying the techniques of Kalman filtering to a regression model with time-varying coefficients of the sort depicted by equations (9.69) and (9.70) when $\Phi = I$ within the former.

A Summary of the Kalman Equations

The algebra associated with the Kalman filter is extensive, and it needs to be summarised. In the table below, which provides a synopsis, we use $\mathcal{I}_t = \{y_1, \ldots, y_t\}$ and $\mathcal{I}_{t-1} = \{y_1, \ldots, y_{t-1}\}$ to denote the information available at times t and $t-1$ respectively. The numbers under which the equations have appeared previously are written on the right.

THE SYSTEM EQUATIONS

$\xi_t = \Phi_t \xi_{t-1} + \nu_t,$	*State Transition*	(9.67)
$y_t = H_t \xi_t + \eta_t,$	*Observation*	(9.68)
$E(\nu_t) = 0, \quad D(\nu_t) = \Psi_t,$	*System Disturbance*	
$E(\eta_t) = 0, \quad D(\eta_t) = \Omega_t.$	*Measurement Error*	

CONDITIONAL EXPECTATIONS

$E(\xi_t \mid \mathcal{I}_{t-1}) = x_{t\|t-1},$	$D(\xi_t \mid \mathcal{I}_{t-1}) = P_{t\|t-1},$	*State Prediction*
$E(\xi_t \mid \mathcal{I}_t) = x_t,$	$D(\xi_t \mid \mathcal{I}_t) = P_t,$	*State Estimate*
$E(y_t \mid \mathcal{I}_{t-1}) = \hat{y}_{t\|t-1},$	$D(y_t \mid \mathcal{I}_{t-1}) = F_t.$	*Observation Prediction*

THE KALMAN FILTER

State Prediction

$x_{t\|t-1} = \Phi_t x_{t-1},$	*State Prediction*	(9.74)
$P_{t\|t-1} = \Phi_t P_{t-1} \Phi_t' + \Psi_t,$	*Prediction Variance*	(9.78)

Observation Prediction

$\hat{y}_{t\|t-1} = H_t x_{t\|t-1},$	*Observation Prediction*	(9.79)
$F_t = H_t P_{t\|t-1} H_t' + \Omega_t,$	*Prediction Variance*	(9.82)

Auxiliary Variables

$e_t = y_t - H_t x_{t\|t-1},$	*Prediction Error*	(9.81)
$K_t = P_{t\|t-1} H_t' F_t^{-1},$	*Kalman Gain*	(9.91)

State Prediction Updating

$x_t = x_{t\|t-1} + K_t e_t,$	*State Estimate*	(9.92)
$P_t = P_{t\|t-1} - K_t F_t^{-1} K_t'.$	*Estimate Variance*	(9.87)

An Alternative Derivation of the Kalman Filter

An alternative derivation of the Kalman filter is available which is based on the calculus of conditional expectations. Consider the jointly distributed random vectors x and y which bear the linear relationship $E(y|x) = \alpha + B'\{x - E(x)\}$. Then the following conditions apply:

(9.93)

$$\text{(i)}\qquad E(y|x) = E(y) + C(y,x)D^{-1}(x)\{x - E(x)\},$$

$$\text{(ii)}\qquad D(y|x) = D(y) - C(y,x)D^{-1}(x)C(x,y),$$

$$\text{(iii)}\qquad E\{E(y|x)\} = E(y),$$

$$\text{(iv)}\qquad D\{E(y|x)\} = C(y,x)D^{-1}(x)C(x,y),$$

$$\text{(v)}\qquad D(y) = D(y|x) + D\{E(y|x)\},$$

$$\text{(vi)}\qquad C\{y - E(y|x), x\} = 0.$$

Here the familiar forms of the conditional expectation and the conditional dispersion are given under (i) and (ii). The result under (iii) follows from (i) when it is recognised that, on the RHS, we have $E\{x - E(x)\} = 0$. To obtain the result under (iv), we may begin by recognising that, on the RHS of (i), only the second term is stochastic. To find the dispersion of this term, we may use the fact that $D[A\{x - E(x)\}] = AD(x)A'$. The result under (v) is simply a combination of (ii) and (iv). Finally, to obtain the result under (vi), which indicates that the error associated with the conditional expectation is uncorrelated with the conditioning variable x, we begin by writing the error as

$$y - E(y|x) = \{y - E(y)\} - C(y,x)D^{-1}(x)\{x - E(x)\}. \tag{9.94}$$

Then, on postmultiplying by x' and taking expectations, we get

$$\begin{aligned} C\{y - E(y|x), x\} &= C(y,x) - C(y,x)D^{-1}(x)D(x) \\ &= 0, \end{aligned} \tag{9.95}$$

which is the desired result.

In applying the results under (9.93) to the task of deriving the equations of the Kalman filter, we must adopt a purely Bayesian interpretation in which the initial state vector ξ_0 is regarded as a random variable. Its mean $x_0 = E(\xi_0)$ and its dispersion matrix $P_0 = D(\xi_0)$ are given in advance.

The initial values x_0 and P_0 give rise to the parameters $x_{1|0} = E(\xi_1|\mathcal{I}_0)$ and $P_{1|0} = D(\xi_1|\mathcal{I}_0)$ of a prior distribution pertaining to the state vector ξ_1 of the first sample period. The task at this stage is to determine the parameters $x_1 = E(\xi_1|\mathcal{I}_1)$ and $P_1 = D(\xi_1|\mathcal{I}_1)$ of a posterior distribution in the light of the information provided by the first observation y_1 which is included in $\mathcal{I}_1$. The task of the tth stage, which stands for all the subsequent stages, is to form the state prediction $x_{t|t-1} = E(\xi_t|\mathcal{I}_{t-1})$ and its dispersion $P_{t|t-1} = D(\xi_t|\mathcal{I}_{t-1})$ and thence to determine $x_t = E(\xi_t|\mathcal{I}_t)$ and $P_t = D(\xi_t|\mathcal{I}_t)$ in the light of the observation y_t.

The first object is to derive the formulae for the state prediction and its dispersion. We use (9.93)(iii) to show that

$$
\begin{aligned}
E(\xi_t|\mathcal{I}_{t-1}) &= E\{E(\xi_t|\xi_{t-1},\mathcal{I}_{t-1})\}\\
&= E\{\Phi_t\xi_{t-1}|\mathcal{I}_{t-1}\}\\
&= \Phi_t x_{t-1}.
\end{aligned}
\tag{9.96}
$$

We can use (9.93)(v) to show that

$$
\begin{aligned}
D(\xi_t|\mathcal{I}_{t-1}) &= D(\xi_t|\xi_{t-1},\mathcal{I}_{t-1}) + D\{E(\xi_t|\xi_{t-1},\mathcal{I}_{t-1})\}\\
&= \Psi_t + D\{\Phi_t E(\xi_{t-1}|\mathcal{I}_{t-1})\}\\
&= \Psi_t + \Phi_t P_{t-1}\Phi_t'.
\end{aligned}
\tag{9.97}
$$

Thus we have

$$
x_{t|t-1} = \Phi_t x_{t-1} \quad \text{and} \quad P_{t|t-1} = \Phi_t P_{t-1}\Phi_t' + \Psi_t,
\tag{9.98}
$$

which are equations that have already appeared under (9.74) and (9.78) respectively.

The next purpose is to find an updated estimate of ξ_t which incorporates the information of y_t. From (9.93)(i), it follows that

$$
E(\xi_t|\mathcal{I}_t) = E(\xi_t|\mathcal{I}_{t-1}) - C(\xi_t, y_t|\mathcal{I}_{t-1})D^{-1}(y_t|\mathcal{I}_{t-1})\{y_t - E(\xi_t|\mathcal{I}_{t-1})\}.
\tag{9.99}
$$

Here there is

$$
\begin{aligned}
C(\xi_t, y_t|\mathcal{I}_{t-1}) &= E\{(\xi_t - x_{t|t-1})(y_t - \hat{y}_{t|t-1})\}\\
&= E\{(\xi_t - x_{t|t-1})(H_t\xi_t + \eta_t - H_t x_{t|t-1})'\}\\
&= P_{t|t-1}H_t'.
\end{aligned}
\tag{9.100}
$$

On substituting this expression into (9.99) and using other definitions which are available in the synopsis, we get the updated state estimate

$$
x_t = x_{t|t-1} + P_{t|t-1}H_t'(H_t P_{t|t-1}H_t' + \Omega_t)^{-1}(y_t - H_t x_{t|t-1}),
\tag{9.101}
$$

which is to be found also under (9.92). From (9.93)(ii), we have

$$
D(\xi_t|\mathcal{I}_t) = D(\xi_t|\mathcal{I}_{t-1}) - C(\xi_t, y_t|\mathcal{I}_{t-1})D^{-1}(y_t|\mathcal{I}_{t-1})C(y_t, \xi_t|\mathcal{I}_{t-1}).
\tag{9.102}
$$

This gives the dispersion matrix for the updated estimate which is to be found under (9.87):

$$
P_t = P_{t|t-1} - P_{t|t-1}H_t'(H_t P_{t|t-1}H_t' + \Omega_t)^{-1}H_t P_{t|t-1}.
\tag{9.103}
$$

Smoothing

In some circumstances, we may wish to improve our estimate x_t of the state vector at time t using information which has arisen subsequently. For the succeeding observations $\{y_{t+1}, y_{t+2}, \ldots\}$ are bound to convey information about the state vector ξ_t which can supplement the information $\mathcal{I}_t = \{y_1, \ldots, y_t\}$ which was available at time t.

The retrospective enhancement of the state estimators using *ex post* information is conventionally described as a process of smoothing. The terminology is somewhat misleading; but we shall adhere to it nevertheless.

There are several ways in which we might effect a process of smoothing. In the first place, there is fixed-point smoothing. This occurs whenever the object is to enhance the estimate of a single state variable ξ_n repeatedly using successive observations. The resulting sequence of estimates is described by

$$(9.104) \qquad \bigl\{x_{n|t} = E(\xi_n|\mathcal{I}_t); t = n+1, n+2, \ldots\bigr\}. \qquad \textit{Fixed-point smoothing}$$

The second mode of smoothing is fixed-lag smoothing. In this case, enhanced estimates of successive state vectors are generated with a fixed lag of, say, n periods:

$$(9.105) \qquad \bigl\{x_{t-n|t} = E(\xi_{t-n}|\mathcal{I}_t); t = n+1, n+2, \ldots\bigr\}. \qquad \textit{Fixed-lag smoothing}$$

Finally, there is fixed-interval smoothing. This is a matter of revising each of the state estimates for a period running from $t = 1$, to $t = n$ once the full set of observations in $\mathcal{I}_n = \{y_1, \ldots, y_n\}$ has become available. The sequence of revised estimates is

$$(9.106) \qquad \bigl\{x_{n-t|n} = E(\xi_t|\mathcal{I}_n); t = 1, 2, \ldots, n-1\bigr\}. \qquad \textit{Fixed-interval smoothing}$$

Here, instead of $x_{t|n}$, we have taken $x_{n-t|n}$ as the generic element which gives the sequence in reverse order. This it to reflect the fact that, with most algorithms, the smoothed estimates are generated by running backwards through the initial set of estimates.

There is also a variant of fixed-interval smoothing which we shall describe as *intermittent smoothing*. For it transpires that, if the fixed-interval smoothing operation is repeated periodically to take account of new data, then some use can be made of the products of the previous smoothing operation.

For each mode of smoothing, there is an appropriate recursive formula. We shall derive these formulae, in the first instance, from a general expression for the expectation of the state vector ξ_t conditional upon the information contained in the set of innovations $\{e_1, \ldots, e_n\}$ which, as we shall show in the next section, is identical to the information contained in the observations $\{y_1, \ldots, y_n\}$.

Innovations and the Information Set

The task of this section is to establish that the information of $\mathcal{I}_t = \{y_1, \ldots, y_t\}$ is also conveyed by the prediction errors or innovations $\{e_1, \ldots, e_t\}$ and that the latter are mutually uncorrelated random variables. For this purpose, it will helpful to define some additional matrix quantities:

(9.107) $$M_t = \Phi_t K_{t-1} \quad \text{and}$$

(9.108) $$\Lambda_t = \Phi_t(I - K_{t-1}H_{t-1}).$$

We begin by demonstrating that each error e_t is a linear function of $y_1, \ldots, y_t$. From equations (9.92), (9.81) and (9.74), which are to be found in the synopsis, we obtain the equation $x_{t|t-1} = \Lambda_t x_{t-1|t-2} + M_t y_{t-1}$. Repeated back-substitution gives

(9.109) $$x_{t|t-1} = \sum_{j=1}^{t-1} \Lambda_{t,j+2} M_{j+1} y_j + \Lambda_{t,2} x_{1|0},$$

where $\Lambda_{t,j+2} = \Lambda_t \Lambda_{t-1} \cdots \Lambda_{j+2}$ is a product of matrices which specialises to $\Lambda_{t,t} = \Lambda_t$ and to $\Lambda_{t,t+1} = I$. It follows that

(9.110) $$\begin{aligned} e_t &= y_t - H_t x_{t|t-1} \\ &= y_t - H_t \sum_{j=1}^{t-1} \Lambda_{t,j+2} M_{j+1} y_j - H_t \Lambda_{t,2} x_{1|0}. \end{aligned}$$

Next, we demonstrate that each y_t is a linear function of $e_1, \ldots, e_t$. By back-substitution in the equation $x_{t|t-1} = \Phi_t x_{t-1|t-2} + M_t e_{t-1}$ obtained from (9.74) and (9.92), we get

(9.111) $$x_{t|t-1} = \sum_{j=1}^{t-1} \Phi_{t,j+2} M_{j+1} e_j + \Phi_{t,2} x_{1|0},$$

wherein $\Phi_{t,j+2} = \Phi_t \Phi_{t-1} \cdots \Phi_{j+2}$ is a product of matrices which specialises to $\Phi_{t,t} = \Phi_t$ and to $\Phi_{t,t+1} = I$. It follows that

(9.112) $$\begin{aligned} y_t &= e_t + H_t x_{t|t-1} \\ &= e_t + H_t \sum_{j=1}^{t-1} \Phi_{t,j+2} M_{j+1} e_j + H_t \Phi_{t,2} x_{1|0}. \end{aligned}$$

Given that there is a one-to-one linear relationship between the observations and the prediction errors, it follows that we can represent the information set in terms of either. Thus we have $\mathcal{I}_{t-1} = \{e_1, \ldots, e_{t-1}\}$; and, given that $e_t = y_t - E(y_t|\mathcal{I}_{t-1})$, it follows from (9.93)(vi) that e_t is uncorrelated with the preceding errors $e_1, \ldots, e_{t-1}$. The result indicates that the prediction errors are mutually uncorrelated.

Conditional Expectations and Dispersions of the State Vector

Given that the sequence $e_1, \ldots, e_n$ of Kalman-filter innovations are mutually independent vectors with zero expectations, it follows from (9.93)(i) that, for any indices m and $n > m$,

$$E(\xi_t|\mathcal{I}_n) = E(\xi_t|\mathcal{I}_m) + \sum_{j=m+1}^{n} C(\xi_t, e_j)D^{-1}(e_j)e_j. \tag{9.113}$$

In a similar way, we see from equation (9.93)(ii) that the dispersion matrix satisfies

$$D(\xi_t|\mathcal{I}_n) = D(\xi_t|\mathcal{I}_m) - \sum_{j=m+1}^{n} C(\xi_t, e_j)D^{-1}(e_j)C(e_j, \xi_t). \tag{9.114}$$

The task of evaluating the expressions under (9.113) and (9.114) is to find the generic covariance $C(\xi_t, e_k)$. For this purpose, we must develop a recursive formula which represents e_k in terms of $\xi_t - E(\xi_t|\mathcal{I}_{t-1})$ and in terms of the state disturbances and observation errors which occur from time t.

Consider the expression for the innovation

$$\begin{aligned} e_k &= y_k - H_k x_{k|k-1} \\ &= H_k(\xi_k - x_{k|k-1}) + \eta_k. \end{aligned} \tag{9.115}$$

Here the term $\xi_k - x_{k|k-1}$ follows a recursion which is indicated by the equation

$$\xi_k - x_{k|k-1} = \Lambda_k(\xi_{k-1} - x_{k-1|k-2}) + (\nu_k - M_k\eta_{k-1}). \tag{9.116}$$

The latter comes from subtracting from the transition equation (9.67) the equation $x_{t|t-1} = \Lambda_t x_{t-1|t-2} + M_t(H_{t-1}\xi_{t-1} + \eta_{t-1})$, obtained by substituting the observation equation (9.68) into (9.90) and putting the result, lagged one period, into (9.74). By running the recursion from time k back to time t, we may deduce that

$$\xi_k - x_{k|k-1} = \Lambda_{k,t+1}(\xi_t - x_{t|t-1}) + \sum_{j=t}^{k-1} \Lambda_{k,j+2}(\nu_{j+1} - M_{j+1}\eta_j), \tag{9.117}$$

wherein $\Lambda_{k,k+1} = I$ and $\Lambda_{k,k} = \Lambda_k$. It follows from (9.115) and (9.117) that, when $k \geq t$,

$$\begin{aligned} C(\xi_t, e_k) &= E\{\xi_t(\xi_t - x_{t|t-1})\Lambda'_{k,t+1}H'_k\} \\ &= P_{t|t-1}\Lambda'_{k,t+1}H'_k. \end{aligned} \tag{9.118}$$

Using the identity $\Phi_{t+1}P_t = \Lambda_{t+1}P_{t|t-1}$ which comes via (9.87), we get for $k > t$

$$C(\xi_t, e_k) = P_t\Phi'_{t+1}\Lambda'_{k,t+2}H'_k. \tag{9.119}$$

Next we note that

$$C(\xi_{t+1}, e_k) = P_{t+1|t}\Lambda'_{k,t+2}H'_k. \tag{9.120}$$

It follows, from comparing (9.119) and (9.120), that

$$C(\xi_t, e_k) = P_t \Phi'_{t+1} P^{-1}_{t+1|t} C(\xi_{t+1}, e_k). \tag{9.121}$$

If we substitute the expression under (9.118) into the formula of (9.113) where $m \geq t-1$, and if we set $D^{-1}(e_j) = F_j^{-1}$, then we get

$$\begin{aligned} E(\xi_t|\mathcal{I}_n) &= E(\xi_t|\mathcal{I}_m) + \sum_{j=m+1}^{n} C(\xi_t, e_j) D^{-1}(e_j) e_j \\ &= E(\xi_t|\mathcal{I}_m) + \sum_{j=m+1}^{n} P_{t|t-1} \Lambda'_{j,t+1} H'_j F_j^{-1} e_j \\ &= E(\xi_t|\mathcal{I}_m) + P_{t|t-1} \Lambda'_{m+1,t+1} \sum_{j=m+1}^{n} \Lambda'_{j,m+2} H'_j F_j^{-1} e_j. \end{aligned} \tag{9.122}$$

An expression for the dispersion matrix is found in a similar way:

$$\begin{aligned} D(\xi_t|\mathcal{I}_n) &= D(\xi_t|\mathcal{I}_m) \\ & -P_{t|t-1} \Lambda'_{m+1,t+1} \left\{ \sum_{j=m+1}^{n} \Lambda'_{j,m+2} H'_j F_j^{-1} H_j \Lambda_{j,m+2} \right\} \Lambda_{m+1,t+1} P_{t|t-1}. \end{aligned} \tag{9.123}$$

Notice that the sums in the two final expressions may be accumulated using recursions running backwards in time of the form

$$\begin{aligned} q_t &= \sum_{j=t}^{n} \Lambda'_{j,t+1} H'_j F_j^{-1} e_j \\ &= H'_t F_t^{-1} e_t + \Lambda'_{t+1} q_{t+1} \end{aligned} \tag{9.124}$$

and

$$\begin{aligned} Q_t &= \sum_{j=t}^{n} \Lambda'_{j,t+1} H'_j F_j^{-1} H_j \Lambda_{j,t+1} \\ &= H'_t F_t^{-1} H_t + \Lambda'_{t+1} Q_{t+1} \Lambda_{t+1}. \end{aligned} \tag{9.125}$$

These recursions are initiated with $q_n = H'_n F_n^{-1} e_n$ and $Q_n = H'_n F_n^{-1} H_n$.

The Classical Smoothing Algorithms

An account of the classical smoothing algorithms is to be found in the book by Anderson and Moore [12] which has become a standard reference for the Kalman filter.

Anderson and Moore have adopted a method for deriving the filtering equations which depends upon an augmented state-transition equation wherein the enlarged state vector contains a sequence of the state vectors from the original transition equation. This approach is common to several authors including Willman [526], who deals with fixed-point smoothing, Premier and Vacroux [408], who treat fixed-lag smoothing and Farooq and Mahalanabis [181], who treat fixed-interval smoothing. It seems that an approach via the calculus of conditional expectations is more direct.

The fixed-point smoother. Of the classical smoothing algorithms, the fixed-point smoothing equations are the easiest to derive. The task is as follows: given $x_{t|n} = E(\xi_t|e_1,\ldots,e_n)$, we must find an expression for $x_{t|n+1} = E(\xi_t|e_1,\ldots,e_{n+1})$ with $n \geq t$. That is to say, we must enhance the estimate of ξ_t by incorporating the extra information which is afforded by the new innovation e_{n+1}. The formula is simply

$$E(\xi_t|\mathcal{I}_{n+1}) = E(\xi_t|\mathcal{I}_n) + C(\xi_t, e_{n+1})D^{-1}(e_{n+1})e_{n+1}. \tag{9.126}$$

Now, (9.118) gives

$$\begin{aligned} C(\xi_t, e_n) &= P_{t|t-1}\Lambda'_{n,t+1}H'_n \\ &= L_n H'_n \end{aligned} \tag{9.127}$$

and

$$\begin{aligned} C(\xi_t, e_{n+1}) &= P_{t|t-1}\Lambda'_{n+1,t+1}H'_{n+1} \\ &= L_n\Lambda'_{n+1}H'_{n+1}. \end{aligned} \tag{9.128}$$

Therefore, we may write the fixed-point algorithm as

$$\begin{aligned} &E(\xi_t|\mathcal{I}_{n+1}) = E(\xi_t|\mathcal{I}_n) + L_{n+1}H'_{n+1}F^{-1}_{n+1}e_{n+1} \\ &\text{where} \quad L_{n+1} = L_n\Lambda'_{n+1} \quad \text{and} \quad L_t = P_{t|t-1}. \end{aligned} \tag{9.129}$$

The accompanying dispersion matrix can be calculated from

$$D(\xi_t|\mathcal{I}_{n+1}) = D(\xi_t|\mathcal{I}_n) - L_{n+1}H'_{n+1}F^{-1}_{n+1}H_{n+1}L'_{n+1}. \tag{9.130}$$

The fixed-point smoother is initiated with values for $E(\xi_t|\mathcal{I}_t)$, $D(\xi_t|\mathcal{I}_t)$ and $L_t = P_{t|t-1}$, which are provided by the Kalman filter. From these initial quantities, a sequence of enhanced estimates of ξ_t is calculated recursively using subsequent observations. The values of e_{n+1}, F_{n+1} and K_n, needed in computing (9.129) and (9.130), are also provided by the Kalman filter, which runs concurrently with the smoother.

The fixed-interval smoother. The next version of the smoothing equation to be derived is the fixed-interval form. Consider using the identity of (9.121) to rewrite equation (9.113), with m set to t, as

$$E(\xi_t|\mathcal{I}_n) = E(\xi_t|\mathcal{I}_t) + P_t\Phi'_{t+1}P^{-1}_{t+1|t}\sum_{j=t+1}^{n} C(\xi_{t+1}, e_j)D^{-1}(e_j)e_j. \tag{9.131}$$

Now

$$(9.132) \qquad E(\xi_{t+1}|\mathcal{I}_n) = E(\xi_{t+1}|\mathcal{I}_t) + \sum_{j=t+1}^{n} C(\xi_{t+1}, e_j)D^{-1}(e_j)e_j;$$

so it follows that equation (9.131) can be rewritten in turn as

$$(9.133) \quad E(\xi_t|\mathcal{I}_n) = E(\xi_t|\mathcal{I}_t) + P_t\Phi'_{t+1}P^{-1}_{t+1|t}\left\{E(\xi_{t+1}|\mathcal{I}_n) - E(\xi_{t+1}|\mathcal{I}_t)\right\}.$$

This is the formula for the fixed-interval smoother.

A similar strategy is adopted in the derivation of the dispersion of the smoothed estimate. According to (9.114), we have

$$(9.134) \qquad D(\xi_t|\mathcal{I}_n) = D(\xi_t|\mathcal{I}_t) - \sum_{j=t+1}^{n} C(\xi_t, e_j)D^{-1}(e_j)C(e_j, \xi_t)$$

and

$$(9.135) \quad D(\xi_{t+1}|\mathcal{I}_n) = D(\xi_{t+1}|\mathcal{I}_t) - \sum_{j=t+1}^{n} C(\xi_{t+1}, e_j)D^{-1}(e_j)C(e_j, \xi_{t+1}).$$

Using the identity of (9.121) in (9.134) and taking the result from (9.135) enables us to write

$$(9.136) \qquad P_{t|n} = P_t - P_t\Phi'_{t+1}P^{-1}_{t+1|t}\{P_{t+1|t} - P_{t+1|n}\}P^{-1}_{t+1|t}\Phi_{t+1}P_t.$$

An interpretation. Consider $E(\xi_t|\mathcal{I}_n)$, and let us represent the information set, at first, by

$$(9.137) \quad \mathcal{I}_n = \{\mathcal{I}_t, h_{t+1}, e_{t+2}, \ldots, e_n\} \quad \text{where} \quad h_{t+1} = \xi_{t+1} - E(\xi_{t+1}|\mathcal{I}_t).$$

We may begin by finding

$$(9.138) \qquad E(\xi_t|\mathcal{I}_t, h_{t+1}) = E(\xi_t|\mathcal{I}_t) + C(\xi_t, h_{t+1}|\mathcal{I}_t)D^{-1}(h_{t+1}|\mathcal{I}_t)h_{t+1}.$$

Here we have

$$(9.139) \qquad \begin{aligned} C(\xi_t, h_{t+1}|\mathcal{I}_t) &= E\{\xi_t(\xi_t - x_t)'\Phi'_{t+1} + \xi_t\nu'_t|\mathcal{I}_t\} = P_t\Phi'_{t+1} \quad \text{and} \\ D(h_{t+1}|\mathcal{I}_t) &= P_{t+1|t}. \end{aligned}$$

It follows that

$$(9.140) \qquad E(\xi_t|\mathcal{I}_t, h_{t+1}) = E(\xi_t|\mathcal{I}_t) + P_t\Phi'_{t+1}P^{-1}_{t+1|t}\{\xi_{t+1} - E(\xi_{t+1}|\mathcal{I}_t)\}.$$

Of course, the value of ξ_{t+1} in the RHS of this equation is not observable. However, if we take the expectation of the equation conditional upon all of the information in the set $\mathcal{I}_n = \{e_1, \ldots, e_n\}$, then ξ_{t+1} is replaced by $E(\xi_{t+1}|\mathcal{I}_n)$ and we get the formula under (9.133). This interpretation was published by Ansley and Kohn [24]. It highlights the notion that the information which is used in enhancing the estimate of ξ_t is contained entirely within the smoothed estimate of ξ_{t+1}.

The intermittent smoother. Consider the case where smoothing is intermittent with m sample points accumulating between successive smoothing operations. Then it is possible to use the estimates arising from the previous smoothing operation.

Imagine that the operation is performed when $n = jm$ points are available. Then, for $t > (j-1)m$, the smoothed estimate of the state vector ξ_t is given by the ordinary fixed-interval smoothing formula found under (9.133). For $t \leq (j-1)m$, the appropriate formula is

(9.141)
$$E(\xi_t|\mathcal{I}_n) = E(\xi_t|\mathcal{I}_{(j-1)m}) + P_t\Phi'_{t+1}P^{-1}_{t+1|t}\{E(\xi_{t+1}|\mathcal{I}_n) - E(\xi_{t+1}|\mathcal{I}_{(j-1)m})\}.$$

Here $E(\xi_t|\mathcal{I}_{(j-1)m})$ is being used in place of $E(\xi_t|\mathcal{I}_t)$. The advantage of the algorithm is that it does not require the values of unsmoothed estimates to be held in memory when smoothed estimates are available.

A limiting case of the intermittent smoothing algorithm arises when the smoothing operation is performed each time a new observation is registered. Then the formula becomes

$$(9.142)\quad E(\xi_t|\mathcal{I}_n) = E(\xi_t|\mathcal{I}_{n-1}) + P_t\Phi'_{t+1}P^{-1}_{t+1|t}\{E(\xi_{t+1}|\mathcal{I}_n) - E(\xi_{t+1}|\mathcal{I}_{n-1})\}.$$

The formula is attributable to Chow [106] who provided a somewhat lengthy derivation. Chow proposed this algorithm for the purpose of ordinary fixed-interval smoothing, for which it is clearly inefficient.

The fixed-lag smoother. The task is to move from the smoothed estimate of ξ_{n-t} made at time n to the estimate of ξ_{n+1-t} once the new information in the prediction error e_{n+1} has become available. Equation (9.93)(i) indicates that

$$(9.143)\quad E(\xi_{n+1-t}|\mathcal{I}_{n+1}) = E(\xi_{n+1-t}|\mathcal{I}_n) + C(\xi_{n+1-t}, e_{n+1})D^{-1}(e_{n+1})e_{n+1},$$

which is the formula for the smoothed estimate, whilst the corresponding formula for the dispersion matrix is

(9.144)
$$D(\xi_{n+1-t}|\mathcal{I}_{n+1}) = D(\xi_{n+1-t}|\mathcal{I}_n) - C(\xi_{n+1-t}, e_{n+1})D^{-1}(e_{n+1})C(e_{n+1}, \xi_{n+1-t}).$$

To evaluate (9.143), we must first find the value of $E(\xi_{n+1-t}|\mathcal{I}_n)$ from the value of $E(\xi_{n-t}|\mathcal{I}_n)$. On setting $t = k$ in the fixed-interval formula under (9.133), and rearranging the result, we get

$$(9.145)\quad E(\xi_{k+1}|\mathcal{I}_n) = E(\xi_{k+1}|\mathcal{I}_k) + P_{k+1|k}\Phi'^{-1}_{k+1}P^{-1}_k\{E(\xi_k|\mathcal{I}_n) - E(\xi_k|\mathcal{I}_k)\}.$$

To obtain the desired result, we simply set $k = n - t$, which gives

$$(9.146)\quad \begin{aligned} E(\xi_{n+1-t}|\mathcal{I}_n) &= E(\xi_{n+1-t}|\mathcal{I}_{n-t}) \\ &+P_{n+1-t|n-t}\Phi'^{-1}_{n+1-t}P^{-1}_{n-t}\{E(\xi_{n-t}|\mathcal{I}_n) - E(\xi_{n-t}|\mathcal{I}_{n-t})\}. \end{aligned}$$

The formula for the smoothed estimate also comprises

$$C(\xi_{n+1-t}, e_{n+1}) = P_{n+1-t|n-t}\Lambda'_{n+1,n+2-t}H'_{n+1}. \tag{9.147}$$

If Λ_{n+1-t} is nonsingular, then $\Lambda_{n+1,n+2-t} = \Lambda_{n+1}\{\Lambda_{n,n+1-t}\}\Lambda^{-1}_{n+1-t}$; and thus we may profit from the calculations entailed in finding the previous smoothed estimate which will have generated the matrix product in the parentheses.

In evaluating the formula (9.144) for the dispersion of the smoothed estimates, we may use the following expression for $D(\xi_{n+1-t}|\mathcal{I}_n) = P_{n+1-t|n}$:

$$\begin{aligned} P_{n+1-t|n} &= P_{n+1-t|n-t} \\ &-P_{n+1-t|n-t}\Phi'^{-1}_{n+1-t}P^{-1}_{n-t}(P_{n-t} - P_{n-t|n})P^{-1}_{n-t}\Phi^{-1}_{n+1-t}P_{n+1-t|n-t}. \end{aligned} \tag{9.148}$$

This is demonstrated is the same manner as equation (9.146).

A process of fixed-lag smoothing, with a lag length of t, is initiated with a value for $E(\xi_1|\mathcal{I}_{t+1})$. The latter is provided by running the fixed-point smoothing algorithm for t periods. After time $t+1$, when the $(n+1)$th observation becomes available, $E(\xi_{n+1-t}|\mathcal{I}_n)$ is calculated from $E(\xi_{n-t}|\mathcal{I}_n)$ via equation (9.146). For this purpose, the values of $x_{n+1-t|n-t}$, x_{n-t}, $P_{n+1-t|n-t}$ and P_{n-t} must be available. These are generated by the Kalman filter in the process of calculating e_{n-t}, and they are held in memory for t periods. The next smoothed estimate $E(\xi_{n+1-t}|\mathcal{I}_{n+1})$ is calculated from equation (9.143), for which the values of e_{n+1}, F_{n+1} and K_n are required. These are also provided by the Kalman filter which runs concurrently.

Variants of the Classical Algorithms

The attention which statisticians have paid to the smoothing problem recently has been focused upon fixed-interval smoothing. This mode of smoothing is, perhaps, of less interest to communications engineers than the other modes; which may account for the fact that the statisticians have found scope for improving the algorithms.

Avoiding an inversion. There are some modified versions of the classical fixed-interval smoothing algorithm which avoid the inversion of the matrix $P_{t|t-1}$. In fact, the basis for these has been provided already in a previous section. Thus, by replacing the sums in equations (9.122) and (9.123) by q_{m+1} and Q_{m+1}, which are the products of the recursions under (9.124) and (9.125), we get

$$E(\xi_t|\mathcal{I}_n) = E(\xi_t|\mathcal{I}_m) + P_{t|t-1}\Lambda'_{m+1,t+1}q_{m+1}, \tag{9.149}$$

$$D(\xi_t|\mathcal{I}_n) = D(\xi_t|\mathcal{I}_m) - P_{t|t-1}\Lambda'_{m+1,t+1}Q_{m+1}\Lambda_{m+1,t+1}P_{t|t-1}. \tag{9.150}$$

These expressions are valid for $m \geq t-1$.

Setting $m = t-1$ in (9.149) and (9.150) gives a useful alternative to the classical algorithm for fixed-interval smoothing:

$$x_{t|n} = x_{t|t-1} + P_{t|t-1}q_t, \tag{9.151}$$

$$P_{t|n} = P_{t|t-1} - P_{t|t-1}Q_tP_{t|t-1}. \tag{9.152}$$

We can see that, in moving from q_{t+1} to q_t via equation (9.124), which is the first step towards finding the next smoothed estimate $x_{t-1|n}$, there is no inversion of $P_{t|t-1}$. The equations (9.151) and (9.152) have been derived by De Jong [146].

The connection with the classical smoothing algorithm is easily established. From (9.151), we get $q_{t+1} = P^{-1}_{t+1|t}(x_{t+1|n} - x_{t+1|t})$. By setting $m = t$ in (9.149) and substituting for q_{t+1}, we get

$$\begin{aligned} x_{t|n} &= x_t + P_{t|t-1}\Lambda'_{t+1}P^{-1}_{t+1|t}(x_{t+1|n} - x_{t+1|t}) \\ &= x_t + P_t\Phi'_{t+1}P^{-1}_{t+1|t}(x_{t+1|n} - x_{t+1|t}), \end{aligned} \tag{9.153}$$

where the final equality follows from the identity $\Phi_{t+1}P_t = \Lambda_{t+1}P_{t|t-1}$ already used in (9.119). Equation (9.153) is a repetition of equation (9.133) which belongs to the classical algorithm.

Equation (9.136), which also belongs to the classical algorithm, is obtained by performing similar manipulations with equations (9.150) and (9.152).

Smoothing via state disturbances. Given an initial value for the state vector, a knowledge of the sequence of the state-transition matrices and of the state disturbances in subsequent periods will enable one to infer the values of subsequent state vectors. Therefore the estimation of a sequence of state vectors may be construed as a matter of estimating the state disturbances. The information which is relevant to the estimation of the disturbance ν_t is contained in the prediction errors from time t onwards. Thus

$$E(\nu_t|\mathcal{I}_n) = \sum_{j=t}^{n} C(\nu_t, e_j)D^{-1}(e_j)e_j. \tag{9.154}$$

Here, for $j \geq t$, the generic covariance is given by

$$\begin{aligned} C(\nu_t, e_j) &= E\{\nu_t\nu'_t\Lambda'_{j,t+1}H'_j\} \\ &= \Psi_t\Lambda'_{j,t+1}H'_j, \end{aligned} \tag{9.155}$$

which follows from the expression for e_t which results from substituting (9.117) in (9.155). Putting (9.155) into (9.154) and setting $D^{-1}(e_j) = F^{-1}_j$ gives

$$\begin{aligned} E(\nu_t|\mathcal{I}_n) &= \Psi_t\sum_{j=t}^{n}\Lambda'_{j,t+1}H'_jF^{-1}_je_j \\ &= \Psi_tq_t, \end{aligned} \tag{9.156}$$

where q_t is a sum which may be accumulated using the recursion under (9.124).

By taking the expectation of the transition equation conditional upon all of the information in the fixed sample, we obtain the recursive equation which generates the smoothed estimates of the state vectors:

$$\begin{aligned} x_{t|n} &= \Phi_tx_{t-1|n} + E(\nu_t|\mathcal{I}_n) \\ &= \Phi_tx_{t-1|n} + \Psi_tq_t. \end{aligned} \tag{9.157}$$

The initial value is $x_{0|n} = x_0 + P_0\Phi_1' q_1$. This is obtained by setting $t = 0$ in the equation $x_{t|n} = x_t + P_t\Phi_{t+1}' q_{t+1}$ which comes from (9.153).

Equation (9.157) has been presented recently in a paper by Koopman [299]. A similar approach has been pursued by Mayne [339].

With some effort, a connection can be found between equation (9.157) and equation (9.151) which is its counterpart in the previous algorithm. From (9.74) and (9.92), we get $x_{t|t-1} = \Phi_t(x_{t-1|t-2} + K_{t-1}e_{t-1})$. From (9.78) and (9.87), we get $P_{t|t-1} = \Phi_t P_{t-1|t-2}(I - K_{t-1}H_{t-1})'\Phi_t' + \Psi_t$. Putting these into (9.151) gives

$$x_{t|n} = \Phi_t x_{t-1|t-2} + \Psi_t q_t + \Phi_t(K_{t-1}e_{t-1} + P_{t-1|t-2}\Lambda_t' q_t). \tag{9.158}$$

Equation (9.151) lagged one period also gives an expression for $x_{t-1|t-2}$ in terms of $x_{t-1|n}$:

$$x_{t-1|t-2} = x_{t-1|n} - P_{t-1|t-2}q_{t-1}. \tag{9.159}$$

Using the identity $q_{t-1} = H_{t-1}'F_{t-1}^{-1}e_{t-1} + \Lambda_t' q_t$ and the latter equation, we can rewrite (9.158) as

$$\begin{aligned} x_{t|n} &= \Phi_t x_{t-1|n} + \Psi_t q_t - \Phi_t P_{t-1|t-2}(H_{t-1}'F_{t-1}^{-1}e_{t-1} + \Lambda_t' q_t) \\ &\quad + \Phi_t(K_{t-1}e_{t-1} + P_{t-1|t-2}\Lambda_t' q_t) \\ &= \Phi_t x_{t-1|n} + \Psi_t q_t, \end{aligned} \tag{9.160}$$

where the final equality follows from equation (9.91). This is (9.157) again.

An alternative algorithm exists which also uses estimates of the state disturbances. In contrast to the previous algorithm, it runs backwards in time rather than forwards. The basic equation is

$$x_{t-1|n} = \Phi_t^{-1}x_{t|n} - \Phi_t^{-1}\Psi_t q_t, \tag{9.161}$$

which comes directly from (9.157). The value of q_t is obtained via equation (9.124). However, because we have a backward recursion in (9.161), an alternative recursion for q_t is available, which reduces the number of elements which must be held in memory. A reformulation of equation (9.124) gives

$$\begin{aligned} q_t &= H_t'F_t^{-1}e_t + \Lambda_{t+1}'q_{t+1} \\ &= H_t'F_t^{-1}e_t + (I - K_tH_t)'\Phi_{t+1}'q_{t+1} \\ &= H_t's_t + \Phi_{t+1}'q_{t+1}, \end{aligned} \tag{9.162}$$

where s_t is defined as

$$s_t = F_t^{-1}e_t - K_t'\Phi_{t+1}'q_{t+1}. \tag{9.163}$$

Now, consider the smoothed estimates of the observation errors. Because η_t is independent of $y_1, \ldots, y_{t-1}$, these are given by

$$E(\eta_t|\mathcal{I}_n) = \sum_{j=t}^{n} C(\eta_t, e_j)D^{-1}(e_j)e_j. \tag{9.164}$$

The covariances follow once more from equations (9.115) and (9.117). For $j > t$, we get

$$C(\eta_t, e_j) = -\Omega_t M'_{t+1}\Lambda'_{j,t+2}H'_j, \tag{9.165}$$

whereas, for $j = t$, we have $C(\eta_t, e_t) = \Omega_t$. Substituting these in (9.164) gives

$$\begin{aligned} E(\eta_t|\mathcal{I}_n) &= \Omega_t\Big\{F_t^{-1}e_t - M'_{t+1}\sum_{j=t+1}^{n}\Lambda'_{j,t+2}H'_jF_j^{-1}e_j\Big\} \\ &= \Omega_t\{F_t^{-1}e_t - K'_t\Phi'_{t+1}q_{t+1}\} \\ &= \Omega_t s_t; \end{aligned} \tag{9.166}$$

from which

$$s_t = \Omega_t^{-1}E(\eta_t|\mathcal{I}_n) = \Omega_t^{-1}\{y_t - H_t x_{t|n}\}, \tag{9.167}$$

where the final equality is justified by the observation equation (9.68). Notice that, in order to calculate s_t from this expression, we need $x_{t|n}$, which is available only because we are using a backward smoothing algorithm. Thus s_t is calculated from (9.167) using the previous smoothed estimate. Then it is substituted in (9.162) to obtain q_t. Finally, the smoothed estimate of the state vector is obtained from equation (9.161). Whittle [517] has derived this algorithm by maximising a log-likelihood function.

Multi-step Prediction

Consider a sequence of predictions into the future which are made at time t, and let us juxtapose with these predictions the expressions for the corresponding values of the true state variables. Then, on the assumption that $\Phi_t = \Phi$ is a constant matrix, we have

$$\begin{aligned} x_{t+1|t} &= \Phi x_t, & \xi_{t+1} &= \Phi\xi_t + \nu_{t+1}, \\ x_{t+2|t} &= \Phi^2 x_t, & \xi_{t+2} &= \Phi^2\xi_t + \Phi\nu_{t+1} + \nu_{t+2}, \\ x_{t+3|t} &= \Phi^3 x_t, & \xi_{t+3} &= \Phi^3\xi_t + \Phi^2\nu_{t+1} + \Phi\nu_{t+2} + \nu_{t+3}, \\ &\vdots & &\vdots \\ x_{t+n|t} &= \Phi^n x_t, & \xi_{t+n} &= \Phi^n\xi_t + \sum_{j=0}^{n-1}\Phi^j\nu_{t+n-j}. \end{aligned} \tag{9.168}$$

It follows that the error in predicting n periods into the future is given by

$$x_{t+n|t} - \xi_{t+n} = \Phi^n(x_t - \xi_t) - \sum_{j=0}^{n-1}\Phi^j\nu_{t+n-j}. \tag{9.169}$$

The vectors ν_{t+j} are statistically independent and have dispersion matrices which are denoted by $D(\nu_{t+j}) = \Psi_{t+j}$. Therefore the dispersion of the error in predicting the state is just

$$P_{t+n|t} = \Phi^n P_t(\Phi^n)' + \sum_{j=0}^{n-1} \Phi^j \Psi_{t+n-j}(\Phi^j)'. \tag{9.170}$$

Bibliography

[12] Anderson, B.D.O., and Moore, J.B., (1979), *Optimal Filtering*, Prentice-Hall, Englewood Cliffs, New Jersey.

[24] Ansley, C.F., and R. Kohn, (1982), A Geometrical Derivation of the Fixed Interval Smoothing Equations, *Biometrika*, **69**, 486–7.

[25] Ansley, C.F., and R. Kohn, (1985), Estimation, Filtering and Smoothing in State Space Models with Incompletely Specified Initial Conditions, *The Annals of Statistics*, **13**, 1286–1316.

[50] Bertrand, J., (1855), *Méthode des Moindres Carrés: Mémoires sur la Combinaison des Observations par C.F. Gauss*, translation into French of *Theoria combinationis observationum erroribus minimis obnoxiae, by C.F. Gauss*, Mallet-Bachelier, Paris.

[82] Brown, R.L., J. Durbin and J.M. Evans, (1975), Techniques for Testing the Constancy of Regression Relationships over Time, *Journal of the Royal Statistical Society, Series B*, **37**, 149–163.

[87] Buja, A., T. Hastie and R. Tibshirani, (1989), Linear Smoothers and Additive Models, *The Annals of Statistics*, **17**, 453–555.

[106] Chow, G.C., (1983), *Econometrics*, McGraw–Hill, New York.

[144] De Jong, P., (1988), The Likelihood for a State Space Model, *Biometrika*, **75**, 165–169.

[145] De Jong, P., (1988), A Cross-Validation Filter for Time Series Models, *Biometrika*, **75**, 594–600.

[146] De Jong, P., (1989), Smoothing and Interpolation with the State-Space Model, *Journal of the American Statistical Association*, 84, 1085–1088.

[147] De Jong, P., (1991), Stable Algorithms for the State Space Model, *Journal of Time Series Analysis*, **12**, 143–157.

[148] De Jong, P., (1991), The Diffuse Kalman Filter, *The Annals of Statistics*, **19**, 1073–1083.

[163] Duncan, D.B., and S.D. Horn, (1972), Linear Dynamic Recursive Estimation from the Viewpoint of Regression Analysis, *Journal of the American Statistical Association*, **67**, 815–821.

[178] Farebrother, R.W., (1990), Mnemonics for the Kalman Filter Covariance, *Statische Hefte (Statistical Papers)*, **31**, 281–284.

[181] Farooq, M., and A.K. Mahalanabis, (1971), A Note on the Maximum Likelihood State Estimation of Linear Discrete Systems with Multiple Time Delays, *IEEE Transactions on Automatic Control*, **AC-16**, 105–106.

[204] Gauss, C.F., 1777–1855, (1809), *Theoria Motus Corporum Celestium*, English translation by C.H. Davis (1857). Reprinted 1963, Dover Publications, New York.

[205] Gauss, C.F., 1777–1855, (1821, 1823, 1826), *Theoria combinationis observationum erroribus minimis obnoxiae, (Theory of the combination of observations least subject to error)*, French translation by J. Bertrand (1855), *Méthode de Moindres Carrés: Mémoires sur la combinaison des Observations par C.-F. Gauss*, Mallet–Bachelier, Paris. English translation by G.W. Stewart (1995), Classics in Applied Mathematics no. 11, SIAM Press, Philadelphia.

[225] Gordon, K., and A.F.M. Smith, (1990), Modelling and Monitoring Biomedical Time Series, *Journal of the American Statistical Association*, **85**, 328–337.

[241] Hannan, E.J., and M. Deistler, (1988), *The Statistical Theory of Linear Systems*, John Wiley and Co., New York.

[247] Harvey, A.C., (1989), *Forecasting, Structural Time Series Models and the Kalman Filter*, Cambridge University Press, Cambridge.

[281] Kalman, R.E., (1960), A New Approach to Linear Filtering and Prediction Problems, *Transactions of the American Society of Mechanical Engineers (ASME), Journal of Basic Engineering*, **82**, 35–45.

[282] Kalman, R.E., and R.S. Bucy, (1961), New Results in Linear Filtering and Prediction Theory, *Transactions of the American Society of Mechanical Engineers (ASME), Journal of Basic Engineering*, **83**, 95–107.

[296] Kohn, R., and C.F. Ansley, (1989), A Fast Algorithm for Signal Extraction, Influence and Cross-Validation in State Space Models, *Biometrika*, **76**, 65–79.

[299] Koopman, S.J., (1990), *Efficient Smoothing Algorithms for Time Series Models*, Discussion Paper of the Department of Statistics, The London London School School of Economics.

[339] Mayne, D.Q., (1966), A Solution of the Smoothing Problem for Linear Dynamic Systems, *Automatica*, **4**, 73–92.

[341] Mehra, R.K., (1979), Kalman Filters and their Applications to Forecasting, *TIMS Studies in Management Sciences*, **12**, 75–94.

[395] Plackett, R.L., (1950), Some Theorems in Least Squares, *Biometrika*, **37**, 149–157.

[408] Premier, R., and A.G. Vacroux, (1971), On Smoothing in Linear Discrete Systems with Time Delays, *International Journal of Control*, **13**, 299–303.

[441] Schneider, W., (1988), Analytical Uses of Kalman Filtering in Econometrics: A Survey, *Statistiche Hefte*, **29**, 3–33.

[517] Whittle, P., (1991), Likelihood and Cost as Path Integrals, *Journal of the Royal Statistical Society, Series B*, **53**, 505–538.

[525] Willems, J.C., (1978), Recursive Filtering, *Statistica Neerlandica*, **32**, 1–39.

[526] Willman, W.W., (1969), On the Linear Smoothing Problem, *IEEE Transactions on Automatic Control*, **AC-14**, 116–117.

[537] Young, P., (1984), *Recursive Estimation and Time-Series Analysis*, Springer Verlag, Berlin.

CHAPTER 10

Estimation of Polynomial Trends

In many time series, broad movements can be discerned which evolve more gradually than do the other motions which are evident. These gradual changes are described as trends. The changes which are of a transitory nature are described as fluctuations.

In some cases, the trend should be regarded as nothing more than the accumulated effect of the fluctuations. In other cases, we feel that the trends and the fluctuations reflect different sorts of influences, and we are inclined to decompose the time series into the corresponding components.

It may be possible to capture the salient features of a trend with a polynomial function of a low degree; and, sometimes, a simple physical analogy suggests why this is appropriate. A body which possesses a high degree of inertia will tend to travel in a straight line with constant velocity and it will not be diverted much by minor impacts. The resulting motion will give rise to a linear time-trend. A body which is subject to a constant force which is independent of its velocity will be accelerated at a constant rate; and, therefore, it will follow a quadratic time-trend. If the objects which are observed are attached, in ways which are more or less flexible, to such inexorable motions, then, in the long run, their underlying trends will be strongly expressed.

Even when there is no theory to specify the trend as a particular function of time, a polynomial may, nevertheless, stand in place of a more complicated yet unknown function. In the early parts of this chapter, we shall deal, at length, with the means of estimating polynomial time-trends.

In other classes of phenomena, a polynomial has insufficient flexibility for capturing the trend. Some trends are due to gradual and continuous processes, such as the processes of biological growth, which are affected by the events which occur during the course of their evolution. Such trends may be extracted from the data via a process of smoothing. The classical methods of data-smoothing rely on weighted averages and other filtering devices. In the latter half of this chapter, we shall present an alternative method of trend estimation which makes use of functions which are constructed from polynomial segments. In the following chapter, we shall develop this method further when we present the so-called smoothing spline which can serve the same purpose as a smoothing filter.

Polynomial Regression

The topic of polynomial regression may be approached by considering the problem of approximating a function whose values at the points $x_0, \ldots, x_n$ are known only via the observations $y_0, \ldots, y_n$ which are subject to error. If we choose to represent the underlying function by a polynomial in x of degree $q \leq n$, then we are led to a model in the form of

$$y_t = \beta_0 + \beta_1 x_t + \beta_2 x_t^2 + \cdots + \beta_q x_t^q + \varepsilon_t; \quad t = 0, \ldots, n. \tag{10.1}$$

In matrix notation, this becomes

$$y = X\beta + \varepsilon, \tag{10.2}$$

where $y = [y_0, \ldots, y_n]'$, $X = [x_t^j]$, $\beta = [\beta_0, \ldots, \beta_q]'$ and $\varepsilon = [\varepsilon_0, \ldots, \varepsilon_n]'$. This is just a case of the usual regression model with the peculiar feature that the independent variables are powers of x. Therefore the problem of determining the coefficients $\beta_0, \ldots, \beta_q$ appears to fall well within the ambit of the regression procedures which are described in Chapter 8. Such procedures are often adequate when the degree of the polynomial in no greater than three or four. However, as the degree increases, they are beset by worsening problems of numerical instability.

In the case of a procedure which determines the parameter vector β by solving the normal equations $T^{-1}X'X\beta = T^{-1}X'y$, where $T = n+1$, the instability may be attributed to the ill-conditioned nature of the moment matrix $X'X/T$. The generic element in the $(i+1)$th row and $(j+1)$th column of this matrix has the value of $T^{-1}\sum_t x_t^{i+j}$. A simple statistical analogy suggests that, if the sample points $x_0, \ldots, x_n$ are distributed uniformly in the interval $[0, 1]$ and if n is large, then the value of the element can be approximated by

$$\int_0^1 x^{i+j} dx = \frac{1}{i+j+1}. \tag{10.3}$$

Therefore the moment matrix as a whole can be approximated by

$$\begin{bmatrix} 1 & \frac{1}{2} & \frac{1}{3} & \cdots & \frac{1}{q+1} \\ \frac{1}{2} & \frac{1}{3} & \frac{1}{4} & \cdots & \frac{1}{q+2} \\ \frac{1}{3} & \frac{1}{4} & \frac{1}{5} & \cdots & \frac{1}{q+3} \\ \vdots & \vdots & \vdots & & \vdots \\ \frac{1}{q+1} & \frac{1}{q+2} & \frac{1}{q+3} & \cdots & \frac{1}{2q+1} \end{bmatrix}. \tag{10.4}$$

This is an instance of the notoriously ill-conditioned Hilbert matrix whose inverse has elements with values which are very large and which increase rapidly with the order of the matrix. It follows that any rounding error associated with the floating-point representation of an element of the matrix X is liable to be greatly magnified in the process of inverting the moment matrix. This makes it impossible to compute an accurate solution to the normal equations. Later, we shall see that, in practice, when x is not confined to the interval $[0, 1]$, the numerical problems can be far more severe that the foregoing example suggests that they are.

To avoid the numerical problems which can arise from the attempt to fit the equation under (10.1), the powers $1, x, x^2, \ldots, x^q$ may be replaced by a set of polynomials $\phi_0(x), \phi_1(x), \phi_2(x), \ldots, \phi_q(x)$ which are mutually orthogonal. Thus equation (10.1) may be replaced by an equation

$$\begin{aligned} y_t &= \alpha_0\phi_0(x_t) + \alpha_1\phi_1(x_t) + \cdots + \alpha_q\phi_q(x_t) + \varepsilon_t \\ &= \alpha_0\phi_{t0} + \alpha_1\phi_{t1} + \cdots + \alpha_q\phi_{tq} + \varepsilon_t \end{aligned} \tag{10.5}$$

wherein the values $\phi_{tj} = \phi_j(x_t); t = 0, \ldots, n$, generated by the polynomial functions at the points $x_0, x_1, \ldots, x_n$, are subject to the orthogonality conditions

$$\sum_t \phi_{tj}\phi_{tk} = \begin{cases} 0, & \text{if } j \neq k; \\ \lambda_j, & \text{if } j = k. \end{cases} \tag{10.6}$$

Let $\alpha = [\alpha_0, \alpha_1, \ldots, \alpha_q]'$ be the vector of the coefficients of (10.5) and let $\Phi = [\phi_{tj}]$ be the matrix of the values of the polynomials. Then the regression equation of (10.2) may be rewritten as

$$y = \Phi\alpha + \varepsilon \quad \text{with} \quad \Phi = XR^{-1} \quad \text{and} \quad \alpha = R\beta, \tag{10.7}$$

where R is a nonsingular matrix of order $q + 1$. Given that $\Phi'\Phi = \Lambda$ is a diagonal matrix, it follows that the ordinary least-squares estimating equations $\alpha = (\Phi'\Phi)^{-1}\Phi'y = \Lambda^{-1}\Phi'y$ resolve themselves into a set of $q + 1$ simple regression equations of the form

$$\alpha_j = \frac{\sum_t \phi_{tj}y_t}{\sum_t \phi_{tj}^2}; \quad j = 0, 1, \ldots, q. \tag{10.8}$$

The orthogonalisation of the vectors of observations on an arbitrary set of functions can be accomplished using the Gram–Schmidt procedure. Two versions of this procedure will be described in the following sections. However, a recursive method for generating an orthogonal set of polynomials is also available; and this will be used in a procedure dedicated to the task of fitting polynomial regressions which will be presented in a later section.

The Gram–Schmidt Orthogonalisation Procedure

The polynomial functions $\phi_0(x), \phi_1(x), \ldots, \phi_q(x)$ are said to be linearly independent over the domain $\{x_0, x_1, \ldots, x_n\}$, where $n \geq q$, if the condition that

$$\mu_0\phi_0(x_t) + \mu_1\phi_1(x_t) + \cdots + \mu_q\phi_q(x_t) = 0 \quad \text{for} \quad t = 0, 1, \ldots, n \tag{10.9}$$

can be satisfied only by setting $\mu_0 = \mu_1 = \cdots = \mu_q = 0$. If the polynomials are linearly independent over a restricted domain, then they will be linearly dependent over an enlarged domain which includes the restricted domain.

The sets of polynomials $p_0(x), \ldots, p_q(x)$ and $\phi_0(x), \ldots, \phi_q(x)$ are equivalent over the real line if there exist sets of scalars $\lambda_{i0}, \ldots, \lambda_{iq}$ and $\mu_{j0}, \ldots, \mu_{jq}$ such that

$$p_i(x) = \sum_j \lambda_{ij}\phi_j(x) \quad \text{and} \quad \phi_j(x) = \sum_i \mu_{ji}p_i(x) \tag{10.10}$$

for all x and for all $i, j = 0, \ldots, q$.

An arbitrary set of linearly independent polynomials can be transformed to an equivalent orthogonal set by subjecting it to the Gram–Schmidt procedure. A convenient choice for these linearly independent polynomials $p_0(x), p_1(x), p_2(x), \ldots, p_q(x)$ is the set of powers $1, x, x^2, \ldots, x^q$ whose use in the regression equation was rejected on the grounds of numerical instability.

We shall use $p_j = [p_{0j}, p_{1j}, \ldots, p_{nj}]'$ to denote the vector containing the values of the function $p_j(x)$ at the points $\{x_0, x_1, \ldots, x_n\}$, and we shall use $\phi_j = [\phi_{0j}, \phi_{1j}, \ldots, \phi_{nj}]'$ to denote the analogous vector corresponding to the function $\phi_j(x)$ which belongs to the set of orthogonal polynomials.

The process of orthogonalising the polynomials starts by setting $\phi_0 = p_0$. Then the component of p_1 which is orthogonal to ϕ_0 becomes ϕ_1. The latter may written as

$$\phi_1 = p_1 - r_{01}\phi_0, \tag{10.11}$$

where the scalar r_{01} is determined so as to satisfy the orthogonality condition $\phi_0'\phi_1 = 0$. The condition yields the equation $0 = \phi_0' p_1 - r_{01}\phi_0'\phi_0$; from which it follows that

$$r_{01} = \frac{\phi_0' p_1}{\phi_0'\phi_0}. \tag{10.12}$$

The second step entails finding the component ϕ_2 of p_2 which is orthogonal to both ϕ_0 and ϕ_1. We write

$$\phi_2 = p_2 - r_{02}\phi_0 - r_{12}\phi_1, \tag{10.13}$$

and we use the orthogonality conditions $\phi_0'\phi_2 = \phi_1'\phi_2 = 0$ to determine the coefficients

$$r_{02} = \frac{\phi_0' p_2}{\phi_0'\phi_0}, \qquad r_{12} = \frac{\phi_1' p_2}{\phi_1'\phi_1}. \tag{10.14}$$

The process can be continued through successive steps until the set of vectors $\{p_j\}$ has been exhausted. In the kth step, we set

$$\phi_k = p_k - r_{0k}\phi_0 - r_{1k}\phi_1 - \cdots - r_{k-1,k}\phi_{k-1}; \tag{10.15}$$

and the orthogonality conditions $\phi_i'\phi_j = 0; i, j = 0, 1, \ldots, k, i \neq j$ serve to determine the coefficients

$$r_{0k} = \frac{\phi_0' p_k}{\phi_0'\phi_0}, \quad r_{1k} = \frac{\phi_1' p_k}{\phi_1'\phi_1}, \ldots, r_{k-1,k} = \frac{\phi_{k-1}' p_k}{\phi_{k-1}'\phi_{k-1}}. \tag{10.16}$$

These can be construed as the coefficients of the regression of p_k upon the vectors $\phi_0, \phi_1, \ldots, \phi_{k-1}$. The vector ϕ_k is the residual of this multiple regression.

The orthogonalisation procedure can also be described in the terminology of orthogonal projection which is established at the beginning of Chapter 8. The kth orthogonal vector ϕ_k is simply the projection of p_k onto the orthogonal complement of the subspace spanned by the columns of the matrix $\Phi_{k-1} = [\phi_0, \phi_1, \ldots, \phi_{k-1}]$. Thus $\phi_k = (I - W_{k-1})p_k$, where $W_{k-1} = \Phi_{k-1}(\Phi'_{k-1}\Phi_{k-1})^{-1}\Phi'_{k-1}$. The conditions under (10.6), which declare that the columns of Φ'_{k-1} are mutually orthogonal, indicate that

$$\begin{aligned} W_{k-1}p_k &= \sum_{j=0}^{k-1}\left(\frac{\phi_j\phi'_j}{\phi'_j\phi_j}\right)p_k \\ &= \sum_{j=0}^{k-1}\left(\frac{\phi'_j p_k}{\phi'_j\phi_j}\right)\phi_j = \sum_{j=0}^{k-1} r_{jk}\phi_j; \end{aligned} \tag{10.17}$$

and, by putting the final expression into the equation $\phi_k = p_k - W_{k-1}p_k$, we derive the expression under (10.15).

The sequence of equations

$$\begin{aligned} \phi_0 &= p_0, \\ \phi_1 &= p_1 - r_{01}\phi_0, \\ \phi_2 &= p_2 - r_{02}\phi_0 - r_{12}\phi_1, \\ &\vdots \\ \phi_q &= p_q - r_{0q}\phi_0 - r_{1q}\phi_1 - \cdots - r_{q-1,q}\phi_{q-1}, \end{aligned} \tag{10.18}$$

which summarise the orthogonalisation procedure, can be rearranged as a matrix equation:

$$\begin{bmatrix} p_{00} & p_{01} & \ldots & p_{0q} \\ p_{10} & p_{11} & \ldots & p_{1q} \\ \vdots & \vdots & & \vdots \\ p_{n0} & p_{n1} & \ldots & p_{nq} \end{bmatrix} = \begin{bmatrix} \phi_{00} & \phi_{01} & \ldots & \phi_{0q} \\ \phi_{10} & \phi_{11} & \ldots & \phi_{1q} \\ \vdots & \vdots & & \vdots \\ \phi_{n0} & \phi_{n1} & \ldots & \phi_{nq} \end{bmatrix} \begin{bmatrix} 1 & r_{01} & \ldots & r_{0q} \\ 0 & 1 & \ldots & r_{1q} \\ \vdots & \vdots & \ddots & \vdots \\ 0 & 0 & \ldots & 1 \end{bmatrix}. \tag{10.19}$$

This may be written as $P = \Phi R$; and thus the Gram–Schmidt procedure is recognised as a means of effecting the Q–R decomposition of the matrix P. The Householder procedure, which achieves the same end, generates the orthogonal matrix $Q = [Q_1, Q_2]$ of order $T = n + 1$ as an adjunct of a process by which the $T \times k$ matrix $X = Q_1 R$ is triangulated. That is to say, the matrix X—which is to be compared with P—is transformed into

$$Q'X = \begin{bmatrix} Q'_1 \\ Q'_2 \end{bmatrix} Q_1 R = \begin{bmatrix} R \\ 0 \end{bmatrix}. \tag{10.20}$$

The Gram–Schmidt procedure, which approaches the problem from a different direction, generates an upper-triangular matrix R as a by-product of a process designed to create a matrix Φ comprising $q+1$ orthogonal vectors of order $T = n+1$, which is comparable to the matrix Q_1.

In order to generate a matrix of orthonormal vectors via the Gram–Schmidt procedure, we would need to rescale the column vectors of $\Phi = [\phi_0, \ldots, \phi_q]$. The lengths of these vectors are given by the diagonal elements of the matrix $\Phi'\Phi = \Lambda = \text{diag}[\lambda_0, \ldots, \lambda_q]$. Let $\Lambda^{-1/2} = \text{diag}[1/\sqrt{\lambda_0}, \ldots, 1/\sqrt{\lambda_q}]$. Then $C = \Phi\Lambda^{-1/2}$ is an orthonormal matrix with $C'C = I_{q+1}$.

A Modified Gram–Schmidt Procedure

There is a modified version of the Gram–Schmidt procedure which is superior, in terms of its numerical stability, to the classical version described above.

The classical procedure generates each orthogonal vector ϕ_k in turn in a single step. In the modified procedure, the process of generating ϕ_k is spread over k steps. At the start of the modified procedure, the vector which will become ϕ_k is given the value of $p_k^{(0)} = p_k$. In the first step, this vector is updated to become $p_k^{(1)} = p_k^{(0)} - r_{0k}\phi_0$, where $\phi_0 = p_0$; and, at the same time, the procedure generates $\phi_1 = p_1^{(0)} - r_{01}\phi_0$. By continuing in this way, the procedure ge ıerates, in the kth step, the vector

$$\phi_k = p_k^{(k-1)} - r_{k-1,k}\phi_{k-1}. \tag{10.21}$$

A simple process of back-substitution shows that this vector can be written as

$$\phi_k = p_k^{(0)} - r_{0k}\phi_0 - r_{1k}\phi_1 - \cdots - r_{k-1,k}\phi_{k-1}; \tag{10.22}$$

and this repeats the expression under (10.15) which belongs to the classical procedure. Thus the algebraic equivalence of the two procedures is demonstrated.

To illustrate the modified procedure, we may portray its initial conditions and its first two steps as follows:

$$\begin{array}{lllll}
\textbf{Step 0} & \qquad & \textbf{Step 1} & \qquad & \textbf{Step 2} \\
\phi_0 = p_0 & & \phi_0 & & \phi_0 \\
p_1^{(0)} = p_1 & & \phi_1 = p_1^{(0)} - r_{01}\phi_0 & & \phi_1 \\
\vdots & & p_2^{(1)} = p_2^{(0)} - r_{02}\phi_0 & & \phi_2 = p_2^{(1)} - r_{12}\phi_1 \\
 & & \vdots & & p_3^{(2)} = p_3^{(1)} - r_{13}\phi_1 \\
 & & & & \vdots \\
p_q^{(0)} = p_q & & p_q^{(1)} = p_q^{(0)} - r_{0q}\phi_0 & & p_q^{(2)} = p_q^{(1)} - r_{1q}\phi_1.
\end{array} \tag{10.23}$$

In the modified procedure, the coefficients $r_{k-1,j}$ with $j = k, \ldots, q$, which are generated in the kth step, can be calculated as

$$r_{k-1,j} = \frac{\phi'_{k-1}p_j^{(k-1)}}{\phi'_{k-1}\phi_{k-1}}. \tag{10.24}$$

These coefficients belong to the $(k-1)$th *row* of the matrix R of (10.19). The kth step of the classical procedure, described in the previous section, generates the coefficients of the kth *column* of R. Since the orthogonality conditions which prevail amongst the vectors $\phi_0, \phi_1, \ldots, \phi_{k-1}$ indicate that

$$\begin{aligned}\phi'_{k-1}p_j^{(k-1)} &= \phi'_{k-1}\big(p_j - r_{0j}\phi_0 - r_{1j}\phi_1 - \cdots - r_{k-2,j}\phi_{k-2}\big) \\ &= \phi'_{k-1}p_j,\end{aligned} \tag{10.25}$$

it follows that the formula of (10.24) gives a value for $r_{k-1,j}$ which would be identical to the one generated by the classical procedure, if there were no rounding error. The advantage of using equation (10.24) in the modified procedure is that there is no need to remember the value of the vector p_j, which is lost when the vector is modified in the first step. The superior numerical stability of the modified procedure is due largely to the fact that $p_j^{(k-1)}$ is used in place of p_j in calculating the coefficient $r_{k-1,j}$.

The difference between the classical and the modified Gram–Schmidt procedures can be summarised in a few words. Both procedures take an arbitrary set of linearly independent vectors—the source set—and they transform them one-by-one into the vectors of an orthonormal set—the destination set. In the classical procedure, a source vector remains unaltered until the time comes to transform it and to transfer it to the destination set. In the modified procedure, the destination set and the source set remain mutually orthogonal at all times. As each vector is transferred from the source to the destination, the mutual orthogonality of the two sets is re-established by transforming the entire source set.

The modified Gram–Schmidt procedure is implemented in the following Pascal procedure.

(10.26)
```
procedure GramSchmidt(var phi, r : matrix;
                        n, q : integer);

  var
    t, j, k : integer;
    num, denom : real;

begin
  for k := 1 to q do
    begin {k}
      denom := 0.0;
      for t := 0 to n do
        denom := denom + Sqr(phi[t, k - 1]);
      for j := k to q do
        begin {j}
          num := 0.0;
          for t := 0 to n do
            num := num + phi[t, j] * phi[t, k - 1];
          r[k - 1, j] := num/denom;
          for t := 0 to n do
            phi[t, j] := phi[t, j] - r[k - 1, j] * phi[t, k - 1];
        end; {j}
    end; {k}

end; {GramSchmidt}
```

The variables which are passed to this procedure are the array *phi* which corresponds to the matrix P of (10.19) and the array r which corresponds to an identity matrix. The procedure, which effects the decomposition represented by (10.19), returns an array of orthogonal vectors in *phi* corresponding to the matrix Φ and an array in r corresponding to the upper-triangular matrix R.

Uniqueness of the Gram Polynomials

The so-called Gram polynomials, which are obtained by orthogonalising the powers $p_0(x) = 1, p_1(x) = x, p_2(x) = x^2, \ldots, p_q(x) = x^q$, are uniquely defined. The conditions of orthogonality that serve to define these polynomials are based upon the natural inner product for finite-dimensional vectors which is described as a Euclidian inner product.

The proof of uniqueness rests upon the following result:

(10.27) If $\phi_0(x), \phi_1(x), \ldots, \phi_q(x)$ is a sequence of polynomials which are orthogonal over the set $\{x_0, x_1, \ldots, x_n\}$ and which are indexed by degree—with $\phi_k(x)$ having a degree of k—then $\phi_k(x)$ is orthogonal to every polynomial of degree less than k.

To understand this, recall that the polynomials $\phi_0(x), \phi_1(x), \ldots, \phi_{k-1}(x)$ form an orthonormal basis of the k-dimensional space of all polynomials of degree less than k which are defined over $\{x_0, x_1, \ldots, x_n\}$. Therefore any polynomial $p(x)$ of a degree less than k can be written in the form of $p(x) = \pi_0\phi_0(x) + \pi_1\phi_1(x) + \cdots + \pi_{k-1}\phi_{k-1}(x)$. Hence it follows, by virtue of the orthogonality conditions of (10.6), that $\sum_t p(x_t)\phi_k(x_t) = p'\phi_k = 0$; and this proves the assertion.

The proposition regarding the uniqueness of the polynomials is that

(10.28) If $\phi_0(x), \phi_1(x), \ldots, \phi_q(x)$ and $\theta_0(x), \theta_1(x), \ldots, \theta_q(x)$ are sequences of polynomials which are indexed by degree and which are orthogonal over $\{x_0, x_1, \ldots, x_n\}$, then, for each k, $\phi_k(x)$ and $\theta_k(x)$ are scalar multiples of each other.

To prove this, we express $\theta_k(x)$ in terms of the basis $\phi_0(x), \phi_1(x), \ldots, \phi_k(x)$:

$$\theta_k(x) = r_0\phi_0(x) + r_1\phi_1(x) + \cdots + r_k\phi_k(x). \tag{10.29}$$

By letting $x = x_0, \ldots, x_n$ and by defining the vectors $\theta_k = [\theta_k(x_0), \ldots, \theta_k(x_n)]'$ and $\phi_j = [\phi_j(x_0), \ldots, \phi_j(x_n)]'; j = 0, 1, \ldots, k$, we can obtain the following vector equation:

$$\theta_k = r_0\phi_0 + r_1\phi_1 + \cdots + r_k\phi_k. \tag{10.30}$$

Here the coefficient $r_j = \phi_j'\theta_k/\phi_j'\phi_j$ is found by premultiplying the equation by ϕ_j' and then using the orthogonality conditions $\phi_j'\phi_i = 0$ to show that $\phi_j'\theta_k = \phi_j'\phi_j r_j$. Now, let us recall that $\theta_k(x)$ belongs to a set of polynomials indexed by degree which are orthonormal over the domain $\{x_0, x_1, \ldots, x_n\}$. Then it can be seen that the result under (10.27) implies that, for $j = 0, 1, \ldots, k-1$, there is $\phi_j'\theta_k = 0$ and hence $r_j = 0$. Therefore $\theta_k = r_k\phi_k$ and also $\theta_k(x) = r_k\phi_k(x)$; and this proves the present assertion.

This result implies that, disregarding scalar multiples of the elements, there is only one orthogonal basis of the space of polynomials which is indexed by degree. Thus the basis polynomials would be specified uniquely if it were decreed that they should each be a monic polynomial with unity as the coefficient of the highest power of x.

This result may come as a surprise if one is aware of the many kinds of orthogonal polynomials which are to be found in the literature. However, the variety comes from varying the domain of the polynomials and from varying the definition of the inner product associated with the conditions of orthogonality.

Example 10.1. The specification of the Gram polynomials depends upon the domain of the orthogonality conditions which are entailed in their definition. In Figure 10.1, the domain of definition is the set of points $x_i = 0.25i - 1; i = 0, \ldots, 8$ which range, with intervals of 0.25, from -1 to $+1$.

The Gram polynomials are sometimes described as Legendre polynomials. However, the latter are usually defined as the set of polynomials $P_n(x); n = 0, 1, 2 \ldots,$ which satisfy the differential equation

$$(1 - x^2)\frac{d^2y}{dx^2} - 2x\frac{dy}{dx} + n(n-1)y = 0, \tag{10.31}$$

and which obey the orthogonality relation

$$\int_{-1}^{1} P_m(x)P_n(x)dx = 0 \quad \text{for} \quad m \neq n. \tag{10.32}$$

These functions arise in studies of systems with three-dimensional spherical symmetry; and they are important in quantum mechanics and in other physical applications. See, for example, Arfken [28].

In common with all sets of orthogonal polynomials, the Legendre polynomials satisfy a three-term recurrence relationship which is of great assistance in finding their formulae and in generating their ordinates. Thus, the $(n+1)$th Legendre polynomial is given by

$$P_{n+1} = \frac{2n+1}{n+1}xP_n - \frac{n}{n+1}P_{n-1}. \tag{10.33}$$

This result remains valid in the case of $n = 0$ if, by definition, $P_{-1} = 0$. Taking $P_0(x) = 1$, and applying the formula successively, gives the following sequence:

$$\begin{aligned} P_0(x) &= 1, & P_1(x) &= x, \\ P_2(x) &= \frac{3}{2}x^2 - \frac{1}{2}, & P_3(x) &= \frac{5}{2}x^2 - \frac{3}{2}x, \\ P_4(x) &= \frac{35}{8}x^4 - \frac{15}{4}x^2 + \frac{3}{8}, & P_5(x) &= \frac{68}{8}x^4 - \frac{35}{4}x^3 + \frac{15}{8}x. \end{aligned} \tag{10.34}$$

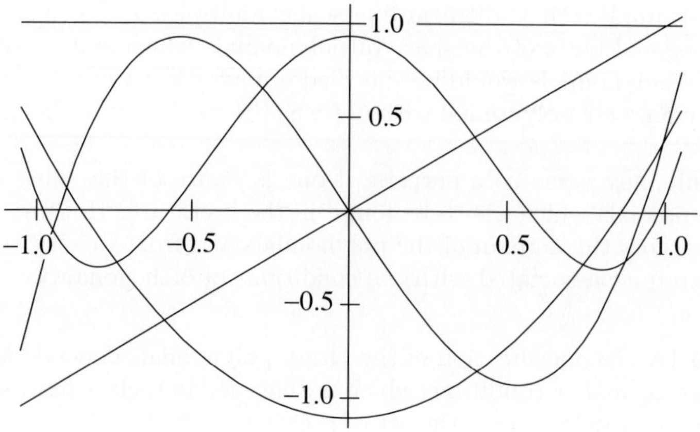

Figure 10.1. The Gram polynomials $\phi_0(x), \ldots, \phi_4(x)$ which are orthogonal over the domain $0.25i - 1; i = 0, \ldots, 8$.

It is easy to envisage that, if we define the Gram polynomials over a set of equally spaced points in the interval $[-1, 1]$ and if we allow the number of points to increase indefinitely, then the Gram polynomials will approximate the Legendre polynomials with ever-increasing accuracy. This suggests that the problems of numerical instability in polynomial regression might be largely overcome by using the Legendre polynomials in place of the powers $1, x, \ldots, x^n$. That is to say, if they are spaced more of less evenly, the data points $x_0, x_1, \ldots, x_n$ may be mapped, via a linear function, into $n + 1$ points on the interval $[-1, 1]$, whence they can be associated with the values taken by the Legendre polynomials at those points.

Recursive Generation of the Polynomials

The Gram–Schmidt procedure has been treated at length because of its theoretical importance and because it provides a method of obtaining the Q–R decomposition of a matrix which is a useful alternative to the Householder method. However, there exists a more efficient method for generating orthogonal polynomials which is based on the so-called three-term recurrence relation, of which an example is given under (10.33). This relation enables the orthogonal polynomial $\phi_{k+1}(x)$ of degree $k + 1$ to be found from the preceding polynomials $\phi_k(x)$ and $\phi_{k-1}(x)$.

To derive the recurrence relation, let us begin by writing the sought-after polynomial as

$$\begin{aligned} \phi_{k+1}(x) &= x\phi_k(x) + \mu_k\phi_k(x) + \mu_{k-1}\phi_{k-1}(x) + \cdots + \mu_0\phi_0(x) \\ &= (x + \mu_k)\phi_k(x) + \mu_{k-1}\phi_{k-1}(x) + \cdots + \mu_0\phi_0(x). \end{aligned} \tag{10.35}$$

This is a linear combination of a set of polynomials $x\phi_k(x), \phi_k(x), \phi_{k-1}(x), \ldots, \phi_0(x)$ whose degrees decline from $k + 1$ to 0. The leading coefficient of the combination is unity because $\phi_{k+1}(x)$ and $x\phi_k(x)$ are both monic polynomials with unity as the coefficient of the term x^{k+1}. Now, if we multiply this equation by $\phi_j(x)$, where $j < k - 1$, and sum over $\{x_0, x_1, \ldots, x_n\}$, then, in consequence of the orthogonality conditions which affect the polynomials $\phi_0(x), \ldots, \phi_{k+1}(x)$ over this domain, we shall find that

$$0 = \sum_t \phi_j(x_t)\{x_t\phi_k(x_t)\} + \mu_j \sum_t \phi_j^2(x_t). \tag{10.36}$$

But, given that $\phi_k(x)$ is orthogonal to all polynomials of degree less than k, including the polynomial $x\phi_j(x)$ which has a degree of $j + 1 < k$, it follows that

$$\sum_t \phi_j(x_t)\{x_t\phi_k(x_t)\} = \sum_t \phi_k(x_t)\{x_t\phi_j(x_t)\} = 0. \tag{10.37}$$

Therefore equation (10.36) implies that $\mu_j = 0$ for all $j < k - 1$. Thus we have demonstrated that the relationship in (10.35) must take the form of

$$\phi_{k+1}(x) = (x + \mu_k)\phi_k(x) + \mu_{k-1}\phi_{k-1}(x). \tag{10.38}$$

In placing this equation in the context of a recursive scheme, it is convenient to rewrite it as

$$\phi_{k+1}(x) = (x - \gamma_{k+1})\phi_k(x) - \delta_{k+1}\phi_{k-1}(x), \tag{10.39}$$

where γ_{k+1} and δ_{k+1} are coefficients which have yet to be determined. This is the three-term recurrence relationship. The initial conditions of the recursion are specified by

$$\phi_0(x) = 1 \quad \text{and} \quad \phi_{-1}(x) = 0. \tag{10.40}$$

Now the values must be found for the coefficients γ_{k+1} and δ_{k+1} which will ensure that $\phi_{k+1}(x)$ is orthogonal to its predecessors. For this purpose, we multiply equation (10.39) by $\phi_{k-1}(x)$ and sum over x to get

$$0 = \sum_t x_t\phi_{k-1}(x_t)\phi_k(x_t) - \delta_{k+1}\sum_t \phi_{k-1}^2(x_t). \tag{10.41}$$

But, with k in place of $k + 1$, equation (10.39) becomes

$$\phi_k(x) = (x - \gamma_k)\phi_{k-1}(x) - \delta_k\phi_{k-2}(x); \tag{10.42}$$

and, when this is multiplied by $\phi_k(x)$ and then summed over x, it is found that

$$\sum_t x_t\phi_{k-1}(x_t)\phi_k(x_t) = \sum_t \phi_k^2(x_t). \tag{10.43}$$

Substituting the latter into (10.41) and rearranging the result gives

(10.44) $$\delta_{k+1} = \frac{\sum_t \phi_k^2(x_t)}{\sum_t \phi_{k-1}^2(x_t)}.$$

Finally, when equation (10.39) is a multiplied by $\phi_k(x)$ and summed over x, we get

(10.45) $$0 = \sum_t x_t \phi_k^2(x_t) - \gamma_{k+1} \sum_t \phi_k^2(x_t),$$

which gives

(10.46) $$\gamma_{k+1} = \frac{\sum_t x_t \phi_k^2(x_t)}{\sum_t \phi_k^2(x_t)}.$$

The Polynomial Regression Procedure

The three-term recurrence is used in a Pascal procedure for fitting a polynomial in x to a data series $(x_0, y_0), \ldots, (x_n y_n)$. The fitted polynomial equation has the form

(10.47) $$\begin{aligned} y_t &= \alpha_0 \phi_0(x_t) + \alpha_1 \phi_1(x_t) + \cdots + \alpha_q \phi_q(x_t) + e_t \\ &= \beta(x_t) + e_t, \end{aligned}$$

and the regression coefficients are given by

(10.48) $$\alpha_k = \frac{\sum_t \phi_k(x_t) y_t}{\sum_t \phi_k^2(x_t)} = \frac{\sum_t \phi_k(x_t) e_t^{(k-1)}}{\sum_t \phi_k^2(x_t)},$$

where

(10.49) $$e_t^{(k-1)} = y_t - \alpha_0 \phi_0(x_t) - \alpha_1 \phi_1(x_t) - \cdots - \alpha_{k-1} \phi_{k-1}(x_t)$$

represents the residual of y_t after fitting a polynomial of degree $k - 1$ to the data.

(10.50)
```
procedure PolyRegress(x, y : vector;
                      var alpha, gamma, delta, poly : vector;
                      q, n : integer);

  var
    phi, philag : vector;
    denom, lagdenom, temp : real;
    i, k, t, Tcap : integer;

begin {PolyRegress}
  Tcap := n + 1;
  alpha[0] := 0.0;
  gamma[1] := 0.0;
```

```
for t := 0 to n do
  begin
    alpha[0] := alpha[0] + y[t];
    gamma[1] := gamma[1] + x[t];
  end;

alpha[0] := alpha[0]/Tcap;
gamma[1] := gamma[1]/Tcap;
lagdenom := Tcap;

for t := 0 to n do
  begin
    philag[t] := 1.0;
    phi[t] := x[t] - gamma[1];
    poly[t] := alpha[0];
  end;

for k := 1 to q do
  begin {k}

    alpha[k] := 0.0;
    gamma[k + 1] := 0.0;
    denom := 0.0;

    for t := 0 to n do
      begin {t}
        alpha[k] := alpha[k] + y[t] * phi[t];
        denom := denom + Sqr(phi[t]);
        gamma[k + 1] := gamma[k + 1] + Sqr(phi[t]) * x[t];
      end; {t}

    alpha[k] := alpha[k]/denom;
    gamma[k + 1] := gamma[k + 1]/denom;
    delta[k + 1] := denom/lagdenom;
    lagdenom := denom;

    for t := 0 to n do
      begin {t}
        poly[t] := poly[t] + alpha[k] * phi[t];
        temp := phi[t];
        phi[t] := (x[t] - gamma[k + 1]) * phi[t]
                    -delta[k + 1] * philag[t];
        philag[t] := temp;
      end; {t}

  end; {k}

end; {PolyRegress}
```

We might wish to examine the values of the coefficients of the power-series representation of the fitted polynomial:

(10.51) $$\beta(x) = \beta_0 + \beta_1 x_t + \beta_2 x_t^2 + \cdots + \beta_q x_t^q.$$

The following procedure is provided for this purpose:

(10.52)
```
procedure OrthoToPower(alpha, gamma, delta : vector;
                        var beta : vector;
                        q : integer);

  var
    phiplus : real;
    phi, philag : vector;
    i, k : integer;

begin
  phi[-1] := 0;
  phi[0] := -gamma[1];
  phi[1] := 1;
  philag[0] := 1;
  philag[1] := 0;
  beta[0] := alpha[0];

{Find the power-form coefficients}
  for k := 1 to q do
    begin {k}
      beta[k] := 0.0;
      for i := k downto 0 do
        begin
          beta[i] := beta[i] + alpha[k] * phi[i];
          phiplus := phi[i - 1] - gamma[k + 1] * phi[i]
                         -delta[k + 1] * philag[i];
          philag[i] := phi[i];
          phi[i] := phiplus;
        end;
      phi[k + 1] := 1;
      philag[k + 1] := 0;
    end; {k}

end; {OrthoToPower}
```

There should be a means of generating the value of the fitted polynomial $\beta(x)$ at an arbitrary point in its domain which is not one of the data points. Such a facility might be used in plotting the function accurately.

The means of expressing the polynomial as a function of rising powers of its argument x is already available; and so one way of generating the requisite value is to use Horner's method of nested multiplication given under (4.11). However, an alternative method is available which avoids the conversion to power-series form and which depends upon the three-term recurrence relationship.

The fitted polynomial is

$$\beta(x) = \alpha_0\phi_0(x) + \alpha_1\phi_1(x) + \cdots + \alpha_q\phi_q(x), \tag{10.53}$$

wherein $\phi_0(x), \phi_1(x), \ldots, \phi_q(x)$ are the orthogonal polynomials. The three-term recurrence relationship of (10.39) may be used to eliminate the orthogonal polynomials from this expression successively, beginning with the polynomial of highest degree q. The recurrence relationship indicates that

$$\phi_q(x) = (x - \gamma_q)\phi_{q-1}(x) - \delta_q\phi_{q-2}(x). \tag{10.54}$$

Putting this into (10.53) gives

$$\begin{aligned}\beta(x) = \alpha_0\phi_0(x) + \alpha_1\phi_1(x) + \cdots + (\alpha_{q-2} - \alpha_q\delta_q)\phi_{q-2}(x)\\ + \{\alpha_{q-1} + \alpha_q(x - \gamma_q)\}\phi_{q-1}(x);\end{aligned} \tag{10.55}$$

and, on defining

$$d_q = \alpha_q \quad \text{and} \quad d_{q-1} = \alpha_{q-1} + d_q(x - \gamma_q), \tag{10.56}$$

the latter becomes

$$\beta(x) = \alpha_0\phi_0(x) + \alpha_1\phi_1(x) + \cdots + (\alpha_{q-2} - d_q\delta_q)\phi_{q-2}(x) + d_{q-1}\phi_{q-1}(x). \tag{10.57}$$

Now the three-term recurrence can be used again to give

$$\phi_{q-1}(x) = (x - \gamma_{q-1})\phi_{q-2}(x) - \delta_{q-1}\phi_{q-3}(x) \tag{10.58}$$

and, when this is substituted into (10.57), we get

$$\beta(x) = \alpha_0\phi_0(x) + \alpha_1\phi_1(x) + \cdots + (\alpha_{q-3} - d_{q-1}\delta_{q-1})\phi_{q-3}(x) + d_{q-2}\phi_{q-2}(x), \tag{10.59}$$

wherein

$$d_{q-2} = \alpha_{q-2} + d_{q-1}(x - \gamma_{q-1}) - d_q\delta_q. \tag{10.60}$$

The process, of which we have given the first two steps, can be continued down to $d_0 = \beta(x)$ using

$$d_j = \alpha_j + d_{j+1}(x - \gamma_{j+1}) - d_{j+2}\delta_{j+2}; \qquad j = q-2, \ldots, 0. \tag{10.61}$$

The following function evaluates the polynomial $\beta(x)$ at the point x using the relationships (10.56) and (10.61):

(10.62)

```
function PolyOrdinate(x : real;
                      alpha, gamma, delta : vector;
                      q : integer) : real;

var
  d : vector;
  i, j : integer;

begin
  d[q] := alpha[q];
  d[q - 1] := alpha[q - 1] + d[q] * (x - gamma[q]);
  for j := q - 2 downto 0 do
    d[j] := alpha[j] + d[j + 1] * (x - gamma[j + 1])
              -d[j + 2] * delta[j + 2];
  PolyOrdinate := d[0];
end; {PolyOrdinate}
```

Example 10.2. Table 10.1, which is due to Tintner [482, p. 195], gives the consumption of meat per head in the United States in each year from 1919 to 1941.

The raw data present acute difficulties when attempts are made to fit the power-series form of a polynomial directly by means of an ordinary regression algorithm. Attempts which have been made by the author to fit polynomials to these data using the appropriate options for polynomial regression in statistical packages which are sold commercially have usually met with failure. In one notable case, the program was unable to fit a cubic polynomial on account of "data singularity".

The procedures *PolyRegress* of (10.50) and *OrthoToPower* of (10.52) give rise to the following estimated coefficients for a cubic polynomial:

$$\begin{aligned} \alpha_0 &= 166.191,28 & \beta_0 &= -158,774,192.000,00 \\ \alpha_1 &= -0.379,05 & \beta_1 &= 246,945.765,62 \\ \alpha_2 &= 0.073,57 & \beta_2 &= -128.025,85 \\ \alpha_3 &= 0.022,12 & \beta_3 &= 0.022,12. \end{aligned} \tag{10.63}$$

Table 10.1. Annual consumption of meat in the United States 1919–1941 measured in pounds per capita.

1919	171.5	1929	163.0	1939	165.4
1920	167.0	1930	162.1	1940	174.7
1921	164.5	1931	160.2	1941	178.7
1922	169.3	1932	161.2		
1923	179.4	1933	165.8		
1924	179.2	1934	163.5		
1925	172.6	1935	146.7		
1926	170.5	1936	160.2		
1927	168.6	1937	156.8		
1928	164.7	1938	156.8		

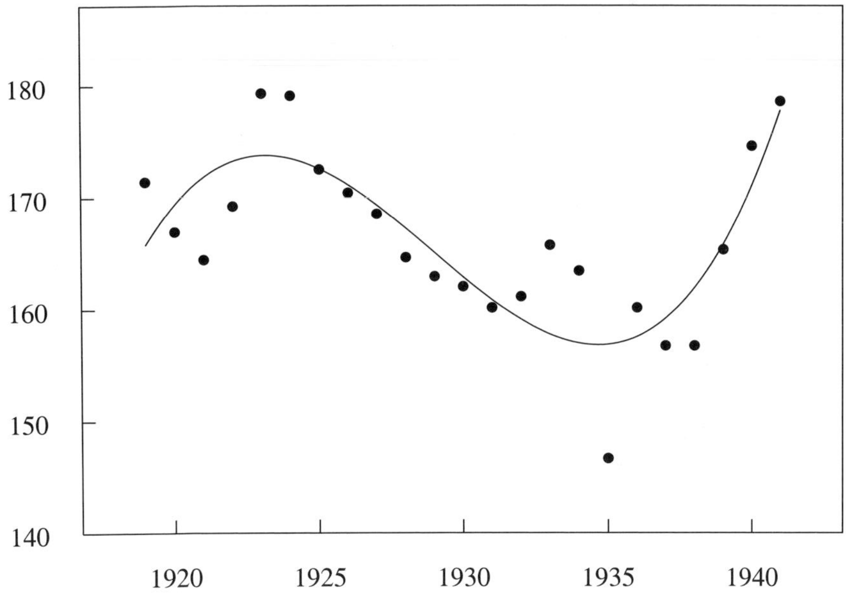

Figure 10.2. A cubic function fitted to the data on meat consumption in the United States, 1919–1941.

The α coefficients relate to the regression with orthogonal polynomials whilst the β coefficients relate to the corresponding power-series regression.

One of the reasons for the failure of the ordinary regression algorithm is the disparity in the sizes of the β coefficients; for it is clear that such values cannot coexist without the danger of serious rounding errors. If an ordinary regression algorithm is to be used for fitting a polynomial to such data, then the data must first be put into deviation form and the elements of the cross-product matrix must be scaled. Such steps are taken by the *GaussianRegression* procedure of (8.56). By putting the data in deviation form, the difficulties caused by the large absolute value of the intercept coefficient β_0 can be overcome.

The appropriate degree for the polynomial may be determined by the formal methods of statistical hypothesis testing described in Chapter 6. These methods have been discussed at length by Anderson [16] who also uses Tintner's data to provide an example. However, in this example, there is no difficulty in recognising that the appropriate polynomial is a cubic. The residual sum of squares from fitting polynomials with degrees up to five are as follows:

$$\begin{aligned} S_0 &= 1,369.538 & S_3 &= 452.833 \\ S_1 &= 1,224.413 & S_4 &= 430.152 \\ S_2 &= 1,032.400 & S_5 &= 430.151. \end{aligned} \tag{10.64}$$

The goodness of fit improves very slowly when the degree of the polynomial is raised beyond three. Figure 10.2 depicts a cubic polynomial which has been fitted to the data.

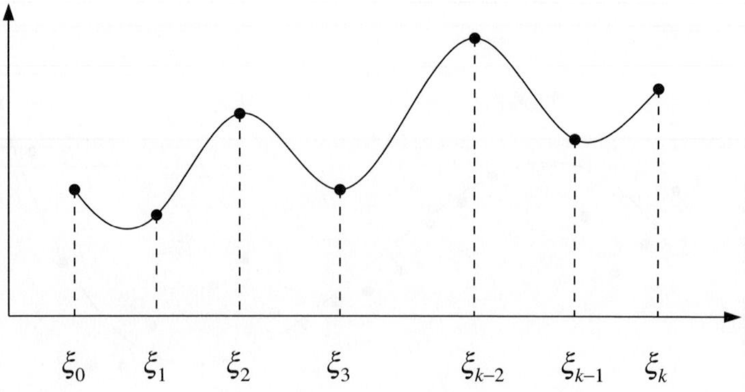

Figure 10.3. A cubic spline with knots at $\xi_0, \ldots, \xi_k$

Grafted Polynomials

A polynomial curve has certain characteristics which may make it inappropriate as a means of modelling a trend. The most notable of these is its behaviour beyond the range of the data. The branches of the polynomial tend to plus or minus infinity at an increasing rate. Moreover, the degree of the polynomial has only to be changed by one, and one of the branches will change its direction. Often, the most radical form of extrapolation which we are prepared to consider is a linear trend; and we may envisage upper and lower bounds for the quantity being extrapolated. In such cases, an extrapolation based on a polynomial is clearly at odds with our preconceptions.

There is another, more fundamental, reason why a polynomial, as much as any other analytic function, may be inappropriate for modelling a trend. This is the inability of such a function to adapt to local variations in a trend without endowing them with global consequences. An analytic function has a Taylor series expansion which is valid for all points in its domain. The expansion may be defined in terms of the derivatives of the function at an arbitrary point; and, from the knowledge of these derivatives, the value of the function at any other point may be inferred.

One way of modelling the local characteristics of a trend without prejudicing its global characteristics is to use a segmented curve. In many applications, it has been found that a curve with cubic polynomial segments is appropriate. The segments must be joined in a way which avoids evident discontinuities; and, in practice, the requirement is usually for continuous first-order and second-order derivatives (see Figure 10.3).

(10.65) The function $S(x)$ is a piecewise cubic polynomial on the interval $[x_0, x_n]$ if it is continuous and if there exists a set of points $\{\xi_i; i = 0, 1, \ldots, k\}$ of strictly increasing values with $x_0 = \xi_0 < \xi_1 < \cdots < \xi_k = x_n$ such that $S(x)$ is a polynomial of degree three at most on each of the sub-intervals $[\xi_{i-1}, \xi_i]; i = 1, \ldots, k$. If $S(x)$ has continuous derivatives up to the second order, then it is a cubic spline.

The property of second-order continuity is denoted by writing $S(x) \in \mathcal{C}^2$. The

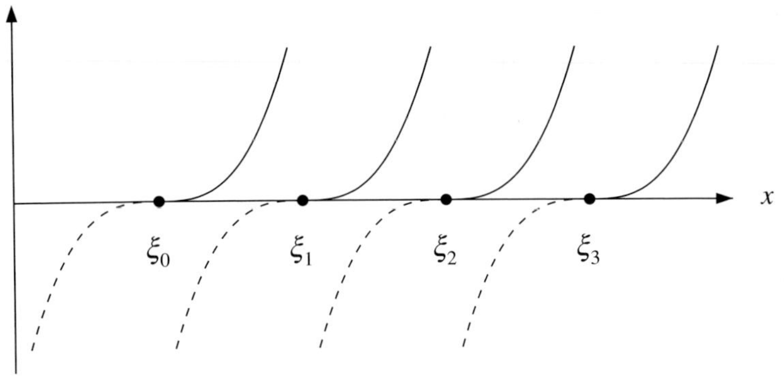

Figure 10.4. The truncated cubic power functions $(x-\xi_i)^3_+; i=0,\ldots,3$.

points $\xi_0,\ldots,\xi_k$ are described as knots whilst the coordinates $\{\xi_0, S(\xi_0)\},\ldots,\{\xi_k, S(\xi_k)\}$, which include the endpoints, are described as nodes. The joints or meeting points of the segments are $\{\xi_1, S(\xi_1)\},\ldots,\{\xi_{k-1}, S(\xi_{k-1})\}$.

If $S(x)$ is a cubic spline with joints at $\xi_1,\ldots,\xi_{k-1}$, then it can be written in the form of

(10.66)
$$\begin{aligned} S(x) &= \sum_{j=0}^{3} c_j x^j + \sum_{j=1}^{k-1} d_j (x-\xi_j)^3_+ \\ &= \sum_{j=-3}^{k-1} d_j (x-\xi_j)^3_+, \end{aligned}$$

where

(10.67)
$$(x-\xi_j)_+ = \max[0, x-\xi_j].$$

The "+" notation means simply that $u_+ = u$ if $u \geq 0$ and $u_+ = 0$ if $u \leq 0$. It is known, amongst engineers, as the Heaviside notation. The function $(x-\xi_j)^p_+$ is described as a truncated power. In Figure 10.4, the continuous curves represent truncated cubic power functions. The discontinuous curves are the parts which are discarded.

In a special case, which will be treated in the next chapter, the knots $\{\xi_i; i = 0,1,\ldots,k\}$ coincide with the data points $x_0,\ldots,x_n$. For the present, we shall consider cases where the segments in $S(x)$ are fewer that $n-1$ and where the placement of the joints of the segments is at the discretion of the investigator. We shall describe $S(x)$ as a grafted polynomial curve whenever $k < n$.

The following extended example examines more closely the formulation of (10.66); and it demonstrates the second-order continuity of functions which are expressed in this way:

Example 10.3. Let $P(x)$ and $Q(x)$ be cubic polynomial functions defined on the interval $[\xi_0, \xi_n]$ which meet at the point $\xi_1 \in [\xi_0, \xi_n]$ where $P(\xi_1) = Q(\xi_1)$. The difference between these functions is also a cubic:

$$(10.68)\qquad \begin{aligned} R(x) &= Q(x) - P(x) \\ &= r_0 + r_1(x-\xi_1) + r_2(x-\xi_1)^2 + r_3(x-\xi_1)^3. \end{aligned}$$

The derivatives of the difference function are

$$(10.69)\qquad \begin{aligned} R'(x) &= r_1 + 2r_2(x-\xi_1) + 3r_3(x-\xi_1)^2, \\ R''(x) &= 2r_2 + 6r_3(x-\xi_1), \\ R'''(x) &= 6r_3. \end{aligned}$$

At the meeting point, or knot, $x = \xi_1$, we find that

$$(10.70)\qquad \begin{aligned} R(\xi_1) &= r_0 = 0, & R''(\xi_1) &= 2r_2, \\ R'(\xi_1) &= r_1, & R'''(\xi_1) &= 6r_3; \end{aligned}$$

and so, if the functions $P(x)$ and $Q(x)$ are to have the same first and second derivatives at this point, then we must have $r_0 = r_1 = r_2 = 0$ and hence

$$(10.71)\qquad R(x) = r_3(x-\xi_1)^3.$$

Now consider the composite function $S(x) \in \mathcal{C}^2$ defined by

$$(10.72)\qquad S(x) = \begin{cases} P(x), & \text{if } x \leq \xi_1; \\ Q(x), & \text{if } x \geq \xi_1. \end{cases}$$

Using the Heaviside notation, this can be written as

$$(10.73)\qquad \begin{aligned} S(x) &= P(x) + \{Q(x) - P(x)\}_+ \\ &= P(x) + R(x)_+ \\ &= P(x) + r_3(x-\xi_1)^3_+, \end{aligned}$$

where $P(x) = \sum_{j=0}^{3} c_j x^j$ is an ordinary cubic function. It should be easy to imagine how a further cubic segment may grafted into the function $S(x)$ at a point $\xi_2 \in [\xi_1, \xi_n]$; and therefore the first expression on the RHS of (10.66) should now be intelligible.

The cubic function $P(x)$ can also be expressed in terms of the Heaviside notation. For this purpose, the points ξ_{-1}, ξ_{-2}, and ξ_{-3} must be defined such that $\xi_{-3} < \xi_{-2} < \xi_{-1} < \xi_0$. When $x \in [\xi_0, \xi_n]$, the four functions $(x-\xi_{-3})^3, \ldots, (x-\xi_0)^3$ form a basis for the set of all cubic functions of x; and, in this domain, the powers and the truncated powers coincide. Therefore, there exists a set of scalars d_{-3}, d_{-2}, d_{-1}, d_0 such that, for $x \in [\xi_0, x_n]$, we have

$$(10.74)\qquad \begin{aligned} P(x) &= \sum_{j=0}^{3} c_j x^j \\ &= \sum_{j=-3}^{0} d_j (x-\xi_j)^3_+. \end{aligned}$$

Putting this into (10.73) and defining $d_1 = r_3$ gives the expression

$$S(x) = \sum_{j=-3}^{1} d_j(x - \xi_j)^3_+, \tag{10.75}$$

which corresponds to the second phase of equation (10.66).

One can approach the problem of fitting a grafted polynomial in the way in which the problem of polynomial regression was approached at first. Thus, given the observations $x_0, x_1, \ldots, x_n$, the coefficients $d_{-3}, d_{-2}, \ldots, d_{k-1}$ in equation (10.66) can be determined by finding the values which minimise the function

$$\sum_{t=0}^{n}\Big\{y_t - S(x_t)\Big\}^2 = \sum_{t=0}^{n}\Big\{y_t - \sum_{j=-3}^{k-1} d_j(x_t - \xi_j)^3_+\Big\}^2. \tag{10.76}$$

The estimator may be expressed in matrix notation. Let $\beta = [d_{-3},, \ldots, d_{k-1}]'$ be the vector of coefficients, and let $X = [x_{tj}]$ be the design matrix whose generic element is $x_{tj} = (x_t - \xi_j)^3_+$. Then the elements of β can be determined in the usual manner via the method of ordinary least-squares regression.

The method of estimation described above, which uses ordinary truncated power functions, is straightforward to implement. However, it is beset by numerical instabilities which get worse as the length of the interval $[x_0, x_n]$ increases. Therefore it is desirable to investigate the use of other sets of basis functions which are better conditioned.

B-Splines

A more sophisticated method of fitting a grafted polynomial uses an alternative set of basis functions which are themselves polynomial splines. These are the so-called B-spline functions, and they allow the grafted polynomial to be written in the form of

$$S(x) = \sum_{j=-3}^{k-1} \lambda_j B_j(x). \tag{10.77}$$

The number $k+3$ of the functions in the basis $B_{-3}, \ldots, B_{k-1}$ is the same as the number of the parameters $c_0, \ldots, c_3, d_1, \ldots, d_{k-1}$ in the representation of $S(x)$ under (10.66).

The concept of the B-splines may be introduced by considering the problem of choosing a set of basis functions $B_j(x); j = -3, \ldots, k-1$ such that each function is zero over a large part of the range $[x_0, x_n]$.

Since we are confining our attention to a grafted polynomial which is a cubic spline, the B-spline functions must also be based on cubic polynomials. The generic cubic B-spline can be written, in terms of the Heaviside notation of the previous section, as

$$(10.78)\qquad \begin{aligned} B_p(x) &= \sum_{i=p}^{q} d_i(x-\xi_i)_+^3 \\ &= \sum_{i=p}^{q} d_i\left(x^3 - 3x^2\xi_i + 3x\xi_i^2 - \xi_i^3\right)_+. \end{aligned}$$

Two features may be observed. The first is that $B_p(x) = 0$ if $x \leq \xi_p$. The second is that $B_p(x) \in \mathcal{C}^2$, which is to say that the function exhibits second-order continuity over the entire range $[x_0, x_n]$. In effect, the function and its first two derivatives rise from zero at the point ξ_p and vary thereafter without discontinuity.

We choose the coefficients $d_p, d_{p+1}, \ldots, d_q$ and, at the same time, we fix the value of q so that $B_p \neq 0$ only if $x \in [\xi_p, \xi_q]$. The interval over which the B-spline is nonzero is described as its support. Since we already have $B_p(x) = 0$ if $x \leq \xi_p$, the problem is to ensure that $B_p(x) = 0$ if $x \geq \xi_q$. In view of (10.78), the necessary and sufficient condition for this is that

$$(10.79)\qquad \sum_{i=p}^{q} d_i\xi_i^k = 0, \quad \text{for} \quad k = 0, \ldots, 3.$$

To guarantee the consistency of these equations, we must take $q \geq p + 4$. To determine the coefficients $d_p, d_{p+1}, \ldots, d_q$ uniquely, it is sufficient to set $q = p + 4$ and to attribute a value of unity to d_p. Then the four remaining coefficients are determined by the system

$$(10.80)\qquad \begin{bmatrix} 1 & 1 & 1 & 1 \\ \xi_{p+1} & \xi_{p+2} & \xi_{p+3} & \xi_{p+4} \\ \xi_{p+1}^2 & \xi_{p+2}^2 & \xi_{p+3}^2 & \xi_{p+4}^2 \\ \xi_{p+1}^3 & \xi_{p+2}^3 & \xi_{p+3}^3 & \xi_{p+4}^3 \end{bmatrix} \begin{bmatrix} d_{p+1} \\ d_{p+2} \\ d_{p+3} \\ d_{p+4} \end{bmatrix} = - \begin{bmatrix} 1 \\ \xi_p \\ \xi_p^2 \\ \xi_p^3 \end{bmatrix}.$$

One should recognise that the values which satisfy the equations of (10.79) will also satisfy the equations

$$(10.81)\qquad \sum_{i=p}^{q} d_i(\xi_i - \theta)^k = 0, \quad \text{for} \quad k = 0, \ldots, 3,$$

wherein θ is an arbitrary constant. This indicates that the coefficients of the B-spline are determined only by the relative positions of the knots ξ_i and are not affected their absolute positions. If the knots are equally spaced, then the B-splines in what is then described as the uniform basis $B_j; j = -3, \ldots, k-1$ will have an identical form. The support of each successive B-spline will be shifted to the right by a constant amount.

In order to construct a basis set comprising $k + 3$ cubic B-splines, some extra support points must be added to the set $\xi_0, \ldots, \xi_k$. Given that the support of each cubic B-spline comprises 5 points, it follows that the original set of points can support only $k-3$ functions. A further 6 points are needed. These should be placed outside the interval $[\xi_0, \xi_k]$ at both ends. If the extra points are denoted by ξ_{-3}, ξ_{-2}, ξ_{-1}, and ξ_{k+1}, ξ_{k+2}, ξ_{k+3}, then we should have

$$\xi_{-3} < \xi_{-2} < \xi_{-1} < \xi_0 \quad \text{and} \quad \xi_k < \xi_{k+1} < \xi_{k+2} < \xi_{k+3}. \tag{10.82}$$

Example 10.4. Let $p=-2$ and $q=2$ in the formula (10.78) and let $\xi_i = i; i = p, \ldots, q$ be equally spaced knots, which gives us the points $-2, 1, 0, 1, 2$. Then, if we set $d_{-2} = 1$, the equations of (10.79) can be written as

$$\begin{bmatrix} 1 & 1 & 1 & 1 \\ -1 & 0 & 1 & 2 \\ 1 & 0 & 1 & 4 \\ -1 & 0 & 1 & 8 \end{bmatrix} \begin{bmatrix} d_{-1} \\ d_0 \\ d_1 \\ d_2 \end{bmatrix} = \begin{bmatrix} -1 \\ 2 \\ -4 \\ 8 \end{bmatrix}. \tag{10.83}$$

The solution gives $d_{-1} = -4$, $d_0 = 6$, $d_1 = -4$, and $d_2 = 1$. When these values are substituted into the equation under (10.78), we find that the basis function $B_p(x)$ is composed of the following segments over the range $[-2, 2]$ which constitutes its support:

$$\begin{aligned} x^3 + 6x^2 + 12x + 8, &\qquad -2 \le x \le -1, \\ -3x^3 - 6x^2 + 4, &\qquad -1 \le x \le 0, \\ 3x^3 - 6x^2 + 4, &\qquad 0 \le x \le 1, \\ -x^3 + 6x^2 - 12x + 8, &\qquad 1 \le x \le 2. \end{aligned} \tag{10.84}$$

Each of these segments can be expressed as a function of a variable $t \in [0, 1]$. Thus, when $t = (x - \xi_i)/(\xi_{i+1} - \xi_i)$, we have $x = \xi_i + t(\xi_{i+1} - \xi_i)$; and this gives the following parametric equations for the segments:

$$\begin{aligned} t^3, &\qquad -2 \le x \le -1, \\ -3t^3 + 3t^2 + 3t + 1, &\qquad -1 \le x \le 0, \\ 3t^3 - 6t^2 + 4, &\qquad 0 \le x \le 1, \\ -t^3 + 3t^2 - 3t + 1, &\qquad 1 \le x \le 2. \end{aligned} \tag{10.85}$$

The segments of $B_p(x)$ are shown in Figure 10.5.

Closed algebraic expressions can be found for the parameters $d_p, \ldots, d_q$ using a result which is given under (4.92). The result indicates that

$$\sum_{i=p}^{q} \frac{\xi_i^k}{\prod_{j \neq i}(\xi_j - \xi_i)} = \delta_{k,(q-p)}, \tag{10.86}$$

where $\delta_{k,(q-p)}$ is Kronecker's delta. This expression is to be compared with the equations under (10.79) which determine the parameters of the cubic B-spline. In

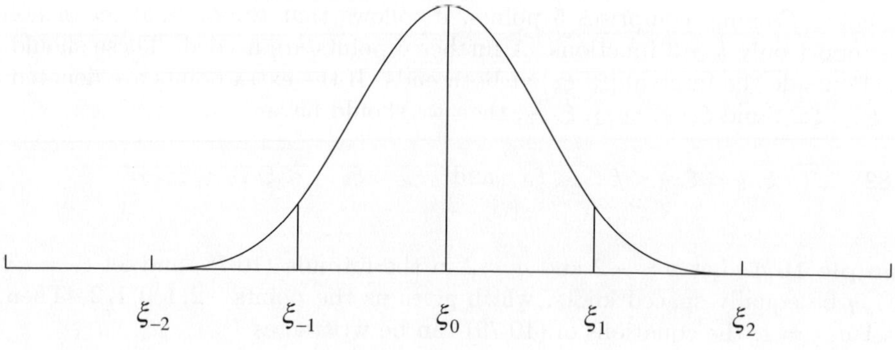

Figure 10.5. A uniform cubic B-spline $B_{-2}(x)$ with the support $[\xi_{-2}, \xi_2]$.

the case where $k = 0, \ldots, 3$ and $q - p = 4$, the appropriate solutions are

$$d_i = \prod_{\substack{j=p \\ j \neq i}}^{p+4} \frac{1}{(\xi_j - \xi_i)}, \quad \text{where} \quad i = p, \ldots, p+4. \tag{10.87}$$

Therefore the cubic B-spline with a support $[\xi_p, \xi_{p+4}]$ can be represented by

$$B_p^3(x) = \sum_{i=p}^{p+4} \left[\prod_{\substack{j=p \\ j \neq i}}^{p+4} \frac{1}{(\xi_j - \xi_i)} \right] (x - \xi_i)_+^3. \tag{10.88}$$

Reference to equation (4.86) shows that this is an expression for the leading coefficient of a polynomial of degree 4 which interpolates the coordinates of the function $(x-\xi_i)_+^3$ at the points $\xi_p, \ldots, \xi_{p+4}$. The expression is therefore synonymous with the fourth-order divided difference of the truncated power function. Indeed, B-splines are commonly defined in terms of divided differences; see, for example, de Boor [138].

Recursive Generation of B-spline Ordinates

The B-spline of degree k with the support $[\xi_p, \xi_{p+k+1}]$ is given by the formula

$$B_p^k(x) = \sum_{i=p}^{p+k+1} \left[\prod_{\substack{j=p \\ j \neq i}}^{p+k+1} \frac{1}{(\xi_j - \xi_i)} \right] (x - \xi_i)_+^k, \tag{10.89}$$

which is evidently a generalisation of (10.88). A recursive scheme for generating such splines is indicated by the following theorem:

(10.90) Let $B_p^{k-1}(x)$ and $B_{p+1}^{k-1}(x)$ be B-splines of degree $k-1$ whose respective supports are the intervals $[\xi_p, \xi_{p+k}]$ and $[\xi_{p+1}, \xi_{p+k+1}]$. Then the B-spline of degree k whose support is the interval $[\xi_p, \xi_{p+k+1}]$ is given by the convex combination

$$B_p^k(x) = \frac{(x - \xi_p)B_p^{k-1}(x) + (\xi_{p+k+1} - x)B_{p+1}^{k-1}(x)}{\xi_{p+k+1} - \xi_p}.$$

To prove the theorem, it is only necessary to only confirm that, if B_p^{k-1} and B_{p+1}^{k-1} fulfil their defining conditions, then the function defined by the RHS of the formula has the correct degree of k, that it is nonzero only over the interval $[\xi_p, \xi_{p+k+1}]$ and that it satisfies the conditions of second-order continuity. This is easily done.

It may also be confirmed directly that the function defined in (10.89) does indeed satisfy the recursion of (10.90). First we can confirm that the RHS of the formula of (10.90) agrees with $B_p^k(x)$ defined in (10.89) in the interval $[\xi_p, \xi_{p+1})$. Now consider the point $\xi_i \in [\xi_{p+1}, \xi_{p+k+1}]$. The recursive formula indicates that, as x passes the point ξ_i, a term is added to the function which takes the form of the polynomial $(x - \xi_i)^{k-1}/(\xi_{p+k+1} - \xi_p)$ multiplied by

(10.91)
$$\begin{aligned}
&(x - \xi_p)\prod_{\substack{j=p\\ j\neq i}}^{p+k}\frac{1}{(\xi_j - \xi_i)} + (\xi_{p+k+1} - x)\prod_{\substack{j=p+1\\ j\neq i}}^{p+k+1}\frac{1}{(\xi_j - \xi_i)}\\
&\qquad = (x - \xi_i)(\xi_{p+k+1} - \xi_p)\prod_{\substack{j=p\\ j\neq i}}^{p+k+1}\frac{1}{(\xi_j - \xi_i)}.
\end{aligned}$$

This agrees with the term which is deduced from the formula of (10.89)

In computing the recursion, we start with the B-spline function of zero degree. The generic function has a constant value over the support $[\xi_p, \xi_{p+1}]$ and a zero value elsewhere. Specifically,

(10.92)
$$\begin{aligned}
B_p^0(x) &= 0 && \text{if} \quad x < \xi_p,\\
B_p^0(x) &= (\xi_{p+1} - \xi_p)^{-1} && \text{if} \quad \xi_p \le x < \xi_{p+1},\\
B_p^0(x) &= 0 && \text{if} \quad \xi_{p+1} \le x.
\end{aligned}$$

The generic linear B-spline is computed as

(10.93)
$$B_p^1(x) = \frac{(x - \xi_p)B_p^0(x) + (\xi_{p+2} - x)B_{p+1}^0(x)}{\xi_{p+2} - \xi_p}.$$

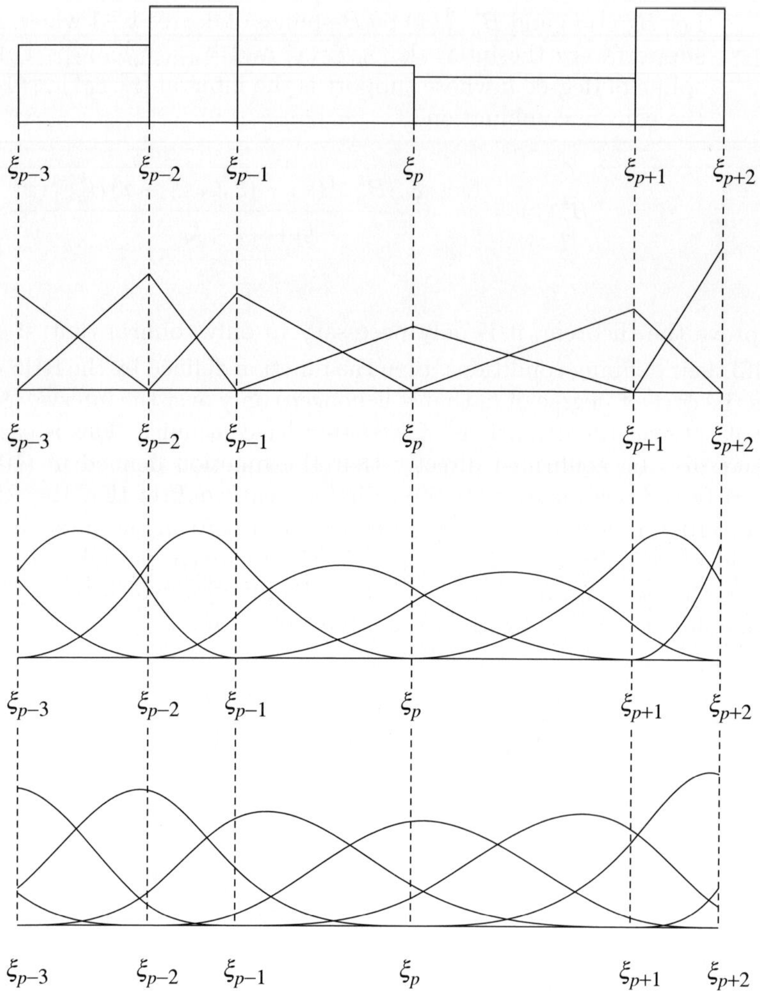

Figure 10.6. The B-splines of degrees $0, \ldots, 3$ generated recursively over the interval $[\xi_{p-3}, \ldots, \xi_{p+2}]$.

This is a tent-shaped function whose support is the interval $[\xi_p, \xi_{p+2}]$. The function rises to a peak at ξ_{p+1}. In particular

$$
(10.94)\qquad
\begin{aligned}
B_p^1(x) &= 0 && \text{if} \quad x \leq \xi_p, \\
B_p^1(x) &= \frac{x - \xi_p}{(\xi_{p+2} - \xi_p)(\xi_{p+1} - \xi_p)} && \text{if} \quad \xi_p \leq x \leq \xi_{p+1}, \\
B_p^1(x) &= \frac{\xi_{p+2} - x}{(\xi_{p+2} - \xi_p)(\xi_{p+2} - \xi_{p+1})} && \text{if} \quad \xi_{p+1} \leq x \leq \xi_{p+2}, \\
B_p^1(x) &= 0 && \text{if} \quad \xi_{p+2} \leq x.
\end{aligned}
$$

The quadratic B-spline, which is computed as a convex combination of linear splines, is a bell-shaped function. Finally, the cubic spline, which is formed from quadratic splines, is also a bell-shaped function, as was shown in Figures 10.5 and 10.6.

There is an alternative version of the recursive scheme which uses the formula

$$N_p^k(x) = \frac{(x - \xi_p)}{(\xi_{p+k} - \xi_p)} N_p^{k-1}(x) + \frac{(\xi_{p+k+1} - x)}{(\xi_{p+k+1} - \xi_{p+1})} N_{p+1}^{k-1}(x), \tag{10.95}$$

where

$$N_p^k = (\xi_{p+k+1} - \xi_p) B_p^k(x) \tag{10.96}$$

is simply a rescaled version of the B-spline defined in (10.90). Notice that the two denominators on the RHS of (10.95) correspond to the lengths of the supports of the constituent B-splines $N_p^{k-1}(x)$ and $N_{p+1}^{k-1}(x)$ of degree $k - 1$, whereas the common denominator in the definition under (10.90) is the length of the support of the B-spline $B_p^k(x)$ on the LHS.

The initial conditions for the algorithm of (10.95), which replace those of (10.92), are

$$N_p^0(x) = \begin{cases} 1, & \text{if } \xi_p \leq x < \xi_{p+1}; \\ 0, & \text{otherwise.} \end{cases} \tag{10.97}$$

A facility is required for evaluating the ordinates of the cubic B-spline at an arbitrary point x. Give that a cubic B-spline is nonzero over a support consisting of four consecutive intervals, it follows that there are only four such splines which are not automatically zero at x. The aim is to evaluate each of these functions; and a simple recursive algorithm based on (10.90) is available for the purpose.

Imagine that $x \in [\xi_p, \xi_{p+1}]$. Then there is one B-spline of zero degree which is nonzero in the interval, there are two of the first degree which are nonzero, there are three of the second degree and four of the third degree. The following diagram illustrates the relationship amongst these splines:

$$\begin{array}{ccccccc} & & & B_p^0 & & & \\ & & \swarrow & & \searrow & & \\ & & B_{p-1}^1 & & B_p^1 & & \\ & \swarrow & & \searrow\ \swarrow & & \searrow & \\ & B_{p-2}^2 & & B_{p-1}^2 & & B_p^2 & \\ \swarrow & & \searrow\ \swarrow & & \searrow\ \swarrow & & \searrow \\ B_{p-3}^3 & & B_{p-2}^3 & & B_{p-1}^3 & & B_p^3 \end{array} \tag{10.98}$$

In algebraic terms, the relationships are

$$
\begin{aligned}
B^1_{p-1} &= (1-\lambda^1_1)B^0_p,\\
B^1_p &= \lambda^1_0 B^0_p,\\
B^2_{p-2} &= (1-\lambda^2_2)B^1_{p-1},\\
B^2_{p-1} &= \lambda^2_1 B^1_{p-1} + (1-\lambda^2_1)B^1_p,\\
B^2_p &= \lambda^2_0 B^1_p,\\
B^3_{p-3} &= (1-\lambda^3_3)B^2_{p-2},\\
B^3_{p-2} &= \lambda^3_2 B^2_{p-2} + (1-\lambda^3_2)B^2_{p-1},\\
B^3_{p-1} &= \lambda^3_1 B^2_{p-1} + (1-\lambda^3_1)B^2_p,\\
B^3_p &= \lambda^3_0 B^2_p.
\end{aligned}
\tag{10.99}
$$

Here λ^k_i and $1-\lambda^k_i$ are the weights of the convex combinations defined in (10.90). Notice that the elements on the borders of the triangular array comprise a zero-valued function as one of the elements of this combination.

The following procedure generates the ordinates of the four cubic B-splines $B^3_{p-i}; i = 0,\ldots,3$ which are nonzero at the point $x \in [\xi_p, \xi_{p+1}]$; and it returns the values under the index $i = 0,\ldots,3$, in the array b. In fact, the algorithm generates in turn each of the elements in the display of (10.99) running from top to bottom. However, because the elements are successively overwritten, only the final four emerge from the procedure.

(10.100)
```
procedure BSplineOrdinates(p : integer;
                           x : real;
                           xi : vector;
                           var b : vector);

  var
    k, j : integer;
    lambda : real;

begin
  b[-1] := 0;
  b[0] := 1/(xi[p + 1] - xi[p]);
  b[1] := 0;

  for k := 1 to 3 do
    begin {k}
      for j := k downto 0 do
        begin {j}
          lambda := (x - xi[p - j])/(xi[p - j + k + 1] - xi[p - j]);
          b[j] := lambda * b[j] + (1 - lambda) * b[j - 1];
        end; {j}
```

```
        b[k + 1] := 0;
      end; {k}

  end; {BSplineOrdinates}
```

Another task which arises is that of finding the coefficients of the polynomial segments which constitute the cubic B-splines. These coefficients may be useful in plotting the individual B-splines or in plotting a curve which is a linear combination of the B-splines. The task is accomplished by the following procedure which builds upon the previous one:

(10.101)

```
procedure BSplineCoefficients(p : integer;
                              xi : vector;
                              mode : string;
                              var c : matrix);

  var
    i, j, k : integer;
    denom : real;

begin
  for i := 0 to 1 do
    for j := -1 to 1 do
      c[i, j] := 0;
  c[0, 0] := 1/(xi[p + 1] - xi[p]);
  for k := 1 to 3 do {degree}
    begin {k}
      c[k, k] := 0;
      for j := k downto 0 do {the splines}
        begin {j}
          for i := k downto 0 do {spline coefficients}
            begin {i}
              denom := (xi[p - j + k + 1] - xi[p - j]);
              c[i, j] := c[i - 1, j] - xi[p - j] * c[i, j];
              c[i, j] := c[i, j] - c[i - 1, j - 1];
              c[i, j] := c[i, j] + xi[p - j + k + 1] * c[i, j - 1];
              c[i, j] := c[i, j]/denom;
              c[i, k + 1] := 0;
            end; {i}
          c[k + 1, j] := 0;
        end; {j}
      c[k + 1, -1] := 0;
    end; {k}

  if mode = 'local' then
    begin {Find shifted-form coefficients}
      for j := 0 to 3 do
        for k := 0 to 2 do
```

for $i := 1$ **to** $3 - k$ **do**
    $c[3 - i, j] := c[3 - i, j] + c[3 - i + 1, j] * xi[p];$
end; $\{Shifted\text{-}form\ coefficients\}$

end; $\{BSplineCoefficients\}$

The output of this procedure is governed by the parameter *mode*. If this string is set to *'global'*, then the procedure finds the coefficients c_{ij} of the power representations $P^3_{p-j} = \sum_{i=0}^{3} c_{ij}x^i$ of the four cubic B-spline segments $B^3_{p-j}; j := 0, \ldots, 3$ which are nonzero in the interval $[\xi_p, \xi_{p+1}]$. If the string is set to *'local'*, then the coefficients of the shifted power forms $S^3_{p-j} = \sum_{i=0}^{3} c_{ij}(x - \xi_p)^i$, which are centred on ξ_p, are delivered instead. The final segment of the code, which effects the recentring, is copied from the procedure *ShiftedForm* of (4.16).

The recentred polynomials are in the same form as the segments of the cubic splines which are to be described in the next chapter. Therefore their coefficients can be supplied directly to the procedure *SplineToBezier* of (11.50) which generates the corresponding Bézier control points which are the arguments of the **curveto** command of the PostScript graphics language.

Regression with B-Splines

Imagine that a choice has been made of a sequence of knots $\xi_0, \ldots, \xi_k$ which serves to demarcate the k segments of a piecewise cubic polynomial $S(x)$. The piecewise polynomial is to be fitted to the data points $(x_0, y_0), \ldots, (x_n, y_n)$ of which the abscissae form a sequence of increasing values which fall in the interval $[\xi_0, \xi_k]$ spanned by the knots.

When six supplementary knots $\xi_{-3}, \xi_{-2}, \xi_{-1}$, and $\xi_{k+1}, \xi_{k+2}, \xi_{k+3}$ are added to the beginning and the end of the sequence of knots, there is sufficient support for a total of $k + 3$ B-spline functions. Then, each $x_i \in [x_0, x_n]$ will fall in some interval $[\xi_j, \xi_{j+1}]$ over which four functions, $B_j(x), B_{j-1}(x), B_{j-2}(x), B_{j-3}(x)$, are defined. The values of these functions at the data point x_i, which are generated by the procedure *BSplineOrdinates* of (10.100), may be recorded as four consecutive nonzero elements in the ith row of a matrix X of order $(n + 1) \times (k + 3)$.

The structure of the matrix X is typified by the following display:

$$\begin{bmatrix} * & * & * & * & 0 & 0 & \ldots & 0 \\ * & * & * & * & 0 & 0 & \ldots & 0 \\ * & * & * & * & 0 & 0 & \ldots & 0 \\ 0 & * & * & * & * & 0 & \ldots & 0 \\ 0 & * & * & * & * & 0 & \ldots & 0 \\ 0 & 0 & * & * & * & * & \ldots & 0 \\ 0 & 0 & * & * & * & * & \ldots & 0 \\ \vdots & \vdots & & \vdots & \vdots & \vdots & & \vdots \\ 0 & 0 & \ldots & 0 & * & * & * & * \\ 0 & 0 & \ldots & 0 & * & * & * & * \end{bmatrix}. \tag{10.102}$$

The matrix X together with the vector $y = [y_0, \ldots, y_n]'$ of the corresponding ordinates of the data points may be subjected to a procedure for calculating a vector of $k+3$ regression parameters $\beta = [\lambda_{-3}, \ldots, \lambda_{k-1}]'$. The fitted polynomial is then the function

$$S(x) = \sum_{j=-3}^{k-1} \lambda_j B_j(x). \tag{10.103}$$

It can be seen, in view of the structure of the matrix X, that $X'X$ is a symmetric matrix of seven diagonal bands. This fact may be exploited in creating a specialised procedure for solving the normal equations $X'X\beta = X'y$ of the least-squares regression. Indeed, if the number of knots is large, then it may be necessary to use such a procedure in the interests of conserving computer memory and of saving time. Procedures of this nature have been provided by de Boor [138].

The placement of the knots ought to be at the discretion of the user. Placing a cluster of knots in a certain area allows the curve $S(x)$ to follow the local features of the data closely. Conversely, a wider spacing of the knots over the region will result in a smoother curve. However, If the knots are numerous, it may be necessary to use an automatic procedure for placing them in appropriate positions along the x-axis. Failing this, they should be placed at regular intervals. In that case, some of the essential advantages of using B-splines in constructing grafted polynomial curves are lost; and it becomes appropriate to adopt the alternative procedure of estimating a least-squares smoothing spline in which the knots coincide with the data points. The cubic smoothing spline is the subject of the next chapter.

Bibliography

[16] Anderson, T.W., (1971), *The Statistical Analysis of Time Series*, John Wiley and Sons, Chichester.

[28] Arfken, G., (1985), *Mathematical Methods for Physicists, Third Edition*, Academic Press, San Diego, California.

[107] Chu, C-K., and J.S. Marron, (1991), Choosing a Kernel Regression Estimator, *Statistical Science*, **6**, 404–436.

[138] de Boor, C., (1978), *A Practical Guide to Splines*, Springer Verlag, New York.

[174] Eubank, R.A., (1988), *Spline Smoothing and Nonparametric Regression*, Marcel Dekker, New York and Basel.

[335] Martin, R.D., (1979), Approximate Conditional Mean Type Smoothers and Interpolators, in *Smoothing Techniques for Curve Estimation,* T. Gasser and M. Rosenblatt (eds.), Springer Verlag, Berlin.

[367] Nychka, D., (1991), Choosing a Range for the Amount of Smoothing in a Nonparametric Regression, *Journal of the American Statistical Association*, **86**, 653–664.

[396] Plass, M., and M. Stone, (1983), Curve-Fitting with Piecewise Parametric Cubics, *Computer Graphics*, **17**, 229–238.

[482] Tintner, G., (1952), *Econometrics*, John Wiley and Sons, New York.

CHAPTER 11

Smoothing with Cubic Splines

A spline function is a curve constructed from polynomial segments which are subject to conditions of continuity at their joints. In this chapter, we shall develop the algorithm of the cubic smoothing spline and we shall justify its use in estimating trends in time series.

Considerable effort has been devoted over several decades to developing the mathematics of spline functions. Much of the interest is due to the importance of splines in industrial design. In statistics, smoothing splines have been used in fitting curves to data ever since workable algorithms first became available in the late 1960s—see Schoenberg [442] and Reinsch [423]. However, many statisticians have felt concern at the apparently arbitrary nature of this device.

The difficulty is in finding an objective criterion for choosing the value of the parameter which governs the trade-off between the smoothness of the curve and its closeness to the data points. At one extreme, where the smoothness is all that matters, the spline degenerates to the straight line of an ordinary linear least-squares regression. At the other extreme, it becomes the interpolating spline which passes through each of the data points. It appears to be a matter of judgment where in the spectrum between these two extremes the most appropriate curve should lie.

One attempt at overcoming this arbitrariness has led to the criterion of cross-validation. Here the underlying notion is that the degree of smoothness should be chosen so as to make the spline the best possible predictor of any points in the data set to which it has not been fitted. Instead of reserving a collection of data points for the sole purpose of making this choice, it has been proposed that each of the available points should be left out in turn while the spline is fitted to the remainder. For each omitted point, there is then a corresponding error of prediction; and the optimal degree of smoothing is that which results in the minimum sum of squares of the prediction errors.

To find the optimal degree of smoothing by the criterion of cross-validation can require an enormous amount of computing. An alternative procedure, which has emerged more recently, is based on the notion that the spline with an appropriate smoothing parameter represents the optimal predictor of the path of a certain stochastic differential equation of which the observations are affected by noise. This is a startling result, and it provides a strong justification for the practice of representing trends with splines. The optimal degree of smoothing now becomes a function of the parameters of the underlying stochastic differential equation and of the parameters of the noise process; and therefore the element of judgment in

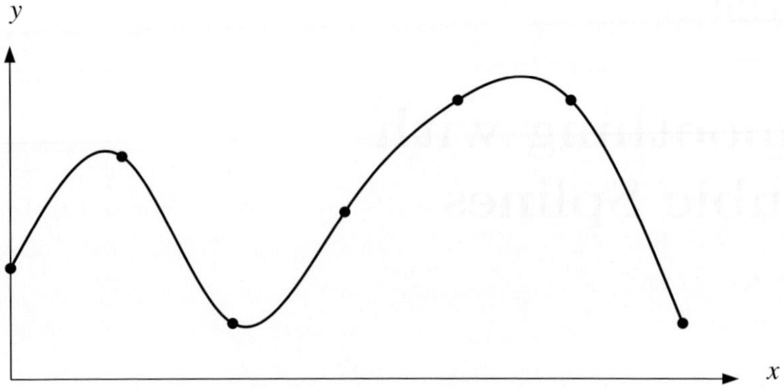

Figure 11.1. A cubic spline.

fitting the curve is eliminated.

We shall begin this chapter by establishing the algorithm for an ordinary interpolating spline. Thereafter, we shall give a detailed exposition of the classical smoothing spline of which the degree of smoothness is a matter of choice. In the final section, we shall give an account of a model-based method of determining an optimal degree of smoothing.

It should be emphasised that a model-based procedure for determining the degree of smoothing will prove superior to a judgmental procedure only if the model has been appropriately specified. The specification of a model is itself a matter of judgment.

Cubic Spline Interpolation

Imagine that we are given a set of coordinates $(x_0, y_0), (x_1, y_1), \ldots, (x_n, y_n)$ of the function $y = y(x)$, where the values of x are in ascending order. The object is to bridge the gap between adjacent points (x_i, y_i), (x_{i+1}, y_{i+1}) using the cubic functions $S_i; i = 0, \ldots, n-1$ so as to piece together a curve with continuous first and second derivatives. Such a curve, which is described as a cubic spline, is the mathematical equivalent of a draughtsman's spline which is a thin strip of flexible wood used for drawing curves in engineering work. The junctions of the cubic segments, which correspond to the points at which the draughtsman's spline would be fixed, are known as knots or nodes (see Figure 11.1).

The function S_i can be expressed as

$$S_i(x) = a_i(x - x_i)^3 + b_i(x - x_i)^2 + c_i(x - x_i) + d_i, \tag{11.1}$$

where x ranges from x_i to x_{i+1}.

The first and second derivatives of this function are

$$\begin{aligned} S_i'(x) &= 3a_i(x - x_i)^2 + 2b_i(x - x_i) + c_i \quad \text{and} \\ S_i''(x) &= 6a_i(x - x_i) + 2b_i. \end{aligned} \tag{11.2}$$

The condition that the adjacent functions S_{i-1} and S_i for $i = 1, \ldots, n$ should meet at the point (x_i, y_i) is expressed in the equation

$$\begin{aligned} &S_{i-1}(x_i) = S_i(x_i) = y_i \quad \text{or, equivalently,} \\ &a_{i-1}h_{i-1}^3 + b_{i-1}h_{i-1}^2 + c_{i-1}h_{i-1} + d_{i-1} = d_i = y_i, \end{aligned} \tag{11.3}$$

where $h_{i-1} = x_i - x_{i-1}$. The condition that the first derivatives should be equal at the junction is expressed in the equation

$$\begin{aligned} &S'_{i-1}(x_i) = S'_i(x_i) \quad \text{or, equivalently,} \\ &3a_{i-1}h_{i-1}^2 + 2b_{i-1}h_{i-1} + c_{i-1} = c_i; \end{aligned} \tag{11.4}$$

and the condition that the second derivatives should be equal is expressed as

$$\begin{aligned} &S''_{i-1}(x_i) = S''_i(x_i) \quad \text{or, equivalently,} \\ &6a_{i-1}h_{i-1} + 2b_{i-1} = 2b_i. \end{aligned} \tag{11.5}$$

It is also necessary to specify the conditions which prevail at the endpoints (x_0, y_0) and (x_n, y_n). The first derivatives of the cubic functions at these points can be set to the values of the corresponding derivatives of $y = y(x)$ thus:

$$\begin{aligned} S'_0(x_0) &= c_0 \\ &= y'(x_0) \end{aligned} \quad \text{and} \quad \begin{aligned} S'_{n-1}(x_n) &= c_n \\ &= y'(x_n). \end{aligned} \tag{11.6}$$

This is described as clamping the spline. By clamping the spline, additional information about the function $y = y(x)$ is introduced; and this should result in a better approximation. However, extra information of an equivalent nature can often be obtained by assessing the function at additional points close to the ends.

If the ends are left free, then the conditions

$$\begin{aligned} S''_0(x_0) &= 2b_0 \\ &= 0 \end{aligned} \quad \text{and} \quad \begin{aligned} S''_{n-1}(x_n) &= 2b_n \\ &= 0 \end{aligned} \tag{11.7}$$

will prevail. These imply that the spline is linear when it passes through the endpoints. The latter conditions may be used when the information about the first derivatives of the function $y = y(x)$ is hard to come by.

We shall begin by treating the case of the natural spline which has free ends. In this case, the values of b_0 and b_n are known, and we can begin by determining the remaining second-degree parameters $b_1, \ldots, b_{n-1}$ from the data values $y_0, \ldots, y_n$ and from the conditions of continuity. Once the values for the second-degree parameters have been found, the values can be determined of the remaining parameters of the cubic segments.

Consider, therefore, the following four conditions relating to the ith segment:

$$\begin{aligned} &\text{(i)} \quad S_i(x_i) = y_i, && \text{(ii)} \quad S_i(x_{i+1}) = y_{i+1}, \\ &\text{(iii)} \quad S''_i(x_i) = 2b_i, && \text{(iv)} \quad S''_i(x_{i+1}) = 2b_{i+1}. \end{aligned} \tag{11.8}$$

If b_i and b_{i+1} were known in advance, as they would be in the case of $n = 1$, then these conditions would serve to specify uniquely the four parameters of S_i. In the case of $n > 1$, the conditions of first-order continuity provide the necessary link between the segments which enables the parameters $b_1, \ldots, b_{n-1}$ to be determined simultaneously.

The first of the four conditions specifies that

$$d_i = y_i. \tag{11.9}$$

The second condition specifies that $a_i h_i^3 + b_i h_i^2 + c_i h_i + d_i = y_{i+1}$, whence it follows that

$$c_i = \frac{y_{i+1} - y_i}{h_i} - a_i h_i^2 - b_i h_i. \tag{11.10}$$

The third condition may be regarded as an identity. The fourth condition specifies that $6a_i h_i + 2b_i = 2b_{i+1}$, which gives

$$a_i = \frac{b_{i+1} - b_i}{3h_i}. \tag{11.11}$$

Putting this into (11.10) gives

$$c_i = \frac{(y_{i+1} - y_i)}{h_i} - \frac{1}{3}(b_{i+1} + 2b_i)h_i; \tag{11.12}$$

and now the parameters of the ith segment are expressed in terms of the second-order parameters b_{i+1}, b_i and the data values y_{i+1}, y_i.

The condition $S'_{i-1}(x_i) = S'_i(x_i)$ of first-order continuity, which is to be found under (11.4), can now be rewritten with the help of equations (11.11) and (11.12) to give

$$b_{i-1}h_{i-1} + 2b_i(h_{i-1} + h_i) + b_{i+1}h_i = \frac{3}{h_i}(y_{i+1} - y_i) - \frac{3}{h_{i-1}}(y_i - y_{i-1}). \tag{11.13}$$

By letting i run from 1 to $n - 1$ in this equation and by taking account of the end conditions $b_0 = b_n = 0$, a tridiagonal system of $n - 1$ equations is generated in the form of

$$\begin{bmatrix} p_1 & h_1 & 0 & \ldots & 0 & 0 \\ h_1 & p_2 & h_2 & \ldots & 0 & 0 \\ 0 & h_2 & p_3 & \ldots & 0 & 0 \\ \vdots & \vdots & \vdots & \ddots & \vdots & \vdots \\ 0 & 0 & 0 & \ldots & p_{n-2} & h_{n-2} \\ 0 & 0 & 0 & \ldots & h_{n-2} & p_{n-1} \end{bmatrix} \begin{bmatrix} b_1 \\ b_2 \\ b_3 \\ \vdots \\ b_{n-2} \\ b_{n-1} \end{bmatrix} = \begin{bmatrix} q_1 \\ q_2 \\ q_3 \\ \vdots \\ q_{n-2} \\ q_{n-1} \end{bmatrix}, \tag{11.14}$$

where

$$\begin{aligned} p_i &= 2(h_{i-1} + h_i) = 2(x_{i+1} - x_{i-1}) \quad \text{and} \\ q_i &= \frac{3}{h_i}(y_{i+1} - y_i) - \frac{3}{h_{i-1}}(y_i - y_{i-1}). \end{aligned} \tag{11.15}$$

These can be reduced by Gaussian elimination to a bidiagonal system

$$(11.16)\qquad \begin{bmatrix} p'_1 & h_1 & 0 & \dots & 0 & 0 \\ 0 & p'_2 & h_2 & \dots & 0 & 0 \\ 0 & 0 & p'_3 & \dots & 0 & 0 \\ \vdots & \vdots & \vdots & \ddots & \vdots & \vdots \\ 0 & 0 & 0 & \dots & p'_{n-2} & h_{n-2} \\ 0 & 0 & 0 & \dots & 0 & p'_{n-1} \end{bmatrix} \begin{bmatrix} b_1 \\ b_2 \\ b_3 \\ \vdots \\ b_{n-2} \\ b_{n-1} \end{bmatrix} = \begin{bmatrix} q'_1 \\ q'_2 \\ q'_3 \\ \vdots \\ q'_n - 2 \\ q'_{n-1} \end{bmatrix},$$

which may be solved by back-substitution to obtain the values $b_1, \dots, b_{n-1}$. The values of $a_i; i = 0, \dots, n-1$ can be obtained from (11.11). The value of c_0 can be obtained from (11.12) and then the remaining values $c_i; i = 1, \dots, n-1$ can be generated by a recursion based on the equation

$$(11.17)\qquad c_i = (b_i + b_{i-1})h_{i-1} + c_{i-1},$$

which comes from substituting into equation (11.4) the expression for a_{i-1} given by (11.11).

The following procedure calculates the parameters of a cubic spline of which the ends have been left free in accordance with the conditions under (11.7):

(11.18)
```
procedure CubicSplines(var S : SplineVec;
                         n : integer);

  var
    i : integer;
    h, p, q, b : vector;

  begin {CubicSplines}

    h[0] := S[1].x - S[0].x;
    for i := 1 to n - 1 do
      begin
        h[i] := S[i + 1].x - S[i].x;
        p[i] := 2 * (S[i + 1].x - S[i - 1].x);
        q[i] := 3 * (S[i + 1].y - S[i].y)/h[i]
                  -3 * (S[i].y - S[i - 1].y)/h[i - 1];
      end;

    {Gaussian elimination}
    for i := 2 to n - 1 do
      begin
        p[i] := p[i] - h[i - 1] * h[i - 1]/p[i - 1];
        q[i] := q[i] - q[i - 1] * h[i - 1]/p[i - 1];
      end;

    {Back-substitution}
```

```
    b[n] := 0;
    b[n - 1] := q[n - 1]/p[n - 1];
    for i := 2 to n - 1 do
      b[n - i] := (q[n - i] - h[n - i] * b[n - i + 1])/p[n - i];

  {Spline parameters}
    S[0].a := b[1]/(3 * h[0]);
    S[0].b := 0;
    S[0].c := (S[1].y - S[0].y)/h[0] - b[1] * h[0]/3;
    S[0].d := S[0].y;
    S[n].b := 0;

    for i := 1 to n - 1 do
      begin
        S[i].a := (b[i + 1] - b[i])/(3 * h[i]);
        S[i].b := b[i];
        S[i].c := (b[i] + b[i - 1]) * h[i - 1] + S[i - 1].c;
        S[i].d := S[i].y;
      end;

  end; {CubicSplines}
```

The procedure must be placed in an environment containing the following type statements:

(11.19)
```
  type
    SplineParameters = record
        a, b, c, d, x, y : real
      end;
    SplineVec = array[0..dim] of SplineParameters;
```

At the beginning of the procedure, the record $S[i]$ contains only the values of x_i and y_i which are held as $S[i].x$ and $S[i].y$ respectively. At the conclusion of the procedure, the parameters a_i, b_i, c_i, d_i of the ith cubic segment are held in $S[i].a$, $S[i].b$, $S[i].c$ and $S[i].d$ respectively.

Now let us consider the case where the ends of the spline are clamped. Then the values of c_0 and c_n are known, and we may begin by determining the remaining first-degree parameters $c_1, \ldots, c_{n-1}$ from the data points $y_0, \ldots, y_n$ and from the continuity conditions. Consider, therefore, the following four conditions relating to the ith segment:

(11.20)
$$\begin{aligned} &\text{(i)} \quad S_i(x_i) = y_i, &\qquad &\text{(ii)} \quad S_i(x_{i+1}) = y_{i+1}, \\ &\text{(iii)} \quad S_i'(x_i) = c_i, &\qquad &\text{(iv)} \quad S_i'(x_{i+1}) = c_{i+1}. \end{aligned}$$

If c_i and c_{i+1} were known in advance, as they would be in the case of $n = 1$, then these four conditions would serve to specify the parameters of the segment. The first and second conditions, which are the same as in the case of the natural spline, lead to equation (11.10). The third condition is an identity, whilst the fourth condition specifies that

$$c_{i+1} = 3a_i h_i^2 + 2b_i h_i + c_i. \tag{11.21}$$

The equations (11.10) and (11.21) can be solved simultaneously to give

$$a_i = \frac{1}{h_i^2}(c_i + c_{i+1}) + \frac{2}{h_i^3}(y_i - y_{i+1}) \tag{11.22}$$

and

$$b_i = \frac{3}{h_i^2}(y_{i+1} - y_i) - \frac{1}{h_i}(c_{i+1} + 2c_i). \tag{11.23}$$

The condition $S''_{i-1}(x_i) = S''_i(x_i)$ of second-order continuity, which is to be found under (11.5), can now be rewritten with the help of equations (11.22) and (11.23) to give

$$\frac{c_{i-1}}{h_{i-1}} + 2\Big(\frac{1}{h_{i-1}} + \frac{1}{h_i}\Big)c_i + \frac{c_{i+1}}{h_i} = \frac{3}{h_{i-1}^2}(y_i - y_{i-1}) + \frac{3}{h_i^2}(y_{i+1} - y_i). \tag{11.24}$$

This is similar to the expression under (11.13); and, by letting i run from 1 to $n-1$, the following system of equations is generated:

$$\begin{bmatrix} f_1 & h_1^{-1} & 0 & \ldots & 0 & 0 \\ h_1^{-1} & f_2 & h_2^{-1} & \ldots & 0 & 0 \\ 0 & h_2^{-1} & f_3 & \ldots & 0 & 0 \\ \vdots & \vdots & \vdots & \ddots & \vdots & \vdots \\ 0 & 0 & 0 & \ldots & f_{n-2} & h_{n-2}^{-1} \\ 0 & 0 & 0 & \ldots & h_{n-2}^{-1} & f_{n-1} \end{bmatrix} \begin{bmatrix} c_1 \\ c_2 \\ c_3 \\ \vdots \\ c_{n-2} \\ c_{n-1} \end{bmatrix} = \begin{bmatrix} g_1 - c_0 h_0^{-1} \\ g_2 \\ g_3 \\ \vdots \\ g_{n-2} \\ g_{n-1} - c_n h_{n-1}^{-1} \end{bmatrix}, \tag{11.25}$$

where

$$\begin{aligned} f_i &= 2(h_{i-1}^{-1} - h_i^{-1}) \quad \text{and} \\ g_i &= \frac{3}{h_{i-1}^2}(y_i - y_{i-1}) + \frac{3}{h_i^2}(y_{i+1} - y_i). \end{aligned} \tag{11.26}$$

These may be solved for the values $c_1, \ldots, c_{n-1}$ in the same manner as the equations under (11.14) are solved for $b_1, \ldots, b_{n-1}$, by reducing the tridiagonal matrix to a bidiagonal matrix and then using a process of back-substitution. The values $a_0, \ldots, a_{n-1}$ may then be obtained from equation (11.22). The value b_0 can be obtained from (11.23), and then the remaining values $b_1, \ldots, b_n$ can be generated using a recursion based on the equation

$$b_i = b_{i-1} + 3a_{i-1}h_{i-1}, \tag{11.27}$$

which comes from (11.5).

The alternative procedure for calculating the clamped spline requires the system of (11.14) to be extended so as to accommodate the additional information concerning the values of the first derivatives at the endpoints. Since the conditions under (11.7) no longer prevail, there are two more parameters to be determined.

The value of the derivative at x_0 affects the parameters of the spline via the equation

$$y_0' = c_0 = \frac{y_1 - y_0}{h_0} - \frac{1}{3}(b_1 + 2b_0)h_0, \tag{11.28}$$

which comes from combining the first condition under (11.6) with the equation under (11.12). This becomes

$$p_0 b_0 + h_0 b_1 = q_0 \tag{11.29}$$

when we define

$$p_0 = 2h_0 \quad \text{and} \quad q_0 = \frac{3}{h_0}(y_1 - y_0) - 3y_0'. \tag{11.30}$$

The value of the derivative at x_n affects the parameters of the spline via the equation

$$y_n' = c_n = 3a_{n-1}h_{n-1}^2 + 2b_{n-1}h_{n-1} + c_{n-1}, \tag{11.31}$$

which comes from combining the second condition under (11.6) with the condition under (11.4). Using (11.11) and (11.12), this can be rewritten as

$$y_n' - \frac{(y_n - y_{n-1})}{h_{n-1}} = \frac{2}{3}b_n h_{n-1} + \frac{1}{3}b_{n-1}h_{n-1} \tag{11.32}$$

which becomes

$$h_{n-1}b_{n-1} + p_n b_n = q_n \tag{11.33}$$

when we define

$$p_n = 2h_n \quad \text{and} \quad q_n = 3y_n' - \frac{3}{h_{n-1}}(y_n - y_{n-1}). \tag{11.34}$$

The extended system can now be written as

$$\begin{bmatrix} p_0 & h_0 & 0 & \dots & 0 & 0 \\ h_0 & p_1 & h_1 & \dots & 0 & 0 \\ 0 & h_1 & p_2 & \dots & 0 & 0 \\ \vdots & \vdots & \vdots & \ddots & \vdots & \vdots \\ 0 & 0 & 0 & \dots & p_{n-1} & h_{n-1} \\ 0 & 0 & 0 & \dots & h_{n-1} & p_n \end{bmatrix} \begin{bmatrix} b_0 \\ b_1 \\ b_2 \\ \vdots \\ b_{n-1} \\ b_n \end{bmatrix} = \begin{bmatrix} q_0 \\ q_1 \\ q_2 \\ \vdots \\ q_{n-1} \\ q_n \end{bmatrix}. \tag{11.35}$$

Cubic Splines and Bézier Curves

Parametric cubic splines have been much used in the past in ship-building and aircraft design and they have been used, to a lesser extent, in the design of car bodies. However, their suitability to an iterative design process is limited by the fact that, if the location of one of the knots is altered, then the whole spline must be recalculated. In recent years, cubic splines have been replaced increasingly in computer-aided design applications by the so-called cubic Bézier curve.

A testimony to the versatility of cubic Bézier curves is the fact that the PostScript [3] page-description language, which has been used in constructing the letter forms on these pages, employs Bézier curve segments exclusively in constructing curved paths, including very close approximations to circles.

The usual parametrisation of a Bézier curve differs from the parametrisation of the cubic polynomial to be found under (11.1). Therefore, in order to make use of the Bézier function provided by a PostScript-compatible printer, we need to establish the correspondence between the two sets of parameters. The Bézier function greatly facilitates the plotting of functions which can be represented exactly or approximately by cubic segments.

The curve-drawing method of Bézier [53], [54] is based on a classical method of approximation known as the Bernstein polynomial approximation. Let $f(t)$ with $t \in [0, 1]$ be an arbitrary real-valued function taking the values $f_k = f(t_k)$ at the points $t_k = k/n; k = 0, \ldots, n$ which are equally spaced in the interval $[0, 1]$. Then the Bernstein polynomial of degree n is defined by

$$B_n(t) = \sum_{k=0}^{n} f_k \frac{n!}{k!(n-k)!} t^k (1-t)^{n-k}. \tag{11.36}$$

The coefficients in this sum are just the terms of the expansion of the binomial

$$\{t + (1-t)\}^n = \sum_{k=0}^{n} \frac{n!}{k!(n-k)!} t^k (1-t)^{n-k}; \tag{11.37}$$

from which it can be seen that the sum of the coefficients is unity.

Bernstein (1912) [49] used this polynomial in a classic proof of the Weierstrass approximation theorem [508] which asserts that, for any $\epsilon > 0$, there exists a polynomial $P_n(t)$ of some degree $n = n(\epsilon)$ such that $|f(t) - P_n(t)| < \epsilon$. The consequence of Bernstein's proof is that $B_n(t)$ converges uniformly to $f(t)$ in $[0, 1]$ as $n \to \infty$.

The restriction of the functions to the interval $[0, 1]$ is unessential to the theorem. To see how it may be relieved, consider a continuous monotonic transformation $x = x(t)$ defined over an interval bounded by $x_0 = x(0)$ and $x_1 = x(1)$. The inverse function $t = t(x)$ exists; and, if $f(x) = f\{t(x)\}$ and $B_n(x) = B_n\{t(x)\}$, then $B_n(x)$ converges uniformly to $f(x)$ as $B_n(t)$ converges to $f(t)$.

Whilst the Bernstein polynomials lead to an elegant constructive proof of the Weierstrass theorem, they do not in themselves provide useful polynomial approximations. One reason for their inadequacy is the slowness of their convergence to $f(t)$, which means that, in order to obtain a good approximation, a polynomial

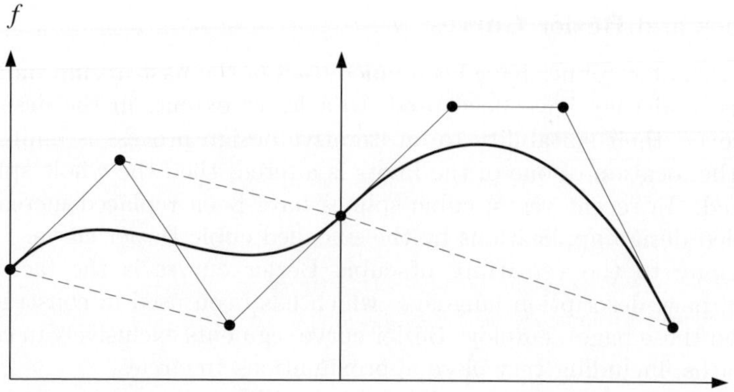

Figure 11.2. Adjacent cubic Bézier segments linked by a condition of first-order continuity. The solid lines link the Bézier control points in sequence. The dotted lines complete the boundaries of the convex hulls.

of a high degree is required. However, a high-degree polynomial is liable to have undesirable ripples. Until the advent of computer-aided design and the discoveries of Bézier, the Bernstein polynomials were regarded as little more that an adjunct to a proof of the Weierstrass theorem—see Achieser [2]. The approach of Bézier in designing a smooth curve is to use Bernstein polynomials of low degree to construct short segments which are linked by conditions of first-order continuity. An ordered set of $n+1$ points $(t_k, f_k); k = 0, \ldots, t_n$ which serves to define a Bernstein polynomial of degree n also defines a convex hull containing the path of the polynomial (see Figure 11.2).

This path, which is known as the Bézier curve, has two important features. On the one hand, it passes through the endpoints $(t_0, f_0), (t_n, f_n)$ which define the boundaries of a segment. This can be seen by setting $t = 0$ and $t = 1$ in (11.36) to give

$$B_n(0) = f_0 \quad \text{and} \quad B_n(1) = f_n. \tag{11.38}$$

On the other hand, the slopes of the vectors which are tangent to the Bézier curve at the endpoints are equal to the slopes of the adjacent sides of the polygon which forms the convex hull. Thus, maintaining assumption that the $n+1$ points $t_0, \ldots, t_n$ are equally spaced in the interval $[0, 1]$, we have

$$\begin{aligned} B_n'(0) &= n(f_0 - f_1) \\ &= \frac{f_0 - f_1}{t_0 - t_1} \end{aligned} \quad \text{and} \quad \begin{aligned} B_n'(1) &= n(f_n - f_{n-1}) \\ &= \frac{f_n - f_{n-1}}{t_n - t_{n-1}}. \end{aligned} \tag{11.39}$$

If the endpoints of the Bézier curve are regarded as fixed, then the intermediate points $(t_1, f_1), \ldots, (t_{n-1}, f_{n-1})$ may be adjusted in an interactive manner to make the Bézier curve conform to whatever shape is desired.

Example 11.1. Consider the cubic Bernstein polynomial

$$\begin{aligned} B_3(t) &= f_0(1-t)^3 + 3f_1t(1-t)^2 + 3f_2t^2(1-t) + f_3t^3 \\ &= \alpha t^3 + \beta t^2 + \gamma t + \delta. \end{aligned} \tag{11.40}$$

Equating the coefficients of the powers of t shows that

$$\begin{aligned} \alpha &= f_3 - 3f_2 + 3f_1 - f_0, \\ \beta &= 3f_2 - 6f_1 + 3f_0, \\ \gamma &= 3f_1 - 3f_0, \\ \delta &= f_0. \end{aligned} \tag{11.41}$$

Differentiating $B_3(t)$ with respect to t gives

$$B_3'(t) = -3f_0(1-t)^2 + 3f_1t(1-4t+3t^2) + 3f_2(2t-3t^2) + 3f_3t^2, \tag{11.42}$$

from which it can be see that the conditions under (11.39) are satisfied:

$$B_3'(0) = 3(f_0 - f_1) \quad \text{and} \quad B_3'(1) = 3(f_3 - f_2). \tag{11.43}$$

In order to exploit the Bézier command which is available in the PostScript language, we need to define the relationship between the ordinates f_0, f_1, f_2, f_3 of the four control points of a cubic Bézier curve and the four parameters a, b, c, d of the representation of a cubic polynomial which is to be found under (11.1).

Let us imagine that the Bézier function $B_3(t)$ ranges from f_0 to f_3 as t ranges from $t_0 = 0$ to $t_3 = 1$, and let

$$S(x) = a(x-x_0)^3 + b(x-x_0)^2 + c(x-x_0) + d \tag{11.44}$$

be a segment of the cubic spline which spans the gap between two points which are (x_0, y_0) and (x_1, y_1) with $y_0 = f_0$ and $y_1 = f_3$. Then, if we define

$$\begin{aligned} x(t) &= (x_1 - x_0)t + x_0 \\ &= ht + x_0, \end{aligned} \tag{11.45}$$

we can identify the function $S(x)$ with the function $B(x) = B\{t(x)\}$. Thus, on taking $B(t) = \alpha t^3 + \beta t^2 + \gamma t + \delta$ and putting $t(x) = (x - x_0)/h$, with $h = x_1 - x_0$, in place of t, we get

$$\begin{aligned} S(x) &= \frac{\alpha}{h^3}(x-x_0)^3 + \frac{\beta}{h^2}(x-x_0)^2 + \frac{\gamma}{h}(x-x_0) + \delta \\ &= a(x-x_0)^3 + b(x-x_0)^2 + c(x-x_0) + d. \end{aligned} \tag{11.46}$$

The mapping from the ordinates of the Bézier control points to the parameters $\alpha, \beta, \gamma, \delta$ is given by

$$\begin{bmatrix} -1 & 3 & -3 & 1 \\ 3 & -6 & 3 & 0 \\ -3 & 3 & 0 & 0 \\ 1 & 0 & 0 & 0 \end{bmatrix} \begin{bmatrix} f_0 \\ f_1 \\ f_2 \\ f_3 \end{bmatrix} = \begin{bmatrix} \alpha \\ \beta \\ \gamma \\ \delta \end{bmatrix} = \begin{bmatrix} ah^3 \\ bh^2 \\ ch \\ d \end{bmatrix}. \tag{11.47}$$

The inverse of mapping is given by

$$(11.48)\qquad \begin{bmatrix} f_0, \\ f_1 \\ f_2 \\ f_3 \end{bmatrix} = \begin{bmatrix} 0 & 0 & 0 & 1 \\ 0 & 0 & 1/3 & 1 \\ 0 & 1/3 & 2/3 & 1 \\ 1 & 1 & 1 & 1 \end{bmatrix} \begin{bmatrix} ah^3 \\ bh^2 \\ ch \\ d \end{bmatrix}.$$

The PostScript Bézier command is the **curveto** command which takes as its arguments the values $z_1, f_1, z_2, f_2, z_3, f_3$ and adds a cubic Bézier segment to the current path between the current point (z_0, f_0) and the point (z_3, f_3) using (z_1, f_1) and (z_2, f_2) as the control points. Then (z_3, f_3) becomes the new current point. The **curveto** function is based upon a pair of parametric cubic equations:

$$(11.49)\qquad \begin{aligned} z(t) &= a_z t^3 + b_z t^2 + c_z t + z_0, \\ y(t) &= a_y t^3 + b_y t^2 + c_y t + f_0. \end{aligned}$$

The parameters a_z, b_z, c_z are obtained from the abscissae z_0, z_1, z_2, z_3 via the transformation of (11.47) which is used to obtain a_y, b_y, c_y from f_0, f_1, f_2, f_3.

The parametric equation $z = z(t)$ enables the t-axis to be expanded, contracted and even folded back on itself. There is therefore no requirement that values z_0, z_1, z_2, z_3 should be equally spaced. More significantly, curves may be plotted which do not correspond to single-valued functions of z. For our own purposes, the function reduces to $z(t) = ht + z_0$ with $h = z_3 - z_0$, where $z_0 = x_i$ and $z_3 = x_{i+1}$ are the values of adjacent knots of a spline curve.

The conversion of the parameters of the cubic function under (11.1) to the parameters of the cubic Bézier curve may be accomplished using the following procedure.

(11.50)
```
procedure SplineToBezier(S : SplineVec;
                         var B : BezierVec;
                         n : integer);

  var
    i : iinteger;
    h, delt : real;

begin {SplineToBezier}
  for i := 0 to n - 1 do
    begin {i}
      h := S[i + 1].x - S[i].x;
      delt := h/3;
      with B[i], S[i] do
        begin {with}
          z0 := x;
          z1 := z0 + delt;
          z2 := z1 + delt;
          z3 := z2 + delt;
```

```
            f0 := d;
            f1 := f0 + c * h/3;
            f2 := f1 + (c + b * h) * h/3;
            f3 := f0 + (c + (b + a * h) * h) * h
          end; {with}
      end; {i}

end; {SplineToBezier}
```

The *BezierVec* type is defined in the following statements which must be included in the program which calls the procedure:

(11.51)
```
type
   BezierPoints = record
       z0, f0, z1, f1, z2, f2, z3, f3 : real
   end;
   BezierVec = array[0..dim] of BezierPoints;
```

The Minimum-Norm Property of Splines

The draughtsman's spline assumes a shape which minimises the potential energy due to the bending strain. The strain energy is approximately proportional to the integral of the square of the second derivative along the path of the spline; and therefore the minimisation of the potential energy leads to a property of minimum curvature. It can be demonstrated that the cubic spline has a similar property, which justifies us in likening it to the draughtsman's spline.

Let $f(x) \in \mathcal{C}^2$ be any function defined over the interval $[x_0, x_n]$ which has a continuous second-order derivative. Then a measure of the curvature of the function is provided by the squared norm

$$\|f\|^2 = \int_{x_0}^{x_n} \{f''(x)\}^2 dx. \tag{11.52}$$

This differs from the ideal measure of curvature which would be the line integral of $\{f''(x)\}^2$ along the path of the function. Thus the squared norm provides only a rough approximation to the potential energy of the draughtsman's spline.

Our object is to show that, amongst all functions $f(x) \in \mathcal{C}^2$, which pass through the points $(x_i, y_i); i = 0, \ldots, n$, it is the spline function which minimises the squared norm.

Let the spline be denoted by $S(x)$, where $x \in [x_0, x_n]$, and let the ith segment continue to be expressed as

$$S_i(x) = a_i(x - x_i)^3 + b_i(x - x_i)^2 + c_i(x - x_i) + d_i, \tag{11.53}$$

where $x \in [x_i, x_{i+1}]$. The derivatives of this function are

$$\begin{aligned} S_i'(x) &= 3a_i(x - x_i)^2 + 2b_i(x - x_i) + c_i, \\ S_i''(x) &= 6a_i(x - x_i) + 2b_i, \\ S_i'''(x) &= 6a_i. \end{aligned} \tag{11.54}$$

The minimum-norm property of the cubic spline can be established using the following result:

(11.55) Let $f(x) \in \mathcal{C}^2$ be a function defined on the interval $[x_0, x_n]$ which passes through the points $(x_i, y_i); i = 0, \ldots, n$ which are the knots of the spline function $S(x)$. Then

$$\|f - S\|^2 = \|f\|^2 - \|S\|^2 - 2\Big[S''(x)\{f'(x) - S'(x)\}\Big]_{x_0}^{x_n}.$$

Proof. By definition, there is

$$\begin{aligned}\|f - S\|^2 &= \|f\|^2 - 2\int_{x_0}^{x_n} f''(x)S''(x)dx + \|S\|^2 \\ &= \|f\|^2 - 2\int_{x_0}^{x_n} S''(x)\{f''(x) - S''(x)\}dx - \|S\|^2.\end{aligned} \tag{11.56}$$

Within this expression, it is found, through integrating by parts, that

$$\begin{aligned}\int_{x_0}^{x_n} S''(x)\{f''(x) - S''(x)\}dx &= \Big[S''(x)\{f'(x) - S'(x)\}\Big]_{x_0}^{x_n} \\ &\quad - \int_{x_0}^{x_n} S'''(x)\{f'(x) - S'(x)\}dx.\end{aligned} \tag{11.57}$$

Since $S(x)$ consists of the cubic segments $S_i(x); i = 0, \ldots, n-1$, it follows that the third derivative $S'''(x)$ is constant in each open interval (x_i, x_{i+1}), with a value of $S_i'''(x) = 6a_i$. Therefore

$$\begin{aligned}\int_{x_0}^{x_n} S'''(x)\{f'(x) - S'(x)\}dx &= \sum_{i=0}^{n-1}\int_{x_i}^{x_{i+1}} 6a_i\{f'(x) - S'(x)\}dx \\ &= \sum_{i=0}^{n-1} 6a_i\Big[f(x) - S(x)\Big]_{x_i}^{x_{i+1}} = 0,\end{aligned} \tag{11.58}$$

since $f(x) = S(x)$ at x_i and x_{i+1}; and hence (11.57) becomes

$$\int_{x_0}^{x_n} S''(x)\{f''(x) - S''(x)\}dx = \Big[S''(x)\{f'(x) - S'(x)\}\Big]_{x_0}^{x_n}. \tag{11.59}$$

Putting this into (11.56) gives the result which we wish to prove.

Now consider the case of the natural spline which satisfies the conditions $S''(x_0) = 0$ and $S''(x_n) = 0$. Putting these into the equality of (11.55) reduces it to

$$\|f - S\|^2 = \|f\|^2 - \|S\|^2, \tag{11.60}$$

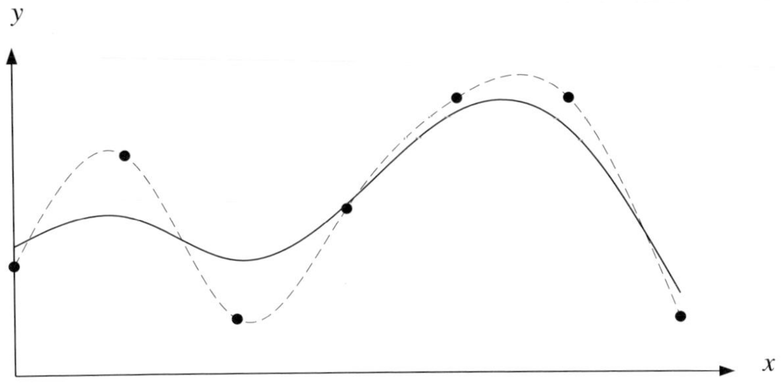

Figure 11.3. An cubic interpolating spline—the dotted path—and a cubic smoothing spline—the continuous path. Here, the parameter of the convex combination of equation (11.62) has been given the value of $\mu = 0.5$.

which demonstrates that $\|f\|^2 \geq \|S\|^2$. In the case of a clamped spline with $S'(x_0) = f'(x_0)$ and $S'(x_n) = f'(x_n)$, the equality of (11.55) is also reduced to that of (11.60). Thus it can be seen that, in either case, the cubic spline has the minimum-norm property.

Smoothing Splines

The interpolating spline provides a useful way of approximating a smooth function $f(x) \in \mathcal{C}^2$ only when the data points lie along the path of the function or very close to it. If the data are scattered at random in the vicinity of the path, then an interpolating polynomial, which is bound to follow the same random fluctuations, will belie the nature of the underlying function. Therefore, in the interests of smoothness, we may wish to allow the spline to depart from the data points (see Figure 11.3).

We may imagine that the ordinates of the data are given by the equation

$$y_i = f(x_i) + \eta_i, \tag{11.61}$$

where $\eta_i; i = 0, \ldots, n$ form a sequence of independently distributed random variables with $V(\eta_i) = \sigma_i^2$. In that case, we can attempt to reconstitute the function $f(x)$ by constructing a spline function $S(x)$ which minimises the value of

$$L = \mu \sum_{i=0}^{n} \left(\frac{y_i - S_i}{\sigma_i} \right)^2 + (1 - \mu) \int_{x_0}^{x_n} \{S''(x)\}^2 dx, \tag{11.62}$$

wherein $S_i = S(x_i)$.

The parameter $\mu \in [0, 1]$ reflects the relative importance which we give to the conflicting objectives of remaining close to the data, on the one hand, and of obtaining a smooth curve, on the other hand. Notice that a linear function satisfies the equation

$$\int_{x_0}^{x_n}\{S''(x)\}^2dx=0, \tag{11.63}$$

which suggests that, in the limiting case, where $\mu=0$ and where smoothness is all that matters, the spline function $S(x)$ will become a straight line. At the other extreme, where $\mu=1$ and where the closeness of the spline to the data is the only concern, we will obtain an interpolating spline which passes exactly through the data points.

Given the piecewise nature of the spline, the integral in the second term on the RHS of (11.62) can be written as

$$\int_{x_0}^{x_n}\{S''(x)\}^2dx=\sum_{i=0}^{n-1}\int_{x_i}^{x_{i+1}}\{S_i''(x)\}^2dx. \tag{11.64}$$

Since the spline is composed of cubic segments, the second derivative in any interval $[x_i,x_{i+1}]$ is a linear function which changes from $2b_i$ at x_i to $2b_{i+1}$ at x_{i+1}. Therefore we have

$$\begin{aligned}\int_{x_i}^{x_{i+1}}\{S_i''(x)\}^2dx&=4\int_0^{h_i}\left\{b_i\left(1-\frac{x}{h_i}\right)+b_{i+1}\frac{x}{h_i}\right\}^2dx\\&=\frac{4h_i}{3}\left(b_i^2+b_ib_{i+1}+b_{i+1}^2\right),\end{aligned} \tag{11.65}$$

where $h_i=x_{i+1}-x_i$; and the criterion function can be rewritten as

$$L=\sum_{i=0}^{n}\left(\frac{y_i-d_i}{\sigma_i}\right)^2+2\lambda\sum_{i=0}^{n-1}h_i\left(b_i^2+b_ib_{i+1}+b_{i+1}^2\right), \tag{11.66}$$

wherein $d_i=S_i(x_i)$ and $\lambda=2(1-\mu)/3\mu$, which is the so-called smoothing parameter.

We shall treat the case of the natural spline which passes through the knots $(x_i,d_i);i=0,\ldots,n$ and which satisfies the end conditions $S''(x_0)=2b_0=0$ and $S''(x_n)=2b_n=0$. The additional feature of the problem of fitting a smoothing spline, compared with that of fitting an interpolating spline, is the need to determine the ordinates $d_i;i=0,\ldots,n$ which are no longer provided by the data values $y_i;i=0,\ldots,n$.

We can concentrate upon the problem of determining the parameters $b_i,d_i;i=0,\ldots,n$ if we eliminate the remaining parameters $a_i,c_i;i=1,\ldots,n-1$. Consider, therefore, the ith cubic segment which spans the gap between the knots (x_i,d_i) and (x_{i+1},d_{i+1}) and which is subject to the following conditions:

$$\begin{aligned}&\text{(i)}\quad S_i(x_i)=d_i, &&\text{(ii)}\quad S_i(x_{i+1})=d_{i+1},\\&\text{(iii)}\quad S_i''(x_i)=2b_i, &&\text{(iv)}\quad S_i''(x_{i+1})=2b_{i+1}.\end{aligned} \tag{11.67}$$

The first condition may be regarded as an identity. The second condition, which specifies that $a_i h_i^3 + b_i h_i^2 + c_i h_i + d_i = d_{i+1}$, gives

$$c_i = \frac{d_{i+1} - d_i}{h_i} - a_i h_i^2 + b_i h_i. \tag{11.68}$$

The third condition is again an identity, whilst the fourth condition, which specifies that $2b_{i+1} = 6a_i h_i + 2b_i$, gives

$$a_i = \frac{b_{i+1} - b_i}{3h_i}. \tag{11.69}$$

Putting this into (11.68) gives

$$c_i = \frac{d_{i+1} - d_i}{h_i} - \frac{1}{3}(b_{i+1} - 2b_i)h_i. \tag{11.70}$$

Here we have expressions for a_i and c_i which are in terms of b_{i+1}, b_i and d_{i+1}, d_i. To determine the latter parameters, we must use the conditions of first-order continuity to link the segments. The condition $S'_{i-1}(x_i) = S'_i(x_i)$ specifies that

$$3a_{i-1}h_{i-1}^2 + 2b_{i-1}h_{i-1} + c_{i-1} = c_i. \tag{11.71}$$

On replacing a_{i-1} and c_{i-1} by expressions derived from (11.69) and (11.70) and rearranging the result, we get

$$b_{i-1}h_{i-1} + 2b_i(h_{i-1} + h_i) + b_{i+1}h_i = \frac{3}{h_i}(d_{i+1} - d_i) - \frac{3}{h_{i-1}}(d_i - d_{i-1}), \tag{11.72}$$

where $h_i = x_{i+1} - x_i$ and $h_{i-1} = x_i - x_{i-1}$. This is similar to the condition under (11.13). By letting i run from 1 to $n-1$ and taking account of the end conditions $b_0 = b_n = 0$, we can obtain the following matrix system:

$$\begin{bmatrix} p_1 & h_1 & 0 & \ldots & 0 & 0 \\ h_1 & p_2 & h_2 & \ldots & 0 & 0 \\ 0 & h_2 & p_3 & \ldots & 0 & 0 \\ \vdots & \vdots & \vdots & \ddots & \vdots & \vdots \\ 0 & 0 & 0 & \ldots & p_{n-2} & h_{n-2} \\ 0 & 0 & 0 & \ldots & h_{n-2} & p_{n-1} \end{bmatrix} \begin{bmatrix} b_1 \\ b_2 \\ b_3 \\ \vdots \\ b_{n-2} \\ b_{n-1} \end{bmatrix}$$

$$= \begin{bmatrix} r_0 & f_1 & r_1 & 0 & \ldots & 0 & 0 \\ 0 & r_1 & f_2 & r_2 & \ldots & 0 & 0 \\ \vdots & \vdots & \vdots & \vdots & \ddots & \vdots & \vdots \\ 0 & 0 & 0 & 0 & \ldots & r_{n-2} & 0 \\ 0 & 0 & 0 & 0 & \ldots & f_{n-1} & r_{n-1} \end{bmatrix} \begin{bmatrix} d_0 \\ d_1 \\ d_2 \\ d_3 \\ \vdots \\ d_{n-1} \\ d_n \end{bmatrix}, \tag{11.73}$$

where

$$(11.74)\qquad \begin{aligned} p_i &= 2(h_{i-1}+h_i), \\ r_i &= \frac{3}{h_i} \quad \text{and} \\ f_i &= -\left(\frac{3}{h_{i-1}}+\frac{3}{h_i}\right) = -(r_{i-1}+r_i). \end{aligned}$$

The matrix equation can be written in a summary notation as

$$(11.75)\qquad Mb = Q'd.$$

This notation can also be used to write the criterion function of (11.66) as

$$(11.76)\qquad L = (y-d)'\Sigma^{-1}(y-d) + \lambda b'Mb,$$

where $\Sigma = \text{diag}\{\sigma_0,\ldots,\sigma_n\}$. Using $b = M^{-1}Q'd$, which comes from (11.75), enables us to write the function solely in terms of the vector d which contains the ordinates of the knots:

$$(11.77)\qquad L(d) = (y-d)'\Sigma^{-1}(y-d) + \lambda d'QM^{-1}Q'd.$$

The optimal values of the ordinates are those which minimise the function $L(d)$. Differentiating with respect to d and setting the result to zero gives

$$(11.78)\qquad -(y-d)'\Sigma^{-1} + \lambda d'QM^{-1}Q' = 0,$$

which is the first-order condition for minimisation. This gives

$$(11.79)\qquad \begin{aligned} \Sigma^{-1}(y-d) &= \lambda QM^{-1}Q'd \\ &= \lambda Qb. \end{aligned}$$

When this equation is premultiplied by $Q'\Sigma$ and rearranged with further help from the identity $Mb = Q'd$ of (11.75), we get

$$(11.80)\qquad \big(M + \lambda Q'\Sigma Q\big)b = Q'y.$$

Once this has been solved for b, the value of d can be obtained from equation (11.79). Thus

$$(11.81)\qquad d = y - \lambda\Sigma Qb.$$

The value of the criterion function is given by

$$(11.82)\qquad L = (y-d)'\Sigma^{-1}(y-d) = \lambda^2 b'Q'\Sigma Qb.$$

The matrix $A = M + \lambda Q'\Sigma Q$ of equation (11.80) is symmetric with five diagonal bands; and the structure of the matrix may be exploited in deriving a specialised procedure for solving the equation. The procedure is as follows. First the factorisation $A = LDL'$ is found, where L is a lower-triangular matrix and D is a diagonal matrix. Then the system $LDL'b = Q'y$ is cast in the form of $Lx = Q'y$ and solved for x whose elements are stored in place of those of $Q'y$. Finally, $L'b = D^{-1}x$ is solved for b whose elements replace those of x.

The procedure *Quincunx*, which effects this solution, takes as arguments the vectors u, v and w which are, respectively, the diagonal, the first supradiagonal and the second supradiagonal of the banded matrix $A = M + \lambda Q'\Sigma Q$. The vector $Q'y$ on the RHS of the equation (11.80) is placed in the array q which contains the solution vector b on the completion of the procedure.

(11.83)

```
procedure Quincunx(n : integer;
                        var u, v, w, q : vector);

  var
    j : integer;

begin {Quincunx}
{Factorisation}
  u[-1] := 0;
  u[0] := 0;
  v[0] := 0;
  w[-1] := 0;
  w[0] := 0;
  for j := 1 to n - 1 do
    begin
      u[j] := u[j] - u[j - 2] * Sqr(w[j - 2]) - u[j - 1] * Sqr(v[j - 1]);
      v[j] := (v[j] - u[j - 1] * v[j - 1] * w[j - 1])/u[j];
      w[j] := w[j]/u[j];
    end;

{Forward-substitution}
  q[0] := 0;
  q[-1] := 0;
  for j := 1 to n - 1 do
    q[j] := q[j] - v[j - 1] * q[j - 1] - w[j - 2] * q[j - 2];
  for j := 1 to n - 1 do
    q[j] := q[j]/u[j];

{Back-substitution}
  q[n + 1] := 0;
  q[n] := 0;
  for j := n - 1 downto 1 do
    q[j] := q[j] - v[j] * q[j + 1] - w[j] * q[j + 2];

end; {Quincunx}
```

The procedure which calculates the smoothing spline may be envisaged as a generalisation of the procedure *CubicSplines* of (11.18) which calculates an interpolating spline. In fact, by setting $\lambda = 0$, we obtain the interpolating spline. Figure 11.4, which uses the same data on the consumption of meat as Figure 10.2, gives an example of the effects of varying the smoothing parameter.

The *SmoothingSpline* procedure is wasteful of computer memory, since there is no need to store the contents of the vectors r and f which have been included in the code only for reasons of clarity. At any stage of the iteration of the index j, only two consecutive elements from each of these vectors are called for; and one of these elements may be calculated concurrently. However, the waste of memory is of little concern unless one envisages applying the procedure to a very long run of data. In that case, it should be straightforward to modify the procedure.

(11.84)
```
procedure SmoothingSpline(var S : SplineVec;
                          sigma : vector;
                          lambda : real;
                          n : integer);

  var
    h, r, f, p, q, u, v, w : vector;
    i, j : integer;

begin {SmoothingSpline}

  h[0] := S[1].x - S[0].x;
  r[0] := 3/h[0];
  for i := 1 to n - 1 do
    begin
      h[i] := S[i + 1].x - S[i].x;
      r[i] := 3/h[i];
      f[i] := -(r[i - 1] + r[i]);
      p[i] := 2 * (S[i + 1].x - S[i - 1].x);
      q[i] := 3 * (S[i + 1].y - S[i].y)/h[i]
            -3 * (S[i].y - S[i - 1].y)/h[i - 1];
    end;

    r[n] := 0;
    f[n] := 0;

  for i := 1 to n - 1 do
    begin
      u[i] := Sqr(r[i - 1]) * sigma[i - 1]
            +Sqr(f[i]) * sigma[i] + Sqr(r[i]) * sigma[i + 1];
      u[i] := lambda * u[i] + p[i];
      v[i] := f[i] * r[i] * sigma[i] + r[i] * f[i + 1] * sigma[i + 1];
      v[i] := lambda * v[i] + h[i];
      w[i] := lambda * r[i] * r[i + 1] * sigma[i + 1];
```

```
      end;

    Quincunx(n, u, v, w, q);

  {Spline parameters}
    S[0].d := S[0].y − lambda ∗ r[0] ∗ q[1] ∗ sigma[0];
    S[1].d := S[1].y − lambda ∗ (f[1] ∗ q[1] + r[1] ∗ q[2]) ∗ sigma[0];
    S[0].a := q[1]/(3 ∗ h[0]);
    S[0].b := 0;
    S[0].c := (S[1].d − S[0].d)/h[0] − q[1] ∗ h[0]/3;
    r[0] := 0;

    for j := 1 to n − 1 do
      begin
        S[j].a := (q[j + 1] − q[j])/(3 ∗ h[j]);
        S[j].b := q[j];
        S[j].c := (q[j] + q[j − 1]) ∗ h[j − 1] + S[j − 1].c;
        S[j].d := r[j − 1] ∗ q[j − 1] + f[j] ∗ q[j] + r[j] ∗ q[j + 1];
        S[j].d := S[j].y − lambda ∗ S[j].d ∗ sigma[j];
      end;
    S[n].d := S[n].y − lambda ∗ r[n − 1] ∗ q[n − 1] ∗ sigma[n];

  end; {SmoothingSpline}
```

A Stochastic Model for the Smoothing Spline

A disadvantage of the smoothing spline is the extent to which the choice of the value for the smoothing parameter remains a matter of judgment. One way of avoiding such judgments is to adopt an appropriate model of the process which has generated the data to which the spline is to be fitted. Then the value of the smoothing parameter may be determined in the process of fitting the model.

Since the smoothing spline is a continuous function, it is natural to imagine that the process underlying the data is also continuous. A model which is likely to prove appropriate to many circumstances is a so-called integrated Wiener process which is the continuous analogue of the familiar discrete-time unit-root autoregressive processes commonly described as a random walk. To the continuous process, a discrete process is added which represents a set of random errors of observation. Thus, the estimation of the trend becomes a matter of signal extraction.

A Wiener process $Z(t)$ consists of an accumulation of independently distributed stochastic increments. The path of $Z(t)$ is continuous almost everywhere and differentiable almost nowhere. If $dZ(t)$ stands for the increment of the process in the infinitesimal interval dt, and if $Z(a)$ is the value of the function at time a, then the value at time $\tau > a$ is given by

$$Z(\tau) = Z(a) + \int_a^\tau dZ(t). \tag{11.85}$$

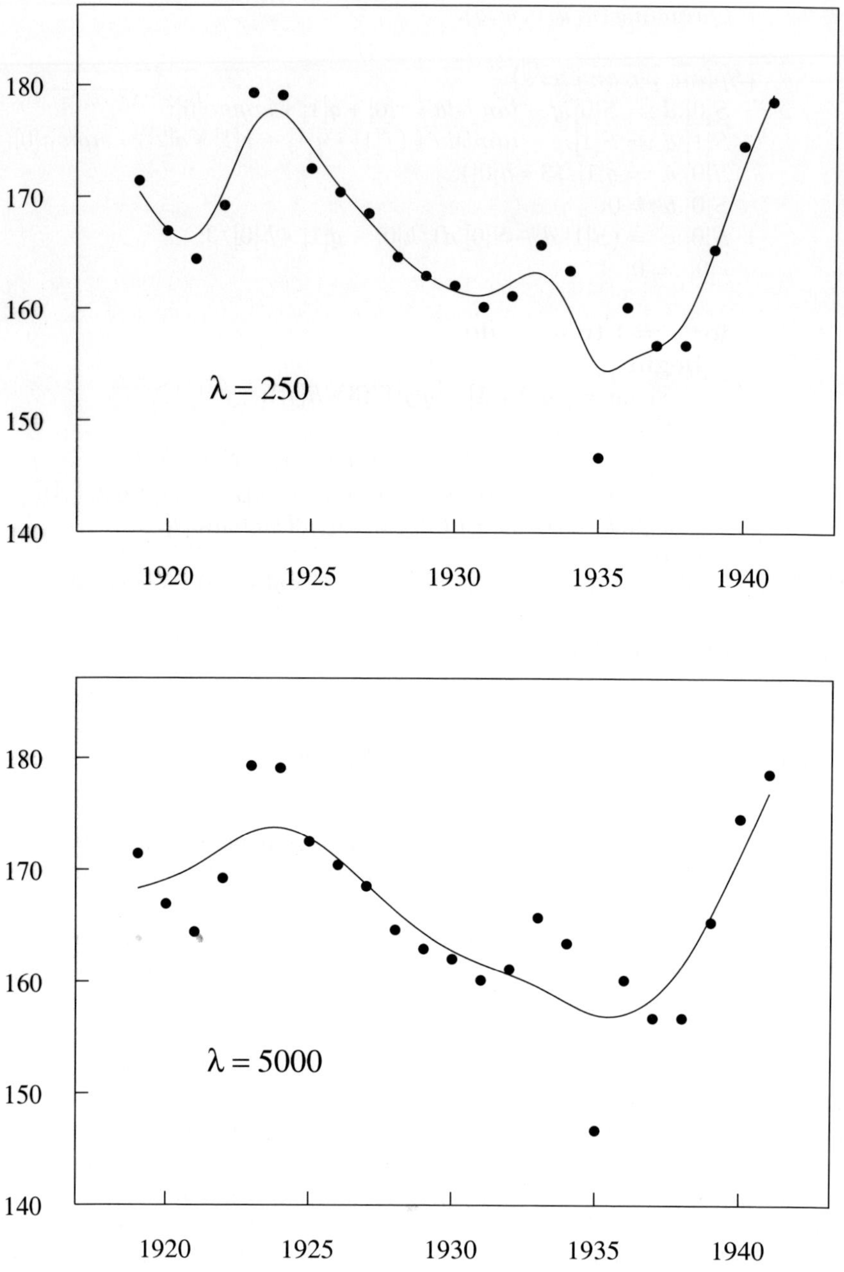

Figure 11.4. Cubic smoothing splines fitted to data on meat consumption in the United States, 1919–1941.

Moreover, it is assumed that the change in the value of the function over any finite interval $(a, \tau]$ is a random variable with a zero expectation:

$$E\{Z(\tau) - Z(a)\} = 0. \tag{11.86}$$

Let us write $ds \cap dt = \emptyset$ whenever ds and dt represent nonoverlapping intervals. Then the conditions affecting the increments may be expressed by writing

$$E\{dZ(s)dZ(t)\} = \begin{cases} 0, & \text{if } ds \cap dt = \emptyset; \\ \sigma^2 dt, & \text{if } ds = dt. \end{cases} \tag{11.87}$$

These conditions imply that the variance of the change over the interval $(a, \tau]$ is proportional to the length of the interval. Thus

$$\begin{aligned} V\{Z(\tau) - Z(a)\} &= \int_{s=a}^{\tau} \int_{t=a}^{\tau} E\{dZ(s)dZ(t)\} \\ &= \int_{t=a}^{\tau} \sigma^2 dt = \sigma^2(\tau - a). \end{aligned} \tag{11.88}$$

The definite integrals of the Wiener process may be defined also in terms of the increments. The value of the first integral at time τ is given by

$$\begin{aligned} Z^{(1)}(\tau) &= Z^{(1)}(a) + \int_a^{\tau} Z(t)dt \\ &= Z^{(1)}(a) + Z(a)(\tau - a) + \int_a^{\tau} (\tau - t)dZ(t), \end{aligned} \tag{11.89}$$

The mth integral is

$$Z^{(m)}(\tau) = \sum_{k=0}^{m} Z^{(m-k)}(a)\frac{(\tau - a)^k}{k!} + \int_a^{\tau} \frac{(\tau - t)^m}{m!} dZ(t). \tag{11.90}$$

The covariance of the changes $Z^{(j)}(\tau) - Z^{(j)}(a)$ and $Z^{(k)}(\tau) - Z^{(k)}(a)$ of the jth and the kth integrated processes derived from $Z(t)$ is given by

$$\begin{aligned} C_{(a,\tau)}\{z^{(j)}, z^{(k)}\} &= \int_{s=a}^{\tau} \int_{t=a}^{\tau} \frac{(\tau - s)^j(\tau - t)^k}{j!k!} E\{dZ(s)dZ(t)\} \\ &= \sigma^2 \int_a^{\tau} \frac{(\tau - t)^j(\tau - t)^k}{j!k!} dt = \sigma^2 \frac{(\tau - a)^{j+k+1}}{(j + k + 1)j!k!}. \end{aligned} \tag{11.91}$$

The simplest stochastic model which can give rise to the smoothing spline is one in which the generic observation is depicted as the sum of a trend component described by an integrated Wiener process and a random error taken from a discrete white-noise sequence. We may imagine that the observations $y_0, y_1, \ldots, y_n$ are made at the times $t_0, t_1, \ldots, t_n$. The interval between t_{i+1} and t_i is $h_i = t_{i+1} - t_i$ which, for the sake of generality, is allowed to vary. These points in time replace the abscissae $x_0, x_1, \ldots, x_n$ which have, hitherto, formed part of our observations.

In order to conform to the existing notation, we define

$$(11.92)\qquad c_i = Z(t_i) \quad \text{and} \quad d_i = Z^{(1)}(t_i)$$

to be, respectively, the slope of the trend component and its level at time t_i, where $Z(t_i)$ and $Z^{(1)}(t_i)$ are described by equations (11.85) and (11.89). Also, we define

$$(11.93)\qquad \varepsilon_{i+1} = \int_{t_i}^{t_{i+1}} dZ(t) \quad \text{and} \quad \nu_{i+1} = \int_{t_i}^{t_{i+1}} (t_{i+1} - t)dZ(t).$$

Then the integrated Wiener process of (11.89), which is the model of the underlying trend, can be written is state-space form as follows:

$$(11.94)\qquad \begin{bmatrix} d_{i+1} \\ c_{i+1} \end{bmatrix} = \begin{bmatrix} 1 & h_i \\ 0 & 1 \end{bmatrix} \begin{bmatrix} d_i \\ c_i \end{bmatrix} + \begin{bmatrix} \nu_{i+1} \\ \varepsilon_{i+1} \end{bmatrix},$$

whilst the equation of the corresponding observation is

$$(11.95)\qquad y_{i+1} = \begin{bmatrix} 1 & 0 \end{bmatrix} \begin{bmatrix} d_{i+1} \\ c_{i+1} \end{bmatrix} + \eta_{i+1}.$$

Using the result under (11.91), we find that the dispersion matrix for the state disturbances is

$$(11.96)\qquad D\begin{bmatrix} \nu_{i+1} \\ \varepsilon_{i+1} \end{bmatrix} = \sigma_\eta^2 \phi \begin{bmatrix} \frac{1}{3}h_i^3 & \frac{1}{2}h_i^2 \\ \frac{1}{2}h_i^2 & h_i \end{bmatrix},$$

where $\sigma_\eta^2\phi = \sigma_\varepsilon^2$ is the variance of the Wiener process expressed as the product of the variance σ_η^2 of the observations errors and of the signal-to-noise ratio $\phi = \sigma_\varepsilon^2/\sigma_\eta^2$.

The estimation of the model according to the criterion of maximum likelihood could be accomplished by a straightforward application of the Kalman filter which serves to generate the prediction errors whose sum of squares is the major element of the criterion function. In fact, when it has been concentrated in respect of σ_η^2, the criterion function has the signal-to-noise ratio ϕ as its sole argument. Once the minimising value of ϕ has been determined, the definitive smoothed estimates of the state parameters c_i, d_i for $i = 0, \ldots, n$ may be obtained via one of the algorithms presented in Chapter 9. The values should coincide with those which would be generated by the *SmoothingSpline* algorithm given the appropriate value of the smoothing parameter.

The advantage of the *SmoothingSpline* algorithm of (11.84) is that it automatically generates the remaining parameters of the cubic segments which bridge the gaps between the points (t_i, d_i) and which serve, thereby, to estimate the underlying trend.

In order to estimate the path of the trend on the basis of the postulated Wiener process, it is necessary to represent the values which lie between the adjacent points (t_i, d_i), (t_{i+1}, d_{i+1}) by an interpolated function whose first derivatives at the two points are given by c_i and c_{i+1}. It has been demonstrated by Craven and Wahba [130] that the curve which represents the minimum-mean-square-error estimator of a trend generated by an integrated Wiener process is indeed a smoothing spline.

The practical details of constructing the spline within the framework of a Wiener process have been set forth by Wecker and Ansley [506]. A lucid exposition, which we shall follow here, has been provided recently by de Vos and Steyn [149].

The problem of estimating the intermediate value of the trend between the times t_i and t_{i+1} of two adjacent observations is that of finding its expectation conditional upon the values $\xi_i = (c_i, d_i)$ and $\xi_{i+1} = (c_{i+1}, d_{i+1})$. Let $t \in (t_i, t_{i+1}]$ be the date of the intermediate values c_t and d_t; and let us define the following quantities which represent the stochastic increments which accumulate over the sub-intervals $(t_i, t]$ and $(t, t_{i+1}]$:

(11.97)
$$\begin{aligned} \varepsilon_t &= \int_{t_1}^{t} dZ(\tau), & \bar{\varepsilon}_t &= \int_{t}^{t_{i+1}} dZ(\tau), \\ \nu_t &= \int_{t_1}^{t} (t - t_i) dZ(\tau), & \bar{\nu}_t &= \int_{t}^{t_{i+1}} (t_{i+1} - t) dZ(\tau). \end{aligned}$$

In these terms, the stochastic increments over the entire interval $(t_i, t_{i+1}]$ are given by

(11.98)
$$\begin{bmatrix} \nu_{i+1} \\ \varepsilon_{i+1} \end{bmatrix} = \begin{bmatrix} 1 & (t_{i+1} - t) \\ 0 & 1 \end{bmatrix} \begin{bmatrix} \nu_t \\ \varepsilon_t \end{bmatrix} + \begin{bmatrix} \bar{\nu}_t \\ \bar{\varepsilon}_t \end{bmatrix},$$

which is a variant of equation (11.94).

The values of the slope and the level of the Wiener process at time t can be given in terms of two of the quantities under (11.97) as follows:

(11.99)
$$\begin{aligned} c_t &= c_i + \varepsilon_t \quad \text{and} \\ d_t &= d_i + (t - t_i)c_i + \nu_t. \end{aligned}$$

After the rest of the interval from t to t_{i+1} has been covered, the slope and the level become

(11.100)
$$\begin{aligned} c_{i+1} &= c_t + \bar{\varepsilon}_t \quad \text{and} \\ d_{i+1} &= d_t + (t_{i+1} - t)c_t + \bar{\nu}_t, \end{aligned}$$

which entail the remaining quantities under (11.97). Substituting for c_t and d_t in these expressions gives

(11.101)
$$\begin{aligned} c_{i+1} &= c_i + \varepsilon_t + \bar{\varepsilon}_t \quad \text{and} \\ d_{i+1} &= d_i + h_i c_i + (t_{i+1} - t)\varepsilon_t + \nu_t + \bar{\nu}_t, \end{aligned}$$

wherein $(t_{i+1} - t)\varepsilon_t + \nu_t + \bar{\nu}_t = \nu_{i+1}$ is an expression which is also provided by equation (11.98).

The equations of (11.99) and (11.101) enable us to evaluate the joint moments of d_t, d_{i+1} and c_{i+1} conditional upon the values c_i and d_i. Thus, with reference to the result under (11.91), we find that

$$C(d_t, c_{i+1}) = C(\nu_t, \varepsilon_t) = \frac{1}{2}(t - t_i)^2 \tag{11.102}$$

and that

$$\begin{aligned} C(d_t, d_{i+1}) &= (t_{i+1} - t)C(\nu_t, \varepsilon_t) + V(\nu_t) \\ &= \frac{1}{2}(t_{i+1} - t)(t - t_i)^2 + \frac{1}{3}(t - t_i)^3. \end{aligned} \tag{11.103}$$

The conditional expectation of the intermediate trend value d_t is given by the regression equation

$$E(d_t|\mathcal{I}_{i+1}) = E(d_t|\mathcal{I}_i) + C(d_t, \xi_{i+1})D(\xi_{i+1})^{-1}(\xi_{i+1} - E\{\xi_{i+1}|\mathcal{I}_i\}), \tag{11.104}$$

where $\xi_{i+1} = (d_{i+1}, c_{i+1})$, and where $\mathcal{I}_i$ and $\mathcal{I}_{i+1}$ represent the information available at t_i and t_{i+1} which is conveyed, in fact, by the values of ξ_i and ξ_{i+1}.

On the RHS of the expression there is

$$\begin{aligned} E(d_t|\mathcal{I}_i) &= d_i + (t - t_i)c_i \quad \text{and} \\ \xi_{i+1} - E\{\xi_{i+1}|\mathcal{I}_i\} &= \begin{bmatrix} d_{i+1} - d_i - h_i c_i \\ c_{i+1} - c_i \end{bmatrix}. \end{aligned} \tag{11.105}$$

Of the remaining terms on the RHS, the elements of the vector $C(d_t, \xi_{i+1}) = [C(d_t, d_{i+1}), C(d_t, c_{i+1})]$ are found under (11.102) and (11.103), whilst the dispersion matrix $D(\xi_{i+1}) = D[\nu_{i+1}, \varepsilon_{i+1}]$ is to be found under (11.96).

Detailed computation shows that the regression equation of (11.104) is a cubic function of t of the form

$$f(t) = a_i(t - t_i)^3 + b_i(t - t_i)^2 + c_i(t - t_i) + d_i \tag{11.106}$$

wherein

$$a_i = \frac{1}{h_i^2}(c_i + c_{i+1}) + \frac{2}{h_i^3}(d_i - d_{i+1}) \tag{11.107}$$

and

$$b_i = \frac{3}{h_i^2}(d_{i+1} - d_i) - \frac{1}{h_i}(c_{i+1} + 2c_i). \tag{11.108}$$

The expressions for a_i and b_i could be obtained from those under (11.22) and (11.23) simply by substituting d_{i+1} and d_i for y_{i+1} and y_i respectively. The latter expressions relate to a segment of an interpolating spline of which the ends have been clamped.

The mere fact that the estimate of the stochastic trend between the points (t_i, d_i) and (t_{i+1}, d_{i+1}) has the same form as a segment of a spline does not establish that the estimated trend function as a whole is equivalent to a smoothing spline. It has to be shown, in addition, that the knots of the segmented trend curve are identical to those which would be generated by a smoothing spline for a particular value of the smoothing parameter. A demonstration of this result, which is on an abstract level, has been provided by Wahba [498].

We shall be able to demonstrate the result, more easily, at a later stage when, in Chapter 19, we derive anew the part of the algorithm of the smoothing spline which generates the knots. Then we shall concentrate solely on the problem of estimating the trend values at the points of observation. For this purpose, we shall rely upon a discrete-time model to depict the trend values. In the appendix to this chapter, we demonstrate the necessary result, which is that the exact observations of an integrated Wiener process, taken at equally spaced intervals, follow an ordinary integrated moving-average IMA$(2,1)$ discrete-time process.

Appendix: The Wiener Process and the IMA Process

The values at the times $t_0, t_1, \ldots, t_n$ which are generated by an integrated Wiener process follow a Markov process which is depicted in equation (11.94). Our object is to show that this can also be expressed as an integrated moving-average (IMA) process $\xi(t)$. Then it will become a straightforward matter to apply the Wiener–Kolmogorov theory of signal extraction to the problem of estimating the values of $\xi(t)$ from a sequence of observations $y(t) = \xi(t) + \eta(t)$ which are afflicted by a white-noise error process $\eta(t)$.

In Chapter 19, we shall present a finite-sample version of the Wiener–Kolmogorov filter which shares its algorithm with the Reinsch smoothing spline. The filter will be based upon the IMA process which will be revealed in this appendix.

On the assumption that observations are equally spaced at unit intervals, which is the assumption that $h_t = 1$ for all t, the model of equation (11.94), which depicts the values of an integrated Wiener process, can be written is state-space form as follows:

$$\begin{bmatrix} \xi_t \\ \zeta_t \end{bmatrix} = \begin{bmatrix} 1 & 1 \\ 0 & 1 \end{bmatrix} \begin{bmatrix} \xi_{t-1} \\ \zeta_{t-1} \end{bmatrix} + \begin{bmatrix} \nu_t \\ \varepsilon_t \end{bmatrix}. \tag{11.109}$$

Here ν_t and ε_t are from mutually correlated white-noise processes. According to the result under (11.96), the dispersion matrix for these state disturbances is

$$D\begin{bmatrix} \nu_t \\ \varepsilon_t \end{bmatrix} = \sigma_\varepsilon^2 \begin{bmatrix} \frac{1}{3} & \frac{1}{2} \\ \frac{1}{2} & 1 \end{bmatrix}, \tag{11.110}$$

where σ_ε^2 is the variance of the Wiener process.

The discrete-time processes entailed in equation (11.109) can be written as

$$\begin{aligned} \nabla\xi(t) &= \xi(t) - \xi(t-1) = \zeta(t-1) + \nu(t) \qquad \text{and} \\ \nabla\zeta(t) &= \zeta(t) - \zeta(t-1) = \varepsilon(t). \end{aligned} \tag{11.111}$$

Applying the difference operator a second time to the first of these and substituting for $\nabla\zeta(t-1) = \varepsilon(t-1)$ gives

$$\begin{aligned}\nabla^2\xi(t) &= \nabla\zeta(t-1) + \nabla\nu(t)\\ &= \varepsilon(t-1) + \nu(t) - \nu(t-1).\end{aligned} \tag{11.112}$$

On the RHS of this equation is a sum of stationary stochastic processes which can be expressed as an ordinary first-order moving-average process. Thus

$$\varepsilon(t-1) + \nu(t) - \nu(t-1) = \eta(t) + \mu\eta(t-1), \tag{11.113}$$

where $\eta(t)$ is a white-noise process with $V\{\eta(t)\} = \sigma^2_\eta$.

The parameters of the latter process may be inferred from its autocovariances which arise from a combination of the autocovariances of $\varepsilon(t)$ and $\nu(t)$. The variance γ_0 of the MA process is given by the sum of the elements of the matrix

$$E\begin{bmatrix} \nu_t^2 & -\nu_t\nu_{t-1} & \nu_t\varepsilon_{t-1} \\ -\nu_{t-1}\nu_t & \nu_{t-1}^2 & -\nu_{t-1}\varepsilon_{t-1} \\ \varepsilon_{t-1}\nu_t & -\varepsilon_{t-1}\nu_{t-1} & \varepsilon_{t-1}^2 \end{bmatrix} = \sigma^2_\varepsilon \begin{bmatrix} \frac{1}{3} & 0 & 0 \\ 0 & \frac{1}{3} & -\frac{1}{2} \\ 0 & -\frac{1}{2} & 1 \end{bmatrix}. \tag{11.114}$$

Thus it is found that $\gamma_0 = 4\sigma^2_\varepsilon/6$. The first autocovariance γ_1 of the MA process is given by the sum of the elements of the matrix

$$E\begin{bmatrix} \nu_t\nu_{t-1} & -\nu_t\nu_{t-2} & \nu_t\varepsilon_{t-2} \\ -\nu_{t-1}^2 & \nu_{t-1}\nu_{t-2} & -\nu_{t-1}\varepsilon_{t-2} \\ \varepsilon_{t-1}\nu_{t-1} & -\varepsilon_{t-1}\nu_{t-2} & \varepsilon_{t-1}\varepsilon_{t-2} \end{bmatrix} = \sigma^2_\varepsilon \begin{bmatrix} 0 & 0 & 0 \\ -\frac{1}{3} & 0 & 0 \\ \frac{1}{2} & 0 & 0 \end{bmatrix}. \tag{11.115}$$

Thus $\gamma_1 = \sigma^2_\varepsilon/6$. The values of the moving-average parameters are found by solving the equations

$$\gamma_0 = \frac{2\sigma^2_\varepsilon}{3} = \sigma^2_\eta(1+\mu^2) \quad \text{and} \quad \gamma_1 = \frac{\sigma^2_\varepsilon}{6} = \sigma^2_\eta\mu. \tag{11.116}$$

There are two solution for μ; and we should take the one which fulfils the condition of invertibility: $\mu = 2 - \sqrt{3}$.

Bibliography

[2] Achieser, N.I., (1956), *Theory of Approximation*, Frederick Ungar Publishing Co, New York.

[3] Adobe Systems Incorporated, (1985), *PostScript Language, Reference Manual*, Addison Wesley, Reading, Massachusetts.

[49] Bernstein, S.N., (1912), Demonstration du théorème de Weierstrass fondeé sur le calcul de probabilité, *Proceedings of the Mathematical Society of Kharkov*, **X111.**

[53] Bézier, P., (1966), Définition Numérique des Courbes et Surfaces I, *Automatisme*, **11**, 625–632.

[54] Bézier, P., (1967), Définition Numérique des Courbes et Surfaces II, *Automatisme*, **12**, 17–21.

[68] Bohm, W., (1981), Generating the Bézier Points of B-spline Curves and Surfaces, *Computer-Aided Design*, **13**, 365–366.

[117] Cohen, Elaine, and R.F. Riesenfeld, (1982), General Matrix Representations for Bézier and B-Spline Curves, *Computers in Industry*, **3**, 9–15.

[130] Craven, P., and Grace Wahba, (1979), Smoothing Noisy Data with Spline Functions, *Numerische Mathematik*, **31**, 377–403.

[138] de Boor, C., (1978), *A Practical Guide to Splines*, Springer Verlag, New York.

[149] De Vos, A.F., and I.J. Steyn, (1990), *Stochastic Nonlinearity: A Firm Basis for the Flexible Functional Form*, Research Memorandum 1990-13, Department of Econometrics, Vrije Universiteit Amsterdam.

[174] Eubank, R.A., (1988), *Spline Smoothing and Nonparametric Regression*, Marcel Dekker, New York and Basel.

[180] Farin, G., (1988), *Curves and Surfaces for Computer Aided Geometric Design: A Practical Guide*, Academic Press, San Diego, California.

[196] Friedman, J.H., (1991), Multivariate Adaptive Regression Splines, *The Annals of Statistics*, **19**, 1–141.

[203] Gasser, T., and M. Rosenblatt, (eds.), (1979), *Smoothing Techniques for Curve Estimation*, Springer Verlag, Berlin.

[264] Hutchinson, M.F., and F.R. de Hoog, (1985), Smoothing Noisy Data with Spline Functions, *Numerische Mathematik*, **47**, 99–106.

[290] Kimeldorf, G.S., and Grace Wahba, (1970), A Correspondence between Bayesian Estimation on Stochastic Processes and Smoothing Splines, *The Annals of Mathematical Statistics*, **41**, 495–502.

[386] Pavlidis, T., (1983), Curve Fitting with Conic Splines, *ACM Transactions on Graphics*, **2**, 1–31.

[423] Reinsch, C.H., (1976), Smoothing by Spline Functions, *Numerische Mathematik*, **10**, 177–183.

[442] Schoenberg, I.J., (1964), Spline Functions and the Problem of Graduation, *Proceedings of the National Academy of Sciences*, **52**, 947–950.

[456] Silverman, B.W., (1984), Spline Smoothing: The Equivalent Variable Kernel Method, *The Annals of Statistics*, **12**, 898–916.

[457] Silverman, B.W., (1985), Some Effects of the Spline Smoothing Approach to Non-parametric Regression Curve Fitting, *Journal of the Royal Statistical Society, Series B*, **47**, 1–52.

[469] Smith, P.L., (1979), Splines as a Useful and Convenient Statistical Tool, *The American Statistician*, **33**, 57–62.

[477] Su, Bu-qing, and Liu Ding-yuan, (1989), *Computational Geometry, Curve and Surface Modelling*, Academic Press, San Diego, California.

[498] Wahba, Grace, (1978), Improper Priors, Spline Smoothing and the Problem of Guarding against Model Errors in Regression, *Journal of the Royal Statistical Society, Series B*, **40**, 364–372.

[500] Wahba, Grace, (1983), Bayesian Confidence Intervals for the Cross Validated Smoothing Spline, *Journal of the Royal Statistical Society, Series B*, **45**, 133–150.

[501] Wahba, Grace, (1985), A Comparison of the GCV and GML for Choosing the Smoothing Parameter in the Generalised Spline Smoothing Problem, *The Annals of Statistics*, **13**, 1378–1402.

[506] Wecker, W.P., and C.F. Ansley, (1983), The Signal Extraction Approach to Nonlinear Regression and Spline Smoothing, *Journal of the American Statistical Association*, **78**, 81–89.

[507] Wegman, E.J., and I.W. Wright, (1983), Splines in Statistics, *Journal of the American Statistical Society*, **78**, 351–365.

[508] Weierstrass, K., (1885), *Über die analytische Darstellbarkheit sogenannter willhülicher Functionen einer reellen Veränderlichen*, Berliner Berichte.

[531] Wold, S., (1974), Spline Functions in Data Analysis, *Technometrics*, **16**, 1–11.

CHAPTER 12

Unconstrained Optimisation

A usual way of estimating the parameters of a statistical model is to seek the values which maximise or minimise a criterion function such as a likelihood function or a sum of squares of prediction errors. If the criterion function is a quadratic function of the unknown parameters, then the first-order conditions for its optimisation will give rise to a set of linear estimating equations which are easily solved to obtain the estimates.

If the criterion function is not a quadratic, then we cannot expect the first-order conditions to have an analytic or closed-form solution. There are two ways of overcoming this difficulty. Either one may endeavour to solve the nonlinear estimating equations by iterative methods, or else one may use iterative techniques to find the values which optimise the criterion function. In this chapter, we shall pursue the latter approach.

In a formal sense, the two approaches are equivalent. In practice, however, they can be quite different. An approach which is aimed at solving the first-order conditions can take account of the particular features of the problem at hand. An optimisation approach, in contrast, must rely upon a general theory of nonlinear functions. If an optimisation technique is to be widely applicable, then it must be capable of dealing with all manner of contingencies; and, therefore, robust methods tend to be complex.

In view of the complexity of modern optimisation techniques, and in view of the likelihood that naive implementations of the techniques will run into problems, some authorities—with Gill *et al.* [211, p. 5] amongst them—have sought to discourage the typical user from writing their own programs. They have suggested that, instead, one should use algorithms selected from high-quality mathematical software libraries. This opinion is too unremitting; for, if a relatively simple and well-understood technique is applied, and if its performance is closely monitored, then the dangers can be averted. Such monitoring is barely possible, even if it is unnecessary, when library routines are used; and the consequence is that some of the more interesting features of the problem at hand may be missed.

Conditions of Optimality

In discussing optimisation methods, we shall consider only the minimisation of functions, since a problem of maximisation can be solved by minimising the negative of the function in question. The functions are assumed to be continuous and smooth, which means that they must be twice-differentiable.

We should begin by giving a precise definition of the minimum of a multivariate function.

(12.1) A point θ^* is said to be a strict minimum of the function $S(\theta)$ if $S(\theta^*) < S(\theta^* + d)$ for all d in a convex set $\mathcal{B} = \{d : 0 < \|d\| \leq \epsilon\}$. The point is said to be a weak minimum of $S(\theta)$ if $S(\theta^*) \leq S(\theta^* + d)$ for all $d \in \mathcal{B}$.

In effect, the point θ^* is a strict minimum if the value of S increases locally in all directions departing from θ^*, whereas it is a weak minimum if the function decreases in none of the directions and may increase in some. In general, a function may exhibit these properties at a number of points which are described as local minima. If there is a unique point at which the function is lowest, then this is called a global minimum.

It is not possible to demonstrate that an analytic function has a global minimum without a complete knowledge of its derivatives of all orders. The conditions which are sufficient for the existence of a local minimum are modest by comparison.

(12.2) The function $S(\theta) \in \mathcal{C}^2$ has a strict minimum at the point θ^* if and only if $\gamma = \partial S/\partial\theta = 0$ at θ^* and the Hessian matrix $H = \partial(\partial S/\partial\theta)'/\partial\theta$ is positive definite in a neighbourhood of θ^* such that $d'Hd > 0$ for any $d \neq 0$.

Proof. We may set $d = \lambda p$, where λ is a scalar and p is a vector. Then the mean-value theorem indicates that

$$S(\theta^* + \lambda p) = S(\theta^*) + \lambda\gamma'(\theta^*)p + \frac{1}{2}\lambda^2 p' H(\theta^* + \kappa\lambda p)p, \tag{12.3}$$

where the κ satisfies $0 \leq \kappa \leq 1$.

The condition $S(\theta^* + \lambda p) \geq S(\theta^*)$ implies that $\gamma' p + \frac{1}{2}\lambda p' H p \geq 0$; and letting $\lambda \rightarrow 0$ shows that $\gamma' p \geq 0$.

The condition $S(\theta^* - \lambda p) \geq S(\theta^*)$ also holds when $|\lambda|$ is small enough. According to the mean-value theorem, this condition implies that $-\gamma' p + \frac{1}{2}\lambda p' H p \geq 0$. Letting $\lambda \rightarrow 0$ shows that $-\gamma' p \geq 0$; and this can be reconciled with the previous inequality only if $\gamma' p = 0$ which implies that $\gamma = 0$, since p is arbitrary. This is a necessary condition for a minimum.

Now, if $\gamma(\theta^*) = 0$, then the inequality $S(\theta^* + d) \geq S(\theta^*)$ holds for all $d = \lambda p$ in a neighbourhood of zero if and only if $d'Hd \geq 0$. Therefore, $\gamma = 0$ and $d'Hd \geq 0$ are jointly the necessary and sufficient conditions for a minimum. If $d'Hd > 0$, then there is a strict minimum.

The multivariate minimisation methods discussed in this chapter are iterative methods which begin with an initial vector θ_0 and proceed to generate a sequence $\theta_1, \ldots, \theta_z$ ending in a value which minimises the function $S(\theta)$ approximately.

The $(r+1)$th element of the sequence is found from its predecessor θ_r according to the updating formula

$$\theta_{r+1} = \theta_r + \lambda_r p_r. \tag{12.4}$$

This embodies the direction vector p_r and the step-adjustment scalar λ_r. The decrement of the objective function for a unit change in λ, evaluated at θ_r, is

$$\frac{\partial S(\theta_r)}{\partial\lambda} = \frac{\partial S(\theta_r)}{\partial\theta}\frac{\partial\theta}{\partial\lambda} = \gamma_r' p_r, \tag{12.5}$$

whilst the so-called directional derivative $\gamma_r' p_r / \|p_r\|$ is obtained by normalising the length of the direction vector.

Since our object is to devise methods which are applicable to any continuous twice-differentiable objective function, the form of this function will not be specified in any of the computer code which we shall present. Instead, we shall use a generic function call of the form *Funct(lambda, theta, pvec, n)* wherein *lambda* is the step-adjustment scalar, *theta* is the value of the function's argument from the end of the previous iteration, *pvec* is the new direction vector and n is the order of *theta* and *pvec*.

As an example of the function call, which is intended only for illustrative purposes, we may consider the following:

(12.6)

```
function Funct(lambda : real;
               theta, pvec : vector;
               n : integer) : real;

  var
    i : integer;

begin
  if lambda <> 0 then
    for i := 1 to n do
      theta[i] := theta[i] + lambda * pvec[i];
  Funct := 3 * Sqr(theta[1] - 1) + 2 * Sqr(theta[2] - 2)
                                  + Sqr(theta[3] - 3);
end;
```

Notice that, by setting $lambda = 0$ or $n = 0$, the function can be evaluated at *theta* instead of $theta + lambda * pvec$.

A means of evaluating $\gamma(\theta)$, the gradient vector of the function at the point θ, is often required. If γ is not available in an analytic form, then it may have to be determined by numerical means. In the present case, the gradient vector is provided by the following procedure:

(12.7)

```
procedure gradient(var gamma : vector;
                   theta : vector;
                   n : integer);

begin
  gamma[1] := 6 * (theta[1] - 1);
  gamma[2] := 4 * (theta[2] - 2);
  gamma[3] := 2 * (theta[3] - 3);
end;
```

The typical optimisation procedure has two main elements. The first is a routine for determining the direction vector p_r in equation (12.4). The second is a routine which determines the value of λ_r. Usually, the value of λ_r corresponds approximately to the minimum of the objective function along the line $\theta = \theta_r +$

λp_r. Finding such a value is a matter of univariate minimisation. Procedures for univariate minimisation are presented in the sections which follow immediately.

Univariate Search

Imagine that it is known that the interval $[a, b]$ contains a unique local minimum of the univariate function $f(x)$. To find a smaller interval which contains the minimum, we should have to evaluate the function at no fewer than two additional points in $[a, b]$; for if the function were evaluated only at $x_1 \in [a, b]$ there would be no way of telling which of the sub-intervals $[a, x_1]$ and $[x_1, b]$ contains the minimum.

Therefore, let us assume that the function has been evaluated at the two points $x_1, x_2 \in [a, b]$, where $x_1 < x_2$. Then there are three possibilities. If $f(x_1) > f(x_2)$, then the minimum must lie in $[x_1, b]$ since it may be excluded from $[a, x_1]$ on the grounds that $f(x)$ is declining monotonically as x increases from a to x_1. Likewise, if $f(x_1) < f(x_2)$, then the minimum must lie in $[a, x_2]$ since it may be excluded from $[x_2, b]$, where $f(x)$ is rising monotonically. Finally, if $f(x_1) = f(x_2)$, then the minimum must lie in $[x_1, x_2]$. In practice, however, this is such an unlikely event that we can afford to ignore it by taking either of the wider sub-intervals $[a, x_2]$, $[x_1, b]$ instead (see Figure 12.1).

We have to decide how to locate the points x_1, x_2 within $[a, b]$. It is desirable that the two overlapping sub-intervals $[a, x_2]$, $[x_1, b]$ which might contain the minimum should be of equal length. If it transpires, after evaluating the function at the points x_1, x_2, that the minimum lies within $[x_1, b]$, then we should place the next point x_3 in such a way as ensure that next two sub-intervals which might contain the minimum, namely $[x_1, x_3]$ and $[x_2, b]$, are also of equal length.

Let us denote the length of the interval $[x, y]$ by $I = |x, y|$. Then, with reference to Figure 12.1, it can be seen that, if the requirements of equal sub-intervals are fulfilled, then

$$\begin{aligned} I_0 &= |a, b| = |a, x_2| + |x_2, b|, \\ I_1 &= |x_1, b| = |a, x_2|, \\ I_2 &= |x_1, x_3| = |x_2, b|; \end{aligned} \tag{12.8}$$

and it follows that $I_0 = I_1 + I_2$. In general, the relationship

$$I_j = I_{j+1} + I_{j+2} \tag{12.9}$$

should hold between the interval sizes of successive iterations of a search procedure. This prescription is not enough to determine the interval length completely. To do so, we may impose the further requirement that the ratio of successive lengths is constant such that

$$I_{j+1} = \kappa I_j \quad \text{and} \quad I_{j+2} = \kappa I_{j+1} = \kappa^2 I_j. \tag{12.10}$$

Combining the two requirements gives

$$I_j = \kappa I_j + \kappa^2 I_j. \tag{12.11}$$

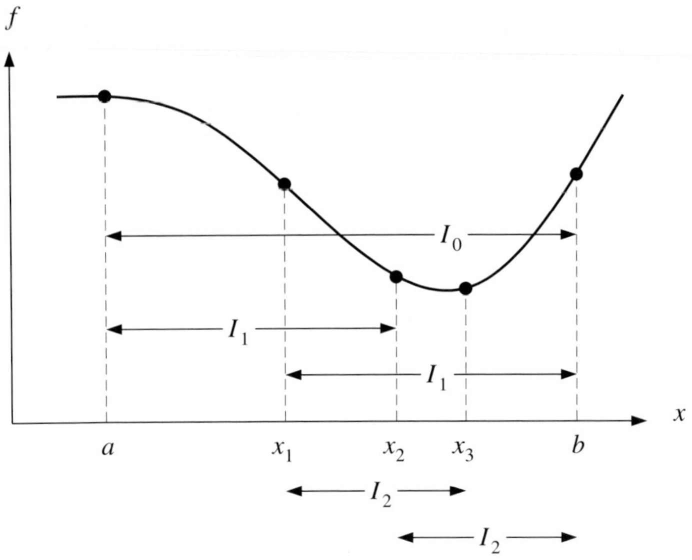

Figure 12.1. Two overlapping sub-intervals whose union overlays the minimum should be equal length, and the lengths of successive sub-intervals within a nested sequence should bear a constant ratio.

Solving the resulting quadratic equation $\kappa^2 + k - 1 = 0$ gives

$$\kappa = \frac{\sqrt{5} - 1}{2} \simeq 0.618. \tag{12.12}$$

A rectangle with the proportions 1 : 0.618 was described in classical architecture as a golden section, and it was said to be one of the most aesthetically pleasing of geometric forms. The search procedure in which the successive intervals have this ratio is called a golden-section search.

(12.13)

```
procedure GoldenSearch(function Funct(x : real) : real;
                       var a, b : real;
                       limit : integer;
                       tolerance : real);

var
  x1, x2, f1, f2, kappa : real;
  iteration : integer;

begin
  kappa := (Sqrt(5.0) − 1)/2.0;
  iteration := 0;
  x1 := b − kappa * (b − a);
  x2 := a + kappa * (b − a);
```

```
f1 := Funct(x1);
f2 := Funct(x2);

while (b - a > tolerance) and (iteration < limit) do
  begin
    if f1 > f2 then
      begin
        a := x1;
        x1 := x2;
        x2 := a + kappa * (b - a);
        f1 := f2;
        f2 := Funct(x2);
      end
    else if f1 <= f2 then
      begin
        b := x2;
        x2 := x1;
        x1 := b - kappa * (b - a);
        f2 := f1;
        f1 := Funct(x1);
      end;

    iteration := iteration + 1;
  end; {while}

end; {GoldenSearch}
```

In this procedure, which is liable to be applied only to problems of univariate minimisation, we have replaced the generic function call of (12.6) by the simpler function call *Funct*(x). The latter is passed to the *GoldenSearch* procedure as a formal parameter; and, therefore, the heading of the procedure contains the full declaration **function** *Funct*$(x : real) : real$. When the procedure is called, the name of an actual function must be supplied. The actual function must have the same number and types of parameters as the formal function which is found in the heading of the procedure; and it must deliver a result of the same type.

The purpose of this construction is to enable the procedure *GoldenSearch* to be applied to several functions which may coexist within the same block of a program. The device of passing a function as a parameter will be used in all of the optimisation procedures which are to presented in this chapter.

Quadratic Interpolation

In the process of searching the interval $[a, b]$, we acquire information which can help us to discern the shape of the function $f(x)$ which we desire to minimise. Such information might lead us more rapidly to an accurate assessment of where the minimum is located than does a process of successive bracketing based on a series of sub-intervals of predetermined sizes.

Recall that, if the minimum is to be located in a sub-interval of $[a, b]$, then the function $f(x)$ must be evaluated at least at two internal points $c, z \in [a, b]$. Imagine that we already have one internal point c and that we wish to locate a second point. We might suppose that $f(x)$ could be usefully approximated over $[a, b]$ by the quadratic $g(x)$ which agrees with $f(x)$ at the points a, b and c, where $f(a) = f_a$, $f(b) = f_b$ and $f(c) = f_c$. In that case, it would be reasonable to locate z at the point where the approximating quadratic function has its minimum.

By evaluating $f(x)$ at c and z, we can find a sub-interval containing the minimum in the manner of the *GoldenSearch* procedure of (12.13):

(12.14)
```
if (z < c) and (fz <= fc) then
  begin {discard [c, b]}
    b := c;
    c := z;
  end;

if (c < z) and (fz <= fc) then
  begin {discard [a, c]}
    a := c;
    c := z;
  end;

if (z < c) and (fz > fc) then
  a := z; {discard [a, z]}

if (c < z) and (fz > fc) then
  b := z; {discard [z, b]}
```

The code of the algorithm which selects the sub-interval is complicated slightly by the fact that we can no longer tell in advance whether $z < c$ or $z > c$.

The quadratic function $g(x)$ which approximates $f(x)$ over $[a, b]$ can be found by the method of Lagrangean interpolation. Thus we have

$$g(x) = f_a\frac{(x-b)(x-c)}{(a-b)(a-c)} + f_b\frac{(x-a)(x-c)}{(b-a)(b-c)} + f_c\frac{(x-a)(x-b)}{(c-a)(c-b)}. \tag{12.15}$$

This can also be written as

$$g(x) = px^2 + qx + r, \tag{12.16}$$

where

$$p = \frac{f_a(b-c) + f_b(c-a) + f_c(a-b)}{(a-b)(a-c)(b-c)} \tag{12.17}$$

and

$$-q = \frac{f_a(b^2-c^2) + f_b(c^2-a^2) + f_c(a^2-b^2)}{(a-b)(a-c)(b-c)}. \tag{12.18}$$

The value which minimises the quadratic function $g(x)$ is

$$z = -\frac{q}{2p}. \tag{12.19}$$

The parameters p and q are calculated by the following procedure:

(12.20)
```
procedure Quadratic(var p, q : real;
                        a, b, c, fa, fb, fc : real);

  const
    epsilon = 10E - 10;

  var
    num, denom : real;

begin
  num := fa * (b * b - c * c);
  num := num + fb * (c * c - a * a);
  num := num + fc * (a * a - b * b);
  denom := (a - b) * (a - c) * (b - c);
  p := 0;
  q := 0;

  if (Abs(num) > epsilon) and (Abs(denom) > epsilon) then
    begin {if}
      q := -num/denom;
      num := fa * (b - c) + fb * (c - a) + fc * (a - b);
      p := num/denom;
    end; {if}

end; {Quadratic}
```

If the approximating quadratic for the jth iteration is constructed using the points a_j, b_j and c_j and their corresponding function values, then we can be assured that its minimum z_j will lie in $[a_j, b_j]$, which is the current interval of uncertainty.

In spite of this assurance, the procedure for quadratic interpolation, as it stands, is not to be recommended. To understand how it can fail, consider the case where, on the jth iteration, we find that $z_j < c_j$ and $f(z_j) < f(c_j)$. The response of the algorithm is to set $a_{j+1} = a_j$, $c_{j+1} = z_j$ and $b_{j+1} = c_j$ and to discard the sub-interval $[c_j, b_j]$. Imagine also that $f(a_j)$ greatly exceeds $f(c_j)$ and that $f(c_j)$ exceeds $f(z_j)$ by very little. Then it is likely that the $(j+1)$th iteration will be a replica of the jth iteration. Thus we may find that $z_{j+1} < c_{j+1}$ and $f(z_{j+1}) < f(c_{j+1})$. Once more, the response of the algorithm will be to discard a sub-interval $[c_{j+1}, b_{j+1}]$ on the RHS of the interval of uncertainty $[a_{j+1}, b_{j+1}]$. Successive iterations may continue in like fashion with ever smaller sub-intervals being subtracted from the RHS; and thus the progress in diminishing the interval of uncertainty will continue to worsen.

It is clear that the main cause of this problem is the retention throughout these iterations of the point on the LHS which has the highest function value. This is an inappropriate point to use in constructing the approximating quadratic function. The problem can be overcome if we are prepared to keep a record of some of the points which have been most recently evaluated. From amongst such points, we can find three which will enable us to construct an appropriate quadratic approximation over the current interval of uncertainty. Clearly, c_j is one of these points since $f(c_j)$ is the lowest function value to be found so far. The second point is w_j which has the next lowest function value, and the third point is v_j which is where the second lowest function value was formerly located.

It is possible for v_j and w_j to coincide with a_j and b_j. In that case, the minimum of the approximating quadratic will be found in $[a_j, b_j]$ and it will serve as the updating point z_j. If v_j and w_j do not coincide with a_j and b_j, then it is possible that the minimum of the quadratic will lie outside the interval $[a_j, b_j]$. In that case, z_j must be determined by another means. We can imitate the strategy of the *GoldenSearch* procedure of (12.13) by taking

$$(12.21)\qquad z_j = \begin{cases} c_j - (1-\kappa)(c_j - a_j), & \text{if } c_j \geq (a_j + b_j)/2; \\ c_j + (1-\kappa)(b_j - c_j), & \text{if } c_j < (a_j + b_j)/2. \end{cases}$$

This places z_j in whichever is the longer of the intervals $[a_j, c_j]$ and $[c_j, b_j]$. The resulting algorithm, which is due to Brent [74], is presented below.

(12.22)

```
procedure QuadraticSearch(function Funct(lambda : real;
                              theta, pvec : vector;
                              n : integer) : real;
                          var a, b, c, fa, fb, fc : real;
                          theta, pvec : vector;
                          n : integer);

  var
    z, kappa, p, q, v, w, fm, fz, fv, fw : real;
    termination : boolean;

begin
  kappa := (Sqrt(5.0) - 1)/2.0;
  v := a;
  w := b;
  fv := fa;
  fw := fb;

  repeat {until termination}
    Quadratic(p, q, v, w, c, fv, fw, fc);

    if (p <> 0) then
      z := -q/(2 * p);
```

```
if (z < a) or (b < z) or (p = 0) then
  if c >= (a + b)/2 then
    z := c - (1 - kappa) * (c - a)
  else
    z := c + (1 - kappa) * (b - c);
fz := Funct(z, theta, pvec, n);

if (fz <= fc) then
  begin

    if (z < c) then
      begin {discard [c, b]}
        b := c;
        fb := fc;
      end;

    if (c < z) then
      begin {discard [a, c]}
        a := c;
        fa := fc;
      end;

    v := w;
    w := c;
    c := z;
    fv := fw;
    fw := fc;
    fc := fz;

  end{fz  <= fc}

else if (fz > fc) then
  begin

    if (z < c) then
      begin {discard [a, z]}
        a := z;
        fa := fz;
      end;

    if (c < z) then
      begin {discard [z, b]}
        b := z;
        fb := fz
      end;

    if (fz <= fw) or (w = c) then
```

```
        begin
          v := w;
          w := z;
          fv := fw;
          fw := fz;
        end
      else if (fz <= fv) or (v = c) or (v = w) then
        begin
          v := z;
          fv := fz;
        end;

    end; {fz > fc}

  termination := Check('weak', a, b, c, fa, fb, fc, fw);
until termination;

end; {QuadraticSearch}
```

It remains to specify the function which determines, at the end of each of its iterations, whether or not the *QuadraticSearch* procedure should be terminated. There are a variety of criteria which can be used to test for the convergence of a sequence $\{f_i = f(x_i)\}$ when $f(x)$ is a continuous differentiable function. Amongst these are

$$|x_i - x_{i-1}| \leq \epsilon \quad \text{and} \\ |f_i - f_{i-1}| \leq \epsilon. \tag{12.23}$$

Also, it is wise to impose a predetermined limit on the number of iterations.

In the context of a multivariate optimisation procedure, we usually conduct a line search aimed at minimising the function in the direction which is being pursued in the current iteration. Since there may be many iterations, and since accurate line searches are expensive to carry out, we might wish to weaken the criterion of convergence considerably and to look only for an adequate decrease in the value of the function. However, merely requiring a decrease in the function in each iteration does not ensure the convergence to the minimum; for a sequence of ever-decreasing decrements could tend to limiting value which is greater than the minimum.

To show how we can guarantee a significant reduction in the value of $f(x)$, let us consider the case where the updating point z lies in the interval $[a_0, b_0]$, where b_0 is such that $f(b_0) = f(a_0)$ (see Figure 12.2). Then we have to ensure that z is close neither to a_0 nor to b_0. We can distance z from b_0 by requiring that

$$f(z) < f(a_0) + \rho(z - a_0)f'(a_0), \tag{12.24}$$

where ρ is some value obeying the condition $0 \leq \rho \leq \frac{1}{2}$ and $f'(a_0)$ is the first derivative of $f(x)$ evaluated at a_0. The function on the RHS represents a line $\mathcal{L}_1$ through a_0 which descends less steeply that does $f(x)$ at that point. We can

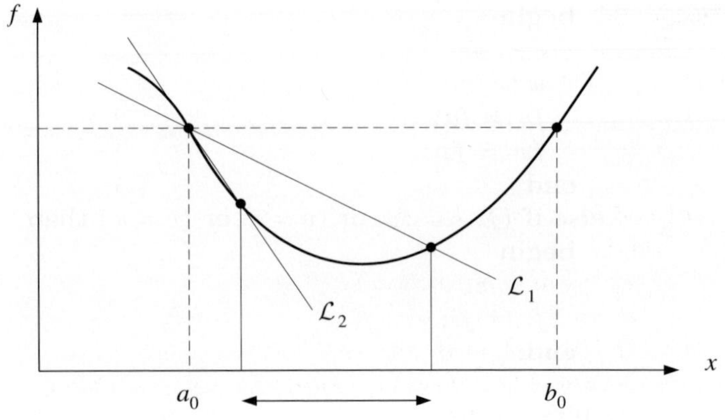

Figure 12.2. The set of acceptable points is indicated by the span of the double-headed arrow.

distance z from a_0 by requiring that

$$f(z) > f(a_0) + (1 - \rho)(z - a_0)f'(a_0). \tag{12.25}$$

Here the function on the RHS represents a line $\mathcal{L}_2$ through a_0 which descends more steeply than $\mathcal{L}_1$. The abscissae of the intersections of $\mathcal{L}_1$ and $\mathcal{L}_2$ with the curve of the function $f(x)$ define an interval containing the set of acceptable points for z. As $\rho \to \frac{1}{2}$, the lines come together and the interval between the points of intersection vanishes. As $\rho \to 0$, the interval approaches that of $[a_0, b_0]$.

Unfortunately, the conditions under (12.24) and (12.25) depend upon the values of the derivative of f at the point a_0, and this might not be available. However, we may replace $f'(a_0)$ by $g'(a_0)$ which is the derivative at a_0 of the approximating quadratic.

The following Pascal function *Check* incorporates both the strong conditions for convergence found under (12.23) above and the weak conditions of termination given by (12.24) and (12.25). If the string *'strong'* is assigned to the parameter *mode* in the statement which invokes *Check*, then the conditions of strong convergence are imposed. If another string is assigned, then the procedure will terminate when the weak conditions are fulfilled.

(12.26)
```
function Check(mode : string;
                a, b, c, fa, fb, fc, fw : real) : boolean;

  const
    xtol = 0.001;
    ftol = 0.0001;
    rho = 0.25;

  var
    p, q, dv : real;
```

```
begin
  Check := false;

  if ((b - a) < xtol) or (Abs(fc - fw) < ftol) then
    Check := true

  else if mode <> 'strong' then
    begin
      Quadratic(p, q, a, b, c, fa, fb, fc);
      dv := 2 * p * a + q;
      if (fc < fa + rho * (c - a) * dv)
          and (fc > fa + (1 - rho) * (c - a) * dv) then
        Check := true
    end;

  end; {Check};
```

Bracketing the Minimum

Both of the methods of univariate minimisation which we have presented require that we have an interval $[a, b]$ which is known to contain a minimum of the function. There has been a tacit assumption that there is only one minimum within the interval, but this is unnecessary. When the univariate minimisation is part of a line-search procedure within a multivariate minimisation, the interval has to be located by a preliminary computation. We shall describe a method which makes use of the existing facility for interpolating a quadratic function through three points.

Imagine that we are given only the point a which is supposed to be reasonably close to a minimum of the function $f(x)$. Then, by evaluating $f(x)$ at a and at an adjacent point $c > a$, we can tell in which direction the function is decreasing. If $f_a > f_c$, then the function declines to the right and the additional point b may be placed to the right of c to form an interval $[a, b]$. If $f_a < f_c$, then the function declines to the left. In that case, we relabel the original point as c and we place an interval $[a, b]$ around it.

Let us assume that we are confronted by the case where $f_a > f_c$, and let us ignore the other case which is its mirror image. If $f_b > f_c$ as well, then we have an interval which brackets a minimum and there is nothing more to be done. If we are not so fortunate, then the next step is to interpolate a quadratic function $g(x)$ through the points (a, f_a), (c, f_c) and (b, f_b). The sign of the second derivative $g''(x) = 2p$ indicates whether $g(x)$ is convex or concave.

If $g(x)$ is convex or linear, which is when $p \leq 0$, then it has no minimum, and we should not use it as an approximation for $f(x)$. However, we do know that $f(x)$ has a minimum to the right of c. Therefore, we should establish a new interval by extending to the right as far as we are prepared to go and by discarding $[a, c]$ on the left.

If $g(x)$ is concave, then it has a minimum z which lies either in $[a, b]$ or to the right of b. If $z \in [a, b]$ and if it transpires that $f_z < f_a, f_b$, then we have an interval which brackets a minimum. If, on the other hand, $f_z \geq f_a$ or $f_z \geq f_b$, then we have failed to find a bracketing interval and we should move b to the right.

If z falls to the right of b, which means that $f_a > f_c > f_b$, then we should expand the interval by moving b to the right. We can use the value of z to provide a new value for b provided that it is at an acceptable distance from the old value. At the same time, we can discard $[a, c]$ on the left.

At the conclusion of these evaluations, we have a new set of points a, c, b together with their corresponding function values; and we are in a position either to accept these values or to embark upon a new round.

The method which we have described is due to Powell [406]. It is incorporated in the *LineSearch* procedure which follows. The latter forms a shell around the procedure *QuadraticSearch* of (12.22) which is invoked once an interval $[a, b]$ has been found which brackets a minimum. Thus *LineSearch*, together with its subsidiary procedures, forms a fully fledged method of univariate minimisation. In fact, we shall use it in the context of a multivariate optimisation procedure where it will serve to locate the minima along predetermined directions of search. To use the procedure for ordinary univariate minimisation, we need only replace the generic multivariate function call of (12.6) by a univariate function call *Funct*(z).

(12.27)
```
procedure LineSearch(function Funct(lambda : real;
                               theta, pvec : vector;
                               n : integer) : real;
                         var a : real;
                         theta, pvec : vector;
                         n : integer);

  var
    b, c, z, p, q, fa, fb, fc, fz, step, maxstep : real;

begin
  step := 0.15;
  maxStep := 0.3;
  c := a + step;
  fa := Funct(a, theta, pvec, n);
  fc := Funct(c, theta, pvec, n);

  if fc < fa then {the function declines to the right}
    begin
      b := c + step;
      fb := Funct(b, theta, pvec, n);
    end

  else {if fa <= fc, then the function declines to the left}
    begin
      b := c;
      c := a;
      a := a - step;
      fb := fc;
      fc := fa;
```

```
    fa := Funct(a, theta, pvec, n);
  end;

while (fc > fb) or (fc > fa) do
  begin {while}

    Quadratic(p, q, a, b, c, fa, fb, fc);

    if p > 0 then {the quadratic is concave}
      begin {p > 0}
        z := -q/(2 * p);
        if Abs(z - a) < step then
          z := a - step;
        if Abs(b - z) < step then
          z := b + step;
        if (z < a - maxStep) then
          z := a - maxStep;
        if (b + maxStep < z) then
          z := b + maxStep;
      end; {p > 0}

    if p <= 0 then {the quadratic is convex or linear}
      begin {p <= 0}
        if fa > fc then
          z := b + maxStep;
        if fa < fc then
          z := a - maxStep;
      end; {p <= 0}

    fz := Funct(z, theta, pvec, n);

    if (fc < fa) and (b < z) then
      begin {extend b to the right and discard [a, c]}
        a := c;
        c := b;
        b := z;
        fa := fc;
        fc := fb;
        fb := fz;
      end

    else if (fa <= fc) and (z < a) then
      begin {extend a to the left and discard [c, b]}
        b := c;
        c := a;
        a := z;
        fb := fc;
```

```
                fc := fa;
                fa := fz;
              end

            else if (a < z) and (z < b) then
              begin
                Writeln('z falls in the interval [a, b]');
                if (fz < fa) and (fz < fb) then
                  begin
                    Writeln('fz is less than fa, fb : set fc := fz');
                    c := z;
                    fc := fz;
                  end
                else if (fc < fa) then
                  b := b + step
                else if (fa <= fc) then
                  a := a - step
              end;

        end; {while}

      QuadraticSearch(Funct, a, b, c, fa, fb, fc, theta, pvec, n);
      a := c;

    end; {LineSearch}
```

Unconstrained Optimisation via Quadratic Approximations

In this section, we shall consider iterative methods for finding the minimum value of a multivariate function $S = S(\theta)$ by seeking the values which satisfy the equation $\gamma = (\partial S/\partial\theta)' = 0$ and which render the Hessian matrix $H = \partial(\partial S/\partial\theta)'/\partial\theta$ positive definite.

Recall that the $(r+1)$th approximation to the minimising value is found from its predecessor according to the formula

$$\theta_{r+1} = \theta_r + d_r = \theta_r + \lambda_r p_r, \tag{12.28}$$

where p_r is the vector specifying the direction of the search and λ_r is the step-adjustment scalar. The so-called gradient methods, to which we shall now confine our attention, conform to the scheme

$$\theta_{r+1} = \theta_r - \lambda_r Q_r \gamma_r, \tag{12.29}$$

whereby the $(r+1)$th approximation to the solution is obtained by subtracting from the rth approximation a vector of adjustments based on the gradient vector γ_r determined at the point θ_r. The vector of adjustments also depends upon a direction matrix Q_r and the step-adjustment scalar λ_r. Both Q and λ are also liable to be determined in the current iteration, although, in some applications, λ remains fixed.

The choice of Q is influenced by the requirement that any step which is confined to the immediate neighbourhood of θ_r ought to result in a decreased value for S. To discover what this requirement entails, let us consider the linear approximation to the function $S = S(\theta)$ in the neighbourhood of θ_r which is provided by the first two terms of a Taylor-series expansion. The approximating function can be written as

$$S_l = S_r + \gamma_r' d_r, \tag{12.30}$$

where $S_r = S(\theta_r)$ is the value of the function at θ_r and $d_r = \theta - \theta_r$ is the step which takes us away from θ_r. According to (12.29), we have

$$d_r = -\lambda_r Q_r \gamma_r. \tag{12.31}$$

On substituting this into (12.30) and on rearranging the resulting expression, we find that

$$S_l - S_r = -\lambda_r \gamma_r' Q_r \gamma_r. \tag{12.32}$$

Assuming that $\lambda_r > 0$, the quantity $S_l - S_r$ will be negative for any value of γ if and only if Q_r is positive definite; and this constitutes the condition that a reduction in the value of S will always result from minimal steps with λ close to zero. However, the larger the step from θ_r, the less accurate will be the linear approximation of (12.30); and, if we step too far, we may find that the actual value of $S = S(\theta)$ has increased even though the linear approximation suggests a decrease.

The choices of λ and Q are made usually with reference to a quadratic approximation of the function $S = S(\theta)$ in the neighbourhood of θ_r which is based on the first three terms of the Taylor-series expansion. The approximating function can be written as

$$S_q = S_r + \gamma_r' d_r + \frac{1}{2} d_r' H_r d_r, \tag{12.33}$$

where $H_r = \partial\{\partial S(\theta_r)/\partial\theta\}'\partial\theta$ is the value of the Hessian at θ_r.

The quadratic function S_q provides the simplest model of S which can be used for testing the performance of the proposed algorithms. Also, their performance in more complicated circumstances is apt to be described in terms of their response to the invalidity of the quadratic approximation to S.

The Method of Steepest Descent

The gradient vector points in the direction of the maximum increase in $S(\theta)$ in departing from θ_r. To see this, we may refer to equation (12.30) which shows that, when d_r is small, the change in the value of the function is given by

$$\begin{aligned} k_{r+1} &= S_l - S_r \\ &= \gamma_r' d_r. \end{aligned} \tag{12.34}$$

This can also be expressed as

$$k_{r+1} = \|\gamma_r\| \, \|d_r\| \cos\theta, \tag{12.35}$$

where $\cos\theta$ is the angle between the vectors γ_r and d_r. For a fixed step length $\|d_r\|$ and a given gradient vector γ_r, this quantity is maximised by setting $\theta = 0$, which implies that the maximising step d_r lies in the direction of γ_r. The minimising step lies in the direction of $-\gamma_r$.

The optimisation procedure known as the method of steepest descent is the result of setting $Q = I$ in equation (12.29) to give the following algorithm:

$$\theta_{r+1} = \theta_r - \lambda_r \gamma_r. \tag{12.36}$$

If the step length λ_r is determined so as to maximise the change in the value of the function, then there is a simple consequence for the sequence of directions which are pursued by the procedure. Let the direction of the $(r+1)$th step be given by the vector p_r. Then the value of the function along this axis is given by $S(\theta_r + \lambda p_r)$ and the derivative is given by $\gamma(\theta_r + \lambda p_r)$. The first-order condition for a minimum is therefore

$$\begin{aligned} \frac{\partial S}{\partial \lambda} &= \gamma'(\theta_r + \lambda p_r) \frac{\partial}{\partial \lambda}(\theta_r + \lambda p_r) \\ &= \gamma'_{r+1} p_r = 0, \end{aligned} \tag{12.37}$$

where $\gamma_{r+1} = \gamma(\theta_{r+1})$ denotes the value of the gradient vector at the minimising point θ_{r+1}. This condition implies that, at the end of the step, the direction of the next departure, which is given by $-\gamma_{r+1} = p_{r+1}$, is at right angles to the previous direction, which was given by p_r.

An advantage of the method of steepest descent is that the requirement that Q should be positive definite is invariably fulfilled. A disadvantage is that it takes no account of the global structure of the function to be minimised. An adverse consequence of this shortsightedness can be illustrated by considering the case where $S = S(\theta)$ is a concave quadratic function giving rise to elongated elliptical contours over the plane. It is possible that, from certain starting points, the direction of steepest ascent will be almost at right angles from the direction in which the minimum lies (see Figure 12.3).

The fact that the steepest-descent property is only a local property means that, in descending a narrow valley, the procedure is liable to follow a zig-zag path to the minimum with small steps and frequent changes of direction.

The Newton–Raphson Method

The Newton–Raphson method makes full use of the quadratic approximation of S in choosing the direction vector. On putting $d_r = \theta - \theta_r$ into (12.33), we get the quadratic approximating function in the form of

$$S_q = S_r + \gamma'_r(\theta - \theta_r) + \frac{1}{2}(\theta - \theta_r)' H_r (\theta - \theta_r). \tag{12.38}$$

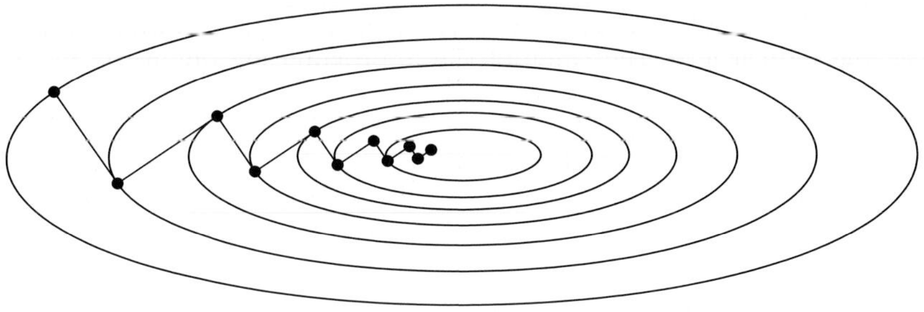

Figure 12.3. In descending a narrow valley with elliptical contours, the method of steepest descent is liable to follow a zig-zag path.

By differentiating S_q in respect of θ and setting the result to zero, we obtain the condition

$$0 = \gamma_r' + (\theta - \theta_r)' H_r. \tag{12.39}$$

The value which minimises the function is therefore

$$\theta_{r+1} = \theta_r - H_r^{-1}\gamma_r; \tag{12.40}$$

and this expression describes the Newton–Raphson algorithm. If the function to be minimised is indeed a concave quadratic, then the Newton–Raphson procedure will attain the minimum in a single step. Notice also that, if $H = I$, then the method coincides with the method of steepest descent with $\lambda = 1$. In the case of $H = I$, the contours of the quadratic function are circular.

The disadvantages of the Newton–Raphson procedure arise when the value of the Hessian matrix at θ_r is not positive definite. In that case, the step from θ_r to θ_{r+1} is liable to be in a direction which is away from the minimum value. This hazard can be illustrated by a simple diagram which relates to the problem of finding the minimum of a function defined over the real line. The problems only arise when the approximation θ_r is remote from the true minimum of the function. Of course, in the neighbourhood of the minimising value, the function is concave; and, provided that the initial approximation θ_0, with which the iterations begin, is within this neighbourhood, the Newton–Raphson procedure is likely to perform well.

A Modified Newton Procedure

In an attempt to overcome the problems which can beset the Newton–Raphson procedure when θ_r is remote from the minimising value, we may adopt two expedients. The first of these is to replace the Hessian matrix H_r, whenever it fails to be positive definite, by some alternative matrix Q. The second expedient is to limit the size of the step length $\|d_r\| = \sqrt{\{(\theta_{r+1} - \theta_r)'(\theta_{r+1} - \theta_r)\}}$. These expedients should guarantee that each iteration of the algorithm leads to a reduction in the value of S.

Imagine that we have limited the step length to $\|d_r\| = l$, and let us seek to maximise the reduction in the value of the approximating quadratic function S_q subject to this restriction. Then, given that

$$S_q - S_r = \gamma_r' d_r + \frac{1}{2} d_r' H_r d_r, \tag{12.41}$$

it follows that we can seek this optimal improvement by evaluating the Lagrangean expression

$$L = \gamma_r' d_r + \frac{1}{2} d_r' H_r d_r + \frac{1}{2}\kappa(d_r' d_r - l^2). \tag{12.42}$$

By differentiating L with respect to d_r and setting the result to zero, we obtain the condition

$$\begin{aligned} 0 &= \gamma_r' + d_r' H_r + \kappa d_r' \\ &= \gamma_r' + (\theta - \theta_r)'(H_r + \kappa I). \end{aligned} \tag{12.43}$$

The value which ensures the maximum change in the criterion function S_q is therefore given by

$$\theta_{r+1} = \theta_r - (H_r + \kappa I)^{-1}\gamma_r; \tag{12.44}$$

and, provided that the matrix $H_r + \kappa I$ is positive definite, the change will be a decrease. We can easily prove the following:

(12.45) The matrix $H + \kappa I$ is positive definite if and only if $\kappa + \mu_s > 0$, where μ_s is smallest latent root of H.

Proof. Since H is a symmetric matrix, there exist an orthonormal matrix C, such that $C'C = I$, which gives $C'HC = M$, where $M = \text{diag}(\mu_1, \ldots, \mu_n)$ contains the latent roots of H. The matrix C also provides the matrix containing the latent roots of $H + \kappa I$ in the form of $C'(H + \kappa I)C = M + \kappa C'C = M + \kappa I$. For $H + \kappa I$ to be positive definite, it is necessary and sufficient that $\mu_i + \kappa > 0$ for all i; and this proves the result.

This result might encourage us to adopt the following modified Newton method which has been described by Goldfeld *et al.* [219]:

(12.46) (i) Find the smallest latent root μ_s of H_r and test whether $\mu_s > 0$.

(ii) If $\mu_s \leq 0$, then set $\kappa_r = \epsilon - \mu_s$ for some small value ϵ and proceed to calculate

$$\theta_{r+1} = \theta_r - (H_r + \kappa_r I)^{-1}\gamma_r.$$

(iii) If $\mu_s > 0$, then set $\kappa_r = 0$ and proceed to calculate

$$\theta_{r+1} = \theta_r - H_r^{-1}\gamma_r.$$

The scalar $\mu_\circ$ is both the smallest latent root of H and the largest root of H^{-1}. It is easily calculated from H^{-1} by the power method which entails only a succession of matrix multiplications.

To investigate the question of the step length, let us write

$$\begin{aligned}\|d_r\|^2 &= (\theta_{r+1} - \theta_r)'(\theta_{r+1} - \theta_r)\\ &= \gamma_r'(H_r + \kappa_r I)^{-2}\gamma_r \qquad (12.47)\\ &= \gamma_r' C'(M_r^{-2} + \kappa_r^{-2} I)C\gamma_r.\end{aligned}$$

From the final expression, it is clear that the step length is a declining function of κ_r. Therefore, a simple interpretation of the procedure defined in (12.46) is available. For, when $\kappa = 0$, we have a Newton–Raphson procedure; and, as the value of κ increases, we have a procedure whose behaviour increasingly resembles that of the method of steepest descent. Also, as the value of κ increases, the size of the step length decreases.

The Minimisation of a Sum of Squares

In statistics, we often encounter the kind of optimisation problem which requires us to minimise a sum-of-squares function

$$S(\theta) = \varepsilon'(\theta)\varepsilon(\theta), \qquad (12.48)$$

wherein $\varepsilon(\theta)$ is a vector of residuals which is a nonlinear function of a vector θ. The value of θ corresponding to the minimum of the function commonly represents the least-squares estimate of the parameters of a statistical model. Such problems may be approached using the Newton–Raphson method which we have described in the previous section. However, the specialised nature of the function $S(\theta)$ allows us to pursue a method which avoids the trouble of finding its second-order derivatives and which has other advantages as well. This is the Gauss–Newton method, and it depends upon a linear approximation of the function $\varepsilon = \varepsilon(\theta)$. In the neighbourhood of θ_r, the approximating function is

$$e = \varepsilon(\theta_r) + \frac{\partial \varepsilon(\theta_r)}{\partial \theta}(\theta - \theta_r), \qquad (12.49)$$

where $\partial\varepsilon(\theta_r)/\partial\theta$ stands for the first derivative of $\varepsilon(\theta)$ evaluated at $\theta = \theta_r$. This gives rise, in turn, to an approximation to S in the form of

$$\begin{aligned}S_g &= \varepsilon'(\theta_r)\varepsilon(\theta_r) + (\theta - \theta_r)'\left\{\frac{\partial \varepsilon(\theta_r)}{\partial \theta}\right\}'\left\{\frac{\partial \varepsilon(\theta_r)}{\partial \theta}\right\}(\theta - \theta_r)\\ &\quad + 2\varepsilon'(\theta_r)\frac{\partial \varepsilon(\theta_r)}{\partial \theta}(\theta - \theta_r).\end{aligned} \qquad (12.50)$$

By differentiating S_g in respect of θ and setting the result to zero, we obtain the condition

$$0 = 2(\theta - \theta_r)'\left\{\frac{\partial \varepsilon(\theta_r)}{\partial \theta}\right\}'\left\{\frac{\partial \varepsilon(\theta_r)}{\partial \theta}\right\} + 2\varepsilon'(\theta_r)\frac{\partial \varepsilon(\theta_r)}{\partial \theta}. \qquad (12.51)$$

The value which minimises the function S_g is therefore

$$\theta_{r+1} = \theta_r - \left[\left\{\frac{\partial\varepsilon(\theta_r)}{\partial\theta}\right\}'\left\{\frac{\partial\varepsilon(\theta_r)}{\partial\theta}\right\}\right]^{-1}\left\{\frac{\partial\varepsilon(\theta_r)}{\partial\theta}\right\}'\varepsilon(\theta_r). \tag{12.52}$$

This equation represents the algorithm of the Gauss–Newton procedure, and it provides the formula by which we can find the $(r+1)$th approximation to the value which minimises sum of squares once we have the rth approximation. Since the gradient of the function $S = S(\theta)$ is given by $\gamma = \{\partial\varepsilon(\theta)/\partial\theta\}'\varepsilon(\theta)$, it is clear that the Gauss–Newton method is a gradient method of the sort which is represented by equation (12.29). Moreover, the affinity of the Gauss–Newton and the Newton–Raphson methods is confirmed when we recognise that the direction matrix in (12.52) is simply an approximation to the Hessian matrix of the sum-of-squares function which is

$$\frac{\partial(\partial S/\partial\theta)'}{\partial\theta} = 2\left[\left(\frac{\partial\varepsilon}{\partial\theta}\right)'\left(\frac{\partial\varepsilon}{\partial\theta}\right) + \sum_t \varepsilon_t\left\{\frac{\partial(\partial\varepsilon_t/\partial\theta)'}{\partial\theta}\right\}'\right]. \tag{12.53}$$

The direction matrix of the Gauss–Newton procedure is always positive semi-definite; and, in this respect, the procedure has an advantage over the Newton–Raphson procedure.

Quadratic Convergence

There is a need for alternatives to the Newton–Raphson procedure and its variants whenever the second derivatives of the function $S(\theta)$ are unavailable or are too laborious to compute. If there is no analytic expression for these derivatives, then it may prove impractical to evaluate them by numerical means.

The method of steepest descent is one method which dispenses with second derivatives, but it is ruled out of consideration by its rate of convergence which is liable to be very slow, even when $S(\theta)$ is a quadratic function. A standard by which to judge the rate of convergence is indicated by the fact that, if $S(\theta)$ is a quadratic function in n arguments, then, using only information provided by the gradient vector, we should be able to reach the overall minimum in, at most, n steps.

To understand this result, let us imagine, to begin with, that the quadratic function has spherical contours. Then

$$S(\theta) = S(\theta_0) + \gamma'(\theta_0)(\theta - \theta_0) + \frac{1}{2}(\theta - \theta_0)'(\theta - \theta_0); \tag{12.54}$$

and, to find the minimum, we might search in turn along each of the directions defined by the vectors $e_1, \ldots, e_n$ which are comprised by the identity matrix I_n. Such a procedure would amount to a series of partial minimisations of the function in respect of each its arguments within the vector θ.

The minimum could also be found by searching in the mutually orthogonal directions specified by the vectors of a matrix $Q = [q_0, \ldots, q_{n-1}]$ such that $Q'Q =$

$\mathrm{diag}(f_0, \ldots, f_{n-1})$. In that case, the vector θ, which is reached after taking n steps away from θ_0, can be represented by

$$\theta = \theta_0 + \sum_{i=0}^{n-1} \lambda_i q_i = \theta_0 + Q\lambda, \tag{12.55}$$

where $\lambda = [\lambda_0, \ldots, \lambda_{n-1}]$. Setting $\theta - \theta_0 = Q\lambda$ in (12.54) gives

$$\begin{aligned} S(\theta) &= S(\theta_0) + \gamma_0' Q\lambda + \frac{1}{2}\lambda' Q' Q\lambda \\ &= S(\theta_0) + \sum_{i=0}^{n-1} \left(\gamma_0' q_i \lambda_i + \frac{1}{2}\lambda_i^2 f_i^2\right), \end{aligned} \tag{12.56}$$

where $\gamma_0 = \gamma(\theta_0)$. This sum comprises n simple quadratic functions, each with a unique argument λ_i. Therefore, it can be minimised in n steps by minimising each of the constituent quadratic terms individually.

A similar search procedure can be used to find the minimum of an arbitrary quadratic function. Such a function takes the form of

$$S(\theta) = S(\theta_0) + \gamma'(\theta_0)(\theta - \theta_0) + \frac{1}{2}(\theta - \theta_0)' H(\theta - \theta_0). \tag{12.57}$$

Now, instead of searching in directions which are mutually orthogonal in the ordinary sense, we must search in directions which are orthogonal or conjugate in respect of a metric defined by the Hessian matrix H. An appropriate set of directions is specified by any matrix $P = [p_0, \ldots, p_{n-1}]$ which serves to reduce H to a diagonal matrix: $P'HP = D = \mathrm{diag}(d_0, \ldots, d_{n-1})$. Let us therefore express the argument of the function as

$$\theta = \theta_0 + \sum_{i=0}^{n-1} \lambda_i p_i = \theta_0 + P\lambda. \tag{12.58}$$

Setting $\theta - \theta_0 = P\lambda$ in (12.57) gives

$$\begin{aligned} S(\theta) &= S(\theta_0) + \gamma_0' P\lambda + \frac{1}{2}\lambda' P' H P\lambda \\ &= S(\theta_0) + \sum_{i=0}^{n-1} \left(\gamma_0' p_i \lambda_i + \frac{1}{2}\lambda_i^2 d_i^2\right). \end{aligned} \tag{12.59}$$

In effect, we have reduced the problem to one which is akin to the former problem by transforming the coordinate system in such a way as to decouple the arguments which are now the elements of $\lambda = P^{-1}(\theta - \theta_0)$. Once more, it follows that the function can be minimised in n steps.

The proposition concerning the termination of the search procedure, which defines the concept of quadratic convergence, may be stated formally as follows:

(12.60) If the minimum of an arbitrary concave quadratic function in the form of $S(\theta)$ of (12.57) is sought by locating the exact minima along a sequence of directions $p_0, \ldots, p_{n-1}$ which are mutually conjugate with respect to its positive-definite Hessian matrix H, such that $p_i'Hp_j = 0$ for all $i \neq j$, then the minimum of the function will be found in n searches at most.

Although we have proved this already by constructive arguments, we shall prove it again more formally since, in so doing, we can develop some algebra which will be useful in the sequel. Therefore, consider differentiating the function $S(\theta)$ with respect to θ to obtain

$$\gamma(\theta) = \gamma(\theta_0) + H(\theta - \theta_0). \tag{12.61}$$

This gives $\gamma(\theta_j) = \gamma(\theta_0) + H(\theta_j - \theta_0)$ and $\gamma(\theta_{j+1}) = \gamma(\theta_0) + H(\theta_{j+1} - \theta_0)$ and, by taking one from the other, we find that

$$\begin{aligned} \gamma_{j+1} - \gamma_j &= H(\theta_{j+1} - \theta_j) \\ &= \lambda_j H p_j, \end{aligned} \tag{12.62}$$

where, for simplicity, we have written $\gamma_{j+1} = \gamma(\theta_{j+1})$ and $\gamma_j = \gamma(\theta_j)$. Here p_j is the direction vector for the step from θ_j to θ_{j+1} and λ_j is the length of the step.

Now consider the identity

$$\gamma_k = \gamma_{i+1} + \sum_{j=i+1}^{k-1} (\gamma_{j+1} - \gamma_j). \tag{12.63}$$

Premultiplying by p_i' gives

$$\begin{aligned} p_i'\gamma_k &= p_i'\gamma_{i+1} + \sum_{j=i+1}^{k-1} p_i'(\gamma_{j+1} - \gamma_j) \\ &= \sum_{j=i+1}^{k-1} p_i'(\gamma_{j+1} - \gamma_j). \end{aligned} \tag{12.64}$$

Here the second equality follows from the condition $p_i'\gamma_{i+1} = 0$ of (12.37) which indicates that the gradient γ_{i+1} at the point θ_{i+1}, which is where the minimum is found in the $(i+1)$th search, is orthogonal to the direction of the search. On putting the expression from (12.62) into (12.64), we find, in view of the conditions of conjugacy, that

$$\begin{aligned} p_i'\gamma_k &= \sum_{j=i+1}^{k-1} \lambda_j p_i' H p_j \\ &= 0 \quad \text{when} \quad k > i. \end{aligned} \tag{12.65}$$

This means that the gradient vector γ_k at the end of the kth search is orthogonal not just to the direction of that search, as in (12.37), but also to the directions of all previous searches. It follows that, after n searches, we have

$$\gamma_n'[p_0, \ldots, p_{n-1}] = 0. \tag{12.66}$$

Since $[p_0, \ldots, p_{n-1}] = P$ is a matrix of full rank, this can only imply that $\gamma_n = 0$, which is to say that θ_n is a stationary point. Finally, given that H is a positive-definite matrix, it follows that the stationary point corresponds to a minimum of the function.

There is one particular set of conjugate directions which also fulfils an ordinary condition of orthogonality. These are the characteristic vectors of the Hessian matrix H which are defined by the conditions

$$Hp_i = \lambda_i p_i; \quad i = 0, \ldots, n-1, \tag{12.67}$$

and which therefore fulfil the conditions

$$\lambda_i p_i' p_j = \lambda_j p_i' p_j = p_i' H p_j. \tag{12.68}$$

If $\lambda_i \neq \lambda_j$, then the only way in which this equality can be maintained is if $p_i'p_j = 0$; which is to say that characteristic vectors corresponding to distinct roots are orthogonal. Thus

$$p_i'p_j = 0 \quad \text{and} \quad p_i'Hp_j = 0, \quad \text{when} \quad i \neq j. \tag{12.69}$$

Of course, it would be impractical to pursue a method of optimisation which depends upon finding the characteristic vectors of the Hessian matrix; for this would require far too much computation.

The Conjugate Gradient Method

In view of the definition of conjugacy, it might seem that we should need to know the matrix H in order to implement an algorithm for searching along mutually conjugate directions. However, if we knew the value of this matrix, then we should be able to implement a Newton method which, doubtless, we would prefer on the grounds that it should enable us to locate the minimum of a quadratic function in a single step.

It transpires that we can find the conjugate directions without knowing H. To show this, let us recall the condition under (12.62). Premultiplying by p_i' gives

$$\begin{aligned} p_i'(\gamma_{j+1} - \gamma_j) &= \lambda p_i' H p_j \\ &= 0 \qquad \text{when} \quad i \neq j. \end{aligned} \tag{12.70}$$

Here is a version of the condition of conjugacy which depends only on the gradient vectors; and this is what makes it practical to use a conjugate search procedure.

We can endeavour to determine the search direction p_r for the $(r+1)$th stage by finding the component of γ_r which is conjugate with respect to all of $p_0,\ldots,p_{r-1}$. Consider, therefore, an expression of the form

$$p_r = -\gamma_r + \sum_{j=0}^{r-1}\beta_j p_j. \tag{12.71}$$

Now, the conditions of conjugacy give

$$\begin{aligned} p_r'Hp_i &= -\gamma_r'Hp_i + \sum_{j=0}^{r-1}\beta_j p_j'Hp_i \\ &= -\gamma_r'Hp_i + \beta_i p_i'Hp_i = 0. \end{aligned} \tag{12.72}$$

It follows that

$$\beta_i = \frac{\gamma_r'Hp_i}{p_i'Hp_i} = \frac{\gamma_r'(\gamma_{i+1}-\gamma_i)}{p_i'(\gamma_{i+1}-\gamma_i)}, \tag{12.73}$$

where the second equality comes from (12.62). Next, we can show that this term β_i is zero-valued unless $i = r-1$. For this, we premultiply an equation in the form of (12.71) by γ_k' , where $k > i$, to give

$$\gamma_k'p_i = -\gamma_k'\gamma_i + \sum_{j=0}^{i-1}\beta_j\gamma_k'p_j. \tag{12.74}$$

The condition $\gamma_k'p_i = 0$ of (12.65), which characterises a conjugate search procedure, enables us to set terms on both sides of the equation to zero, leaving only $0 = \gamma_k'\gamma_i$; which indicates, in general, that

$$\gamma_i'\gamma_j = 0 \quad \text{for} \quad i \neq j. \tag{12.75}$$

It follows, from equation (12.73), that $\beta_i = 0$ for $i = 0,\ldots,r-2$. Therefore, equation (12.71) becomes

$$p_r = -\gamma_r + \beta_{r-1}p_{r-1} \qquad \text{where} \qquad \beta_{r-1} = \frac{\gamma_r'(\gamma_r-\gamma_{r-1})}{p_r'(\gamma_r-\gamma_{r-1})}. \tag{12.76}$$

This serves to define the conjugate gradient procedure.

It is notable that the sequences of the gradient vectors γ_i and the direction vectors p_i, which are generated by the conjugate gradient procedure in the case of a quadratic function, obey conditions of orthogonality and conjugacy, respectively, which are similar to the conditions under (12.69) which pertain to the characteristic vectors of the Hessian matrix.

We can simplify both the numerator and the denominator of the expression for β_{r-1} under (12.76). For the denominator, we have

$$\begin{aligned} (\gamma_r-\gamma_{r-1})'p_{r-1} &= \gamma_r'p_{r-1} - \gamma_{r-1}'p_{r-1} \\ &= -\gamma_{r-1}'(-\gamma_{r-1} + \beta_{r-2}p_{r-2}) \\ &= \gamma_{r-1}'\gamma_{r-1}, \end{aligned} \tag{12.77}$$

where the second equality comes from using (12.65) and (12.76) and the third from using (12.65) again. In view of (12.75), the numerator can be simplified to give

$$(12.78) \qquad (\gamma_r - \gamma_{r-1})'\gamma_r = \gamma_r'\gamma_r.$$

Thus we can express β_{r-1} as

$$(12.79) \qquad \beta_{r-1} = \frac{(\gamma_r - \gamma_{r-1})'\gamma_r}{\gamma_{r-1}'\gamma_{r-1}}$$

and as

$$(12.80) \qquad \beta_{r-1} = \frac{\gamma_r'\gamma_r}{\gamma_{r-1}'\gamma_{r-1}}.$$

Although these alternative expressions for β_{r-1} are equivalent in the context of a quadratic function, they will differ in value when the function to be minimised is only approximately quadratic. There are arguments in favour of the use of either of these formulae. However, we shall adopt the expression under (12.80) which has been advocated by Fletcher and Reeves [189].

The algorithm of the conjugate gradient procedure is specified by the equations (12.76) and (12.80) together with the updating equation

$$(12.81) \qquad \theta_{r+1} = \theta_r + \lambda_r p_r$$

which defines the $(r+1)$th step. The value of λ_r is determined so as to provide a good approximation to the minimum of the objective function along the line defined by the equation $\theta = \theta_r + \lambda p_r$. The choice of a starting value θ_0 is, of course, arbitrary, whilst the initial direction p_0 may be provided by the gradient vector $\gamma_0 = \gamma(\theta_0)$.

When the conjugate gradient procedure is applied to a nonquadratic function, it cannot be expected that a minimum will be attained in n steps or minor iterations. Usually, several cycles of minor iterations are required, each of which begins by resetting the direction vector to the value of the current gradient vector. A cycle of minor iterations may be described as a major iteration.

A procedure which implements the conjugate gradient method is presented below. The brevity of the code belies the fact that a major part of the algorithm consists of the *LineSearch* procedure of (12.27) for determining the value of λ which has been presented in an earlier section. However, a line search forms an essential part of any multivariate optimisation procedure which is intended to be robust.

(12.82)
```
procedure ConjugateGradients(function Funct(lambda : real;
                                 theta, pvec : vector;
                                 n : integer) : real;
                               var theta : vector;
                               n : integer);

    const
```

```
    tolerance = 0.01;

  var
    gamma, gammaLag, pvec : vector;
    num, denom, beta, lambda : real;
    i, j : integer;

  function SqrNorm(gamma : vector;
                         n : integer) : real;

    var
      i : integer;
      s : real;

  begin
    s := 0.0;
    for i := 1 to n do
      s := s + Sqr(gamma[i]);
    SqrNorm := s;
  end; {SqrNorm}
begin {ConjugateGradients}
  for i := 1 to n do
    pvec[i] := 0;

  repeat {major iterations}
    Gradient(Funct, gamma, theta, n);
    beta := 0;

    for j := 0 to n - 1 do
      begin {minor iterations}

        for i := 1 to n do
          pvec[i] := -gamma[i] + beta * pvec[i];
        lambda := 0.0;
        LineSearch(Funct, lambda, theta, pvec, n);
        for i := 1 to n do
          theta[i] := theta[i] + lambda * pvec[i];

        for i := 1 to n do
          gammaLag[i] := gamma[i];
        Gradient(Funct, gamma, theta, n);
        num := SqrNorm(gamma, n);
        denom := SqrNorm(gammaLag, n);
        beta := num/denom;

      end; {minor iterations}
```

until $Sqrt(num)/n < tolerance$;
end; $\{ConjugateGradients\}$

Numerical Approximations to the Gradient

It may not always be practical, or even possible, to obtain the derivatives of the objective function by analytic means. In some circumstances, we might think of replacing them by numerical approximations obtained by finite-difference methods. Formally, such numerical derivatives are based either on a linear or a quadratic approximation to the function at the point in question.

Consider the Taylor-series expansion of $S(\theta)$ about the point $S(\theta+he_j)$, where e_j stands for the jth column of the identity matrix I_n and h is a small increment. Then

$$\begin{aligned} S(\theta + he_j) &\simeq S(\theta) + he_j'\gamma(\theta) + \frac{1}{2}h^2 e_j' H(\theta) e_j \\ &= S(\theta) + h\gamma_j(\theta) + \frac{1}{2}h^2 H_{jj}(\theta), \end{aligned} \tag{12.83}$$

and, likewise,

$$S(\theta - he_j) \simeq S(\theta) - h\gamma_j(\theta) + \frac{1}{2}h^2 H_{jj}(\theta). \tag{12.84}$$

The forward-difference approximation, which is exact for a linear function, is

$$\gamma_j(\theta) \simeq \frac{S(\theta + he_j) - S(\theta)}{h}. \tag{12.85}$$

This is obtained from (12.83) by suppressing the term containing the second derivative H_{jj}. The central-difference approximation, which is exact for a quadratic function, is

$$\gamma_j(\theta) \simeq \frac{S(\theta + he_j) - S(\theta - he_j)}{2h}. \tag{12.86}$$

This is obtained by subtracting (12.84) from (12.83). The approximation amounts to the derivative, at the point θ, of the quadratic function which passes through the coordinates $\{\theta - he_j, S(\theta - he_j)\}$, $\{\theta, S(\theta)\}$ and $\{\theta + he_j, S(\theta + he_j)\}$. The enhanced accuracy of the central-difference approximation is purchased, of course, at the cost of increasing the number of function evaluations which are necessary in computing the derivatives.

Further accuracy might be purchased by increasing the degree of the approximating polynomial, leading to yet more function evaluations. However, it is doubtful whether even the extra computation entailed by the central-difference approximation is justified if all that is being sought are the first derivatives. For the purpose of the derivatives is to provide the directions for the line searches; and, so long as an adequate search procedure is available, great accuracy is not required in these directions.

Another problem is to choose the step size h to be sufficiently small so that the truncation errors incurred by the approximations are minor, but not so small as invite cancellation errors resulting from the subtraction, in the numerator, of two virtually equal function values. The choice of h can become the subject of a sophisticated computation, see for example Gill *et al.* [211, p. 127], but it depends largely on the precision of the computer.

In the following procedure for computing the forward-difference approximations, a rather simple criterion for choosing h is employed which also depends upon the size of the element θ_j.

(12.87)

```
procedure FdGradient(function Funct(lambda : real;
                            theta, pvec : vector;
                            n : integer) : real;
                        var gamma : vector;
                        theta : vector;
                        n : integer);

  var
    i, j : integer;
    epsilon, lambda, stepSize, ftheta, fjstep : real;
    hvec : vector;

  begin
    epsilon := 0.5E - 4;
    for i := 1 to n do
      hvec[i] := 0.0;
    lambda := 1;
    ftheta := Funct(lambda, theta, hvec, n);

    for j := 1 to n do
      begin {j}
        stepSize := Rmax(epsilon * Abs(theta[j]), epsilon);
        hvec[j] := stepSize;
        fjstep := Funct(lambda, theta, hvec, n);
        gamma[j] := (fjstep - ftheta)/stepSize;
        hvec[j] := 0.0;
      end; {j}

  end; {FdGradient}
```

Quasi-Newton Methods

If we evaluate the gradient of the quadratic function $S_q = S_q(\theta)$ at two points θ_r and θ_{r+1}, then we obtain complete information about the curvature of the function along the line passing through these points. If we evaluate the gradient at $n + 1$ points, where n is the number of elements of the argument θ, then we should have enough information to reconstitute the Hessian matrix of the quadratic function.

To understand this result, consider the condition under (12.62), which is characteristic of a quadratic function. This may be written as

$$(12.88)\quad q_r = Hd_r, \qquad \text{where} \qquad q_r = \gamma_{r+1} - \gamma_r \quad \text{and} \quad d_r = \theta_{r+1} - \theta_r.$$

Imagine that the gradient is evaluated at the starting point θ_0 of an iterative procedure for optimising the quadratic function, and thereafter at n successive points $\theta_1, \ldots, \theta_n$. Then, from (12.88), we could form the equation

$$(12.89)\qquad Q = [q_0, \ldots, q_{n-1}] = H[d_0, \ldots, d_{n-1}] = HD.$$

If the line segments d_r joining consecutive points θ_r and θ_{r+1} correspond to a set of linearly independent vectors, then we should be able to invert the matrix D so as to obtain the Hessian matrix

$$(12.90)\qquad H = QD^{-1}.$$

A powerful idea springs to mind: an approximation to the curvature of a nonlinear function can be computed without evaluating the Hessian matrix in its entirety at any one point. If the function were quadratic, then, after n steps, we should have gathered enough information to form the Hessian matrix. At that stage, we should be in a position to apply the Newton procedure and, if we had not already arrived, one more step would bring us to the minimum of the quadratic function.

Our aim should be to form an ever-improving approximation M to the Hessian matrix H as each step provides additional gradient information. A reasonable criterion is that, at the end of the rth step, a quadratic function based upon the updated Hessian approximation M_{r+1} should have the same curvature in the direction of that step as the function S_q which is based upon the true matrix H. The direction is given by p_r or $d_r = \lambda p_r$; and, since $Hd_r = q_r$, we require that

$$(12.91)\qquad M_{r+1}d_r = q_r.$$

This is the so-called quasi-Newton condition.

There is also a question of the directions in which the steps are to be taken. Here it seems reasonable to imitate the Newton–Raphson algorithm so that, in place of equation (12.40), we have

$$(12.92)\quad \theta_{r+1} = \theta_r - \lambda_r M_r^{-1}\gamma_r \qquad \text{or, equivalently,} \qquad d_r = -\lambda_r M_r^{-1}\gamma_r.$$

The latter equation suggests that, instead of approximating the Hessian matrix, we might chose to approximate the inverse of the Hessian matrix by a matrix W. Then, given that $d_r = H^{-1}q_r$, the corresponding quasi-Newton condition for the rth step would be the requirement that

$$(12.93)\qquad d_r = W_{r+1}q_r.$$

Let us concentrate, for a start, on approximating the Hessian matrix rather than its inverse, and let the updated approximation be

$$M_{r+1} = M_r + vv', \tag{12.94}$$

in which a symmetric matrix vv' of rank one is added to M_r. In that case, the quasi-Newton condition becomes

$$(M_r + vv')d_r = q_r, \quad \text{or} \quad vv'd_r = q_r - M_r d_r, \tag{12.95}$$

which indicates that v must be proportional to $q_r - M_r d_r$. The constant of proportionality is $v'd_r$ and its square is $(v'd_r)^2 = d_r'(q_r - M_r d_r)$. Therefore, the updated approximation to the Hessian matrix is

$$M_{r+1} = M_r + \frac{(q_r - M_r d_r)(q_r - M_r d_r)'}{d_r'(q_r - M_r d_r)}. \tag{12.96}$$

This is the so-called symmetric rank-one update—and it is clearly uniquely determined. For a starting value, we can take any symmetric positive-definite matrix; but, in the absence of any prior information, it seems reasonable to set $M_0 = I$.

It can be shown that, if it does not fail by becoming undefined, and if $d_1, \ldots, d_n$ are linearly independent, then the rank-one update method reaches the minimum of a quadratic function in $n + 1$ steps at most. Also $M_{n+1} = H$, which is to say that the exact value of the Hessian matrix is recovered at the end.

There are two problems which can affect the performance of the rank-one update algorithm, even when it is applied to a quadratic function. The first is that there is no guarantee that the updated matrix M_{r+1} will retain the property of positive definiteness. The second is that the denominator term $d_r'(q_r - M_r d_r)$ may come dangerously close to zero, which can affect the numerical stability of the algorithm.

Rank-Two Updating of the Hessian Matrix

A more satisfactory quasi-Newton algorithm is one which is based upon an updating matrix of rank two. The idea is to update the approximation to the Hessian matrix by incorporating the gradient information provided by q_r and by removing the previous version of the same information which is contained in the vector $M_r d_r$. The updated matrix is in the form of

$$M_{r+1} = M_r + \alpha q_r q_r' - \beta M_r d_r d_r' M_r, \tag{12.97}$$

where α and β are scaling factors. The quasi-Newton condition of (12.91) requires that

$$M_{r+1}d_r = M_r d_r + \{\alpha q_r' d_r\}q_r - \{\beta d_r' M_r d_r\}M_r d_r = q_r; \tag{12.98}$$

and, if we choose to set $\alpha q_r' d_r = \beta d_r' M_r d_r = 1$, then α and β are determined, and the updated matrix becomes

$$M_{r+1} = M_r + \frac{q_r q_r'}{q_r' d_r} - \frac{M_r d_r d_r' M_r}{d_r' M_r d_r}. \tag{12.99}$$

This corresponds to the formula which was proposed severally by Broyden [83], Fletcher [187], Goldfarb [218] and Shanno [449] in 1970 and which is known as the *BFGS* formula. Once more, it seems reasonable to take $M_0 = I$ as the starting value for the sequence of Hessian approximations. The direction vectors for the procedure continue to be specified by equation (12.92).

There are two points regarding the *BFGS* procedure which need to be established. The first is that, in normal circumstances, we may be assured that the matrix M_{r+1} will be positive definite if M_r is positive definite. The second point is that, if accurate line searches are conducted in each step of the procedure, and if the function $S = S(\theta)$ is quadratic, then the outcome of n steps will be a matrix $M_n = H$ equal to the Hessian matrix.

To demonstrate that M_{r+1} is liable to be positive definite, let z be an arbitrary vector and consider the quantity

$$z'M_{r+1}z = z'M_rz + \frac{(z'q_r)^2}{q_r'd_r} - \frac{(z'M_rd_r)^2}{d_r'M_rd_r}. \tag{12.100}$$

Since it is positive definite by assumption, M_r can be factorised as $M_r = G'G$ for some matrix G. Therefore, if we define $a = Gz$ and $b = Gd_r$, then $a'a = z'M_rz$, $b'b = d_r'M_rd_r$ and $a'b = z'M_rd_r$, and we have

$$z'M_{r+1}z = a'a - \frac{(a'b)^2}{b'b} + \frac{(z'q_r)^2}{q_r'd_r}. \tag{12.101}$$

Now, the Cauchy–Schwarz inequality asserts that $(a'b)^2 < (a'a)(b'b)$ for any two vectors a, b which are not collinear. Therefore, the sum of the first and second terms of the RHS of (12.101) is positive; and we shall have $z'M_{r+1}z > 0$ provided that $q_r'd_r = (\gamma_{r+1} - \gamma_r)'d_r > 0$. The latter is simply the condition that the steepness of the gradient along the direction of the rth step diminishes in passing from the point θ_r to the point θ_{r+1}. This must happen if the function $S(\theta)$ which is to be minimised is a quadratic; and it is virtually guaranteed to happen in all other circumstances where the quadratic approximation used in our line-search algorithm is valid.

We can demonstrate that $H = M_n$ by showing that $Q = M_nD$, because this is identical to the equation $Q = HD$ under (12.89) which yields $H = QD^{-1}$. What has to be shown is that

$$q_j = M_{r+1}d_j \qquad \text{whenever} \qquad r \geq j. \tag{12.102}$$

The latter condition, which is described as the inheritance condition, means that the gradient information which is obtained in the jth step, and which is used to form the Hessian approximation M_{j+1}, is preserved in all subsequent approximations M_{r+1}.

When $r = j$, the inheritance condition, which is then just the quasi-Newton condition, is satisfied by construction. When $r > j$, we must consider the expression

$$M_{r+1}d_j = M_rd_j + \frac{q_r}{q_r'd_r}\{q_r'd_j\} - \frac{M_rd_r}{d_r'M_rd_r}\{d_r'M_rd_j\}. \tag{12.103}$$

We can see that the inheritance condition of (12.102) is fulfilled if the previous inheritance condition $q_j = M_r d_j$ is granted and if

$$q_r' d_j = d_r' H d_j = d_r' M_r d_j = 0 \quad \text{whenever} \quad r > j. \tag{12.104}$$

The latter are described as the conditions of conjugacy.

To establish the conditions under (12.104), we proceed by induction. We begin by noting the quasi-Newton condition $q_0 = Hd_0 = M_1 d_0$ which must be fulfilled by the first revision M_1 of the Hessian approximation. The condition implies that

$$d_0 = M_1^{-1} H d_0. \tag{12.105}$$

We proceed to deduce the first of the conditions of conjugacy, which is that $d_1' H d_0 = d_1' M_1 d_0 = 0$. Consider the expression

$$d_1 = \lambda_1 p_1 = -\lambda_1 M_1^{-1} \gamma_1, \tag{12.106}$$

which comes from (12.92). In view of (12.105), this gives

$$\begin{aligned} d_1' H d_0 &= -\lambda_1 \gamma_1' M_1^{-1} H d_0 \\ &= -\lambda_1 \gamma_1' d_0 = -\lambda_1 \lambda_0 \gamma_1' p_0. \end{aligned} \tag{12.107}$$

But, if exact line searches are used, then the gradient γ_1, at the end of the first search, will be orthogonal to the direction of the search. Therefore, $\gamma_1' p_0 = 0$ and, since $M_1 d_0 = H d_0 = q_0$ in consequence of the quasi-Newton condition, it follows, on setting (12.107) to zero, that

$$q_1' d_0 = d_1' H d_0 = d_1' M_1 d_0 = 0. \tag{12.108}$$

This condition of conjugacy is used together with the quasi-Newton condition in demonstrating the following inheritance condition:

$$\begin{aligned} M_2 d_0 &= M_1 d_0 + \frac{q_1}{q_1' d_1} \{q_1' d_0\} - \frac{M_1 d_1}{d_1' M_1 d_1} \{d_1' M_1 d_0\} \\ &= q_0 + \frac{q_1}{q_1' d_1} \{d_1' H d_0\} - \frac{M_1 d_1}{d_1' M_1 d_1} \{d_1' H d_0\} \\ &= q_0. \end{aligned} \tag{12.109}$$

At this stage, we may observe that we have another quasi-Newton condition in the form of $q_1 = H d_1 = M_2 d_1$.

The process of establishing the inheritance conditions proceeds in the manner indicated. To establish the conditions in their full generality, we may assume that $q_j = M_r d_j$ for some r and for all $j < r$, and that the conjugacy conditions $d_i' H d_j = d_i' M_i d_j = 0$ hold for $j < i < r$. Then the object is to deduce the further conjugacy condition $d_r' H d_j = d_r' M_r d_j = 0$ and to show that $q_j = M_{r+1} d_j$.

From (12.61), we may obtain the expressions $\gamma_r = \gamma_0 + H(\theta_r - \theta_0)$ and $\gamma_{j+1} = \gamma_0 + H(\theta_{j+1} - \theta_0)$. Taking one from the other gives

$$\begin{aligned} \gamma_r &= \gamma_{j+1} + H(\theta_r - \theta_{j+1}) \\ &= \gamma_{j+1} + H(d_{j+1} + d_{j+2} + \cdots + d_{r-1}). \end{aligned} \tag{12.110}$$

Transposing and postmultiplying by d_j gives

$$\gamma_r' d_j = \gamma_{j+1}' d_j + \{d_{j+1}' H d_j + d_{j+2}' H d_j + \cdots + d_{r-1}' H d_j\} = 0, \tag{12.111}$$

which follows by the fact that every term on the RHS is equal to zero. Here the condition $\gamma_{j+1}' d_j = 0$ is a property of the line search procedure which finds an exact minimum in the jth stage. The conditions $d_i' H d_j$ for $i = j+1, \ldots, r-1$ are the conjugacy conditions, which are true by assumption.

Now use the condition $q_j = M_r d_j = H d_j$ to write

$$d_j = M_r^{-1} H d_j; \tag{12.112}$$

and also recall that, in the jth stage, we have $d_r = -\lambda_r M_r^{-1} \gamma_r$ by virtue of (12.92). It follows from (12.111) that

$$\begin{aligned} 0 &= \gamma_r' M_r^{-1} H d_j \\ &= \frac{1}{\lambda_r} d_r' H d_j. \end{aligned} \tag{12.113}$$

When this is joined with the condition $q_j = H d_j = M_r d_j$, which is true by assumption, we obtain the conjugacy condition of (12.104). Then we have all the conditions which are required to ensure that equation (12.103) reduces to

$$q_j = M_{r+1} d_j. \tag{12.114}$$

This completes the proof.

In implementing the *BFGS* procedure, we need to generate the direction vector $p_r = M_r^{-1} \gamma_r$ which is to be found in equation (12.92). There are two possibilities here. The first is to obtain p_r simply by solving the equation $M_r p_r = \gamma_r$ in each iteration of the procedure. The second possibility is to generate the sequence of approximations W_r to the inverse of the Hessian matrix H^{-1} in place of the matrices M_r, which are the approximations to H. Then $p_r = W_r \gamma_r$, and the essential equation of the *BFGS* procedure becomes

$$\theta_{r+1} = \theta_r - \lambda_r W_r \gamma_r. \tag{12.115}$$

It can be shown by some algebraic manipulation, which is straightforward but tedious, that, if $W_r = M_r^{-1}$, then the inverse of the matrix M_{r+1} of (12.99) is given by

$$\begin{aligned} W_{r+1} &= W_r + \frac{d_r d_r'}{d_r' q_r} - \frac{W_r q_r q_r' W_r}{q_r' W_r q_r} + q_r' W_r q_r h_r h_r' \\ &= W_r + \alpha d_r d_r' - \beta W_r q_r q_r' W_r + \delta h_r h_r', \end{aligned} \tag{12.116}$$

where

$$\alpha = \frac{1}{d_r' q_r}, \qquad \beta = \frac{1}{q_r' W_r q_r}, \qquad \delta = q_r' W_r q_r, \tag{12.117}$$

and

$$(12.118)\qquad \begin{aligned} h_r &= \frac{d_r}{d_r' q_r} + \frac{W_r}{q_r' W_r q_r} \\ &= \alpha d_r + \beta W_r q_r. \end{aligned}$$

An implementation of the *BFGS* procedure which uses W instead of M, and which is based on equations (12.115)–(12.118), is presented below.

(12.119)

```
procedure BFGS(function Funct(lambda : real;
                        theta, pvec : vector;
                        n : integer) : real;
                    var theta : vector;
                    n : integer);

  const
    epsilon = 1.0E − 5;
    tolerance = 1.0E − 5;

  var
    failure, convergence : boolean;
    i, j, iterations : integer;
    lambda, S, Slag, alpha, beta, delta : real;
    pvec, qvec, dvec, hvec : vector;
    gamma, gammaLag, Wq : vector;
    W : matrix;

begin {BFGS : Broyden, Fletcher, Goldfarb, Shanno}

{Initialise the matrix W}
  for i := 1 to n do
    begin {i}
      pvec[i] := 0.0
      for j := 1 to n do
        begin {j}
          if i = j then
            W[i, i] := 1.0
          else
            W[i, j] := 0.0;
        end; {j}
    end; {i}

  failure := false;
  convergence := false;
  iterations := 0;
  S := Funct(0, theta, pvec, n);
  Gradient(Funct, gamma, theta, n);
```

```
repeat {major iterations}
  iterations := iterations + 1;

{Calculate the direction vector and step length}
    for i := 1 to n do
      begin {i}
        pvec[i] := 0.0;
        for j := 1 to n do
          pvec[i] := pvec[i] - W[i, j] * gamma[j]
      end; {i}

    lambda := 0.0;
    LineSearch(Funct, lambda, theta, pvec, n);
    for i := 1 to n do
      begin {i}
        dvec[i] := lambda * pvec[i];
        theta[i] := theta[i] + dvec[i];
      end; {i}

{New function value and gradient}
    Slag := S;
    for i := 1 to n do
      gammaLag[i] := gamma[i];
    S := Funct(0, theta, pvec, n);
    Gradient(Funct, gamma, theta, n);

    if (Abs(Slag - S) <= tolerance * Abs(S) + epsilon) then
      convergence := true;
    if iterations > 200 then
      failure := true;

    if not (failure or convergence) then
      begin {Approximation W of the Hessian inverse }

  {Calculate elements for updating W}
        for i := 1 to n do
          begin {i}
            qvec[i] := gamma[i] - gammaLag[i];
            Wq[i] := 0.0;
          end; {i}
        for i := 1 to n do
          for j := 1 to n do
            Wq[i] := Wq[i] + W[i, j] * qvec[j];

        alpha := 0.0;
        delta := 0.0;
```

```
            for i := 1 to n do
              begin
                alpha := alpha + qvec[i] * dvec[i];
                delta := delta + qvec[i] * Wq[i]
              end;
            alpha := 1.0/alpha;
            beta := 1.0/delta;
            for i := 1 to n do
              hvec[i] := alpha * dvec[i] - beta * Wq[i];

      {Update the matrix W}
            for i := 1 to n do
              for j := 1 to n do
                begin {i, j}
                  W[i, j] := W[i, j] + alpha * dvec[i] * dvec[j];
                  W[i, j] := W[i, j] - beta * Wq[i] * Wq[j];
                  W[i, j] := W[i, j] + delta * hvec[i] * hvec[j];
                end; {i, j}

          end; {Approximation of the Hessian inverse}

    until failure or convergence;
  {end of iterations}
    if failure then
      Writeln('no convergence in 200 iterations');

  end; {BFGS}
```

Bibliography

[39] Beale, E.M.L., (1988), *Introduction to Optimisation*, John Wiley and Sons, Chichester.

[74] Brent, R.P., (1973), *Algorithms for Minimisation without Derivatives*, Prentice-Hall, Englewood Cliffs, New Jersey.

[83] Broyden, C.G., (1970), The Convergence of a Class of Double-Rank Minimisation Algorithms, *Journal of the Institute of Mathematics and its Applications*, **6**, 76–90.

[84] Brundy, D.B., and G.R. Garside, (1978), *Optimisation Methods in Pascal*, Edward Arnold, Baltimore.

[154] Dennis, J.E., and R.B. Schnabel, (1983), *Numerical Methods of Unconstrained Optimisation*, Prentice-Hall, Englewood Cliffs, New Jersey.

[187] Fletcher, R., (1970), A New Approach to Variable Metric Algorithms, *Computer Journal*, **13**, 317–322.

[188] Fletcher, R., (1987), *Practical Methods of Optimisation, Second Edition*, John Wiley and Sons, New York.

[189] Fletcher, R., and C.M. Reeves, (1964), Function Minimisation by Conjugate Gradients, *The Computer Journal*, **7**, 149–154.

[211] Gill, P.E., W. Murray and M.H. Wright, (1981), *Practical Optimisation*, Academic Press, London.

[218] Goldfarb, D., (1970), A Family of Variable Metric Methods Derived by Variational Means, *Mathematics of Computation*, **24**, 23–26.

[219] Goldfeld, S.M., R.E. Quandt and H.F. Trotter, (1966), Minimisation by Quadratic Hill Climbing, *Econometrica*, **34**, 541–551.

[288] Kennedy, W.J., and J.E. Gentle, (1980), *Statistical Computing*, Marcel Dekker, New York.

[324] Luenberger, D.G., (1969), *Optimisation by Vector Space Methods*, John Wiley and Sons, New York.

[391] Phillips, C., and B. Cornelius, (1986), *Computational Numerical Methods*, Ellis Horwood, Chichester, England.

[406] Powell, M.J.D., (1964), An Efficient Method of Finding the Minimum of a Function of Several Variables without Calculating the Derivatives, *Computer Journal*, **7**, 155–162.

[416] Quandt, R.E., (1983), Computational Problems and Methods, Chapter 12 in *Handbook of Econometrics*, Z. Griliches and M.D. Intriligator (eds.), North-Holland Publishing Co., Amsterdam.

[449] Shanno, D.F, (1970), Conditioning of Quasi-Newton Methods for Function Minimisation, *Mathematics of Computation*, **24**, 647–657.

[532] Wolfe, M.A., (1978), *Numerical Methods for Unconstrained Optimisation*, Van Nostrand Reinhold, New York.

Fourier Methods

CHAPTER 13

Fourier Series and Fourier Integrals

A Fourier method is a technique for analysing a mathematical function, which may be continuous or discrete, by representing it as a combination of trigonometrical functions, or, equivalently, as a combination of complex exponentials. Such a Fourier combination is either a weighted sum or a weighted integral of the trigonometrical functions. The weighting function, whether it be a sequence or a continuous function, is what is known as the Fourier transform.

In the case of a Fourier sum, or a Fourier series as it is liable to be called when the summation is infinite, the weighting function defines a discrete spectrum. In the case of a Fourier integral, the weighting function defines a spectral density function.

The Fourier transforms which are described in this chapter and the next can be assigned to one of four categories which are the product of a pair of dichotomies. On the one hand, there is the question of whether the function to be transformed is a continuous function defined over the real line or a discrete function defined on the set of integers. If the original function is discrete, then its Fourier transform will be a periodic function. Otherwise, if the original function is continuous, then the transform will be aperiodic. On the other hand is the question of whether the original function is periodic or aperiodic. If it is periodic, then the Fourier transform will be a sequence. If it is aperiodic, then the Fourier transform will be a continuous function.

The result is a fourfold classification given in Table 13.1. A discrete periodic function has a discrete periodic transform—*the discrete Fourier transform.* A

Table 13.1. The classes of Fourier transforms*

	Periodic	**Aperiodic**
Continuous	Discrete aperiodic *Fourier series*	Continuous aperiodic *Fourier integral*
Discrete	Discrete periodic *Discrete FT*	Continuous periodic *Discrete-time FT*

* The class of the Fourier transform depends upon the nature of the function which is transformed. This function may be discrete or continuous and it may be periodic or aperiodic. The names and the natures of corresponding transforms are shown in the cells of the table.

continuous aperiodic function has a continuous aperiodic transform—*the Fourier integral*. The other two cases are mixed: a continuous periodic function transforming to a discrete aperiodic sequence and vice versa—*the Fourier series* and the *discrete-time Fourier transform*.

There is no extra complication in taking account of functions which are defined only over an interval of the real line or over a finite set of consecutive integers. Such functions can be treated as if they represent just one cycle of a periodic function. The idea here is that there is no significance in what happens to the function outside its domain of definition; and therefore to imagine that it replicates itself in each successive interval is an acceptable fiction which happens to facilitate the analysis.

This chapter is subdivided in accordance with the classification which we have outlined above. The first section of the chapter deals with the classical mode of Fourier analysis where a continuous periodic function is transformed into a discrete Fourier series. The continuous function, which is assigned to the time domain, represents a time-varying quantity. The coefficients of its Fourier transform are in the frequency domain.

The second section, which deals with the discrete-time Fourier transform, concerns the transformation of an aperiodic sequence in the time domain into a continuous periodic function of the frequency variable. The time-domain sequence represents either a signal or a sequence of filter coefficients—which is the same thing as the output response of the filter to a unit-impulse input. Compared with the classical mode, the roles of the frequency domain and the time domain are interchanged. Now the direct transform from the time domain to the frequency domain is comparable to the inverse transform of the classical mode of analysis, which is from the frequency domain to the time domain. By exploiting this comparison, we can save ourselves the trouble of repeating some of the analysis of the first section.

The third section deals with continuous aperiodic functions. Such functions do not have Fourier-series expansions any more than do aperiodic sequences. Nevertheless, if they fulfil certain conditions of convergence, they too have a Fourier representation, which is now in terms of integrals. In this case, the weighting function is a continuous aperiodic function. Therefore, the original function and its Fourier transform are of the same nature.

A well-known application of the Fourier integral transformation is in mathematical physics, where a finite wave train is resolved into sinusoidal waves. The inverse relationship between the dispersion of the wave train and that of its Fourier transform is expressed in the famous uncertainty principle of Heisenberg.

In the context of time-series analysis and signal processing, the question arises as to how we should analyse a continuous function or "analogue" function for which only a set of discrete observations are available. The sampling theorem indicates that there are conditions under which the analogue-to-discrete conversion entails no loss of information.

The remaining species of Fourier transform is the discrete Fourier transform which entails a one-to-one mapping between a finite (or periodic) sequence in the time domain and a finite (or periodic) sequence in the frequency domain. This transform is of great practical importance since it provides the discrete approximations which allow us to evaluate all manner of Fourier transforms with a digital computer. The discrete Fourier transform and the fast Fourier transform, which is an efficient means of computing it, are dealt with at length in two succeeding chapters. The chapter on the discrete Fourier transform offers an alternative point of entry into the realms of Fourier analysis to any reader who is, at first, daunted by the complicated structure of the present chapter.

Fourier Series

A trigonometrical series is one which can be expressed in the form of

$$(13.1) \qquad x(t) = \alpha_0 + \sum_{j=1}^{\infty} \alpha_j \cos(j\omega t) + \sum_{j=1}^{\infty} \beta_j \sin(j\omega t),$$

where the coefficients α_j and β_j are assumed to be real and where t is any real number.

The series is assumed to converge for all values of ωt under consideration. The frequencies $\{\omega, 2\omega, 3\omega, \ldots\}$, which are multiples of the fundamental frequency ω, are said to constitute an harmonic sequence. The corresponding sine and cosine functions complete an integral number of cycles in a space of $T = 2\pi/\omega$ units which is the fundamental period. The functions $\cos(\omega t)$ and $\sin(\omega t)$ complete just one cycle in this period. Thus, $x(t)$ is a periodic function with $x(t) = x(t+T)$ for all t; and it follows that it is completely defined in terms of the values which it assumes over the interval $(0, T]$. Thus, in talking of a function defined on $(0, T]$, we are talking equally of a periodic function.

The alternative trigonometrical expression for the series is

$$(13.2) \qquad x(t) = \alpha_0 + \sum_{j=1}^{\infty} \rho_j \cos(j\omega t - \theta_j).$$

Here $\rho_j \cos(j\omega t - \theta_j)$ is described as the jth harmonic component of $x(t)$. The amplitude of this component is ρ_j whilst its phase displacement, measured in radians, is θ_j. The effect of the phase displacement is to delay, by $\theta_j/(j\omega)$ time periods, the peak of the cosine function, which would occur, otherwise, at $t = 0$.

In view of the identity $\cos(j\omega t - \theta_j) = \cos\theta_j \cos(j\omega t) + \sin\theta_j \sin(j\omega t)$, it follows, from comparing (13.1) and (13.2), that

$$(13.3) \qquad \alpha_j = \rho_j \cos\theta_j \quad \text{and} \quad \beta_j = \rho_j \sin\theta_j.$$

Therefore

$$(13.4) \qquad \rho_j^2 = \alpha_j^2 + \beta_j^2 \quad \text{and} \quad \theta_j = \tan^{-1}(\beta_j/\alpha_j).$$

It is often more convenient to write the series in terms of complex exponentials. According to Euler's equations,

$$(13.5) \quad \cos(j\omega t) = \frac{1}{2}\left(e^{ij\omega t} + e^{-ij\omega t}\right) \quad \text{and} \quad \sin(j\omega t) = \frac{-i}{2}\left(e^{ij\omega t} - e^{-ij\omega t}\right),$$

where $i = \sqrt{-1}$. Therefore, equation (13.1) can be expressed as

$$x(t) = \alpha_0 + \sum_{j=1}^{\infty} \frac{\alpha_j + i\beta_j}{2} e^{-i\omega_j t} + \sum_{j=1}^{\infty} \frac{\alpha_j - i\beta_j}{2} e^{i\omega j t}. \tag{13.6}$$

If all the terms are gathered under a single summation sign, then this can be written as

$$x(t) = \sum_{j=-\infty}^{\infty} \xi_j e^{i\omega j t}, \tag{13.7}$$

where

$$\xi_0 = \alpha_0, \quad \xi_j = \frac{\alpha_j - i\beta_j}{2} \quad \text{and} \quad \xi_{-j} = \xi_j^* = \frac{\alpha_j + i\beta_j}{2}. \tag{13.8}$$

The condition $\xi_{-j} = \xi_j^*$ guarantees that the function $x(t)$ defined in (13.7) will be real-valued. In developing the theory of the Fourier transform, it is often appropriate to set aside such conditions and to proceed as if the subjects of the transform were complex-valued. This enhances the symmetry of the relationship of the Fourier transform and its inverse, thereby simplifying the theory.

A fundamental question which arises in Fourier analysis is whether it is always possible to represent a prescribed function $x(t)$ over the interval from $t = 0$ to $t = 2\pi/\omega$ by a trigonometrical series. The question, which was much debated by Fourier's contemporaries, was answered by Dirichlet who, in 1829, gave sufficient conditions for the convergence of the series. The result of Dirichlet states that, if $x(t)$ is bounded in the interval $(0, T]$, where $T = 2\pi/\omega$, in the sense that

$$\int_0^T |x(t)| dt < \infty, \tag{13.9}$$

and if it has a only a finite number of discontinuities and only a finite number of maxima and minima in the interval, then there exists a Fourier series which converges at any point t to the sum

$$\frac{1}{2}\{x(t^+) + x(t^-)\}, \tag{13.10}$$

where $x(t^+)$ is the value of x as t is approached from the right $x(t^-)$ is the value of x as t is approached from the left. The endpoints of the interval can be included in this prescription by using the fact that $x(t + T) = x(t)$. The condition that there be only a finite number of discontinuities means that, for all but a finite set of points, $x(t^+)$ and $x(t^-)$ will be equal. A function which fulfils this condition will be described as piecewise continuous.

Dirichlet's conditions, which are sufficient rather than necessary, are somewhat restrictive. Nevertheless, it is often declared in textbooks that they are valid for most functions which arise in mathematical physics. This assertion is out of date; and, as we shall see in a later chapter, the conditions are certainly not fulfilled by the functions which are the subject of Wiener's generalised harmonic analysis [522] and which are entailed in the spectral representation of a stationary stochastic process. However, such generalisations of Fourier analysis are sought not so much by weakening Dirichlet's conditions as by generalising the concept of integration.

To obtain the coefficients of the trigonometrical Fourier series, we make use of the orthogonality conditions which prevail amongst the harmonic components. They are as follows:

$$\int_0^T \cos(j\omega t)\cos(k\omega t)dt = \begin{cases} 0, & \text{if } j \neq k; \\ T/2, & \text{if } j = k > 0; \\ T, & \text{if } j = k = 0; \end{cases} \tag{13.11}$$

$$\int_0^T \sin(j\omega t)\sin(k\omega t)dt = \begin{cases} 0, & \text{if } j \neq k; \\ T/2, & \text{if } j = k > 0; \end{cases}$$

$$\int_0^T \cos(j\omega t)\sin(k\omega t)dt = 0 \quad \text{for all} \quad j, k.$$

Here the range of the integration, which is over one complete cycle of the fundamental harmonic component, can be replaced be any interval from an arbitrary point t_0 to a point $t_1 = t_0 + T$, where T is the length of the period. These results, which are proved in the appendix, become transparent when the integrands are rewritten with the help of the compound-angle formulae of trigonometry. They may be used to show that the coefficients of the trigonometrical Fourier series of equation (13.1) are

$$\begin{aligned} \alpha_0 &= \frac{1}{T}\int_0^T x(t)dt, \\ \alpha_j &= \frac{2}{T}\int_0^T x(t)\cos(\omega jt)dt \quad \text{for} \quad j > 0, \\ \beta_j &= \frac{2}{T}\int_0^T x(t)\sin(\omega jt)dt \quad \text{for} \quad j > 0. \end{aligned} \tag{13.12}$$

The results follow immediately on replacing $x(t)$ by the expression under (13.1).

The coefficients of the complex exponential series of (13.7) may be obtained from the above results using the definitions of (13.8). They can also be obtained more readily from (13.7) itself using the orthogonality conditions for the complex exponentials. The conditions indicate that

$$\int_0^T e^{i\omega(j-k)t}dt = \begin{cases} 0, & \text{if } j \neq k; \\ T, & \text{if } j = k. \end{cases} \tag{13.13}$$

Thus, when the Fourier transform and its inverse are displayed together, we have

$$\xi_j = \frac{\omega}{2\pi}\int_{-\pi/\omega}^{\pi/\omega} x(t)e^{-i\omega jt}dt = \frac{1}{T}\int_{-T/2}^{T/2} x(t)e^{-i\omega jt}dt, \tag{13.14}$$

$$x(t) = \sum_{j=-\infty}^{\infty} \xi_j e^{i\omega jt}. \tag{13.15}$$

Here, we have written $T = 2\pi/\omega$ and we have disposed the limits of integration symmetrically about $t = 0$ in order to facilitate a subsequent comparison.

To demonstrate that the function $x(t)$ is indeed the inverse of the sequence $\{\xi_j\}$, let us substitute equation (13.15) into equation (13.14) with the object of obtaining an identity. When $1/T$ is put in place of $\omega/(2\pi)$, and when the integral is taken from 0 to T, this gives

$$\begin{aligned}\xi_j &= \frac{1}{T}\int_0^T x(t)e^{-i\omega jt}dt = \frac{1}{T}\int_0^T \Big\{\sum_k \xi_k e^{i\omega kt}\Big\} e^{-i\omega jt}dt \\ &= \frac{1}{T}\sum_k \xi_k \Big\{\int_0^T e^{i\omega(k-j)t}dt\Big\} = \xi_j,\end{aligned} \tag{13.16}$$

where the final equality is by virtue of the orthogonality conditions of (13.13).

There are advantages and disadvantages in both the exponential and trigonometric forms of the Fourier series. The exponential form provides a neater set of expressions, but this is sometimes at the cost of concealing interesting details.

Example 13.1. Consider a periodic square wave $x(t) = x(t+T)$, with a (fundamental) frequency of $\omega = 2\pi/T$, which is defined by

$$x(t) = \begin{cases} 1, & \text{if } |t| < \tau; \\ 0, & \text{if } \tau < |t| < T/2. \end{cases} \tag{13.17}$$

In finding the coefficients of the Fourier series, it is convenient, in view of the symmetry of $x(t)$ about $t = 0$, to perform the integration over the interval $-T/2 \geq t > T/2$ (see Figure 13.1). Setting $j = 0$ in (13.14) and taking $x(t)$ from (13.17) gives

$$\xi_0 = \frac{1}{T}\int_{-\tau}^{\tau} dt = \frac{2\tau}{T} = \frac{\tau\omega}{\pi}. \tag{13.18}$$

For $j \neq 0$, we find that

$$\begin{aligned}\xi_j = \frac{1}{T}\int_{-\tau}^{\tau} e^{-i\omega jt}dt &= \frac{2}{j\omega T}\left\{\frac{e^{i\omega j\tau} - e^{-i\omega j\tau}}{2i}\right\} \\ &= \frac{2\sin j\omega\tau}{j\omega T} = \frac{\sin j\omega\tau}{j\pi},\end{aligned} \tag{13.19}$$

where the final equality comes from setting setting $\omega T = 2\pi$.

If $x(t)$ assumes the alternate values of 1 and 0 with equal duration, then $\tau = T/4$ and $\omega\tau = \pi/2$, and we get

$$\xi_0 = \frac{1}{2}, \qquad \xi_j = \frac{\sin(\pi j/2)}{j\pi} = \begin{cases} \pm 1/(j\pi), & \text{if } j \text{ is odd;} \\ 0, & \text{if } j \text{ is even.} \end{cases} \tag{13.20}$$

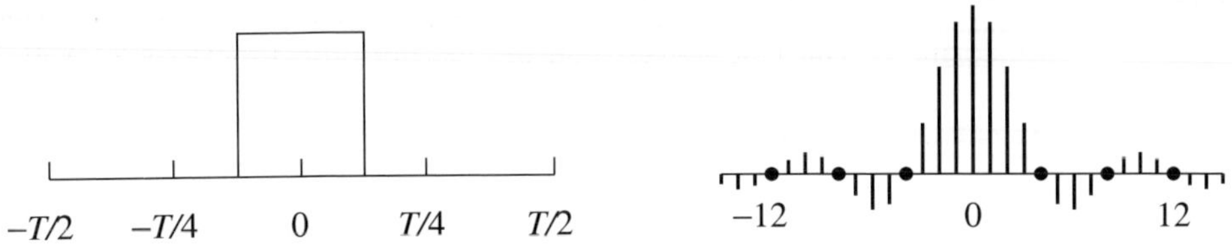

Figure 13.1. A periodic square wave and its Fourier coefficients.

Also $\xi_j = \xi_{-j}$. Therefore the equation under (13.15) can be written as

$$\begin{aligned} x(t) &= \frac{1}{2} + \left\{ \frac{e^{i\omega t} + e^{-i\omega t}}{\pi} - \frac{e^{i\omega 3t} + e^{-i\omega 3t}}{3\pi} + \frac{e^{i\omega 5t} + e^{-i\omega 5t}}{5\pi} - \cdots \right\} \\ &= \frac{1}{2} + \frac{2}{\pi}\left\{ \cos(t) - \frac{1}{3}\cos(3t) + \frac{1}{5}\cos(5t) - \cdots \right\}. \end{aligned} \tag{13.21}$$

The expression following the second equality can be subsumed equally under equations (13.1) and (13.2).

This example shows two important features which are the consequence of the fact that the original real-valued function is even, or symmetric, about $t = 0$. First, there is no phase displacement amongst the harmonic components of equation (13.2). Secondly, the coefficients of the complex exponentials are all real-valued.

If the original function had been real-valued and odd, as in the case where $y(t) = 1$ if $0 < t < T/2$ and $y(t) = 0$ if $T/2 < t < T$, then the coefficients of the complex exponentials would have taken the form of $\xi_j = -i\beta_j/2$ with $\xi_{-j} = -\xi_j$. Then the expression under (13.1) would be in terms of sine functions only; which is to say that each of the harmonic components of the cosine expression under (13.2) would incorporate a phase displacement of π radians.

Convolution

Let $f(t)$ and $g(t)$ be two continuous functions defined over the interval $(0, T]$ which are bounded in the sense of (13.9). Then their convolution product is defined by

$$f(t) * g(t) = \frac{1}{T}\int_0^T f(\tau)g(t-\tau)d\tau = \frac{1}{T}\int_0^T f(t-\tau)g(\tau)d\tau. \tag{13.22}$$

Likewise, if $\phi(j) = \{\phi_j\}$ and $\gamma(j) = \{\gamma_j\}$ are two sequences which are absolutely summable with $\sum|\phi_j| < \infty$ and $\sum|\gamma_j| < \infty$, then their convolution product $\xi(j) = \phi(j) * \gamma(j)$ is the sequence whose elements are

$$\xi_j = \sum_{\tau=-\infty}^{\infty} \phi_\tau \gamma_{j-\tau} = \sum_{\tau=-\infty}^{\infty} \phi_{j-\tau}\gamma_\tau. \tag{13.23}$$

A basic theorem states that a convolution in the time domain corresponds to a multiplication in the frequency domain:

(13.24) Let the sequences $\gamma(j) = \{\gamma_j\}$ and $\phi(j) = \{\phi_j\}$ represent the Fourier transforms of the (piecewise) continuous periodic functions $g(t)$ and $f(t)$ respectively. Then the Fourier transform of the convolution $f(t) * g(t)$ is the sequence $\phi(j)\gamma(j) = \{\gamma_j\phi_j\}$.

Proof. The functions $f(t)$ and $g(t)$ are related to their Fourier transforms $\{\phi_j\}$ and $\{\gamma_j\}$ by equations in the forms of (13.14) and (13.15). Therefore

$$\begin{aligned}\frac{1}{T}\int_0^T f(\tau)g(t-\tau)d\tau &= \frac{1}{T}\int_0^T f(\tau)\Big\{\sum_{j=-\infty}^{\infty}\gamma_j e^{i\omega j(t-\tau)}\Big\}d\tau\\ &= \sum_{j=-\infty}^{\infty}\gamma_j e^{i\omega jt}\Big\{\frac{1}{T}\int_0^T f(\tau)e^{-i\omega j\tau}d\tau\Big\}\\ &= \sum_{j=-\infty}^{\infty}\gamma_j\phi_j e^{i\omega jt}.\end{aligned} \tag{13.25}$$

Thus it can be see that the convolution $f(t) * g(t)$ is the Fourier transform of the sequence $\phi(j)\gamma(j) = \{\gamma_j\phi_j\}$.

There is an analogous theorem which asserts that a convolution in the frequency domain corresponds to a multiplication in the time domain, which is also described as the modulation of one function of time by another:

(13.26) Let the sequences $\gamma(j) = \{\gamma_j\}$ and $\phi(j) = \{\phi_j\}$ represent the Fourier transforms of the (piecewise) continuous periodic functions $g(t)$ and $f(t)$. Then their convolution $\gamma(j) * \phi(j)$ is the Fourier transform of the modulation product $g(t)f(t)$.

Proof. We may model the proof on the previous proof by interchanging the roles of integration and summation. Thus the convolution of the sequences is

$$\begin{aligned}\sum_\tau \phi_\tau\gamma_{j-\tau} &= \sum_\tau \phi_\tau\Big\{\frac{1}{T}\int_0^T g(t)e^{-i\omega(j-\tau)t}dt\Big\}\\ &= \frac{1}{T}\int_0^T g(t)e^{-i\omega jt}\Big\{\sum_\tau \phi_\tau e^{i\omega\tau t}\Big\}dt\\ &= \frac{1}{T}\int_0^T g(t)f(t)e^{-i\omega jt}dt;\end{aligned} \tag{13.27}$$

and this is just the Fourier transform of the product $g(t)f(t)$.

Closely related to these theorems on convolution are Parseval's relationships:

(13.28) Let $f(t)$ and $g(t)$ be complex-valued piecewise-continuous periodic functions whose Fourier-series expansions have the coefficients ϕ_j and γ_j, and let $g^*(t)$ be the complex conjugate of $g(t)$. Then

$$\frac{1}{T}\int_0^T f(t)g^*(t)dt = \sum_{j=-\infty}^{\infty}\phi_j\gamma_j^*.$$

This is demonstrated by writing

$$\begin{aligned}\int_0^T f(t)g^*(t)dt &= \int_0^T f(t)\Big\{\sum_{j=-\infty}^{\infty}\gamma_j^* e^{-i\omega jt}\Big\}dt \\ &= \sum_{j=-\infty}^{\infty}\gamma_j^*\Big\{\int_0^T f(t)e^{-i\omega jt}dt\Big\} \\ &= T\sum_{j=-\infty}^{\infty}\gamma_j^*\phi_j.\end{aligned} \tag{13.29}$$

It follows that, when $f(t) = g(t) = x(t)$, we get

$$\frac{1}{T}\int_0^T |x(t)|^2 dt = \sum_{j=-\infty}^{\infty}|\xi_j|^2. \tag{13.30}$$

This is Parseval's equation for a Fourier series. If $x(t)$ is, in fact, a real-valued time-varying periodic function, then this can be interpreted to mean that the power of the signal may be determined either from the original function or from its transform.

Notice that, upon setting $t = 0$ in the expression

$$\frac{1}{T}\int_0^T f(\tau)g(t-\tau)d\tau = \sum_j \phi_j\gamma_j e^{i\omega jt}, \tag{13.31}$$

which comes from (13.25), we get

$$\frac{1}{T}\int_0^T f(\tau)g(-\tau)d\tau = \sum_j \phi_j\gamma_j. \tag{13.32}$$

This is equivalent to the equation of (13.28) when $\gamma_j^* = \gamma_j$ is real-valued, for which it is necessary and sufficient that $g^*(\tau) = g(-\tau)$. The latter is the analogue of the condition $\xi_j^* = \xi_{-j}$ of (13.8) which ensures that the function $x(t)$ from the time domain is real-valued.

It may be useful to have a summary of the results of this section:

(13.33) Let the correspondence between a continuous periodic function $x(t)$ and its Fourier transform $\xi(j)$ be denoted by $x(t) \longleftrightarrow \xi(j)$. Likewise, let $f(t) \longleftrightarrow \phi(j)$ and $g(t) \longleftrightarrow \gamma(j)$. Then the following conditions apply:

Convolution and Modulation

(i) $f(t) * g(t) \longleftrightarrow \gamma(j)\phi(j)$,

(ii) $f(t)g(t) \longleftrightarrow \gamma(j) * \phi(j)$,

Parseval's Theorem

$$\text{(iii)}\quad \int_0^T f(t)g^*(t)dt = T\sum_{j=-\infty}^{\infty}\gamma_j^*\phi_j,$$

$$\text{(iv)}\quad \int_0^T |x(t)|^2 dt = T\sum_{j=-\infty}^{\infty}|\xi_j|^2.$$

Example 13.2. If $x(t) = \rho\cos(\omega t)$ represents a voltage applied across a resistance of one ohm, then the power dissipated will be $\rho^2/2$ watts. The result is not affected by a shift in the phase of the cycle. Thus, by using the compound-angle formula (13.126)(c) from the appendix of this chapter, it can be shown that

$$\text{(13.34)}\qquad \frac{1}{T}\int_0^T \big\{\rho\cos(\omega t-\theta)\big\}^2 dt = \frac{\rho^2}{2T}\int_0^T \big\{1+\cos(2\omega t - 2\theta)\big\}dt.$$

Since the integral of the cosine term is over two complete cycles, its value is zero; and thus the expression as a whole is evaluated as $\rho^2/2$. Any of the expressions from the identities

$$\text{(13.35)}\qquad \rho\cos(\omega t-\theta) = \alpha\cos\omega t + \beta\sin\omega t = \xi e^{i\omega t} + \xi^* e^{-i\omega t},$$

where $2\xi = \rho e^{-i\theta}$, may be used in place of the integrand. It follows that

$$\text{(13.36)}\qquad \frac{\rho^2}{2} = \frac{\alpha^2+\beta^2}{2} = 2\xi\xi^*.$$

The same result can be obtained via Parseval's equation (13.30). This becomes $\rho^2/2 = \sum_{j=-1}^{1}|\xi_j|^2$, where $|\xi_{-1}| = |\xi_1| = \rho/2$ and $|\xi_0| = 0$. The latter condition reflects the fact that there is no D.C. component in the current.

Fourier Approximations

Let

$$\text{(13.37)}\qquad f(t) = \sum_j \phi_j e^{i\omega jt} \quad\text{and}\quad g(t) = \sum_j \gamma_j e^{i\omega jt}$$

be two functions defined on the interval $(0, T]$, and let the summations be finite or infinite. Then a measure of the squared distance between the functions is given by

$$\text{(13.38)}\qquad \|f(t)-g(t)\|^2 = \|f(t)\|^2 + \|g(t)\|^2 - 2\langle f(t), g(t)\rangle,$$

where

$$\text{(13.39)}\qquad \begin{aligned} \|f(t)\|^2 &= T\sum_j |\phi_j|^2,\\ \|g(t)\|^2 &= T\sum_j |\gamma_j|^2,\\ \langle f(t), g(t)\rangle &= T\sum_j \phi_j^*\gamma_j. \end{aligned}$$

When $g(t) = f(t)$, there is $\langle f(t), f(t)\rangle = \|f(t)\|^2$.

Suppose that we are given the function $f(t)$ and that we wish to approximate it in the interval $(0, T]$ by a function $g_n(t)$ of the type defined in (13.37) with a finite summation which runs from $-n$ to n:

$$g_n(t) = \gamma_0 + \sum_{j=1}^{n} \left\{\gamma_j e^{i\omega jt} + \gamma_{-j} e^{-i\omega jt}\right\}. \tag{13.40}$$

When $\gamma_{-j} = \gamma_j^*$, which is to say that $\gamma_j^{re} = \gamma_{-j}^{re}$ and $\gamma_j^{im} = -\gamma_{-j}^{im}$, this function $g_n(t)$ is real-valued, and it becomes an ordinary trigonometrical polynomial of degree n of the form

$$\begin{aligned} g_n(t) &= \gamma_0 + \sum_{j=1}^{n} \left\{\gamma_j^{re}\left(e^{i\omega jt} + e^{-i\omega jt}\right) + i\gamma_j^{im}\left(e^{i\omega jt} - e^{-i\omega jt}\right)\right\} \\ &= \gamma_0 + 2\sum_{j=1}^{n} \left\{\gamma_j^{re} \cos(\omega jt) - \gamma_j^{im} \sin(\omega jt)\right\}. \end{aligned} \tag{13.41}$$

Let $f_n(t)$ be the function defined in the same manner as $g_n(t)$ but with the leading coefficients $\phi_{-n}, \ldots, \phi_0, \ldots, \phi_n$ of $f(t)$ in place of $\gamma_{-n}, \ldots, \gamma_0, \ldots, \gamma_n$. Then $f_n(t)$ is the best approximation of its type, and it can be asserted that

$$\|f(t) - f_n(t)\|^2 \leq \|f(t) - g_n(t)\|^2. \tag{13.42}$$

This implies, in particular, that, amongst all of the trigonometrical polynomials of degree n, the one which gives the best approximation to a given real-valued function $f(t)$ is the so-called Fourier polynomial which is based on a partial sum comprising terms of the Fourier expansion of $f(t)$ up to the nth order.

The result depends upon the fact that f_n represents the minimum-distance projection of f onto the linear subspace spanned by the set of functions $e^{\pm i\omega jt}$; $j = 0, \ldots, n$ which constitute a basis for functions of the type g_n.

To prove this result, we can employ the method which has been used in proving the minimum-distance property of an ordinary-least squares regression. Consider

$$\|f - g_n\|^2 = \|f\|^2 + \|g_n\|^2 - 2\langle f, g_n\rangle \tag{13.43}$$

and

$$\begin{aligned} \|f - f_n\|^2 &= \|f\|^2 + \|f_n\|^2 - 2\langle f, f_n\rangle \\ &= \|f\|^2 - \|f_n\|^2, \end{aligned} \tag{13.44}$$

where the second equality of (13.44) follows from $\langle f, f_n\rangle = \langle f_n, f_n\rangle = \|f_n\|^2$. Taking one from the other gives

$$\begin{aligned} \|f - g_n\|^2 - \|f - f_n\|^2 &= \|g_n\|^2 + \|f_n\|^2 - 2\langle f, g_n\rangle \\ &= \|g_n\|^2 + \|f_n\|^2 - 2\langle f_n, g_n\rangle \\ &= \|g_n - f_n\|^2. \end{aligned} \tag{13.45}$$

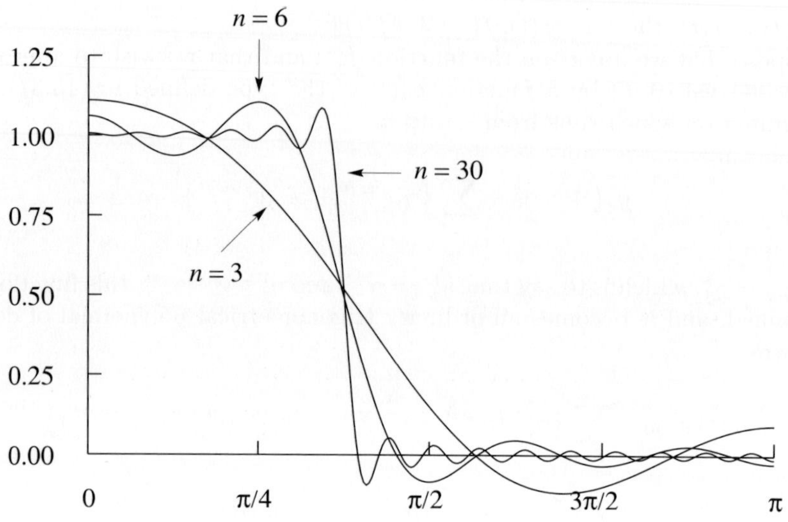

Figure 13.2. The Fourier approximation of a square wave.

On rearranging the expression, we obtain the inequality of (13.42), in view of the nonnegativity of the squares.

From the construction of f_n, it follows that

$$\|f - f_n\|^2 = T \sum_{j=n+1}^{\infty} \left\{ |\phi_j|^2 + |\phi_{-j}|^2 \right\}. \tag{13.46}$$

This comprises the tails of the convergent series

$$\int_0^T |f(t)|^2 dt = T \sum_{j=-\infty}^{\infty} |\phi_j|^2 \tag{13.47}$$

which is given by Parseval's equation. It follows that the quantity in (13.46) can be made smaller than any preassigned number by ensuring that n is large enough. This result is known as the mean-square convergence of the Fourier approximation.

If f is, in fact, a continuous periodic function, then its Fourier series expansion converges to the value of the function at every point in the interval $(0, T]$. The consequence is that any continuous periodic function can be uniformly approximated by trigonometrical polynomials. This is the trigonometrical form of the famous Weierstrass approximation theorem [508], of which the following is a formal statement:

(13.48) Let $f(t)$ be a continuous real-valued function on the interval $[0, T]$ and suppose that $f(T) = f(0)$. Then, given any real number $\epsilon > 0$, there exists a trigonometrical polynomial $f_n(t)$ such that $|f(t) - f_n(t)| < \epsilon$ for all $t \in [0, T]$.

It should be emphasised that this result of uniform pointwise convergence does not apply unless f is strictly continuous. If f is only piecewise-continuous, then only the mean-square convergence of f_n to f is assured.

Some important practical consequence of the failure of a Fourier series to converge uniformly to the values of a piecewise continuous function were discovered at the end of the nineteenth century. The American physicist Michelson had constructed a mechanical apparatus—the Michelson–Stratton Harmonic Analyser—which could be used both for the Fourier analysis of a function and for its synthesis from a set of trigonometrical components. He discovered that the apparatus was capable of synthesising a square wave successfully everywhere except at the points of discontinuity where the series approximation was liable to overshoot the value of the function. Michelson published his finding in 1898 in a letter to *Nature* [346]. An explanation of it was provided by the mathematical physicist J.W. Gibbs in two further letters, [208], [209]; and Michelson's discovery came to be known as Gibbs' Phenomenon.

This phenomenon is illustrated in Figure 13.2 where it is apparent that not all of the oscillations in the partial sums are decreasing at a uniform rate as n increases. Instead, the oscillations which are adjacent to the point of discontinuity are tending to a limiting amplitude which is about 9% of the jump. However, as n increases, the width of these end-oscillations becomes vanishingly small; and thus the mean-square convergence of the Fourier series is assured.

Discrete-Time Fourier Transform

If $x(t) = \{x_t; t = 0 \pm 1, \pm 2, \ldots\}$ is an absolutely summable sequence of real values such that

$$\sum_{t=-\infty}^{\infty} |x_t| < \infty, \tag{13.49}$$

then its elements may be expressed as

$$\begin{aligned} x_t &= \frac{1}{2\pi}\int_0^{\pi} \alpha(\omega)\cos(\omega t)d\omega + \frac{1}{2\pi}\int_0^{\pi} \beta(\omega)\sin(\omega t)d\omega \\ &= \frac{1}{2\pi}\int_0^{\pi} \left\{\frac{\alpha(\omega) - i\beta(\omega)}{2}\right\} e^{i\omega t}d\omega + \frac{1}{2\pi}\int_0^{\pi} \left\{\frac{\alpha(\omega) + i\beta(\omega)}{2}\right\} e^{-i\omega t}d\omega, \end{aligned} \tag{13.50}$$

where $\alpha(\omega) = \alpha(-\omega)$ and $\beta(-\omega) = -\beta(\omega)$ are periodic functions of ω which are an even function and an odd function respectively.

We have written the expression this manner simply for the purpose of comparing it with the trigonometrical expression for the Fourier series which is given under (13.1). On defining

$$\xi(\omega) = \frac{\alpha(\omega) - i\beta(\omega)}{2} \quad \text{and} \quad \xi(-\omega) = \xi^*(\omega) = \frac{\alpha(\omega) + i\beta(\omega)}{2}, \tag{13.51}$$

the expression under (13.50) together with its inverse may be written as

$$x_t = \frac{1}{2\pi}\int_{-\pi}^{\pi} \xi(\omega)e^{i\omega t}d\omega, \tag{13.52}$$

(13.53) $$\xi(\omega) = \sum_{t=-\infty}^{\infty} x_t e^{-i\omega t}.$$

Equation (13.53) represents the so-called discrete-time Fourier transform of the temporal sequence $x(t)$. This alternative form makes two points apparent. The first is that, if we allow the variables on both sides of the Fourier transforms to be complex-valued, then the equations under (13.52) and (13.53) bear a dual relationship with the equations under (13.14) and (13.15). The latter equations define the Fourier transform of a continuous periodic function and its inverse.

To affirm this point, we may change the variable of integration in the expression for ξ_j under (13.14) from t to $\psi = -\omega t$. Then the two equations become

(13.54) $$\xi_j = \frac{1}{2\pi}\int_{-\pi}^{\pi} x(\psi)e^{i\psi j}d\psi,$$

(13.55) $$x(\psi) = \sum_{j=-\infty}^{\infty} \xi_j e^{-i\psi j};$$

and these forms, which should be quickly forgotten for fear of confusing the notation, are isomorphic with the equations under (13.52) and (13.53).

The second point is that the expression for $\xi(\omega)$ under (13.53) is simply the z-transform of the sequence $x(t) = \{x_t\}$ which has been specialised by setting $z = e^{-i\omega}$ or, equivalently, by constraining z to lie on the unit circle. Moreover, if we change the variable of integration in the expression for x_t under (13.52) from ω to $z = e^{i\omega}$, then we get

(13.56) $$x_t = \frac{1}{2\pi i}\oint \xi(z)z^t\frac{dz}{z},$$

which is an expression for the generic coefficient of the Laurent series $\xi(z) = \sum x_t z^t$ which is in accordance with the formula given under (3.94).

Symmetry Properties of the Fourier Transform

The basic properties of the sine and cosine functions which underlie the Fourier transform give rise to certain symmetry conditions which are useful in understanding the frequency-response functions of linear systems. These properties, which we shall develop in the context of the discrete-time Fourier transform, are common to all the species of Fourier transforms. Thus they also lead to useful simplifications in the computing of the discrete Fourier transform of a finite sequence.

To demonstrate the symmetry conditions, it is appropriate to expand equations (13.52) and (13.53) so as to make the trigonometrical functions explicit as well as to reveal the real and imaginary components. First, we may write equation (13.53) as

$$
\begin{aligned}
\xi(\omega) &= \sum_{t=-\infty}^{\infty} \{x_t^{re} + ix_t^{im}\}\{\cos(\omega t) - i\sin(\omega t)\} \\
&= \sum_{t=-\infty}^{\infty} \{x_t^{re}\cos(\omega t) + x_t^{im}\sin(\omega t)\} \\
&\qquad -i\sum_{t=-\infty}^{\infty} \{x_t^{re}\sin(\omega t) - x_t^{im}\cos(\omega t)\} \\
&= \frac{1}{2}\{\alpha(\omega) - i\beta(\omega)\}.
\end{aligned} \tag{13.57}
$$

Here there is no presumption that $\alpha(\omega)$ is an even function or that $\beta(\omega)$ is odd. When there is no presumption that x_t is real, equation (13.52) can be written likewise as

$$
\begin{aligned}
x_t &= \frac{1}{4\pi}\int_{-\pi}^{\pi} \{\alpha(\omega) - i\beta(\omega)\}\{\cos(\omega t) + i\sin(\omega t)\}d\omega \\
&= \frac{1}{4\pi}\int_{-\pi}^{\pi} \{\alpha(\omega)\cos(\omega t) + \beta(\omega)\sin(\omega t)\}d\omega \\
&\qquad +i\frac{1}{4\pi}\int_{-\pi}^{\pi} \{\alpha(\omega)\sin(\omega t) - \beta(\omega)\cos(\omega t)\}d\omega \\
&= x_t^{re} + ix_t^{im}.
\end{aligned} \tag{13.58}
$$

This is a generalisation of (13.50).

Consider setting $x_t^{re} = x_t$ and $x_t^{im} = 0$ in (13.57), which is the case when $x(t) = \{x_t\}$ is a real-valued sequence. Then the equation becomes

$$
\begin{aligned}
\xi(\omega) &= \sum_{j=-\infty}^{\infty} \{x_t\cos(\omega t) - ix_t\sin(\omega t)\} \\
&= \frac{1}{2}\{\alpha(\omega) - i\beta(\omega)\};
\end{aligned} \tag{13.59}
$$

and, in view of the properties of the trigonometrical functions, it can be seen that $x(t)$ has a Fourier transform of which the real part $\alpha(\omega) = \alpha(-\omega)$ is now an even function and the imaginary part $\beta(\omega) = -\beta(-\omega)$ is now an odd function. When the latter conditions are applied to equation (13.58), it can be seen that the imaginary term vanishes, since the integral over $(-\pi, 0]$ cancels with the integral over $(0, \pi]$. Thus equation (13.58) is reduced to equation (13.50).

Other useful symmetry properties of a similar nature can also be deduced directly from the equations (13.57) and (13.58). They are presented in Table 13.2. It is possible to deduce the symmetry conditions in a more abstract manner by referring to some fundamental relationships which are give below and which may themselves be confirmed by reference to equations (13.52) and (13.53):

Table 13.2. Symmetry properties of the Fourier transform

Time domain $x(t)$	Frequency domain $\xi(\omega)$
Real	Real even, imaginary odd
Imaginary	Real odd, imaginary even
Real even, imaginary odd	Real
Real odd, imaginary even	Imaginary
Real and even	Real and even
Real and odd	Imaginary and odd
Imaginary and even	Imaginary and even
Complex and even	Complex and even
Complex and odd	Complex and odd

(13.60) Let $x(t)$ be a complex-valued sequence, and let $\xi(\omega)$ be its Fourier transform. Then the following relationships exist:

$$\text{(i)}\quad x(t) \longleftrightarrow \xi(\omega), \qquad \text{(ii)}\quad x(-t) \longleftrightarrow \xi(-\omega),$$
$$\text{(iii)}\quad x^*(t) \longleftrightarrow \xi^*(-\omega), \qquad \text{(iv)}\quad x^*(-t) \longleftrightarrow \xi^*(\omega).$$

Using the result under (iii), we can readily prove the first of the symmetry conditions of Table 13.2:

(13.61) If $\xi(\omega) = \{\alpha(\omega) - i\beta(\omega)\}/2$ is the Fourier transform of a real-valued sequence $x(t)$, then $\alpha(\omega) = \alpha(-\omega)$ is an even function and $\beta(\omega) = -\beta(-\omega)$ is an odd function.

Proof. If $x(t)$ is real-valued, then $x(t) = x^*(t)$ and therefore $\xi(\omega) = \xi^*(-\omega)$, or, equivalently, $\alpha(\omega) - i\beta(\omega) = \alpha(-\omega) + i\beta(-\omega)$. Equating separately the real and imaginary parts of this equation proves the result.

From here it is a small step to prove that, if $x(t)$ is real and even, then $\xi(\omega)$ is real and even. For, with $x(t) = x(-t)$ and hence $\xi(\omega) = \xi(-\omega)$, it follows that $\alpha(\omega) = \alpha(-\omega)$ and $\beta(\omega) = \beta(-\omega)$. But if $x(t)$ is real, then $\beta(\omega) = -\beta(-\omega)$; and the two conditions on $\beta(\omega)$ can be reconciled only if $\beta(\omega) = 0$.

The other results in Table 13.2 are proved in the same manner with equal facility.

The Frequency Response of a Discrete-Time System

The Fourier series and the discrete-time Fourier transform bear a dual relationship to each other which makes them mathematically identical. (See Table 13.1). Nevertheless, in the context of signal processing, they have quite different interpretations, since one of them transforms a continuous periodic signal from the time domain into a discrete aperiodic sequence of the frequency domain, whereas the other transforms an aperiodic time-domain sequence into a continuous periodic frequency-domain function.

The differences are heightened if the time-domain functions are restricted to be real-valued; for given that, in general, the frequency-domain transforms will be complex, this specialisation destroys the duality of the two Fourier transforms.

One of the principal uses of the discrete-time Fourier transform in signal processing is in describing the effects upon arbitrary signals of discrete-time transfer functions which are also called filters. In the case of a linear filter, a filtered sequence $y(t)$ is derived from an input sequence $x(t)$ via a relationship of the form

$$y(t) = \psi(L)x(t) = \sum_j \psi_j x(t-j). \tag{13.62}$$

Here, $\psi(L) = \sum \psi L^j$ is, in general, a two-sided transfer function which comprises both positive and negative powers of the lag operator L.

A natural way of characterising a filter is to specify the sequence $\psi(j) = \{\psi_j\}$ of the filter coefficients. However, a filter may also be characterised usefully by describing its effect upon a variety of standardised input signals. The simplest of these signals is the unit impulse specified by

$$\delta(t) = \begin{cases} 1, & \text{if } t = 0; \\ 0, & \text{if } t \neq 0. \end{cases} \tag{13.63}$$

Setting $x(t) = \delta(t)$ in (13.62) gives

$$y(t) = \psi(L)\delta(t) = \sum_j \psi_j \delta(t-j) = \psi(t). \tag{13.64}$$

For a given value of t, the equation yields the filter coefficient ψ_t. As a whole, the sequence $y(t) = \psi(t)$, which is described as the impulse response of the filter, is synonymous with the sequence of filter coefficients.

It is also appropriate to consider the response to the complex exponential input $x(t) = e^{i\omega t} = \cos(\omega t) + i\sin(\omega t)$. The equation

$$\begin{aligned} y(t) = \psi(L)e^{i\omega t} &= \sum_j \psi_j e^{i\omega(t-j)} \\ &= \Big\{\sum_j \psi_j e^{-i\omega j}\Big\}e^{i\omega t} = \psi(\omega)e^{i\omega t} \end{aligned} \tag{13.65}$$

indicates that the output sequence is formed by multiplying the input sequence by a complex-valued function

$$\psi(\omega) = \sum_j \psi_j e^{-i\omega j}. \tag{13.66}$$

This function, which is the discrete-time Fourier transform of the impulse-response sequence $\psi(j) = \{\psi_j\}$, is called the frequency-response function of the filter. Given that the transfer function $\psi(L)$ satisfies the BIBO stability condition, which is that $\sum_j |\psi_j| < \infty$, it follows that the condition of convergence given under (13.49) is satisfied—which guarantees the existence of the transform.

To reveal the effect of the filter upon the complex exponential input, the frequency response function may be cast in the form of

$$\psi(\omega) = |\psi(\omega)|e^{-i\theta(\omega)} = |\psi(\omega)|\left[\cos\{\theta(\omega)\} - i\sin\{\theta(\omega)\}\right]. \tag{13.67}$$

Then the output response of the filter to the complex-exponential input, given under (13.65), becomes

$$\begin{aligned} y(t) &= |\psi(\omega)|e^{i\{\omega t-\theta(\omega)\}} \\ &= |\psi(\omega)|\left[\cos\{\omega t - \theta(\omega)\} + i\sin\{\omega t - \theta(\omega)\}\right]. \end{aligned} \tag{13.68}$$

This indicates that, on passing through the filter, a sinusoidal input at frequency ω will have its amplitude altered by a factor of $|\psi(\omega)|$, described as the gain effect, and it will have its phase displaced by a amount equal to $\theta(\omega)$ radians, described as the phase effect.

An alternative approach to the derivation of the frequency response is to apply a transformation directly to equation (13.64) which depicts the impulse response. The fact that the Fourier transform is a linear operator indicates that the transform of a sum is a sum of transforms. The Fourier transform of the sequence $\delta(t-j)$, which is zero-valued apart from a unit impulse at $t = j$, is $e^{-i\omega j}$. Therefore the transform of the sum on the RHS of equation (13.64) is the sum of transformed impulses found under (13.66).

Example 13.3. Consider an ideal lowpass filter with a frequency response defined by

$$\psi(\omega) = \begin{cases} 1, & \text{if } |\omega| < \omega_c; \\ 0, & \text{if } \omega_c < |\omega| \leq \pi. \end{cases} \tag{13.69}$$

This function is periodic, but one may imagine that the frequencies of the input signal do not exceed π. The coefficients of the filter are found by means of the (inverse) Fourier transform of (13.52). Thus

$$\psi_j = \frac{1}{2\pi}\int_{-\omega_c}^{\omega_c} e^{i\omega j}d\omega = \frac{\sin\omega_c j}{\pi j}. \tag{13.70}$$

The integral has appeared already under (13.19) in connection with a previous example. In particular, the sequence of filter coefficients has a profile which is the same as the sequence of Fourier coefficients presented in Figure 13.1. This result illustrates the duality of the Fourier series and the discrete-time Fourier transform.

Since this particular frequency-response function $\psi(\omega)$ is real-valued and equal everywhere to its modulus $|\psi(\omega)|$, there is no phase effect. Such is the case whenever the filter coefficients are symmetric about the point $j = 0$.

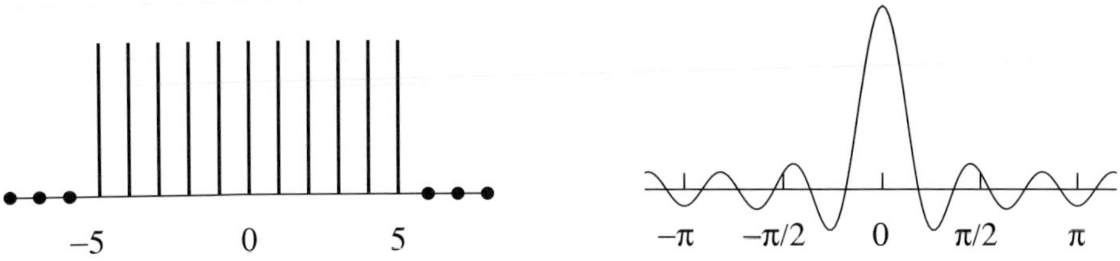

Figure 13.3. The Fourier transform of a uniform sequence yields the Dirichlet kernel.

The coefficients of the filter are nonzero for $j < 0$, and therefore the filter is not available for real-time signal processing where filters must look backwards in time. Also, the coefficients constitute an infinite sequence which must be truncated before filter can be implemented. Nevertheless, the ideal filter does represent a standard which practical filters are often designed to emulate; and something close to it is available for the processing of digitally recorded sound when there is not a real-time constraint and when the sampling rate of the recording is high. However, such filter designs are not particularly favoured in sound engineering because, even at relatively high sampling rates, the impulse response of the filter appears to be under-damped, which corresponds, in sound, to the phenomenon of ringing.

Example 13.4. As a further example of the frequency response, consider the simple n-point moving-average operator

$$\psi(L) = n^{-1}(I + L + \cdots + L^{n-1}). \tag{13.71}$$

The frequency response of this filter is given by

$$\psi(\omega) = \frac{1}{n}\sum_{j=0}^{n-1} e^{-i\omega j} = \begin{cases} 1, & \text{if } \omega j = 0; \\ \dfrac{1}{n}\dfrac{1 - e^{-i\omega n}}{1 - e^{-i\omega}}, & \text{otherwise.} \end{cases} \tag{13.72}$$

Multiplying top and bottom of $n\psi(\omega)$ by $\exp(i\omega n/2)$ gives

$$n\psi(\omega) = \frac{e^{i\omega n/2} - e^{-i\omega n/2}}{e^{i\omega(n-1)/2}(e^{i\omega/2} - e^{-i\omega/2})} = \frac{\sin(\omega n/2)}{\sin(\omega/2)} e^{-i\omega(n-1)/2}. \tag{13.73}$$

The ratio of the sine functions divided by n gives the gain of the filter, whilst the phase effect is $\theta(\omega) = (n-1)\omega/2$. This is a linear effect which indicates that each harmonic component in the input signal is delayed by $\theta(\omega)/\omega = (n-1)/2$ periods.

Imagine that n is an odd integer, and define a symmetric two-sided moving-average filter by

$$\begin{aligned} \kappa(L) &= L^{(1-n)/2} + \cdots + L^{-1} + I + L + \cdots + L^{(n-1)/2} \\ &= nL^{(1-n)/2}\psi(L). \end{aligned} \tag{13.74}$$

Then the frequency response becomes

$$\kappa(\omega) = n\psi(\omega)e^{i\omega(n-1)/2} = \frac{\sin(\omega n/2)}{\sin(\omega/2)}, \tag{13.75}$$

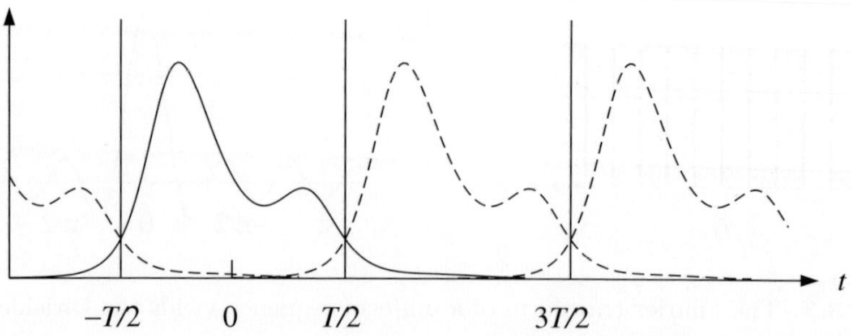

Figure 13.4. A periodic function may be constructed by replicating, in successive intervals, a segment of a function which is absolutely integrable over the real line.

and there is no longer any phase effect. It transpires, from this example, that the factor $e^{-i\omega(n-1)/2}$ is the frequency-domain equivalent of the operator $L^{(n-1)/2}$. The function defined is (13.75) is known as the Dirichlet kernel—see, for example, Titchmarsh [484, p. 402]. The function is illustrated in Figure 13.3

The Fourier Integral

Continuous aperiodic signals do not have Fourier series expansions, and it is necessary to represent them in terms of an nondenumerable infinity of sinusoidal components which are gathered under an integral.

The Fourier integral can be developed from the Fourier series by allowing the period T of the function $x(t) = x(t+T)$ to increase indefinitely whilst maintaining the condition of (13.9) that the function is absolutely integrable over the interval $(-T/2, T/2]$. One may imagine that the periodic function $x(t)$ was defined, in the first place, by replicating, in successive intervals, a segment of a function which is absolutely integrable over the real line. (See Figure 13.4). Then the process of increasing T is a matter of extending this segment in both directions.

Consider rewriting equation of (13.1) as

$$x(t) = \frac{1}{2\pi}\sum_{j=0}^{\infty}\Big\{dA(\omega j)\cos(j\omega t) + dB(\omega j)\sin(j\omega t)\Big\}, \tag{13.76}$$

where $dA(\omega j) = 2\pi\alpha_j$ stands for the increments of a step function or "staircase" function A at the points of discontinuity ωj where it rises vertically. These points are integer multiples of the fundamental frequency $\omega = 2\pi/T$. In the limit, as $T \to \infty$, the intervals between the successive harmonic frequencies, which are the treads of the staircase, vanish and the increments $dA(\omega j)$ also become vanishingly small. Then the summation is replaced by integration and the expression becomes

$$\begin{aligned} x(t) &= \frac{1}{2\pi}\int_0^{\infty}\cos(\omega t)dA(\omega) + \frac{1}{2\pi}\int_0^{\infty}\sin(\omega t)dB(\omega) \\ &= \frac{1}{2\pi}\int_0^{\infty}\alpha(\omega)\cos(\omega t)d\omega + \frac{1}{2\pi}\int_0^{\infty}\beta(\omega)\sin(\omega t)d\omega. \end{aligned} \tag{13.77}$$

Here ω no longer stands for the fundamental frequency. Instead, it now represents the continuous variable of integration. Also, $A(\omega)$ and $B(\omega)$ have become continuous functions which have the derivatives $\alpha(\omega)$ and $\beta(\omega)$ respectively.

The Fourier integral and its inverse may be expressed in terms of the complex exponential function. Thus

(13.78) $$x(t) = \frac{1}{2\pi}\int_{-\infty}^{\infty} \xi(\omega)e^{i\omega t}d\omega,$$

(13.79) $$\xi(\omega) = \int_{-\infty}^{\infty} x(t)e^{-i\omega t}dt,$$

where $\xi(\omega) = \{\alpha(\omega) + i\beta(\omega)\}/2$. A condition which is sufficient for the existence of the Fourier integral is that of absolute integrability which corresponds to the condition under (13.9):

(13.80) $$\int_{-\infty}^{\infty} |x(t)|dt < \infty.$$

Signals which are absolutely integrable have a finite energy.

But for the factor $1/2\pi$, the Fourier integral and its inverse are symmetrically related to each other; and a perfect symmetry could be achieved by sharing the factor equally between the Fourier transform and its inverse, which is the practice of some authors. Also, the factor disappears if ordinary frequency $f = \omega/2\pi$ is used in place of angular frequency ω as an argument. However, to use f in place of ω poses an impediment to complex analysis. The duality of the relationship between the Fourier integral and its inverse can be expressed by writing

(13.81) $$x(t) \longleftrightarrow \xi(\omega) \qquad \text{and} \qquad \xi(t) \longleftrightarrow 2\pi x(-\omega).$$

The properties of the Fourier integral are the natural analogues of those of the Fourier series and of the discrete-time Fourier transform; and once this is recognised there is no need to restate these results. Nevertheless, to illustrate the proposition, let us take the example of the convolution operations.

Let $g(t)$ and $\gamma(\omega)$ be a pair of Fourier functions which have the same forms as $x(t)$ of (13.78) and $\xi(\omega)$ of (13.79) respectively. Then the time-domain convolution of $x(t)$ and $g(t)$ leads to the identity

(13.82) $$g(t) * x(t) = \int_{\tau} g(\tau)x(t-\tau)d\tau = \frac{1}{2\pi}\int_{\omega} \gamma(\omega)\xi(\omega)e^{i\omega t}d\omega.$$

This is the analogue of the equation under (13.25). On the other hand, the frequency-domain convolution of $\gamma(\omega)$ and $\xi(\omega)$ leads to the identity

(13.83) $$\gamma(\omega) * \xi(\omega) = \int_{\lambda} \gamma(\lambda)\xi(\omega-\lambda)d\lambda = 2\pi\int_{t} g(t)x(t)e^{-i\lambda t}dt,$$

which is the analogue of the result under (13.27). The summary of these results, which corresponds to the statement under (13.33), is as follows:

(13.84) Let the correspondence between a continuous aperiodic function $x(t)$ and its Fourier transform $\xi(\omega)$ be denoted by $x(t) \longleftrightarrow \xi(\omega)$. Likewise, let $g(t) \longleftrightarrow \gamma(j)$. Then the following conditions apply:

$$\text{(i) } x(t) * g(t) \longleftrightarrow \gamma(\omega)\xi(\omega),$$

$$\text{(ii) } x(t)g(t) \longleftrightarrow \frac{1}{2\pi}\gamma(\omega) * \xi(\omega).$$

Example 13.5. Consider an isolated rectangular pulse defined by

$$x(t) = \begin{cases} 1, & \text{if } |t| \leq \tau; \\ 0, & \text{if } \tau < |t|. \end{cases} \tag{13.85}$$

The condition of (13.80) is certainly fulfilled, and therefore there is a Fourier transform which is the so-called sinc function:

$$\xi(\omega) = \int_{-\tau}^{\tau} e^{-i\omega t} dt = \frac{2\sin\omega\tau}{\omega}. \tag{13.86}$$

This is to be compared with equation (13.19) which defines the jth coefficient ξ_j of the Fourier expansion of a periodic square wave. Multiplying the latter by T gives

$$T\xi_j = \frac{2\sin j\omega\tau}{j\omega}. \tag{13.87}$$

The periodic wave approaches the single rectangular pulse as its period T becomes large relative to 2τ, which is the width of the pulse. Also, the Fourier coefficients of the wave scaled by T become more densely spaced under their envelope, which is the same function as the Fourier transform of the pulse.

The Uncertainty Relationship

There is an inverse relationship between the dispersion of a function and the range of the frequencies which are present in its transform. Thus one finds that, the shorter is the duration of a transient signal, the wider is the spread of the frequencies in its transform.

In electrical engineering, this notion finds expression in the so-called bandwidth theorem. In mathematical physics, an analogous relationship between the spatial dispersion of a wave train and its frequency dispersion is the basis of the uncertainty principle of Heisenberg.

To illustrate the relationship, we may consider a Gaussian or normal distribution. This is defined in terms of the random variable x by

$$f(x) = \frac{1}{\sqrt{2\pi\sigma^2}} \exp\left\{ -\frac{1}{2\sigma^2}(x-\mu)^2 \right\}. \tag{13.88}$$

The Fourier transform of $f(x)$, which in known in mathematical statistics as the characteristic function of the normal distribution, is given by

$$\begin{aligned}\phi(\omega) &= \frac{1}{\sqrt{2\pi\sigma^2}}\int_{-\infty}^{\infty} \exp\left\{i\omega x - \frac{1}{2\sigma^2}(x-\mu)^2\right\}dx\\ &= \exp\left\{i\omega\mu - \frac{1}{2}\sigma^2\omega^2\right\}\frac{1}{\sqrt{2\pi\sigma^2}}\int_{-\infty}^{\infty} \exp\left\{-\frac{1}{2\sigma^2}(x-\mu-i\sigma^2\omega)^2\right\}dx.\end{aligned} \tag{13.89}$$

The integral here is that of the function $\exp\{z^2/(2\sigma^2)\}$, where z is a complex variable which runs along a line parallel to the real axis. This can be shown to be equal to the integral of the corresponding real function which has the value of $\sigma\sqrt{2\pi}$. Therefore the characteristic function is

$$\phi(\omega) = \exp\left\{i\omega\mu - \frac{1}{2}\sigma^2\omega^2\right\}. \tag{13.90}$$

The characteristic function is so-called because it completely characterises the distribution. The parameters of the distribution are the mean μ and the variance σ^2 which measures the dispersion of x. The distribution is symmetric about the value μ; and if, $\mu = 0$, then $\phi(\omega)$ is real-valued, as we are led to expect from the symmetry properties of the Fourier transform.

The inverse relationship between the dispersions of $f(x)$ and $\phi(\omega)$ is manifest from the comparison of the two functions which, apart from a scalar factor, have the same form when $\mu = 0$. Thus, if the dispersion of $f(x)$ is represented by σ, then that of $\phi(\omega)$ is directly related to σ^{-1}.

The measure of dispersion which is used in mathematical statistics, and which is based on the presumption that the function is nonnegative, is inappropriate for measuring the width of an oscillatory signal or a wave train. In such cases, the usual measure of dispersion of x is

$$\Delta_x^2 = \int_{-\infty}^{\infty} x^2|f(x)|^2dx \bigg/ \int_{-\infty}^{\infty} |f(x)|^2dx. \tag{13.91}$$

The dispersion Δ_ω^2 in the frequency domain is defined likewise.

In quantum mechanics, a particle is also depicted as a De Broglie wave. Schrödinger's wave function $\psi(x)$ serves to define the spatial extent of the wave train, and its dispersion Δ_x is liable to be interpreted as a measure of the uncertainty of our knowledge of the particle's position.

The formulation of De Broglie [141] also relates the momentum ρ of a particle to its wavelength $\lambda = 1/\omega$ according to the formula $\rho = h/\lambda$, where h is Planck's constant. Thus, the spread of momentum is $\Delta_\rho = h\Delta_\omega$; and the position–momentum uncertainty principle states that

$$\Delta_x\Delta_\rho \geq \frac{h}{4\pi}. \tag{13.92}$$

It can be shown that the Gaussian wave train is the only one which leads to an equality in this relationship.

The Delta Function

For completeness, it is useful to extend the definition of the Fourier integral so that it can be applied to periodic as well as to aperiodic signals. Also, we should like to subsume the case of a discrete-time process under the general theory of the Fourier integral.

The problem in attempting to define the Fourier transform of a periodic function is that the function fails to satisfy the condition under (13.80). A natural interpretation of this feature is that a continuous periodic function represents a process which, over time, dissipates an indefinite amount of energy.

The problem in attempting to define the Fourier transform of a discrete-time process is, in a sense, the diametric opposite of the previous problem: it is that the process possesses negligible energy. It transpires that there is a species of duality between the two problems, both of which can be treated with the help of Dirac's delta function [160].

The Dirac delta function is a mathematical idealisation of the mechanical concept of an impulse which is defined as an indefinitely large force acting instantaneously so as to impart a finite momentum to the body on which it impinges. The unit impulse in the time domain is a delta function which fulfils the following conditions:

$$\begin{aligned} &\delta(t) = 0 \quad \text{for all} \quad t \neq 0, \\ &\int_{-\infty}^{\infty} \delta(t)dt = 1. \end{aligned} \tag{13.93}$$

A consequence of these two properties is that $\delta(t)$ must be infinite at $t = 0$. A frequency-domain impulse function $\delta(\omega)$ may also be defined which fulfils conditions which are derived from those of (13.93) simply by replacing the time-domain argument t by the frequency-domain argument ω.

The delta function $\delta(t)$ may be approximated by any number of functions which integrate to unity and which have a minimal dispersion about the point $t = 0$. An easy way of conceptualising the function is to consider a rectangle of unit area defined on the interval $[0, \Delta]$. The height of the rectangle, which is $1/\Delta$, increases indefinitely as $\Delta \rightarrow 0$; and, in the limit, we obtain Dirac's function. An alternative approach is to consider the normal density function of statistics whose expression is to be found under (13.88). The integral of the function over the real line is unity, and, as the variance σ^2 tends to zero, it too becomes an increasingly accurate representation of a unit impulse.

An essential property of the Dirac delta is the so-called *sifting property* whereby

$$\begin{aligned} f(\tau) &= \int_{-\infty}^{\infty} f(t)\delta(t-\tau)dt \\ &= \int_{-\infty}^{\infty} f(t)\delta(\tau-t)dt. \end{aligned} \tag{13.94}$$

This equation, which has a discrete-time counterpart in equation (13.64), may be explained by representing the delta function in terms of a unit-area rectangle defined over the interval $[\tau, \tau + \Delta]$. The value of the integral is approximated by the area of the rectangle times the average of the values which the function attains over the interval. As $\Delta \to 0$, this average tends to the value of the function at the point τ, which is $f_\tau = f(\tau)$. Notice that it also follows from the conditions under (13.93) that

$$f(t)\delta(t-\tau) = f(\tau)\delta(t-\tau). \tag{13.95}$$

The Fourier transform of the Dirac function $\delta(t-\tau)$ can be obtained directly from the formula of (13.79) by setting $x(t) = \delta(t-\tau)$. From (13.94), it follows that

$$\xi(\omega) = \int_{-\infty}^{\infty} e^{-i\omega t}\delta(t-\tau)dt = e^{-i\omega\tau}. \tag{13.96}$$

When $\tau = 0$, we have $\xi(\omega) = 1$; which is to say that the transform of the impulse is a constant function which is dispersed over the entire real line. In effect, every frequency is needed in order to synthesise the impulse.

In contemplating the idea that an impulse comprises all of the frequencies, one is reminded of the fact that, to investigate the harmonics of a bell, all that is required—in order to excite the resonant frequencies—is a sharp stroke of the clapper, which is an impulse in other words.

It is now possible to write down the frequency-domain expression for a signal which comprises a train of impulses, each separated by a unit time interval and each weighted by a value from the sequence $\{x_k\}$:

$$f(t) = \sum_k x_k \delta(t-k). \tag{13.97}$$

Equation (13.96) indicates that the Fourier transform of $f(t)$ is just

$$\phi(\omega) = \sum_t x_t e^{-i\omega t}; \tag{13.98}$$

and we notice that this corresponds precisely to the Fourier transform of the discrete-time sequence $x(k) = \{x_k\}$ already given under (13.53). In effect, we have managed to subsume the case of discrete aperiodic functions under the theory of the Fourier integral.

The expression under (13.97), which is for a continuous-time signal, is indistinguishable from an expression in terms of the discrete-time unit impulse function such as the one appearing under (13.64). The fault—if it is one—lies in our use of the same symbol to denote both the discrete-time and the continuous-time impulse.

Now let us consider the problem of applying the Fourier transform to a signal which is a continuous periodic function of t. We may tackle this case in a seemingly roundabout way by considering a function $x(t)$ whose Fourier transform $\xi(\omega)$ is a single frequency-domain impulse at $\omega = \omega_0$ with an area of 2π:

$$\xi(\omega) = 2\pi\delta(\omega - \omega_0). \tag{13.99}$$

On applying the inverse Fourier transform which is defined under (13.78), we find that this transform of (13.99) belongs to the function

$$
\begin{aligned}
x(t) &= \frac{1}{2\pi}\int_{-\infty}^{\infty} 2\pi\delta(\omega-\omega_0)e^{i\omega t}d\omega \\
&= e^{i\omega_0 t}.
\end{aligned} \tag{13.100}
$$

A comparison with equation (13.96), which gives the frequency-domain representation of a time-domain impulse, suggests that time-domain and frequency-domain impulses can be treated in ways which are mathematically equivalent.

Now imagine that $\xi(\omega)$ is a linear combination of a set of impulses equally spaced in frequency of the form

$$
\xi(\omega) = \sum_{j=-\infty}^{\infty} \xi_j 2\pi\delta(\omega - j\omega_0). \tag{13.101}
$$

Then the inverse Fourier transform yields

$$
x(t) = \sum_{j=-\infty}^{\infty} \xi_j e^{i\omega_0 jt}. \tag{13.102}
$$

This corresponds exactly to the Fourier-series representation of a periodic signal which is given under (13.7). Thus, in effect, we have subsumed the case of continuous periodic functions under the theory of the Fourier integral.

It may be observed that, in the absence of the factor 2π, the relationship $f(t) \longleftrightarrow \phi(\omega)$, which connects the functions of (13.97) and (13.98), would be identical to the relationship $\xi(\omega) \longleftrightarrow x(t)$, which connects those of (13.101) and (13.102). This is a consequence of the duality of the Fourier integral transform and its inverse which is expressed in (13.81).

It might be helpful to summarise the results of this section and to reveal two further implications.

(13.103) Let $\delta(t)$ and $\delta(\omega)$ represent Dirac's delta function in the time domain and in the frequency domain respectively. Then the following conditions apply:

(i) $\delta(t) \longleftrightarrow 1$,

(ii) $1 \longleftrightarrow 2\pi\delta(\omega)$,

(iii) $\delta(t-\tau) \longleftrightarrow e^{i\omega\tau}$,

(iv) $e^{i\omega_0 t} \longleftrightarrow 2\pi\delta(\omega-\omega_0)$,

(v) $\cos(\omega_0 t) \longleftrightarrow \pi\big\{\delta(\omega-\omega_0)+\delta(\omega-\omega_0)\big\}$,

(vi) $\sin(\omega_0 t) \longleftrightarrow i\pi\big\{\delta(\omega-\omega_0)-\delta(\omega-\omega_0)\big\}$.

The last two of these results are nothing but alternative renditions of the Euler equations, seen previously under (13.5).

Impulse Trains

In describing the periodic sampling of a continuous-time signal, as we shall in the next section, it is useful consider a train of impulses separated by a time period of T. This is represented by the function

$$g(t) = \sum_{j=-\infty}^{\infty} \delta(t - jT) \tag{13.104}$$

which is both periodic and discrete. The periodic nature of this function indicates that it can be expanded as a Fourier series

$$g(t) = \sum_{j-\infty}^{\infty} \gamma_j e^{i\omega jt}. \tag{13.105}$$

The coefficients of this expansion may be determined, according to the formula of (13.14), by integrating over just one cycle. Thus

$$\gamma_j = \frac{1}{T}\int_0^T \delta(t)e^{-i\omega_0 jt}dt = \frac{1}{T}, \tag{13.106}$$

wherein $\omega_0 = 2\pi/T$ represents the fundamental frequency. On setting $\gamma_j = T^{-1}$ for all j in the Fourier-series expression for $g(t)$ and invoking the result under (13.103)(iv), it is found that the Fourier transform of the continuous-time impulse train $g(t)$ is the function

$$\begin{aligned}\gamma(\omega) &= \frac{2\pi}{T}\sum_{j=-\infty}^{\infty} \delta\left(\omega - j\frac{2\pi}{T}\right)\\ &= \omega_0 \sum_{j=-\infty}^{\infty} \delta(\omega - j\omega_0).\end{aligned} \tag{13.107}$$

Thus it transpires that a periodic impulse train $g(t)$ in the time domain corresponds to a periodic impulse train $\gamma(\omega)$ in the frequency domain. Notice that there is an inverse relationship between the length T of the sampling interval in the time domain and the length $2\pi/T$ of the corresponding interval between the frequency-domain pulses.

Example 13.6. The impulse train in the time domain may be compared with the discrete-time sequence $\{\delta_t = 1; t = 0, \pm 1, \pm 2, \ldots\}$. The latter may be regarded as a limiting case, as $M \to \infty$, of the rectangular window sequence defined by

$$\kappa_t = \begin{cases} 1, & \text{if } |t| \leq M; \\ 0, & \text{if } |t| > M, \end{cases} \tag{13.108}$$

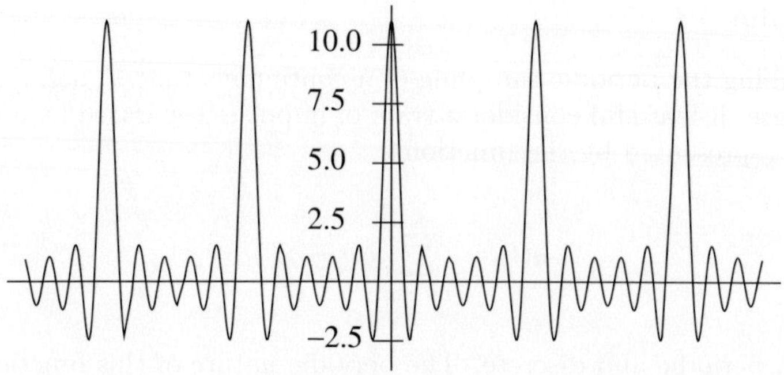

Figure 13.5. The Dirichlet function with $n = 11$, plotted over five cycles.

which comprises $n = 2M + 1$ consecutive units centred on $t = 0$. The Fourier transform of the window sequence is the Dirichlet kernel

$$\kappa(\omega) = \sum_{t=-M}^{M} e^{-i\omega t} = 1 + 2\sum_{t=1}^{M} \cos(\omega t). \tag{13.109}$$

This can also be expressed, in the manner of (13.75), as a ratio of sine functions, except in the case where $\omega = 2\pi j$ for any integer j. It can be observed that, by sampling the Dirichlet function at the points $\omega = 2\pi j/n$, one obtains a discrete impulse train in the frequency domain which takes a value of n at every nth point and which takes zero values on the other points. Thus

$$\kappa\left(\frac{2\pi}{n}j\right) = \begin{cases} n, & \text{for } j \bmod n = 0; \\ 0, & \text{otherwise.} \end{cases} \tag{13.110}$$

These features are readily discernible from Figure 13.5.

It is interesting to witness how the Dirichlet function approximates an impulse train in the frequency domain with ever-increasing accuracy as the length n of the window function increases. As n increases, the main lobe becomes more prominent, whereas the side lobes are attenuated.

The Sampling Theorem

Under certain conditions, a continuous-time signal can be completely represented by a sequence of values which have been sampled at equally spaced intervals. This fact, which is expressed in the so-called sampling theorem, is the basis of much of the modern signal-processing technology. In many applications, an analogue signal is converted to a digital signal to enable it to be processed numerically. After the processing, the signal is restored to analogue form. The sampling theorem indicates that these steps may be carried out without the loss of information. A familiar example of such processing is provided by the sound recordings which are available on compact discs.

The mathematical representation of the sampling process depends upon the periodic impulse train or sampling function $g(t)$ which is already defined under (13.104). The period T is the sampling interval, whilst the fundamental frequency of this function, which is $\omega_0 = 2\pi/T$, is the sampling frequency.

The activity of sampling may be depicted as a process of amplitude modulation wherein the impulse train $g(t)$ is the carrier signal and the sampled function $x(t)$ is the modulating signal. In the time domain, the modulated signal is described by the following *multiplication* of $g(t)$ and $x(t)$:

(13.111)
$$\begin{aligned} x_s(t) &= x(t)g(t) \\ &= \sum_{j=-\infty}^{\infty} x(jT)\delta(t - jT). \end{aligned}$$

The Fourier transform $\xi_s(\omega)$ of $x_s(t)$ is the *convolution* of the transforms of $x(t)$ and $g(t)$ which are denoted by $\xi(\omega)$ and $\gamma(\omega)$ respectively. Thus, from (13.83),

(13.112)
$$\begin{aligned} \xi_s(\omega) &= \int_{-\infty}^{\infty} x_s(t)e^{-i\omega t}dt \\ &= \frac{1}{2\pi}\int_{-\infty}^{\infty} \gamma(\lambda)\xi(\omega - \lambda)d\lambda. \end{aligned}$$

On substituting the expression for $\gamma(\lambda)$ found under (13.107), we get

(13.113)
$$\begin{aligned} \xi_s(\omega) &= \frac{\omega_0}{2\pi}\int_{-\infty}^{\infty} \xi(\omega - \lambda)\left\{\sum_{j=-\infty}^{\infty} \delta(\lambda - j\omega_0)\right\}d\lambda \\ &= \frac{1}{T}\sum_{j=-\infty}^{\infty}\left\{\int_{-\infty}^{\infty} \xi(\omega - \lambda)\delta(\lambda - j\omega_0)d\lambda\right\} \\ &= \frac{1}{T}\sum_{j=-\infty}^{\infty} \xi(\omega - j\omega_0). \end{aligned}$$

The final expression indicates that $\xi_s(\omega)$, which is the Fourier transform of the sampled signal $x_s(t)$, is a periodic function consisting repeated copies of the transform $\xi(\omega)$ of the original continuous-time signal $x(t)$. Each copy is shifted by an integral multiple of the sampling frequency before being superimposed.

Imagine that $x(t)$ is a band-limited signal whose frequency components are confined to the interval $[0, \omega_c]$, which is to say that the function $\xi(\omega)$ is nonzero only over the interval $[-\omega_c, \omega_c]$. If

(13.114)
$$\frac{2\pi}{T} = \omega_0 > 2\omega_c,$$

then the successive copies of $\xi(\omega)$ will not overlap; and therefore the properties of $\xi(\omega)$, and hence those of $x(t)$, can be deduced from those displayed by $\xi_s(\omega)$ over the interval $[0, \omega_0]$. In principle, the original signal could be recovered by passing its sampled version through an ideal lowpass filter which transmits all components of frequency less that ω_0 and rejects all others.

If, on the contrary, the sampling frequency is such that $\omega_0 < 2\omega_c$, then the resulting overlapping of the copies of $\xi(\omega)$ will imply that the spectrum of the sampled signal is no longer simply related to that of the original; and no linear filtering operation can be expected to recover the original signal from its sampled version. The effect of the overlap is to confound the components of the original process which have frequencies greater that π/T with those of frequencies lower than π/T; and this is described as the aliasing error.

The foregoing results are expressed in the famous samplin; theorem which is summarised as follows:

(13.115) Let $x(t)$ be a continuous-time signal with a transform $\xi(\omega)$ which is zero-valued for all $\omega > \omega_c$. Then $x(t)$ can be recovered from its samples provided that the sampling rate $\omega_0 = 2\pi/T$ exceeds $2\omega_c$.

An alternative way of expressing this result is to declare that the rate of sampling sets an upper limit to the frequencies which can be detected in an underlying process. Thus, when the sampling provides one observation in T seconds, the highest frequency which can be detected has a value of π/T radians per second. This is the so-called Nyquist frequency.

The sampling theorem was known in various forms long before is application to signal processing. The essence of the sampling theorem is contained, for example, in Whittaker's [516] tract on functional interpolation. The theorem has also been attributed to Nyquist [368] and to Shannon [450].

The Frequency Response of a Continuous-Time System

The Fourier integral may be used in clarifying some concepts associated with continuous-time systems which correspond closely to those which have been developed in a previous section in connection to discrete-time systems.

A linear system which transforms a continuous-time signal $x(t)$ into an output $y(t)$ may be described by an equation in the form of

$$y(t) = \varphi(D)x(t), \tag{13.116}$$

wherein $\varphi(D)$ is a polynomial or a rational function in the operator D which finds the derivative of a function of t. Equation (13.116) is the continuous-time analogue of the discrete-time equation (13.62). It is assumed that the system is causal and stable. The causal nature of the system implies that only positive powers of D are present in the series expansion of the operator. The necessary and sufficient condition for stability is that all of the poles of the function $\varphi(z)$ must lie in the left half of the complex plane.

If $x(t)$, which is the system's input, possesses a Fourier representation in the form of (13.78) then $y(t)$, which is its output, will be given by

$$\begin{aligned} y(t) &= \varphi(D)\left\{\frac{1}{2\pi}\int_{-\infty}^{\infty}\xi(\omega)e^{i\omega t}d\omega\right\} \\ &= \frac{1}{2\pi}\int_{-\infty}^{\infty}\varphi(D)e^{i\omega t}\xi(\omega)d\omega \\ &= \frac{1}{2\pi}\int_{-\infty}^{\infty}\varphi(i\omega)\xi(\omega)e^{i\omega t}d\omega. \end{aligned} \tag{13.117}$$

Here the final equality depends upon the result that $\varphi(D)e^{i\omega t} = \varphi(i\omega)e^{i\omega t}$ which is to be found under (5.69)(i). The Fourier representation of the output is therefore

$$y(t) = \frac{1}{2\pi}\int_{-\infty}^{\infty}\upsilon(\omega)e^{i\omega t}d\omega = \frac{1}{2\pi}\int_{-\infty}^{\infty}\phi(\omega)\xi(\omega)e^{i\omega t}d\omega. \tag{13.118}$$

Here $\phi(\omega) = \varphi(i\omega)$ is the complex-valued frequency response function; and the only purpose in changing the notation is to suppress the imaginary number i.

The continuous-time frequency-response function can be written in the same manner the discrete-time result of (13.67):

$$\phi(\omega) = |\phi(\omega)|e^{i\theta(\omega)}. \tag{13.119}$$

Here the RHS incorporates the gain $|\phi(\omega)|$ of the filter $\varphi(D)$ and its phase $\theta(\omega)$.

Now consider the matter of the impulse response of the filter. This is

$$f(t) = \varphi(D)\delta(t), \tag{13.120}$$

where $\delta(t)$ is the impulse in the time domain which satisfies the conditions under (13.93). These indicate that the Fourier transform of the impulse function $x(t) = \delta(t)$ is just the constant function $\xi(\omega) = 1$. On substituting the latter into equation (13.118) and putting $y(t) = f(t)$, we get

$$f(t) = \frac{1}{2\pi}\int_{-\infty}^{\infty}\phi(\omega)e^{i\omega t}d\omega. \tag{13.121}$$

This indicates that the impulse response function $f(t)$ is the (inverse) Fourier transform of the frequency response function $\phi(\omega)$. Thus $f(t)$ and $\phi(\omega)$ are a Fourier pair.

A further relationship is established when the function $y(t)$ of (13.118) is expressed as a convolution of functions in the time domain. In the manner of equation (13.82), we may write

$$f(t) * x(t) = \int_{\tau} f(\tau)x(t-\tau)d\tau = \frac{1}{2\pi}\int_{\omega}\phi(\omega)\xi(\omega)e^{i\omega t}d\omega. \tag{13.122}$$

Therefore

$$y(t) = \int_{-\infty}^{\infty} f(\tau)x(t-\tau)d\tau. \tag{13.123}$$

Some additional terminology is also used in describing the attributes of a linear system. Thus, the operator $\varphi(D)$ is often described as the system function. Also, the Laplace transform of the impulse-response function, which is given by

$$\varphi(s) = \int_{0}^{\infty} e^{-st} f(t)dt, \tag{13.124}$$

where $s = \alpha + i\omega$ with $\alpha > 0$, is usually described as the transfer function. The advantage of using the Laplace transform in place of the Fourier transform is that it enables the foregoing analysis to be applied to cases where the input signal $x(t)$ does not fulfil the finite-energy condition of (13.80).

Appendix of Trigonometry

The addition theorems or compound-angle theorems are familiar from elementary trigonometry where they are proved by geometric means. See, for example, Abbot [1]. The theorems are as follows:

$$\begin{aligned} &(a) \quad \cos(A+B) = \cos A\cos B - \sin A\sin B,\\ &(b) \quad \cos(A-B) = \cos A\cos B + \sin A\sin B,\\ &(c) \quad \sin(A+B) = \sin A\cos B + \cos A\sin B,\\ &(d) \quad \sin(A-B) = \sin A\cos B - \cos A\sin B. \end{aligned} \tag{13.125}$$

We can also prove these using Euler's equations from (13.5). Consider, for example, the first equation (a). We have

$$\begin{aligned} \cos(A+B) &= \frac{1}{2}\big\{ \exp\big(i[A+B]\big) + \exp\big(-i[A+B]\big)\big\}\\ &= \frac{1}{2}\big\{ \exp(iA)\exp(iB) + \exp(-iA)\exp(-iB)\big\}\\ &= \frac{1}{2}\big\{(\cos A + i\sin A)(\cos B + i\sin B)\\ &\qquad\qquad + (\cos A - i\sin A)(\cos B - i\sin B)\big\}\\ &= \cos A\cos B - \sin A\sin B. \end{aligned}$$

The other relationships are established likewise.

From the addition theorems, we can directly establish the following sum–product transformations:

$$\begin{aligned} &(a) \quad \sin A\cos B = \frac{1}{2}\big\{ \sin(A+B) + \sin(A-B)\big\},\\ &(b) \quad \cos A\sin B = \frac{1}{2}\big\{ \sin(A+B) - \sin(A-B)\big\},\\ &(c) \quad \cos A\cos B = \frac{1}{2}\big\{ \cos(A+B) + \cos(A-B)\big\},\\ &(d) \quad \sin A\sin B = \frac{1}{2}\big\{ \cos(A-B) - \cos(A+B)\big\}. \end{aligned} \tag{13.126}$$

Orthogonality Conditions

In Fourier analysis, we make use of certain fundamental orthogonality conditions which prevail amongst harmonically related trigonometrical functions. They are as follows:

$$(13.127)\quad \begin{aligned} &(a) \quad \int_0^{2\pi} \cos(jx)\cos(kx)dx = \begin{cases} 0, & \text{if } j \neq k; \\ \pi, & \text{if } j = k > 0; \\ 2\pi, & \text{if } j = k = 0; \end{cases} \\ &(b) \quad \int_0^{2\pi} \sin(jx)\sin(kx)dx = \begin{cases} 0, & \text{if } j \neq k; \\ \pi, & \text{if } j = k > 0; \end{cases} \\ &(c) \quad \int_0^{2\pi} \cos(jx)\sin(kx)dx = 0, \quad \text{for all} \quad j, k. \end{aligned}$$

To prove the results in (13.127)(a), we may use (13.126)(c) to rewrite the integral as

$$\int_0^{2\pi} \cos(jx)\cos(kx)dx = \frac{1}{2}\int_0^{2\pi} \Big\{ \cos([j+k]x) + \cos([j-k]x) \Big\} dx.$$

If $j \neq k$, then both the cosine terms complete an integral number of cycles over the range $[0, 2\pi]$; and therefore they integrate to zero. If $j = k > 0$, then the second cosine becomes unity, and therefore it integrates to 2π over the range $[0, 2\pi]$, whilst the first cosine term integrates to zero. If $j = k = 0$, then both cosine terms become unity and both have integrals of 2π.

The results under (13.127)(b) and (13.127)(c) are also established easily using the relevant results from (13.126).

Bibliography

[1] Abbott, P., (1940), *Teach Yourself Trigonometry*, Teach Yourself Books, English Universities Press, London.

[72] Bracewell, R.N., (1986), *The Fourier Transform and its Applications, Second Edition*, MGraw-Hill, New York.

[98] Carslaw, H.S., (1925), A Historical Note on Gibb's Phenomenon in Fourier Series and Integrals, *Bulletin of the American Mathematical Society*, **31**, 420–424.

[100] Champeney, D.C., (1973), *Fourier Transforms and their Physical Applications*, Academic Press, New York.

[140] De Broglie, L., (1953), *The Revolution in Physics*, (translated by R.W. Niemeyer), Noonday Press, New York.

[141] De Broglie, L., (1955), *Physics and Microphysics*, (translated by M. Davidson), Hutchinson.

[160] Dirac, P.A.M., (1958), *The Principles of Quantum Mechanics, Fourth Edition*, Oxford University Press, Oxford.

[170] Dym, H., and H.P. McKean, (1972), *Fourier Series and Integrals*, Academic Press, New York.

[195] French, A.P., and E.F. Taylor, (1978), *An Introduction to Quantum Physics*, Van Nostrand Reinhold, Wokingham.

[208] Gibbs, J.W., (1898), Fourier's Series: A Letter to the Editor, *Nature*, December 29 1898, **59**, 200.

[209] Gibbs, J.W., (1899), Fourier's Series: A Letter to the Editor, *Nature*, April 27 1899, **58**, 606.

[308] Lanczos, C., (1961), *Linear Differential Operators*, Van Nostrand Co., London.

[309] Lanczos, C., (1966), *Discourse on Fourier Series*, Oliver and Boyd, Edinburgh and London.

[317] Lighthill, M.J., (1958), *Introduction to Fourier Analysis and Generalised Functions*, Cambridge University Press, Cambridge..

[334] Marshall, A.G., and F.R. Verdun, (1990), *Fourier Transforms in NMR, Optical and Mass Spectrometry*, Elsevier Science Publishers, Amsterdam.

[346] Michelson, A.A., (1898), Fourier's Series: A Letter to the Editor, *Nature*, October 6 1898, **58**, 544–545.

[368] Nyquist, H., (1928), Certain Topics in Telegraph Transmission Theory, *AIEE Journal*, **47**, 214–216.

[377] Papoulis, A., (1962), *The Fourier Integral and its Applications*, McGraw-Hill, New York.

[428] Robinson, P.D., (1968), *Fourier and Laplace Transforms*, Routledge Kegan and Paul, London.

[450] Shannon, C.E., (1949), Communication in the Presence of Noise, *Proceedings of the IRE*, **37**, 10–21.

[471] Sneddon, I.N., (1961), *Fourier Series*, Routledge Kegan and Paul, London.

[483] Titchmarsh, E.C., (1939), *Theory of Functions, Second Edition*, Oxford University Press, London.

[484] Titchmarsh, E.C., (1948), *Introduction to the Theory of Fourier Integrals, Second Edition*, Oxford, Clarendon Press.

[508] Weierstrass, K., (1885), *Über die analytische Darstellbarkheit sogenannter willhülicher Functionen einer reellen Veränderlichen*, Berliner Berichte.

[516] Whittaker, J.M., (1945), *Interpolatory Function Theory*, Cambridge Tracts in Mathematics and Mathematical Physics no. 33.

[522] Wiener, N., (1930), Generalised Harmonic Analysis, *Acta Mathematica*, **55**, 117–258.

CHAPTER 14

The Discrete Fourier Transform

In the previous chapter, a classification was established for the Fourier representations of four classes of time-domain functions which depends upon whether the functions are discrete or continuous and on whether they are periodic or aperiodic. The functions which are both discrete and periodic were set aside for a separate and extended treatment which is provided by the present chapter.

A discrete periodic function may be represented completely by a finite sequence which contains one or more complete cycles of the function; and it makes sense to concentrate one's attention upon a single cycle. Equally, any finite sequence may be treated as if it were a single cycle of a periodic sequence. Therefore, in general, periodic sequences and finite sequences are to be regarded as synonymous for the purposes of Fourier analysis; and their Fourier representation entails the so-called discrete Fourier transform (DFT).

The fact that the data for time-series analysis is in the form of finite sequences implies that the DFT is of prime importance. However, finite sequences are also used to approximate other classes of functions in order to make them amenable to digital computation; and this enhances its importance. The efficient algorithm for computing the DFT is called the fast Fourier transform (FFT). This is treated at length in the next chapter

We shall maintain two ways of looking at the DFT. The first is as a device for uncovering hidden periodicities in real-valued data sequences. The business of finding the coefficients of the Fourier representation of a finite sequence can be interpreted as a regression analysis wherein the explanatory variables are the values taken by a variety of sine and cosine functions throughout the sample period. The advantage of this viewpoint is that it provides a context within which to develop an understanding of the DFT which also encompasses the periodogram and the power spectrum.

The second way in which we shall look at the DFT is as a computational realisation of the other kinds of Fourier transform. From this point of view, it is more appropriate to work with the complex-exponential representation of the trigonometrical functions and, ultimately, to replace the real-valued data sequence, which is the subject of the analysis in the first instance, by a complex-valued sequence. One of our concerns is to assess the accuracy of the approximations which are entailed in using the DFT. The accuracy may be affected by the problems of aliasing and leakage which will be discussed at length in the sequel.

Trigonometrical Representation of the DFT

Imagine a sequence $y_t; t = 0, 1, \ldots, T-1$ which comprises T consecutive values of a time series sampled at equal intervals from a continuous process. The object is to uncover any underlying periodic motions which might be present in the time series.

There are two approaches which might be followed. The first of these is to perform a regression analysis in which the explanatory variables are the ordinates of a limited number of sine and cosine functions with carefully chosen frequencies aimed at approximating the frequencies of the underlying motions. The chosen frequencies might be adjusted in a systematic manner until the most appropriate values have been found. A thorough account of the relevant procedure, described as trigonometrical regression, has been provided by Bloomfield [67].

A second approach to uncovering the hidden motions is to employ a maximal set of linearly independent trigonometric functions, with frequencies which are equally spaced throughout the feasible range, with the aim of constructing a net which is sufficiently fine to capture the hidden motions. This approach leads to the Fourier decomposition of the sequence.

In the Fourier decomposition of the sequence $y_t; t = 0, 1, \ldots, T-1$, the values are expressed as

$$\begin{aligned} y_t &= \sum_{j=0}^{n} \left(\alpha_j \cos \omega_j t + \beta_j \sin \omega_j t\right) \\ &= \sum_{j=0}^{n} \left(\alpha_j c_{jt} + \beta_j s_{jt}\right), \end{aligned} \tag{14.1}$$

where $\omega_j = 2\pi j/T$ is a so-called Fourier frequency. The set of scalars α_j, β_j; $j = 0, 1, \ldots, n$ are called the Fourier coefficients. The value of n, which is the limit of the summation, is given by

$$n = \begin{cases} \frac{1}{2}T, & \text{if } T \text{ is even;} \\ \frac{1}{2}(T-1), & \text{if } T \text{ is odd.} \end{cases} \tag{14.2}$$

We shall consider the two cases in succession.

If T is even, then $\omega_j = 0, 2\pi/T, \ldots, \pi$ as $j = 0, 1, \ldots, T/2$. Thus the expression in (14.1) seems to entail $T+2$ functions. However, for integral values of t, it transpires that

$$\begin{aligned} c_0(t) &= \cos 0 = 1, \\ s_0(t) &= \sin 0 = 0, \\ c_n(t) &= \cos(\pi t) = (-1)^t, \\ s_n(t) &= \sin(\pi t) = 0; \end{aligned} \tag{14.3}$$

so, in fact, there are only T nonzero functions; and the expression can be written as

$$y_t = \alpha_0 + \sum_{j=1}^{n-1}(\alpha_j \cos\omega_j t + \beta_j \sin\omega_j t) + \alpha_n(-1)^t. \tag{14.4}$$

If T is odd, then $\omega_j = 0, 2\pi/T, \ldots, \pi(T-1)/T$, and the expression becomes

$$y_t = \alpha_0 + \sum_{j=1}^{n}(\alpha_j \cos\omega_j t + \beta_j \sin\omega_j t); \tag{14.5}$$

so, if the constant function is counted, then, again, T functions are entailed.

The angular velocity $\omega_j = 2\pi j/T$ relates to a pair of trigonometrical functions, $\sin(\omega_j t)$ and $\cos(\omega_j t)$, which accomplish j cycles in the T periods which are spanned by the data. The lowest of the angular velocities is $\omega_1 = 2\pi/T$, which relates to a cosine function which takes T periods to complete one cycle. The highest of the velocities is $\omega_n = \pi$ which relates to a function which takes two periods to complete a cycle. The function $c_n(t) = \cos(\pi t) = (-1)^t$ is present in equation (14.1) only if T is even.

The velocity π corresponds to the so-called Nyquist frequency $f_n = 1/2$. To detect, within the process which has generated the data, any component whose frequency exceeds this value would require a greater acuity of observation than is afforded by the sampling interval. It follows that the effects within the process of the components whose frequencies are greater than the Nyquist value are liable to be confounded with the effects of those whose frequencies fall below it.

Imagine, for example, that the process contains a component which is a pure cosine wave of unit amplitude and zero phase for which the angular frequency ω is in excess of the Nyquist value. Suppose that ω obeys the condition $\pi < \omega < 2\pi$, and define $\omega_* = 2\pi - \omega$. The frequency $\omega_* < \pi$ is below the Nyquist value. For all values of $t = 0, \ldots, T-1$, the following identity holds:

$$\begin{aligned} \cos(\omega t) &= \cos(2\pi t - \omega_* t) \\ &= \cos(2\pi)\cos(\omega_* t) + \sin(2\pi)\sin(\omega_* t) \\ &= \cos(\omega_* t). \end{aligned} \tag{14.6}$$

Therefore the components at the frequencies ω and ω_* are observationally indistinguishable. The pseudo-frequency $\omega_* < \pi$ is described as the alias of $\omega > \pi$. Figure 14.1 provides a graphical illustration of the phenomenon of aliasing.

This result is nothing more than a restatement of the sampling theorem which is to be found under (13.115). In that context, the concern was to determine a rate of sampling sufficient to detect the highest frequency which is present in a continuous-time process. In the present context, the rate of sampling is assumed to be fixed and the concern is to specify the highest detectable frequency—which is the Nyquist frequency.

Example 14.1. For an illustration of the problem of aliasing, let us imagine that a person observes the sea level at 6 a.m. and 6 p.m. each day. He should notice

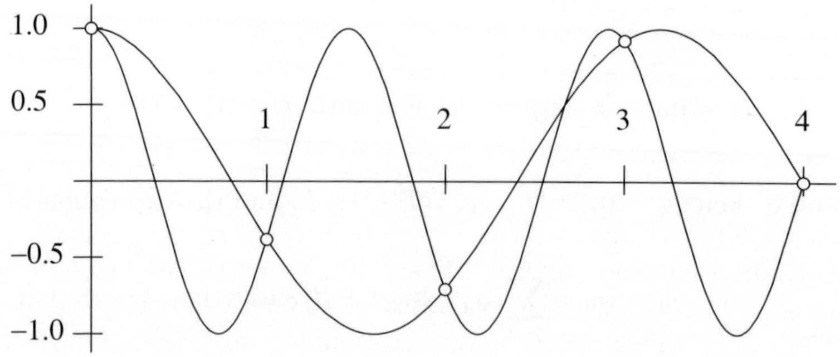

Figure 14.1. The values of the function $\cos\{(11/8)\pi t\}$ coincide with those of its alias $\cos\{(5/8)\pi t\}$ at the integer points $\{t = 0, \pm 1, \pm 2, \ldots\}$.

a very gradual recession and advance of the water level, the frequency of the cycle being $f = 1/28$, which amounts to one tide in 14 days. In fact, the true frequency is $f = 1 - 1/28$, which gives 27 tides in 14 days. Observing the sea level every six hours should enable him to infer the correct frequency.

The nature of the relationship between the sequence of observations and the sequence of Fourier coefficients is clarified by writing the T instances of equation (14.1) in a matrix format. In the case where T is even, these become

$$\begin{bmatrix} y_0 \\ y_1 \\ y_2 \\ \vdots \\ y_{T-2} \\ y_{T-1} \end{bmatrix} = \begin{bmatrix} 1 & 1 & 0 & \ldots & 1 \\ 1 & c_{11} & s_{11} & \ldots & c_{n1} \\ 1 & c_{12} & s_{12} & \ldots & c_{n2} \\ \vdots & \vdots & \vdots & & \vdots \\ 1 & c_{1,T-2} & s_{1,T-2} & \ldots & c_{n,T-2} \\ 1 & c_{1,T-1} & s_{1,T-1} & \ldots & c_{n,T-1} \end{bmatrix} \begin{bmatrix} \alpha_0 \\ \alpha_1 \\ \beta_1 \\ \vdots \\ \beta_{n-1} \\ \alpha_n \end{bmatrix}. \tag{14.7}$$

The equations can be written more easily in a summary notation by separating the sine functions from the cosine functions. Define

$$n_c = \begin{cases} \frac{1}{2}T, & \text{if } T \text{ is even,} \\ \frac{1}{2}(T-1), & \text{if } T \text{ is odd,} \end{cases} \qquad n_s = \begin{cases} \frac{1}{2}T - 1, & \text{if } T \text{ is even,} \\ \frac{1}{2}(T-1), & \text{if } T \text{ is odd.} \end{cases} \tag{14.8}$$

Then equation (14.1) can be written as

$$\begin{aligned} y_t &= \sum_{j=0}^{n_c} \alpha_j \cos \omega_j t + \sum_{j=1}^{n_s} \beta_j \sin \omega_j t \\ &= \sum_{j=0}^{n_c} \alpha_j c_{jt} + \sum_{j=1}^{n_s} \beta_j s_{jt}. \end{aligned} \tag{14.9}$$

In matrix notation, the set of equations for $t = 0, 1, \ldots, T-1$ can be rendered as

$$y = \begin{bmatrix} C & S \end{bmatrix} \begin{bmatrix} \alpha \\ \beta \end{bmatrix}, \tag{14.10}$$

where y is the vector on the LHS of (14.7), where

$$\text{(14.11)}\qquad \begin{aligned} C &= [c_{jt}]; \quad j = 0, 1, \ldots, n_c, \\ S &= [s_{jt}]; \quad j = 1, 2, \ldots, n_s, \end{aligned}$$

and where $\alpha = [\alpha_0, \ldots, \alpha_{n_c}]'$ and $\beta = [\beta_1, \ldots, \beta_{n_s}]'$ are the Fourier coefficients.

The matrix $[C, S]$ of the mapping from the Fourier coefficients to the data values is square and nonsingular; and, in the light of this fact, the mere existence of the Fourier representation under (14.1) seems unremarkable. The Fourier coefficients are recoverable from the data values by solving the equations. This may be done easily in a way which does not require a matrix inversion; and, in fact, the coefficients may be found one at a time.

Determination of the Fourier Coefficients

For heuristic purposes, one can imagine calculating the Fourier coefficients using an ordinary regression procedure to fit equation (14.1) to the data. In this case, there are no regression residuals, for the reason that we are "estimating" a total of T coefficients from T data points; and this is a matter of solving a set of T linear equations in T unknowns.

A reason for not using a multiple regression procedure is that, in this case, the vectors of "explanatory" variables are mutually orthogonal. Therefore T applications of a univariate regression procedure would be sufficient for the purpose.

Let $c_j = [c_{0j}, \ldots, c_{T-1,j}]'$ and $s_j = [s_{0j}, \ldots, s_{T-1,j}]'$ represent vectors of T values of the generic functions $\cos(\omega_j t)$ and $\sin(\omega_j t)$ respectively. Also observe that $s_0 = [0, 0, \ldots, 0]'$, that $c_0 = [1, 1, \ldots, 1]'$ is the summation vector and that the vector associated with the Nyquist frequency, in the case where $T = 2n$ is even, is $c_n = [1, -1, \ldots, 1, -1]'$. Amongst these vectors the following orthogonality conditions hold:

$$\text{(14.12)}\qquad \begin{aligned} c_i'c_j &= 0 \quad \text{if} \quad i \neq j, \\ s_i's_j &= 0 \quad \text{if} \quad i \neq j, \\ c_i's_j &= 0 \quad \text{for all} \quad i, j. \end{aligned}$$

In addition, there are some sums of squares which can be taken into account in computing the coefficients of the Fourier decomposition:

$$\text{(14.13)}\qquad \begin{aligned} c_0'c_0 &= T, \\ s_0's_0 &= 0, \\ \left.\begin{aligned} c_j'c_j &= \tfrac{1}{2}T, \\ s_j's_j &= \tfrac{1}{2}T. \end{aligned}\right\} &\quad \text{for} \quad j = 1, \ldots, n-1. \end{aligned}$$

For $j = n$, there are

$$\text{(14.14)}\qquad \begin{aligned} \left.\begin{aligned} s_n's_n &= \tfrac{1}{2}T, \\ c_n'c_n &= \tfrac{1}{2}T, \end{aligned}\right\} &\quad \text{if} \quad 2n = T - 1, \\ \left.\begin{aligned} s_n's_n &= 0, \\ c_n'c_n &= T, \end{aligned}\right\} &\quad \text{if} \quad 2n = T; \end{aligned}$$

which correspond, respectively, to the cases where T is odd and T is even. These various results are established in an appendix to the chapter.

The "regression" formulae for the Fourier coefficients can now be given. First there is

$$\alpha_0 = (i'i)^{-1}i'y = \frac{1}{T}\sum_t y_t = \bar{y}. \tag{14.15}$$

Then, for $j = 1, \ldots, n-1$, there are

$$\alpha_j = (c_j'c_j)^{-1}c_j'y = \frac{2}{T}\sum_t y_t \cos\omega_j t, \tag{14.16}$$

and

$$\beta_j = (s_j's_j)^{-1}s_j'y = \frac{2}{T}\sum_t y_t \sin\omega_j t. \tag{14.17}$$

Finally, if $T = 2n-1$ is odd, then the formulae above serve for the case where $j = n$. If $T = 2n$ is even, then there is no coefficient β_n and there is

$$\alpha_n = (c_n'c_n)^{-1}c_n'y = \frac{1}{T}\sum_t (-1)^t y_t. \tag{14.18}$$

By pursuing the analogy of multiple regression, it can be seen, in view of the orthogonality relationships, that there is a complete decomposition of the sum of squares of the elements of the vector y which is given by

$$y'y = \alpha_0^2 i'i + \sum_{j=1}^{n_c}\alpha_j^2 c_j'c_j + \sum_{j=1}^{n_s}\beta_j^2 s_j's_j. \tag{14.19}$$

The formulae of this section can be rendered compactly in a matrix notation. Consider the matrix $[C, S]$ of equation (14.10). The conditions under (14.12), (14.13) and (14.14) indicate that this consists of a set of orthogonal vectors. Hence

$$\begin{bmatrix} C'C & C'S \\ S'C & S'S \end{bmatrix} = \begin{bmatrix} \Lambda_c & 0 \\ 0 & \Lambda_s \end{bmatrix}, \tag{14.20}$$

where Λ_c and Λ_s are diagonal matrices. Also

$$\begin{aligned} I &= C(C'C)^{-1}C' + S(S'S)^{-1}S' \\ &= C\Lambda_c^{-1}C' + S\Lambda_s^{-1}S'. \end{aligned} \tag{14.21}$$

The vectors of Fourier coefficients can be rendered as

$$\begin{aligned} \alpha &= (C'C)^{-1}C'y = \Lambda_c^{-1}C'y \quad \text{and} \\ \beta &= (S'S)^{-1}S'y = \Lambda_s^{-1}S'y. \end{aligned} \tag{14.22}$$

The decomposition of (14.19) is then

$$\begin{aligned} y'y &= y'C(C'C)^{-1}C'y + y'S(S'S)^{-1}S'y \\ &= \alpha'\Lambda_c\alpha + \beta'\Lambda_s\beta. \end{aligned} \tag{14.23}$$

The Periodogram and Hidden Periodicities

The variance of the sample $y_0, y_1, \ldots, y_{T-1}$ is given by

$$\begin{aligned}\frac{1}{T}(y'y - i'i\alpha_0^2) &= \frac{1}{T}\sum_{t=0}^{T} y_t^2 - \bar{y}^2 \\ &= \frac{1}{T}\sum_{t=0}^{T}(y_t - \bar{y})^2.\end{aligned} \tag{14.24}$$

This can be used in rearranging equation (14.19). Also, the inner products $c_j'c_j$ and $s_j's_j$ within equation (14.19) can be replaced using the results under (14.13) and (14.14). If T is even, then it will be found that

$$\begin{aligned}\frac{1}{T}\sum_{t=0}^{T}(y_t - \bar{y})^2 &= \frac{1}{2}\sum_{j=1}^{n-1}(\alpha_j^2 + \beta_j^2) + \alpha_n^2 \\ &= \frac{1}{2}\sum_{j=1}^{n-1}\rho_j^2 + \alpha_n^2,\end{aligned} \tag{14.25}$$

whereas, if T is odd, the result will be

$$\begin{aligned}\frac{1}{T}\sum_{t=0}^{T}(y_t - \bar{y})^2 &= \frac{1}{2}\sum_{j=1}^{n}(\alpha_j^2 + \beta_j^2) \\ &= \frac{1}{2}\sum_{j=1}^{n}\rho_j^2.\end{aligned} \tag{14.26}$$

Here $\rho_j = \sqrt{(\alpha_j^2 + \beta_j^2)}$ is the amplitude of the jth harmonic component

$$\begin{aligned}\rho_j\cos(\omega_j t - \theta_j) &= \rho_j\cos\theta_j\cos(\omega_j t) + \rho_j\sin\theta_j\sin(\omega_j t) \\ &= \alpha_j\cos(\omega_j t) + \beta_j\sin(\omega_j t).\end{aligned} \tag{14.27}$$

The RHS of the equations (14.25) and (14.26) stands for the variance of the sample $y_0, \ldots, y_{T-1}$. Thus $\rho_j^2/2$ is the contribution of the jth harmonic to the sample variance; and, indeed, the two equations are tantamount to a classical statistical analysis of variance.

For a further interpretation of equation (14.27), one can imagine that the mean-adjusted data points $y_t - \bar{y}; t = 0, 1, \ldots, T-1$ represent a set of observations on a continuous periodic process of period T. As is established in the Example 13.2, the power which is attributable to the jth harmonic component of the process is

$$\frac{1}{T}\int_0^T \{\rho_j\cos(\omega_j t - \theta_j)\}^2 dt = \frac{1}{2}\rho_j^2. \tag{14.28}$$

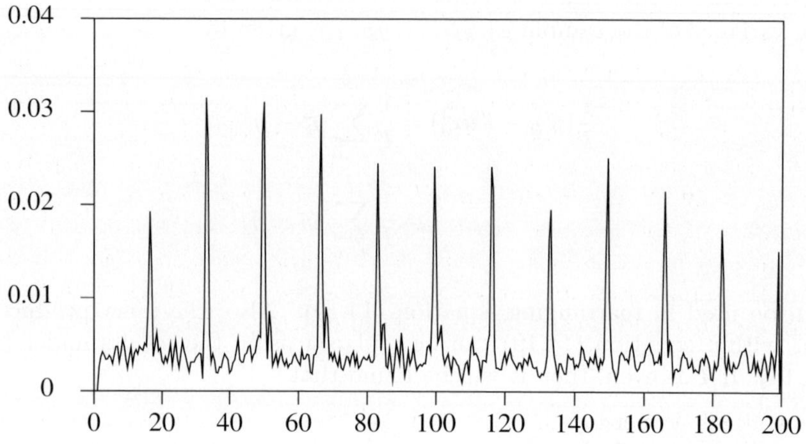

Figure 14.2. The power spectrum of the vibrations transduced from the casing of an electric motor in the process of a routine maintenance inspection. The units of the horizontal axis are hertz. The first peak at 16.6 hertz corresponds to a shaft rotation speed of 1000 rpm. The prominence of its successive harmonics corresponds to the rattling of a loose shaft. (By courtesy of Chris Ward, VISTECH Ltd.)

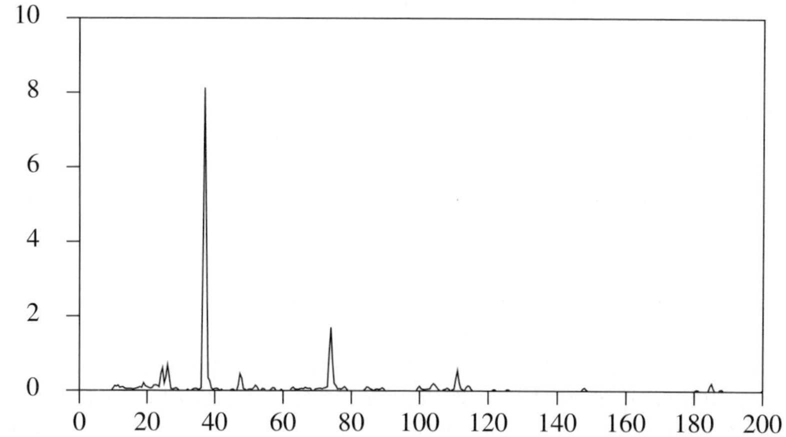

Figure 14.3. A velocity spectrum transduced from a fan installation. The units of the horizontal axis are hertz and the units on the vertical axis are mm/sec. The prominence of the highest peak at the RMS level of 8.3mm/sec indicates that the installation is in an unstable condition. Further measurements indicated a serious problem with the flexibility of the mounting frame. (By courtesy of Chris Ward, VISTECH Ltd.)

Under this construction, the sample variance stands for the power of the periodic process and the assemblage of values $\rho_1^2, \ldots, \rho_n^2$ (with $\rho_n^2 = 2\alpha_n^2$ in the case that T is even) represents its power spectrum.

The process in question may be thought of as a fluctuating voltage signal associated with an electric current. In that case, the mean adjustment represents the removal of a D.C. component. If the current were taken from the mains supply, then, of course, there would be a dominant periodic component at 50 Hz (as in Europe) or at 60 Hz (as in the United States) or thereabouts. In the idealised representation of the mains current, no other components are present; and the mains frequency is apt to be described as the fundamental frequency. If the time span T of the sample were an integral multiple of the period of the mains frequency, then the sample would consist of a number of exact repetitions of a unique cycle.

In a set of empirical observations, one might expect to detect other frequency components in addition to the mains component; and the measured cycles would not be identical. In performing a Fourier analysis of such a record, one proceeds as if the entire record of T observations represents just one cycle of a periodic process; and then it is the frequency of this notional process which is liable to be described as the fundamental frequency.

The case must also be considered where the sample span of T periods does not coincide with with an integral multiple of the period of the mains frequency. Then none of the harmonic Fourier frequencies $\omega_j = 2\pi j/T$ will coincide with the nonharmonic mains frequency; and the power of the mains frequency will be attributed, in various measures, to all of the Fourier frequencies. The nearest of the Fourier frequencies will acquire the largest portion of the power; but, if none of them is close, then an undesirable phenomenon of leakage will occur whereby the power is widely dispersed. This phenomenon will be analysed in a later section.

In a statistical Fourier analysis, the number of the Fourier frequencies increases at the same rate as the sample size T. Therefore, if the variance of the sample remains finite, and if there are no regular harmonic components in the process generating the data, then we can expect the proportion of the variance attributed to the individual frequencies to decline as the sample size increases. If there is such a regular component within the process, then we can expect the proportion of the variance attributable to it to converge to a finite value as the sample size increases.

In order provide a graphical representation of the decomposition of the sample variance, the squared amplitudes of the harmonic components are scaled by a factor of T. The graph of the function $I(\omega_j) = (T/2)(\alpha_j^2 + \beta_j^2)$ is known as the periodogram. We should note that authors differ widely in their choice of a scalar factor to apply to $\alpha_j^2 + \beta_j^2$ in defining the periodogram, albeit that the factor is usually proportional to T.

There are many impressive examples where the estimation of the periodogram has revealed the presence of regular harmonic components in a data series which might otherwise have passed undetected. One of the best-known examples concerns the analysis of the brightness or magnitude of the star T. Ursa Major. It was shown by Whittaker and Robinson [515] in 1924 that this series could be described almost completely in terms of two trigonometrical functions with periods of 24 and 29 days. Figures 14.2 and 14.3 illustrate a common industrial application of periodogram analysis.

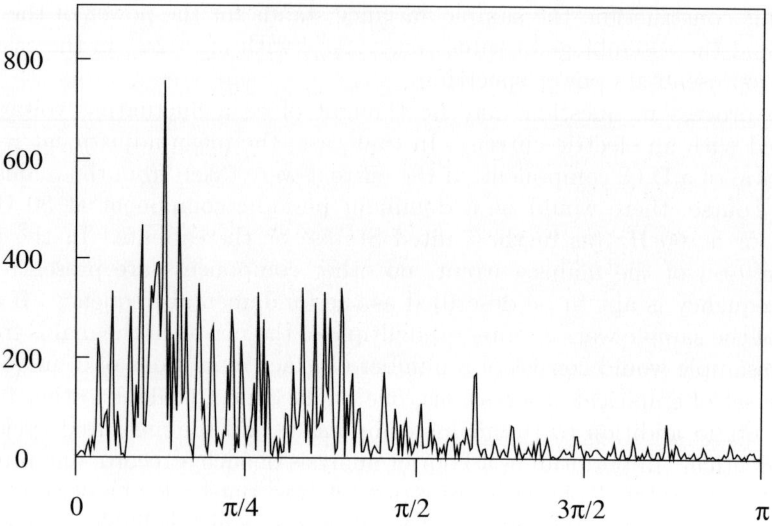

Figure 14.4. The periodogram of the Beveridge's trend-free wheat price index 1500–1869.

The attempts to discover underlying components in economic time-series have been less successful. One application of periodogram analysis which was a notorious failure was its use by William Beveridge [51], [52] in 1921 and 1922 to analyse a long series of European wheat prices. The periodogram had so many peaks that at least twenty possible hidden periodicities could be picked out, and this seemed to be many more than could be accounted for by plausible explanations within the realms of economic history (see Figure 14.4).

Such findings seem to diminish the importance of periodogram analysis in econometrics. However, the fundamental importance of the periodogram is established once it is recognised that it represents nothing less than the Fourier transform of the sequence of empirical autocovariances.

The Periodogram and the Empirical Autocovariances

The periodogram of the sample $y_0, \ldots, y_{T-1}$ is the function

$$\begin{aligned} I(\omega_j) &= \frac{2}{T}\left\{\left[\sum_t y_t \cos(\omega_j t)\right]^2 + \left[\sum_t y_t \sin(\omega_j t)\right]^2\right\} \\ &= \frac{T}{2}\left\{\alpha^2(\omega_j) + \beta^2(\omega_j)\right\}. \end{aligned} \tag{14.29}$$

The ordinates of the periodogram are just the values $\rho_j^2/2$ scaled by a factor of T. This scaling ensures that the magnitude of the ordinates will not diminish as T increases.

The estimate of the autocovariance at lag τ is

(14.30) $$c_\tau = \frac{1}{T}\sum_{t=\tau}^{T-1}(y_t - \bar{y})(y_{t-\tau} - \bar{y}).$$

There is a fundamental theorem which relates the periodogram to the estimated autocovariance function:

(14.31) The *Wiener–Khintchine Theorem* states that

$$\begin{aligned} I(\omega_j) &= 2\left\{c_0 + 2\sum_{\tau=1}^{T-1} c_\tau \cos(\omega_j\tau)\right\} \\ &= 2\left\{c_0 + \sum_{\tau=1}^{T-1}(c_\tau + c_{T-\tau})\cos(\omega_j\tau)\right\}. \end{aligned}$$

Proof. First consider

(14.32) $$\begin{aligned} I(\omega_j) &= \frac{T}{2}\{\alpha^2(\omega_j) + \beta^2(\omega_j)\} \\ &= \frac{T}{2}\{\alpha(\omega_j) - i\beta(\omega_j)\}\{\alpha(\omega_j) + i\beta(\omega_j)\}. \end{aligned}$$

Now

(14.33) $$\begin{aligned} \alpha(\omega_j) - i\beta(\omega_j) &= \frac{2}{T}\sum_t y_t\{\cos(\omega_j t) - i\sin(\omega_j t)\} \\ &= \frac{2}{T}\sum_t (y_t - \bar{y})\{\cos(\omega_j t) - i\sin(\omega_j t)\} \\ &= \frac{2}{T}\sum_t (y_t - \bar{y})e^{-i\omega_j t}, \end{aligned}$$

where the second equality follows from the identity $\sum_t \cos(\omega_j t) = \sum_t \sin(\omega_j t) = 0$, which is given in the appendix under (14.78). Likewise, it can be shown that

(14.34) $$\alpha(\omega_j) + i\beta(\omega_j) = \frac{2}{T}\sum_s (y_s - \bar{y})e^{i\omega_j s},$$

On recombining the conjugate complex numbers and setting $t - s = \tau$, we get

(14.35) $$\begin{aligned} I(\omega_j) &= \frac{2}{T}\sum_{t=0}^{T-1}\sum_{s=0}^{T-1}(y_t - \bar{y})(y_s - \bar{y})e^{-i\omega_j(t-s)} \\ &= 2\sum_{\tau=1-T}^{T-1} c_\tau e^{-i\omega_j\tau} \\ &= 2\left\{c_0 + 2\sum_{\tau=1}^{T-1} c_\tau \cos(\omega_j\tau)\right\}, \end{aligned}$$

which gives the first expression on the RHS of (14.31). Now consider the identity

$$\begin{aligned} \cos(\omega_j\tau) &= \cos(2\pi j\tau/T) \\ &= \cos(2\pi j[T-\tau]/T) \\ &= \cos(\omega_j[T-\tau]). \end{aligned} \tag{14.36}$$

By using this in the final expression of (14.35), the second expression on the RHS of (14.31) can be derived.

The Exponential Form of the Fourier Transform

A concise and elegant formulation of the Fourier representation of a finite sequence is achieved by expressing the trigonometrical functions in terms of complex exponential functions.

Consider again the equation which expresses the value of y_t as a Fourier sum of sines and cosines. If T is even, then $n = T/2$; and the equation takes the form of

$$y_t = \alpha_0 + \sum_{j=1}^{n-1}(\alpha_j \cos\omega_j t + \beta_j \sin\omega_j t) + \alpha_n(-1)^t, \tag{14.37}$$

which was given previously under (14.4). According to Euler's equations,

$$\cos\omega_j t = \frac{1}{2}\left(e^{i\omega_j t} + e^{-i\omega_j t}\right) \quad \text{and} \quad \sin\omega_j t = \frac{-i}{2}\left(e^{i\omega_j t} - e^{-i\omega_j t}\right). \tag{14.38}$$

It follows that

$$\begin{aligned} y_t &= \alpha_0 + \sum_{j=1}^{n-1}\left\{\frac{\alpha_j + i\beta_j}{2}e^{-i\omega_j t} + \frac{\alpha_j - i\beta_j}{2}e^{i\omega_j t}\right\} + \alpha_n(-1)^t \\ &= \sum_{j=1}^{n-1}\zeta_j^* e^{-i\omega_j t} + \zeta_0 + \sum_{j=1}^{n-1}\zeta_j e^{i\omega_j t} + \zeta_n(-1)^t, \end{aligned} \tag{14.39}$$

where we have defined

$$\begin{aligned} \zeta_j &= \frac{\alpha_j - i\beta_j}{2}, & \zeta_j^* &= \frac{\alpha_j + i\beta_j}{2}, \\ \zeta_0 &= \alpha_0 \quad \text{and} & \zeta_n &= \alpha_n. \end{aligned} \tag{14.40}$$

Equation (14.39) can be written more compactly by gathering its terms under a single summation sign. For this purpose, a set of negative frequencies $\omega_{-j} = -\omega_j$ are defined for $j = 1, \ldots, n$. Then, setting $\zeta_{-j} = \zeta_j^*$ for $j = 1, \ldots, n$ gives

$$y_t = \sum_{j=1-n}^{n}\zeta_j e^{i\omega_j t}, \quad \text{where} \quad n = \frac{T}{2} \quad \text{and} \quad T \quad \text{is even.} \tag{14.41}$$

In comparing this expression with the equation (14.39), it should be recognised that the terms associated with ζ_0 and ζ_n are $\exp\{i\omega_0 t\} = 1$ and $\exp\{i\omega_n t\} = \exp\{i\pi t\} = (-1)^t$ respectively.

When T is odd, there is $n = (T-1)/2$; and equation (14.37) is replaced by

(14.42) $$y_t = \alpha_0 + \sum_{j=1}^{n}(\alpha_j \cos\omega_j t + \beta_j \sin\omega_j t),$$

which has been given previously under (14.5). On substituting into this the expressions for $\cos\omega_j t$ and $\sin\omega_j t$ from (14.38), and by using the definitions

(14.43) $$\zeta_n = \frac{\alpha_n - i\beta_n}{2}, \qquad \zeta_{-n} = \frac{\alpha_n + i\beta_n}{2},$$

we find that we can write

(14.44) $$y_t = \sum_{j=-n}^{n} \zeta_j e^{i\omega_j t}, \quad \text{where} \quad n = \frac{T-1}{2} \quad \text{and} \quad T \quad \text{is odd}.$$

If defining negative frequencies seems to conflict with intuition, then an alternative way of compacting the expressions is available which extends the range of the positive frequencies. Consider the fact that the exponential function $\exp(-i\omega_j) = \exp(-i2\pi j/T)$ is T-periodic in the index j. This implies that

(14.45) $$\exp(i\omega_{-j}) = \exp(i\omega_{T-j}).$$

By using this identity and by taking $\zeta_{T-j} = \zeta_{-j} = \zeta_j^*$, we can rewrite both equation (14.41) and equation (14.44) as

(14.46) $$y_t = \sum_{j=0}^{T-1} \zeta_j e^{i\omega_j t}.$$

In this sum, there are frequencies which exceed the Nyquist value; and this might also conflict with intuition. However, such problems with intuition arise only if we expect the complex exponential functions to behave in the same way as the real trigonometrical functions which are their parents.

The equations of (14.46) may be written in a matrix format similar to that of equation (14.7) which represents the Fourier transform in terms of trigonometrical functions rather than complex exponentials. For this purpose, we adopt the traditional notation $W_T = \exp(-2\pi/T)$ so that the exponential expression within equation (14.46) becomes

(14.47) $$\begin{aligned}\exp(i\omega_j t) &= \exp(2\pi j t/T)\\ &= \left(W_T^{-j}\right)^t.\end{aligned}$$

Then, if the subscript T in W_T is suppressed, the T instances of equation (14.46) with $t = 0, \ldots, T-1$ can be written as

$$\begin{bmatrix} y_0 \\ y_1 \\ y_2 \\ \vdots \\ y_{T-1} \end{bmatrix} = \begin{bmatrix} 1 & 1 & 1 & \ldots & 1 \\ 1 & W^{-1} & W^{-2} & \ldots & W^{1-T} \\ 1 & W^{-2} & W^{-4} & \ldots & W^{2(1-T)} \\ \vdots & \vdots & \vdots & & \vdots \\ 1 & W^{1-T} & W^{2(1-T)} & \ldots & W^{-(T-1)^2} \end{bmatrix} \begin{bmatrix} \zeta_0 \\ \zeta_1 \\ \zeta_2 \\ \vdots \\ \zeta_{T-1} \end{bmatrix}. \tag{14.48}$$

A way can also be found of representing the value of ζ_j in terms of the values $y_t; t = 0, \ldots, T-1$ which is as parsimonious as the expression under (14.46). Consider

$$\zeta_j = \frac{1}{T}\left\{\sum_t y_t \cos\omega_j t - i\sum_t y_t \sin\omega_j t\right\} \tag{14.49}$$

which comes from substituting the expressions

$$\alpha_j = \frac{2}{T}\sum_t y_t \cos\omega_j t \quad \text{and} \quad \beta_j = \frac{2}{T}\sum_t y_t \sin\omega_j t, \tag{14.50}$$

given previously under (14.16) and (14.17), into the expression for ζ_j given under (14.40). By using Euler's equations, this can be rewritten as

$$\zeta_j = \frac{1}{T}\sum_{t=0}^{T-1} y_t e^{-i\omega_j t}. \tag{14.51}$$

The latter represents the inverse of the transformation in (14.46). The relationship is confirmed by writing

$$\begin{aligned} \zeta_j &= \frac{1}{T}\sum_{t=0}^{T-1}\left\{\sum_{k=0}^{T-1}\zeta_k e^{i\omega_k t}\right\} e^{-i\omega_j t} \\ &= \frac{1}{T}\sum_{k=0}^{T-1}\zeta_k\left\{\sum_{t=0}^{T-1} e^{i(\omega_k - \omega_j)t}\right\} \end{aligned} \tag{14.52}$$

and by recognising that

$$\sum_{t=0}^{T-1} e^{i(\omega_k - \omega_j)t} = T\delta_{kj}$$

takes the value of T if $k = j$ and the value of zero otherwise.

In terms of the notation under (14.47), the equation of (14.51) for $t = 0, \ldots, T-1$ can be written as

$$\begin{bmatrix} \zeta_0 \\ \zeta_1 \\ \zeta_2 \\ \vdots \\ \zeta_{T-1} \end{bmatrix} = \frac{1}{T}\begin{bmatrix} 1 & 1 & 1 & \ldots & 1 \\ 1 & W^{1} & W^{2} & \ldots & W^{T-1} \\ 1 & W^{2} & W^{4} & \ldots & W^{2(T-1)} \\ \vdots & \vdots & \vdots & & \vdots \\ 1 & W^{T-1} & W^{2(T-1)} & \ldots & W^{(T-1)^2} \end{bmatrix} \begin{bmatrix} y_0 \\ y_1 \\ y_2 \\ \vdots \\ y_{T-1} \end{bmatrix}. \tag{14.53}$$

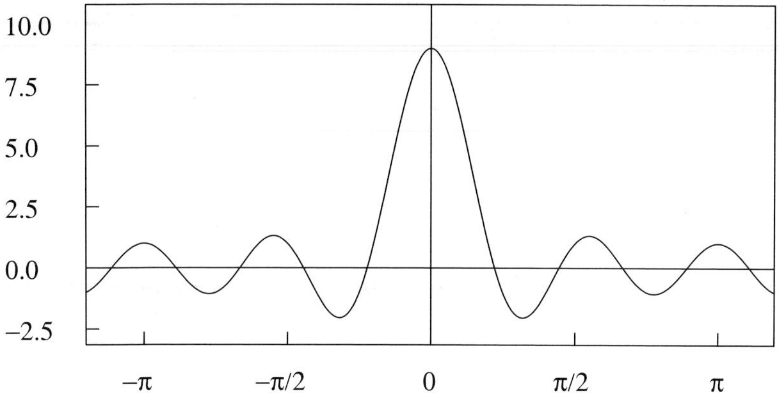

Figure 14.5. The function $\sin(T\lambda/2)/\{T\sin(\lambda/2)\}$ has zeros at the values $\lambda = 2\pi j/T$ where j/T are fractional values.

Leakage from Nonharmonic Frequencies

The complex exponential representation of the Fourier transform facilitates an understanding of the phenomenon of leakage. This occurs whenever the signal which is to be analysed contains a nonharmonic component which does not complete an integral number of cycles in the time spanned by the data. Imagine that the component in question is a simple cosine function

$$\cos(\bar{\omega}t - \theta) = \frac{1}{2}\Big[\exp\big\{i(\bar{\omega}t - \theta)\big\} + \exp\big\{-i(\bar{\omega}t - \theta)\big\}\Big]. \tag{14.54}$$

In that case, it is sufficient to consider the Fourier transform of $\exp(i\bar{\omega}t)$ alone, since that of $\exp(i\theta)$ represents only a scale factor. The transform of the sequence $\{y_t = \exp(i\bar{\omega}t); t = 0, \ldots, T-1\}$ is a sequence whose jth coefficient is given by equation (14.51). By means of the manipulations which have given rise to equation (13.73), an expression for the coefficient is found in the form of

$$\begin{aligned}\zeta_j &= \frac{1}{T}\sum_{t=0}^{T-1}\exp\big\{i(\bar{\omega} - \omega_j)t\big\}\\ &= \frac{\sin\{T(\bar{\omega} - \omega_j)/2\}}{T\sin\{(\bar{\omega} - \omega_j)/2\}}\exp\left\{\frac{i(T-1)(\bar{\omega} - \omega_j)}{2}\right\}.\end{aligned} \tag{14.55}$$

Here the first equality delivers a value of unity if $\bar{\omega} = \omega_j$, whilst the second equality presupposes that $\bar{\omega} \neq \omega_j$.

Imagine that $\bar{\omega} = \omega_k = 2\pi k/T$, which is to say that the frequency of the component coincides with a Fourier frequency. Then, if $j = k$, the transform takes the value of unity, whereas, if $j \neq k$, it takes the value of zero. Therefore the power of the component is attributed entirely to the appropriate Fourier frequency.

If $\bar{\omega}$ is not a Fourier frequency, then the transform will assume nonzero values at other Fourier frequencies in addition to the Fourier frequency which is closest to $\bar{\omega}$. The appearance of nonzero terms in the transform over the entire range of Fourier frequencies is described as the phenomenon of leakage; and its effect is to give a misleading impression of the dispersion of the power of a signal which may, in truth, be concentrated on a single frequency or on a limited number of frequencies.

The function $\delta(\lambda) = \sin(T\lambda/2)/\{T\sin(\lambda/2)\}$, which is entailed in equation (14.55), is an instance of the Dirichlet kernel. (Its graph is plotted in Figure 14.5.) Its argument is $\lambda = \omega - \omega_j$; and its zeros are the nonzero values $\lambda_j = 2\pi j/T$, other than integer multiples of 2π, which are separated by the distance between adjacent Fourier frequencies. The ordinates of the function at the points $\lambda = \bar{\omega} - \omega_j$ give the amplitudes of the components entailed in the Fourier decomposition of the sequence $\exp(i\bar{\omega}t)$; and, as the value of $|\bar{\omega} - \omega_j|$ increases towards π, these amplitudes decline. The amplitudes of all but the leading components are smaller the closer $\bar{\omega}$ is to a Fourier frequency and the larger is the value of T. Also, as T increases, the bandwidth spanned by adjacent Fourier frequencies diminishes, which means that the phenomenon of leakage becomes more localised.

The Fourier Transform and the z-Transform

It may be helpful to present some of the foregoing results in the notation of the z-transform. Consider the z-transforms of the data sequence $y_t; t = 0, 1, \ldots, T-1$ and of the sequence of the corresponding Fourier coefficients $\zeta_j; t = 0, 1, \ldots, T-1$. Recall that the Fourier transform of a real sequence is a complex sequence in which the real part is even and the imaginary part is odd. This implies that $\zeta_{T-j} = \zeta_j^*$. The two z-transforms are

$$\zeta(z) = \sum_{j=0}^{T-1} \zeta_j z^j, \tag{14.56}$$

$$y(z) = \sum_{t=0}^{T-1} y_t z^t. \tag{14.57}$$

Now set $z = W^{-t}$ in (14.56), and set $z = W^j$ in (14.57) and divide the latter by T. The results are

$$y_t = \zeta(W^{-t}) = \sum_{j=0}^{T-1} \zeta_j (W^{-t})^j, \tag{14.58}$$

$$\zeta_j = \frac{1}{T} y(W^j) = \frac{1}{T} \sum_{t=0}^{T-1} y_t (W^j)^t. \tag{14.59}$$

These are just the generic expressions for the equations comprised by the matrix systems of (14.48) and (14.53), respectively, which stand for the discrete Fourier transform and its inverse.

Now consider the two autocovariance generating functions defined by

$$c(z) = \sum_{\tau=1-T}^{T-1} c_\tau z^\tau = c_0 + \sum_{\tau=1}^{T-1} \left(c_\tau z^\tau + c_{T-\tau} z^{\tau-T}\right), \tag{14.60}$$

$$c^{\circ}(z) = \sum_{\tau=0}^{T-1} c^{\circ}_{\tau} z^{\tau} = c_0 + \sum_{\tau=1}^{T-1} (c_{\tau} + c_{T-\tau}) z^{\tau}. \tag{14.61}$$

The first of these is the generating function for the *ordinary autocovariances* $\{c_{\tau}; \tau = 0, \pm 1, \ldots, \pm T-1\}$, whereas the second is the generating function for the *circular autocovariances* $\{c^{\circ}_{\tau}; \tau = 0, 1, \ldots, T-1\}$ which are defined by $c^{\circ}_0 = c_0$ and $c^{\circ}_{\tau} = c_{\tau} + c_{T-\tau}; \tau = 1, \ldots, T-1$. The ordinary autocovariances satisfy the condition that $c_{\tau} = c_{-\tau}$, whereas the circular autocovariances, which are elements of a periodic sequence of period T, satisfy the condition that $c^{\circ}_{\tau} = c^{\circ}_{T-\tau}$. When $z = W^j$, the argument is T-periodic such that $(W^j)^{\tau - T} = (W^j)^{\tau}$. It follows that

$$\sum_{\tau=1-T}^{T-1} c_{\tau}(W^j)^{\tau} = \sum_{\tau=0}^{T-1} c^{\circ}_{\tau}(W^j)^{\tau}. \tag{14.62}$$

Now observe that, when $j \neq 0$, the condition $\sum_t (W^j)^t = 0$ holds. Therefore equation (14.59) can be written as

$$\zeta_j = \frac{1}{T} \sum_{t=0}^{T-1} (y_t - \bar{y})(W^j)^t. \tag{14.63}$$

It follows that

$$\begin{aligned} T\zeta_j\zeta_j^* &= \frac{1}{T}\left\{\sum_{t=0}^{T-1}(y_t - \bar{y})(W^j)^t\right\}\left\{\sum_{s=0}^{T-1}(y_s - \bar{y})(W^j)^{-s}\right\} \\ &= \sum_{\tau=1-T}^{T-1}\left\{\frac{1}{T}\sum_{t=\tau}^{T-1}(y_t - \bar{y})(y_{t-\tau} - \bar{y})\right\}(W^j)^{\tau} \\ &= \sum_{t=1-T}^{T-1} c_{\tau}(W^j)^{\tau}, \end{aligned} \tag{14.64}$$

wherein $\tau = t - s$.

Equation (14.64) is the basis of the Wiener–Khintchine theorem of (14.31) which relates the sequence of periodogram ordinates $\{I_j\}$ to the sequence of autocovariances $\{c_{\tau}\}$. This relationship is demonstrated anew by observing, in reference to (14.29) and (14.40), that

$$I_j = \frac{T}{2}\{\alpha_j^2 + \beta_j^2\} = 2T\zeta_j\zeta_j^*. \tag{14.65}$$

There are altogether T distinct ordinary autocovariances in the set $\{c_{\tau}; \tau = 0, 1, \ldots, T-1\}$, whereas there are only n ordinates of the periodogram $\{I_j, j = 0, 1, \ldots, n\}$ corresponding to n distinct frequency values, where n is defined by (14.2). Therefore the ordinary autocovariances cannot be recovered from the

$$
\begin{array}{ccc}
\{c^\circ_\tau\} & \xleftrightarrow{DFT} & \{I_j\} \\
\uparrow \text{acv.} & & \uparrow \text{mod.} \\
\{y_t\} & \xleftrightarrow{DFT} & \{\zeta_j\}
\end{array}
$$

Figure 14.6. The relationship between time-domain sequences and the frequency-domain sequences.

periodogram. However, consider an extended periodogram sequence $\{I_j; j = 0, \ldots T-1\}$ defined over T points, instead of only n points, which is real-valued and even with $I_{T-j} = I_j$. Then it is manifest that there is a one-to-one relationship between the periodogram ordinates and the sequence of *circular* autocovariances which can be expressed in terms of a discrete Fourier transform and its inverse defined on T points:

$$
(14.66) \qquad I_j = 2\sum_{\tau=0}^{T-1} c^\circ_\tau \left(W^j\right)^\tau,
$$

$$
(14.67) \qquad c^\circ_\tau = \frac{1}{2T}\sum_{j=0}^{T-1} I_j \left(W^{-\tau}\right)^j.
$$

As we have already observed, the circular autocovariances $\{c^\circ_j; j = 0, \ldots, T-1\}$ also constitute a sequence which is real and even. Thus the two sequences fulfil one of the symmetry relationships between Fourier pairs which is detailed in Table 13.2.

The relationships amongst the various sequences which have been considered in this section are depicted in Figure 14.6.

The Classes of Fourier Transforms

The nature of a time-domain function or signal determines the nature of its Fourier transform, which is a function in the frequency domain. The signal is classified according to whether it is discrete or continuous—i.e. its continuity attribute—and according to whether it is periodic or aperiodic—i.e. its periodicity attribute. Its Fourier transform is classified likewise.

The manner in which the class of the signal corresponds to that of its transform is indicated in Table 13.1 which is to be found at the start of the previous chapter. The fourfold classification of the table is based upon the time-domain attributes of the signal. The frequency-domain attributes of the corresponding transforms, which are revealed in the course of Chapter 13, are represented by annotations within the body of the table.

The information of the table is reproduced in Table 14.1. Here, the Fourier transformations are classified in terms of their twofold attributes in both domains. The class of a transformation is specified equally by a pair of time-domain attributes or by a pair of frequency-domain attributes. It is also specified by declaring the periodicity attributes of the transformation in both domains or by declaring its continuity attributes. The various Fourier transforms are illustrated in Figure 14.7.

The Fourier integral: Ref. (13.78), (13.79)

$$x(t) = \frac{1}{2\pi}\int_{-\infty}^{\infty} \xi(\omega)e^{i\omega t}d\omega \quad \longleftrightarrow \quad \xi(\omega) = \int_{-\infty}^{\infty} x(t)e^{-i\omega t}dt$$

The classical Fourier series: Ref. (13.14), (13.15)

$$x(t) = \sum_{j=-\infty}^{\infty} \xi_j e^{i\omega jt} \quad \longleftrightarrow \quad \xi_j = \frac{1}{T}\int_0^T x(t)e^{-i\omega jt}dt$$

The discrete-time Fourier transform: Ref. (13.52), (13.53)

$$x_t = \frac{1}{2\pi}\int_{-\pi}^{\pi} \xi(\omega)e^{i\omega t}d\omega \quad \longleftrightarrow \quad \xi(\omega) = \sum_{t=-\infty}^{\infty} x_t e^{-i\omega t}$$

The discrete Fourier transform: Ref. (14.46), (14.51)

$$x_t = \sum_{j=0}^{T-1} \xi_j e^{i\omega_j t} \quad \longleftrightarrow \quad \xi_j = \frac{1}{T}\sum_{t=0}^{T-1} x_t e^{-i\omega_j t}$$

Figure 14.7. The classes of the Fourier transforms.

Table 14.1. The classes of Fourier transformations*

	Aperiodic in frequency Continuous in time	Periodic in frequency Discrete in time
Aperiodic in time **Continuous in frequency**	*Fourier integral*	*Discrete-time FT*
Periodic in time **Discrete in frequency**	*Fourier series*	*Discrete FT*

* Each cell of the table contains the name of a transform which is the product of a Fourier transformation mapping from the time domain to the frequency domain. The nature of the transform is determined by the nature of the signal (i.e., the time-domain function)—which is continuous or discrete, and periodic or aperiodic.

We may regard the signals which are continuous and aperiodic as the most general class of functions within the time domain. Such functions are defined over the entire domain $\{t \in (-\infty, \infty)\}$. A time-limited function is one which takes nonzero values only over a finite interval of this domain. The transform of a continuous aperiodic signal, which is obtained via the *Fourier integral,* is also continuous and aperiodic; and it is defined over the entire frequency domain $\{\omega \in (-\infty, \infty)\}$. The transform is said to be band-limited if it is nonzero only over a finite frequency interval. A time-limited signal cannot have a band-limited transform.

Whereas the Fourier integral is the dominant transformation in a theoretical perspective, it is the *discrete Fourier transform* (DFT), together with its realisation via the fast Fourier transform (FFT), which is used in practice for computing all classes of Fourier transforms. In order to reduce a time-domain function which is continuous and aperiodic, and which would be subject to the Fourier integral, to one which is discrete and finite and which is amenable to the DFT, two modifications are necessary. The first modification entails a process of time-domain sampling which converts the function from a continuous one to a discrete one. The second modification is the truncation of the signal which serves to restrict it to a finite interval. In the following sections, we shall examine the effects of these two modifications.

Sampling in the Time Domain

By the process of time-domain sampling, a continuous function of time is reduced to a discrete sequence. A temporal sequence possesses a *discrete-time Fourier transform,* which is a continuous periodic function. Thus the act of sampling a signal is accompanied by the repeated copying of its transform at regular frequency intervals so as to create a periodic function. The length of the frequency-domain intervals is inversely related to the frequency of the time-domain sampling. If the sampling is not sufficiently rapid, then the copying of the transform at close intervals along the frequency axis will lead to an overlapping which constitutes the problem of aliasing. These results concerning time-domain sampling have been established already in the previous chapter under the heading of the sampling theorem—see (13.115). Nevertheless, it seems worthwhile to reiterate them here.

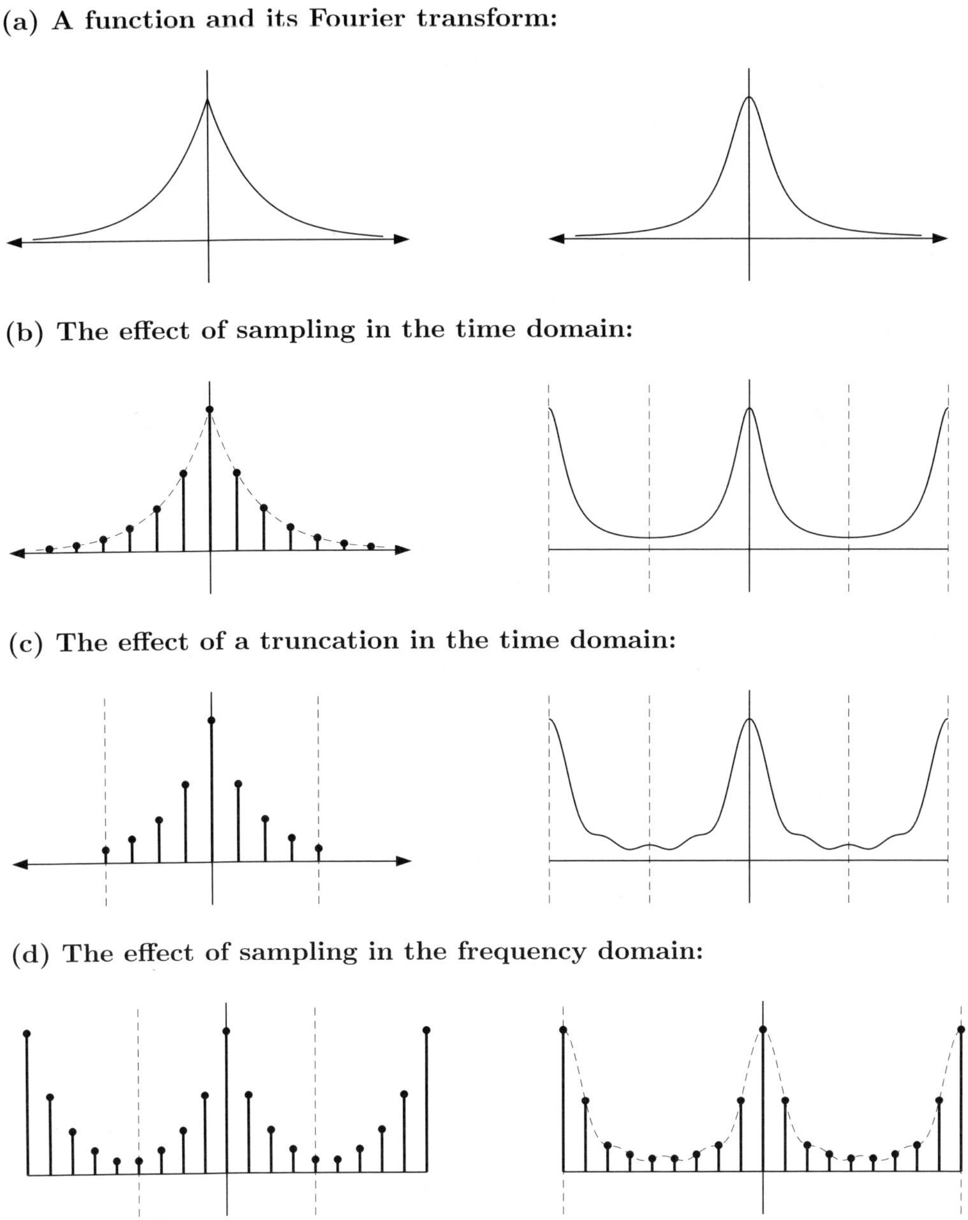

Figure 14.8. The effects of sampling and truncation. The time-domain functions appear on the left and their Fourier transforms, which are functions in the frequency domain, appear on the right.

In the previous chapter, a representation of a sampled signal was given which is based upon the notion of an impulse train. The impulse train is represented by

$$g(t) = \sum_{j=-\infty}^{\infty} \delta(t - jT_0), \tag{14.68}$$

where $\delta(t)$ is Dirac's delta function and where T_0 is the interval between the pulses. The Fourier transform of a time-domain impulse train is an impulse train in the frequency domain which is specified by

$$\gamma(\omega) = \omega_0 \sum_{j=-\infty}^{\infty} \delta(\omega - j\omega_0), \tag{14.69}$$

where $\omega_0 = 2\pi/T_0$ is the so-called sampling frequency, which is the distance in radians which separates successive pulses. The sampled version of the signal $x(t)$ is the modulated signal

$$x_s(t) = x(t)g(t) = \sum_{j=-\infty}^{\infty} x(t)\delta(t - jT_0), \tag{14.70}$$

which is obtained by multiplying together the original signal and the impulse train.

The multiplication of two signals in the time domain corresponds to the convolution of their transforms in the frequency domain. Thus, if $x(t)$ and $\xi(\omega)$ are the signal and its Fourier transform, then the relationship between the sampled signal and its transform is represented by writing

$$x_s(t) = x(t)g(t) \longleftrightarrow \gamma(\omega) * \xi(\omega) = \xi_s(\omega). \tag{14.71}$$

Using this relationship, it has been shown under (13.113) that the Fourier transform of the sampled signal is given by

$$\xi_s(\omega) = \frac{1}{T_0} \sum_{j=-\infty}^{\infty} \xi(\omega - j\omega_0). \tag{14.72}$$

Thus it transpires that the transform of the sampled signal is a periodic function $\xi_s(\omega)$ constructed from copies of the original function $\xi(\omega)$ superimposed at equal intervals separated by ω_0 radians. These copies will overlap each other unless $\xi(\omega) = 0$ for all $|\omega| \geq \frac{1}{2}\omega_0$—see Figure 14.8(b). The consequence is that the values of $\xi_s(\omega)$ over the interval $[-\omega_0, \omega_0]$ will represent a sum of the values of the original function plus the values of the tails of the shifted functions functions $\xi(\omega \pm j\omega_0); j = \{0, 1, 2, \ldots\}$ which extend into this interval.

In virtually all time-series applications, we can afford to set $T_0 = 1$; which is to say that the sampling interval can be taken as the unit interval. In that case, the sampling frequency is 2π; and the condition that no aliasing occurs is the condition that the signal contains no components with frequencies equal to or in excess of the Nyquist frequency of π.

Truncation in the Time Domain

The second modification which is applied to the continuous-time signal in order to make it amenable to digital computation is that of truncation. The truncation severs the tails of the function which lie outside the finite interval; and, if these tails form a significant part of the original function, then the effect upon the transform of the function is also liable to be significant.

Recall that there is an inverse relationship between the dispersion of a signal in the time domain and the dispersion of its transform in the frequency domain. Imagine that the signal $x_s(t) = g(t)x(t)$ which is to be processed is a infinite sequence which has been sampled from a continuous function $x(t)$ with a band-limited transform. The truncated version of the signal, which is time-limited, cannot have a band-limited transform. In effect, the truncation of the signal must be accompanied by the spreading of its transform over the entire frequency domain. This extension of the dispersion of the transform is described as the problem of leakage.

The truncation of a continuous-time signal is achieved formally by multiplying the signal by a window function defined by

$$(14.73) \qquad k(t) = \begin{cases} 1, & \text{if } t \in [a, a+T); \\ 0, & \text{otherwise,} \end{cases}$$

where $T = nT_0$ is the length of a period which spans n sample points. The truncation of the sampled signal $x_s(t) = x(t)g(t)$ will give rise to the function $x_k(t) = x(t)g(t)k(t)$. Notice that the factors on the RHS of this expression can be written in any order. The implication is that, in mathematical terms, the order in which the operations of windowing and sampling are performed is immaterial.

The window function, which can be regarded as a rectangular pulse, has the sinc function of (13.86) as its Fourier transform. Therefore the Fourier transform of the windowed or truncated version of the sampled signal $x_s(t)$ is the following convolution:

$$(14.74) \qquad \xi_k(\omega) = \frac{1}{\pi\omega}\int_{-\infty}^{\infty} \xi_s(\lambda)\text{sinc}\left(\frac{T}{2}\omega - \lambda\right) d\lambda.$$

At first sight, it appears that, in general, an expression such as this requires to be evaluated by numerical integration. The difficulty stems from the nature of the quasi-continuous function $x_s(t)$ which has been obtained from the original signal $x(t)$ by means of Dirac's delta function. When $x_s(t)$ is replaced by an ordinary sequence $\{x_t; t = 0 \pm 1 \pm 2, \ldots\}$, the difficulty vanishes. In that case, the function $\xi_k(\omega)$ can be replaced by the readily accessible discrete-time Fourier transform of the sequence $\{x_t k_t\}$, wherein k_t takes the value of unity for $t = 0, 1, \ldots, n-1$ and the value of zero elsewhere.

The Fourier transform of the sequence $\{x_t k_t\}$, whose elements are the products of those of $\{x_t\}$ and $\{k_t\}$, is just the (frequency-domain) convolution of the respective Fourier transforms, $\xi_s(\omega)$ and $\kappa(\omega)$:

$$(14.75) \qquad \xi_k(\omega) = \frac{1}{2\pi}\int_{-\pi}^{\pi} \xi_s(\lambda)\kappa(\omega - \lambda) d\lambda.$$

The function $\kappa(\omega)$ is, in fact, the Dirichlet kernel defined under (13.75). The number n of sample points falling within the truncation interval is the number of consecutive nonzero values of $\{k_t\}$. When n is large, the aperiodic sinc function, which is found in equation (14.74), is closely approximated, over the appropriate interval, by a single cycle of the periodic Dirichlet kernel.

The effect of the convolution of (14.75) is to smear and spread the profile of $\xi_s(\omega)$ so that any peaks which it might posses will lose some of their prominence. The convolution will also induce ripples in the profile—see Figure 14.8(c). The classical example of these effects is provided by Gibbs' phenomenon which has been described in the previous chapter and which is represented by Figure 13.2. In its original context, the phenomenon arises from the truncation of the sequence of the coefficients entailed in the Fourier-series representation of a periodic square wave. To apply the example to the present context, we need only interchange the frequency domain and the time domain so that the Fourier coefficients are re-interpreted as a sequence of points in the time domain and their transform becomes a periodic function in the frequency domain.

We may note that, in the limit as $n \to \infty$, the Dirichlet kernel becomes Dirac's delta function. In that case, it follows from the so-called *sifting property* of the delta function—see (13.94)—that the convolution would deliver the function $\xi_s(\omega)$ unimpaired.

Sampling in the Frequency Domain

The effect of applying the operations of sampling and truncation to a continuous signal $x(t)$ is to deliver a discrete-time signal sequence $\{x_t\}$, which is nonzero only over a finite interval, and which, in theory, possesses a continuous periodic transform $\xi_k(\omega)$ which is described as a discrete-time Fourier transform (DTFT)—see Figure 14.8(c). However, the discrete Fourier transformation or DFT, which is to be applied in practice to the signal sequence, has been represented as a mapping from one finite (or periodic) sequence in the time domain to another equivalent finite (or periodic) sequence in the frequency domain.

This apparent discrepancy is easily resolved by applying a notional process of frequency-domain sampling to the continuous Fourier transform $\xi_k(\omega)$ so as to obtain a discrete version $\xi_r(\omega_j)$—see Figure 14.8(d). An accompanying effect of this sampling process is to generate the periodic extension $x_r(t)$ of the finite signal sequence $\{x_t\}$. Now there exists a one-to-one relationship $x_r(t) \longleftrightarrow \xi_r(\omega_j)$ between two discrete periodic sequences each comprising the same number of elements per cycle; and such a relationship is the object of the DFT, which is depicted under (14.48), and its inverse, which is depicted under (14.53).

To demonstrate that frequency sampling need have no substantive effect, let us consider expressing the sampled transform as

$$\xi_r(\omega) = \gamma(\omega)\xi_k(\omega), \tag{14.76}$$

where $\gamma(\omega)$ is the impulse train in the frequency domain specified by (14.69). The time-domain function $x_r(t)$ corresponding to $\xi_r(\omega)$ is obtained from the convolution of $x(t)$, which is the ordinary extension of $\{x_t\}$ defined according to (2.1), with the time-domain impulse train $g(t)$ of (14.68). Thus

$$\begin{aligned}x_r(t) &= \int_\tau x(t-\tau)\Big\{\sum_{j=-\infty}^{\infty}\delta(\tau - jT_0)\Big\}d\tau \\ &= \sum_{j=-\infty}^{\infty}\Big\{\int_{-\infty}^{\infty} x(t-\tau)\delta(\tau - jT_0)d\tau\Big\} \\ &= \sum_{j=-\infty}^{\infty} x(t - jT_0).\end{aligned} \tag{14.77}$$

The final expression indicates that $x_r(t)$, is a periodic function consisting of repeated copies of $x(t)$ separated by a time interval of T_0. Imagine that $T_0 = T$ and that the signal is time limited to the interval $[0, T)$. Then it follows that there will be no aliasing effect in the time domain in consequence of the notional frequency-domain sampling which is at intervals of $2\pi/T$; and therefore $x_r(t)$ is simply the periodic extension of $\{x_t\}$ defined according to (2.2).

Appendix: Harmonic Cycles

If a trigonometrical function completes an integral number of cycles in T periods, then the sum of its ordinates at the points $t = 0, 1, \ldots, T-1$ is zero. We state this more formally as follows:

(14.78) Let $\omega_j = 2\pi j/T$ where $j \in \{0, 1, \ldots, T/2\}$, if T is even, and $j \in \{0, 1, \ldots, (T-1)/2\}$, if T is odd. Then

$$\sum_{t=0}^{T-1}\cos(\omega_j t) = \sum_{t=0}^{T-1}\sin(\omega_j t) = 0.$$

Proof. We have

$$\begin{aligned}\sum_{t=0}^{T-1}\cos(\omega_j t) &= \frac{1}{2}\sum_{t=0}^{T-1}\{\exp(i\omega_j t) + \exp(-i\omega_j t)\} \\ &= \frac{1}{2}\sum_{t=0}^{T-1}\exp(i2\pi jt/T) + \frac{1}{2}\sum_{t=0}^{T-1}\exp(-i2\pi jt/T).\end{aligned} \tag{14.79}$$

By using the formula $1 + \lambda + \cdots + \lambda^{T-1} = (1-\lambda^T)/(1-\lambda)$, we find that

$$\sum_{t=0}^{T-1}\exp(i2\pi jt/T) = \frac{1-\exp(i2\pi j)}{1-\exp(i2\pi j/T)}.$$

But Euler's equation indicates that $\exp(i2\pi j) = \cos(2\pi j) + i\sin(2\pi j) = 1$, so the numerator in the expression above is zero, and hence $\sum_t \exp(i2\pi j/T) = 0$. By similar means, it can be show that $\sum_t \exp(-i2\pi j/T) = 0$; and, therefore, it follows that $\sum_t \cos(\omega_j t) = 0$.

An analogous proof shows that $\sum_t \sin(\omega_j t) = 0$.

The proposition of (14.78) is used to establish the orthogonality conditions affecting functions with an integral number of cycles.

(14.80) Let $\omega_j = 2\pi j/T$ and $\psi_k = 2\pi k/T$ where $j, k \in 0, 1, \ldots, T/2$ if T is even and $j, k \in 0, 1, \ldots, (T-1)/2$ if T is odd. Then

$$\text{(a)}\qquad \sum_{t=0}^{T-1} \cos(\omega_j t)\cos(\psi_k t) = 0 \qquad \text{if } j \neq k,$$

$$\sum_{t=0}^{T-1} \cos^2(\omega_j t) = T/2,$$

$$\text{(b)}\qquad \sum_{t=0}^{T-1} \sin(\omega_j t)\sin(\psi_k t) = 0 \qquad \text{if } j \neq k,$$

$$\sum_{t=0}^{T-1} \sin^2(\omega_j t) = T/2,$$

$$\text{(c)}\qquad \sum_{t=0}^{T-1} \cos(\omega_j t)\sin(\psi_k t) = 0 \qquad \text{if } j \neq k.$$

Proof. From the formula $\cos A \cos B = \frac{1}{2}\{\cos(A+B) + \cos(A-B)\}$, we have

$$\begin{aligned}\sum_{t=0}^{T-1} \cos(\omega_j t)\cos(\psi_k t) &= \frac{1}{2}\sum \{\cos([\omega_j + \psi_k]t) + \cos([\omega_j - \psi_k]t)\} \\ &= \frac{1}{2}\sum_{t=0}^{T-1} \{\cos(2\pi[j+k]t/T) + \cos(2\pi[j-k]t/T)\}.\end{aligned} \tag{14.81}$$

We find, in consequence of (14.78), that if $j \neq k$, then both terms on the RHS vanish, which gives the first part of (a). If $j = k$, then $\cos(2\pi[j-k]t/T) = \cos 0 = 1$ and so, whilst the first term vanishes, the second terms yields the value of T under summation. This gives the second part of (a).

The proofs of (b) and (c) follow along similar lines once the relevant sum–product relationships of (13.126) have been invoked.

Bibliography

[51] Beveridge, W.H., (1921), Weather and Harvest Cycles, *The Economic Journal*, **31**, 429–45.

[52] Beveridge, W.H., (1922), Wheat Prices and Rainfall in Western Europe, *Journal of the Royal Statistical Society*, **85**, 412–47.

[67] Bloomfield, P., (1976), *Fourier Analysis of Time Series: An Introduction*, John Wiley and Sons, New York.

[110] Cizek, V., (1986), *Discrete Fourier Transforms and their Applications*, Adam Hilger, Bristol.

[481] Tiao, G.C., and M.R. Grupe, (1980), Hidden Periodic Autoregressive-Moving Average Models in Time Series Data, *Biometrika*, **67**, 365–73.

[515] Whittaker, E., and G. Robinson, (1944), *The Calculus of Observation: A Treatise on Numerical Mathematics, Fourth Edition*, Blakie and Son, London.

CHAPTER 15

The Fast Fourier Transform

In the late 1960s, an efficient method for computing the discrete Fourier transform became available which has revolutionised many fields of applied science and engineering where, hitherto, problems of computing had posed a serious obstacle to progress. The algorithm, which became known as the fast Fourier transform or as the FFT for short, has been attributed primarily to J.W. Cooley and J.W. Tukey [125] who presented a version of it in a seminal paper of 1965. Subsequent enquiry (see Cooley *et al.* [126]) has shown that the principles behind the algorithm have been understood by others at earlier dates. Nevertheless, there is no doubt that the paper of Cooley and Tukey acted as the catalyst for one of the most important advances in applied mathematics of this century.

Basic Concepts

To assist our understanding of the nature of the economies which can result from the use of the fast Fourier transform, we shall begin by showing a matrix formulation of the transformation.

The discrete Fourier transform represents a one-to-one mapping from the sequence of data values $y_t; t = 0, \ldots, T-1$, which are to be regarded as complex numbers, to the sequence of Fourier coefficients $\zeta_j; j = 0, \ldots, T-1$ which are, likewise, complex. The equation of the transform can be written as

$$\zeta_j = \frac{1}{T}\sum_{t=0}^{T-1} y_t e^{-i\omega_j t}; \quad \omega_j = 2\pi j/T, \tag{15.1}$$

wherein ω_j is a so-called Fourier frequency. The complex multiplications which are entailed by this expression take the form of

$$\begin{aligned} y_t e^{-i\omega_j t} &= (y_t^{re} + iy_t^{im})(\cos\omega_j t - i\sin\omega_j t) \\ &= (y_t^{re}\cos\omega_j t + y_t^{im}\sin\omega_j t) + i(y_t^{im}\cos\omega_j t - y_t^{re}\sin\omega_j t). \end{aligned} \tag{15.2}$$

Equation (15.2) can be rewritten conveniently in terms of W_T, which is the first of the Tth roots of unity when these are taken in the clockwise order:

$$\zeta_j = \frac{1}{T}\sum_{t=0}^{T-1} y_t W_T^{jt}; \quad W_T = \exp(-i2\pi/T). \tag{15.3}$$

Now consider, as an illustration, the case where $T = 6$. Then, if the subscript on W_T is suppressed, the T instances of equation (15.3) can be written as

$$(15.4)\qquad T\begin{bmatrix}\zeta_0\\ \zeta_1\\ \zeta_2\\ \zeta_3\\ \zeta_4\\ \zeta_5\end{bmatrix}=\begin{bmatrix}W^0 & W^0 & W^0 & W^0 & W^0 & W^0\\ W^0 & W^1 & W^2 & W^3 & W^4 & W^5\\ W^0 & W^2 & W^4 & W^6 & W^8 & W^{10}\\ W^0 & W^3 & W^6 & W^9 & W^{12} & W^{15}\\ W^0 & W^4 & W^8 & W^{12} & W^{16} & W^{20}\\ W^0 & W^5 & W^{10} & W^{15} & W^{20} & W^{25}\end{bmatrix}\begin{bmatrix}y_0\\ y_1\\ y_2\\ y_3\\ y_4\\ y_5\end{bmatrix}.$$

After a cursory inspection of this equation, it might be imagined that to compute the mapping from the vector $y = [y_0, y_1, \ldots, y_{T-1}]'$ to the vector $\zeta = [\zeta_0, \zeta_1, \ldots, \zeta_{T-1}]'$ would require a total of $T^2 = 36$ complex multiplications and $T(T-1) = 30$ complex additions. However, a more careful inspection shows that some of the multiplications are repeated several times.

To begin, it may be observed that $W^\alpha = \exp(-i2\pi\alpha/T)$ is a T-periodic function of the integer sequence $\{\alpha = 0, 1, 2, \ldots\}$. Thus, if $q = \alpha \text{ div } T$ and $r = \alpha \bmod T$ are, respectively, the quotient and the remainder from the division of α by T, then $\alpha = Tq + r$ and $W^\alpha = (W^T)^q W^r = W^r$, where $r = 1, 2, \ldots, T-1$ and $W^T = 1$.

The next point to observe is that the identity $\exp(\pm i\pi) = -1$ implies that

$$(15.5)\qquad \begin{aligned}W^{q+T/2} &= \exp(-i2\pi q/T)\exp(-i\pi)\\ &= -W^q.\end{aligned}$$

This property, which is described as the half-wave symmetry of the periodic function, further reduces the number of distinct values of W^α which need to be considered when T is even.

The consequence of the T-periodicity of W^α for the example under (15.4), where $T = 6$, is that only the values $W^0 = 1, W^1, \ldots, W^5$ are present. The half-symmetry of W^α implies that $W^3 = -1$, $W^4 = -W$ and $W^5 = -W^2$. Therefore, the matrix in (15.4) can be written as

$$(15.6)\qquad \begin{bmatrix}1 & 1 & 1 & 1 & 1 & 1\\ 1 & W^1 & W^2 & W^3 & W^4 & W^5\\ 1 & W^2 & W^4 & 1 & W^2 & W^4\\ 1 & W^3 & 1 & W^3 & 1 & W^3\\ 1 & W^4 & W^2 & 1 & W^4 & W^2\\ 1 & W^5 & W^4 & W^3 & W^2 & W\end{bmatrix}=\begin{bmatrix}1 & 1 & 1 & 1 & 1 & 1\\ 1 & W & W^2 & -1 & -W & -W^2\\ 1 & W^2 & -W & 1 & W^2 & -W\\ 1 & -1 & 1 & -1 & 1 & -1\\ 1 & -W & W^2 & 1 & -W & W^2\\ 1 & -W^2 & -W & -1 & W^2 & W\end{bmatrix}.$$

On putting the matrix on the RHS in place of the matrix in equation (15.4), it becomes clear that the elements ζ_0, ζ_1, ζ_2 are formed from the same components as the elements ζ_3, ζ_4, ζ_5 respectively.

A much greater potential for reducing the number of operations in computing the transform comes from the next step, which is to factorise the matrix in (15.4). It is readily confirmed that, ignoring the half-symmetry of W^α, the matrix can be written as the following product:

$$(15.7)\qquad \begin{bmatrix} 1 & 0 & 1 & 0 & 1 & 0 \\ 0 & 1 & 0 & W & 0 & W^2 \\ 1 & 0 & W^2 & 0 & W^4 & 0 \\ 0 & 1 & 0 & W^3 & 0 & 1 \\ 1 & 0 & W^4 & 0 & W^2 & 0 \\ 0 & 1 & 0 & W^5 & 0 & W^4 \end{bmatrix} \begin{bmatrix} 1 & 0 & 0 & 1 & 0 & 0 \\ 1 & 0 & 0 & W^3 & 0 & 0 \\ 0 & 1 & 0 & 0 & 1 & 0 \\ 0 & 1 & 0 & 0 & W^3 & 0 \\ 0 & 0 & 1 & 0 & 0 & 1 \\ 0 & 0 & 1 & 0 & 0 & W^3 \end{bmatrix}.$$

Here the leading matrix contains $P = 3$ elements in each row whilst the following matrix contains $Q = 2$ in each row. The factorisation enables the transformation of y into ζ to be performed in two steps which together involve less computation than does the single step depicted in (15.4).

To obtain a rough measure of the economies which can be achieved by using the two-step procedure, let us ignore any economies which may result from the periodic nature of the function W^{jt}. Then it can be can seen that, if $T = PQ$, the number of operations entailed by the two-step procedure increases, as a function of the length T of the data sequence, at the same rate as $T(P + Q)$. This is to be compared with the number of operations in the single-step procedure which increases at the rate of T^2. Of course, with $T = PQ = 3 \times 2$, the difference between $T^2 = 36$ and $T(P + Q) = 30$ is not great. However, when $T = 143 = 13 \times 11$, the difference between $T^2 = 20,449$ and $T(P + Q) = 3,432$ is already beginning to show.

The principle of factorisation can be exploited more fully in a multistep procedure. Thus, if the sample size is a highly composite number which can be factorised as $T = N_1N_2 \cdots N_g$, then the transformation from y to ζ can be accomplished in g steps which entail, altogether, a number of operations which is proportional to $T(N_1+N_2+\cdots+N_g)$. A special case arises when $T = 2^g$. Then the number of operations entailed in a g-step procedure is proportional to $2Tg$; and this increases with T at the same rate as $T \log T$. Apart from allowing a dramatic economy in the number of computational operations, the cases with $T = 2^g$ can be treated in a comparatively simple computer program. The consequence is that, in many applications, the length of the data sequence is contrived, wherever possible, to be a power of 2.

There are two ways of ensuring that a data sequence has a length of $T = 2^g$. The first way is to truncate a sequence whose original length is T' so as to retain $T = 2^g$ observations. The other way is to supplement the T' observations with zeros. This is described as padding the sequence.

An effect of padding the sequence will be to change the definition of the Fourier frequencies. In certain applications, where the frequencies are of no intrinsic interest, padding is the appropriate recourse. An example is when the FFT is used for convoluting two sequences. In other applications, such as in the calculation of the periodogram, padding may be unacceptable; for it may cause a severe spectral leakage. Then, either the loss of data which is occasioned by truncation must be accepted, or else a version of the FFT must be used which can cope with numbers T with arbitrary factors. Usually, we are prepared to suffer a small loss of data in pursuit of a number which is sufficiently composite to allow for a reasonably efficient exploitation of the FFT algorithm. Numbers tend to become increasingly composite as they increase in size; and it is relevant to recall that, according to the prime number theorem, the relative frequency of the prime numbers in the set $\{1, \ldots, T\}$ tends to $1/\ln T$ as T increases.

The following procedure represents a simple but effective way of finding the factors of $T = N_1 N_2 \cdots N_g$. The procedure attempts to arrange the factors in the form of palindrome with $N_{1+l} = N_{g-l}$. This is not always possible; and the value taken by the Boolean variable *palindrome* at the end of the operations indicates whether or not the object has been achieved.

(15.8)
```
procedure PrimeFactors(Tcap : integer;
                       var g : integer;
                       var N : ivector;
                       var palindrome : boolean);

  var
    i, p, T, first, last, store : integer;

  function Next(p : integer) : integer;
  begin
    if p = 2 then
      Next := 3
    else
      Next := p + 2
  end; {Next}

begin {PrimeFactors}
  palindrome := false;
  g := 1;
  p := 2;
  T := Tcap;

{Find the prime factors}
  while Sqr(p) <= T do
    begin {while}
      if T mod p = 0 then
        begin
          T := T div p;
          N[g] := p;
          g := g + 1
        end
      else
        p := Next(p);
    end; {while}
  N[g] := T;

  first := 1;
  last := g;
```

```
{Rearrange the factors}
  while (N[first] <= N[first + 1]) do
    begin {while}
      store := N[first];
      for i := first to last - 1 do
        N[i] := N[i + 1];
      N[last] := store;
      if (N[first] = N[last]) then
        first := first + 1;
      last := g + 1 - first;
    end; {while}
  if (last - first <= 0) and (g <> 1) then
    palindrome := true

end; {PrimeFactors}
```

In the following sections, we shall begin by considering, in detail, the case where T has two factors. This is for illustrative purposes. We shall proceed to develop carefully an algorithm for the case where T has arbitrary factors. We shall end by presenting an algorithm for the case where $T = 2^g$.

The Two-Factor Case

The basic algebraic features of the fast Fourier transform can be seen by considering the case where $T = PQ$ has just two factors. Then the indices t and j, both of which run from 0 to $T - 1$, can be expressed as

$$\begin{aligned} &t = Pr + s \qquad \text{and} \qquad j = l + Qk, \\ &\text{where} \quad s, k = 0, 1, \ldots, P - 1 \quad \text{and} \\ &\qquad\qquad r, l = 0, 1, \ldots, Q - 1. \end{aligned} \tag{15.9}$$

The indices r, s are the digits in a mixed-base or mixed-radix representation of the number t. When $P = Q = 10$, then, of course, r and s stand for "tens" and "units" respectively. The indices k and l of the expression for j can be characterised in the same way.

The new indices can be used notionally to arrange the elements of $\{y_t; t = 0, \ldots, T - 1\}$ and $\{\zeta_j; j = 0, \ldots, T - 1\}$ in two matrix arrays of orders $Q \times P$ whose respective elements are $y(r, s) = y_t$ and $\zeta(l, k) = \zeta_j$. With the new notation, equation (15.3) can be rewritten as

$$\begin{aligned} \zeta(l, k) &= \frac{1}{T} \sum_{s=0}^{P-1} \sum_{r=0}^{Q-1} y(r, s) W_T^{(Pr+s)(l+Qk)} \\ &= \frac{1}{T} \sum_{s=0}^{P-1} \left\{ \sum_{r=0}^{Q-1} y(r, s) W_T^{(Pr+s)l} \right\} W_T^{Q(Pr+s)k}. \end{aligned} \tag{15.10}$$

In the outer sum of this expression, there is the factor

$$W_T^{Q(Pr+s)k} = \left(W_T^Q\right)^{Prk} \left(W_T^Q\right)^{sk}. \tag{15.11}$$

Given that $W_T^Q = \exp(-2\pi/P) = W_P$, it follows that

$$(15.12)\qquad \begin{aligned} \left(W_T^Q\right)^{sk} &= W_P^{sk} \quad \text{and} \\ \left(W_T^Q\right)^{Prk} &= W_P^{Prk} = 1, \quad \text{for all} \quad k. \end{aligned}$$

Therefore,

$$(15.13)\qquad W_T^{Q(Pr+s)k} = W_P^{sk}.$$

By applying the same principles to the factor in the inner sum, it is found that

$$(15.14)\qquad W_T^{(Pr+s)l} = W_Q^{rl} W_T^{sl}.$$

When the results of (15.13) and (15.14) are substituted into equation (15.10), the latter can be written as

$$(15.15)\qquad \begin{aligned} \zeta(l,k) &= \frac{1}{T}\sum_{s=0}^{P-1}\left[\left\{\sum_{r=0}^{Q-1} y(r,s)W_Q^{rl}\right\}W_T^{sl}\right]W_P^{sk} \\ &= \frac{1}{T}\sum_{s=0}^{P-1}\left[\xi(l,s)W_T^{sl}\right]W_P^{sk} \\ &= \frac{1}{T}\sum_{s=0}^{P-1}\varphi(l,s)W_P^{sk}. \end{aligned}$$

The factor W_T^{sl} which is associated with $\xi(l,s)$ in this equation has been called the twiddle factor by Gentleman and Sande [206].

The elements $\zeta(l,k)$, which are the end-products of the transformation, are found by proceeding through four stages. The first stage generates

$$(15.16)\qquad \xi(l,s) = \sum_{r=0}^{Q-1} y(r,s)W_Q^{rl}$$

for $l = 0,1,\ldots,Q-1$ and $s = 0,1,\ldots,P-1$. For each value of s, the elements $\xi(l,s); l = 0,1,\ldots,Q-1$ entail the same Q data points $y(r,s); r = 0,1,\ldots,Q-1$, which are contained in the sth column of the notional $Q \times P$ data matrix. For any two distinct values of s, say p and q, the elements $\xi(l,p)$ and $\xi(l,q)$ entail disjoint sets of data points from different columns of the notional matrix. Therefore, once a group of Q elements $\xi(l,s); l = 0,1,\ldots,Q-1$ has been computed, they can be written in place of the data elements $y(r,s); r = 0,1,\ldots,Q-1$ from which they have been derived and for which there is no further need. In effect, the sequence $y_t; t = 0,1,\ldots,T-1$ can be sorted into a set of P subsequences of length Q, each of which can be transformed separately and then returned to its place of origin.

To illustrate this stage, let us reconsider the example where $T = PQ = 6$ with $P = 3$ and $Q = 2$. Then there is

$$(15.17)\qquad \begin{bmatrix} \xi(0,0) \\ \xi(0,1) \\ \xi(0,2) \\ \xi(1,0) \\ \xi(1,1) \\ \xi(1,2) \end{bmatrix} = \begin{bmatrix} 1 & 0 & 0 & 1 & 0 & 0 \\ 0 & 1 & 0 & 0 & 1 & 0 \\ 0 & 0 & 1 & 0 & 0 & 1 \\ 1 & 0 & 0 & W^3 & 0 & 0 \\ 0 & 1 & 0 & 0 & W^3 & 0 \\ 0 & 0 & 1 & 0 & 0 & W^3 \end{bmatrix} \begin{bmatrix} y(0,0) \\ y(0,1) \\ y(0,2) \\ y(1,0) \\ y(1,1) \\ y(1,2) \end{bmatrix}.$$

By careful inspection, three separate transformations can be picked out, each of which is a Fourier transform in microcosm:

$$(15.18)\qquad \begin{bmatrix} \xi(0,s) \\ \xi(1,s) \end{bmatrix} = \begin{bmatrix} 1 & 1 \\ 1 & W^3 \end{bmatrix} \begin{bmatrix} y(0,s) \\ y(1,s) \end{bmatrix}; \quad s = 0, 1, 2.$$

For each value of s, the elements $\xi(0,s)$ and $\xi(1,s)$ are calculated, and then they are written in place of the elements $y(0,s)$ and $y(1,s)$ which are no longer needed. We may recall that the half-wave symmetry of the function W^α implies that $W^3 = -1$. Therefore, $\xi(0,s)$ and $\xi(1,s)$ are respectively the sum and the difference of $y(0,s)$ and $y(1,s)$.

The second stage of the procedure is to compute

$$(15.19)\qquad \varphi(l,s) = \xi(l,s)W_T^{sl}$$

for each value of l and s. This is a straightforward matter of using the so-called twiddle factor to scale each element of the vector ξ. The mapping from ξ to φ is therefore associated with a diagonal matrix. In the example, the mapping is given by

$$(15.20)\qquad \begin{bmatrix} \varphi(0,0) \\ \varphi(0,1) \\ \varphi(0,2) \\ \varphi(1,0) \\ \varphi(1,1) \\ \varphi(1,2) \end{bmatrix} = \begin{bmatrix} 1 & 0 & 0 & 0 & 0 & 0 \\ 0 & 1 & 0 & 0 & 0 & 0 \\ 0 & 0 & 1 & 0 & 0 & 0 \\ 0 & 0 & 0 & 1 & 0 & 0 \\ 0 & 0 & 0 & 0 & W & 0 \\ 0 & 0 & 0 & 0 & 0 & W^2 \end{bmatrix} \begin{bmatrix} \xi(0,0) \\ \xi(0,1) \\ \xi(0,2) \\ \xi(1,0) \\ \xi(1,1) \\ \xi(1,2) \end{bmatrix}.$$

The third stage of the procedure is compute the transformation

$$(15.21)\qquad \zeta(l,k) = \frac{1}{T}\sum_{s=0}^{P-1} \varphi(l,s)W_P^{sk}.$$

Here we can use the methods which have already been applied to equation (15.16). Thus, for a given value of l, each of the elements $\zeta(l,k); k = 0, 1, \ldots, P-1$ are computed from the same P elements $\varphi(l,s); s = 0, 1, \ldots, P-1$; and then they are written in place of the latter. In the example, this stage of the procedure takes the form of

$$(15.22)\qquad T\begin{bmatrix} \zeta(0,0) \\ \zeta(0,1) \\ \zeta(0,2) \\ \zeta(1,0) \\ \zeta(1,1) \\ \zeta(1,2) \end{bmatrix} = \begin{bmatrix} 1 & 1 & 1 & 0 & 0 & 0 \\ 1 & W^2 & W^4 & 0 & 0 & 0 \\ 1 & W^4 & W^2 & 0 & 0 & 0 \\ 0 & 0 & 0 & 1 & 1 & 1 \\ 0 & 0 & 0 & 1 & W^2 & W^4 \\ 0 & 0 & 0 & 1 & W^4 & W^2 \end{bmatrix} \begin{bmatrix} \varphi(0,0) \\ \varphi(0,1) \\ \varphi(0,2) \\ \varphi(1,0) \\ \varphi(1,1) \\ \varphi(1,2) \end{bmatrix}.$$

Table 15.1. The correspondence between the digits of t and j for the case where $P = 3$ and $Q = 2$.

$r = l$	$s = k$	$t = Pr + s$	$j = l + Qk$
0	0	0	0
0	1	1	2
0	2	2	4
1	0	3	1
1	1	4	3
1	2	5	5

The fourth and final stage of the procedure is necessitated by the fact that the elements $\zeta_j = \zeta(l, k)$ have been obtained in an order which differs from the natural order of the index j. For, as a result of the strategy of overwriting, which is designed to minimise the use of the computer's memory, the elements $\zeta_j = \zeta(l, k)$ are to be found in the places which were originally occupied by the data values $y_t = y(r, s)$. This means that the indices l and k vary in step with the indices r and s respectively. Given that $t = Pr + s$ and $j = l + Qk$, it is clear that setting $l = r$ and $k = s$ implies that, in general, $t \neq j$; and so it is not possible for both t and j to follow the natural order.

The scrambling of the index j which occurs in the example is shown in Table 15.1, which gives the corresponding values of t and j.

The order of the elements $\zeta_j = \zeta(l, k)$ can be unscrambled by a permutation transformation. In the example, the transformation takes the following form:

$$
(15.23) \qquad \begin{bmatrix} \zeta(0,0) \\ \zeta(1,0) \\ \zeta(0,1) \\ \zeta(1,1) \\ \zeta(0,2) \\ \zeta(1,2) \end{bmatrix} = \begin{bmatrix} 1 & 0 & 0 & 0 & 0 & 0 \\ 0 & 0 & 0 & 1 & 0 & 0 \\ 0 & 1 & 0 & 0 & 0 & 0 \\ 0 & 0 & 0 & 0 & 1 & 0 \\ 0 & 0 & 1 & 0 & 0 & 0 \\ 0 & 0 & 0 & 0 & 0 & 1 \end{bmatrix} \begin{bmatrix} \zeta(0,0) \\ \zeta(0,1) \\ \zeta(0,2) \\ \zeta(1,0) \\ \zeta(1,1) \\ \zeta(1,2) \end{bmatrix}.
$$

The FFT for Arbitrary Factors

The fast Fourier transform achieves its greatest efficiency when the sample size T is a highly composite number with many factors which is written as

$$
(15.24) \qquad T = \prod_{l=1}^{g} N_l.
$$

In that case, the Fourier transform, which is given by the matrix equation $\zeta = T^{-1}Wy$, can be accomplished in g stages with T/N_l transforms of order N_l in the lth stage. The g-fold decomposition of the transformation matrix W can be written as

$$
(15.25) \qquad W = PF_gF_{g-1}\cdots F_2F_1,
$$

where P is a permutation matrix which restores the elements $\zeta_j; j = 0, \ldots, T-1$ to the natural order. The individual transformations F_l can be decomposed as

$$F_l = R_l W_l, \tag{15.26}$$

where R_l is a diagonal matrix containing the twiddle factors. The virtue of this further decomposition is that the T/N_l submatrices into which W_l can be partitioned are now identical in form.

The easiest way of expounding the algebra of the multifactor Fourier transform is to concentrate on the details of the three-factor case; for the algebra remains simple, whilst the generalisation to the multifactor case follows immediately.

Let us therefore imagine that

$$T = N_1 N_2 N_3, \tag{15.27}$$

and let us define

$$\begin{aligned} P_1 &= N_2 N_3, & Q_1 &= 1, \\ P_2 &= N_3, & Q_2 &= N_1, \\ P_3 &= 1, & Q_3 &= N_2 N_1. \end{aligned} \tag{15.28}$$

Then the indices t and j can be expressed as

$$\begin{aligned} t &= t_1 N_2 N_3 + t_2 N_3 + t_3 \\ &= t_1 P_1 + t_2 P_2 + t_3 P_3, \\ j &= j_1 + j_2 N_1 + j_3 N_2 N_1 \\ &= j_1 Q_1 + j_2 Q_2 + j_3 Q_3, \end{aligned} \tag{15.29}$$

where

$$\begin{aligned} t_1, j_1 &\in \{0, 1, \ldots, N_1 - 1\}, \\ t_2, j_2 &\in \{0, 1, \ldots, N_2 - 1\}, \\ t_3, j_3 &\in \{0, 1, \ldots, N_3 - 1\}. \end{aligned} \tag{15.30}$$

More generally, when $T = N_1 N_2 \cdots N_g$, there are

$$\begin{aligned} &t = t_1 P_1 + t_2 P_2 + \cdots + t_g P_g \\ &\quad \text{with} \quad P_i = N_{i+1} N_{i+2} \cdots N_g \quad \text{and} \quad P_g = 1 \end{aligned} \tag{15.31}$$

and

$$\begin{aligned} &j = j_1 Q_1 + j_2 Q_2 + \cdots + j_g Q_g \\ &\quad \text{with} \quad Q_i = N_1 N_2 \cdots N_{i-1} \quad \text{and} \quad Q_1 = 1, \end{aligned} \tag{15.32}$$

where $t_i, j_i \in \{0, 1, \ldots, N_i - 1\}$.

Using the expressions from (15.29), the equation (15.3) for the Fourier coefficients, which was written as $\zeta_j = T^{-1}\sum_t y_t(W_T)^{jt}$, can now be expressed, for the three-factor case, as

$$\begin{aligned}\zeta(j_1,j_2,j_3) &= \frac{1}{T}\sum_{t_1}\sum_{t_2}\sum_{t_3} y(t_1,t_2,t_3)W_T\uparrow[tj]\\ &= \frac{1}{T}\sum_{t_1}\sum_{t_2}\sum_{t_3} y(t_1,t_2,t_3)W_T\uparrow\left[\Big(\sum_m t_mP_m\Big)\Big(\sum_l j_lQ_l\Big)\right],\end{aligned}\tag{15.33}$$

where the upward arrow signifies exponentiation. Now, in view of (15.28), it can be seen that, if $m < l$, then $P_mQ_l = \alpha T$ is an integer multiple of T; and, given that W_T^{α} is a T-periodic function of α, it follows that $W_T^{\alpha T} = 1$. This result, which holds for any number of factors, leads to the following identity:

$$\begin{aligned}W_T\uparrow[tj] &= W_T\uparrow\left[\Big(\sum_{m=1}^{g} t_mP_m\Big)\Big(\sum_{l=1}^{g} j_lQ_l\Big)\right]\\ &= W_T\uparrow\left[\sum_{l=1}^{g} j_l\Big(\sum_{m=l}^{g} t_mP_m\Big)Q_l\right]\\ &= \prod_{l=1}^{g}\left\{W_T\uparrow\left[j_l\Big(\sum_{m=l+1}^{g} t_mP_m\Big)Q_l\right]W_{N_l}\uparrow[j_lt_l]\right\}.\end{aligned}\tag{15.34}$$

The second equality comes from discarding, from the sum in the exponent, all elements with $m < l$. The final identity is obtained by extracting the factor

$$W_T\uparrow[j_lt_lP_lQ_l] = W_{N_l}\uparrow[j_lt_l],\tag{15.35}$$

wherein $W_T\uparrow[P_lQ_l] = W_{N_l}$. This serves to isolate the twiddle factor

$$W_T\uparrow\left[j_l\Big(\sum_{m=l+1}^{g} t_mP_m\Big)Q_l\right].\tag{15.36}$$

With $g = 3$, the three factors of $W_T\uparrow[tj]$ are

$$\begin{aligned}f_1 &= W_T\uparrow[j_1(t_1P_1+t_2P_2+t_3P_3)Q_1]\\ &= W_T\uparrow[j_1(t_2P_2+t_3P_3)Q_1]W_{N_1}\uparrow[j_1t_1],\\ f_2 &= W_T\uparrow[j_2(t_2P_2+t_3P_3)Q_2]\\ &= W_T\uparrow[j_2t_3P_3Q_2]W_{N_2}\uparrow[j_2t_2],\\ f_3 &= W_T\uparrow[j_3t_3P_3Q_3] = W_{N_3}\uparrow[j_3t_3].\end{aligned}\tag{15.37}$$

The factors f_1 and f_2 are decomposed here into a twiddle factor and a residual factor.

Substituting the decomposition $W_T\uparrow[tj] = f_1f_2f_3$ into equation (15.33) gives

$$T\zeta(j_1,j_2,j_3) = \frac{1}{T}\sum_{t_3}\left\{\sum_{t_2}\left\{\sum_{t_1} y(t_1,t_2,t_3)f_1\right\}f_2\right\}f_3.\tag{15.38}$$

This expression represents the transformation from $y(t_1, t_2, t_3)$ to $\zeta(j_1, j_2, j_3)$ as the product of three successive transformations which can be represented, in turn, as follows:

$$(15.39)\quad \begin{aligned}
\zeta_1(j_1, t_2, t_3) &= \Big\{\sum_{t_1} y(t_1, t_2, t_3) W_{N_1} \uparrow [j_1 t_1]\Big\} W_T \uparrow [j_1(t_2 P_2 + t_3 P_3) Q_1] \\
&= \xi_1(j_1, t_2, t_3) W_T \uparrow [j_1(t_2 P_2 + t_3 P_3) Q_1], \\
\zeta_2(j_1, j_2, t_3) &= \Big\{\sum_{t_2} \zeta_1(j_1, t_2, t_3) W_{N_2} \uparrow [j_2 t_2]\Big\} W_T \uparrow [j_2 t_3 P_3 Q_2] \\
&= \xi_2(j_1, j_2, t_3) W_T \uparrow [j_2 t_3 P_3 Q_2], \\
\zeta_3(j_1, j_2, j_3) &= \sum_{t_3} \zeta_2(j_1, j_2, t_3) W_{N_3} \uparrow [j_3 t_3].
\end{aligned}$$

The algorithm of the three-factor case can be generalised easily to the case of g factors. The lth stage of the g-factor algorithm is represented by

$$(15.40)\quad \begin{aligned}
&\zeta_l(j_1, \ldots, j_l, t_{l+1}, \ldots, t_g) \\
&= \Big\{\sum_{t_l} \zeta_{l-1}(j_1, \ldots, j_{l-1}, t_l, \ldots, t_g) W_{N_l} \uparrow [j_l t_l]\Big\} W_T \uparrow \Big[\Big(\sum_{m=l+1}^{g} t_m P_m\Big) j_l Q_l\Big] \\
&= \xi_l(j_1, \ldots, j_{l-1}, t_l, \ldots, t_g) W_T \uparrow \Big[\Big(\sum_{m=l+1}^{g} t_m P_m\Big) j_l Q_l\Big].
\end{aligned}$$

In the lth stage of the algorithm, the elements of ζ_{l-1} are transformed in self-contained groups or subsequences of N_l at a time. Each group of elements is subjected to the same transformation which is a Fourier transform on a small scale. The resulting elements of ξ_l are multiplied by their corresponding twiddle factors to convert them to the elements of ζ_l. This can be done at any time from the moment that an element is available until the beginning of the next stage of the algorithm. However, as we shall discover, there is an advantage in performing the twiddle as soon as the elements of a subsequence of ζ_{l-1} have been transformed into the corresponding elements of ξ_l.

Locating the Subsequences

In the lth stage of the algorithm for computing the FFT, the vector ζ_{l-1} is divided into a T/N_l disjoint subsequences, each of length N_l, which are transformed separately. To locate the elements of a given subsequence, the indices $j_1, \ldots, j_{l-1}$ and $t_{l+1}, \ldots, t_g$ must be held constant while the value of the index t_l is increased from 0 to $N_l - 1$. To pass from one subsequence to the next, the other indices are varied in their natural order. In fact, there is no need to vary $j_1, \ldots, j_{l-1}$ and $t_{l+1}, \ldots, t_g$ explicitly; for they can be replaced by the composite indices $a = j_1 P_1 + \cdots + j_{l-1} P_{l-1}$ and $c = t_{l+1} P_{l+1} + \cdots + t_g P_g$, wherein $P_m = N_{m+1} \cdots N_g$. It will be recognised that c is a value which is already incorporated in the expression for the twiddle factor given under (15.36).

For an example of how the subsequences may be located in practice, let us consider the case where $T = N_1N_2N_3$ so that, according to (15.29),

$$\begin{aligned} t &= t_1N_2N_3 + t_2N_3 + t_3 \\ &= a + b + c, \end{aligned} \tag{15.41}$$

with

$$\begin{aligned} a &= t_1N_2N_3 = t_1P_1, \\ b &= t_2N_3 = t_2P_2, \\ c &= t_3 = t_3P_3. \end{aligned} \tag{15.42}$$

Let us imagine that the purpose is to calculate the elements

$$\xi_2(j_1, j_2, t_3) = \sum_{t_2} \zeta_1(j_1, t_2, t_3)W_{N_2} \uparrow [j_2t_2] \tag{15.43}$$

from the second stage of the transformation under (15.39). Then the indices should be varied by gearing them to each other in a such a way that a single increment of t_3 follows after a complete cycle of t_2 and an increment of $j_1 = t_1$ follows after a cycle of t_3. This can be accomplished by the following algorithm in which the indices i, j and k are proxies for t_1, t_2 and t_3 respectively:

(15.44)
```
begin
  a := 0;
  b := 0;
  c := 0;
  t := 0;

  for i := 0 to N1 - 1 do
    begin {i}
      for k := 0 to N3 - 1 do
        begin {k}
          for j := 0 to N2 - 1 do
            begin {j}
              t := a + b + c;
              Writeln(t : 10);
              b := b + N3
            end; {j}
          b := 0;
          c := c + 1;
        end; {k}
      c := 0;
      a := a + N2 * N3;
    end; {i}
end;
```

Table 15.2. The index t in natural order and permuted order.

Natural order				Permuted order			
t_1	t_2	t_3	t	t_1	t_2	t_3	t
0	0	0	0	0	0	0	0
0	0	1	1	0	1	0	2
0	1	0	2	0	2	0	4
0	1	1	3	0	0	1	1
0	2	0	4	0	1	1	3
0	2	1	5	0	2	1	5
1	0	0	6	1	0	0	6
1	0	1	7	1	1	0	8
1	1	0	8	1	2	0	10
1	1	1	9	1	0	1	7
1	2	0	10	1	1	1	9
1	2	1	11	1	2	1	11

An example of the output of this algorithm is given in Table 15.2 which shows, for the case where $N_1 = 2$, $N_2 = 3$ and $N_3 = 2$, the effect of varying the indices in their natural order and in the order of the algorithm.

To apply the algorithm to the lth stage of a multifactor transformation with $T = (N_1 \cdots N_{l-1})N_l(N_{l+1} \cdots N_g) = Q_l N_l P_l$, we need only replace N_1, N_2 and N_3 by Q_l, N_l and P_l respectively.

The Core of the Mixed-Radix Algorithm

In effect, the fragment under (15.44) provides a basic framework for the mixed-radix FFT algorithm. In place of the inner *Writeln* statement indexed by j, three operations must be inserted.

The first operation is to make a copy of the designated subsequence. The second operation is to effect the transformation from $\zeta_{l-1}(j_1, \ldots, j_{l-1}, t_l, \ldots, t_g)$ to $\xi_l(j_1, \ldots, j_{l-1}, j_l, \ldots, t_g)$ for each value of the index $t_l = j_l$. In the third operation, each element of the transformed subsequence is scaled by the relevant twiddle factor so as to give $\zeta_l(j_1, \ldots, j_{l-1}, j_l, \ldots, t_g)$.

The iterations of the indices i and k carry the algorithm from one subsequence to the next. The index l corresponds to the stages of the FFT. At the beginning of each stage, the matrix is constructed of the transformation which is to be applied identically to each of the subsequences. Altogether, there are g stages, which is the number of factors in T; and, in the final stage, the twiddle factors are all unity, which eliminates the scaling operations.

The values Q_l, and P_l, which are generated at the outset of the lth stage, are written into the vectors Q and P which are empty when passed to the procedure.

(15.45) **procedure** *MixedRadixCore*(**var** *yReal*, *yImag* : *vector*;
var N, P, Q : *ivector*;
Tcap, g : *integer*);

```
  const
    twopi = 6.283185;

  type
    Wvector = array[0..23] of real;
    Wmatrix = array[0..23] of Wvector;

  var
    a, b, c, t, i, j, k, l, r : integer;
    W, Wl, theta : real;
    yR, yI : Wvector;
    cosine, sine : Wmatrix;

begin {MixedRadixCore}

  W := twopi/Tcap;
  Q[0] := 1;
  N[0] := 1;
  P[0] := Tcap;
  b := 0;
  c := 0;

  for l := 1 to g do
   begin {l : this is the major loop}
    a := 0;
    t := 0;
    Q[l] := Q[l − 1] ∗ N[l − 1];
    P[l] := P[l − 1] div N[l];

{Construct the transformation matrix}
    Wl := twopi/N[l];
    for j := 0 to N[l] − 1 do
     for r := j to N[l] − 1 do
      begin {r, j}
       theta := Wl ∗ ((j ∗ r) mod N[l]);
       cosine[j, r] := Cos(theta);
       sine[j, r] := Sin(theta);
       cosine[r, j] := cosine[j, r];
       sine[r, j] := sine[j, r];
      end; {r, j}

    for i := 0 to Q[l] − 1 do
     begin {i}
      for k := 0 to P[l] − 1 do
       begin {k}
    {subsequences are indexed by i, k jointly}
```

```
{Copy a subsequence of the data}
        for j := 0 to N[l] - 1 do
         begin {j}
          t := a + b + c;
          yR[j] := yReal[t];
          yI[j] := yImag[t];
          b := b + P[l];
         end; {j}
        b := 0;

{Transform the subsequence}
        for j := 0 to N[l] - 1 do
         begin {j}
          t := a + b + c;
          yReal[t] := 0.0;
          yImag[t] := 0.0;
          for r := 0 to N[l] - 1 do
          begin {r}
          yReal[t] := yReal[t] + yR[r] * cosine[j, r]
                                + yI[r] * sine[j, r];
          yImag[t] := yImag[t] + yI[r] * cosine[j, r]
                                 - yR[r] * sine[j, r];
          end; {r}
          b := b + P[l];
         end; {j}
        b := 0;

{Scale the subsequence by the twiddle factors}
        if l < g then
         begin {if}
          for j := 0 to N[l] - 1 do
          begin {j : twiddle factors}
          t := a + b + c;
          theta := W * ((j * c * Q[l]) mod Tcap);
          yR[0] := yReal[t];
          yI[0] := yImag[t];
          yReal[t] := yR[0] * Cos(theta) + yI[0] * Sin(theta);
          yImag[t] := yI[0] * Cos(theta) - yR[0] * Sin(theta);
          b := b + P[l];
          end; {j : twiddle factors}
          b := 0;
         end; {if}

        c := c + 1;
       end; {k}
      c := 0;
      a := a + N[l] * P[l];
```

```
        end; {i}
      end; {l : the major loop}

    end; {Mixed Radix Core}
```

Unscrambling

On the completion of the gth stage of the transformation, a set of Fourier coefficients ζ_j is obtained which is indexed by a sequence $j = j(t)$ which is a permutation of the natural sequence of $t = 0, 1, \ldots, T-1$. Access to the coefficients in their natural order is obtained by computing the inverse function $t = t(j)$ for $j = 1, \ldots, T-1$. This enables the Fourier coefficient ζ_j to be recovered from the scrambled array which is indexed by t. Therefore, a correspondence must be established between $t = t_1P_1 + \cdots + t_gP_g$ and $j = j_1Q_1 + \cdots + j_gQ_g$ when $j_i = t_i$ for all i.

To begin, the digits of $j = (j_1, \ldots, j_g)$ must be found. Given that Q_i divides Q_j if and only if $i \leq j$, it follows that, if $j = j_1Q_1 + \cdots + j_lQ_l + \cdots + j_gQ_g$, then

$$j \bmod Q_{l+1} = j_1Q_1 + \cdots + j_lQ_l. \tag{15.46}$$

Therefore,

$$j_l = (j \bmod Q_{l+1}) \operatorname{div} Q_l \tag{15.47}$$

is provided by the integral part from the division of $j \bmod Q_{l+1}$ by Q_l. Next, if $t_i = j_i$ for all i, it follows that

$$\begin{aligned} t &= j_1P_1 + \cdots + j_{g-1}P_{g-1} + j_gP_g \\ &= \{(j \bmod Q_2) \operatorname{div} Q_1\}P_1 + \cdots \\ &\quad + \{(j \bmod Q_g) \operatorname{div} Q_{g-1}\}P_{g-1} + (j \operatorname{div} Q_g)P_g. \end{aligned} \tag{15.48}$$

This expression is incorporated in the following algorithm:

(15.49)
```
function tOfj(j, g : integer;
                    P, Q : ivector) : integer;

  var
    i, t : integer;

begin
  t := (j div Q[g]) * P[g];
  for i := g - 1 downto 1 do
    t := t + ((j mod Q[i + 1]) div Q[i]) * P[i];
  tOfj := t
end; {tOfj}
```

Table 15.3. The reordering of the index $j = j_1 + j_2N_1 + j_2N_2N_1$ when $N_1 = 2$, $N_2 = 2$, and $N_3 = 3$. In this example, the reordering is accomplished by three permutation cycles. The presence of two bracketed elements in a column signifies the completion of a permutation cycle.

j_3 t_3	j_2 t_2	j_1 t_1	t	$j = j(t)$	Successive reordering						
0	0	0	0	0	0	0	0	0	0	0	0
1	0	0	1	4	[1]	1	1	1	1	1	1
2	0	0	2	8	8	8	[2]	2	2	2	2
0	1	0	3	2	2	2	[ ]	[3]	3	3	3
1	1	0	4	6	6	[4]	4	4	4	4	4
2	1	0	5	10	10	10	10	10	10	[5]	5
0	0	1	6	1	[ ]	[6]	6	6	6	6	6
1	0	1	7	5	5	5	5	5	5	[ ]	[7]
2	0	1	8	9	9	9	9	9	[8]	8	8
0	1	1	9	3	3	3	3	[ ]	[9]	9	9
1	1	1	10	7	7	7	7	7	7	7	[10]
2	1	1	11	11	11	11	11	11	11	11	11

The ease with which the elements ζ_j can be accessed in their natural order diminishes the incentive to reorder the sequence in which they are stored. It is important that any method which is used for reordering the elements should not waste computer memory.

An effective way of reordering the elements of the scrambled vector is to pursue a chain of replacements. This begins with a misplaced element which is removed from its location and set aside. Then the empty place is filled with its appropriate element whose removal from another location will create another empty space. This space is filled, in turn, with its appropriate element. The process continues until an empty space occurs which can be filled with the element which was put aside at the start. Such a chain of replacements is known as a permutation cycle. Table 15.3 shows the effect of using three permutation cycles to reorder the numbers $j = j(t)$ in the case where $N_1 = 2$, $N_2 = 2$ and $N_3 = 3$.

A problem with this procedure is the difficulty of knowing when to initiate a permutation cycle. If we were to begin a new cycle with each successive element of the scrambled vector, then we should be reordering the same elements several times; and there would be no guarantee that, at the end, we should succeed in unscrambling the vector. We might be tempted, therefore, to keep a record to show which of the elements have been restored already to their proper places. However, since such a record would require T entries, the object of conserving computer memory would be defeated.

The proper recourse is to test each permutation cycle before applying it to the elements of the vector. Imagine that the previous permutation cycle has begun and ended at the $(r-1)$th location. Then we should proceed to test the cycle which begins at the rth location. If this cycle leads to a location preceding the rth, then it should be abandoned on the grounds that the same cycle, albeit with a different starting point, has been applied already to the vector. If the cycle is completed without leading to any location preceding the rth location, then it can be applied safely.

If the factorisation $T = N_1N_2 \cdots N_g$ is in the form of a palindrome with $N_{1+l} = N_{g-l}$, then there is a straightforward correspondence between $t = t_1P_1 + \cdots + t_gP_g$ and $j = j_1Q_1 + \cdots + j_gQ_g$ when $t_i = j_i$ for all i. It can be seen, with reference to the example under (15.28), that, in this case, $P_1 = Q_g, \ldots, P_g = Q_1$. It follows that $t = t(j) = j_1Q_g + \cdots + j_gQ_1$, which is to say that the value of t is obtained by reversing the order of the digits in the mixed-base representation of the number j. The consequence is that the elements of the scrambled vector ζ can be reordered through pairwise interchanges; and time is not wasted in testing permutation cycles. To secure this advantage, the procedure *PrimeFactors*, given under (15.8), arranges the factors of T in the form of a palindrome whenever possible.

(15.50)
```
procedure ReOrder(P, Q : ivector;
                  Tcap, g : integer;
                  var yImag, yReal : vector);

var
  r, t, j, Pcount : integer;


procedure Pcycle(r : integer;
                 var yImag, yReal : vector;
                 var Pcount : integer);

  var
    j, t : integer;
    Rstore, Istore : real;

begin {Pcycle}
  j := r;
  t := tOfj(j, g, P, Q);
  if t = j then
    Pcount := Pcount + 1
  else
    begin {else}
      Rstore := yReal[j];
      Istore := yImag[j];
      repeat {t}
        yReal[j] := yReal[t];
        yimag[j] := yImag[t];
        Pcount := Pcount + 1;
        j := t;
        t := tOfj(j, g, P, Q);
```

```
            until t = r;
            yReal[j] := Rstore;
            yimag[j] := Istore;
            Pcount := Pcount + 1;
         end; {else}
      end; {Pcycle}

   begin {ReOrder}
      r := 1;
      Pcount := 1;
      Pcycle(r, yImag, yReal, Pcount);

      repeat {r}
         r := r + 1;
         j := r;
         repeat {j}
            t := tOfj(j, g, P, Q);
            j := t;
         until j <= r;
         if j = r then
            Pcycle(r, yImag, yReal, Pcount);
      until (r = Tcap - 1) or (Pcount = Tcap - 1);

   end; {ReOrder}
```

The Shell of the Mixed-Radix Procedure

Now the operations of factorisation, transformation and unscrambling can be gathered together to form a coherent program for the mixed-radix FFT. To ensure that the efficiencies inherent in the FFT are exploited sufficiently, it is worth sacrificing a few points of data in order to obtain a number $T = N_1 \cdots N_g$ for the sample size which is reasonably composite. The amount of data which needs to be sacrificed is usually small.

To obtain the composite number, we shall impose the condition that T must have at least three factors and that the value of none of them should exceed 23. These prescriptions are somewhat arbitrary, and they can be varied easily. The restriction on the size of the prime factors also defines the maximum order of the transformation matrix which is constructed at the beginning of each stage of the FFT. The effects of these restrictions are illustrated in Table 15.4, which shows a short sequence of numbers which fulfil the conditions.

In order to find a number which satisfies the conditions, only one or two data points need to be sacrificed in most cases; and in no case covered by the table is it necessary to discard more than five points.

It is interesting to see that the palindromes are relatively frequent. One should recall that the process of unscrambling the sequence of Fourier coefficients is greatly assisted when the factors of T can be arranged in this form.

Table 15.4. The numbers $T = N_1 \cdots N_g$ in the interval $300 \geq T > 240$ which fulfil the conditions $\max(N_i) \leq 23$ and $g \geq 3$. The palindromes with $N_{1+i} = N_{g-i}$ are marked with asterisks.

$300 = 2 \times 5 \times 3 \times 5 \times 2$ *	$270 = 3 \times 2 \times 3 \times 5 \times 3$
$297 = 3 \times 3 \times 11 \times 3$	$266 = 2 \times 7 \times 19$
$294 = 7 \times 2 \times 3 \times 7$	$264 = 2 \times 2 \times 3 \times 11 \times 2$
$288 = 2 \times 2 \times 3 \times 2 \times 3 \times 2 \times 2$ *	$260 = 2 \times 5 \times 13 \times 2$
$286 = 2 \times 11 \times 13$	$256 = 2 \times 2 \times 2 \times 2 \times 2 \times 2 \times 2 \times 2$ *
$285 = 3 \times 5 \times 19$	$255 = 3 \times 5 \times 17$
$280 = 2 \times 2 \times 5 \times 7 \times 2$	$252 = 2 \times 3 \times 7 \times 3 \times 2$ *
$276 = 2 \times 3 \times 23 \times 2$	$250 = 5 \times 2 \times 5 \times 5$
$275 = 5 \times 11 \times 5$ *	$245 = 7 \times 5 \times 7$ *
$273 = 3 \times 7 \times 13$	$243 = 3 \times 3 \times 3 \times 3 \times 3$ *
$272 = 2 \times 2 \times 17 \times 2 \times 2$ *	$242 = 11 \times 2 \times 11$ *

The following Pascal procedure, which represents the shell of the mixed-radix FFT, incorporates a routine which searches for an appropriate value for T:

(15.51)
```
procedure MixedRadixFFT(var yReal, yImag : vector;
                        var Tcap, g : integer;
                        inverse : boolean);

  var
    t, i, Nmax : integer;
    Q, N, P : ivector;
    palindrome : boolean;

begin {MixedRadixFFT}
  g := 0;
  Tcap := Tcap + 1;

  repeat {Find a composite number for T}
      Tcap := Tcap - 1;
      PrimeFactors(Tcap, g, N, palindrome);
      Nmax := 0;
      for i := 1 to g do
        if N[i] > Nmax then
          Nmax := N[i];
    until; (g >= 3) and (Nmax <= 23)

  for t := 0 to Tcap - 1 do
    begin {t}
      if inverse = true then
        yImag[t] := -Imag[t]
```

```
        else
          begin {else}
            yReal[t] := yReal[t]/Tcap;
            yImag[t] := Imag[t]/Tcap
          end; {else}
      end; {t}

    MixedRadixCore(yReal, yImag, N, P, Q, Tcap, g);
    ReOrder(P, Q, Tcap, g, yImag, yReal);

  end; {MixedRadixFFT}
```

Amongst the arguments of the procedure is the Boolean variable *inverse*. If the value of this variable is *false*, then the direct form of the FFT, indicated by equation (15.1), is calculated. The scale factor T^{-1} is therefore applied to the data prior to its transformation. If the value of *inverse* is *true*, then the inverse form of the FFT is calculated. In that case, the scale factor is omitted and the sign of the imaginary part of the data sequence is reversed.

The Base-2 Fast Fourier Transform

When the data is available rapidly and in abundance, there should be little hesitation in sacrificing some of it in order to benefit from the speed and efficiency of an FFT procedure which caters specifically to the case when the sample size T is a power of 2.

The advantages of the condition $T = 2^g$, which are usually exploited by such algorithms, are the highly composite nature of T and the ease with which the sequence of Fourier coefficients can be unscrambled by pairwise interchanges. A further advantage, which is available if a twiddle-factor version of the FFT algorithm is used, is the avoidance of several of the multiplications involving trigonometrical functions. The evaluation of trigonometrical functions is a relatively time-consuming business; and most of the hard-wired implementations of the FFT incorporate permanent trigonometrical tables to help in speeding the process.

When $T = 2^g$, there are $T/2 = T/N_l$ two-point subsequences to be transformed in the lth stage of the procedure. If the twiddle-factor algorithm is used, then each subsequence is subjected to the same transformation. Moreover, when $N_l = 2$ for all l, the transformation is the same in every stage. The matrix of this transformation takes the form of

$$(15.52)\qquad \begin{bmatrix} 1 & 1 \\ 1 & W_{N_l} \end{bmatrix}.$$

With $N_l = 2$ for all l, there is $W_{N_l} = \exp(-i2\pi/N_l) = \exp(-i\pi) = -1$. Therefore, the transformation involves nothing more than the sum and the difference of two elements. The only recourse to trigonometrical functions is in forming the twiddle factor.

These features are readily discernible in the code of the base-2 FFT procedure which follows. In constructing the procedure, the real and imaginary components of the input have been placed in a single array which takes the form of

$$[y_0^{re}, y_1^{re}, \ldots, y_{T-1}^{re}, y_0^{im}, y_1^{im}, \ldots, y_{T-1}^{im}]. \tag{15.53}$$

Thus y_t^{re} and y_t^{im} are coded as $y[t]$ and $y[Tcap + t]$ respectively.

(15.54)
```
procedure Base2FFT(var y : longVector;
                Tcap, g : integer);

  const
    twopi = 6.283185;

  var
    a, c, t, i, j, k, l, P, Q : integer;
    W, theta, sine, cosine : real;
    yR, yI : real;

  function BitReverse(j, g : integer) : integer;

    var
      t, tl, r, l : integer;

  begin
    t := 0;
    r := j;
    for l := 1 to g do
      begin
        tl := r mod 2;
        t := t * 2 + tl;
        r := r div 2;
      end;
    BitReverse := t
  end; {BitReverse}

begin {Base2FFT}
  W := twopi/Tcap;
  P := Tcap;
  c := 0;

  for l := 1 to g do
    begin {l : this is the major loop}
      a := 0;
      t := 0;
      if l = 1 then
        Q := 1
      else
        Q := Q * 2;
```

```
    P := P div 2;

    for i := 0 to Q - 1 do
      begin {i}
        for k := 0 to P - 1 do
          begin {k : transform the subsequence}
            t := a + c;
            yR := y[t];
            yI := y[t + Tcap];
            y[t] := y[t] + y[t + P];
            y[t + Tcap] := y[t + Tcap] + y[t + P + Tcap];
            y[t + P] := yR - y[t + P];
            y[t + P + Tcap] := yI - y[t + P + Tcap];

            if l < g then
              begin {twiddle}
                theta := W * ((c * Q) mod Tcap);
                cosine := Cos(theta);
                sine := Sin(theta);
                yR := y[t + P];
                yI := y[t + P + Tcap];
                y[t + P] := yR * cosine + yI * sine;
                y[t + P + Tcap] := yI * cosine - yR * sine;
              end; {twiddle}

            c := c + 1;
          end; {k}
        c := 0;
        a := a + 2 * P;
      end; {i}
  end; {l : the major loop}

  for j := 1 to Tcap - 2 do
    begin {t : unscramble the vector}
      t := BitReverse(j, g);
      if t > j then
        begin {t}
          yR := y[t];
          yI := y[t + Tcap];
          y[t] := y[j];
          y[t + Tcap] := y[j + Tcap];
          y[j] := yR;
          y[j + Tcap] := yI;
        end; {t}
    end; {t : the vector is unscrambled}

end; {Base2FFT}
```

The above procedure can be seen as a specialisation of the mixed-radix FFT procedure presented in earlier sections. The specialisation extends to the algorithm for unscrambling the Fourier coefficients. Thus, in place of the function *tOfj*, the function *BitReverse* is used which maps from $j = j_1 + j_2 2 + \cdots + j_g 2^{g-1}$ to $t = j_1 2^{g-1} + j_2 2^{g-2} + \cdots + j_g$. This function exploits the fact that factors of $T = 2^g$ form a palindrome. The need for a lengthy evaluation of each permutation cycle, before applying it to the elements of the coefficient vector or discarding it, is avoided. Now, all of the permutation cycles are of length 2, which is to say that they involve nothing more than pairwise interchanges. The tth and jth elements of the vector of coefficients are interchanged if and only if it is found that $t(j) = t > j$.

FFT Algorithms for Real Data

So far, in our exposition of the FFT, the data sequence $y(t) = y^{re}(t) + iy^{im}(t)$ has been regarded as complex-valued. However, measurements taken in the real world are invariably real-valued, and so the task of transforming a sequence which is purely real often arises in practice. However, even when the elements of the imaginary vector are set to zero, the algorithm still performs the multiplications involving $y^{im}(t)$, and this is inefficient. There are several strategies which may be adopted which are aimed at overcoming such inefficiencies.

Let us begin by imagining that the object is to obtain the Fourier transforms $\phi(j)$ and $\delta(j)$ of two real sequences $f(t)$ and $d(t)$. This can be done via a single application of the FFT. The procedure is to construct a synthetic complex sequence

$$y(t) = f(t) + id(t) \tag{15.55}$$

and to recover $\phi(j)$ and $\delta(j)$ from its Fourier transform which is

$$\zeta(j) = \phi(j) + i\delta(j). \tag{15.56}$$

Here $\phi(j)$ and $\delta(j)$ are complex-valued sequences. However, their sum can be recast as the sum of a purely real sequence and a purely imaginary sequence. Thus

$$\begin{aligned}\zeta(j) &= \zeta^{re}(j) + i\zeta^{im}(j)\\ &= \{\zeta_e^{re}(j) + \zeta_o^{re}(j)\} + i\{\zeta_e^{im}(j) + \zeta_o^{im}(j)\},\end{aligned} \tag{15.57}$$

where

$$\begin{aligned}\zeta_e^{re}(j) &= \frac{1}{2}\{\zeta^{re}(j) + \zeta^{re}(T-j)\},\\ \zeta_e^{im}(j) &= \frac{1}{2}\{\zeta^{im}(j) + \zeta^{im}(T-j)\}\end{aligned} \tag{15.58}$$

are even functions, whilst

$$\begin{aligned}\zeta_o^{re}(j) &= \frac{1}{2}\{\zeta^{re}(j) - \zeta^{re}(T-j)\},\\ \zeta_o^{im}(j) &= \frac{1}{2}\{\zeta^{im}(j) - \zeta^{im}(T-j)\}\end{aligned} \tag{15.59}$$

are odd functions. Now, it is easy to see, with reference to equations (15.1) and (15.2), that, if $f(t)$ is purely real, then its Fourier transform $\phi(j)$ must have a real part which is even and an imaginary part which is odd. Conversely, if $id(t)$ is purely imaginary, then its Fourier transform $i\delta(j)$ must have a real part which is odd and an imaginary part which is even. Therefore, in comparing (15.56) with (15.57), it can be seen that

$$\phi(j) = \zeta_e^{re}(j) + i\zeta_o^{im}(j), \tag{15.60}$$

and that

$$\begin{aligned} i\delta(j) &= \zeta_o^{re}(j) + i\zeta_e^{im}(j) \quad \text{or, equivalently,} \\ \delta(j) &= \zeta_e^{im}(j) - i\zeta_o^{re}(j). \end{aligned} \tag{15.61}$$

These identities show how $\phi(j)$ and $\delta(j)$ can be recovered from $\zeta(j)$.

It should be recognised that the T-periodicity of the Fourier transform implies that $\zeta^{re}(T) = \zeta^{re}(0)$ and $\zeta^{im}(T) = \zeta^{im}(0)$, whence the definitions under (15.58) and (15.59) imply that $\zeta_e^{re}(0) = \zeta^{re}(0)$, $\zeta_e^{im}(0) = \zeta^{im}(0)$ and $\zeta_o^{re}(0) = \zeta_o^{im}(0) = 0$. Therefore, (15.60) and (15.61) imply that

$$\begin{aligned} \phi(0) &= \zeta^{re}(0), \\ \delta(0) &= \zeta^{im}(0); \end{aligned} \tag{15.62}$$

and these conditions are reflected in the procedure below which treats $\phi(0)$ and $\delta(0)$ separately from the remaining elements of the vectors. The procedure returns ϕ_t^{re} and ϕ_t^{im} as $f[t]$ and $f[Tcap+t]$ respectively. Likewise δ_t^{re} and δ_t^{im} are returned as $d[t]$ and $d[Tcap+t]$ respectively.

(15.63)
```
procedure TwoRealFFTs(var f, d : longVector
                      Tcap, g : integer);

  var
    y : longVector;
    t, Ncap : integer;

begin {TwoRealFFTs}

  Ncap := 2 * Tcap;
  for t := 0 to Tcap - 1 do
    begin
      y[t] := f[t];
      y[Tcap + t] := d[t]
    end;

  Base2FFT(y, Tcap, g);
```

```
f[0] := y[0];
f[Tcap] := 0;
d[0] := y[Tcap];
d[Tcap] := 0;

for t := 1 to Tcap - 1 do
  begin
    f[t] := (y[t] + y[Tcap - t])/2;
    f[Tcap + t] := (y[Tcap + t] - y[Ncap - t])/2;
    d[t] := (y[Tcap + t] + y[Ncap - t])/2;
    d[Tcap + t] := (y[Tcap - t] - y[t])/2
  end;

end; {TwoRealFFTs}
```

FFT for a Single Real-valued Sequence

The foregoing technique can be extended in order to transform a single real-valued sequence in an efficient manner. Given the real sequence $x(t) = \{x_0, x_1, \ldots, x_{N-1}\}$, where $N = 2T$, a synthetic complex-valued sequence $y(t)$ can be constructed by assigning the even-numbered elements of $x(t)$ to the real part and the odd-numbered elements to the imaginary part. Thus

$$\begin{aligned} y(t) &= x(2t) + ix(2t+1) \\ &= f(t) + id(t), \end{aligned} \tag{15.64}$$

where $t = 0, 1, \ldots, T-1$. The transform of $y(t)$ is denoted $\zeta(j) = \phi(j) + i\delta(j)$ as before, with $\phi(j)$ and $\delta(j)$ as the transforms of $f(t)$ and $d(t)$ respectively. To see how the transform of $x(t)$ can be constructed from $\phi(j)$ and $\delta(j)$, we may consider writing it in the following form:

$$\begin{aligned} T\xi(j) &= \sum_{t=0}^{N-1} x(t)W_N^{jt} \\ &= \sum_{t=0}^{T-1} x(2t)W_N^{j(2t)} + \sum_{t=0}^{T-1} x(2t+1)W_N^{j(2t+1)} \\ &= \sum_{t=0}^{T-1} x(2t)W_T^{jt} + W_N^j \sum_{t=0}^{T-1} x(2t+1)W_T^{jt} \\ &= T\big\{\phi(j) + W_N^j\delta(j)\big\}. \end{aligned} \tag{15.65}$$

Here, we have used $W_N^{2t} = \exp(-i2\pi\{2t\}/N) = \exp(-i2\pi t/T) = W_T^t$. Now, as in the previous instance, there are the identities

$$\begin{aligned} \phi(j) &= \zeta_e^{re}(j) + i\zeta_o^{im}(j), \\ \delta(j) &= \zeta_e^{im}(j) - i\zeta_o^{re}(j), \end{aligned} \tag{15.66}$$

which arise from the fact that $f(t)$ and $d(t)$ are real-valued sequences. On substituting these into the equation $\xi(j) = \phi(j) + W_N^j\delta(j)$ of (15.65) and using $W_N^j = \exp(-i\pi j/T) = \cos(\theta_j) - i\sin(\theta_j)$, where $\theta_j = \pi j/T$, we find that the real and imaginary parts of $\xi(j) = \xi^{re}(j) + i\xi^{im}(j)$ are given by

$$(15.67)\qquad \begin{aligned} \xi^{re}(j) &= \zeta_e^{re}(j) + \cos(\theta_j)\zeta_e^{im}(j) - \sin(\theta_j)\zeta_o^{re}(j), \\ \xi^{im}(j) &= \zeta_o^{im}(j) - \cos(\theta_j)\zeta_o^{re}(j) - \sin(\theta_j)\zeta_e^{im}(j), \end{aligned}$$

where the index j runs from 0 to $T-1$. The remainder of the sequence $\xi(j)$ for $j = T, \ldots, N-1$ is obtained from the condition that $\xi(N-j) = \xi^{re}(j) - i\xi^{im}(j)$ which arises from the fact that $x(t)$ itself is a real-valued sequence whose transform $\xi(j)$ has a real part which is even and an imaginary part which is odd. However, there is no need to recover the second half of the sequence since it is composed of values which are already present in the first half.

A further economy in computation arises from the half-wave symmetries whereby $\cos(\theta_j) = -\cos(\theta_{T-j})$ and $\sin(\theta_j) = \sin(\theta_{T-j})$. These, in conjunction with the fact that $\zeta_e^{re}(T-j) = \zeta_e^{re}(j)$ and $\zeta_o^{im}(T-j) = -\zeta_o^{im}(j)$, imply that

$$(15.68)\qquad \begin{aligned} \xi^{re}(T-j) &= \zeta_e^{re}(j) - \cos(\theta_j)\zeta_e^{im}(j) + \sin(\theta_j)\zeta_o^{re}(j), \\ \xi^{im}(T-j) &= -\zeta_o^{im}(j) - \cos(\theta_j)\zeta_o^{re}(j) - \sin(\theta_j)\zeta_e^{im}(j). \end{aligned}$$

Thus, the components of $\xi(T-j)$ are composed of the same elements as those of $\xi(j)$ and, therefore, they can be calculated at the same time.

Finally, by taking account of the T-periodicity of $\zeta^{re}(j)$ and $\zeta^{im}(j)$, it can be seen that the definitions under (15.58) imply that

$$(15.69)\qquad \begin{aligned} \zeta_e^{re}(0) &= \zeta_e^{re}(T) = \zeta^{re}(0), \\ \zeta_e^{im}(0) &= \zeta_e^{im}(T) = \zeta^{im}(0). \end{aligned}$$

Likewise, the conditions under (15.59) imply that

$$(15.70)\qquad \begin{aligned} \zeta_o^{re}(0) &= \zeta_o^{re}(T) = 0, \\ \zeta_o^{im}(0) &= \zeta_o^{im}(T) = 0. \end{aligned}$$

Therefore, since $\cos(\theta_0) = 1$ and $\cos(\theta_T) = -1$, it follows from (15.67) that

$$(15.71)\qquad \begin{aligned} \xi^{re}(0) &= \zeta^{re}(0) + \zeta^{im}(0), \\ \xi^{im}(0) &= 0, \\ \xi^{re}(T) &= \zeta^{re}(0) - \zeta^{im}(0), \\ \xi^{im}(T) &= 0. \end{aligned}$$

In constructing a procedure, attempts should be made to conserve computer storage. Thus, instead of creating the vector

$$(15.72)\qquad \begin{aligned} y &= [f_0, f_1, \ldots, f_{T-1}, d_0, d_1, \ldots, d_{T-1}] \\ &= [x_0, x_2, \ldots, x_{N-2}, x_1, x_3, \ldots, x_{N-1}] \end{aligned}$$

in a new array, one might wish to contain it in the space originally occupied by the data vector

$$x = [x_0, x_1, \ldots, x_{T-1}, x_T, x_{T+1}, \ldots, x_{N-1}]. \tag{15.73}$$

The process of rearranging the elements of x to form the vector y can be accomplished by the following procedure which is an adapted form of the procedure *ReOrder* of (15.50) which was used for unscrambling the product of the mixed-radix FFT:

(15.74)
```
procedure OddSort(Ncap : integer;
                  var y : longVector);

  var
    Tcap, t, j, k : integer;
    store : real;

begin
  Tcap := Ncap div 2;

  for j := 1 to Tcap − 1 do
    begin {j}
      k := j;
      if j > 1 then
        repeat {Test the cycle}
          k := (2 ∗ k) mod (2 ∗ Tcap − 1);
        until k <= j;
      if k = j then {ReOrder}
        begin {if}
          store := y[j];
          t := j;
          repeat
            k := t;
            t := (2 ∗ k) mod (2 ∗ Tcap − 1);
            y[k] := y[t];
          until t = j;
          y[k] := store;
        end; {if}
    end; {j}

end; {OddSort}
```

Further scope for saving computer storage derives from a fact already remarked upon, which is that the Fourier transform $\xi(j) = \xi^{re}(j) + i\xi^{im}(j); j \in \{0, 1, \ldots, N-1\}$ of the real-valued data sequence $x(t); t \in \{0, 1, \ldots, N-1\}$ contains only $N = 2T$ distinct elements. Therefore, in place of

$$\begin{aligned}\xi = \; & [\xi_0^{re}, \xi_1^{re}, \ldots, \xi_{T-1}^{re}, \xi_T^{re}, \xi_{T-1}^{re}, \ldots, \xi_2^{re}, \xi_1^{re}] \\ & + i[0, \xi_1^{im}, \ldots, \xi_{T-1}^{im}, 0, -\xi_{T-1}^{im}, \ldots, -\xi_2^{im}, -\xi_1^{im}],\end{aligned} \tag{15.75}$$

we need only record the vector

$$[\xi_0^{re}, \xi_1^{re}, \ldots, \xi_{T-1}^{re}, \xi_T^{re}, \xi_1^{im}, \ldots, \xi_{T-1}^{im}]. \tag{15.76}$$

Moreover, the latter may be created in the space occupied originally by the data vector. Only a small amount of extra space needs to be set aside to facilitate the work in hand.

The resulting procedure is as follows:

(15.77)
```
procedure CompactRealFFT(var x : longVector;
                              Ncap, g : integer);

  const
    pi = 3.1415926;

  var
    t, Tcap : integer;
    xReven, xRodd, xIeven, xIodd, store : real;
    theta, increment, sine, cosine : real;

begin {RealFFT}
  Tcap := Ncap div 2;
  increment := pi/Tcap;
  theta := 0;

  OddSort(Ncap, x);

  g := g - 1;
  Base2FFT(x, Tcap, g);

  for t := 1 to Tcap div 2 do
    begin
      theta := theta + increment;
      cosine := Cos(theta);
      sine := Sin(theta);

      xReven := (x[t] + x[Tcap - t])/2;
      xRodd := (x[t] - x[Tcap - t])/2;
      xIeven := (x[t + Tcap] + x[Ncap - t])/2;
      xIodd := (x[t + Tcap] - x[Ncap - t])/2;

      x[t] := xReven + cosine * xIeven - sine * xRodd;
      x[Tcap - t] := xReven - cosine * xIeven + sine * xRodd;
      x[t + Tcap] := xIodd - cosine * xRodd - sine * xIeven;
      x[Ncap - t] := -xIodd - cosine * xRodd - sine * xIeven;
    end;
  store := x[0];
```

$x[0] := x[0] + x[Tcap];$
$x[Tcap] := store - x[Tcap];$

end; {*RealFFT*}

Bibliography

[47] Bergland, G.D., (1969), *A Guided Tour of the Fast Fourier Transform*, IEEE Spectrum.

[72] Bracewell, R.N., (1986), *The Hartley Transform*, Oxford University Press, New York.

[75] Brigham, E.O., (1988), *The Fast Fourier Transform and its Applications*, Prentice-Hall, Englewood Cliffs, New Jersey.

[125] Cooley, J.W., and J.W. Tukey, (1965), An Algorithm for the Machine Calculation of Complex Fourier Series, *Mathematics of Computation*, **19**, 297–301.

[126] Cooley, J.W., P.A.W. Lewis and P.D. Welch, (1967), Historical Notes on the Fast Fourier Transform, *IEEE Transactions on Audio and Electroacoustics*, **AU-15**, 76–79.

[139] de Boor, C., (1980), FFT as Nested Multiplications with a Twist, *SIAM Journal on Scientific and Statistical Computing*, **1**, 173–178.

[206] Gentleman, W.M., and G. Sande, (1966), Fast Fourier Transforms—for Profit and Fun, in *1966 Fall Joint Computer Conference, American Federation of Information Processing Societies (AFIPS) Conference Proceedings*, **29**, 563–78.

[226] Granger, C.W.J., (1966), The Typical Spectral Shape of an Economic Variable, *Econometrica*, **34**, 150–161.

[252] Heideman, M.T., D.H. Johnson and C.S. Burrus, (1984), Gauss and the History of the Fast Fourier Transform, *IEEE ASSP Magazine*, **1**, 14–21.

[350] Monro, D.M., (1975), Complex Discrete Fast Fourier Transform, Algorithm AS 83, *Applied Statistics*, **24**, 153–160.

[351] Monro, D.M., (1976), Real Discrete Fast Fourier Transform, Algorithm AS 97, *Applied Statistics*, **25**, 166–172.

[462] Singleton, R.C., (1967), A Method for Computing the Fast Fourier Transform with Auxiliary Memory and Limited High-Speed Storage, *IEEE Transactions on Audio and Electroacoustics*, **AU-15**, 91–97.

[466] Singleton, R.C., (1969), An Algorithm for Computing the Mixed Radix Fast Fourier Transform, *IEEE Transactions on Audio and Electroacoustics*, **AU-17**, 93–103.

Time-Series Models

CHAPTER 16

Linear Filters

The theory of linear filtering was a major concern of electrical engineers long before it became practical to use digital computers to process the rapid sequences which come from sampling continuous-time signals at high frequencies. This helps to explain why much of the theory of digital signal processing still makes reference to the theory of analogue filters. In fact, some very effective digital filters are obtained simply by translating analogue designs into digital terms; and many engineers derive their intuition in matters of linear filtering by thinking in terms of the electrical circuits of analogue filters.

Digital filters fall into two classes which are know as the finite impulse-response (FIR) filters and the infinite impulse-response (IIR) filters. For linear filters, these classes correspond, respectively, to finite polynomial lag operators and rational lag operators. In analogue signal processing, the natural device is an IIR filter which is implemented in simple LCR circuits comprising resistors capacitors and inductors. An FIR filter is not easily implemented in an analogue electrical circuit; and this accounts for the preponderance of IIR filtering techniques in analogue signal processing.

In this chapter, we shall deal with FIR filters and IIR filters in succession. Some of the results relating to FIR filters will be directly relevant to the problem of smoothing the periodogram of a stationary time series, which is the subject of Chapter 23. Some of the results relating to IIR filters will be invoked in Chapter 19 in connection with the problem of signal extraction.

Frequency Response and Transfer Functions

Whenever we form a linear combination of successive elements of a discrete-time signal $x(t)$, we are performing an operation which is described as linear filtering. Such an operation can be represented by the equation

$$y(t) = \psi(L)x(t) = \sum_j \psi_j x(t-j) \tag{16.1}$$

wherein $\psi(L) = \{\cdots + \psi_{-1}L^{-1} + \psi_0 + \psi_1 L + \cdots\}$ is described as the linear filter.

The sequence $\{\psi_j\}$ of the filter's coefficients constitutes its response, on the output side, to an input in the form of a unit impulse. If the sequence is finite, then $\psi(L)$ is described as a moving-average filter or as a finite impulse-response (FIR) filter. When the filter produces an impulse response of an indefinite duration, it is called an infinite impulse-response (IIR) filter. The filter is said to be causal or backward-looking if none of its coefficients is associated with a negative power of L. In that case, the filter is available for real-time signal processing.

A practical filter, which is constructed from a limited number of components of hardware or software, must be capable of being expressed in terms of a finite number of parameters. Therefore, linear IIR filters which are causal correspond invariably to recursive structures of the form

$$\gamma(L)y(t) = \delta(L)x(t), \tag{16.2}$$

wherein $\gamma(L) = \gamma_0 + \gamma_1 L + \cdots + \gamma_g L^g$ and $\delta(L) = \delta_0 + \delta L + \cdots + \delta_d L^d$ are finite-degree polynomials of the lag operator. The leading coefficient of $\gamma(L)$ may be set to unity without loss of generality; and thus $y(t)$ in equation (16.2) becomes a function not only of past and present inputs but also of past outputs, which are described as feedback.

The recursive equation may be assimilated to the equation under (16.1) by writing it in rational form:

$$y(t) = \frac{\delta(L)}{\gamma(L)}x(t) = \psi(L)x(t). \tag{16.3}$$

On the condition that the filter is stable, the expression $\psi(L)$ stands for the series expansion of the ratio of the polynomials.

The input signal $x(t)$ may be regarded as a sequences of impulses. Therefore, the effect of the filter upon the signal is completely summarised by its response to a unit impulse. This provides the so-called time-domain description of the filter. However, the signal may be represented equally as a Fourier combination of trigonometrical or complex-exponential functions. Thus, for example, if $x(t)$ is generated by a stationary stochastic process, then it may be represented by the Fourier–Stieltjes integral

$$x(t) = \int_{-\pi}^{\pi} e^{i\omega t} dZ(\omega), \tag{16.4}$$

wherein the $dZ(\omega)$ are the discontinuous increments of a cumulative weighting function. (See Chapter 18).

The effects of a linear filter upon the cyclical components which constitute a signal are twofold. First, the filtering is liable to alter the amplitude of any component. This effect, which will vary according to the frequency of the component, is described as the gain of the filter. Secondly, the filter will translate the components along the time axis; and, for any component of a given frequency, there will be a corresponding displacement in terms of an alteration of the phase angle.

A linear filtering operation does not alter the frequencies of the components which constitute the signal; so that, if a simple sinusoidal component with a given frequency is passed through a linear filter, then the output will be a sinusoid of the same frequency.

To understand these results, let us consider the case of an elementary complex exponential function which is one of the components subsumed under the integral of (16.4):

$$x(t) = e^{i\omega t} = \cos(\omega t) + i\sin(\omega t). \tag{16.5}$$

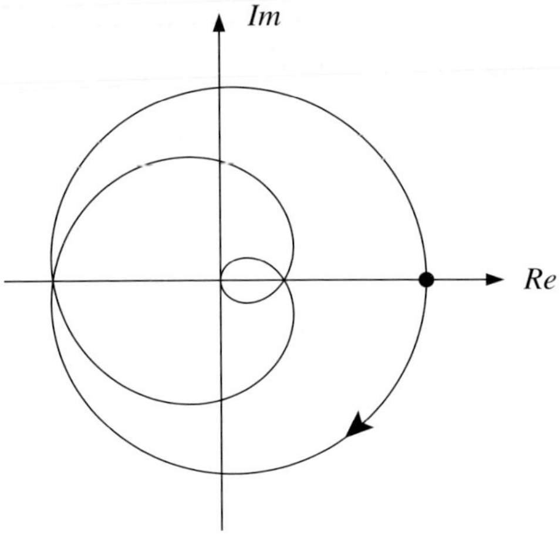

Figure 16.1. The path described in the complex plane by the frequency-response function $\psi(\omega)$ of the differencing filter $\psi(L) = L^2 - L^3$ as ω increases from $-\pi$ to π. The corresponding gain and phase functions are given by Figures 16.2 and 16.3. The trajectory originates in the point on the real axis marked by a dot. It traverses the origin when $\omega = 0$ and it returns to the dot when $\omega = \pi$.

When this is passed through the filter, it becomes

$$(16.6)\qquad \begin{aligned} y(t) = \psi(L)e^{i\omega t} &= \sum_j \psi_j e^{i\omega(t-j)} \\ &= \Big\{\sum_j \psi_j e^{-i\omega j}\Big\}e^{i\omega t} = \psi(\omega)e^{i\omega t}. \end{aligned}$$

Thus the effects of the filter are summarised by the complex-valued frequency-response function

$$(16.7)\qquad \psi(\omega) = \sum_j \psi_j e^{-i\omega j}.$$

This is just the discrete-time Fourier transform of the sequence of filter coefficients. Equally, it is the z-transform $\psi(z^{-1}) = \sum \psi_j z^{-1}$ evaluated at the points $z = e^{i\omega}$. Notice that this transform is in terms of z^{-1} rather than z.

There is a one-to-one correspondence between the coefficients and their Fourier transform. Therefore, the frequency-domain description of the filter, which is provided by the frequency-response function, is equivalent to the time-domain description, which is provided by the impulse-response function.

As ω progresses from $-\pi$ to π, or, equally, as $z = e^{i\omega}$ travels around the unit circle, the frequency-response function defines a trajectory in the complex plane which becomes a closed contour when ω reaches π. The points on the trajectory are characterised by their polar coordinates. These are the modulus $|\psi(\omega)|$, which is the length of the radius vector joining $\psi(\omega)$ to the origin, and the argument $\mathrm{Arg}\{\psi(\omega)\} = -\theta(\omega)$ which is the (anticlockwise) angle in radians which the radius makes with the positive real axis.

To demonstrate the effects of the filter in these terms, consider writing the frequency-response function in polar form:

$$\psi(\omega) = |\psi(\omega)|e^{-i\theta(\omega)} = |\psi(\omega)|\left[\cos\{\theta(\omega)\} - i\sin\{\theta(\omega)\}\right]. \tag{16.8}$$

Then the final expression under (16.6) becomes

$$\begin{aligned} y(t) &= |\psi(\omega)|e^{i\{\omega t-\theta(\omega)\}} \\ &= |\psi(\omega)|\left[\cos\{\omega t - \theta(\omega)\} + i\sin\{\omega t - \theta(\omega)\}\right]. \end{aligned} \tag{16.9}$$

This shows that the amplitude of the input signal, which is the complex exponential function $x(t) = e^{i\omega t}$ of (16.5), is multiplied by the gain $|\psi(\omega)|$, due to the modulus of the frequency-response function. It also shows that the phase of the signal is displaced by an angle of $\theta(\omega)$, due to the argument of the function.

If the phase shift is divided by ω, then it becomes the phase delay, which is a measure of the time delay experienced by the signal $x(t)$ in passing through the filter:

$$\tau(\omega) = \frac{\theta(\omega)}{\omega}. \tag{16.10}$$

The group delay of the filter is the derivative of the phase function with respect to ω:

$$\tilde{\tau}(\omega) = \frac{d\theta(\omega)}{d\omega}. \tag{16.11}$$

The gain and the phase are periodic functions of the angular frequency ω which are completely characterised by the values which they take over the interval $(-\pi, \pi]$. Moreover, the functions are often plotted only for values of ω in the interval $[0, \pi]$ in the knowledge that, for real-valued impulse responses, the gain $|\psi(-\omega)| = |\psi(\omega)|$ is an even function and the phase $\theta(-\omega) = -\theta(\omega)$ is an odd function.

Whereas the frequency response $\psi(\omega)$ is an analytic function, the corresponding argument and modulus are not. The slope of the modulus may become discontinuous when the modulus itself assumes a value of zero. The argument also is liable to show discontinuities. It increases by jumps of π at the points where $|\psi(\omega)|$ has a zero value and a discontinuous slope. It also jumps to $-\pi$ whenever it looks set to exceed a value of π. These features can be understood in reference to a diagram of the trajectory of $\psi(\omega)$. (See Figure 16.1 and the related diagrams of Figures 16.2 and 16.3.) A jump in $\theta(\omega)$ from π to $-\pi$ occurs whenever the trajectory crosses the real axis to the left of the origin. A jump of π occurs whenever the trajectory passes through the origin, which happens whenever $z = e^{i\omega}$ coincides with a zero of $\psi(z^{-1})$ located on the unit circle.

These various discontinuities are consequences of definitions which may be amended easily. A trigonometrical function which has a lag of $\theta^* \in [\pi, 2\pi]$ is indistinguishable, in effect, from one which has a lead of $2\pi - \theta^* = \theta \in [0, \pi]$; and, according to the definition of the Arg function, when a lag exceeds π it is

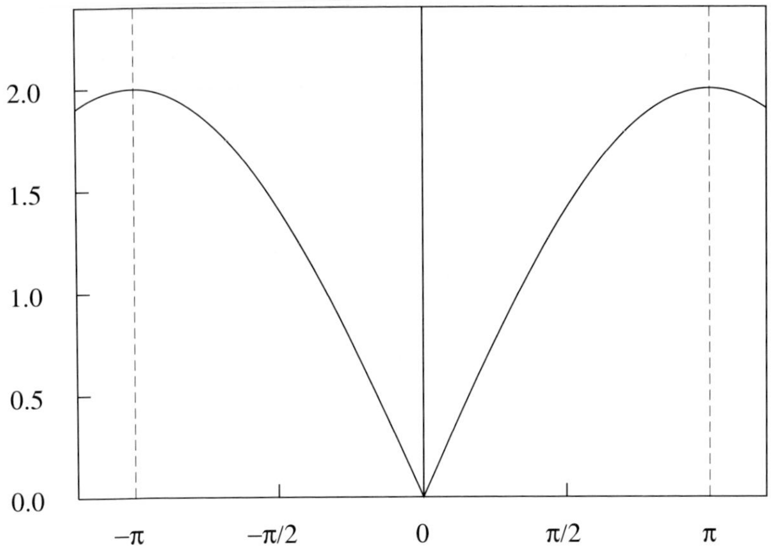

Figure 16.2. The gain $|\psi(\omega)|$ of the differencing filter $\psi(L) = L^2 - L^3$.

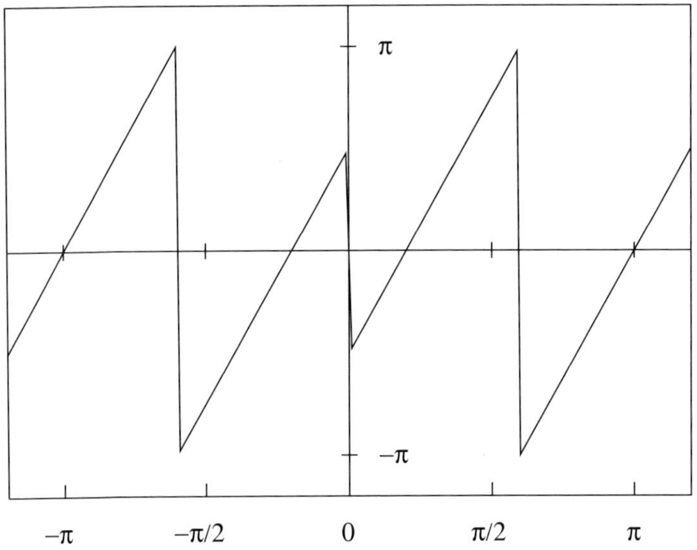

Figure 16.3. The phase $-\text{Arg}\{\psi(\omega)\}$ of the differencing filter $\psi(L) = L^2 - L^3$.

instantly commuted into a lead. The corresponding jump of 2π would not occur if the phase lag were allowed to accumulate continuously to reflect the fact that a linear filter imposes virtually the same delay upon components of the signal with adjacent frequencies. The jumps in $\theta(\omega)$ of π, which occur when the trajectory of $\psi(\omega)$ passes through the origin, are the consequence of not allowing the modulus to change its sign. The sign change of $-1 = e^{\pm i\pi}$ is bound to be absorbed by the phase function. If the appropriate sign were applied instead to the modulus, then there would be no phase jump.

Example 16.1. Consider the application of a four-point moving average to an quarterly economic time series—the object being to remove the seasonal component from the series. Imagine that the data is processed with a three-month delay. Then, if $x(t)$ is the quarterly series, the deseasonalised series, which is published at time t, will be given by

$$y(t) = \frac{1}{4}\{x(t-1) + \cdots + x(t-4)\}. \tag{16.12}$$

The z-transform of the corresponding filter is

$$\begin{aligned}\psi(z^{-1}) &= \frac{1}{4}\left(z^{-1} + z^{-2} + z^{-3} + z^{-4}\right)\\ &= \frac{1}{4z^4}(z+1)(z+i)(z-i).\end{aligned} \tag{16.13}$$

The factorisation indicates that $\psi(z^{-1})$ has roots of $z = -1$ and $z = \pm i$. These correspond to points on the unit circle at the angles $\omega = \pm\pi$ and $\omega = \pm\pi/2$. Figure 16.4 shows that the gain is zero at the latter frequencies and unity at the zero frequency. The zero gain at $\omega = \pm\pi/2$ corresponds to the removal of a four-period (i.e. annual) cycle, whilst the unit gain at zero frequency indicates that the trend in the series is preserved.

To understand the phase function which is depicted in Figure 16.5, consider writing the frequency-response function as

$$\begin{aligned}\psi(\omega) &= e^{-i\omega}\frac{1}{4}\left(1 + e^{-i\omega} + e^{-i2\omega} + e^{-i3\omega}\right)\\ &= e^{-i5\omega/2}\frac{1}{4}\left(e^{i3\omega/2} + e^{i\omega/2} + e^{-i\omega/2} + e^{-i3\omega/2}\right)\\ &= \frac{1}{2}\left\{\cos\left(\tfrac{1}{2}\omega\right) + \cos\left(\tfrac{3}{2}\omega\right)\right\}e^{-i5\omega/2}.\end{aligned} \tag{16.14}$$

The final expression is the product of a real-valued factor and a complex exponential factor. The real-valued factor is the amplitude function of which the modulus $|\psi(\omega)|$ is the absolute value. The exponential term bears an exponent which corresponds to a continuous version of the phase function from which the jumps have been eliminated. This function indicates that the filter imposes a delay of $2\frac{1}{2}$ periods upon the deseasonalised series which is uniform over the range of frequencies.

The linearity of the phase effect is apparent in Figure 16.5. There are four jumps in the phase function $\theta(\omega)$ in the interval $[0, \pi]$. The first jump of 2π radians occurs at $\omega = (2/5)\pi$, which is where the value of $5\omega/2$ equals π. The second jump, which is of π radians, occurs at $\omega = \pi$ where the real-valued amplitude function changes sign. The remaining jumps have similar explanations.

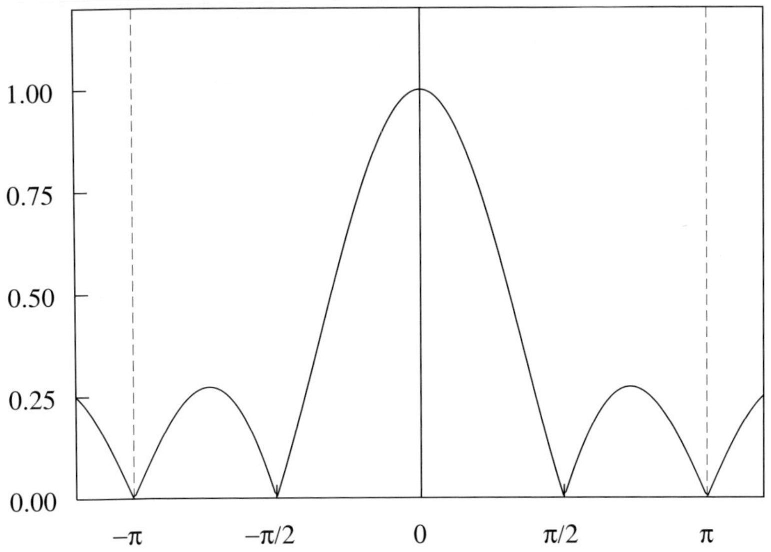

Figure 16.4. The gain of the deseasonalising filter $\psi(L) = \frac{1}{4}(L + \cdots + L^4)$.

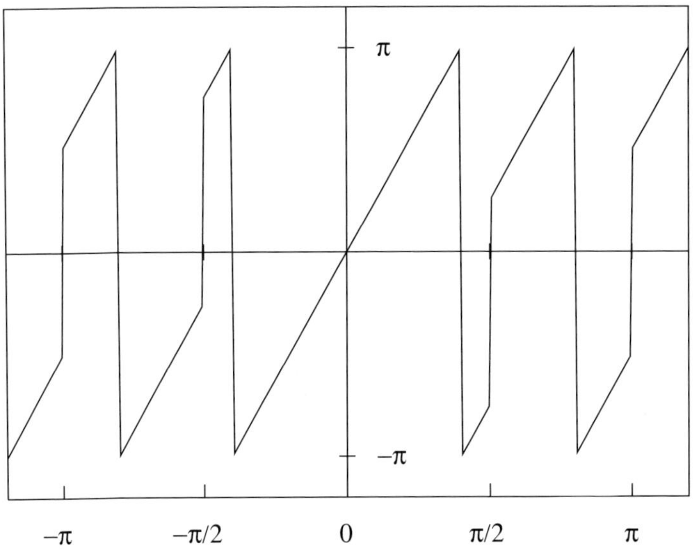

Figure 16.5. The phase effect of the deseasonalising filter $\psi(L) = \frac{1}{4}(L + \cdots + L^4)$.

Computing the Gain and Phase Functions

To provide more explicit expressions for the gain and the phase displacement, which will help in computing, consider decomposing $\psi(\omega)$ into its real and imaginary components. Then (16.7) becomes

$$\psi(\omega) = \psi^{re}(\omega) + i\psi^{im}(\omega) \tag{16.15}$$

with

$$\psi^{re}(\omega) = \sum_j \psi_j \cos(\omega j) \quad \text{and} \quad \psi^{im}(\omega) = -\sum_j \psi_j \sin(\omega j), \tag{16.16}$$

The squared-gain of the filter is given by

$$|\psi(\omega)|^2 = \{\psi^{re}(\omega)\}^2 + \{\psi^{im}(\omega)\}^2, \tag{16.17}$$

whilst its phase displacement is given by

$$\theta(\omega) = -\text{Arg}\{\psi^{re}(\omega) + i\psi^{im}(\omega)\}. \tag{16.18}$$

In general, the filter $\psi(L) = \delta(L)/\gamma(L)$ is expressible as the ratio of two polynomials in the lag operator. There is no need to generate the sequence $\{\psi_j\}$ of filter coefficients by expanding the rational function. Instead, the real and imaginary parts of $\psi(\omega)$ may be obtained via the equation

$$\begin{aligned}\psi(\omega) &= \frac{\delta^{re}(\omega) + i\delta^{im}(\omega)}{\gamma^{re}(\omega) + i\gamma^{im}(\omega)} = \frac{\{\delta^{re} + i\delta^{im}\}\{\gamma^{re} - i\gamma^{im}\}}{\{\gamma^{re}\}^2 + \{\gamma^{im}\}^2} \\ &= \frac{\{\delta^{re}\gamma^{re} + \delta^{im}\gamma^{im}\} + i\{\delta^{im}\gamma^{re} - \delta^{re}\gamma^{im}\}}{\{\gamma^{re}\}^2 + \{\gamma^{im}\}^2}.\end{aligned} \tag{16.19}$$

The following Pascal procedure calculates both the gain and the phase for a given value of the ω.

(16.20)

```
procedure GainAndPhase(var gain, phase : real;
            delta, gamma : vector;
            omega : real;
            d, g : integer);

  var
    i, j : integer;
    gamm, delt, psi : complex;
    numerator, denominator : real;

begin
  delt.re := 0.0;
  delt.im := 0.0;
```

```
    for j := 0 to d do
      begin
        delt.re := delt.re + delta[j] * Cos(omega * j);
        delt.im := delt.im + delta[j] * Sin(omega * j);
      end;

    gamm.re := 0.0;
    gamm.im := 0.0;
    for j := 0 to g do
      begin
        gamm.re := gamm.re + gamma[j] * Cos(omega * j);
        gamm.im := gamm.im + gamma[j] * Sin(omega * j);
      end;

    psi.re := delt.re * gamm.re + delt.im * gamm.im;
    psi.im := delt.im * gamm.re - delt.re * gamm.im;
    numerator := Sqr(psi.re) + Sqr(psi.im);
    denominator := Sqr(gamm.re) + Sqr(gamm.im);
    gain := Sqrt(numerator)/denominator;
    phase := -Arg(psi);

  end; {GainAndPhase}
```

The procedure uses a special function for calculating $-\theta(\omega) = \text{Arg}\{\psi^{re}(\omega) + i\psi^{im}(\omega)\}$. The value of Arg$(z)$ is simply the angle which the vector $z = \alpha + i\beta$ makes with the positive real axis. In FORTRAN, this would be found using the function ATAN2 which takes as its arguments the real numbers α and β. In Pascal, we have to make use of the function arctan(x) which finds the angle in radians corresponding to a positive or negative tangent $x = \beta/\alpha$. The difference between arctan(x) and Arg(z) is illustrated in Figure 16.6.

There is some variation amongst texts of signal processing and elsewhere in the definition of the arctan function. Thus, for example, Oppenheim and Schafer [372, p. 215] regard it as synonymous with the Arg function. We shall adopt the definition of arctan which is common in programming languages. Moreover, we shall persist in using $\tan^{-1}(\beta/\alpha)$ to denote a function which is synonymous with Arg$(\alpha + i\beta)$ for all $\beta \geq 0$.

The Arg function, which is defined for all α and β, is coded in Pascal as follows:

(16.21)
```
  function Arg(psi : complex) : real;

    var
      theta : real;

  begin
    if psi.re = 0 then
      theta := Sign(psi.im) * pi/2;
    if psi.re <> 0 then
```

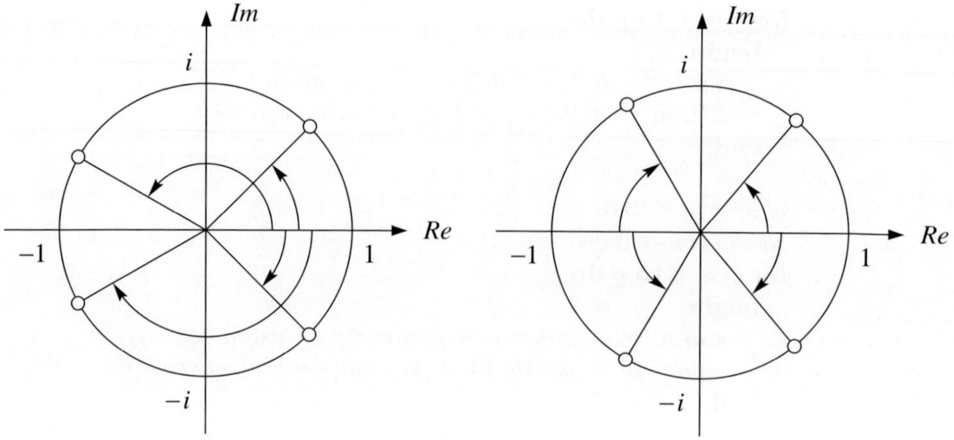

Figure 16.6. The function $\text{Arg}(\alpha + i\beta)$ finds the angle in radians which $z = \alpha + i\beta$ makes with the positive real axis. The function $\arctan(\beta/\alpha)$ finds the angle in radians corresponding to a positive or negative tangent.

```
    theta := ArcTan(psi.im/psi.re);
  if psi.re < 0 then
    theta := theta + Sign(psi.im) * pi;
  Arg := theta;
end;
```

An alternative way of computing the squared gain is available if there is no interest in calculating the phase. Consider the equation

$$
\begin{aligned}
|\psi(\omega)|^2 &= \Big(\sum_j \psi_j e^{-i\omega j}\Big)\Big(\sum_k \psi_k e^{i\omega k}\Big) \\
&= \sum_j \sum_k \psi_j \psi_k e^{-i\omega(j-k)} \\
&= \sum_\tau \Big(\sum_j \psi_j \psi_{j-\tau}\Big) e^{-i\omega\tau}; \quad \tau = j - k.
\end{aligned}
\tag{16.22}
$$

On defining

$$\phi_\tau = \sum_j \psi_j \psi_{j-\tau}, \tag{16.23}$$

this can be written as

$$|\psi(\omega)|^2 = \sum_\tau \phi_\tau e^{-i\omega\tau} = \phi_0 + 2\sum_{\tau>0} \phi_\tau \cos(\omega\tau), \tag{16.24}$$

where the final expression depends upon the conditions $\phi_\tau = \phi_{-\tau}$ and the identity $(e^{i\omega\tau} + e^{-i\omega\tau}) = 2\cos(\omega\tau)$ which is from the familiar Euler equation. The squared gain is calculated from (16.23) and (16.24).

The same expressions for the squared gain can be derived by expanding equation (16.17):

$$
\begin{aligned}
|\psi(\omega)|^2 &= \Big\{\sum_j \psi_j \cos(\omega j)\Big\}^2 + \Big\{\sum_j \psi_j \sin(\omega j)\Big\}^2 \\
&= \Big\{\sum_j\sum_k \psi_j\psi_k \cos(\omega j)\cos(\omega k)\Big\} + \Big\{\sum_j\sum_k \psi_j\psi_k \sin(\omega j)\sin(\omega k)\Big\}.
\end{aligned} \tag{16.25}
$$

The identity $\cos(A)\cos(B) + \sin(A)\sin(B) = \cos(A - B)$ indicates that this may be written as

$$
\begin{aligned}
|\psi(\omega)|^2 &= \sum_j\sum_k \psi_j\psi_k \cos(\omega[j-k]) \\
&= \sum_\tau \phi_\tau \cos(\omega\tau).
\end{aligned} \tag{16.26}
$$

In the final expression, the sum is over positive and negative values of τ. Since $\phi_\tau = \phi_{-\tau}$ and since the cosine is an even function, the final expression may be rewritten as the one-sided sum of (16.24).

In the case of an FIR filter, these formulae lend themselves to more rapid computation than do those which are incorporated in the Pascal procedure. However, they do not assist us in computing the phase.

The interest in the phase effect of linear filtering varies greatly amongst practical applications. Whereas it is usually necessary to take account of the phase effect in communications engineering, the effect is of minor interest in the analysis of random mechanical vibrations.

The Poles and Zeros of the Filter

The characteristics of a linear filter $\psi(L) = \delta(L)/\gamma(L)$, which are manifested in its frequency-response function, can be explained in terms of the location in the complex plane of the poles and zeros of $\psi(z^{-1}) = \delta(z^{-1})/\gamma(z^{-1})$ which include the roots of the constituent polynomials $\gamma(z^{-1})$ and $\delta(z^{-1})$. Consider, therefore, the expression

$$
\psi(z^{-1}) = z^{g-d}\frac{\delta_0 z^d + \delta_1 z^{d-1} + \cdots + \delta_d}{\gamma_0 z^g + \gamma_1 z^{g-1} + \cdots + \gamma_g}. \tag{16.27}
$$

This stands for a causal or backward-looking filter. In fact, the restriction of causality is unnecessary, and the action of the filter can be shifted in time without affecting its essential properties. Such a shift would be represented by multiplying the filter by a power of z. There would be no effect upon the gain of the filter, whilst the effect upon the phase would be linear, in the sense that each component of a signal, regardless of its frequency, would be advanced (if the power were positive) or delayed (if the power were negative) by the same amount of time.

The numerator and denominator of $\psi(z^{-1})$ may be factorised to give

$$(16.28) \qquad \psi(z^{-1}) = z^{g-d}\frac{\delta_0}{\gamma_0}\frac{(z-\mu_1)(z-\mu_2)\cdots(z-\mu_d)}{(z-\kappa_1)(z-\kappa_2)\cdots(z-\kappa_g)},$$

where $\mu_1, \mu_2, \ldots, \mu_d$ are zeros of $\psi(z^{-1})$ and $\kappa_1, \kappa_2, \ldots, \kappa_g$ are poles. The term z^{g-d} contributes a further g zeros and d poles at the origin. If these do not cancel completely, then they will leave, as a remainder, a positive or negative power of z whose phase-shifting effect has been mentioned above.

The BIBO stability condition requires that $\psi(z^{-1})$ must be finite-valued for all z with $|z| \geq 1$, for which it is necessary and sufficient that $|\kappa_j| < 1$ for all $j = 1, \ldots, g$.

The effect of the filter can be assessed by plotting its poles and zeros on an Argand diagram. The frequency-response function is simply the set of the values which are assumed by the complex function $\psi(z^{-1})$ as z travels around the unit circle; and, at any point on the circle, we can assess the contribution which each pole and zero makes to the gain and phase of the filter. Setting $z = e^{i\omega}$ in (16.28), which places z on the circumference of the unit circle, gives

$$(16.29) \qquad \psi(e^{-i\omega}) = e^{i(g-d)\omega}\frac{\delta_0}{\gamma_0}\frac{(e^{i\omega}-\mu_1)(e^{i\omega}-\mu_2)\cdots(e^{i\omega}-\mu_d)}{(e^{i\omega}-\kappa_1)(e^{i\omega}-\kappa_2)\cdots(e^{i\omega}-\kappa_g)}.$$

The generic factors in this expression can be written in polar form as

$$(16.30) \qquad \begin{aligned} e^{i\omega} - \mu_j &= |e^{i\omega} - \mu_j| e^{i\phi_j(\omega)} \\ &= \rho_j(\omega) e^{i\phi_j(\omega)} \end{aligned} \quad \text{and} \quad \begin{aligned} e^{i\omega} - \kappa_j &= |e^{i\omega} - \kappa_j| e^{i\varphi_j(\omega)} \\ &= \lambda_j(\omega) e^{i\varphi_j(\omega)}. \end{aligned}$$

When the frequency-response function as a whole is written in polar form, it becomes $\psi(\omega) = |\psi(\omega)|e^{-i\theta(\omega)}$, with

$$(16.31) \qquad \begin{aligned} |\psi(e^{-i\omega})| &= \left|\frac{\delta_0}{\gamma_0}\right| \frac{|e^{i\omega}-\mu_1|\;|e^{i\omega}-\mu_2|\cdots|e^{i\omega}-\mu_d|}{|e^{i\omega}-\kappa_1|\;|e^{i\omega}-\kappa_2|\cdots|e^{i\omega}-\kappa_g|} \\ &= \left|\frac{\delta_0}{\gamma_0}\right| \frac{\prod \rho_j(\omega)}{\prod \lambda_j(\omega)} \end{aligned}$$

and

$$(16.32) \qquad \theta(\omega) = (d-g)\omega - \{\phi_1(\omega) + \cdots + \phi_d(\omega)\} + \{\varphi_1(\omega) + \cdots + \varphi_g(\omega)\}.$$

The value of $\lambda_j = |e^{i\omega} - \kappa_j|$ is simply the distance from the pole κ_j to the point $z = e^{i\omega}$ on the unit circle whose radius makes an angle of ω with the positive real axis. It can be seen that the value of λ_j is minimised when $\omega = \text{Arg}(\kappa_j)$ and maximised when $\omega = \pi + \text{Arg}(\kappa_j)$. Since λ_j is a factor in the denominator of the function $|\psi(\omega)|$, it follows that the pole κ_j makes its greatest contribution to the gain of the filter when $\omega = \text{Arg}(\kappa_j)$ and its least contribution when $\omega = \pi + \text{Arg}(\kappa_j)$. Moreover, if κ_j is very close to the unit circle, then its contribution to the gain at $\omega = \text{Arg}(\kappa_j)$

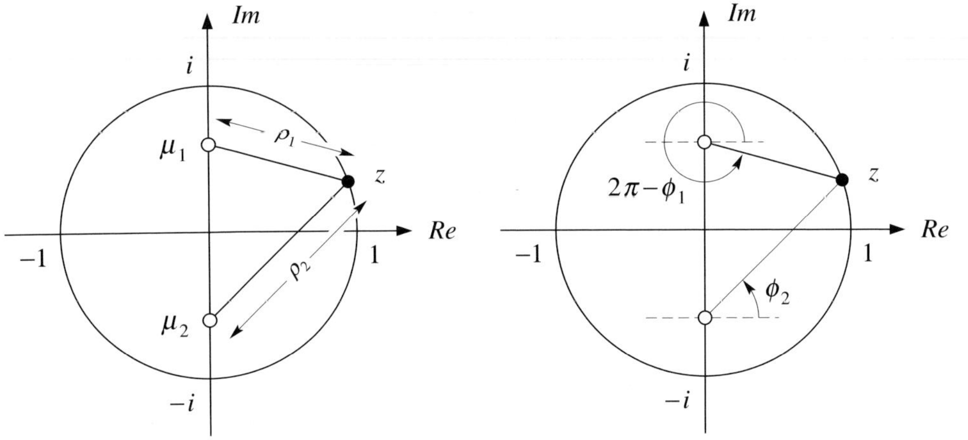

Figure 16.7. The pair of conjugate zeros $\mu_j, j = 1, 2$ are evaluated at the point $z = e^{i\omega}$ on the unit circle by finding the moduli $\rho_j = |z - \mu_j|$ and the arguments $\phi_j = \text{Arg}(z - \mu_j)$ of the corresponding factors.

will be very large. The effect of the zeros upon the gain of the filter is the opposite of the effect of the poles. In particular, a zero μ_j which lies on the perimeter of the unit circle will cause the gain of the filter to become zero at the frequency value which coincides with the zero's argument—that is to say, when $\omega = \text{Arg}(\mu_j)$.

The effect of the placement of the poles and zeros upon the phase of the filter may also be discerned from the Argand diagram. Thus it can be seen that the value of the derivative $d\text{Arg}(e^{i\omega} - \mu_j)/d\omega$ is maximised an the point on the circle where $\omega = \text{Arg}(\mu_j)$. Moreover, the value of the maximum increases with the diminution of $|e^{i\omega} - \mu_j|$, which is the distance between the root and the point on the circle.

The results of this section may be related to the argument principle which is stated under (3.127). According to the principle, the number of times the trajectory of a function $f(z)$ encircles the origin as $z = e^{i\omega}$ travels around the unit circle is equal to the number N of the zeros of $f(z)$ which lie within the circle less the number P of the poles which lie within the circle.

In applying the principle in the present context, one must observe the fact that the polynomials comprised by the rational function $\psi(z^{-1})$ are in terms of negative powers of z. For, on factorising the numerator polynomial $\delta(z^{-1}) = \delta_0 + \delta_1 z^{-1} + \cdots + \delta_d z^{-d}$, it is found that $\delta(z^{-1}) = \delta_0 \prod_j (1 - \mu_j/z) = z^{-d}\delta_0 \prod_j (z - \mu)$, which indicates that the polynomial contributes d poles as well as d zeros to the rational function.

Example 16.2. Consider the FIR filter

$$\psi(L) = (1 - \mu_1 L)(1 - \mu_2 L) = 1 - (\mu_1 + \mu_2)L + \mu_1\mu_2 L^2. \tag{16.33}$$

With z^{-1} in place of L, this becomes

$$\psi(z^{-1}) = (1 - \mu_1 z^{-1})(1 - \mu_2 z^{-1}) = \frac{(z - \mu_1)(z - \mu_2)}{z^2}, \tag{16.34}$$

from which it can be seen that there is a double pole located at zero and two zeros at the points μ_1 and μ_2. We shall assume for, the sake of generality, that the zeros are conjugate complex numbers $\mu_1 = \mu = \rho e^{i\theta}$ and $\mu_2 = \mu^* = \rho e^{-i\theta}$. Then, with $z = e^{i\omega}$, the first factor $1 - \mu_1 z$ becomes

$$\begin{aligned}1 - \rho e^{i\theta}e^{-i\omega} &= 1 - \rho\{\cos(\theta - \omega) + i\sin(\theta - \omega)\}\\ &= 1 - \rho\cos(\omega - \theta) + i\rho\sin(\omega - \theta).\end{aligned} \tag{16.35}$$

The case where μ is real can be accommodated simply by setting $\theta = 0$.

The contribution of this factor to the squared gain of the filter is

$$\begin{aligned}|1 - \rho e^{i\theta}e^{-i\omega}|^2 &= \{1 - \rho e^{i(\theta-\omega)}\}\{1 - \rho e^{-i(\theta-\omega)}\}\\ &= 1 - \rho\{e^{i(\theta-\omega)} + e^{-i(\theta-\omega)}\} + \rho^2\\ &= 1 - 2\rho\cos(\theta - \omega) + \rho^2;\end{aligned} \tag{16.36}$$

and it is manifest that, as the geometry of Figure 16.7 suggests, this is minimised when $\omega = \theta$, which is when the cosine takes the value of 1. Notice also that the same result would be obtained from $z - \mu_1$, which is to say the extra factor z^{-1}, which corresponds to a pole at zero, has no effect upon the gain of the filter. The contribution of the conjugate factor $1 - \rho e^{-i\theta}e^{-i\omega}$ is

$$|1 - \rho e^{-i\theta}e^{-i\omega}|^2 = 1 - 2\rho\cos(\omega + \theta) + \rho^2. \tag{16.37}$$

The contribution to the phase of the first factor is $-\mathrm{Arg}\big(1 - \rho e^{i(\theta-\omega)}\big)$. It can be seen, in reference to the expression for a complex exponential under (16.5), that $\mathrm{Arg}(1 - \rho e^{i\theta-\omega}) = 0$ when $\omega = \theta$. For a given value of ρ, the variations in the value of θ have the effect simply of shifting the phase curve along the horizontal axis. The combined contributions to the phase of the factor $1 - \rho e^{i\phi}e^{-i\omega}$ and its conjugate $1 - \rho e^{-i\phi}e^{-i\omega}$ are plotted in Figure 16.8.

It is sometimes useful to have an explicit expression for the group delay attributable to the factor of the filter polynomial. It can be shown that

$$-\frac{d}{d\omega}\mathrm{Arg}\{1 - \rho e^{i(\theta-\omega)}\} = \frac{\rho^2 - \rho\cos(\omega - \theta)}{1 - 2\rho\cos(\omega - \theta) + \rho^2}. \tag{16.38}$$

Example 16.3. Consider the second-order IIR filter

$$\psi(L) = \frac{1}{(1 - \kappa L)(1 - \kappa^* L)},$$

with $\kappa = \lambda e^{i\varphi}$, where $|\lambda| < 1$ to ensure stability. The effects of this filter can be inferred from the effects of the corresponding IIR filter of which it represents the inverse. That is to say, at any particular frequency, the product of the gain of the FIR filter with that of the inverse IIR filter will be unity. Therefore, a peak in the gain of the IIR filter corresponds to a trough in the gain of the FIR filter. The phase effect of the IIR filter is simply the negative of the phase effect of the inverse FIR filter.

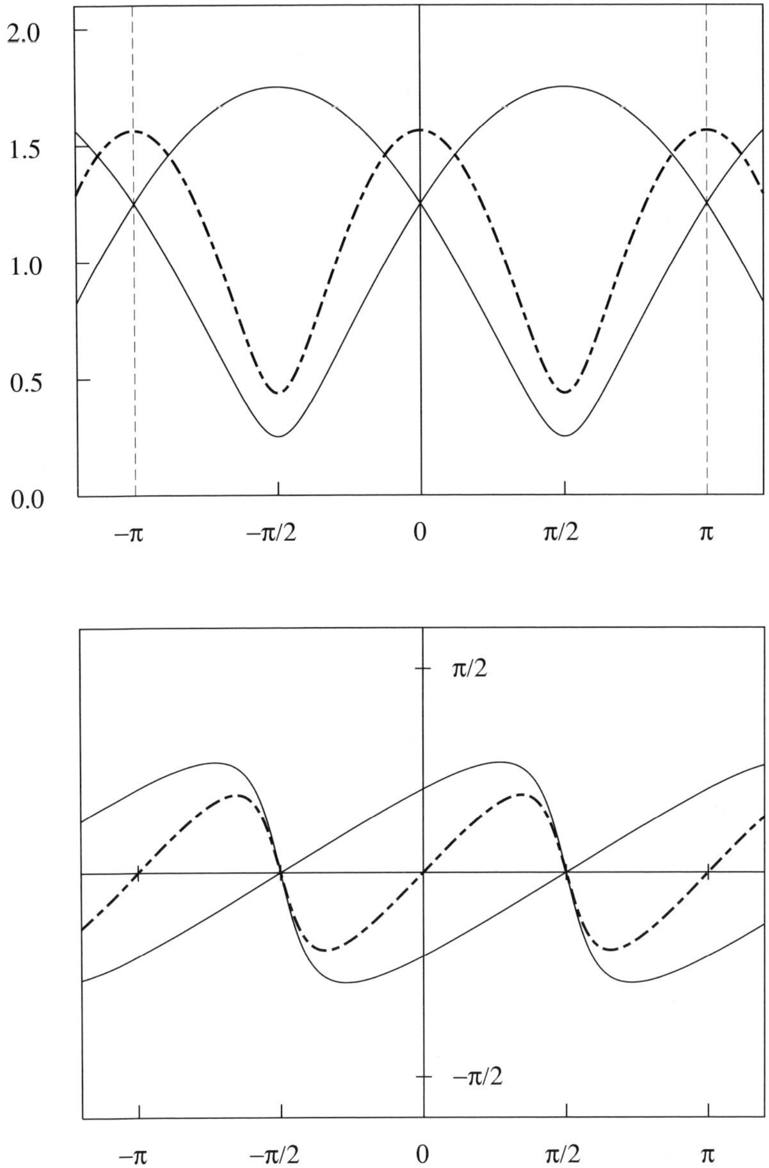

Figure 16.8. The gain and the phase effect of the second-order FIR filter $(1-\mu L)(1-\mu^* L)$ where $\mu = \rho e^{i\theta}$ with $\rho = 0.5$ and $\theta = \pi/2$. The contributions of the individual zeros are represented by the continuous lines. Their joint contribution is represented by the dashed lines.

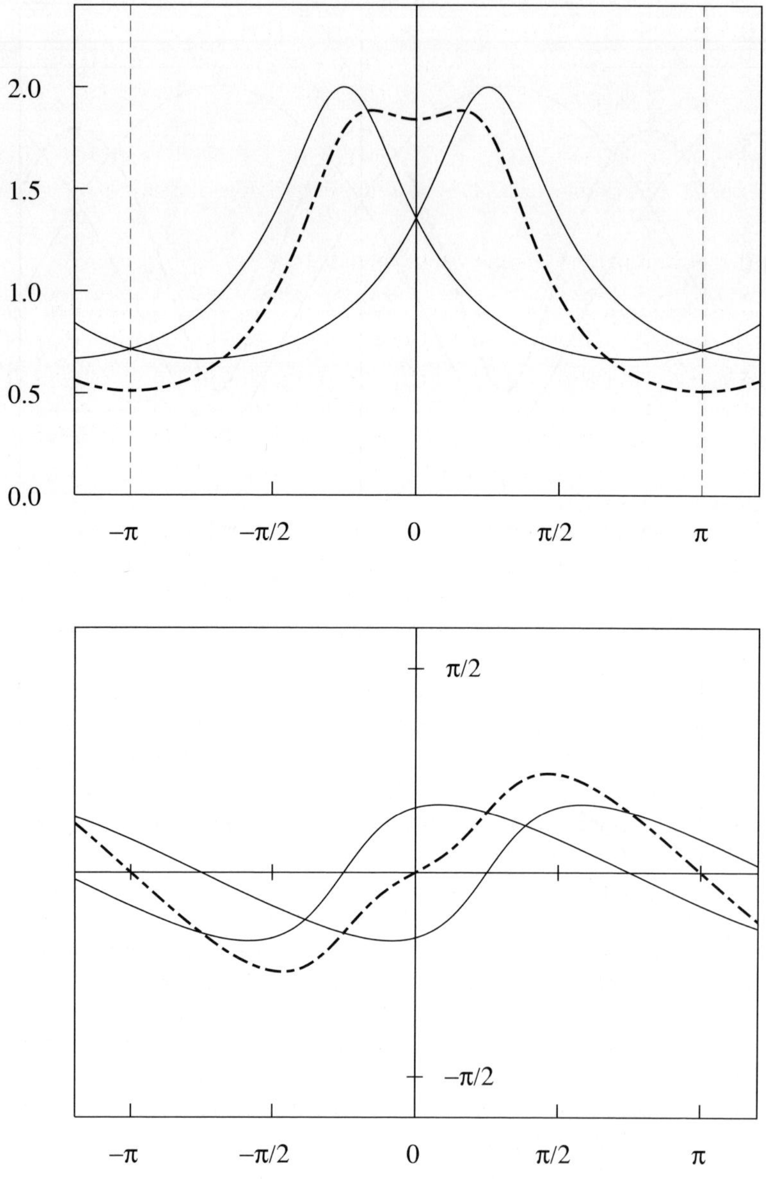

Figure 16.9. The gain and the phase effect of the second-order IIR filter $\{(1-\kappa L)(1-\kappa^* L)\}^{-1}$ where $\kappa = \lambda e^{i\varphi}$ with $\lambda = 0.5$ and $\varphi = \pi/4$. The contributions of the individual poles are represented by the continuous lines. Their joint contribution is represented by the dashed lines.

The magnitude of the gain at the frequency ω of the second-order FIR filter will be the reciprocal of the products of the lengths of the vectors $z-\kappa$ and $z-\kappa^*$, where $z=e^{i\omega}$ is the point on the unit circle which makes an angle of ω with the positive real axis. If κ is close to the unit circle, then its contribution to the gain at the frequency $\omega=\text{Arg}(\kappa)$ will dominate the contribution of the conjugate pole κ^*. In that case, there will be a peak in the gain at a frequency close to the value of $\text{Arg}(\kappa)$.

In Figure 16.9, where κ is remote from the unit circle, the gain has a peak in the interval $[0,\pi]$ at a frequency which is significantly below the value of $\pi/4=\text{Arg}(\kappa)$.

Inverse Filtering and Minimum-Phase Filters

The question sometimes arises of whether it is possible to recover the original form of a signal after it has been subjected to a process of linear filtering. The question has practical importance since it relates to the problem of removing distortions from observed signals which have been introduced by measuring instruments or by the physical medium through which the signal has been transmitted. If these effects can be modelled by a linear filter, then the issue is a matter of whether or not the distorting filter is invertible.

A filter may noninvertible for two reasons. On the one hand, the filter may impose an irreversible delay upon the signal. This, in itself, implies no disadvantage unless there is a requirement for rapid processing in real time. On the other hand, the filter may destroy part of the information in the original signal which becomes irrecoverable.

A filter which has both of these effects is to be found in Example 16.1 which concerns a method for processing a quarterly economic time series in order to remove its seasonal component. This filter, which imposes a real-time delay upon the signal, incorporates several zeros located on the unit circle at various angles. The filter annihilates the components of the original signal whose frequencies correspond to the phase angles of the zeros.

The necessary and sufficient condition for the invertibility of a linear filter is discussed at length in Chapter 17 on linear time-series models, where the invertibility of an autoregressive moving-average filter is investigated. A stable causal filter $\psi(L)=\delta(L)/\gamma(L)$ is invertible if and only if there exists an inverse function $\psi^{-1}(L)=\gamma(L)/\delta(L)$ such that $\psi^{-1}(L)\psi(L)=1$. A necessary and sufficient condition for the existence of $\psi^{-1}(L)$ is that the roots of the numerator polynomial $\delta(z^{-1})=0$, i.e. the zeros of $\psi(z^{-1})$, fall inside the circle. From the presumption that the filter is stable, it follows that the roots of $\gamma(z^{-1})=0$, i.e. the poles of $\psi(z^{-1})$, also fall inside the circle.

A noninvertible filter $\hat{\psi}(L)$ is one for which some of the zeros of the function $\hat{\psi}(z^{-1})$ fall on or outside the unit circle. A filter with zeros outside the unit circle may be converted to an invertible filter with an identical gain simply by replacing these zeros by their reciprocal values and by applying a scale factor, which is the product of the zeros, to the filter as a whole. The factor is, of course, real-valued on the assumption that the complex roots are in conjugate pairs.

Conversely, any noninvertible filter may be depicted as the product of an invertible filter $\psi(L)$ and a noninvertible filter $\alpha(L)$ described as an allpass filter. The rational function $\alpha(z^{-1})$ corresponding to the allpass filter contains poles which cancel some of the zeros of $\psi(z^{-1})$ and it contains a new set of zeros, outside the unit circle, which are the reciprocals of the cancelled zeros. As its name suggests, an allpass filter has no gain effect, but it does have a phase effect which induces delays at all frequencies.

To clarify these matters, consider the factorisation of $\psi(z^{-1})$ given under (16.28), and imagine that this relates to an invertible filter. Then the noninvertible filter, which replaces r of the zeros of $\psi(z^{-1})$ by their reciprocals, may be written as

$$\hat{\psi}(z^{-1}) = \psi(z^{-1})\prod_{j=0}^{r}\mu_j\frac{z-\mu_j^{-1}}{z-\mu_j} = \psi(z^{-1})\alpha(z^{-1}). \tag{16.39}$$

The generic factor of the allpass function $\alpha(z^{-1})$ is

$$\zeta(z^{-1}) = \mu_j\frac{z-\mu_j^{-1}}{z-\mu_j} = -z^{-1}\frac{1-\mu_j z}{1-\mu_j z^{-1}}. \tag{16.40}$$

Setting $z = e^{i\omega}$ gives the frequency response of the factor:

$$\zeta(\omega) = e^{-i\omega}\frac{1-\mu_j e^{i\omega}}{1-\mu_j e^{-i\omega}}. \tag{16.41}$$

From this, it is easy to see that, in the case where μ_j is real, the gain of the factor is $|\zeta(\omega)| = 1$; which is to say that the factor does not affect the gain of the filter. In the case where μ_j is complex, the factor $\zeta(\omega)$ is combined with its conjugate factor $\zeta^*(\omega)$, and the gain of the combination is shown to be unity with the same ease.

In demonstrating the phase effect of the allpass filter, we shall take $\mu_j = \rho_j e^{i\theta}$ to be complex number, for the sake of generality. Observe that $\rho_j = |\mu_j| < 1$, since the original filter $\psi(L)$, to whose numerator the root μ_j belongs, is assumed to be invertible. In this case, we are bound to take the generic factor in conjunction with its conjugate. We find that

$$\begin{aligned}\text{Arg}\{\zeta(\omega)\zeta^*(\omega)\} &= \text{Arg}\left\{\frac{(e^{-i\omega}-\rho e^{i\theta})(e^{-i\omega}-\rho e^{-i\theta})}{(1-\rho e^{i\theta}e^{-i\omega})(1-\rho e^{-i\theta}e^{-i\omega})}\right\}\\ &= -2\omega - 2\text{Arg}\left\{1-\rho^{-i(\omega-\theta)}\right\} - 2\text{Arg}\left\{1-\rho^{-i(\omega+\theta)}\right\}.\end{aligned} \tag{16.42}$$

It transpires that $\text{Arg}\{\zeta(\omega)\zeta^*(\omega)\} \leq 0$ for $\omega \in [0, \pi]$. The same result can be demonstrated for a factor containing a real-valued root. This implies that, via their phase effects, the factors impose delays upon components of all frequencies. Since the phase effect of the allpass filter $\alpha(L)$ is just the sum the phase effects of its factors, the inference is that an allpass filter creates delays at all frequencies.

We are now in a position to give an important syllogism which highlights an essential characteristic of an invertible filter. The first premise is that an allpass filter imposes a delay upon signal components of all frequencies. The second premise is that a noninvertible filter may be represented as the product of an allpass filter and an invertible filter. The conclusion which follows is that an invertible filter imposes the least delay of all the filters which share the same gain. For this reason, an invertible filter is commonly described as a minimum-phase filter. The condition of invertibility is often described as the *miniphase* condition.

Linear-Phase Filters

In designing a filter which selects certain frequencies from signals and rejects others, the gain should be made approximately constant amongst the selected frequencies and the phase effect, if there is one, should be linear so as to impose the same delay at each of the frequencies. A filter with a nonlinear phase effect would distort the shape of the signal; and the frequency components would be advanced in time or retarded to varying extents.

Digital filters with a finite-duration impulse response have the advantage that they can achieve exactly linear phase effects.

An FIR filter which achieves a linear phase effect may be a regarded as the product of a filter which imposes a simple delay and a filter which has no phase effect. That is to say, the frequency-response function of a linear-phase system may be written in the form of

$$\psi(\omega) = \mu(\omega)e^{-i\omega M}, \tag{16.43}$$

where $e^{-i\omega M}$ is the Fourier transform of the function $\delta(t-M)$ which represents a unit impulse delayed by M periods and where $\mu(\omega)$ is a function, with no phase effect, which is either purely real or purely imaginary. To understand the relevance of the latter condition, we may note that, if the trajectory of $\mu(\omega)$ is confined to one or other of the axes of the complex plane, then the only variation in the phase angle will be the jump of π when the origin is traversed. Such a phase jump is a mathematical construct for which there is no corresponding effect in the signal.

The sequence of filter coefficients $\{\mu_j\}$ must be chosen so as to bestow the requisite properties upon the function $\mu(\omega)$ which is its Fourier transform. This is achieved by exploiting some of the symmetry properties of the Fourier transform which are recorded in Table 13.2. The table indicates that, if a sequence of real-valued filter coefficients constitutes an even or symmetric function with $\psi_j = \psi_{-j}$, then the Fourier transform will be real valued, whereas, if the sequence constitutes an odd or antisymmetric function with $-\psi_j = \psi_{-j}$, then Fourier transform will be purely imaginary.

It is easiest to demonstrate the effect of the conditions by considering first the case of the filter

$$\mu(L) = \psi_{-M}L^{-M} + \cdots + \psi_{-1}L^{-1} + \psi_0 + \psi_1 L + \cdots + \psi_M L^M \tag{16.44}$$

which comprises an odd number of $q+1 = 2M+1$ coefficients disposed equally about the central coefficient ψ_0. This is a noncausal filter which looks both backwards and forwards in time; and it is of the sort that can be used only in processing a recorded signal.

If a condition of symmetry is imposed on $\mu(L)$ to the effect that $\psi_j = \psi_{-j}$ for $j = 1, \ldots, M$, then the frequency-response function can be written as

$$\begin{aligned}\mu(\omega) &= \sum_{j=-M}^{M} \psi_j e^{-i\omega j} = \psi_0 + \sum_{j=1}^{M} \psi_j \big(e^{i\omega j} + e^{-i\omega j}\big) \\ &= \psi_0 + 2\sum_{j=1}^{M} \psi_j \cos(\omega j).\end{aligned} \tag{16.45}$$

This function, which is the Fourier transform of an even sequence, is real-valued with $\mu^{re}(\omega) = \mu(\omega)$ and $\mu^{im}(\omega) = 0$. It follows that the phase response is

$$\text{Arg}\{\mu(\omega)\} = \begin{cases} 0, & \text{if } \mu(\omega) > 0; \\ \pm\pi, & \text{if } \mu(\omega) < 0. \end{cases} \tag{16.46}$$

Thus, apart from the jump of π radians which occurs when $\mu(\omega)$ changes sign, there is no phase effect.

Now consider the case where the $q + 1 = 2M + 1$ filter coefficients of (16.44) form an odd or antisymmetric sequence with $\psi_0 = 0$ as the central coefficient. Then $\psi_j = -\psi_{-j}$ for $j = 1, \ldots, M$, and the frequency-response function is

$$\begin{aligned}\mu(\omega) &= \sum_{j=1}^{M} \psi_j \big(e^{i\omega j} - e^{-i\omega j}\big) \\ &= -2i \sum_{j=1}^{M} \psi_j \sin(\omega j).\end{aligned} \tag{16.47}$$

Here $\mu(\omega) = i\mu^{im}(\omega)$, which is the Fourier transform of an odd sequence, is a purely imaginary function of the frequency ω. It follows that the phase response is

$$\text{Arg}\{\mu(\omega)\} = \begin{cases} \pi/2, & \text{if } \mu^{im}(\omega) > 0; \\ \pi/2 \pm \pi, & \text{if } \mu^{im}(\omega) < 0. \end{cases} \tag{16.48}$$

Now let us consider the general case of a causal linear-phase filter

$$\psi(L) = \psi_0 + \psi L + \cdots + \psi_q L^q, \tag{16.49}$$

which may have an odd or an even number of filter coefficients. Let $M = q/2$ be the midpoint of the sequence of coefficients, which will coincide with the central coefficient if q is an even number, and which will fall between two coefficients if q is an odd number. The value of M will also give the length of the least time delay imposed by the filter if it is to be completely causal. The frequency-response function of the filter is

$$\begin{aligned}\psi(\omega) &= \sum_{j=0}^{q} \psi_j e^{-i\omega j} \\ &= e^{-i\omega M} \sum_{j=0}^{q} \psi_j e^{-i\omega(j-M)} = e^{-i\omega M}\mu(\omega).\end{aligned} \tag{16.50}$$

Since $q = 2M$, this can also be written as

$$\begin{aligned}\psi(\omega) &= e^{-i\omega M}\{\psi_0 e^{i\omega M} + \psi_1 e^{i\omega(M-1)} + \cdots \\ &\qquad + \psi_{q-1}e^{-i\omega(M-1)} + \psi_q e^{-i\omega M}\} \\ &= e^{-i\omega M}\{(\psi_0 + \psi_q)\cos(\omega M) + i(\psi_0 - \psi_q)\sin(\omega M) \\ &\qquad + (\psi_1 + \psi_{q-1})\cos(\omega[M-1]) \\ &\qquad + i(\psi_1 - \psi_{q-1})\sin(\omega[M-1]) + \cdots\}.\end{aligned} \tag{16.51}$$

We can proceed to impose some structure upon the set of filter coefficients. There are four cases to consider. The first two depend upon the symmetry conditions $\psi_j = \psi_{q-j}$. When the degree q is even, which is to say that there is an *odd* number of filter coefficients, then we have *Case 1*:

$$\mu(\omega) = \psi_M + 2\sum_{j=0}^{M-1} \psi_j \cos(\omega[M-j]). \tag{16.52}$$

When q is odd and there is an *even* number of coefficients, we have *Case 2*:

$$\mu(\omega) = 2\sum_{j=0}^{(q-1)/2} \psi_j \cos(\omega[M-j]). \tag{16.53}$$

Apart from the indexing of the parameters, these two equations are instances of equation (16.45) which gives rise to the conditions of (16.46). The second pair of cases depend upon the condition of antisymmetry: $\psi_j = -\psi_{q-j}$. When q is even, we have *Case 3*:

$$\mu(\omega) = 2i\sum_{j=0}^{M-1} \psi_j \sin(\omega[M-j]). \tag{16.54}$$

Here, the condition of antisymmetry sets $\psi_M = 0$, since this is the central coefficient of the filter. When q is odd, we have *Case 4*:

$$\mu(\omega) = 2i\sum_{j=0}^{(q-1)/2} \psi_j \sin(\omega[M-j]). \tag{16.55}$$

These equations are instances of equation (16.47) which generates the conditions of (16.48).

Locations of the Zeros of Linear-Phase Filters

The conditions of symmetry and antisymmetry, which are associated with a linear phase effect, impose restrictions on the placement of the zeros of the filter. Consider the primary polynomial $\psi(z)$, obtained as the z-transform of the filter coefficients, together with the corresponding auxiliary polynomial $\psi'(z)$:

$$\begin{aligned}\psi(z) &= \psi_0 + \psi_1 z + \cdots + \psi_{q-1}z^{q-1} + \psi_q z^q, \\ \psi'(z) &= \psi_q + \psi_{q-1}z + \cdots + \psi_1 z^{q-1} + \psi_0 z^q.\end{aligned} \tag{16.56}$$

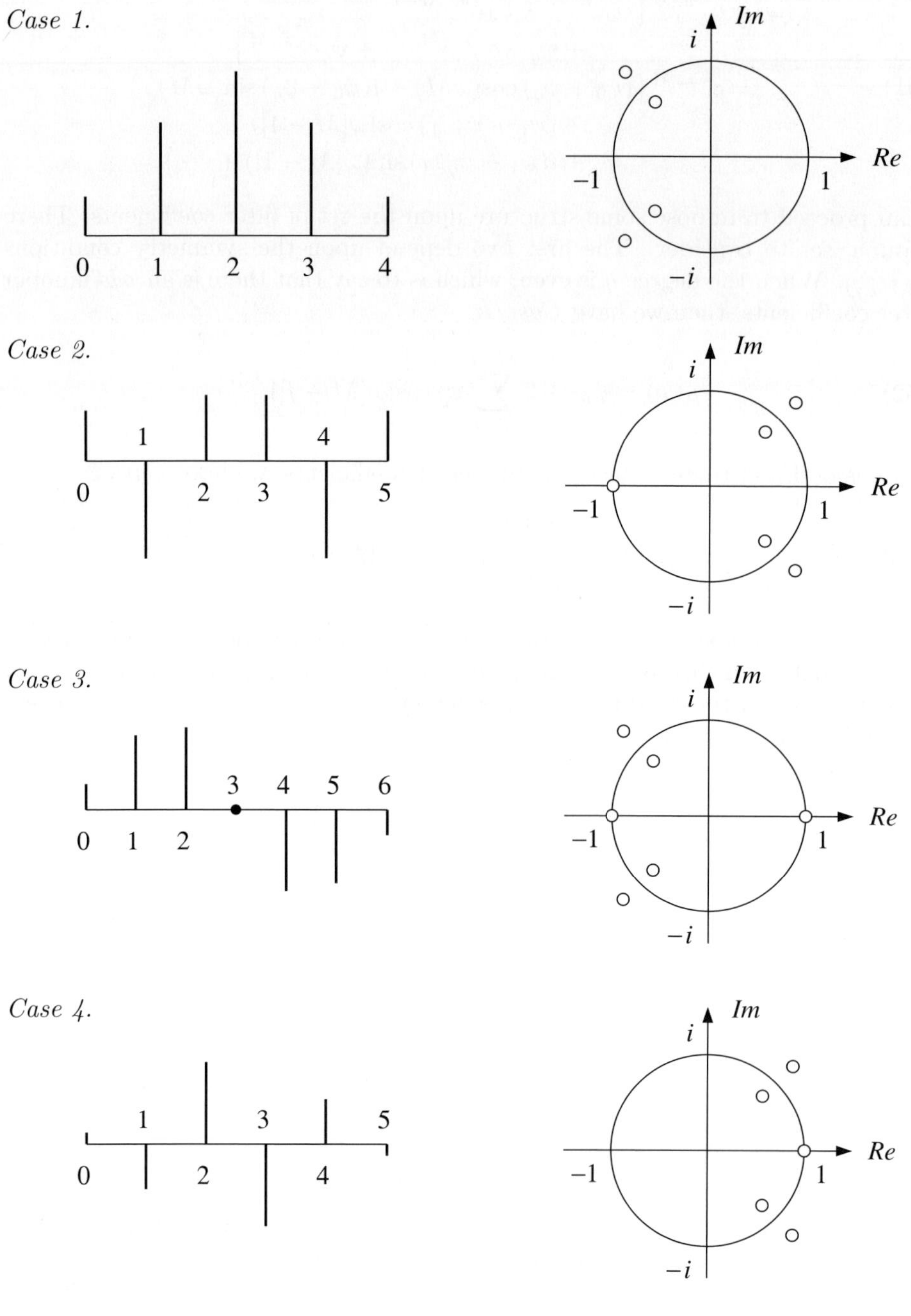

Figure 16.10. The coefficients and zero locations of four linear-phase FIR filters.

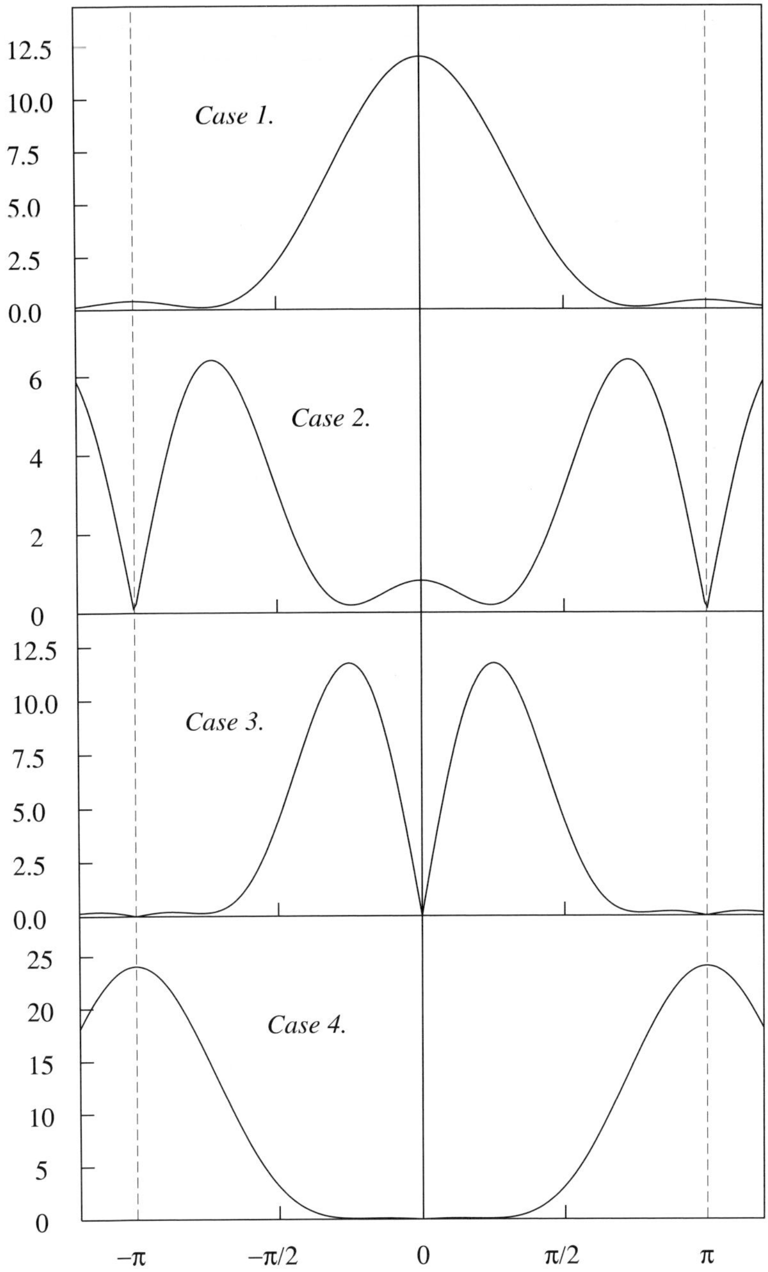

Figure 16.11. The gain of four linear-phase FIR filters.

In this application, it makes no difference whether we consider $\psi(z^{-1})$ or $\psi(z)$, and the latter presents an tidier notation. Notice that $\psi'(z) = z^q\psi(z^{-1})$.

The condition of *symmetry* is simply that $\psi(z) = \psi'(z)$. Now, if λ is a zero of $\psi(z)$ such that $\psi(\lambda) = 0$, then $1/\lambda$ is a zero of $\psi(z^{-1})$ and, therefore, of $\psi'(z)$. But, if $\psi(z) = \psi'(z)$, then it is implied that both the zero λ and its reciprocal $1/\lambda$ are present in the factors of $\psi(z)$. However, if a zero lies on the unit circle, then its reciprocal is its complex conjugate.

The condition of *antisymmetry* is that $\psi(z) = -\psi'(z)$ or, equivalently, that $\psi(z) + \psi'(z) = 0$. Setting $z = 1$ indicates that $\sum \psi_j = 0$, from which it follows that $\lambda = 1$ must be a zero of $\psi(z)$ such that $\psi(z) = (1 - z)\beta(z)$. Now consider

$$\begin{aligned} \psi'(z) &= z^q\psi(z^{-1}) \\ &= z^q\left(1 - \frac{1}{z}\right)\beta(z^{-1}) \\ &= z^q\left(1 - \frac{1}{z}\right)\{z^{1-q}\beta'(z)\} \\ &= (z - 1)\beta'(z). \end{aligned} \tag{16.57}$$

We see immediately that the condition

$$\psi(z) = (1 - z)\beta(z) = -\psi'(z) = (1 - z)\beta'(z) \tag{16.58}$$

implies that $\beta(z) = \beta'(z)$. Therefore, the antisymmetric polynomial $\psi(z) = -\psi'(z)$ is the product of a symmetric polynomial $\beta(z)$ and the antisymmetric factor $(1 - z)$. Given that $(1 - z)^{2n}$ is a symmetric polynomial and that $(1 - z)^{2n+1}$ is an antisymmetric polynomial, we infer that the condition $\psi(z) = -\psi'(z)$ of antisymmetry implies that the zero $\lambda = 1$ must be present in an odd number of the factors of $\psi(z)$. The condition $\psi(z) = \psi'(z)$ of symmetry, on the other hand, implies that, if the zero $\lambda = 1$ is present amongst the factors, then it must occur with an even multiplicity.

A final inference regarding the placement of the zeros concerns a polynomial of odd degree. This must have an odd number of zeros. If the polynomial is either symmetric or antisymmetric, then the only zeros which do not come in pairs or quadruples—which are complex conjugate pairs of reciprocal pairs—are the real-valued zeros which lie on the unit circle. Therefore, if the order q of $\psi(z)$ is odd, then the polynomial must comprise one or other of the zeros $\lambda = 1$, $\lambda = -1$ with an odd multiplicity.

One can be more specific. If $\psi(z)$ is symmetric and of odd degree, then it must contain the zero $\lambda = -1$ with an odd multiplicity. If $\psi(z)$ is antisymmetric and of odd degree, then it must contain the zero $\lambda = 1$ with an odd multiplicity and, if the zero $\lambda = -1$ is present, this must be of even multiplicity. Finally, if $\psi(z)$ is antisymmetric and of even degree then it must contain both $\lambda = 1$ and $\lambda = -1$.

The significance of these results concerning the presence of roots on the unit circle is that they indicate the frequencies at which the filter will have zero gain. Thus, for example, an antisymmetric filter which is bound to have zero gain at zero frequency (on account of the zero $\lambda = 1$) is inappropriate as a lowpass filter. It might be used as a highpass filter; but, in that case, the order q of $\psi(z)$ should be odd to avoid having zero gain at the frequency $\omega = \pi$ (which, in the case of an even q, is due to the presence of the zero $\lambda = -1$).

It may be useful to have a summary of the results of this section:

(16.59) *Case 1:* $q = 2M$ is even, $\psi(z) = \psi'(z)$ is symmetric

(i) $\psi_j = \psi_{q-j}; j = 0, \ldots, M - 1$.

Case 2: q is odd, $\psi(z) = \psi'(z)$ is symmetric

(i) $\psi_j = \psi_{q-j}; j = 0, \ldots, (q-1)/2$,

(ii) $\psi(\omega) = 0$ at $\omega = \pi$.

Case 3: $q = 2M$ is even, $\psi(z) = -\psi'(z)$ is antisymmetric

(i) $\psi_j = -\psi_{q-j}; j = 0, \ldots, M - 1$,

(ii) $\psi_M = 0$,

(iii) $\psi(\omega) = 0$ at $\omega = 0$,

(iv) $\psi(\omega) = 0$ at $\omega = \pi$.

Case 4: q is odd, $\psi(z) = -\psi'(z)$ is antisymmetric

(i) $\psi_j = -\psi_{q-j}; j = 0, \ldots, (q-1)/2$,

(ii) $\psi(\omega) = 0$ at $\omega = 0$.

These four cases are illustrated in Figure 16.10, which displays the coefficients and the zeros of the filters, and in Figure 16.11, which shows the corresponding gain of the filters. The illustration of Case 1 is a prototype of a lowpass filter whereas that of Case 4 is a prototype of a highpass filter.

In fact, the highpass filter can be converted to a lowpass filter and vice versa by reflecting the zeros about the imaginary axis. This is a matter of altering the signs of the real parts of the zeros. It can be inferred from equation (2.73), which expresses the coefficients of a polynomial in terms of its roots, that the effect of this alteration will be to convert the polynomial $\psi_\alpha(z) = \sum_j \psi_j z^j$ into a polynomial $\psi_\beta(z) = \sum_j (-1)^j \psi_j z^j$, which is a matter of reversing the signs of the coefficients associated with odd powers of z.

Practical filters, which are designed to show a gain which is approximately constant over the range of the selected frequencies, and which should show a rapid transition from the passband to the stopband, are likely to require many more coefficients than the numbers which are present in the foregoing examples.

FIR Filter Design by Window Methods

A simple idea for the design of a linear filter is to specify the desired frequency-response function in terms of its gain and phase characteristics and then to attempt to find the corresponding filter coefficients by applying an inverse Fourier transform.

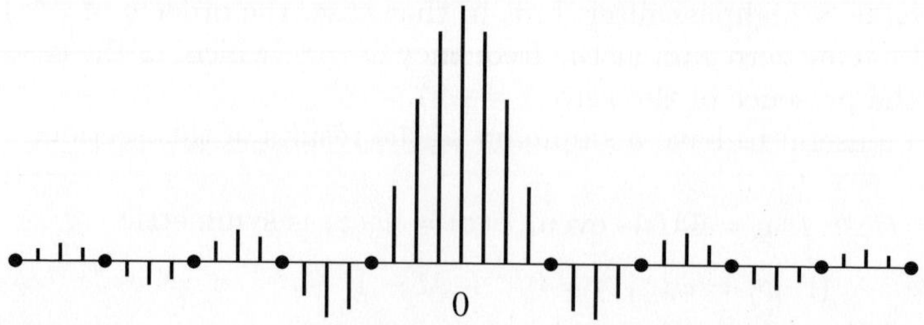

Figure 16.12. The central coefficients of the Fourier transform of a square wave with a jump at $\omega_c = \pi/4$. The sequence of coefficients, which represents the impulse response of an ideal lowpass filter, extends indefinitely in both directions.

Often it is possible to specify the gain of the filter as a nonnegative function $\psi(\omega)$ of the frequency $\omega \in (-\pi, \pi]$ which is real-valued and even. Then, since there is no phase effect to be taken into account, the gain is synonymous with the frequency response of the filter. Therefore, it follows, in view of Table 13.2, that the transform

(16.60) $$\psi_j = \frac{1}{2\pi}\int_{-\pi}^{\pi} \psi(\omega)e^{i\omega j}d\omega$$

gives rise to a sequence of coefficients $\{\psi_j\}$ which is also real-valued and even.

Provided that the sequence is finite with a limited number of elements, say $2M + 1$, then the coefficients may be used directly in constructing a filter of the form $\psi(L) = \psi_{-M}L^{-M} + \cdots + \psi_0 + \cdots + \psi_M L^M$. However, if the filter is to be used in real-time signal processing, then a delay of M periods at least must be imposed to ensure that first nonzero coefficient is associated with a nonnegative power of the lag operator.

This simple prescription in not always a practical one; for the resulting filter coefficients may be very numerous or even infinite in number. (See, for example, Figure 16.12.) In that case, the sequence has to be truncated; after which some further modification of its elements is likely to be desirable.

Lowpass filters are amongst the most common filters in practice, and the ideal lowpass filter will provide a useful model for much of our subsequent discussion. Such a filter is specified by the following frequency-response function:

(16.61) $$\psi(\omega) = \begin{cases} 1, & \text{if } |\omega| < \omega_c; \\ 0, & \text{if } \omega_c < |\omega| \leq \pi. \end{cases}$$

Within the interval $[0, \pi]$, the subinterval $[0, \omega_c)$ is the passband whilst the remainder $(\omega_c, \pi]$ is the stopband.

The coefficients of the lowpass filter are given by the (inverse) Fourier transform

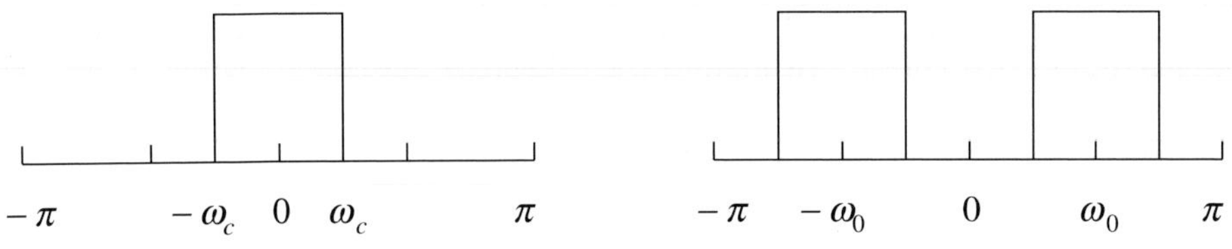

Figure 16.13. Two copies may be made of the passband, which covers the interval $[-\omega_c, \omega_c]$, and the copies shifted so that their centres lie at $-\omega_0$ and at ω_0, whereas, formerly, they were at zero.

of the periodic square wave:

$$\psi_j = \frac{1}{2\pi}\int_{-\omega_c}^{\omega_c} e^{i\omega j}d\omega = \begin{cases} \dfrac{\omega_c}{\pi}, & \text{if } j = 0; \\ \dfrac{\sin(\omega_c j)}{\pi j}, & \text{if } j \neq 0\ . \end{cases} \tag{16.62}$$

However these coefficients constitute a sequence which extends indefinitely in both directions; and, since only a limited number of central coefficients can be taken into account, we are bound to accept approximations to the ideal filter.

It may seem restrictive to concentrate on the case of a lowpass filter. Other related devices such as highpass filters, bandpass filters and bandstop filters are also of interest. However, it is a relatively straightforward matter to transform a lowpass filter into another of these filters by the device of frequency shifting.

Example 16.4. Consider a bandpass filter $\bar{\psi}(L)$ with the specification that

$$\bar{\psi}(\omega) = \begin{cases} 1, & \text{if } |\omega| \in (\omega_1, \omega_2); \\ 0, & \text{if } |\omega| < \omega_1 \text{ and } \omega_2 < |\omega| < \pi. \end{cases} \tag{16.63}$$

By a straightforward evaluation, the coefficients of the filter are found to be

$$\begin{aligned} \bar{\psi}_j &= \frac{1}{2\pi}\int_{-\omega_2}^{-\omega_1} e^{i\omega j}d\omega + \frac{1}{2\pi}\int_{\omega_1}^{\omega_2} e^{i\omega j}d\omega \\ &= \begin{cases} \dfrac{1}{\pi}(\omega_2 - \omega_1), & \text{if } j = 0; \\ \dfrac{1}{\pi j}\{\sin(\omega_2 j) - \sin(\omega_1 j)\}, & \text{if } j \neq 0. \end{cases} \end{aligned} \tag{16.64}$$

Thus, in effect, the coefficients of the bandpass filter are obtained by subtracting the coefficients of the lowpass filter with a cutoff at ω_1 from those of the lowpass filter with a cutoff at ω_2. However, the same results may be obtained by the technique of frequency shifting.

Let $\omega_1 = \omega_0 - \omega_c$ and $\omega_2 = \omega_0 + \omega_c$ where ω_0 is the central frequency of the passband in the bandpass filter of (16.63) and ω_c is the cutoff frequency in the lowpass filter of (16.61). Then, in order to convert the lowpass filter to the bandpass filter, two copies are made of the lowpass passband, which covers the interval $[-\omega_c, \omega_c]$, and the copies are shifted so that their new centres lie at the frequencies $-\omega_0$ and ω_0 (see Figure 16.13).

Now, if the frequency response of the lowpass filter is $\psi(\omega) = \sum \psi_j e^{-i\omega j}$, then the response of one of its shifted copies is

$$\begin{aligned}\psi(\omega - \omega_0) &= \sum_j \psi_j e^{-i(\omega-\omega_0)j} \\ &= \sum_j \{\psi_j e^{i\omega_0 j}\} e^{-i\omega j},\end{aligned} \tag{16.65}$$

and that of the other is $\psi(\omega + \omega_0)$. If follows that the frequency response of the bandpass filter is given by

$$\begin{aligned}\bar{\psi}(\omega) &= \psi(\omega - \omega_0) + \psi(\omega + \omega_0) \\ &= 2\sum_j \left\{\psi_j \left(\frac{e^{i\omega_0 j} + e^{-i\omega_0 j}}{2}\right)\right\} e^{-i\omega j} \\ &= 2\sum_j \{\psi_j \cos(\omega_0 j)\} e^{-i\omega j}.\end{aligned} \tag{16.66}$$

This result may be reconciled easily with the expression for $\bar{\psi}_j$ under (16.64). Thus, from the trigonometrical identities in the appendix of Chapter 13, it can be seen that

$$\begin{aligned}\bar{\psi}_j &= \frac{1}{\pi j}\{\sin(\omega_2 j) - \sin(\omega_1 j)\} \\ &= \frac{1}{\pi j}\left[\sin\{(\omega_0 + \omega_c)j\} - \sin\{(\omega_0 - \omega_c)j\}\right] \\ &= \frac{2}{\pi j}\sin(\omega_c j)\cos(\omega_0 j) = 2\psi_j \cos(\omega_0 j),\end{aligned} \tag{16.67}$$

where the final equality depends upon the definition of the coefficient $\psi_j = \sin(\omega_c j)/(\pi j)$ of the lowpass filter found under (16.62).

A leading example of frequency shifting, which has been mentioned already, concerns the conversion of a lowpass filter to a highpass filter so that the passband over the interval $[0, \omega_c]$ becomes a passband over the interval $[\pi - \omega_c, \pi]$. This is achieved by multiplying each of the lowpass coefficients ψ_j by a factor of $\cos(\pi j) = (-1)^j$, with the effect that the signs of the coefficients with odd-valued indices are reversed. Another evident means of converting a lowpass filter $\psi(\omega)$ to a highpass filter which does *not* entail frequency shifting is by forming $1 - \psi(\omega)$. Then the passband becomes the stopband and vice versa.

Truncating the Filter

Now let us turn our attention to the fact that, in practice, the ideal lowpass filter, which entails an infinite number of coefficients, must be represented by an approximation. The simplest way of approximating the Fourier series $\psi(\omega) = \sum_j \psi_j e^{-i\omega}$ which corresponds to the frequency-response function of the ideal filter is to take a partial sum

$$\psi_M(\omega) = \sum_{j=-M}^{M} \psi_j e^{-i\omega j} \tag{16.68}$$

with j extending to $\pm M$. The coefficients of this sum may be regarded as the products of the Fourier coefficients and the coefficients κ_j of a rectangular window sequence defined by

$$\kappa_j = \begin{cases} 1, & \text{if } |j| \leq M; \\ 0, & \text{if } |j| > M. \end{cases} \tag{16.69}$$

Let $\kappa(\omega)$ be the Fourier transform of the rectangular window. Then, according to a fundamental theorem, the Fourier transform of the sequence $\{\kappa_j\psi_j\}$, whose elements are the products of the elements of $\{\psi_j\}$ and $\{\kappa_j\}$, is just the (frequency-domain) convolution of the respective Fourier transforms, $\psi(\omega)$ and $\kappa(\omega)$:

$$\psi_M(\omega) = \frac{1}{2\pi}\int_{-\pi}^{\pi} \psi(\lambda)\kappa(\omega - \lambda)d\lambda. \tag{16.70}$$

The convolution represents a species of smoothing operation applied to the ideal frequency-response function $\psi(\omega)$. The function $\kappa(\omega)$, which is described as a kernel, represents the weighting function in this operation. To see how the operation distorts the shape of the frequency-response function, we need to reveal the form of the kernel function. The latter may be expressed as

$$\begin{aligned} \kappa(\omega) &= e^{-i\omega M} + \cdots + e^0 + \cdots + e^{i\omega M} \\ &= e^{-i\omega M}(1 + e^{i\omega} + \cdots + e^{i\omega 2M}) \\ &= e^{-i\omega M}\frac{1 - e^{i\omega(2M+1)}}{1 - e^{i\omega}} = \frac{1 - e^{i\omega(2M+1)}}{e^{i\omega M} - e^{i\omega(M+1)}}. \end{aligned} \tag{16.71}$$

On multiplying top and bottom by $-\exp\{-i\omega(2M+1)/2\}$, this becomes

$$\begin{aligned} \kappa(\omega) &= \frac{e^{i\omega(2M+1)/2} - e^{-i\omega(2M+1)/2}}{e^{i\omega/2} - e^{-i\omega/2}} \\ &= \frac{\sin\{\omega(2M+1)/2\}}{\sin(\omega/2)}. \end{aligned} \tag{16.72}$$

This is the Dirichlet kernel which is to be found in the classical proof of the convergence of the Fourier-series expansion of a piecewise-continuous function.

We may note that, in the limit as $M \to \infty$, the Dirichlet kernel in (16.72) becomes Dirac's delta function. In that case, it follows from the so-called *sifting property* of the delta function—see (13.94)—that the convolution would deliver the ideal frequency-response function $\psi(\omega)$ unimpaired. This convergence of the convolution integral is in spite of the fact that the kernel function itself does not converge to a limit with M. (See Lanczos [308, pp. 68–71] for example.)

Now consider the effects of the convolution of the Dirichlet kernel with the ideal frequency-response function. First, because of the dispersion of the kernel, the sharp discontinuity at ω_c in the ideal function is liable to become a gradual transition. A curious phenomenon of overshooting and undershooting also occurs in the vicinity of the discontinuity. In the second place, the side lobes of the kernel function have the effect of transferring some of the power from the passband of the ideal function into the stopband where the desired response is zero. The latter phenomenon is described as leakage.

The main lobe of the Dirichlet function—which is the portion of the function $\kappa(\omega)$ between the points on either side of $\omega = 0$ where it first assumes a zero value—has a width of $8\pi/(2M+1)$. Since this width diminishes as the number of the Fourier coefficients increases, we can expect that the transition from passband to stopband will become sharper as the order of the FIR filter increases. Unfortunately, the effects of the sidelobes do not diminish at the same rate; and, although the width of its oscillation diminishes, the overshoot at the point of discontinuity tends to a value of approximately 9% of the jump. This is an instance of Gibbs' phenomenon which has already been discussed in Chapter 13 on Fourier series.

The leakage which results from the use of a rectangular window is fully evident from the second of the diagrams of Figure 16.14 which displays the real-valued amplitude function $\psi_M(\omega)$ of the filter. The ordinary gain $|\psi_M(\omega)|$ is derived by taking absolute values.

In some applications where the leakage is less evident, and where it needs to be accentuated for the purposes of a graphical representation, it is appropriate to plot the logarithm of the gain. The decibel quantity $20\log_{10}|\psi(\omega)|$ is displayed in the lower diagram of Figure 16.14. Decibels are also an appropriate measure when there is a wide range of intensities amongst the components of a signal or where the quantity measured is a stimulus to a human response. Most human sensations, including vision, hearing and the sensitivity to mechanical vibrations, are logarithmically related to the stimulus, such that, each time the stimulus is doubled, the sensation increases by a constant amount. The intensity of sound, for example, is universally measured in decibels.

To reduce the extent of the leakage and of the overshoot, one may consider alternative window sequences whose transforms have smaller sidelobes. The design of such windows is, to a large extent, an *ad hoc* affair; and, in the past, considerable ingenuity has been devoted to the task.

The simplest suggestion for how to improve the approximation to the ideal lowpass filter is to employ a device due to Fejér which is to be found in classical Fourier analysis. Instead of obtaining approximations merely by terminating the Fourier series to form a partial sum $\psi_M(\omega)$, a new approximation is constructed by taking the arithmetic means of successive partial sums. Thus the Mth-order

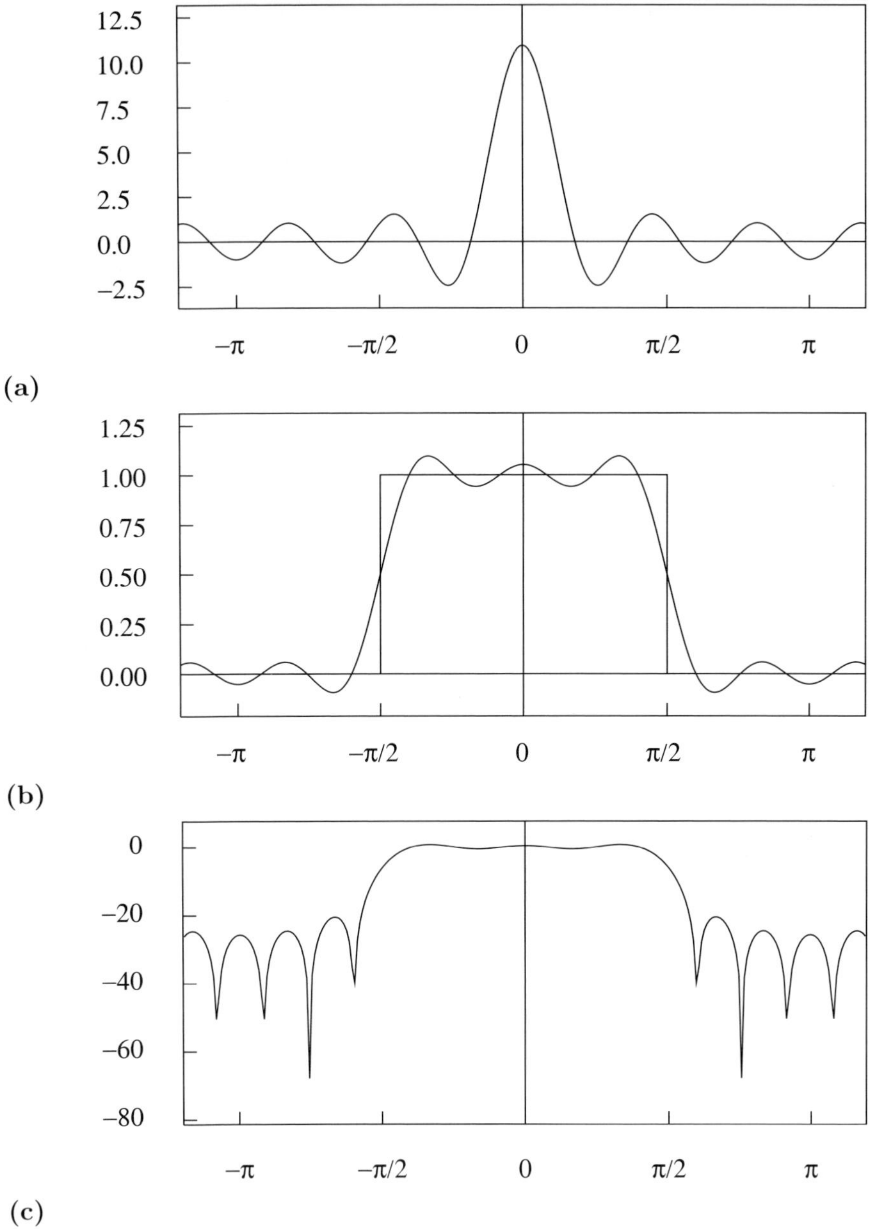

Figure 16.14. The results of applying an 11-point rectangular window to the coefficients of an ideal lowpass filter. In **(a)** is the Dirichlet kernel which is the Fourier transform of the window. In **(b)** is the frequency response of the windowed sequence. In **(c)** is the log magnitude (in decibels) of the frequency response.

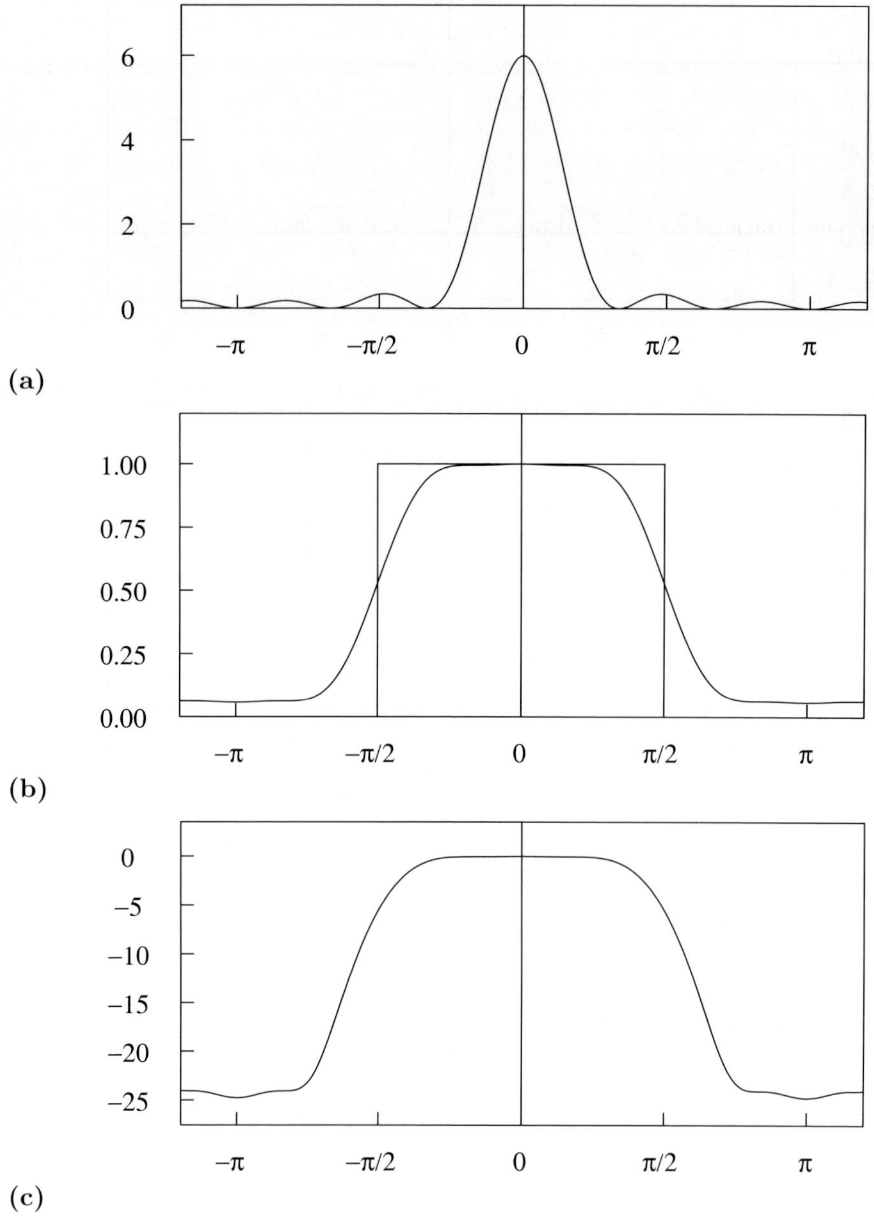

Figure 16.15. The results of applying an 11-point triangular window to the coefficients of an ideal lowpass filter. In **(a)** is the Féjer kernel which is the Fourier transform of the window. In **(b)** is the frequency response of the windowed sequence. In **(c)** is the log magnitude (in decibels) of the frequency response.

approximation is

$$S_M = \frac{1}{M}\{\psi_0(\omega) + \psi_1(\omega) + \cdots + \psi_M(\omega)\}. \tag{16.73}$$

This approximation, which is described as a Cesàro sum, will still converge to $\psi(\omega)$ as $M \to \infty$ if $\psi_M(\omega)$ does so. The new approximation $S_M(\omega)$, which places less emphasis on the higher-order terms, has better convergence properties than the original function $\psi_M(\omega)$. In fact, by using this device, Fejér was able to extend the validity of the Fourier series to a much wider class of functions than those which satisfy the Dirichlet conditions.

It is easy to see that Fejér's function is formed by applying the following triangular weighting function to the sequence of the Fourier coefficients ψ_j:

$$d_j = \begin{cases} 1 - \dfrac{|j|}{M+1}, & \text{if } |j| \leq M; \\ 0, & \text{if } |j| > M. \end{cases} \tag{16.74}$$

This is sometimes known as the Bartlett window. The coefficients of the window are formed from the (time-domain) convolution of two rectangular sequences of $M+1$ units. That is to say,

$$d_j = \frac{1}{M+1}\sum_k m_k m_{j-k}, \tag{16.75}$$

where

$$m_t = \begin{cases} 1, & \text{if } 0 \leq j \leq M+1; \\ 0, & \text{otherwise.} \end{cases} \tag{16.76}$$

The Fourier transform of the rectangular sequence is

$$m(\omega) = \frac{1 - e^{-i\omega(M+1)}}{1 - e^{-i\omega}} = \frac{\sin\{\omega(M+1)/2\}}{\sin(\omega/2)} e^{-i\omega M/2}. \tag{16.77}$$

The Fourier transform of the sequence $\{d_j\}$ is product of $m(\omega)$ with is complex conjugate $m^*(\omega)$. On dividing by $M+1$, we obtain the Fejér kernel

$$\kappa(\omega) = \frac{\sin^2\{\omega(M+1)/2\}}{(M+1)\sin^2(\omega/2)}. \tag{16.78}$$

The effect of using a triangular window in place of a rectangular window in deriving a lowpass filter can be assessed through the comparison of Figures 16.14 and 16.15. It can be see that the problem of Gibbs' phenomenon is considerably alleviated. However, the leakage is not noticeably reduced. Overall, the improvements are not impressive, and they are purchased at the expense of widening the transitional region which falls between the passband and the stopband of the filter. There are several simple window functions which are far more effective than the triangular window.

Cosine Windows

Considerable ingenuity has been exercised in the design of windows and of kernel functions which mitigate the phenomenon of leakage and which ensure a transition from the passband to the stopband which is as rapid as possible. The desirable characteristics of the kernel function are easily described. The function should have minimal sidelobes and the width of the main lobe should be the least possible. Given that there is a limit to the number of filter coefficients, which is the number of coefficients from which the kernel function can be crafted, it is clear that these objectives can be realised only in part and that they are liable to be in conflict.

A family of windows which are both simple in theory and straightforward to apply are based upon cosine functions. The prototype of such windows is the Hanning window, which is named after the Austrian physicist Julius von Hann. The time-domain coefficients of this window have the profile of a cosine bell. The bell is described by the function $\cos(\omega)$ over the interval $[-\pi, \pi]$. Its profile is raised so that that it comes to rest on the horizontal axis, and it is also scaled so that its maximum value is unity. Thus the window coefficients are given by

$$\text{(16.79)} \qquad \kappa_j = \begin{cases} \frac{1}{2}\left\{1 + \cos\left(\frac{\pi j}{M}\right)\right\}, & \text{if } |j| \leq M; \\ 0, & \text{if } |j| \geq M. \end{cases}$$

Notice that the coefficients $\kappa_{\pm M}$ at the ends are zeros so that, in fact, there are only $N = 2M - 1$ nonzero coefficents in this specification.

The kernel function which corresponds to the Hanning window has sidelobes which are much reduced in comparison with those of the kernel functions of the rectangular and triangular windows. Its characteristics are displayed in Figure 16.16 which also shows the frequency response of the filter constructed by applying the Hanning window to the coefficients of the ideal lowpass filter. Since the width of the main lobe of the Hanning kernel is no greater than that of the Fejér kernel which corresponds to the triangular window, the Hanning window is clearly superior.

This reduction in the sidelobes of the Hanning kernel and the consequent reduction in the leakage of the lowpass filter can be attributed to the absence of sharp discontinuities in the profile of the coefficients of the Hanning window. It can also be explained at a more fundamental level by analysing the structure of the kernel.

The analysis begins with the observation that the coefficients κ_j of the Hanning window can be construed as the products of the coefficients β_j of a raised cosine sequence, which extends indefinitely over the set of integers $\{j = 0 \pm 1, \pm 2, \ldots\}$, with the corresponding coefficients γ_j of the rectangular window sequence of (16.64), which are nonzero only for $|j| \leq M$. That is to say, we have

$$\text{(16.80)} \qquad \begin{aligned} \kappa_j &= \frac{1}{2}\left\{1 + \cos\left(\frac{\pi j}{M}\right)\right\}\gamma_j \\ &= \left\{\frac{1}{4}e^{-i\pi j/M} + \frac{1}{2} + \frac{1}{4}e^{i\pi j/M}\right\}\gamma_j = \beta_j\gamma_j. \end{aligned}$$

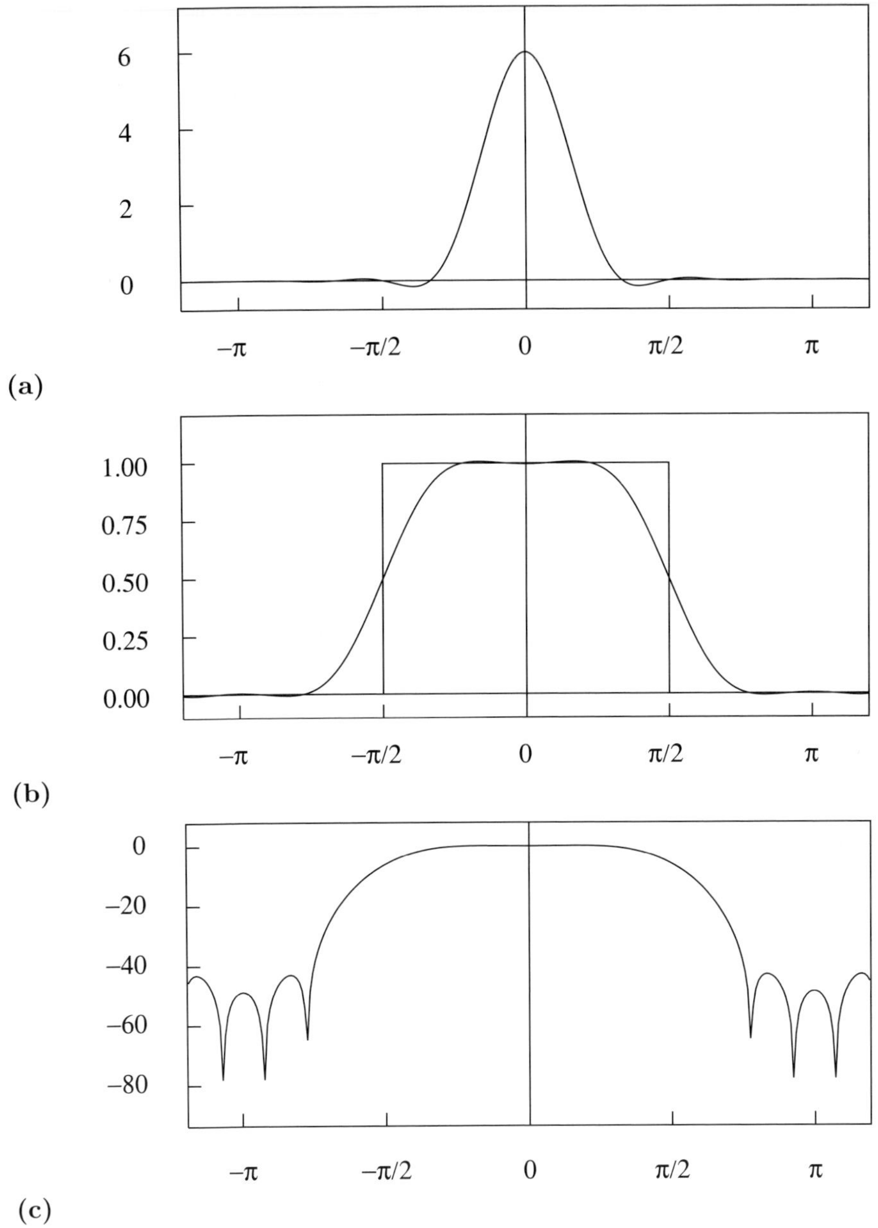

Figure 16.16. The results of applying an 11-point Hannning window to the coefficients of an ideal lowpass filter. In **(a)** is the Hannning kernel which is the Fourier transform of the window. In **(b)** is the frequency response of the windowed sequence. In **(c)** is the log magnitude (in decibels) of the frequency response.

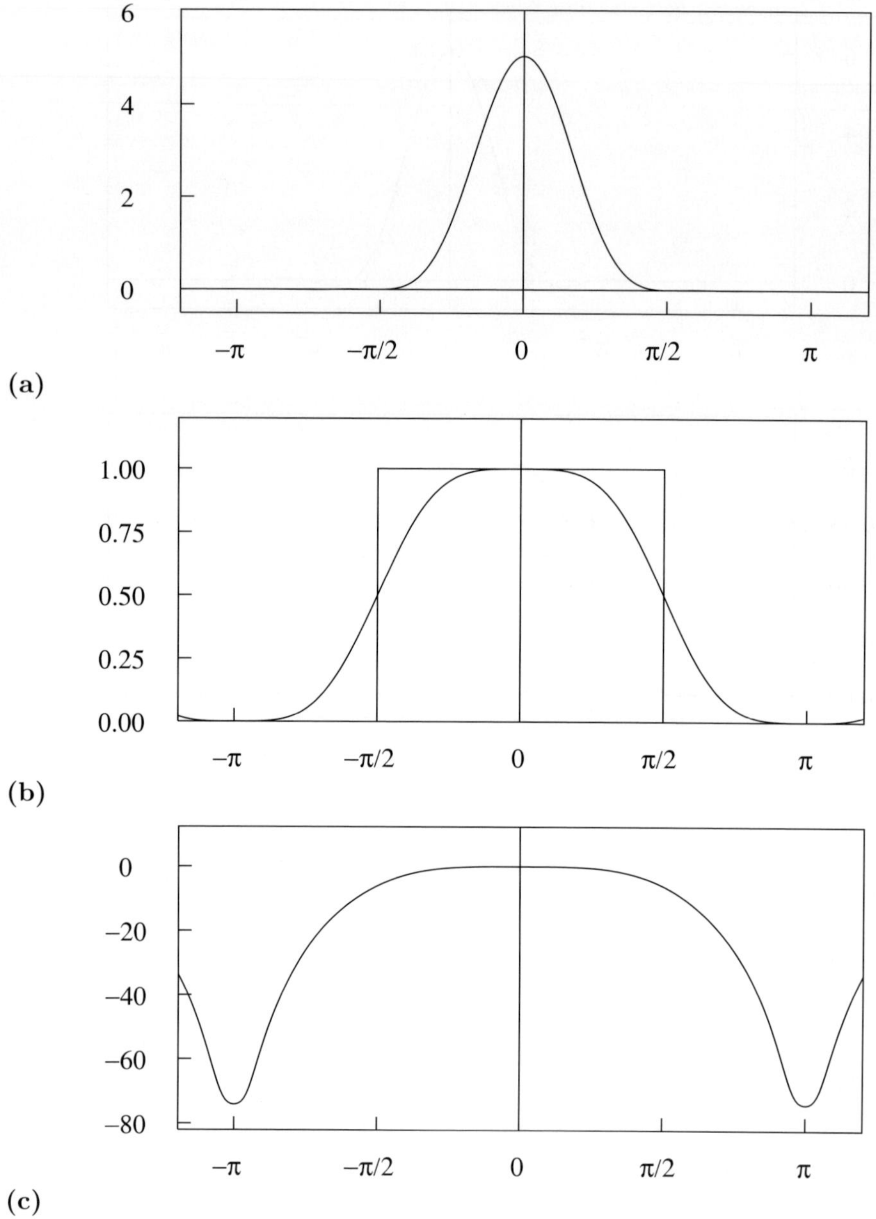

Figure 16.17. The results of applying an 11-point Blackman window to the coefficients of an ideal lowpass filter. In **(a)** is the Blackman kernel which is the Fourier transform of the window. In **(b)** is the frequency response of the windowed sequence. In **(c)** is the log magnitude (in decibels) of the frequency response.

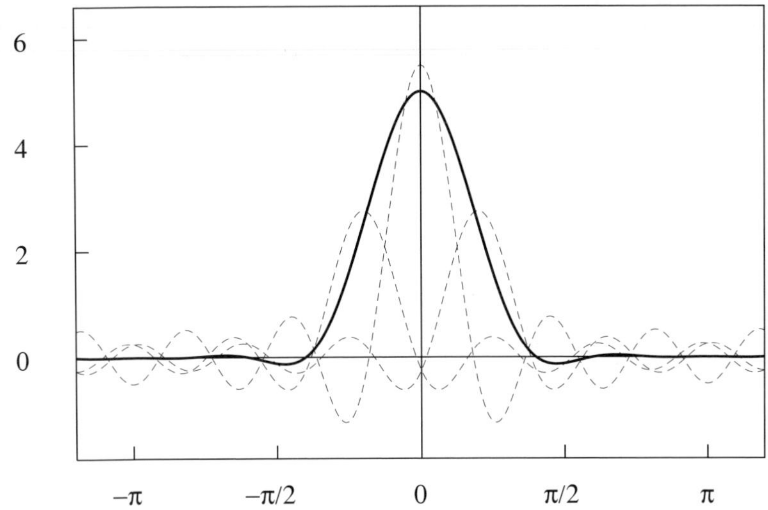

Figure 16.18. The Hanning kernel—the bold line—is the sum of three Dirichlet kernels—the dotted lines—which have been scaled and/or shifted in frequency. The attenuation of the sidelobes of the Hanning kernel is due to the destructive interference of the sidelobes of the Dirichlet kernels.

The kernel is therefore given by the convolution

(16.81) $$\kappa(\omega) = \frac{1}{2\pi}\int_{-\pi}^{\pi} \beta(\lambda)\gamma(\omega - \lambda)d\lambda,$$

where $\beta(\omega)$ and $\gamma(\omega)$ are the Fourier transforms of the sequences $\{\beta_j\}$ and $\{\gamma_j\}$. The function $\gamma(\omega)$ is just the Dirichlet kernel. However, since $\{\beta_j\}$ is composed only of a constant term and a cosine function of an angular frequency of $\omega = \pi/M$, its Fourier transform is just the sum of three Dirac functions:

(16.82) $$\beta(\omega) = 2\pi\left\{\frac{1}{4}\delta\Big(\omega - \frac{\pi}{M}\Big) + \frac{1}{2}\delta(\omega) + \frac{1}{4}\pi\delta\Big(\omega + \frac{\pi}{M}\Big)\right\}.$$

It follows from the sifting property of the Dirac function that the convolution of (16.81) amounts to a sum of three Dirichlet functions which have undergone operations of scaling and shifting. The three scale factors are 0.25, 0.5, 0.25 and the centres of the functions are at the points $\omega = -\pi/M$, $\omega = 0$, $\omega = \pi/M$ which are the locations of the Dirac impulses.

Figure 16.18 shows how the kernel of Figure 16.16 is composed of the three Dirichlet functions. The explanation for the small amplitude of the sidelobes of the Hanning kernel is to be found in the destructive interference of the sidelobes of the constituent Dirichlet functions.

A derivative of the Hanning window is the window due to R.W. Hamming [237]—who also named the Hanning window. The coefficients κ_j of the latter follow a profile which may be described as a raised cosine with a platform:

$$(16.83)\qquad \kappa_j = \begin{cases} 0.54 + 0.46\cos\left(\dfrac{\pi j}{M}\right), & \text{if } |j| \leq M; \\ 0, & \text{if } |j| > M. \end{cases}$$

The corresponding kernel function is therefore a combination of the three Dirichlet functions which form the Hanning kernel, but with weights of 0.23, 0.54, 0.23 instead of 0.25, 0.5, 0.25. These Hamming weights are chosen so as to minimise the maximum of the sidelobe amplitudes of the kernel function. Whereas the sidelobe amplitudes of the Hanning kernel decline with rising frequency, those of the Hamming function display an almost constant amplitude throughout the stopband, which might be regarded as a disadvantage in some applications. The main lobes of the Hamming and the Hanning kernels have the same width.

The final elaboration on the theme of cosine windows is provided by the Blackman window—see Blackman and Tukey [65]. Here the coefficients are given by

$$(16.84)\quad \kappa_j = 0.42 + 0.5\cos\left(\frac{\pi j}{M}\right) + 0.08\cos\left(\frac{2\pi j}{M}\right) \qquad \text{where} \qquad |j| \leq M.$$

The kernel function of the Blackman window is composed of five Dirichlet functions which are shifted and scaled before being added. The weights are 0.04, 0.25, 0.42, 0.25, 0.04 and the corresponding locations are $\omega = -2\pi/M$, $\omega = -\pi/M$, $\omega = 0$, $\omega = \pi/M$, $\omega = 2\pi/M$.

The effects of the Blackman window are shown in Figure 16.17. The width of the main lobe of the Blackman kernel is greater than that of the other kernel functions which we have examined. Therefore, when the Blackman window is applied to the coefficients of the ideal lowpass filter, the result is a filter whose frequency response has a rather gradual transition between the passband and the stopband. However, the Blackman kernel does have the advantage of a much diminished sidelobe amplitude, which implies that the leakage from the passband of the resulting lowpass filter does not extend too far into the higher frequencies.

Design of Recursive IIR Filters

The memory span of an FIR filter is a function of the order of the associated moving-average operator. That is to say, a long memory is usually achieved only at the cost of a large number of coefficients and a heavy burden of computation. A recursive filter is one which bases its memory upon feedback. A single feedback term will endow the filter with an infinite memory span. Therefore, recursive filters can sometimes achieve the same effects as nonrecursive filters with far fewer coefficients and at the cost of much less computation.

A filter with an infinite memory span is also one with a impulse response which continues for ever. It is therefore referred to as an infinite impulse-response (IIR) filter. Since it can be written in the form of (16.3), a recursive filter is also described as a rational filter.

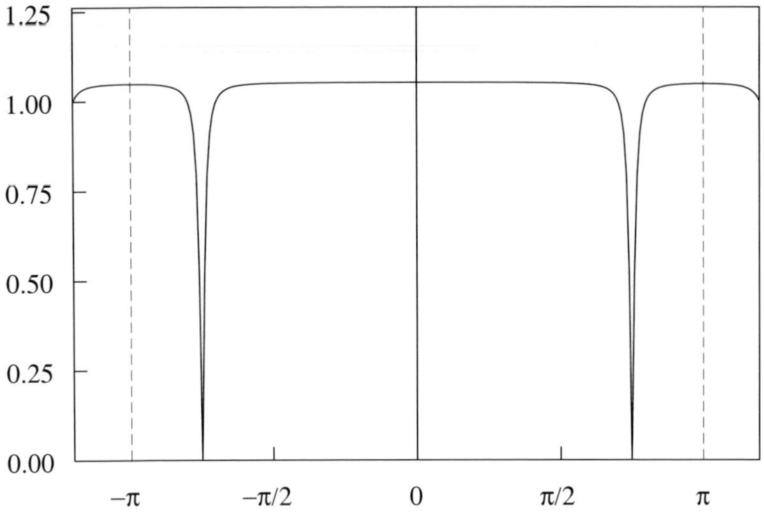

Figure 16.19. The gain of an IIR notch filter designed to eliminate a signal component at the frequency of $3\pi/4$ radians per period.

The rational form of an IIR filter suggest that its advantage also lies in the power of rational functions to provide approximations to arbitrary impulse responses. The added flexibility is due to the presence of parameters in the numerator of the transfer function which are at the disposal of the designer of the filter. A testimony to the power of rational functions in achieving approximations is provided by their widespread use in autoregressive moving-average (ARMA) models and in rational transfer function (RTM) models of the sort which are fitted to stochastic data sequences.

A disadvantage of causal IIR filters is that they cannot achieve a linear phase response. This is a corollary of the evident fact that an infinite impulse response cannot assume a form which is symmetrical with respect to a central point without comprising both values which predate and values which succeed the impulse. In many applications, such as in the digital processing of sound recordings, the nonlinear phase response of causal IIR filters virtually disbars their use. A further impediment is that it is far more difficult to design an IIR filter to achieve a frequency-domain specification than it is to design an FIR filter for the same purpose.

One way of designing an FIR filter is by a judicious placement of its poles and its zeros in view of a clear understanding of the effects which these are likely to have upon the gain of the filter. Perhaps the best-known example of this method is provided by the so-called notch filter, which is aimed at eliminating a signal component at a specific frequency. An example of an unwanted component is provided by the interference in the recordings of sensitive electronic transducers which is caused by the inductance of the alternating current of the main electrical supply. The frequency of the supply is commonly at 50 or 60 Hertz.

The unwanted component can be eliminated by a filter which incorporates a zero on the unit circle, together with its complex conjugate, whose argument corresponds exactly to the frequency in question. However, unless some compensation is provided, the effect of such a zero is likely to be felt over the entire frequency range. Therefore, it must be balanced by a pole in the denominator of the filter's transfer function located at a point in the complex plane near to the zero. The pole should have the same argument as the zero and a modulus which is slightly less than unity.

At any frequency which is remote from the target frequency, the effect of the pole and the zero in the filter will virtually cancel, thereby ensuring that the gain of the filter is close to unity. However, at frequencies close to the target frequency, the effect of the zero, which is on the unit circle, will greatly outweigh the effect of the pole which is at a short distance from it; and a narrow notch will be cut in the gain. This result is illustrated in Figure 16.19. Here the target frequency is $(3/4)\pi$, whilst the moduli of the pole and the zero are 0.95 and 1.0 respectively The resulting second-order filter is given by

$$\psi(L) = \frac{1 - 1.41421L + 1.0L^2}{1 - 1.3435L + 0.9025L^2}. \tag{16.85}$$

IIR Design via Analogue Prototypes

One way of designing an IIR filter, which is often followed, is to translate an analogue design into digital terms. The theory of analogue filters was built up over a period of sixty years or more, and it still represents a valuable resource. Analogue filters are preponderantly of the all-pole variety; and to use them means sacrificing the advantages of a rational transfer function with a free choice of parameters in both numerator and denominator. The fact that designers are willing to make this sacrifice is an indication of the underdeveloped state of the theory of rational transfer-function design. The incentive to develop the theory is diminished with each advance in processor speed which mitigates the disadvantages of computationally burdensome FIR designs.

There are two distinct approaches which may be followed in converting an analogue filter to a digital filter. Both of them aim to convert a stable analogue filter into a stable IIR filter with similar frequency-response characteristics.

The first of approach, which is called the impulse-invariance technique, is to derive the impulse-response function $f(t)$ of the analogue filter and then to design a digital filter whose impulse response is a sampled version of $f(t)$. Such a filter is designed as a set of first-order and second-order sections which are joined in parallel or in series. Each of these sections mimics the response of the corresponding section of the analogue filter. The parallel sections are, in effect, a set of subfilters which take a common input and whose outputs are added together to produce the same output as the series filter. To express the filter in a parallel form, we use the partial-fraction decomposition; and, for this purpose, it is necessary to evaluate the poles of the filter.

The second approach in converting analogue filters to digital form is to use the bilinear Möbius transformation to map the poles of a stable analogue filter, which must lie in the left half of the complex plane, into the corresponding poles of a stable digital filter, which must lie inside the unit circle. This bilinear transformation has been used already in Chapter 5 in describing the relationship between the stability conditions for differential and difference equations. The second approach is both the easier and the more successful of the two; and we shall follow it exclusively in the sequel.

The Butterworth Filter

The simplest of the analogue prototypes is the Butterworth filter. This is defined by the squared magnitude of its frequency response. The frequency response of an analogue filter whose system function is $\varphi(D)$ is obtained by evaluating the function $\varphi(s)$ along the imaginary frequency axis $s = i\Omega$. Therefore, the squared magnitude of the response is

$$|\varphi(s)|^2 = \varphi(s)\varphi(-s), \qquad \text{where} \qquad s = i\Omega. \tag{16.86}$$

In the case of the nth-order lowpass Butterworth filter, this is

$$|\varphi(i\Omega)|^2 = \frac{1}{1 + (\Omega/\Omega_c)^{2n}}. \tag{16.87}$$

Here Ω_c is the cutoff frequency. For $\Omega < \Omega_c$, the series expansion of the function takes the form of

$$|\varphi(i\Omega)|^2 = \left\{1 - (\Omega/\Omega_c)^{2n} + (\Omega/\Omega_c)^{4n} - \cdots\right\}. \tag{16.88}$$

It follows that the derivatives of the function with respect to Ω vanish at $\Omega = 0$ for all orders up to $2n - 1$. For this reason, the frequency response is described as maximally flat at zero; and the feature is clearly apparent in Figure 16.20 which plots the gain $|\varphi(i\Omega)|$ of the filter.

The poles of the analogue Butterworth filter $\varphi(s)$ are a selection of half of the $2n$ values in the s-plane which satisfy the equation

$$1 + \left(\frac{s}{i\Omega_c}\right)^{2n} = 0, \tag{16.89}$$

which comes from replacing Ω by $s/i = -is$ in the denominator of (16.87).

The problem of finding these poles is akin to that of finding the roots of unity. Observe that $-1 = \exp(i\pi) = \exp\{i\pi(2k-1)\}$, where $2k - 1$ stands for any odd-valued integer. It follows that the roots λ_k must satisfy the equation

$$\left(\frac{\lambda_k}{i\Omega_c}\right)^{2n} = -1 = \exp\left\{i\pi(2k-1)\right\}. \tag{16.90}$$

The equation can be simplified by reducing the LHS to $\lambda_k/i\Omega_c$ and by dividing the exponent on the RHS by $2n$. Then, multiplying both sides by $i\Omega_c$ gives the solution in the form of

$$\begin{aligned} \lambda_k &= i\Omega_c \exp\left\{\frac{i\pi}{2}\left(\frac{2k-1}{n}\right)\right\} \\ &= -\Omega_c \sin\left\{\frac{\pi}{2}\left(\frac{2k-1}{n}\right)\right\} + i\Omega_c \cos\left\{\frac{\pi}{2}\left(\frac{2k-1}{n}\right)\right\}, \end{aligned} \tag{16.91}$$

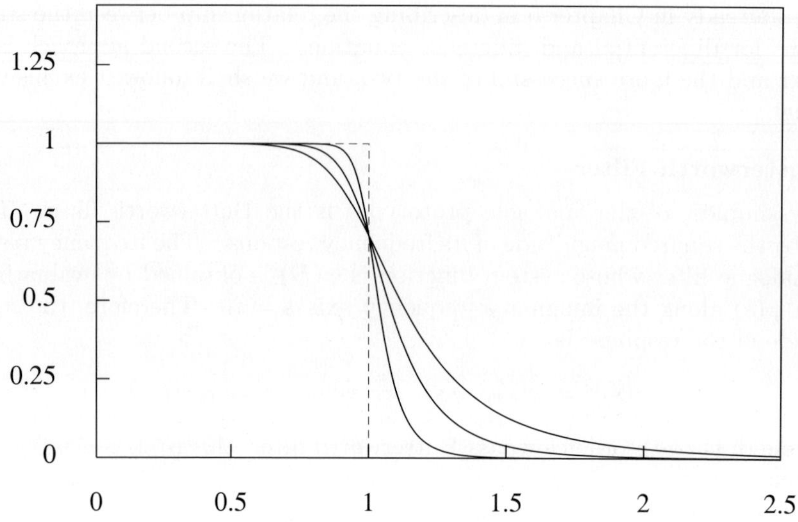

Figure 16.20. The gain of the lowpass Butterworth filter with $n = 5, 8, 16$.

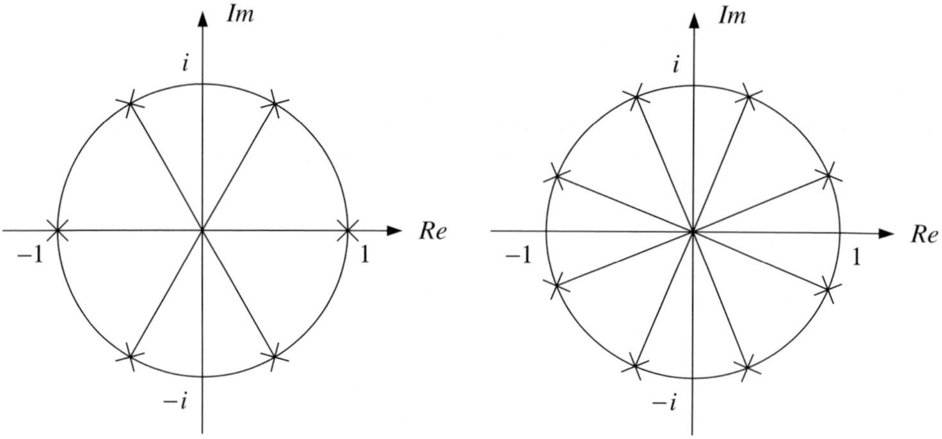

Figure 16.21. The location of the poles of the lowpass Butterworth filters with $n = 3, 4$ in the case where $\Omega_c = 1$.

with $k = 1, 2, \ldots, 2n$. These values are equally spaced on the circumference of a circle of radius Ω_c at angles which increase in steps of π/n (see Figure 16.21). Notice, in particular, that none of them falls on the imaginary axis. Values in the left half-plane, with negative real parts, are obtained when $k = 1, 2, \ldots, n$; and these are the poles of $\varphi(s)$. This selection of poles ensures that the filter is stable. The remaining values become the poles of $\varphi(-s)$.

The object is to derive an expression for the filter which is in terms of the poles. We begin by observing that, if $\lambda_k = \Omega_c \mu_k$ is a root of the equation (16.89), as specified in (16.91), then its inverse is given by

$$\lambda_k^{-1} = \frac{\mu_k^*}{\Omega_c} = \lambda_k^*, \tag{16.92}$$

where λ_k^* is the complex conjugate of λ_k and where $\mu_k = i\exp\{i\pi(2k-1)/2n\}$. This follows from the fact that $\mu_k\mu_k^* = 1$. The appropriate expression for the filter is therefore

$$\varphi(s) = \frac{1}{\prod_{k=1}^{n}(1 - s/\lambda_k)} = \frac{\Omega_c^n}{\prod_{k=1}^{n}(\Omega_c - s\mu_k^*)}. \tag{16.93}$$

Notice that, if μ_k^* is present in the denominator of $\varphi(s)$, then μ_k will be present also. This follows from the fact any pair of conjugate complex numbers must fall in the same half of the complex plane—which is the left half in this case.

When n is even, the denominator of $\varphi(s)$ can be expressed as a product of quadratic factors of the form

$$\begin{aligned}(\Omega_c - s\mu_k)(\Omega_c - s\mu_k^*) &= \Omega_c^2 - 2\Omega_c(\mu_k + \mu_k^*)s + s^2 \\ &= \Omega_c^2 + 2\Omega_c \sin\left\{\frac{\pi}{2}\left(\frac{2k-1}{n}\right)\right\} s + s^2,\end{aligned} \tag{16.94}$$

where $k = 1, 2, \ldots, n/2$. When n is odd, there are $(n-1)/2$ quadratic factors of this form in the denominator together with a single real factor of the form $(\Omega_c + s)$.

The Chebyshev Filter

In common with the Butterworth filter, the Chebyshev filter is defined by a function which specifies the magnitude of its frequency response. In the case of the nth-order lowpass Chebyshev analogue filter with a cutoff frequency of Ω_c, the squared magnitude function is

$$|\varphi(i\Omega)|^2 = \frac{1}{1 + \epsilon^2 T_n^2(\Omega/\Omega_c)}, \tag{16.95}$$

where ϵ is a parameter chosen by the designer and where $T_n(\Omega)$ is the Chebyshev polynomial of degree n.

The ordinates of the Chebyshev polynomial of degree k are given by

$$T_k(\Omega) = \begin{cases} \cos\left\{k\cos^{-1}(\Omega)\right\}, & \text{if } |\Omega| \leq 1; \\ \cosh\left\{k\cosh^{-1}(\Omega)\right\}, & \text{if } |\Omega| > 1, \end{cases} \tag{16.96}$$

where

$$\cosh(\theta) = \frac{e^{\theta} + e^{-\theta}}{2} = \cos(i\theta) \tag{16.97}$$

is the hyperbolic cosine.

Observe that, when $|\Omega| \leq 1$, there is a straightforward relationship between a cosine function $\Omega = \cos(\theta)$ defined on the interval $[0, \pi]$, and its inverse $\theta = \cos^{-1}(\Omega)$. However, when $|\Omega| > 1$, the quantity $\cos^{-1}(\Omega)$ becomes imaginary which is to say that

$$\cos^{-1}(\Omega) = i\theta = i\cosh^{-1}(\Omega). \tag{16.98}$$

In that case,

$$\begin{aligned}\cos\left\{k\cos^{-1}(\Omega)\right\} &= \cos\left\{ik\cosh^{-1}(\Omega)\right\}\\ &= \cosh\left\{k\cosh^{-1}(\Omega)\right\}.\end{aligned} \tag{16.99}$$

Over the interval $[-1, 1]$, the function $T_k(\Omega)$ attains its bounds of ± 1 alternately at the $k+1$ points $\Omega_j = \cos\{\pi(k-j)/k\}$ where $j = 0, \ldots, k$. This is evident from the fact that

$$\begin{aligned}\cos\left\{k\cos^{-1}(\Omega_j)\right\} &= \cos\left\{k\pi(k-j)/k\right\}\\ &= \cos\left\{\pi(k-j)\right\} = (-1)^{k-j}.\end{aligned} \tag{16.100}$$

Another feature of the Chebyshev polynomials is that they are orthogonal over the interval $[-1, 1]$ in terms of a certain inner product, in which respect they resemble the Legendre polynomials.

In common with other orthogonal polynomials, the Chebyshev polynomials obey a three-term recurrence relationship. By setting $A = k\theta$ and $B = \theta$ in the trigonometrical identity $\cos(A + B) = 2\cos(B)\cos(A) - \cos(A - B)$, it is easy to show that

$$T_{k+1}(\Omega) = 2\Omega T_k(\Omega) - T_{k-1}(\Omega). \tag{16.101}$$

The initial values, which are indicated by (16.96), are $T_0(\Omega) = 1$ and $T_1(\Omega) = \Omega$. In fact, the above relationship holds for all values of Ω, as can be shown by using the trigonometrical identity with the hyperbolic cosine in place of the ordinary cosine; and the relationship is of great assistance in rapidly generating the ordinates of the function under (16.95).

The general features of the Chebyshev filter can be explained by observing that the alternation of the function $T_n(\Omega)$ over the interval $[-1, 1]$ between its bounds of ± 1 causes the graph of the gain $|\varphi(i\Omega)|$ to vary between 1 and $1/\sqrt{(1+\epsilon^2)}$. For $|\Omega| > 1$, where the value of $T_n(\Omega)$ increases rapidly with Ω, the gain tends rapidly to zero. The ripples in the passband of the Chebyshev filter are an undesirable feature which is the price that is paid for achieving a sharper cutoff than that of the Butterworth filter of the same order (see Figure 16.22). The sharpness of the cutoff is inversely related to the value of the design parameter ϵ which controls the magnitude of the ripples.

To obtain the poles of the Chebyshev filter, we may set $\Omega = s/i$ within $T_n(\Omega/\Omega_c) = \cos\{n\cos^{-1}(\Omega/\Omega_c)\}$. Then, equating the denominator of (16.95) to

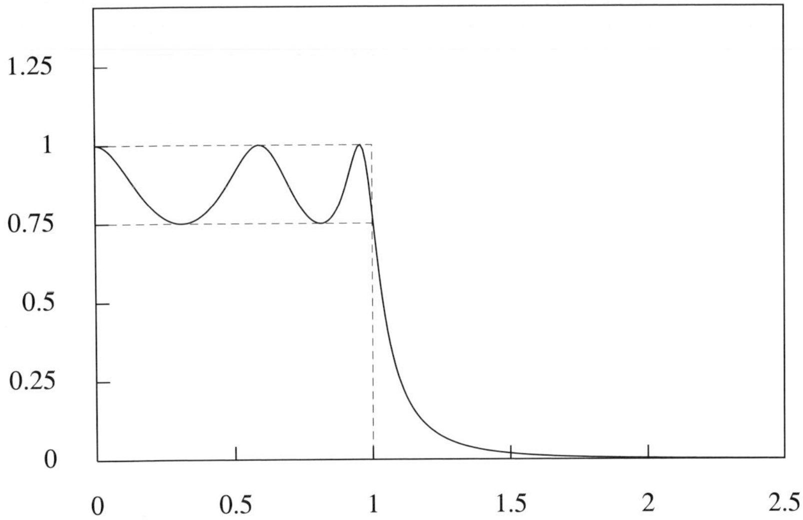

Figure 16.22. The gain of a lowpass Chebyshev Filter for $n = 5$ when the ripple amplitude is 0.25.

zero gives

$$\begin{aligned} 0 &= 1 + \left\{\epsilon T_n\left(\frac{s}{i\Omega_c}\right)\right\}^2 \\ &= 1 + \epsilon^2 \cos^2\left\{n \cos^{-1}\left(\frac{s}{i\Omega_c}\right)\right\}. \end{aligned} \tag{16.102}$$

The poles of the filter are amongst the roots of this equation. Rearranging the equation and taking the square root gives

$$\frac{i}{\epsilon} = \pm\cos\left\{n\cos^{-1}\left(\frac{s}{i\Omega_c}\right)\right\} = \sin\left\{n\cos^{-1}\left(\frac{s}{i\Omega_c}\right) + \frac{\pi}{2}(2k-1)\right\}, \tag{16.103}$$

where k is an integer. Here the second equality reflects the fact that a sine wave is just a cosine wave with a phase lag of $\pi/2$ radians. As k takes successive values, the sign of the term on the RHS will alternate. The equation can be rearranged to give a solution for the pole $s = \lambda_k$ in the form of

$$\begin{aligned} \lambda_k &= i\Omega_c \cos\left\{\frac{1}{n}\sin^{-1}\left(\frac{i}{\epsilon}\right) - \frac{\pi}{2}\left(\frac{2k-1}{n}\right)\right\} \\ &= -\alpha\Omega_c \sin\left\{\frac{\pi}{2}\left(\frac{2k-1}{n}\right)\right\} + i\beta\Omega_c\cos\left\{\frac{\pi}{2}\left(\frac{2k-1}{n}\right)\right\}, \end{aligned} \tag{16.104}$$

where

$$\alpha = -i\sin\left\{\frac{1}{n}\sin^{-1}\left(\frac{i}{\epsilon}\right)\right\} = \sinh\left\{\frac{1}{n}\sinh^{-1}\left(\frac{1}{\epsilon}\right)\right\} \tag{16.105}$$

and

$$\beta = \cos\left\{\frac{1}{n}\sin^{-1}\left(\frac{i}{\epsilon}\right)\right\} = \cosh\left\{\frac{1}{n}\sinh^{-1}\left(\frac{1}{\epsilon}\right)\right\}. \tag{16.106}$$

The second equality of (16.104) depends upon the trigonometrical identity $\cos(A - B) = \cos A \cos B + \sin A \sin B$, whilst that of (16.105) invokes the definition of a hyperbolic sine:

$$\sinh(\theta) = \frac{e^{\theta} - e^{-\theta}}{2} = -i\sin(i\theta). \tag{16.107}$$

The poles with negative real parts, which belong to the function $\varphi(s)$, are obtained from (16.104) by setting $k = 1, 2, \ldots, n$. The poles lie on an ellipse in the s plane. The major axis of the ellipse lies along the imaginary line where the radius has the value of $\beta\Omega_c$. The minor axis lies along the real line where the radius has the value of $\alpha\Omega_c$. Observe that $\alpha^2 - \beta^2 = 1$.

Expressions for α and β can be found which avoid the direct use of the hyperbolic functions. Consider the following identities:

$$\begin{aligned} \sinh^{-1}(\epsilon^{-1}) &= \ln\left(\epsilon^{-1} + \sqrt{\epsilon^{-2} + 1}\right) = \ln(q), \\ q = \exp\left\{\sinh^{-1}(\epsilon^{-1})\right\} &= \epsilon^{-1} + \sqrt{\epsilon^{-2} + 1}. \end{aligned} \tag{16.108}$$

The expression for the inverse of the $\sinh(\theta)$, which we are invoking here, is easily confirmed by direct substitution. In the light of these identities, it is readily confirmed, in reference to the definitions of $\cosh(\theta)$ and $\sinh(\theta)$, that

$$\alpha = \frac{1}{2}\left(q^{1/n} - q^{-1/n}\right) \quad \text{and} \quad \beta = \frac{1}{2}\left(q^{1/n} + q^{-1/n}\right). \tag{16.109}$$

The Bilinear Transformation

Once an analogue filter has been specified in terms of the coefficients or the poles and zeros of its transfer function $\varphi(s)$, it is a relatively straightforward matter to use the bilinear transformation to convert it to a digital filter. The business is simplified if the analogue filter can be decomposed into parallel first-order and second-order sections by use of the partial-fraction decomposition or if it can be decomposed into a cascade or series of low-order filters. This presents no difficulties when the poles of the filter are already known via analytic formulae, as is the case for the Butterworth and Chebyshev filters.

The relevant bilinear transformation is given by the function

$$s(z) = \frac{z - 1}{z + 1}, \tag{16.110}$$

which is a mapping from the z-plane, which contains the poles of the discrete-time digital filter, to the s-plane, which contains the poles of the continuous-time

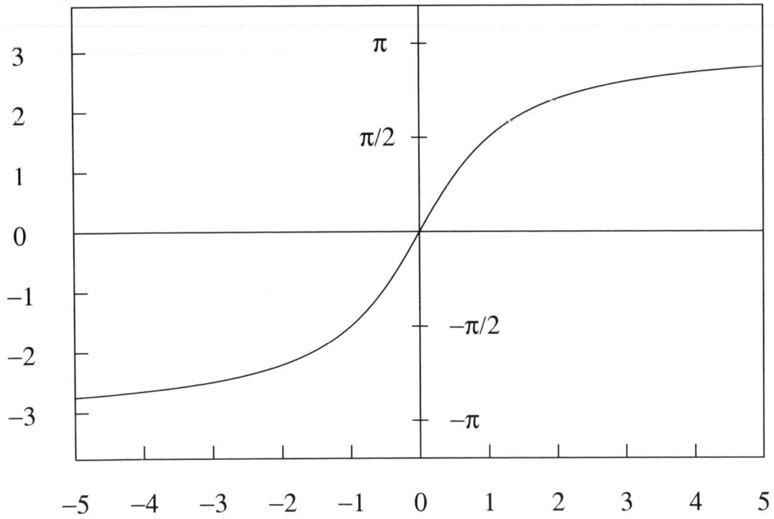

Figure 16.23. The frequency warping of the bilinear transformation whereby the analogue frequencies Ω, which range over the real line, are mapped by the function $\omega = 2\tan^{-1}(\Omega)$ into the digital frequencies ω which fall in the interval $(-\pi, \pi)$.

analogue filter. To understand the effect of the transformation, consider writing $z = \alpha - i\beta$. In these terms, the transformation becomes

$$s = \frac{(\alpha - 1) - i\beta}{(\alpha + 1) - i\beta} = \frac{\{(\alpha^2 + \beta^2) - 1\} - 2i\beta}{(\alpha + 1)^2 + \beta^2}. \tag{16.111}$$

If $s = \sigma + i\Omega$, then it follows, from the expression above, that $\sigma < 0$ if and only if $\alpha^2 + \beta^2 = |z|^2 < 1$. Thus, values within the unit circle on the z-plane are mapped into values which are in the left half of the s-plane, and vice versa.

The squared gain of the digital filter can be found from that of the analogue filter without finding the coefficients of the digital transfer function. Consider setting $z = e^{i\omega}$ in equation (16.110), with ω representing the digital frequency. This gives

$$\begin{aligned} s\big(e^{i\omega}\big) &= \frac{e^{i\omega} - 1}{e^{i\omega} + 1} = \frac{e^{i\omega/2} - e^{-i\omega/2}}{e^{i\omega/2} + e^{-i\omega/2}} \\ &= i\frac{\sin(\omega/2)}{\cos(\omega/2)} = i\tan(\omega/2). \end{aligned} \tag{16.112}$$

If the latter is put in place of the variable $s = i\Omega$ wherever it is found in the function $|\varphi(s)| = \sqrt{\{\varphi(s)\varphi(-s)\}}$, then the result will be the gain or amplitude function of the digital filter.

The function $\Omega = \tan(\omega/2)$, which is obtained from (16.112) by setting $s = i\Omega$, is, in effect, a mapping from the digital frequency $\omega \in [-\pi, \pi]$ to the unbounded analogue frequency Ω. The inverse mapping $\omega = 2\tan^{-1}(\Omega)$, which is illustrated by Figure 16.23, describes the frequency warping which accompanies the conversion of the analogue filter to the digital filter. The mapping is virtually linear within the neighbourhood of zero. However, as Ω increases beyond 0.5, an ever-increasing range of analogue frequencies are mapped into a given range of digital frequencies.

In the process of frequency warping, a disparity is liable to emerge between the cutoff frequencies of the digital and the analogue filters. Since the digital filter is the one which is to be used in the processing of signals, the analogue cutoff frequency Ω_c must be selected so as to give rise to the desired digital value ω_c. Thus the appropriate cutoff frequency for the analogue prototype is given by $\Omega_c = \tan(\omega_c/2)$. The business of choosing an analogue frequency which will map into a designated digital frequency is described as pre-warping the frequency.

When the technique of pre-warping is applied, for example, to the analogue Butterworth filter depicted in (16.87), the squared gain of its digital counterpart becomes

$$(16.113) \qquad \left|\phi(\omega)\right|^2 = \frac{1}{1 + \left\{\dfrac{\tan(\omega/2)}{\tan(\omega_c/2)}\right\}^{2n}}.$$

The Butterworth and Chebyshev Digital Filters

There is some advantage in applying the bilinear transformation directly to the first-order factors of an analogue filter since, in that case, one may easily check that the poles fulfil the conditions of stability. If, for some reason such as numerical rounding error, the poles violate the conditions, then it should be possible to correct their values. Once the poles and zeros of the digital filter have been found, they can be knit together to provide the numerical coefficients of the filter.

In the case of the Butterworth analogue lowpass filter, which is specified by (16.93), and in the case of the analogous Chebyshev filter, the kth first-order factor takes the form of

$$(16.114) \qquad f(s) = \frac{1}{1 - s/\lambda_k} = \frac{\lambda_k}{\lambda_k - s},$$

where λ_k is the analogue pole. Substituting for $s = (z-1)/(z+1)$ in this expression gives the factor of the corresponding digital filter:

$$(16.115) \qquad f(z^{-1}) = \frac{\lambda_k(z+1)}{(\lambda_k - 1)z + (\lambda_k + 1)}.$$

The digital factor has a zero at $z = -1$ and a pole which is given by

$$(16.116) \qquad z = \frac{1 + \lambda_k}{1 - \lambda_k} = \frac{1 + (\lambda_k - \lambda_k^*) - \lambda_k\lambda_k^*}{1 - (\lambda_k + \lambda_k^*) + \lambda_k\lambda_k^*},$$

where the final expression, which has a real-valued denominator, comes from multiplying top and bottom of the second expression by $1 - \lambda_k^*$.

From the expressions for the Butterworth analogue pole found under (16.91), the following components of equation (16.116) can be assembled:

$$\begin{aligned}\lambda_k\lambda_k^* &= \Omega_c^2,\\ \lambda_k - \lambda_k^* &= i2\Omega_c \cos\left\{\frac{\pi}{2}\left(\frac{2k-1}{n}\right)\right\},\\ \lambda_k + \lambda_k^* &= -2\Omega_c \sin\left\{\frac{\pi}{2}\left(\frac{2k-1}{n}\right)\right\}.\end{aligned} \tag{16.117}$$

Here Ω_c is the pre-warped cutoff frequency of the analogue lowpass filter which is given by $\Omega_c = \tan(\omega_c/2)$, where ω_c is the desired digital cutoff frequency. It follows that the poles $\rho_k; k = 1, 2, \ldots, n$ of the Butterworth digital filter are given by

$$\rho_k = \frac{1-\Omega_c^2}{\delta_k} + i\frac{2\Omega_c}{\delta_k}\cos\left\{\frac{\pi}{2}\left(\frac{2k-1}{n}\right)\right\}, \tag{16.118}$$

where

$$\delta_k = 1 + 2\Omega_c \sin\left\{\frac{\pi}{2}\left(\frac{2k-1}{n}\right)\right\} + \Omega_c^2. \tag{16.119}$$

The expression under (16.116) also serves in finding the poles of the Chebyshev digital filter. From the expressions for the Chebyshev analogue pole given under (16.104), the following components are found:

$$\begin{aligned}\lambda_k\lambda_k^* &= \Omega_c^2\alpha^2 \sin\left\{\frac{\pi}{2}\left(\frac{2k-1}{n}\right)\right\} + \Omega_c^2\beta^2 \cos\left\{\frac{\pi}{2}\left(\frac{2k-1}{n}\right)\right\},\\ \lambda_k - \lambda_k^* &= i2\beta\Omega_c \cos\left\{\frac{\pi}{2}\left(\frac{2k-1}{n}\right)\right\},\\ \lambda_k + \lambda_k^* &= -2\Omega_c\alpha \sin\left\{\frac{\pi}{2}\left(\frac{2k-1}{n}\right)\right\}.\end{aligned} \tag{16.120}$$

These can be assembled in expressions which are analogous to those under (16.118) and (16.119). The requisite values for α and β may be determined from the equations under (16.108) and (16.109).

Frequency-Band Transformations

In the preceding sections, the discussion of the techniques for designing IIR filters has concentrated on the case of a lowpass filter with a cutoff point at the analogue frequency of Ω_c. It is straightforward to generalise the results to the cases of lowpass, highpass, bandstop and bandpass filters with arbitrary cutoff frequencies.

There are two ways of deriving the more elaborate types of digital filter. The first way is to transform the lowpass filter to the desired form within the analogue domain and then to convert the result into a digital filter. The second way is to convert the lowpass analogue filter into a lowpass digital filter and then to transform the digital filter to the desired form. The two techniques produce similar results.

The first technique originates in the theory of analogue networks; and the transformations reflect the nature of the analogue hardware. The frequency-band transformations are defined relative to a prototype analogue lowpass filter with a positive-frequency cutoff at $\Omega = 1$ and a negative-frequency cutoff at $\Omega = -1$.

The first of the analogue transformations is one which causes the (positive-frequency) cutoff point to be transposed to the frequency Ω_c. This is achieved by restoring the parameter Ω_c to the places in the filter formulae where it has been suppressed—i.e. set to unity—for the purpose of defining the prototype filter. The change may be denoted by

$$(16.121) \qquad s \longrightarrow \frac{s}{\Omega_c} \quad \text{or} \quad i\Omega \longrightarrow \frac{i\Omega}{\Omega_c}.$$

The second expression reflects the fact that the gain of a filter $\varphi(D)$ is evaluated by setting $s = i\Omega$ in the function $|\varphi(s)|$ and letting Ω range from 0 to ∞.

The second of the transformations maps the lowpass prototype into a highpass filter by letting

$$(16.122) \qquad s \longrightarrow \frac{\Omega_c}{s} \quad \text{or} \quad i\Omega \longrightarrow -i\frac{\Omega_c}{\Omega}.$$

As Ω ranges from 0 to ∞, the function $g(\Omega) = -\Omega_c/\Omega$ ranges from $-\infty$ to 0. Therefore, the effect of putting $g(\Omega)$ in place of Ω is that the negative-frequency passband of the prototype filter, which stands on the interval $[-1, 0]$, becomes a positive-frequency passband on the interval $[\Omega_c, \infty)$. In terms of an analogue network, the change is brought about by replacing inductors by capacitors and capacitors by inductors. See, for example, Van Valkenburg [497].

The next operation is to convert the prototype analogue lowpass filter into a bandpass filter. That is achieved by combining the previous two operations so that

$$(16.123) \qquad s \longrightarrow \beta\left(\frac{s}{\Omega_0} + \frac{\Omega_0}{s}\right) \quad \text{or} \quad i\Omega \longrightarrow i\beta\left(\frac{\Omega}{\Omega_0} - \frac{\Omega_0}{\Omega}\right).$$

Now, as Ω ranges from 0 to ∞, the function $g(\Omega) = \beta\{(\Omega/\Omega_0) - (\Omega_0/\Omega)\}$ ranges from $-\infty$ to ∞. The effect is that the passband of the prototype filter, which stands on the interval $[-1, 1]$ is transposed into the positive half of the frequency range.

There are now two parameters, Ω_0 and β, at the disposal of the designer. Their values are chosen so that $-\Omega_c = -1$, which is the negative-frequency cutoff point of the prototype filter, is mapped into $\Omega_l > 0$, which is the lower limit of the passband, whilst $\Omega_c = 1$, which is the positive-frequency cutoff, is mapped into the $\Omega_u > 0$, which is the upper limit of the passband. Thus Ω_0 and β are determined by the simultaneous solution of the equations

$$(16.124) \qquad -1 = \beta\left(\frac{\Omega_l}{\Omega_0} - \frac{\Omega_0}{\Omega_l}\right) \quad \text{and} \quad 1 = \beta\left(\frac{\Omega_u}{\Omega_0} - \frac{\Omega_0}{\Omega_u}\right);$$

and it is easy to confirm that

$$\Omega_0 = \sqrt{\Omega_l \Omega_u} \quad \text{and} \quad \beta = \frac{\Omega_0}{\Omega_u - \Omega_l}. \tag{16.125}$$

The frequency Ω_0, which is the geometric mean of Ω_l and Ω_u, may be described as the bandpass centre.

The final operation is one which transforms the prototype filter into a bandstop filter. This can be achieved by taking the stopband of a prototype *highpass* filter, which stands on the interval $[-1, 1]$, and transposing it into the positive-frequency interval; and the transformation under (16.123) would serve the purpose. All that is required in order to convert the prototype *lowpass* filter to a bandstop filter is to compound the latter transformation with that of (16.122), which turns a lowpass filter into a highpass filter, so as to give

$$s \longrightarrow \frac{1}{\beta}\left(\frac{s\Omega_0}{s^2 + \Omega_0^2}\right) \quad \text{or} \quad i\Omega \longrightarrow i\frac{1}{\beta}\left(\frac{\Omega\Omega_0}{\Omega^2 + \Omega_0^2}\right), \tag{16.126}$$

where β and Ω_0 are given by the expression under (16.125).

The four cases may be summarised as follows:

$$\begin{aligned}
&\text{lowpass} \longrightarrow \text{lowpass} && s \longrightarrow \frac{s}{\Omega_c},\\
&\text{lowpass} \longrightarrow \text{highpass} && s \longrightarrow \frac{\Omega_c}{s},\\
&\text{lowpass} \longrightarrow \text{bandpass} && s \longrightarrow \frac{s^2 + \Omega_l\Omega_u}{s(\Omega_u - \Omega_l)},\\
&\text{lowpass} \longrightarrow \text{bandstop} && s \longrightarrow \frac{s(\Omega_u - \Omega_l)}{s^2 + \Omega_l\Omega_u}.
\end{aligned} \tag{16.127}$$

Here the expressions for the bandpass and bandstop conversions are obtained by substituting the expressions of β and Ω_0 into equations (16.123) and (16.126) respectively.

When the bandpass and bandstop conversions are applied to the first-order sections of lowpass filters, they will generate second-order sections. It is usually helpful to re-factorise the latter into pairs of first-order sections before converting the analogue filter to a digital filter via the bilinear transformation.

The labour of re-factorising the filter sections can be avoided by effecting the frequency transformations after the prototype lowpass filter has been translated into the digital domain. An exposition of the technique of converting a lowpass digital filter to other forms within the digital domain has been provided by Constantinides [121].

Let β be the cutoff frequency of a prototype digital lowpass filter and let ω_c be a desired cutoff frequency. Also, let ω_u and ω_l be the upper and lower cutoff frequencies of a desired bandpass or bandstop filter. Then the appropriate transformations for converting the prototype are as follows:

Lowpass to lowpass

$$(16.128)\qquad z \longrightarrow \frac{z-\alpha}{1-\alpha z}, \qquad\qquad \alpha = \frac{\sin\{(\beta-\omega_c)/2\}}{\sin\{(\beta+\omega_c)/2\}}.$$

Lowpass to highpass

$$(16.129)\qquad z \longrightarrow -\frac{z+\alpha}{1+\alpha z}, \qquad\qquad \alpha = \frac{\cos\{(\beta+\omega_c)/2\}}{\cos\{(\beta-\omega_c)/2\}}.$$

Lowpass to bandpass

$$(16.130)\quad z \longrightarrow -\frac{z^2 - \dfrac{2\alpha k}{k+1}z + \dfrac{k-1}{k+1}}{\dfrac{k-1}{k+1}z^2 - \dfrac{2\alpha k}{k+1}z + 1}, \qquad \begin{aligned} \alpha &= \frac{\cos\{(\omega_u+\omega_l)/2\}}{\cos\{(\omega_u-\omega_l)/2\}}, \\ k &= \cot\left(\frac{\omega_u-\omega_l}{2}\right)\tan\left(\frac{\beta}{2}\right). \end{aligned}$$

Lowpass to bandstop

$$(16.131)\quad z \longrightarrow \frac{z^2 - \dfrac{2\alpha}{1+k}z + \dfrac{1-k}{1+k}}{\dfrac{1-k}{1+k}z^2 - \dfrac{2\alpha}{1+k}z + 1}, \qquad \begin{aligned} \alpha &= \frac{\cos\{(\omega_u+\omega_l)/2\}}{\cos\{(\omega_u-\omega_l)/2\}}, \\ k &= \tan\left(\frac{\omega_u-\omega_l}{2}\right)\tan\left(\frac{\beta}{2}\right). \end{aligned}$$

These formulae may appear complicated; but, in practice, the transformations can be simplified by a judicious choice of the prototype frequency β.

Consider, for example, the lowpass-to-highpass transformation. If β is chosen such that $\beta+\omega_c=\pi$, then a value of $\alpha=0$ will result and the transformation will become $z \longrightarrow -z$. To see the effect of this, consider the individual poles and zeros of a Butterworth or Chebyshev lowpass filter.

Given that the poles and the zeros of a filter come in conjugate pairs, the change from lowpass to highpass may be accomplished by changing the signs on the real parts of the poles and zeros. From the point of view of the frequency-response diagram, the changes entail the reflection of the graph about the axis of $\omega=\pi/2$. In terms of the Argand diagram, the transformation is achieved via a reflection about the vertical imaginary axis.

For the lowpass-to-bandpass and lowpass-to-bandstop transformations, matters may be simplified by contriving the values of β so as to ensure that $k=1$.

Example 16.5. Consider the conversion of a lowpass filter to a bandpass filter. The expression for the generic first-order section of the prototype lowpass filter, which is given under (16.115), may be rewritten as

$$(16.132)\qquad \frac{\gamma(1+z)}{z-\rho} \quad \text{where} \quad \gamma = \frac{\lambda}{\lambda-1}, \quad \rho = \frac{\lambda+1}{1-\lambda}.$$

Here λ is a pole of the prototype analogue filter whilst ρ is a pole of the corresponding digital filter. The formula for the conversion of the digital prototype is given under (16.130). To simplify matters, let the cutoff frequency of the prototype lowpass filter be set to $\beta = \omega_u - \omega_l$. Then $k = 1$, and the conversion formula becomes

$$z \longrightarrow \frac{z^2 - \alpha z}{\alpha z - 1}. \tag{16.133}$$

The generic second-order segment of the bandpass filter is obtained by making this substitution within equation (16.132) to give

$$\gamma\left\{1 + \frac{z^2 - \alpha z}{\alpha z - 1}\right\}\left\{\frac{z^2 - \alpha z}{\alpha z - 1} - \rho\right\}^{-1} = \gamma\left\{\frac{z^2 - 1}{z^2 - \alpha(1+\rho)z + \rho}\right\}. \tag{16.134}$$

This section has a pair of zeros and a pair of poles given, respectively, by

$$\begin{aligned} z &= \pm 1 \quad \text{and} \\ z &= \frac{\alpha(1+\rho) \pm \sqrt{\alpha^2(1+\rho)^2 - 4\rho}}{2}. \end{aligned} \tag{16.135}$$

The formula for the poles entails complex arithmetic. The placement of the zeros is intuitively intelligible; for it ensures that the gain of the filter is zero both at the zero frequency and at the maximum Nyquist frequency of π.

The squared gain of the filter, which is a function of the digital frequency value ω, may be found without knowing the coefficients of the filter. It is obtained by making the substitution

$$\tan(\omega/2) \longrightarrow \frac{\cos(\omega) - \alpha}{\sin(\omega)} \tag{16.136}$$

within the relevant filter formula. This conversion is derived first by applying the substitution of (16.133) to the formula (16.110) for the bilinear transformation and then by setting $z = e^{i\omega}$, in the manner of (16.112).

The formula for the squared gain of the Butterworth bandpass digital filter, which is derived from (16.113), is

$$|\phi(\omega)|^2 = \frac{1}{1 + \left\{\dfrac{\cos(\omega) - \alpha}{\tan(\omega_c/2)\sin(\omega)}\right\}^{2n}}. \tag{16.137}$$

Bibliography

[46] Bellanger, M., (1989), *Digital Processing of Signals, Second Edition*, John Wiley and Sons, Chichester.

[65] Blackman, R.B., and T.W. Tukey, (1959), *The Measurement of the Power Spectrum from the Point of View of Communications Engineering*, Dover Publications, New York.

[108] Chui, C-K., P.W. Smith and L.Y. Su, (1977), A Minimisation Problem Related to Padé Synthesis of Recursive Digital Filters, in *Padé and Rational Approximations: Theory and Applications*, E.B. Saff and R.S. Vaga (eds.), Academic Press, New York.

[121] Constantinides, A.G., (1970), Spectral Transformations for Digital Filters, *Proceedings of the IEE*, **117**, 1585–1590.

[150] DeFatta, D.J., J.G. Lucas and Hodgkiss, W.S., (1988), *Digital Signal Processing: A System Design Approach*, John Wiley and Sons, New York.

[199] Gabel, R.A., and R.A. Roberts, (1987), *Signals and Linear Systems, Third Edition*, John Wiley and Sons, New York

[237] Hamming, R.W., (1989), *Digital Filters, Third Edition*, Prentice-Hall, Englewood Cliffs, New Jersey.

[280] Kailath, T., (1974), A View of Three Decades of Linear Filtering Theory, *IEEE Transactions on Information Theory*, **IT-20**, 146–181.

[308] Lanczos, C., (1961), *Linear Differential Operators*, Van Nostrand Co., London.

[309] Lanczos, C., (1966), *Discourse on Fourier Series*, Oliver and Boyd, Edinburgh and London.

[371] Oppenheim, A.V., and R.W. Schaffer, (1975), *Digital Signal Processing*, Prentice-Hall, Englewood Cliffs, New Jersey.

[372] Oppenheim, A.V., and R.W. Schaffer, (1989), *Discrete-Time Signal Processing*, Prentice-Hall, Englewood Cliffs, New Jersey.

[373] Oppenheim, A.V., A.S. Willsky and I.Y. Young, (1983), *Signals and Systems*, Prentice-Hall, London.

[380] Parks, T.W., and Burrus, C.S., (1987), *Digital Filter Design*, John Wiley and Sons, New York.

[417] Rabiner, L.R., and B. Gold, (1975), *Theory and Applications of Digital Signal Processing*, Prentice Hall, Englewood Cliffs, New Jersey.

[426] Roberts, R.A., and C.T. Mullis, (1987), *Digital Signal Processing*, Addison-Wesley, Reading, Massachusetts.

[497] van Valkenburg, M.E., (1972), *Network Analysis*, Prentice-Hall, Englewood Cliffs.

CHAPTER 17

Autoregressive and Moving-Average Processes

Amongst the simplest and most widely-used models of stationary stochastic processes are the autoregressive moving-average or ARMA models. A simple autoregressive or AR model is a linear difference equation with constant coefficients in which the forcing function is a white-noise process. A moving-average or MA model is one which expresses an observable stochastic sequence as a linear combination of the current value of a white-noise process and a finite number of lagged values. An ARMA model is a stochastic difference equation in which the forcing function is a moving-average process.

An explanation for the ubiquity of ARMA models begins with the Cramér–Wold theorem which indicates that virtually every stationary stochastic process has a moving-average representation. This important result will be established in a subsequent chapter. The theorem provides only a weak justification for the use of MA models on their own, since the latter embody a limited number of parameters, whereas the theorem establishes the existence of an infinite-order moving-average representation. In practice, finite-order MA models are used only infrequently. Indeed, an AR model has an infinite-order MA representation which is often very poorly approximated by a finite-order MA model with a manageable number of parameters.

The ability of a low-order AR model to approximate an arbitrary stationary process is also limited; and it is wishful thinking to imagine that low-order models are widely applicable. It is true that adequate approximations can be achieved often by high-order AR models; but, in that case, large amounts of data are required in order to find stable values for the estimated parameters.

When the AR and MA components are combined in a mixed model, the ability accurately to approximate the Cramér–Wold representation with a limited number of parameters is greatly increased. In such approximations, both the AR and the MA components of the model have their parts to play. This is the true justification for ARMA models.

Although mixed ARMA models are more serviceable that simple AR and MA models, we shall begin the chapter with extensive treatments of the latter. This is for ease of exposition. Once the theory of the pure models has been developed, that of the mixed models becomes readily accessible by combining established results.

One of the principal concerns of this chapter is to discover the relationship between the parameters of ARMA processes and their autocovariances. We shall show that it is straightforward to find the values of the autocovariances from the values of the parameters. To go in the other direction is only slightly more difficult. The programs which are presented will enable one to move in both directions; and their accuracy may be tested by connecting them in a cycle passing back and forth between the parameters and the autocovariances.

Stationary Stochastic Processes

A temporal stochastic process is a sequence of random variables indexed by a time subscript. We shall denote such a process by $x(t)$ so that the element of the sequence with the time index $t = \tau$ has the value of $x_\tau = x(\tau)$.

Let $\{x_{\tau+1}, x_{\tau+2}, \ldots, x_{\tau+n}\}$ denote n consecutive elements of the sequence. Then the process is said to be strictly stationary if the joint probability distribution of these elements does not depend on τ, regardless of the size of n. This means that any two segments of the sequence of equal length will have identical probability distribution functions. If the moments in question are finite, then stationarity implies that

$$E(x_t) = \mu \quad \text{for all} \quad t \qquad \text{and} \quad C(x_t, x_s) = \gamma_{|t-s|}. \tag{17.1}$$

The second of the conditions indicates that the covariance of any two elements depends only on their temporal separation $|t - s|$. It should be observed that, in a stationary process which is completely characterised by its first and second moments, there is nothing to indicate the direction of time.

Notice that, if the elements of the sequence are normally distributed, then the two conditions of (17.1) are sufficient to guarantee strict stationarity, for the reason that a normal distribution is characterised completely by moments of the first and second orders. On their own, the conditions imply weak or second-order stationarity.

The second condition of (17.1) also implies that $V(x_t) = \gamma_0$ for all t, which is to say that the elements share a common variance. Therefore, the Cauchy–Schwarz inequality serves to show that

$$-1 \leq \frac{C(x_t, x_s)}{\sqrt{V(x_t)V(x_s)}} = \frac{\gamma_{|t-s|}}{\gamma_0} \leq 1, \tag{17.2}$$

which, in view of the nonnegativity of the variance, is expressed more succinctly as the condition that $\gamma_0 \geq |\gamma_\tau|$ for all τ. Thus the scale of the variance dominates that of the ensuing autocovariances.

In graphical representations of the autocovariances, such as those which are to be found in this chapter, it is usually desirable to normalise the scale by dividing each of them by the variance γ_0. This gives rise to the sequence of autocorrelations $\{\rho_\tau = \gamma_\tau/\gamma_0\}$.

The condition that the autocovariances are functions solely of the temporal separation of the elements in question implies that the dispersion matrix of the vector of the n elements $x_0, x_1, \ldots, x_{n-1}$ is a bisymmetric Laurent matrix of the form

$$
(17.3) \qquad \Gamma = \begin{bmatrix} \gamma_0 & \gamma_1 & \gamma_2 & \cdots & \gamma_{n-1} \\ \gamma_1 & \gamma_0 & \gamma_1 & \cdots & \gamma_{n-2} \\ \gamma_2 & \gamma_1 & \gamma_0 & \cdots & \gamma_{n-3} \\ \vdots & \vdots & \vdots & \ddots & \vdots \\ \gamma_{n-1} & \gamma_{n-2} & \gamma_{n-3} & \cdots & \gamma_0 \end{bmatrix}.
$$

This dispersion matrix is positive definite, as is the dispersion matrix of any nondegenerate multivariate distribution. The result follows immediately when one considers an arbitrary quadratic product in the elements of the matrix. The product may be expressed in terms of the elements of the mean-adjusted process $y(t) = x(t) - \mu$ as follows:

$$
(17.4) \qquad \begin{aligned} \sum_{t=0}^{n-1}\sum_{s=0}^{n-1} c_t c_s \gamma_{t-s} &= \sum_{t=0}^{n-1}\sum_{s=0}^{n-1} c_t c_s E(y_t y_s) \\ &= E\left\{\left(\sum_{t=0}^{n-1} c_t y_t\right)^2\right\} > 0. \end{aligned}
$$

The autocovariance generating function is a power series whose coefficients are the autocovariances γ_τ for successive values of τ and whose argument is a complex variable z. This will be denoted by

$$
(17.5) \qquad \begin{aligned} \gamma(z) &= \{\gamma_0 + \gamma_1(z + z^{-1}) + \gamma_2(z^2 + z^{-2}) + \cdots\} \\ &= \sum_\tau \gamma_\tau z^\tau. \end{aligned}
$$

Here it must be noted that the summation is over positive and negative values of τ and that $\gamma_{-\tau} = \gamma_\tau$. The function may be described as the z-transform of the sequence of autocovariances.

The autocovariance generating function $\gamma(z)$ of a stationary process is positive definite in the sense that it fulfils the condition that

$$
(17.6) \qquad 0 < \frac{1}{2\pi i}\oint c(z)\gamma(z)c(z^{-1})\frac{dz}{z},
$$

where the contour of integration is the circumference of the unit circle, and where $c(z) \neq 0$ is any rational function which is analytic within an annulus surrounding the circumference. This integral has the value of the coefficient of $c(z)\gamma(z)c(z^{-1})$ associated with z^0, which is $\sum c_t c_s \gamma_{|t-s|}$. Thus (17.7) is equivalent to the condition of positive definiteness under (17.4) when the order n of the dispersion matrix is indefinitely large.

The dispersion matrix Γ of (17.3) may be derived from the autocovariance generating function by replacing positive powers of the argument z by positive powers of the $n \times n$ matrix L, which has units on the first subdiagonal and zeros elsewhere, and by replacing negative powers of z by powers of L'. The matrix

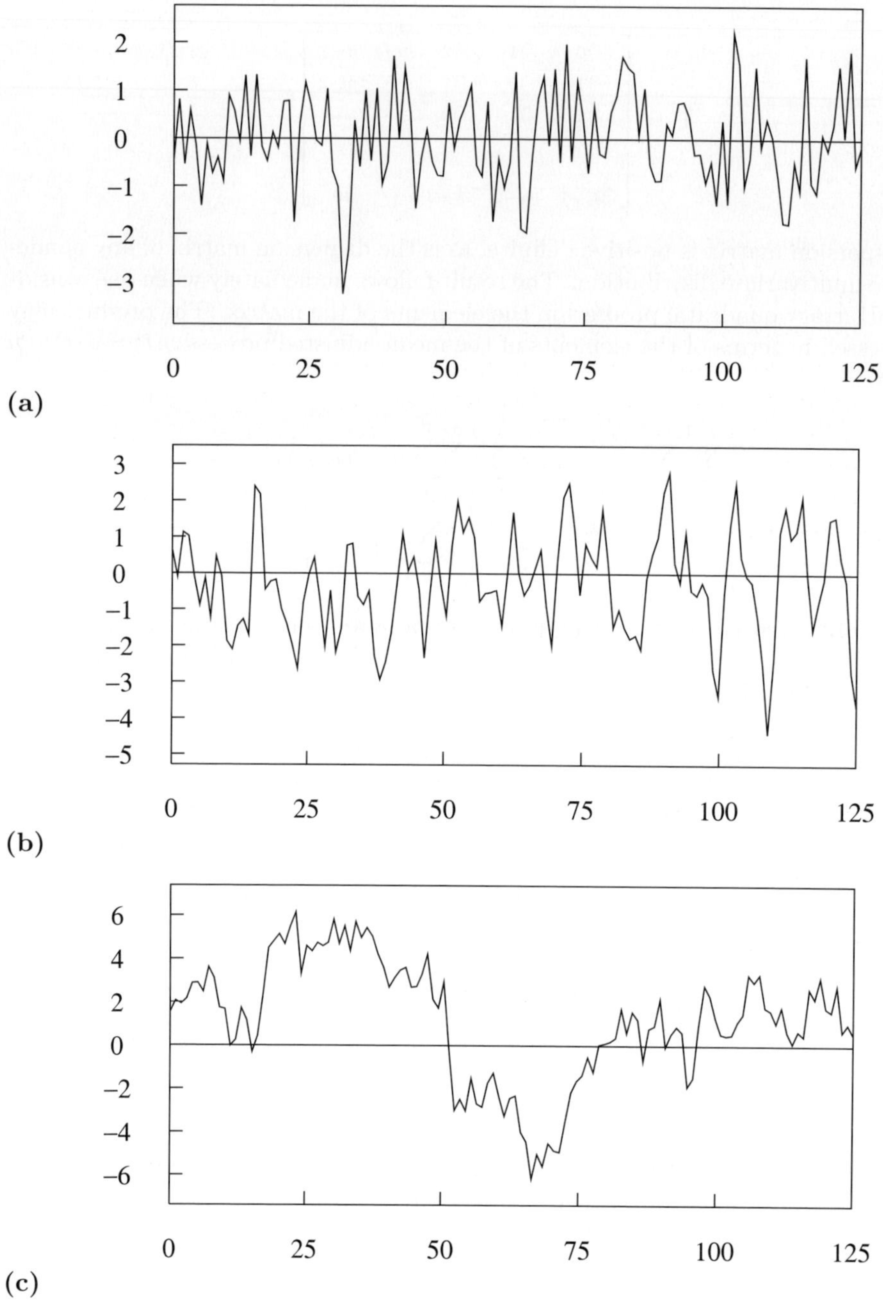

Figure 17.1. The graphs of 125 observations on three simulated series: **(a)** a unit-variance white-noise process $\varepsilon(t)$, **(b)** an MA(1) process $y(t) = (1 + 0.9L)\varepsilon(t)$ and **(c)** an AR(1) process $(1 - 0.9L)y(t) = \varepsilon(t)$.

$L = [e_1, \ldots, e_{n-1}, 0]$, which is nilpotent of degree n, such that $L^j = 0$ for all $j \geq n$, is obtained from the identity matrix $I_n = [e_0, e_1, \ldots, e_{n-1}]$ of order n by deleting the leading vector and by appending a zero vector to the end of the array. Its transpose $L' = [0, e_0, \ldots, e_{n-2}]$ is obtained by deleting the trailing vector of I_n and by appending a zero vector to the front of the array.

In seeking an overview of the algebraic results of this chapter, it may be helpful to bear the autocovariance generating function in mind. In the following chapter, it will give rise to the spectral density function when the locus of z becomes the perimeter of the unit circle.

Given that a finite sequence of observations represents only a segment of a single realisation of the underlying stochastic process $x(t)$, it might be thought that there is little chance of making valid statistical inferences about the parameters of the process. However, if the process is stationary and if the statistical dependencies between widely separated elements are weak, then it is possible to estimate consistently a limited number of autocovariances which express the statistical dependence of proximate elements of the sequence. If an ARMA model is to be fitted to the process, then these autocovariances comprise all the information which is required for the purpose of estimating the parameters.

Moving-Average Processes

The qth-order moving-average process, or MA(q) process, is defined by the equation

$$y(t) = \mu_0\varepsilon(t) + \mu_1\varepsilon(t-1) + \cdots + \mu_q\varepsilon(t-q), \tag{17.7}$$

where $\varepsilon(t)$, which has $E\{\varepsilon(t)\} = 0$, is a white-noise process generating a sequence of independently and identically distributed random variables with zero expectations. The equation may be normalised either by setting $\mu_0 = 1$ or by setting $V\{\varepsilon(t)\} = \sigma_\varepsilon^2 = 1$; and the usual choice is to set $\mu_0 = 1$. The equation may be represented, in a summary notation, by $y(t) = \mu(L)\varepsilon(t)$ where $\mu(L) = \mu_0 + \mu_1 L + \cdots + \mu_q L^q$ is a polynomial in the lag operator.

A moving-average process is stationary by definition, since any two elements y_t and y_s represent the same function of identically distributed sequences $\{\varepsilon_t, \varepsilon_{t-1}, \ldots, \varepsilon_{t-q}\}$ and $\{\varepsilon_s, \varepsilon_{s-1}, \ldots, \varepsilon_{s-q}\}$. The process is often required to be invertible, as well, such that it can be expressed in the form of $\mu^{-1}(L)y(t) = \varepsilon(t)$ where $\mu^{-1}(L) = \psi(L)$ is an power series in L whose coefficients fulfil the condition of absolute summability, which is that $\sum_i |\psi_i| < \infty$. This is an infinite-order autoregressive representation of the process. The representation is available only if all the roots of the equation $\mu(z) = \mu_0 + \mu_1 z + \cdots + \mu_q z^q = 0$ lie outside the unit circle. This conclusion follows from the result under (3.36).

Example 17.1. Consider the first-order moving-average process which is defined by

$$y(t) = \varepsilon(t) - \theta\varepsilon(t-1) = (1 - \theta L)\varepsilon(t). \tag{17.8}$$

The process is illustrated in Figure 17.1. Provided that $|\theta| < 1$, the inverse operator $(1 - \theta L)^{-1}$ can be expanded to give the following autoregressive representation of the process:

$$\begin{aligned}\varepsilon(t) &= (1-\theta L)^{-1}y(t) \\ &= \{y(t)+\theta y(t-1)+\theta^2 y(t-2)+\cdots\}.\end{aligned} \tag{17.9}$$

Imagine that $|\theta|>1$ instead. Then, to obtain an autoregressive representation, we should have to write

$$\begin{aligned}y(t+1) &= \varepsilon(t+1)-\theta\varepsilon(t) \\ &= -\theta\left(1-\frac{F}{\theta}\right)\varepsilon(t),\end{aligned} \tag{17.10}$$

where $F=L^{-1}$ and $F\varepsilon(t)=\varepsilon(t+1)$. This gives

$$\begin{aligned}\varepsilon(t) &= -\frac{1}{\theta}\left(1-\frac{F}{\theta}\right)^{-1}y(t+1) \\ &= -\frac{1}{\theta}\left\{y(t+1)+\frac{y(t+2)}{\theta}+\frac{y(t+3)}{\theta^2}+\cdots\right\}.\end{aligned} \tag{17.11}$$

Often, an expression such as this, which embodies future values of $y(t)$, has no reasonable meaning.

It is straightforward to generate the sequence of autocovariances of an MA(q) process from a knowledge of the parameters of the moving-average process and of the variance of the white-noise process which powers it. Consider

$$\begin{aligned}\gamma_\tau &= E(y_t y_{t-\tau}) \\ &= E\left\{\left(\sum_i \mu_i\varepsilon_{t-i}\right)\left(\sum_j \mu_j\varepsilon_{t-\tau-j}\right)\right\} \\ &= \sum_i\sum_j \mu_i\mu_j E(\varepsilon_{t-i}\varepsilon_{t-\tau-j}).\end{aligned} \tag{17.12}$$

Since $\varepsilon(t)$ is a white-noise sequence of independently and identically distributed random variables with zero expectations, it follows that

$$E(\varepsilon_{t-i}\varepsilon_{t-\tau-j}) = \begin{cases}0, & \text{if } i\neq\tau+j,\\ \sigma_\varepsilon^2, & \text{if } i=\tau+j.\end{cases} \tag{17.13}$$

Therefore,

$$\gamma_\tau = \sigma_\varepsilon^2\sum_{j=0}^{q-\tau}\mu_j\mu_{j+\tau}. \tag{17.14}$$

Now let $\tau=0,1,\ldots,q$. This gives

$$\begin{aligned}\gamma_0 &= \sigma_\varepsilon^2(\mu_0^2+\mu_1^2+\cdots+\mu_q^2), \\ \gamma_1 &= \sigma_\varepsilon^2(\mu_0\mu_1+\mu_1\mu_2+\cdots+\mu_{q-1}\mu_q), \\ &\vdots \\ \gamma_q &= \sigma_\varepsilon^2\mu_0\mu_q.\end{aligned} \tag{17.15}$$

For $\tau > q$, there is $\gamma_\tau = 0$.

The cut-off in the autocorrelation function at lag q, which is liable to be mirrored in its empirical counterpart, provides a means of identifying the order of a moving-average model—see Figure 17.2(b).

Example 17.2. A first-order moving-average process $y(t) = \varepsilon(t) - \theta\varepsilon(t-1)$ has the following autocovariances:

$$\begin{aligned}\gamma_0 &= \sigma_\varepsilon^2(1+\theta^2),\\ \gamma_1 &= -\sigma_\varepsilon^2\theta,\\ \gamma_\tau &= 0 \quad \text{if} \quad \tau > 1.\end{aligned} \tag{17.16}$$

A vector $[y_0, y_1, \ldots, y_{T-1}]$ comprising T consecutive elements from the process has a dispersion matrix of the form

$$\Gamma = \sigma_\varepsilon^2 \begin{bmatrix} 1+\theta^2 & -\theta & 0 & \ldots & 0 \\ -\theta & 1+\theta^2 & -\theta & \ldots & 0 \\ 0 & -\theta & 1+\theta^2 & \ldots & 0 \\ \vdots & \vdots & \vdots & \ddots & \vdots \\ 0 & 0 & 0 & \ldots & 1+\theta^2 \end{bmatrix}. \tag{17.17}$$

In general, the dispersion matrix of a qth-order moving-average process has q subdiagonal and q supradiagonal bands of nonzero elements and zero elements elsewhere.

It should recognised that the equations of (17.16) impose restrictions on the admissible values of the autocovariances of the MA(1) process. The ratio of the autocovariances is

$$\frac{\gamma_1}{\gamma_0} = \frac{-\theta}{1+\theta^2} = \rho; \tag{17.18}$$

and the solution for θ is

$$\theta = \frac{-1 \pm \sqrt{1-4\rho^2}}{2\rho}, \tag{17.19}$$

which is real-valued only if $|\rho| \leq \frac{1}{2}$. The latter is also a necessary condition for the nonnegative-definiteness of the autocovariance function. To demonstrate this, consider a quadratic function of the nth-order dispersion matrix $\Gamma = [\gamma_{|t-s|}]$ in the form of

$$Q = \sum_{t=0}^{n-1}\sum_{s=0}^{n-1} c_t\gamma_{|t-s|}c_s, \qquad \text{where} \qquad c_j = (-1)^j \tag{17.20}$$

and where $\gamma_{|t-s|}$ is defined by (17.16). If $\rho > \frac{1}{2}$, then there would be

$$Q = \sigma_\varepsilon^2(1+\theta^2)\{n - 2(n-1)\rho\} < 0 \qquad \text{for} \qquad n > \frac{2\rho}{2\rho - 1}, \tag{17.21}$$

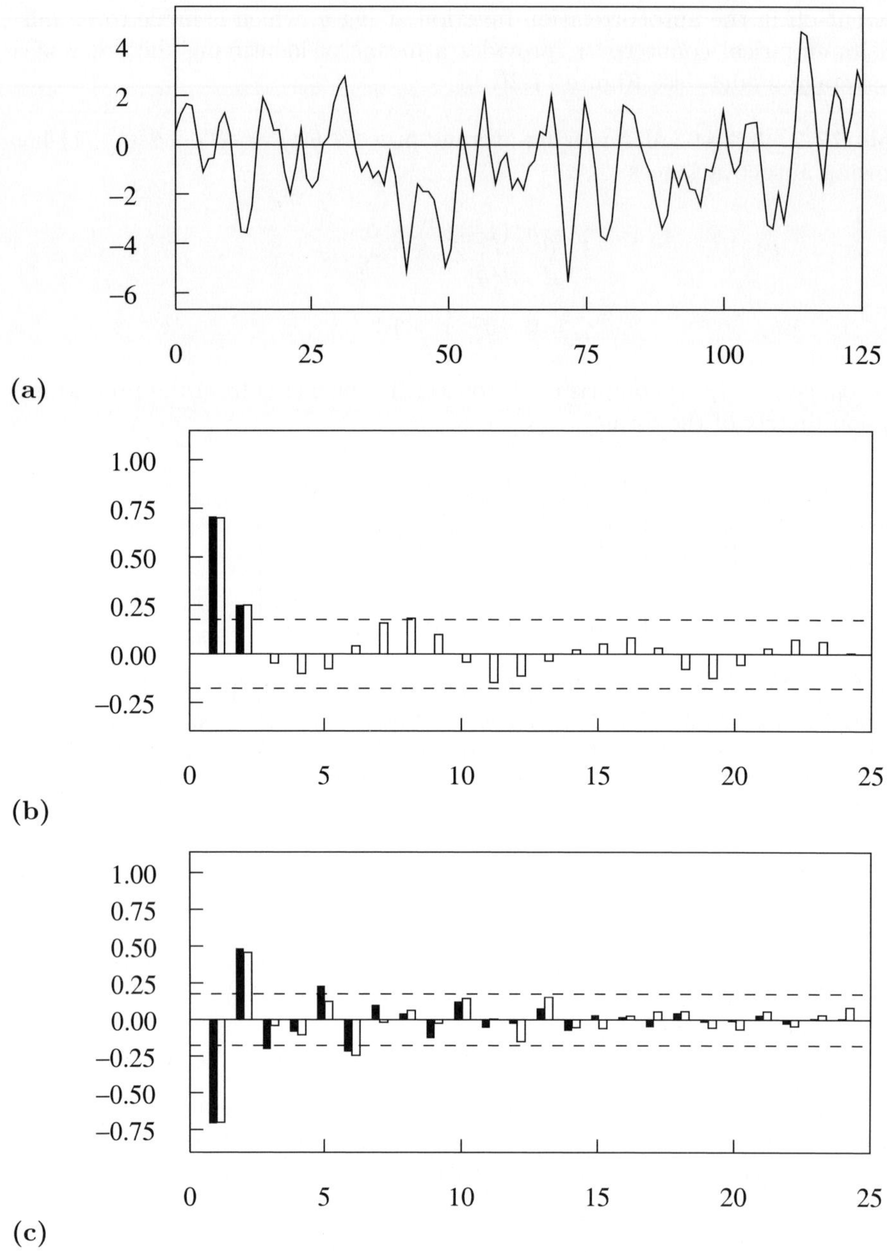

Figure 17.2. **(a)** The graph of 125 observations on a simulated series generated by an MA(2) process $y(t) = (1 + 1.25L + 0.80L^2)\varepsilon(t)$, together with **(b)** the theoretical and empirical autocorrelations and **(c)** the theoretical and empirical partial autocorrelations. The theoretical values correspond to the solid bars.

which is a violation of the condition of nonnegative-definiteness. Therefore, $|\rho| \leq \frac{1}{2}$ is necessary. The fact that it is also a sufficient condition for the nonnegative-definiteness of the autocovariance function will emerge from an example in the following chapter.

The autocovariance generating function of the qth-order moving-average process can be found quite readily. Consider the product

$$\begin{aligned} \mu(z)\mu(z^{-1}) &= \Big(\sum_i \mu_i z^i\Big)\Big(\sum_j \mu_j z^{-j}\Big) \\ &= \sum_i \sum_j \mu_i \mu_j z^{i-j} \\ &= \sum_\tau \Big(\sum_j \mu_j \mu_{j+\tau}\Big) z^\tau, \qquad \tau = i - j. \end{aligned} \tag{17.22}$$

When the expression for the autocovariance of lag τ of a moving-average process given under (17.14) is referred to, it can be see that the autocovariance generating function is just

$$\gamma(z) = \sigma_\varepsilon^2 \mu(z)\mu(z^{-1}). \tag{17.23}$$

The decomposition of $\gamma(z)$ into the factors $\mu(z)$, $\mu(z^{-1})$ and σ_ε^2 is known as the Cramér–Wold factorisation.

Given the correspondence which exists between the autocovariance generating function $\gamma(z)$ of a stationary process and the dispersion matrix Γ of a (finite) sequence of its elements, one might expect to find a straightforward matrix representation of the Cramér–Wold factorisation. Consider, therefore, the lower-triangular Toeplitz matrix $M = \mu(L)$ obtained by replacing the argument z of the polynomial $\mu(z)$ by the $L = [e_1, \ldots, e_{n-1}, 0]$ which is a matrix analogue of the lag operator. Then it transpires that the product $\sigma_\varepsilon^2 MM'$, which is the analogue of $\sigma_\varepsilon^2 \mu(z)\mu(z^{-1})$, is only an approximation to the MA dispersion matrix Γ. An exact expression for Γ, of which the matrix $\sigma_\varepsilon^2 MM'$ is the essential part, is to be found under (22.58).

Another matrix relationship which may also be construed as an analogue of the Cramér–Wold factorisation is the Cholesky decomposition $\Gamma = WDW'$ of the dispersion matrix, wherein W is a lower-triangular matrix which approximates to $M = \mu(L)$ and $D = \text{diag}\{d_0, \ldots, d_{n-1}\}$ is a diagonal matrix whose elements may be construed as successive approximations to σ_ε^2. This representation is discussed in Chapter 19 in connection with the Gram–Schmidt prediction-error algorithm.

Figure 17.2 illustrates the autocorrelation function of a moving-average process, and it shows the relationship of the latter to the partial autocorrelation function which is to be considered in detail later.

Computing the MA Autocovariances

A straightforward procedure which is modelled on the equations in (17.15) can be used to calculate the autocovariances once the values have been provided for the parameters $\mu_0, \mu_1, \ldots, \mu_q$ and σ_ε^2:

(17.24)
```
procedure MACovariances(var mu, gamma : vector;
                        var varEpsilon : real;
                        q : integer);

  var
    i, j : integer;

  begin
    for i := 0 to q do
      begin
        gamma[i] := 0.0;
        for j := 0 to q - i do
          gamma[i] := gamma[i] + mu[j] * mu[j + i];
        gamma[i] := gamma[i] * varEpsilon
      end;
  end; {MACovariances}
```

MA Processes with Common Autocovariances

Several moving-average processes can share the same autocovariance function. Thus, for example, the equations under (17.15) are generated not only by the process $y(t) = \mu_0\varepsilon(t) + \mu_1\varepsilon(t-1) + \cdots + \mu_q\varepsilon(t-q)$ but also by the process $x(t) = \mu_q\varepsilon(t) + \mu_{q-1}\varepsilon(t-1) + \cdots + \mu_0\varepsilon(t-q)$, which is formed by reversing the sequence of coefficients.

If the equation of the original process is $y(t) = \mu(L)\varepsilon(t)$, then the equation of the process with the coefficients in reversed order is $x(t) = \mu'(L)\varepsilon(t)$, wherein the operator $\mu'(L) = L^q\mu(L^{-1})$ is obtained from $\mu(L)$ by inverting all of its roots.

To see the effect of inverting a single root of the operator $\mu(L)$, one may consider the autocovariance function under (17.23) in the light of the factorisation

$$\mu(z) = \mu_0 \prod_{i=1}^{q} \left(1 - \frac{z}{\lambda_i}\right). \tag{17.25}$$

Let λ be a real-valued root of $\mu(z) = 0$. Then λ is inverted by multiplying $\mu(z)$ by $1 - \lambda z$ and dividing it by $1 - z/\lambda$. The result is

$$\tilde{\mu}(z) = \mu(z)\frac{(1 - \lambda z)}{(1 - z\lambda^{-1})}, \tag{17.26}$$

which gives

$$\begin{aligned}\tilde{\mu}(z)\tilde{\mu}(z^{-1}) &= \mu(z)\mu(z^{-1})\frac{(1 - \lambda z)(1 - \lambda z^{-1})}{(1 - z\lambda^{-1})(1 - \{z\lambda\}^{-1})}\\ &= \lambda^2\mu(z)\mu(z^{-1});\end{aligned} \tag{17.27}$$

and it follows that the autocovariance generating function of (17.23) can also be factorised as

$$\gamma(z) = \tilde{\sigma}_\varepsilon^2\tilde{\mu}(z)\tilde{\mu}(z^{-1}), \quad \text{where} \quad \tilde{\sigma}_\varepsilon^2 = \frac{\sigma_\varepsilon^2}{\lambda^2}. \tag{17.28}$$

If λ is a complex-valued root of $\mu(z) = 0$, then inverting it on its own would lead to a complex-valued function $\tilde{\mu}(z)$. For $\tilde{\mu}(z)$ to be real-valued, it is necessary to invert both λ and its complex conjugate λ^* at the same time. In that case, σ_ε^2 must be scaled by a factor of $|\lambda|^{-4}$.

Clearly, an arbitrary selection of the roots of $\mu(z)$ can be inverted in this way without affecting the autocovariance generating function. By taking account of all such inversions, the complete class of the processes which share the common autocovariance function can be defined. Amongst such a class, there is only one process which satisfies the condition of invertibility which requires every root of $\mu(z) = 0$ to lie outside the unit circle.

A particular feature of the invertible model in comparison with others which share the same autocovariances is that the corresponding transfer function entails the minimum time delays in the mapping from $\varepsilon(t)$ to $y(t)$. This is the so-called minimum-phase-delay or "miniphase" property of the invertible model.

Computing the MA Parameters from the Autocovariances

The equations of (17.15) define a mapping from the set of parameters to the sequence of autocovariances. If none of the roots of the polynomial equation $\mu(z) = 0$ lie on the unit circle, then the equations also serve to define an inverse mapping from the autocovariances to a set of parameters which correspond to a unique stationary process which satisfies the condition of invertibility. The equations which must be solved to obtain these parameters can be written, in two alternative ways, as

$$
\begin{aligned}
\begin{bmatrix} \gamma_0 \\ \gamma_1 \\ \vdots \\ \gamma_{q-1} \\ \gamma_q \end{bmatrix}
&= \sigma_\varepsilon^2 \begin{bmatrix} \mu_0 & \mu_1 & \cdots & \mu_{q-1} & \mu_q \\ \mu_1 & \mu_2 & \cdots & \mu_q & 0 \\ \vdots & \vdots & \cdot^{\cdot^{\cdot}} & \vdots & \vdots \\ \mu_{q-1} & \mu_q & \cdots & 0 & 0 \\ \mu_q & 0 & \cdots & 0 & 0 \end{bmatrix}
\begin{bmatrix} \mu_0 \\ \mu_1 \\ \vdots \\ \mu_{q-1} \\ \mu_q \end{bmatrix} \\
&= \sigma_\varepsilon^2 \begin{bmatrix} \mu_0 & \mu_1 & \cdots & \mu_{q-1} & \mu_q \\ 0 & \mu_0 & \cdots & \mu_{q-2} & \mu_{q-1} \\ \vdots & \vdots & \ddots & \vdots & \vdots \\ 0 & 0 & \cdots & \mu_0 & \mu_1 \\ 0 & 0 & \cdots & 0 & \mu_0 \end{bmatrix}
\begin{bmatrix} \mu_0 \\ \mu_1 \\ \vdots \\ \mu_{q-1} \\ \mu_q \end{bmatrix}.
\end{aligned} \tag{17.29}
$$

The equations may be written in summary notation as

$$\gamma = \sigma_\varepsilon^2 M^\# \mu = \sigma_\varepsilon^2 M' \mu, \tag{17.30}$$

where $M^\#$ is a so-called Hankel matrix. The objective is to find a solution, in terms of the vector μ, for the system

$$f(\mu) = \gamma - \sigma_\varepsilon^2 M^\# \mu = 0. \tag{17.31}$$

Since this system is nonlinear, its solution must be found via an iterative procedure.

There are several methods which can be used in solving this system. To reveal one of the simplest methods, let us consider setting $\mu_0 = 1$ and rewriting all but the first of the equations of (17.29) as

$$\begin{bmatrix} \mu_1 \\ \mu_2 \\ \vdots \\ \mu_{q-1} \\ \mu_q \end{bmatrix} = \frac{1}{\sigma_\varepsilon^2} \begin{bmatrix} \gamma_1 \\ \gamma_2 \\ \vdots \\ \gamma_{q-1} \\ \gamma_q \end{bmatrix} - \begin{bmatrix} \mu_2 & \dots & \mu_{q-1} & \mu_q \\ \mu_3 & \dots & \mu_q & 0 \\ \vdots & \cdot^{\cdot^{\cdot}} & \vdots & \vdots \\ \mu_q & \dots & 0 & 0 \\ 0 & \dots & 0 & 0 \end{bmatrix} \begin{bmatrix} \mu_1 \\ \mu_2 \\ \vdots \\ \mu_{q-2} \\ \mu_{q-1} \end{bmatrix}. \tag{17.32}$$

Here the generic equation takes the form of

$$\mu_i = \frac{\gamma_i}{\sigma_\varepsilon^2} - \sum_{j=1}^{q-i} \mu_j \mu_{j+i}, \tag{17.33}$$

whilst the first equation of (17.29), which has been omitted from (17.32), can be rearranged to give

$$\sigma_\varepsilon^2 = \frac{\gamma_0}{1 + \sum \mu_i^2}. \tag{17.34}$$

In solving these equations iteratively, one can start by setting $\mu_1 = \mu_2 = \cdots = \mu_q = 0$ and by setting $\sigma_\varepsilon^2 = \gamma_0$, which, in that case, is the value implied by (17.34). Then the equations of (17.32) can be solved from bottom to top for successive values of μ_i with $i = q, q-1, \ldots, 1$ and with the solutions for $\mu_{i+1}, \mu_{i+2}, \ldots, \mu_q$ replacing the starting values as soon as they become available. The first round of the procedure is completed by finding a revised value for σ_ε^2 from equation (17.34) using the newly found values of $\mu_1, \ldots, \mu_q$. The latter also become the starting values for the next round.

The convergence of this procedure is not particularly rapid; and it is wise to impose a limit on the number of its iterations. Its attractions, in comparison with other procedures, are its simplicity and the speed of its execution. The code for the procedure is as follows:

(17.35)
```
procedure MAParameters(var mu : vector;
                       var varEpsilon : real;
                           gamma : vector;
                           q : integer);

  var
    r, i, j : integer;
    denom, oldVar, temp, test : real;

begin
  r := 0;
  varEpsilon := gamma[0];
  mu[0] := 1;
```

```
for i := 1 to q do
  mu[i] := 0;

repeat {until convergence}

  for i := q downto 1 do
    begin {q}
      temp := gamma[i]/varEpsilon;
      for j := 1 to q - i do
        temp := temp - mu[j] * mu[j + i];
      mu[i] := temp;
    end; {q}

  oldVar := varEpsilon;
  denom := 1;
  for i := 1 to q do
    denom := denom + mu[i] * mu[i];
  varEpsilon := gamma[0]/denom;
  test := (oldVar - varEpsilon)/varEpsilon;
  r := r + 1;
until (Abs(test) < 0.00001) or (r = 99);

end; {MAParameters}
```

A more sophisticated procedure, which is a liable to find the moving-average parameters in fewer iterations, is the procedure of Tunnicliffe–Wilson [527] which depends upon the Newton–Raphson algorithm. In this case, the $(r+1)$th approximation to the solution, which is μ_{r+1}, is obtained from the rth approximation μ_r according to the formula

(17.36) $$\mu_{r+1} = \mu_r - \{Df(\mu_r)\}^{-1} f(\mu_r).$$

Here $f(\mu_r)$ and $Df(\mu_r)$ stand, respectively, for the vector function of (17.31) and its matrix first derivative evaluated at the point $\mu = \mu_r$. It is easily to verify that

(17.37) $$Df(\mu) = -\sigma_\varepsilon^2(M^{\#} + M').$$

Therefore, the algorithm can be written as

(17.38) $$\mu_{r+1} = \mu_r + \left\{\sigma_\varepsilon^2(M^{\#} + M')\right\}_r^{-1}(\gamma - \sigma_\varepsilon^2 M^{\#}\mu)_r,$$

where the subscript on the RHS is to indicate that the elements are to be evaluated at $\mu = \mu_r$.

This iterative procedure for finding the MA parameters requires some starting values. Recall that the equation of the moving average can be normalised either by setting $\sigma_\varepsilon^2 = 1$ or by setting $\mu_0 = 1$. If $\sigma_\varepsilon^2 = 1$ is chosen, then it is reasonable to begin the iterations with $\mu_0 = \gamma_0$ and $\mu_1 = \mu_2 = \cdots = \mu_q = 0$. Once the iterative procedure has converged, the equation can be re-normalised so that $\mu_0 = 1$.

The equations of (17.31) have multiple solutions. Nevertheless, the selection of starting values makes it virtually certain that the iterative procedure will converge upon the parameter values which correspond to the uniquely defined invertible MA model. The following display gives the Pascal code for implementing the procedure:

(17.39)

```
procedure Minit(var mu : vector;
                var varEpsilon : real;
                gamma : vector;
                q : integer);

  var
    d : matrix;
    delta, f : vector;
    i, j, iterations, start, finish : integer;
    tolerance : real;
    convergence : boolean;

begin {Minit}
  tolerance := 1.0E - 5;

  {Initialise the vector mu}
  for i := 0 to q do
    mu[i] := 0.0;
  mu[0] := Sqrt(gamma[0]);
  convergence := false;
  iterations := 0;

  while (convergence = false) and (iterations < 10) do
    begin

  {Form the matrix of derivatives}
      for i := 0 to q do
        for j := 0 to q do
          begin
            d[i, j] := 0.0;
            if (j - i) >= 0 then
              d[i, j] := mu[j - i];
            if (i + j) <= q then
              d[i, j] := d[i, j] + mu[i + j];
          end;

  {Find the function value}
      for j := 0 to q do
        begin
          f[j] := gamma[j];
          for i := 0 to q - j do
            f[j] := f[j] - mu[i] * mu[i + j]
        end;
```

```
    {Find the updating vector}
        LUsolve(0, q + 1, d, delta, f);

    {Update the value of mu}
        for i := 0 to q do
          mu[i] := mu[i] + delta[i];
        iterations := iterations + 1;

    {Check for convergence}
        convergence := CheckDelta(tolerance, q, delta, mu);
      end; {while}

    {Renormalise the results}
    varEpsilon := 1.0;
    for i := 1 to q do
      begin {i}
        mu[i] := mu[i]/mu[0];
        varEpsilon := varEpsilon + mu[i] * mu[i];
      end; {i}
    mu[0] := 1;
    varEpsilon := gamma[0]/varEpsilon;

  end; {Minit}
```

The test of the convergence of the algorithm is conducted by the following function which will be used again in another context:

(17.40)

```
function CheckDelta(tolerance : real;
                       q : integer;
                       var delta, mu : vector) : boolean;

  var
    i, j : integer;
    muNorm, deltaNorm : real;

 begin
   muNorm := 0.0;
   deltaNorm := 0.0;
   for i := 0 to q do
     begin
       muNorm := muNorm + Sqr(mu[i]);
       deltaNorm := deltaNorm + Sqr(delta[i])
     end;
   if (deltaNorm/muNorm) > tolerance then
     CheckDelta := false
   else
     CheckDelta := true;
 end {CheckDelta};
```

Before closing this section, we should mention yet another way in which the MA parameters may be inferred from the autocovariances. This is by using the Gram–Schmidt prediction-error algorithm of Chapter 19 to effect the Cholesky decomposition of the dispersion matrix Γ. This generates a lower-triangular matrix L, with units on its diagonal, and a diagonal matrix D such that $LDL' = \Gamma$. Given that the dispersion matrix Γ of an MA(q) process has q supradiagonal bands and q subdiagonal bands, and zero-valued elements elsewhere, it follows that L is a matrix of q subdiagonal bands. As the order n of Γ and L increases, it will be found that the values of the q nonzero off-diagonal elements in the final row of L, as well as those in other higher-order rows, will converge upon the values of the q MA parameters $\mu_1, \ldots, \mu_q$. This convergence is a reason for regarding the Cholesky decomposition as a matrix analogue of the Cramér–Wold factorisation.

Autoregressive Processes

The pth-order autoregressive process, or AR(p) process $y(t)$, is defined by the equation

$$\alpha_0 y(t) + \alpha_1 y(t-1) + \cdots + \alpha_p y(t-p) = \varepsilon(t). \tag{17.41}$$

This equation is normalised, invariably, by setting $\alpha_0 = 1$, although it would be possible to set $\sigma_\varepsilon^2 = 1$ instead. The equation can be written more concisely as $\alpha(L)y(t) = \varepsilon(t)$, where $\alpha(L) = \alpha_0 + \alpha_1 L + \cdots + \alpha_p L^p$. For the process to be stationary, the roots of the equation $\alpha(z) = \alpha_0 + \alpha_1 z + \cdots + \alpha_p z^p = 0$ must lie outside the unit circle. When this condition is satisfied, the autoregressive process can be represented as an infinite-order moving-average process in the form of $y(t) = \alpha^{-1}(L)\varepsilon(t)$.

Example 17.3. Consider the first-order autoregressive process which is defined by

$$\begin{aligned} \varepsilon(t) &= y(t) - \phi y(t-1) \\ &= (1 - \phi L) y(t). \end{aligned} \tag{17.42}$$

Provided that the process is stationary with $|\phi| < 1$, this can be written in moving-average form as

$$\begin{aligned} y(t) &= (1 - \phi L)^{-1} \varepsilon(t) \\ &= \big\{ \varepsilon(t) + \phi\varepsilon(t-1) + \phi^2 \varepsilon(t-2) + \cdots \big\}. \end{aligned} \tag{17.43}$$

The Autocovariances and the Yule–Walker Equations

Since a stationary autoregressive process is equivalent to an infinite-order moving-average process, its autocovariances can be found using the formula under (17.14), which is applicable to moving-average processes of any order. For the same reason, the autocovariance generating function of the autoregressive process $y(t) = \alpha^{-1}(L)\varepsilon(t)$ is given by

$$\gamma(z) = \frac{\sigma_\varepsilon^2}{\alpha(z)\alpha(z^{-1})}. \tag{17.44}$$

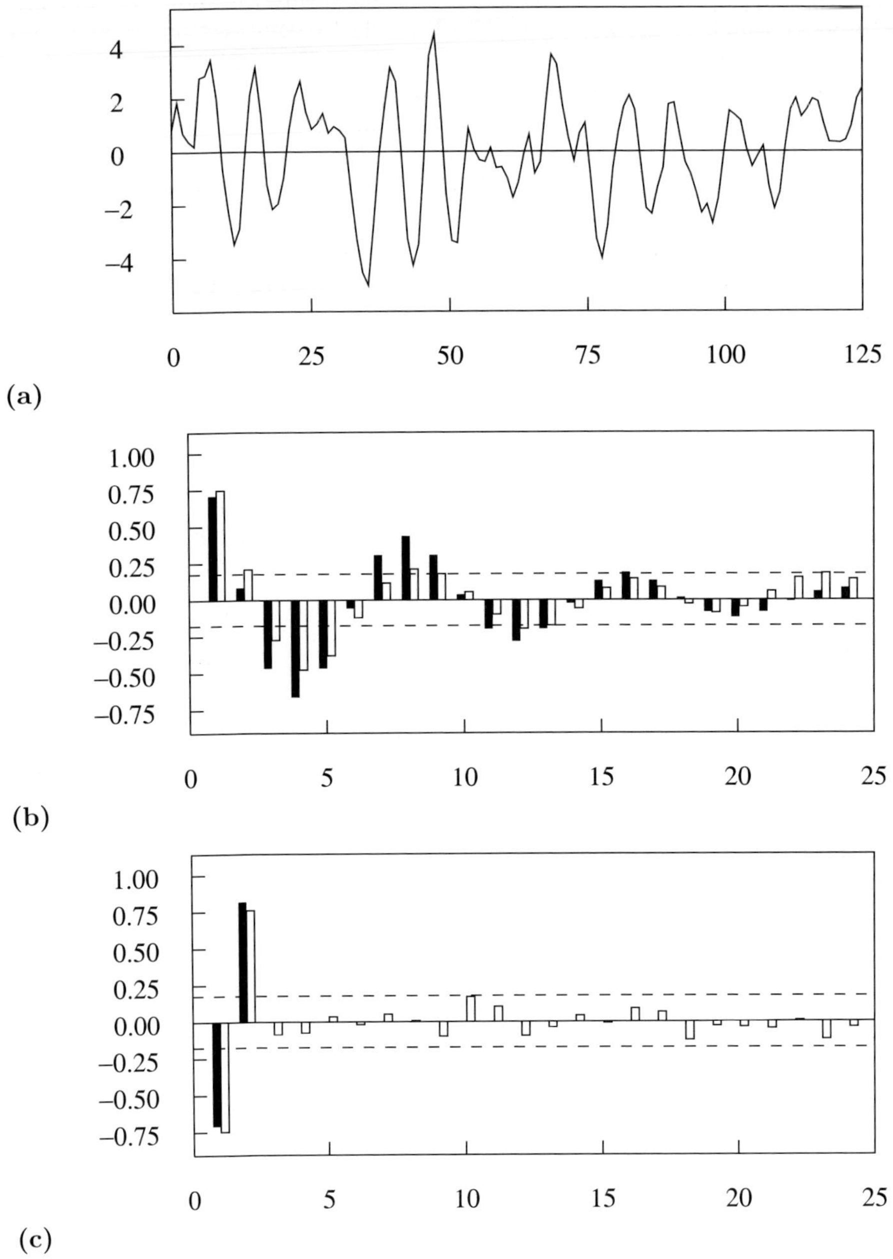

Figure 17.3. **(a)** The graph of 125 observations on a simulated series generated by an AR(2) process $(1 - 0.273L + 0.81L^2)y(t) = \varepsilon(t)$, together with **(b)** the theoretical and empirical autocorrelations and **(c)** the theoretical and empirical partial autocorrelations. The theoretical values correspond to the solid bars.

This is to be compared with the autocovariance generating function of a moving-average process which is given under (17.23). The decomposition of $\gamma(z)$ into $1/\alpha(z)$, $1/\alpha(z^{-1})$ and σ_ε^2 may be described as the Yule–Walker factorisation of the autocovariance generating function.

In view of the correspondence which exists between the autocovariance generating function $\gamma(z)$ of a stationary process and the dispersion matrix Γ, one might look for a matrix analogue of the Yule–Walker factorisation. Therefore, consider the lower-triangular Toeplitz matrix $A = \alpha(L)$ which is derived by replacing the argument z of the polynomial $\alpha(z)$ by the matrix $L = [e_1, \ldots, e_{n-1}, 0]$. Then the form of the autocovariance generating function suggests that the AR dispersion matrix Γ may be approximated by $\sigma_\varepsilon^2 A^{-1}A'^{-1}$. An exact matrix relationship which entails Γ and $A^{-1}A'^{-1}$ is to be found under (22.22).

The Cholesky decomposition of Γ may also be construed as an analogue the Yule–Walker factorisation. This decomposition, which is effected by the Levinson–Durbin algorithm, is presented later in this chapter.

Example 17.4. Consider again the first-order autoregressive process depicted in equations (17.42) and (17.43). From equation (17.43), it follows that

$$(17.45)\qquad \begin{aligned}\gamma_\tau &= E(y_t y_{t-\tau})\\ &= E\left\{\Big(\sum_i \phi^i \varepsilon_{t-i}\Big)\Big(\sum_j \phi^j \varepsilon_{t-\tau-j}\Big)\right\}\\ &= \sum_i\sum_j \phi^i\phi^j E(\varepsilon_{t-i}\varepsilon_{t-\tau-j}).\end{aligned}$$

Because of the absence of correlations amongst the elements of the sequence $\varepsilon(t)$, which is reflected in the conditions under (17.13), this equation becomes

$$(17.46)\qquad \begin{aligned}\gamma_\tau &= \sigma_\varepsilon^2 \sum_j \phi^j \phi^{j+\tau}\\ &= \frac{\sigma_\varepsilon^2 \phi^\tau}{1-\phi^2}.\end{aligned}$$

Therefore, a vector $y = [y_0, y_1, \ldots, y_{T-1}]'$ of T consecutive elements from a first-order autoregressive process has a dispersion matrix of the form

$$(17.47)\qquad D(y) = \frac{\sigma_\varepsilon^2}{1-\phi^2}\begin{bmatrix} 1 & \phi & \phi^2 & \ldots & \phi^{T-1}\\ \phi & 1 & \phi & \ldots & \phi^{T-2}\\ \phi^2 & \phi & 1 & \ldots & \phi^{T-3}\\ \vdots & \vdots & \vdots & \ddots & \vdots\\ \phi^{T-1} & \phi^{T-2} & \phi^{T-3} & \ldots & 1\end{bmatrix}.$$

The method of finding the autocovariances which is exemplified above is algebraically intractable for all but a first-order process. For a practical way of finding the autocovariances of the pth-order process, consider multiplying $\sum_i \alpha_i y_{t-i} = \varepsilon_t$

by $y_{t-\tau}$ and taking expectations to give

$$\sum_i \alpha_i E(y_{t-i}y_{t-\tau}) = E(\varepsilon_t y_{t-\tau}). \tag{17.48}$$

Taking $\alpha_0 = 1$ and writing $y_{t-\tau} = \varepsilon_{t-\tau} - \alpha_1 y_{t-\tau-1} - \cdots - \alpha_p y_{t-\tau-p}$ helps to confirm that

$$E(\varepsilon_t y_{t-\tau}) = \begin{cases} \sigma_\varepsilon^2, & \text{if } \tau = 0, \\ 0, & \text{if } \tau > 0. \end{cases} \tag{17.49}$$

These results depend upon the fact that elements of $\varepsilon(t)$ have no correlation with the elements of $y(t)$ which precede them in time. Therefore, (17.48) becomes

$$\sum_i \alpha_i \gamma_{|\tau-i|} = \begin{cases} \sigma_\varepsilon^2, & \text{if } \tau = 0, \\ 0, & \text{if } \tau > 0. \end{cases} \tag{17.50}$$

The equation $\sum_i \alpha_i \gamma_{\tau-i} = 0$ is a homogeneous difference equation which serves to generate the sequence $\{\gamma_p, \gamma_{p+1}, \ldots\}$ given p starting values $\gamma_0, \gamma_1, \ldots, \gamma_{p-1}$.

The variance of $\varepsilon(t)$ is also given by the following quadratic form:

$$\begin{aligned} \sigma_\varepsilon^2 &= E\Big\{\Big(\sum_i \alpha_i y_{t-i}\Big)\Big(\sum_j \alpha_j y_{t-\tau-j}\Big)\Big\} \\ &= \sum_i \sum_j \alpha_i \alpha_j \gamma_{i-j}. \end{aligned} \tag{17.51}$$

This is the coefficient associated with z^0 on the left-hand side of the identity $\alpha(z)\gamma(z)\alpha(z^{-1}) = \sigma_\varepsilon^2$ which comes directly from (17.44). The identity

$$\begin{aligned} \sum_i \sum_j \alpha_i \alpha_j \gamma_{i-j} &= \sum_j \alpha_j \Big(\sum_i \alpha_i \gamma_{i-j}\Big) \\ &= \sum_i \alpha_i \gamma_i, \end{aligned} \tag{17.52}$$

which relates the alternative expressions for this variance found under (17.50) and (17.51), comes from the fact that $\sum_i \alpha_i \gamma_{i-j} = 0$ for all $j > 0$ and from the normalisation $\alpha_0 = 1$.

By letting $\tau = 0, 1, \ldots, p$ in (17.50), a set of $p + 1$ equations are generated which can be arrayed in matrix form as follows:

$$\begin{bmatrix} \gamma_0 & \gamma_1 & \gamma_2 & \cdots & \gamma_p \\ \gamma_1 & \gamma_0 & \gamma_1 & \cdots & \gamma_{p-1} \\ \gamma_2 & \gamma_1 & \gamma_0 & \cdots & \gamma_{p-2} \\ \vdots & \vdots & \vdots & \ddots & \vdots \\ \gamma_p & \gamma_{p-1} & \gamma_{p-2} & \cdots & \gamma_0 \end{bmatrix} \begin{bmatrix} 1 \\ \alpha_1 \\ \alpha_2 \\ \vdots \\ \alpha_p \end{bmatrix} = \begin{bmatrix} \sigma_\varepsilon^2 \\ 0 \\ 0 \\ \vdots \\ 0 \end{bmatrix}. \tag{17.53}$$

These are called the Yule–Walker equations, and they can be used either for obtaining the values $\gamma_0, \gamma_1, \ldots, \gamma_p$ from the values $\alpha_1, \ldots, \alpha_p, \sigma_\varepsilon^2$ or vice versa.

The Yule–Walker equations may be derived directly from the autocovariance generating function of (17.44), which can be rearranged to give

$$\gamma(z)\alpha(z) = \sigma_\varepsilon^2 \frac{1}{\alpha(z^{-1})}. \tag{17.54}$$

The $p+1$ equations of (17.53) are obtained by equating the coefficients on either side of the equation above which are associated with the $z^0, z, \ldots, z^p$.

The parameters $\alpha_1, \ldots, \alpha_p$ are uniquely determined by the system

$$\begin{bmatrix} \gamma_0 & \gamma_1 & \ldots & \gamma_{p-1} \\ \gamma_1 & \gamma_0 & \ldots & \gamma_{p-2} \\ \vdots & \vdots & \ddots & \vdots \\ \gamma_{p-1} & \gamma_{p-2} & \ldots & \gamma_0 \end{bmatrix} \begin{bmatrix} \alpha_1 \\ \alpha_2 \\ \vdots \\ \alpha_p \end{bmatrix} = - \begin{bmatrix} \gamma_1 \\ \gamma_2 \\ \vdots \\ \gamma_p \end{bmatrix} \tag{17.55}$$

which comprises all but the first equation of (17.53) and which may be described as the normal equations of the regression of $y(t)$ on $y(t-1), \ldots, y(t-p)$. The value of σ_ε^2 is obtained from the first equation of (17.53) using the previously determined values of $\alpha_1, \ldots, \alpha_p$.

To derive the equations which allow the autocovariances to be found from the parameters, let us define $g = \gamma_0/2$ in order to write the LHS of (17.53) as

$$\begin{aligned} &\begin{bmatrix} g & 0 & \ldots & 0 \\ \gamma_1 & g & \ldots & 0 \\ \vdots & \vdots & \ddots & \vdots \\ \gamma_p & \gamma_{p-1} & \ldots & g \end{bmatrix} \begin{bmatrix} 1 \\ \alpha_1 \\ \vdots \\ \alpha_p \end{bmatrix} + \begin{bmatrix} g & \gamma_1 & \ldots & \gamma_p \\ 0 & g & \ldots & \gamma_{p-1} \\ \vdots & \vdots & \ddots & \vdots \\ 0 & 0 & \ldots & g \end{bmatrix} \begin{bmatrix} 1 \\ \alpha_1 \\ \vdots \\ \alpha_p \end{bmatrix} \\ &= \begin{bmatrix} 1 & 0 & \ldots & 0 \\ \alpha_1 & 1 & \ldots & 0 \\ \vdots & \vdots & \ddots & \vdots \\ \alpha_p & \alpha_{p-1} & \ldots & 1 \end{bmatrix} \begin{bmatrix} g \\ \gamma_1 \\ \vdots \\ \gamma_p \end{bmatrix} + \begin{bmatrix} 1 & \ldots & \alpha_{p-1} & \alpha_p \\ \alpha_1 & \ldots & \alpha_p & 0 \\ \vdots & ⋰ & \vdots & \vdots \\ \alpha_p & \ldots & 0 & 0 \end{bmatrix} \begin{bmatrix} g \\ \vdots \\ \gamma_{p-1} \\ \gamma_p \end{bmatrix}. \end{aligned} \tag{17.56}$$

Then, by recombining the matrices, the following alternative version of equation (17.53) is derived:

$$\begin{bmatrix} 1 & \alpha_1 & \ldots & \alpha_{p-1} & \alpha_p \\ \alpha_1 & 1+\alpha_2 & \ldots & \alpha_p & 0 \\ \vdots & \vdots & & \vdots & \vdots \\ \alpha_{p-1} & \alpha_{p-2}+\alpha_p & \ldots & 1 & 0 \\ \alpha_p & \alpha_{p-1} & \ldots & \alpha_1 & 1 \end{bmatrix} \begin{bmatrix} \gamma_0 \\ \gamma_1 \\ \vdots \\ \gamma_{p-1} \\ \gamma_p \end{bmatrix} = \begin{bmatrix} \sigma_\varepsilon^2 \\ 0 \\ \vdots \\ 0 \\ 0 \end{bmatrix}. \tag{17.57}$$

This serves to determine uniquely the set of autocovariances $\gamma_0, \gamma_1, \ldots, \gamma_p$. The set contains one more element than is needed to begin the recursion based on equation (17.50) by which the succeeding autocovariances can be generated.

Example 17.5. For an example of the two uses of the Yule–Walker equations, let us consider the second-order autoregressive process. In that case, there is

$$
\begin{aligned}
\begin{bmatrix} \gamma_0 & \gamma_1 & \gamma_2 \\ \gamma_1 & \gamma_0 & \gamma_1 \\ \gamma_2 & \gamma_1 & \gamma_0 \end{bmatrix}
\begin{bmatrix} 1 \\ \alpha_1 \\ \alpha_2 \end{bmatrix}
&= \begin{bmatrix} \alpha_2 & \alpha_1 & 1 & 0 & 0 \\ 0 & \alpha_2 & \alpha_1 & 1 & 0 \\ 0 & 0 & \alpha_2 & \alpha_1 & 1 \end{bmatrix}
\begin{bmatrix} \gamma_2 \\ \gamma_1 \\ \gamma_0 \\ \gamma_1 \\ \gamma_2 \end{bmatrix} \\
&= \begin{bmatrix} 1 & \alpha_1 & \alpha_2 \\ \alpha_1 & 1+\alpha_2 & 0 \\ \alpha_2 & \alpha_1 & 1 \end{bmatrix}
\begin{bmatrix} \gamma_0 \\ \gamma_1 \\ \gamma_2 \end{bmatrix}
= \begin{bmatrix} \sigma_\varepsilon^2 \\ 0 \\ 0 \end{bmatrix}.
\end{aligned}
\tag{17.58}
$$

Notice how the matrix following the first equality is folded across the axis which divides it vertically to create the matrix which follows the second equality. Pleasing effects of this sort abound in time-series analysis.

If the values for $\gamma_0, \gamma_1, \gamma_2$, are known, then the equations can be solved for the parameter values

$$
\begin{aligned}
\alpha_1 &= \frac{\gamma_0\gamma_1 - \gamma_1\gamma_2}{\gamma_0^2 - \gamma_1^2}, \\
\alpha_2 &= \frac{\gamma_0\gamma_2 - \gamma_1^2}{\gamma_0^2 - \gamma_1^2}.
\end{aligned}
\tag{17.59}
$$

Then $\sigma_\varepsilon^2 = \gamma_0 + \alpha_1\gamma_1 + \alpha_2\gamma_2$ can be found. Conversely, to obtain the first two autocovariances, the second equation of (17.58) may be reduced to

$$
\begin{bmatrix} 1-\alpha_2^2 & \alpha_1(1-\alpha_2) \\ \alpha_1 & 1+\alpha_2 \end{bmatrix}
\begin{bmatrix} \gamma_0 \\ \gamma_1 \end{bmatrix}
= \begin{bmatrix} \sigma_\varepsilon^2 \\ 0 \end{bmatrix}.
\tag{17.60}
$$

Solving this gives

$$
\begin{aligned}
\gamma_0 &= \frac{\sigma_\varepsilon^2(1+\alpha_2)}{(1-\alpha_2)(1+\alpha_2+\alpha_1)(1+\alpha_2-\alpha_1)}, \\
\gamma_1 &= \frac{-\sigma_\varepsilon^2\alpha_1}{(1-\alpha_2)(1+\alpha_2+\alpha_1)(1+\alpha_2-\alpha_1)}.
\end{aligned}
\tag{17.61}
$$

The denominator in these expressions can be parsed in a variety of ways. Given the values of γ_0 and γ_1, we can proceed to find $\gamma_2 = -\alpha_2\gamma_0 - \alpha_1\gamma_1$.

In should be emphasised that the sequence of autocovariances which is generated by solving the system under (17.57) will give rise to a positive-definite dispersion matrix only if the parameters $\alpha_0, \alpha_1, \ldots, \alpha_p$ correspond to a stationary AR process. Conversely, the AR parameters which are obtained by solving equation (17.55) will correspond to a stationary process if and only if the matrix of autocovariances is positive definite. We may express this more succinctly by declaring that

(17.62) The polynomial equation $\alpha(z) = \alpha_0 + \alpha_1 z + \cdots + \alpha_p z^p = 0$, has all of its roots outside the unit circle if and only if the autocovariance function $\gamma(z)$ of the corresponding AR process is positive definite.

Proof. If the roots of $\alpha(z) = 0$ are outside the unit circle, then the AR process is stationary and, by definition, there are well-defined positive-definite dispersion matrices corresponding to segments of $y(t)$ of any length. In particular, the condition of (17.6) is satisifed by the AR autocovariance generating function $\gamma(z) = \{\alpha(z)\alpha(z^{-1})\}^{-1}$. It is only the converse that has to be proved.

Therefore, consider the factorisation $\alpha(z) = \delta(z)\phi(z)$, where $\delta(z) = 1 + \delta_1 z + \delta_2 z^2$ is any of the quadratic factors of $\alpha(z)$. Then the equation in z under (17.54), from which the Yule–Walker equations for the AR(p) process may be derived, can be rewritten as

$$\begin{aligned}\rho(z)\delta(z) &= \{\phi(z^{-1})\gamma(z)\phi(z)\}\delta(z)\\ &= \sigma_\varepsilon^2 \frac{1}{\delta(z^{-1})},\end{aligned} \tag{17.63}$$

where $\rho(z) = \phi(z^{-1})\gamma(z)\phi(z)$ is the autocovariance generating function of the filtered sequence $\phi(L)y(t)$. By equating the coefficients associated with z^0, z^1 and z^2 on both sides, the following equations are derived:

$$\begin{bmatrix}\rho_0 & \rho_1 & \rho_2\\ \rho_1 & \rho_0 & \rho_1\\ \rho_2 & \rho_1 & \rho_0\end{bmatrix}\begin{bmatrix}1\\ \delta_1\\ \delta_2\end{bmatrix} = \begin{bmatrix}1 & \delta_1 & \delta_2\\ \delta_1 & 1+\delta_2 & 0\\ \delta_2 & \delta_1 & 1\end{bmatrix}\begin{bmatrix}\rho_0\\ \rho_1\\ \rho_2\end{bmatrix} = \begin{bmatrix}\sigma_\varepsilon^2\\ 0\\ 0\end{bmatrix}. \tag{17.64}$$

Solving the second equation for the autocovariances gives

$$\begin{aligned}\rho_0 &= \frac{\sigma_\varepsilon^2(1+\delta_2)}{(1-\delta_2)(\{1+\delta_2\}^2 - \delta_1^2)},\\ \rho_1 &= \frac{-\sigma_\varepsilon^2\delta_1}{(1-\delta_2)(\{1+\delta_2\}^2 - \delta_1^2)}.\end{aligned} \tag{17.65}$$

If the function $\gamma(z)$ is positive definite, then so too is $\rho(z)$. The condition that $\rho(z)$ is positive definite implies that $\rho_0 > 0$ and that $\rho_0^2 - \rho_1^2 > 0$. Given that $\sigma_\varepsilon^2 > 0$, it is easy to see that these imply that

$$\begin{aligned}&(1+\delta_2)^2 > \delta_1^2 \quad \text{and}\\ &\frac{1+\delta_2}{1-\delta_2} > 0 \quad \text{or, equivalently,} \quad 1-\delta_2^2 > 0.\end{aligned} \tag{17.66}$$

The latter conditions, which correspond to those listed under (5.148), are necessary and sufficient to ensure that the roots of $\delta(z)$ lie outside the unit circle.

This analysis can be repeated for every other quadratic factor of the polynomial $\alpha(z)$ in order to show that all of the complex roots must lie outside the unit circle in consequence of the positive definite nature of $\rho(z)$. It is easy to show, along similar lines, that the real roots must lie outside the unit circle.

Computing the AR Parameters

To obtain the autoregressive parameters $\alpha_1, \ldots, \alpha_p$ and σ_ε^2 from the autocovariances $\gamma_0, \ldots, \gamma_p$, a straightforward procedure may be devised which solves the Yule–Walker equations. In calling the procedure listed below, one must set $q = 0$. The matrix which is notionally inverted in the process of finding the parameters is, in fact, symmetric as can be seen from (17.55). Therefore, it might be more efficient to employ an inversion algorithm which recognises this symmetry in place of the procedure *LUsolve* of (7.28), which does not. However, there will be an occasion later when it will be necessary to infer the autoregressive parameters from a set of autocovariances which do not give rise to a symmetric matrix; and this is the reason for invoking *LUsolve*.

(17.67)
```
procedure YuleWalker(p, q : integer;
                     gamma : vector;
                     var alpha : vector;
                     var varEpsilon : real);

  var
    a : matrix;
    b : vector;
    i, j, k : integer;

begin {YuleWalker}
  for i := 1 to p do
    begin {i}
      b[i] := -gamma[q + i];
      for j := 1 to p do
        begin {j}
          k := Abs(q + i - j);
          a[i, j] := gamma[k];
        end{j}
    end; {i}

  LUsolve(1, p, a, alpha, b);
  alpha[0] := 1;
  varEpsilon := 0;
  for i := 0 to p do
    varEpsilon := varEpsilon + alpha[i] * gamma[i];

end; {YuleWalker}
```

The autoregressive parameters may be computed, alternatively, by a recursive method which delivers, on the rth iteration, a set of coefficients, denoted by $\alpha_{1(r)}, \ldots, \alpha_{r(r)}$, which are the parameters belonging to the unique autoregressive process of order r which generates the values $\gamma_0, \gamma_1, \ldots, \gamma_r$ as its first r autocovariances.

The sequence $\{\alpha_{r(r)}; r = 1, 2, \ldots\}$ is known as the partial autocorrelation function. If, in fact, the autocovariances have been generated by a process of order p, and if $p < r$, then it will be found that $\alpha_{r(r)} = 0$. If, in place of the true autocovariances, which might be unknown, one were to use a set of empirical values determined from sample data, then this result would no longer hold. Nevertheless, one could expect the corresponding empirical value of $\alpha_{r(r)}$, for any $r > p$, to be close to zero; and this should provide a useful indication of the order of the process underlying the data.

The relationship between the theoretical and the empirical autocorrelation functions is illustrated in Figure 17.3 for the case of $p = 2$. The figure supports the notion that the order of an AR process can be discerned by inspecting the empirical partial autocorrelation function.

The recursive algorithm for generating the sequence of partial autocorrelations was first discovered by Levinson [314] in 1946. It was rediscovered by Durbin [166] in 1960 who gave it an alternative form. It has a startling simplicity.

To derive the algorithm, let us imagine that the values $\alpha_{1(r)}, \ldots, \alpha_{r(r)}$ are already available. Then, by extending the set of rth-order Yule–Walker equations to which these values correspond, we can derive the system

$$\begin{bmatrix} \gamma_0 & \gamma_1 & \cdots & \gamma_r & \gamma_{r+1} \\ \gamma_1 & \gamma_0 & \cdots & \gamma_{r-1} & \gamma_r \\ \vdots & \vdots & \ddots & \vdots & \vdots \\ \gamma_r & \gamma_{r-1} & \cdots & \gamma_0 & \gamma_1 \\ \gamma_{r+1} & \gamma_r & \cdots & \gamma_1 & \gamma_0 \end{bmatrix} \begin{bmatrix} 1 \\ \alpha_{1(r)} \\ \vdots \\ \alpha_{r(r)} \\ 0 \end{bmatrix} = \begin{bmatrix} \sigma^2_{(r)} \\ 0 \\ \vdots \\ 0 \\ g \end{bmatrix}, \tag{17.68}$$

wherein

$$g = \sum_{j=0}^{r} \alpha_{j(r)}\gamma_{r+1-j} \qquad \text{with} \quad \alpha_{0(r)} = 1. \tag{17.69}$$

The system can also be written as

$$\begin{bmatrix} \gamma_0 & \gamma_1 & \cdots & \gamma_r & \gamma_{r+1} \\ \gamma_1 & \gamma_0 & \cdots & \gamma_{r-1} & \gamma_r \\ \vdots & \vdots & \ddots & \vdots & \vdots \\ \gamma_r & \gamma_{r-1} & \cdots & \gamma_0 & \gamma_1 \\ \gamma_{r+1} & \gamma_r & \cdots & \gamma_1 & \gamma_0 \end{bmatrix} \begin{bmatrix} 0 \\ \alpha_{r(r)} \\ \vdots \\ \alpha_{1(r)} \\ 1 \end{bmatrix} = \begin{bmatrix} g \\ 0 \\ \vdots \\ 0 \\ \sigma(r)^2 \end{bmatrix}. \tag{17.70}$$

Now let us combine the two systems of equations under (17.68) and (17.70) to give

$$\begin{bmatrix} \gamma_0 & \gamma_1 & \cdots & \gamma_r & \gamma_{r+1} \\ \gamma_1 & \gamma_0 & \cdots & \gamma_{r-1} & \gamma_r \\ \vdots & \vdots & \ddots & \vdots & \vdots \\ \gamma_r & \gamma_{r-1} & \cdots & \gamma_0 & \gamma_1 \\ \gamma_{r+1} & \gamma_r & \cdots & \gamma_1 & \gamma_0 \end{bmatrix} \begin{bmatrix} 1 \\ \alpha_{1(r)} + c\alpha_{r(r)} \\ \vdots \\ \alpha_{r(r)} + c\alpha_{1(r)} \\ c \end{bmatrix} = \begin{bmatrix} \sigma^2_{(r)} + cg \\ 0 \\ \vdots \\ 0 \\ g + c\sigma^2_{(r)} \end{bmatrix}. \tag{17.71}$$

If the coefficient of the combination, which is described as the reflection coefficient, is taken to be

$$c = -\frac{g}{\sigma^2_{(r)}}, \tag{17.72}$$

then the final element in the vector on the RHS becomes zero and the system becomes the set of Yule–Walker equations of order $r+1$. The solution of the equations, from the last element $\alpha_{r+1(r+1)} = c$ through to the variance term $\sigma^2_{(r+1)}$, is given by

$$\begin{aligned}
\alpha_{r+1(r+1)} &= \frac{-1}{\sigma^2_{(r)}}\left\{\sum_{j=0}^{r} \alpha_{j(r)}\gamma_{r+1-j}\right\},\\
\begin{bmatrix} \alpha_{1(r+1)} \\ \vdots \\ \alpha_{r(r+1)} \end{bmatrix} &= \begin{bmatrix} \alpha_{1(r)} \\ \vdots \\ \alpha_{r(r)} \end{bmatrix} + \alpha_{r+1(r+1)} \begin{bmatrix} \alpha_{r(r)} \\ \vdots \\ \alpha_{1(r)} \end{bmatrix},\\
\sigma^2_{(r+1)} &= \sigma^2_{(r)}\left\{1 - (\alpha_{r+1(r+1)})^2\right\}.
\end{aligned} \tag{17.73}$$

Thus the solution of the Yule–Walker system of order $r+1$ is easily derived from the solution of the system of order r, and there is scope for devising a recursive procedure. The starting values for the recursion are

$$\alpha_{1(1)} = -\gamma_1/\gamma_0 \quad \text{and} \quad \sigma^2_{(1)} = \gamma_0\left\{1 - (\alpha_{1(1)})^2\right\}. \tag{17.74}$$

The Levinson–Durbin algorithm is implemented in the following Pascal procedure:

(17.75)
```
procedure LevinsonDurbin(gamma : vector;
                         p : integer;
                         var alpha, pacv : vector);

var
  r, j, jstop : integer;
  c, g, astore : real;
  sigsqr : vector;

begin {LevinsonDurbin}
  alpha[0] := 1.0;
  pacv[0] := 1.0;
  sigsqr[0] := gamma[0];
  r := 0;

  while r < p do
    begin
      g := 0.0;
      for j := 0 to r do
```

```
        g := g + alpha[j] * gamma[r + 1 - j];
        c := g/sigsqr[r];
        jstop := r div 2;

        for j := 1 to jstop do
          begin {j}
            astore := alpha[j];
            alpha[j] := astore - c * alpha[r + 1 - j];
            alpha[r + 1 - j] := alpha[r + 1 - j] - c * astore;
          end; {j}

        j := jstop + 1;
        if Odd(r) then
          alpha[j] := alpha[j] * (1 - c);
        alpha[r + 1] := -c;
        sigsqr[r + 1] := sigsqr[r] * (1 - Sqr(alpha[r + 1]));
        r := r + 1;
        pacv[r] := -c;

      end; {while}
  end; {LevinsonDurbin}
```

It is interesting to recognise that the Levinson–Durbin algorithm generates the factors of the Cholesky decomposition of the matrix Γ of the autocovariances. Consider the following equation:

(17.76)
$$\begin{bmatrix} \gamma_0 & \gamma_1 & \cdots & \gamma_r \\ \gamma_1 & \gamma_0 & \cdots & \gamma_{r-1} \\ \vdots & \vdots & \ddots & \vdots \\ \gamma_r & \gamma_{r-1} & \cdots & \gamma_0 \end{bmatrix} \begin{bmatrix} 1 & \alpha_{1(1)} & \cdots & \alpha_{r(r)} \\ 0 & 1 & \cdots & \alpha_{r-1(r)} \\ \vdots & \vdots & \ddots & \vdots \\ 0 & 0 & \cdots & 1 \end{bmatrix} = \begin{bmatrix} \gamma_0 & 0 & \cdots & 0 \\ q_{11} & \sigma^2_{(1)} & \cdots & 0 \\ \vdots & \vdots & \ddots & \vdots \\ q_{r1} & q_{r2} & \cdots & \sigma^2_{(r)} \end{bmatrix}.$$

The elements on and above the diagonal of the matrix product are explained by a succession of Yule–Walker equations of increasing order which are in the same inverted form as equation (17.70). Equation (17.76) may be expressed in summary notation as $\Gamma A' = L$.

Now consider premultiplying by A to give $A\Gamma A' = AL$. Since A and L are lower-triangular matrices, it follows that AL is lower-triangular. Also, since A has units on the diagonal, it follows that AL must have the same diagonal elements as L. Finally, $AL = A\Gamma A'$ is symmetric; so the elements both above and below its diagonal must be zeros. Thus $A\Gamma A' = D = \mathrm{diag}\{\gamma_0, \sigma^2_{(1)}, \ldots, \sigma^2_{(r)}\}$; and there is a Cholesky decomposition of Γ in the form of $\Gamma = A^{-1}DA'^{-1}$ together with a corresponding decomposition of its inverse in the form of $\Gamma^{-1} = A'D^{-1}A$.

It may be recalled that, for a matrix such as Γ to be positive definite, it is necessary and sufficient that the elements in the diagonal matrix D of the Cholesky decomposition should all be positive. As we shall see in the next example, it transpires that this is also a necessary and sufficient condition for the stability of an autoregressive process with the parameters $\alpha_{1(r)}, \ldots, \alpha_{r(r)}$; but, in fact, this point has already been established through the proof of the proposition under (17.62).

The Durbin–Levinson algorithm can be construed as a procedure which generates recursively a sequence of polynomials of ever-increasing degrees. The polynomials of the rth degree are

$$\begin{aligned} \alpha_r(z) &= 1 + \alpha_{1(r)}z + \cdots + \alpha_{r-1(r)}z^{r-1} + \alpha_{r(r)}z^r, \quad \text{and} \\ \alpha'_r(z) &= \alpha_{r(r)} + \alpha_{r-1(r)}z + \cdots + \alpha_{1(r)}z^{r-1} + z^r. \end{aligned} \tag{17.77}$$

The polynomials of degree $r+1$, whose coefficients are provided by the equations under (17.73), are

$$\begin{aligned} \alpha_{r+1}(z) &= \alpha_r(z) + zc_{r+1}\alpha'_r(z), \\ \alpha'_{r+1}(z) &= z\alpha'_r(z) + c_{r+1}\alpha_r(z), \end{aligned} \tag{17.78}$$

where $c_{r+1} = \alpha_{r+1(r+1)}$ is the reflection coefficient of (17.72), to which a subscript has been added to identify its order in the recursive scheme.

The Durbin–Levinson procedure can be put into reverse. That is to say, it is possible to devise an algorithm which generates $\alpha_r(z)$ and $\alpha'_r(z)$ from $\alpha_{r+1}(z)$ and $\alpha'_{r+1}(z)$. The equations are

$$\begin{aligned} \alpha_r(z) &= \frac{1}{1-c_{r+1}^2}\{\alpha_{r+1}(z) - c_{r+1}\alpha'_{r+1}(z)\}, \\ \alpha'_r(z) &= \frac{z^{-1}}{1-c_{r+1}^2}\{\alpha'_{r+1}(z) - c_{r+1}\alpha_{r+1}(z)\}. \end{aligned} \tag{17.79}$$

It is remarkable to discover that this inverse algorithm is a straightforward variant of the Schur–Cohn algorithm which serves to determine whether or not a linear difference equation is stable.

Example 17.6. The Schur–Cohn algorithm—see (5.153)—generates a succession of polynomials of decreasing degrees whose leading coefficients are all required to be positive if the roots of the first polynomial in the sequence are to lie outside the unit circle. According to equation (5.152), the polynomials of degree $p-1$ are generated from those of degree p via the following equations:

$$\begin{aligned} f_{p-1}(z) &= \alpha_0 f_p(z) - \alpha_p f'_p(z), \\ f'_{p-1}(z) &= z^{-1}\{\alpha_0 f'_p(z) - \alpha_p f_p(z)\}. \end{aligned} \tag{17.80}$$

Inverting the transformation gives

$$\begin{aligned} f_p(z) &= \frac{1}{\alpha_0^2 - \alpha_p^2}\{\alpha_0 f_{p-1}(z) + z\alpha_p f'_{p-1}(z)\}, \\ f'_p(z) &= \frac{1}{\alpha_0^2 - \alpha_p^2}\{z\alpha_0 f'_{p-1}(z) + \alpha_p f_{p-1}(z)\}. \end{aligned} \tag{17.81}$$

Here equation (17.81) corresponds to equation (17.78) and equation (17.80) corresponds to equation (17.79). There are minor differences of notation which have to be taken into account in making the comparison. Thus, for example, $c_{r+1} = \alpha_{r+1(r+1)}$ of equations (17.78) and (17.79) becomes α_p in equations (17.80) and (17.81). The substantive differences in the two sets of equations are attributable wholly to the normalisation $\alpha_0 = 1$ which is applied to the leading coefficients of the polynomials in the case of the Durbin–Levinson algorithm but not in the case of the Schur–Cohn algorithm as we have presented it.

A necessary condition for the polynomial equation $\alpha_p(z) = \alpha_0 + \alpha_1 z + \cdots + \alpha_p z^p = 0$ to have all if its roots outside the unit circle, which is given under (5.151), is that $\alpha_0^2 - \alpha_p^2 > 0$. The necessary and sufficient condition is that all such products, calculated from the sequence of polynomials $f_p(z), f_{p-1}(z), \ldots, f_1(z)$, should be positive. In terms of the Durbin–Levinson algorithm, the first of these conditions becomes $1 - \alpha_{r+1(r+1)}^2 > 0$ which, according to equation (17.73), implies that

$$\sigma_{(r)}^2\{1 - \alpha_{r+1(r+1)}^2\} = \sigma_{(r+1)}^2 > 0 \tag{17.82}$$

if $\sigma_{(r)}^2 > 0$. A condition which is necessary and sufficient for the roots of $\alpha_{r+1}(z) = 0$ to lie outside the unit circle, which emerges from the Durbin–Levinson algorithm, is that $\gamma_0, \sigma_{(1)}^2, \ldots, \sigma_{(r+1)}^2 > 0$; for, as we have seen, this is equivalent to the condition that the dispersion matrix $\Gamma = [\gamma_{|i-j|}]$ is positive definite.

Autoregressive Moving-Average Processes

The autoregressive moving-average process $y(t)$ of orders p and q, which is known for short as an ARMA(p, q) process, is defined by the equation

$$\begin{aligned}\alpha_0 y(t) + \alpha_1 y(t-1) + \cdots + \alpha_p y(t-p)\\ = \mu_0 \varepsilon(t) + \mu_1 \varepsilon(t-1) + \cdots + \mu_q \varepsilon(t-q).\end{aligned} \tag{17.83}$$

The equation is normalised by setting $\alpha_0 = 1$ and by setting either $\mu_0 = 1$ or $\sigma_\varepsilon^2 = 1$. The equation may also be expressed, in a summary notation, by writing $\alpha(L)y(t) = \mu(L)\varepsilon(t)$. It is assumed that $\alpha(z)$ and $\mu(z)$ have no factors in common.

Provided that the roots of $\alpha(z) = 0$ lie outside the unit circle, the ARMA process can be represented by $y(t) = \alpha^{-1}(L)\mu(L)\varepsilon(t)$, which corresponds to an infinite-order moving-average process. Also, provided the roots of the equation $\mu(z) = 0$ lie outside the unit circle, the process can be represented by the equation $\mu^{-1}(L)\alpha(L)y(t) = \varepsilon(t)$, which corresponds to an infinite-order autoregressive process.

By considering the moving-average form of the process, and by noting the form of the corresponding autocovariance generating function which is given by equation (17.23), it can be recognised that the autocovariance generating function for the autoregressive moving-average process is given by

$$\gamma(z) = \sigma_\varepsilon^2 \frac{\mu(z)\mu(z^{-1})}{\alpha(z)\alpha(z^{-1})}. \tag{17.84}$$

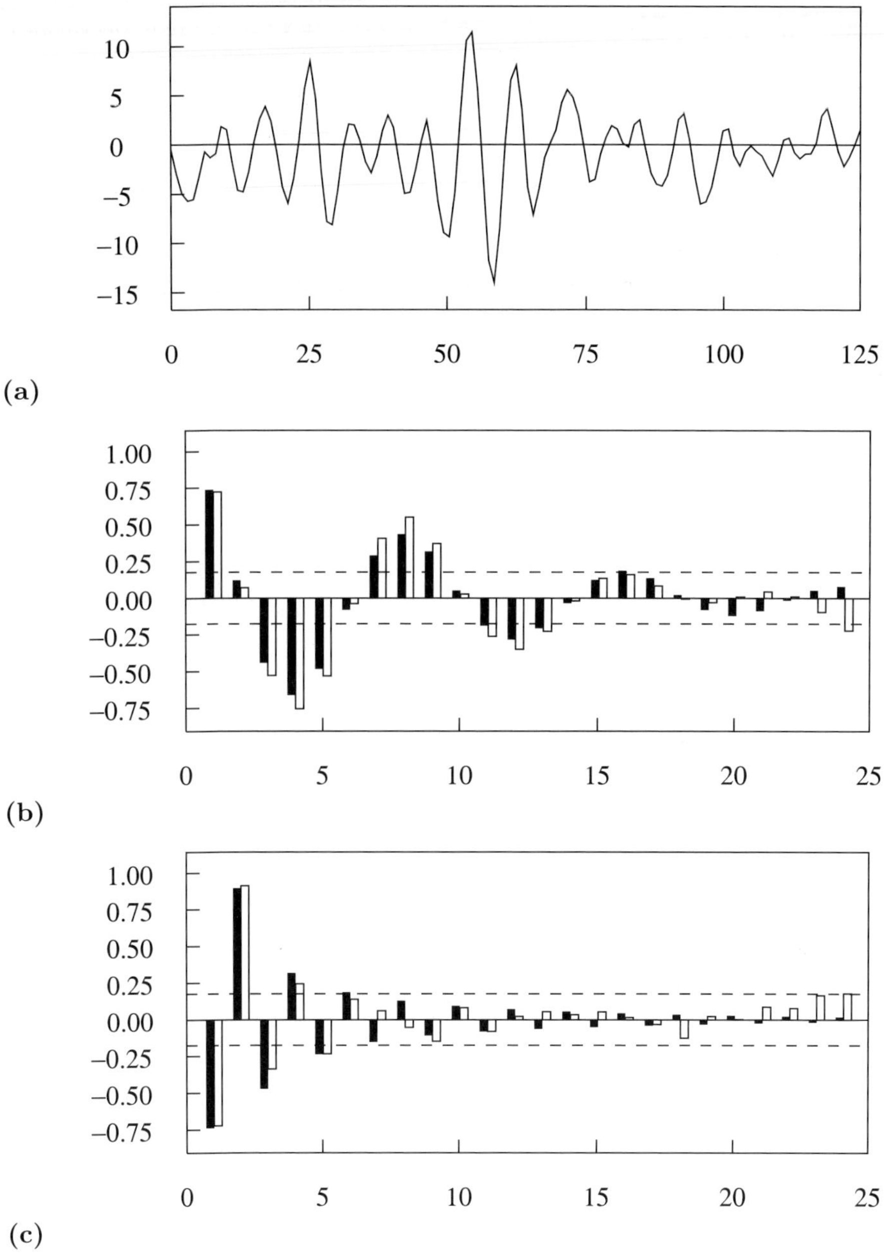

Figure 17.4. **(a)** The graph of 125 observations on a simulated series generated by an ARMA(2, 1) process $(1 - 0.273L + 0.81L^2)y(t) = (1 + 0.9L)\varepsilon(t)$, together with **(b)** the theoretical and empirical autocorrelations and **(c)** the theoretical and empirical partial autocorrelations. The theoretical values correspond to the solid bars.

To find the autocovariances of the ARMA process, we may begin by multiplying the equation $\sum_i \alpha_i y_{t-i} = \sum_i \mu_i \varepsilon_{t-i}$ of (17.83) by $y_{t-\tau}$ and by taking expectations. This gives

$$\sum_{i=0}^{p} \alpha_i \gamma_{\tau-i} = \sum_{i=0}^{q} \mu_i \delta_{i-\tau}, \tag{17.85}$$

where $\gamma_{\tau-i} = E(y_{t-\tau}y_{t-i})$ and $\delta_{i-\tau} = E(y_{t-\tau}\varepsilon_{t-i})$. Since ε_{t-i} is uncorrelated with $y_{t-\tau}$ whenever it postdates to the latter, it follows that $\delta_{i-\tau} = 0$ if $\tau > i$. Since $i = 0, 1, \ldots, q$ in the moving-average operator on the RHS of the equation (17.85), it follows that

$$\sum_{i=0}^{p} \alpha_i \gamma_{\tau-i} = 0 \quad \text{when} \quad \tau > q. \tag{17.86}$$

Given the $q + 1$ nonzero values $\delta_0, \delta_1, \ldots, \delta_q$, and p initial values $\gamma_0, \gamma_1, \ldots, \gamma_{p-1}$ for the autocovariances, the equation above can be solved recursively to obtain the subsequent autocovariances $\{\gamma_p, \gamma_{p+1}, \ldots\}$.

To find the requisite values $\delta_0, \delta_1, \ldots, \delta_q$, consider multiplying the equation $\sum_i \alpha_i y_{t-i} = \sum_i \mu_i \varepsilon_{t-i}$ by $\varepsilon_{t-\tau}$ and taking expectations. This gives

$$\sum_{i=0}^{\tau} \alpha_i \delta_{\tau-i} = \sigma_\varepsilon^2 \mu_\tau, \tag{17.87}$$

where $\delta_{\tau-i} = E(\varepsilon_{t-\tau} y_{t-i})$. Here it should be noted that $\delta_{\tau-i} = 0$ when $i > \tau$. Equation (17.87) can be rewritten as

$$\delta_\tau = \frac{1}{\alpha_0}\left(\mu_\tau \sigma_\varepsilon^2 - \sum_{i=1}^{\tau} \alpha_i \delta_{\tau-i}\right), \tag{17.88}$$

and, by setting $\tau = 0, 1, \ldots, q$, the required values $\delta_0, \delta_1, \ldots, \delta_q$ can be generated recursively.

The schematic aspects of this derivation become clearer when generating functions are used. Multiplying the autocovariance generating function $\gamma(z)$ of (17.84) by $\alpha(z)$ gives

$$\alpha(z)\gamma(z) = \mu(z)\delta(z^{-1}), \tag{17.89}$$

where

$$\delta(z^{-1}) = \sigma_\varepsilon^2 \frac{\mu(z^{-1})}{\alpha(z^{-1})}. \tag{17.90}$$

Equation (17.85) is obtained by equating the coefficients of the same powers of z on the two sides of (17.89). Next, by rearranging the equation defining $\delta(z^{-1})$, it is found that

$$\alpha(z^{-1})\delta(z^{-1}) = \sigma_\varepsilon^2 \mu(z^{-1}). \tag{17.91}$$

This corresponds to equation (17.87). By solving equations (17.89) and (17.91) for $\gamma(z)$, and by eliminating $\delta(z^{-1})$ in the process, the expression for the autocovariance generating function given under (17.84) can be recovered.

An example of the autocorrelation function of an ARMA process is provided by Figure 17.4(b). This relates to an ARMA(2, 1) process which has the same autoregressive operator as the AR(2) process which has given rise to the autocorrelation function depicted in Figure 17.3 (b). The autocorrelation functions of the two processes are barely distinguishable. However, there is a marked difference in the corresponding partial autocorrelation functions which are represented in Figures 14.3(c) and 14.4(c)

The identification of the orders of an ARMA model on the basis if its autocorrelation functions poses a difficult problem.

Example 17.7. Consider the ARMA(2, 2) model which gives the equation

$$\alpha_0 y_t + \alpha_1 y_{t-1} + \alpha_2 y_{t-2} = \mu_0 \varepsilon_t + \mu_1 \varepsilon_{t-1} + \mu_2 \varepsilon_{t-2}. \tag{17.92}$$

On multiplying by y_t, y_{t-1} and y_{t-2} and taking expectations, we get

$$\begin{bmatrix} \gamma_0 & \gamma_1 & \gamma_2 \\ \gamma_1 & \gamma_0 & \gamma_1 \\ \gamma_2 & \gamma_1 & \gamma_0 \end{bmatrix} \begin{bmatrix} \alpha_0 \\ \alpha_1 \\ \alpha_2 \end{bmatrix} = \begin{bmatrix} \delta_0 & \delta_1 & \delta_2 \\ 0 & \delta_0 & \delta_1 \\ 0 & 0 & \delta_0 \end{bmatrix} \begin{bmatrix} \mu_0 \\ \mu_1 \\ \mu_2 \end{bmatrix}. \tag{17.93}$$

On multiplying by $\varepsilon_t, \varepsilon_{t-1}$ and ε_{t-2} and taking expectations, we get

$$\begin{bmatrix} \delta_0 & 0 & 0 \\ \delta_1 & \delta_0 & 0 \\ \delta_2 & \delta_1 & \delta_0 \end{bmatrix} \begin{bmatrix} \alpha_0 \\ \alpha_1 \\ \alpha_2 \end{bmatrix} = \begin{bmatrix} \sigma_\varepsilon^2 & 0 & 0 \\ 0 & \sigma_\varepsilon^2 & 0 \\ 0 & 0 & \sigma_\varepsilon^2 \end{bmatrix} \begin{bmatrix} \mu_0 \\ \mu_1 \\ \mu_2 \end{bmatrix}. \tag{17.94}$$

When the latter equations are written as

$$\begin{bmatrix} \alpha_0 & 0 & 0 \\ \alpha_1 & \alpha_0 & 0 \\ \alpha_2 & \alpha_1 & \alpha_0 \end{bmatrix} \begin{bmatrix} \delta_0 \\ \delta_1 \\ \delta_2 \end{bmatrix} = \sigma_\varepsilon^2 \begin{bmatrix} \mu_0 \\ \mu_1 \\ \mu_2 \end{bmatrix}, \tag{17.95}$$

they can be solved recursively for δ_0, δ_1 and δ_2 on the assumption that the values of the other elements are known. Notice that, when we adopt the normalisation $\alpha_0 = \mu_0 = 1$, we get $\delta_0 = \sigma_\varepsilon^2$. When the equations (17.93) are rewritten as

$$\begin{bmatrix} \alpha_0 & \alpha_1 & \alpha_2 \\ \alpha_1 & \alpha_0 + \alpha_2 & 0 \\ \alpha_2 & \alpha_1 & \alpha_0 \end{bmatrix} \begin{bmatrix} \gamma_0 \\ \gamma_1 \\ \gamma_2 \end{bmatrix} = \begin{bmatrix} \mu_0 & \mu_1 & \mu_2 \\ \mu_1 & \mu_2 & 0 \\ \mu_2 & 0 & 0 \end{bmatrix} \begin{bmatrix} \delta_0 \\ \delta_1 \\ \delta_2 \end{bmatrix}, \tag{17.96}$$

they can be solved for γ_0, γ_1 and γ_2. Thus the starting values are obtained which enable the equation

$$\alpha_0 \gamma_\tau + \alpha_1 \gamma_{\tau-1} + \alpha_2 \gamma_{\tau-2} = 0; \quad \tau > 2 \tag{17.97}$$

to be solved recursively to provide the succeeding values $\{\gamma_3, \gamma_4, \ldots\}$ of the autocovariances.

The methods described in this section for finding the autocovariances of an ARMA process are implemented in the following procedure which can also be used in connection with pure autoregressive and pure moving-average processes:

(17.98)

```
procedure ARMACovariances(alpha, mu : vector;
                          var gamma : vector;
                          var varEpsilon : real;
                          lags, p, q : integer);

  var
    i, j, t, tau, r : integer;
    delta, f : vector;
    a : matrix;

  {Find the delta values}
  begin
    r := Max(p, q);

    for t := 0 to q do
      begin
        delta[t] := mu[t] * varEpsilon;
        tau := Min(t, p);
        for i := 1 to tau do
          delta[t] := delta[t] - delta[t - i] * alpha[i];
        delta[t] := delta[t]/alpha[0]
      end;

    for i := 0 to r do
      for j := 0 to r do
        begin {i, j : form the matrix of alpha values}
          a[i, j] := 0.0;
          if ((i - j) >= 0) and ((i - j) <= p) then
            a[i, j] := alpha[i - j];
          if ((i + j) <= p) and (j > 0) then
            a[i, j] := a[i, j] + alpha[i + j];
        end; {i, j}

    for i := 0 to r do
      begin {i : form the RHS vector}
        f[i] := 0.0;
        for j := i to q do
          f[i] := f[i] + mu[j] * delta[j - i]
      end; {i}

  {Solve for the initial autocovariances}
    LUsolve(0, r + 1, a, gamma, f);

  {Find the succeeding autocovariances}
    for i := r + 1 to lags do
      begin {i}
        gamma[i] := 0.0;
        for j := 1 to p do
          gamma[i] := gamma[i] - alpha[j] * gamma[i - j];
```

```
      gamma[i] := gamma[i]/alpha[0];
    end; {i}

end; {ARMACovariances}
```

Calculating the ARMA Parameters from the Autocovariances

Given a set of autocovariances $\gamma_0, \ldots, \gamma_{p+q}$ from the ARMA(p, q) process, it is possible to infer the values of the parameters. Let us reconsider the equation of (17.85):

$$\sum_{i=0}^{p} \alpha_i \gamma_{i-\tau} = \sum_{i=0}^{q} \mu_i \delta_{i-\tau}. \tag{17.99}$$

By running τ from 0 to $p + q$, the following equations are generated:

$$\left[\begin{array}{cccc} \gamma_0 & \gamma_1 & \ldots & \gamma_p \\ \gamma_1 & \gamma_0 & \ldots & \gamma_{p-1} \\ \vdots & \vdots & & \vdots \\ \gamma_q & \gamma_{q-1} & \ldots & \gamma_{p-q} \\ \hdashline \gamma_{q+1} & \gamma_q & \ldots & \gamma_{p-q-1} \\ \vdots & \vdots & \ddots & \vdots \\ \gamma_{q+p} & \gamma_{q+p+1} & \ldots & \gamma_q \end{array}\right] \begin{bmatrix} \alpha_0 \\ \alpha_1 \\ \vdots \\ \alpha_p \end{bmatrix} = \left[\begin{array}{cccc} \delta_0 & \delta_1 & \ldots & \delta_q \\ 0 & \delta_0 & \ldots & \delta_q - 1 \\ \vdots & \vdots & \ddots & \vdots \\ 0 & 0 & \ldots & \delta_0 \\ \hdashline 0 & 0 & \ldots & 0 \\ \vdots & \vdots & & \vdots \\ 0 & 0 & \ldots & 0 \end{array}\right] \begin{bmatrix} \mu_0 \\ \mu_1 \\ \vdots \\ \mu_p \end{bmatrix}. \tag{17.100}$$

Thus, if we let $\tau = q + 1, \ldots, q + p$, we obtain the system

$$\begin{bmatrix} \gamma_q & \gamma_{q-1} & \cdots & \gamma_{q-p+1} \\ \gamma_{q+1} & \gamma_q & \cdots & \gamma_{q-p} \\ \vdots & \vdots & \ddots & \vdots \\ \gamma_{q+p-1} & \gamma_{q+p-2} & \cdots & \gamma_q \end{bmatrix} \begin{bmatrix} \alpha_1 \\ \alpha_2 \\ \vdots \\ \alpha_p \end{bmatrix} = -\alpha_0 \begin{bmatrix} \gamma_{q+1} \\ \gamma_{q+2} \\ \vdots \\ \gamma_{q+p} \end{bmatrix}; \tag{17.101}$$

and, if the normalisation $\alpha_0 = 1$ is imposed, this can be solved for the parameters $\alpha_1, \ldots, \alpha_p$. The *Yule–Walker* procedure for finding the parameters of an AR(p) process, which was presented under (17.67), has been devised to solve the equations under (17.101) as well as the equation under (17.55).

Now let us reconsider the original equation of the ARMA(p, q) model which is

$$\psi_t = \sum_{i=0}^{p} \alpha_i y_{t-i} = \sum_{i=0}^{q} \mu_i \varepsilon_{t-i}. \tag{17.102}$$

On the one hand, this gives

$$\begin{aligned} E(\psi_t \psi_{t-\tau}) &= E\left\{ \left(\sum_i \mu_i \varepsilon_{t-i} \right) \left(\sum_j \mu_j \varepsilon_{t-\tau-j} \right) \right\} \\ &= \sum_i \sum_j \mu_i \mu_j E(\varepsilon_{t-i} \varepsilon_{t-\tau-j}) \\ &= \sigma_\varepsilon^2 \sum_j \mu_j \mu_{j+\tau}. \end{aligned} \tag{17.103}$$

On the other hand, there is

$$\begin{aligned}E(\psi_t\psi_{t-\tau}) &= E\Big\{\Big(\sum_i \alpha_i y_{t-i}\Big)\Big(\sum_j \alpha_j y_{t-\tau-j}\Big)\Big\}\\ &= \sum_i\sum_j \alpha_i\alpha_j E(y_{t-i}y_{t-\tau-j})\\ &= \sum_i\sum_j \alpha_i\alpha_j \gamma_{\tau+j-i}.\end{aligned} \tag{17.104}$$

Putting these two together gives

$$\sum_i\sum_j \alpha_i\alpha_j\gamma_{\tau+j-i} = \sigma_\varepsilon^2 \sum_j \mu_j\mu_{j+\tau}; \tag{17.105}$$

and, if the elements of the LHS are already known, we can let $\tau = 0, \ldots, q$ in order to generate a set of equations which can be solved for the values of $\mu_1, \ldots, \mu_q$ and σ_ε^2 once the normalisation $\mu_0 = 1$ has been imposed. These parameters of the moving-average component may be obtained via the procedure *Minit* of (17.39) which is also used for obtaining the parameters of a pure MA(q) model.

(17.106)
```
procedure ARMAParameters(p, q : integer;
                         gamma : vector;
                         var alpha, mu : vector;
                         var varEpsilon : real);

  var
    t, i, j, k : integer;
    temp : real;
    a : matrix;
    b, psi : vector;

begin {ARMAParameters}

  YuleWalker(p, q, gamma, alpha, temp);

  for t := 0 to q do
    begin {t}
      psi[t] := 0.0;
      for i := 0 to p do
        for j := 0 to p do
          begin {i, j}
            k := Abs(t + j - i);
            psi[t] := psi[t] + alpha[i] * alpha[j] * gamma[k];
          end; {i, j}
    end; {t}
```

```
if q > 0 then
    Minit(mu, varEpsilon, psi, q)
else
    varEpsilon := temp;

end; {ARMAParameters}
```

Bibliography

[21] Anderson, T.W., and A.A. Takemura, (1986), Why do Noninvertible Estimated Moving Averages Occur?, *Journal of Time Series Analysis*, **7**, 235–25.

[166] Durbin, J., (1960), The Fitting of Time-Series Models, *Revue, Institut International de Statistique*, **28**, 233–243.

[192] Franke, J., (1985), A Levinson–Durbin Recursion for Autoregressive Moving Average Processes, *Biometrika*, **72**, 573–81.

[215] Godolphin, E.J., (1976), On the Cramer–Wold Factorisation, *Biometrika*, **63**, 367–80.

[310] Laurie, D.P., (1980), Efficient Implementation of Wilson's Algorithm for Factorising a Self-Reciprocal Polynomial, *BIT*, **20**, 257–259.

[311] Laurie, D.P., (1982), Cramer–Wold Factorisation: Algorithm AS 175, *Applied Statistics*, **31**, 86–90.

[314] Levinson, N., (1946), The Weiner RMS (Root Mean Square) Error Criterion in Filter Design and Prediction, *Journal of Mathematical Physics*, **25**, 261–278.

[348] Mittnik, S., (1987), Non-Recursive Methods for Computing The Coefficients of the Autoregressive and Moving-Average Representation of Mixed ARMA Processes, *Economic Letters*, **23**, 279–284.

[353] Morettin, P.A., (1984), The Levinson Algorithm and its Applications in Time Series Analysis, *International Statistical Review*, **52**, 83–92.

[375] Pagano, M., (1973), When is an Autoregressive Scheme Stationary?, *Communications in Statistics*, **1**, 533–544.

[376] Pandit, S.M., and S-M. Wu, (1983), *Time Series and System Analysis with Applications*, John Wiley and Sons, New York.

[527] Wilson, G.T., (1969), Factorisation of the Covariance Generating Function of a Pure Moving Average Process, *SIAM Journal of Numerical Analysis*, **6**, 1–7.

CHAPTER 18

Time-Series Analysis in the Frequency Domain

In this chapter, we shall provide an analysis of stationary stochastic processes from the point of view of their spectral representation. This complements the analysis of stationary processes from the point of view of the time domain, which was the subject of the previous chapter. The previous chapter has dealt specifically with linear time-series models of the ARMA variety.

At the outset, we shall make only weak assumptions about the mechanisms which generate the time series; and, therefore, we shall be proceeding at a higher level of generality than hitherto. Nevertheless, a strong justification for linear time-series models models will emerge when we succeed in demonstrating that virtually every stationary stochastic process which has no regular or deterministic components of any significant magnitude can be represented as a moving-average process. This result, which is commonly know as the Cramér–Wold theorem, depends crucially upon the concepts underlying the spectral representation of time series.

The spectral representation is rooted in the basic notion of Fourier analysis which is that well-behaved functions can be approximated over a finite interval, to any degree of accuracy, by a weighted combination of sine and cosine functions whose harmonically rising frequencies are integral multiples of a fundamental frequency. Such linear combinations are described as Fourier sums or Fourier series. Of course, the notion applies to sequences as well; for any number of well-behaved functions may be interpolated through the coordinates of a finite sequence.

We shall approach the Fourier analysis of stochastic processes via the exact Fourier representation of a finite sequence. This is extended to provide a representation of an infinite sequence in terms of an infinity of trigonometrical functions whose frequencies range continuously in the interval $[0, \pi]$. The trigonometrical functions and their weighting functions are gathered under a Fourier–Stieltjes integral. It is remarkable that, whereas a Fourier sum serves only to define a strictly periodic function, a Fourier integral provides the means of representing an aperiodic time series generated by a stationary stochastic process.

The Fourier integral is also used to represent the underlying stochastic process. This is achieved by describing the stochastic processes which generate the weighting functions. There are two such weighting processes, associated respectively with the sine and cosine functions; and the function which defines their common variance is the so-called spectral distribution function whose derivative, when it exists, is the spectral density function or the "spectrum".

The relationship between the spectral density function and the sequence of autocovariances, which is summarised in the Wiener–Khintchine theorem, provides a link between the time-domain and the frequency-domain analyses. The sequence of autocovariances may be obtained from the Fourier transform of the spectral density function and the spectral density function is, conversely, a Fourier transform of the autocovariances. (Figure 18.1 provides a graphical example of the relationship between the autocovariance function of stationary stochastic process and its spectral density function.)

For many practical purposes, it is this relationship between the autocovariances and the spectral density function which is of primary interest; and, in many texts of a practical orientation, the other aspects of frequency-domain analysis are ignored. In contrast to such a straightforward approach is the detailed analytic treatment to be found in some of the classic texts of time-series analysis, amongst which are the works of Doob [162], Wold [530] and Grenander and Rosenblatt [229], which are closely related, both in time and in spirit, to the original pioneering work.

Much of this early work was concerned primarily with stochastic processes in continuous time. Results for discrete-time processes, which have proved, subsequently, to be more important in practice, were often obtained as side-products by other authors at later dates. This feature makes it difficult to give the correct attribution to some of the leading results and even to name them appropriately. The reader who wishes to pursue these issues may consult the above-mentioned texts. The text of Yaglom [536] also contains a number of historical observations which illustrate the perspectives of the Russian school.

Stationarity

Consider two vectors of $n+1$ consecutive elements from the process $y(t)$:

$$(18.1) \qquad [y_t, y_{t+1}, \ldots, y_{t+n}] \quad \text{and} \quad [y_s, y_{s_1}, \ldots, y_{s+n}].$$

Then $y(t) = \{y_t; t = 0, \pm 1, \pm 2, \ldots\}$ is strictly stationary if the joint probability density functions of the two vectors are the same for all values of t and s regardless of the size of n. On the assumption that the first and second-order moments of the distribution are finite, the condition of stationarity implies that all the elements of $y(t)$ have the same expected value and that the covariance between any pair of elements of the sequences is a function only of their temporal separation. Thus,

$$(18.2) \qquad E(y_t) = \mu \quad \text{and} \quad C(y_t, y_s) = \gamma_{|t-s|}.$$

On their own, the conditions of (18.2) constitute the conditions of weak stationarity.

A normal process is completely characterised by its mean and its autocovariances. Therefore, a normal process which satisfies the conditions for weak stationarity is also stationary in the strict sense.

The Filtering of White Noise

A white-noise process is a sequence $\varepsilon(t)$ of uncorrelated random variables with a mean of zero and a common variance σ_ε^2. Thus

$$(18.3) \qquad E(\varepsilon_t) = 0 \quad \text{for all } t, \qquad \text{and} \qquad E(\varepsilon_t \varepsilon_s) = \begin{cases} \sigma_\varepsilon^2, & \text{if } t = s; \\ 0, & \text{if } t \neq s. \end{cases}$$

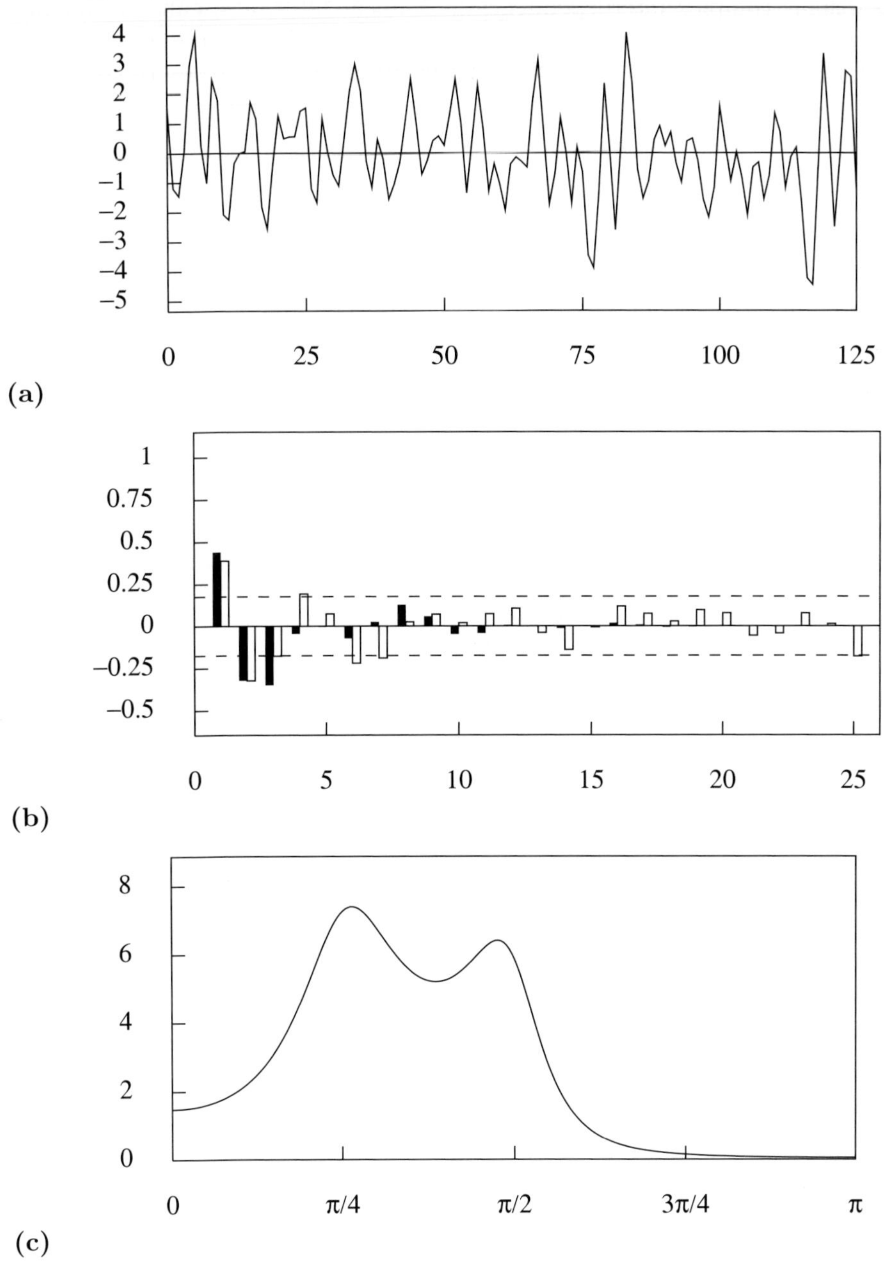

Figure 18.1. **(a)** The graph of 125 points generated by a simulated AR(4) process $(1 - 1.061L + 1.202L^2 - 0.679L^3 + 0.360L^4)y(t) = \varepsilon(t)$, together with **(b)** the theoretical and empirical autocorrelations and **(c)** the spectral density function.

By means of linear filtering, a variety of time series may be constructed whose elements display complex interdependencies. A linear filter, also called a moving-average operator, is a function of the lag operator of the form $\mu(L) = \{\cdots + \mu_{-1}L^{-1} + \mu_0 + \mu_1 L + \cdots\}$. The effect of this filter on $\varepsilon(t)$ is described by the equation

$$\begin{aligned} y(t) &= \mu(L)\varepsilon(t) \\ &= \sum_i \mu_i \varepsilon(t-i). \end{aligned} \tag{18.4}$$

The operator $\mu(L)$ is also be described as the transfer function which maps the input sequence $\varepsilon(t)$ into the output sequence $y(t)$.

In many practical applications, such as in forecasting, one is constrained to employ one-sided or "causal" moving-average operators of the form $\mu(L) = \{\mu_0 + \mu_1 L + \mu_2 L^2 + \cdots\}$. In a practical filtering operation, the order of the operator—i.e. the highest index on a nonzero coefficient—must be finite, or else the coefficients $\{\mu_0, \mu_1, \mu_2, \ldots\}$ must be functions of a limited number of fundamental parameters, as in the case of the expansion of a rational function. In addition, if $y(t)$ is always to remain bounded when $\varepsilon(t)$ is bounded, then it is necessary, and sufficient, that

$$\sum_i |\mu_i| < \infty. \tag{18.5}$$

Given the value of $\sigma_\varepsilon^2 = V\{\varepsilon(t)\}$, the autocovariances of the filtered sequence $y(t) = \mu(L)\varepsilon(t)$ may be determined by evaluating the expression

$$\begin{aligned} \gamma_\tau &= E(y_t y_{t-\tau}) \\ &= E\Big(\sum_i \mu_i \varepsilon_{t-i} \sum_j \mu_j \varepsilon_{t-\tau-j}\Big) \\ &= \sum_i \sum_j \mu_i \mu_j E(\varepsilon_{t-i}\varepsilon_{t-\tau-j}). \end{aligned} \tag{18.6}$$

From equation (18.3), it follows that

$$\gamma_\tau = \sigma_\varepsilon^2 \sum_j \mu_j \mu_{j+\tau}; \tag{18.7}$$

and so the variance of the filtered sequence is

$$\gamma_0 = \sigma_\varepsilon^2 \sum_j \mu_j^2. \tag{18.8}$$

The condition under (18.5) guarantees that these quantities are finite, as is required by the condition of stationarity.

In the subsequent analysis, it will prove helpful to present the results in the notation of the z-transform. The z-transform of a sequence of autocovariances is called the autocovariance generating function. For the moving-average process, this is given by

$$
\begin{aligned}
\gamma(z) &= \sigma_\varepsilon^2 \mu(z)\mu(z^{-1}) \\
&= \sigma_\varepsilon^2 \sum_i \mu_i z^i \sum_j \mu_j z^{-j} \\
&= \sum_\tau \Big\{ \sigma_\varepsilon^2 \sum_j \mu_j \mu_{j+\tau} \Big\} z^\tau \quad ; \quad \tau = i - j \\
&= \sum_{\tau=-\infty}^{\infty} \gamma_\tau z^\tau .
\end{aligned} \tag{18.9}
$$

The final equality is by virtue of equation (18.7).

Cyclical Processes

The present section, which is devoted to cyclical processes, may be regarded as a rehearsal for the development of the spectral representation of a stationary stochastic process.

An elementary cyclical process, which has the same fundamental importance in the analysis of stationary time series as a white-noise process, is the defined by the equation

$$
\begin{aligned}
y(t) &= \alpha \cos(\omega t) + \beta \sin(\omega t) \\
&= \zeta e^{i\omega t} + \zeta^* e^{-i\omega t},
\end{aligned} \tag{18.10}
$$

wherein $\omega \in (0, \pi]$ is an angular frequency and

$$
\zeta = \frac{\alpha + i\beta}{2} \quad \text{and} \quad \zeta^* = \frac{\alpha - i\beta}{2}, \tag{18.11}
$$

are complex-valued conjugate random variables compounded from the real-valued random variables α and β. Such a process becomes strictly periodic when $2\pi/\omega$ is a rational number. Otherwise, it is liable to be described as almost periodic. The autocovariance of the elements y_t and y_s is given by

$$
E(y_t y_s) = E\Big[\zeta^2 e^{i\omega(t+s)} + \zeta\zeta^* \Big\{ e^{i\omega(t-s)} + e^{i\omega(s-t)} \Big\} + \zeta^{*2} e^{-i\omega(t+s)} \Big]. \tag{18.12}
$$

For the process to be stationary, this must be a function of $|t - s|$ only, for which it is necessary that $E(\zeta^2) = E(\zeta^{*2}) = 0$. These conditions imply that

$$
E(\alpha^2) = E(\beta^2) = \sigma^2 \quad \text{and} \quad E(\alpha\beta) = 0. \tag{18.13}
$$

It follows that the autocovariance of (18.12) is

$$
\gamma_\tau = \sigma^2 \cos \omega\tau, \quad \text{where} \quad \tau = t - s. \tag{18.14}
$$

The process $y(t)$ may also be written as

$$y(t) = \rho\cos(\omega t - \theta), \tag{18.15}$$

where

$$\rho^2 = \alpha^2 + \beta^2 \quad \text{and} \quad \theta = \tan^{-1}\left(\frac{\beta}{\alpha}\right) \tag{18.16}$$

are respectively the squared amplitude and the phase displacement. These are a pair of random variables whose values, which are generated at the outset, determine the values of all of the elements of the realisation of $y(t)$. For this reason, one is tempted to describe the process as a deterministic one. If α and β are normally distributed, then ρ^2 is proportional to a chi-square variate of two degrees of freedom whilst θ is uniformly distributed over the interval $(0, 2\pi]$, in consequence of the symmetry of the normal distribution.

A more general cyclical process may be considered which is a sum of uncorrelated elementary processes. This is

$$\begin{aligned} y(t) &= \sum_{j=0}^{n}\{\alpha_j\cos(\omega_j t) + \beta_j\sin(\omega_j t)\} \\ &= \sum_{j=0}^{n}\left\{\zeta_j e^{i\omega_j t} + \zeta_j^* e^{-i\omega_j t}\right\}, \end{aligned} \tag{18.17}$$

wherein

$$\zeta_j = \frac{\alpha_j + i\beta_j}{2} \quad \text{and} \quad \zeta_j^* = \frac{\alpha_j - i\beta_j}{2} \tag{18.18}$$

are complex-valued random variables whose components fulfil the conditions

$$E(\alpha_j^2) = E(\beta_j^2) = \sigma_j^2 \quad \text{and} \quad E(\alpha_j\beta_j) = 0. \tag{18.19}$$

The autocovariance of the elements y_t and y_s is given by

$$\begin{aligned} E(y_t y_s) = \sum_{j=0}^{m}\sum_{k=0}^{m} E\Big[&\zeta_j\zeta_k e^{i(\omega_j t + \omega_k s)} + \zeta_j\zeta_k^* e^{i(\omega_j t - \omega_k s)} \\ &+ \zeta_j^*\zeta_k e^{i(\omega_k s - \omega_j t)} + \zeta_j^*\zeta_k^* e^{-i(\omega_j t + \omega_k s)}\Big]. \end{aligned} \tag{18.20}$$

The condition of stationarity now requires, in addition to the conditions of (18.19), that, whenever $j \neq k$, there is

$$E(\zeta_j\zeta_k) = E(\zeta_j^*\zeta_k^*) = E(\zeta_j^*\zeta_k) = E(\zeta_j\zeta_k^*) = 0. \tag{18.21}$$

The consequence is that the autocovariance of the process at lag $\tau = t - s$ is given by

$$\gamma_\tau = \sum_{j=0}^{n}\sigma_j^2\cos\omega_j\tau. \tag{18.22}$$

Notice that the variance of the process is just

$$\gamma_0 = \sum_{j=0}^{n} \sigma_j^2, \tag{18.23}$$

which is the sum of the variances of the n elementary processes.

Equation (18.23) represents the so-called spectral decomposition of the variance. The spectrum of the process $y(t)$ is an array of vertical lines rising from the horizontal frequency axis and terminating in the points (ω_j, σ_j). The cumulated spectrum, also described as the spectral distribution function, is defined by

$$F(\omega) = \sum_{\omega_j \leq \omega} \sigma_j^2. \tag{18.24}$$

This is a staircase with a step or saltus of σ_j^2 at each frequency value ω_j. These effects are reminiscent of the emission and absorption spectra of chemical elements.

One is liable to regard $y(t)$ as a deterministic process which is masquerading as a stochastic process. There are two features which suggest this notion. First, the process depends upon only a finite number of randomly determined parameters $\zeta_0, \ldots, \zeta_n$, whereas its realisation is an indefinite sequence of numbers. Secondly, the random elements are determined in advance of the realisation of the process. It is as if they were picked out of a hat and recorded in a log before participating in the purely mechanical business of generating the sequence. Such ideas, which might inhibit a proper understanding the spectral representation of a stochastic process, can be held in abeyance until the developments of the following sections have been considered, which concern the spectral representation of stationary processes.

The Fourier Representation of a Sequence

According to the basic result of Fourier analysis, it is always possible to approximate an arbitrary analytic function defined over a finite interval of the real line, to any desired degree of accuracy, by a weighted sum of sine and cosine functions of harmonically increasing frequencies.

Similar results apply in the case of sequences, which may be regarded as functions mapping from the set of integers onto the real line. For a sample of T observations $y_0, \ldots, y_{T-1}$, it is possible to devise an expression in the form

$$y_t = \sum_{j=0}^{n} \left\{ \alpha_j \cos(\omega_j t) + \beta_j \sin(\omega_j t) \right\}, \tag{18.25}$$

wherein $\omega_j = 2\pi j/T$ is a multiple of the fundamental frequency $\omega_1 = 2\pi/T$. Thus, the elements of a finite sequence can be expressed exactly in terms of sines and cosines. This expression is called the Fourier decomposition of y_t, and the set of coefficients $\{\alpha_j, \beta_j; j = 0, 1, \ldots, n\}$ are called the Fourier coefficients.

One should observe that equation (18.25), which is to be regarded for the moment as a device of descriptive statistics, is identical in form to equation (18.17) of the previous section, which relates to a so-called cyclical stochastic process.

When T is even, we have $n = T/2$; and it follows that

$$
\begin{aligned}
\sin(\omega_0 t) &= \sin(0) = 0, \\
\cos(\omega_0 t) &= \cos(0) = 1, \\
\sin(\omega_n t) &= \sin(\pi t) = 0, \\
\cos(\omega_n t) &= \cos(\pi t) = (-1)^t.
\end{aligned} \tag{18.26}
$$

Therefore, equation (18.25) becomes

$$
y_t = \alpha_0 + \sum_{j=1}^{n-1} \{\alpha_j \cos(\omega_j t) + \beta_j \sin(\omega_j t)\} + \alpha_n(-1)^t. \tag{18.27}
$$

When T is odd, we have $n = (T-1)/2$; and then equation (18.25) becomes

$$
y_t = \alpha_0 + \sum_{j=1}^{n} \{\alpha_j \cos(\omega_j t) + \beta_j \sin(\omega_j t)\}. \tag{18.28}
$$

In both cases, there are T nonzero coefficients amongst the set $\{\alpha_j, \beta_j; j = 0, 1, \ldots, n\}$; and the mapping from the sample values to the coefficients constitutes a one-to-one invertible transformation.

In equation (18.27), the frequencies of the trigonometric functions range from $\omega_1 = 2\pi/T$ to $\omega_n = \pi$; whereas, in equation (18.28), they range from $\omega_1 = 2\pi/T$ to $\omega_n = \pi(T-1)/T$. The frequency π is the so-called Nyquist frequency. Although the process generating the data may contain components of frequencies higher than the Nyquist frequency, these will not be detected when it is sampled regularly at unit intervals of time. In fact, the effects on the process of components with frequencies in excess of the Nyquist value will be confounded with those whose frequencies fall below it.

To demonstrate this, consider the case where the process contains a component which is a pure cosine wave of unit amplitude and zero phase whose frequency ω lies in the interval $\pi < \omega < 2\pi$. Let $\omega^* = 2\pi - \omega$. Then

$$
\begin{aligned}
\cos(\omega t) &= \cos\left\{(2\pi - \omega^*)t\right\} \\
&= \cos(2\pi)\cos(\omega^* t) + \sin(2\pi)\sin(\omega^* t) \\
&= \cos(\omega^* t);
\end{aligned} \tag{18.29}
$$

which indicates that ω and ω^* are observationally indistinguishable. Here, $\omega^* < \pi$ is described as the alias of $\omega > \pi$.

The Spectral Representation of a Stationary Process

We shall begin this section by extending the Fourier representation to encompass sequences of indefinite length. We shall proceed to develop a stochastic model of a process which can be deemed to have generated the sequence. It transpires that the model in question is simply a generalisation of the cyclical model of the previous section which arises when the number of its sinusoidal components becomes indefinitely large.

By allowing the value of n in the expression (18.25) to tend to infinity, it is possible to express a sequence of indefinite length in terms of a sum of sine and cosine functions. However, in the limit as $n \to \infty$, the coefficients α_j, β_j tend to vanish; and therefore an alternative representation in terms of differentials is called for.

By writing $\alpha_j = dA(\omega_j)$, $\beta_j = dB(\omega_j)$, where $A(\omega)$, $B(\omega)$ are step functions with discontinuities at the points $\{\omega_j; j = 0, \ldots, n\}$, the expression (18.25) can be rendered as

$$y_t = \sum_j \{\cos(\omega_j t)dA(\omega_j) + \sin(\omega_j t)dB(\omega_j)\}. \tag{18.30}$$

In the limit, as $n \to \infty$, the summation is replaced by an integral to give the expression

$$y(t) = \int_0^\pi \{\cos(\omega t)dA(\omega) + \sin(\omega t)dB(\omega)\}. \tag{18.31}$$

Here, $\cos(\omega t)$ and $\sin(\omega t)$, and therefore $y(t)$, may be regarded as infinite sequences defined over the set $\mathcal{Z} = \{t = 0, \pm 1, \pm 2, \ldots\}$ of all positive and negative integers.

Since $A(\omega)$ and $B(\omega)$ are discontinuous functions for which no derivatives exist, one must avoid using $\alpha(\omega)d\omega$ and $\beta(\omega)d\omega$ in place of $dA(\omega)$ and $dB(\omega)$. Moreover, the integral in equation (18.31) is a Fourier–Stieltjes integral which is more general than the usual Riemann integral.

This representation may be compared with the expression under (13.50) which stands for the Fourier representation of a discrete-time sequence subject to the condition of absolute summability under (13.49). In the case of an indefinite sequence generated by a stationary stochastic process, the condition of summability cannot apply. Therefore the spectral representation of the process lies beyond the realms of ordinary Fourier analysis; and it belongs, instead, to the domain of Wiener's generalised harmonic analysis [522]. For an intuitive understanding of the difference between the ordinary and the generalised Fourier forms, one may compare the "fractal" nature of the generalised functions $A(\omega)$ and $B(\omega)$, which corresponds to the irregularities of the stochastic process, with the more regular nature of the continuous and differentiable functions $\alpha(\omega)$ and $\beta(\omega)$ which are entailed in the Fourier representation of an absolutely summable sequence.

In order to derive a statistical theory for the process which generates $y(t)$, one must make some assumptions concerning the functions $A(\omega)$ and $B(\omega)$. So far, the sequence $y(t)$ has been interpreted as a realisation of a stochastic process. If $y(t)$ is regarded as the stochastic process itself, then the functions $A(\omega)$, $B(\omega)$ must, likewise, be regarded as stochastic processes defined over the interval $(0, \pi]$. A single realisation of these processes now corresponds to a single realisation of the process $y(t)$.

The first assumption to be made is that the functions $A(\omega)$ and $B(\omega)$ represent a pair of stochastic processes of zero mean which are indexed on the continuous parameter ω. Thus

$$E\{dA(\omega)\} = E\{dB(\omega)\} = 0. \tag{18.32}$$

The second and third assumptions are that the two processes are mutually uncorrelated and that nonoverlapping increments within each process are uncorrelated. Thus

$$\begin{aligned} E\{dA(\omega)dB(\lambda)\} &= 0 \quad \text{for all} \quad \omega, \lambda, \\ E\{dA(\omega)dA(\lambda)\} &= 0 \quad \text{if} \quad \omega \neq \lambda, \\ E\{dB(\omega)dB(\lambda)\} &= 0 \quad \text{if} \quad \omega \neq \lambda. \end{aligned} \tag{18.33}$$

The final assumption is that the variance of the increments is given by

$$V\{dA(\omega)\} = V\{dB(\omega)\} = 2dF(\omega). \tag{18.34}$$

The function $F(\omega)$, which is defined initially over the interval $(0, \pi]$, is described as the spectral distribution function. The properties of variances imply that it is a nondecreasing function of ω. In the case where the process $y(t)$ is purely nondeterministic, $F(\omega)$ is a continuous differentiable function. Its derivative $f(\omega)$, which is nonnegative, is described as the spectral density function.

In order to express equation (18.31) in terms of complex exponentials, we may define a pair of conjugate complex stochastic processes:

$$\begin{aligned} dZ(\omega) &= \frac{1}{2}\{dA(\omega) - idB(\omega)\}, \\ dZ^*(\omega) &= \frac{1}{2}\{dA(\omega) + idB(\omega)\}. \end{aligned} \tag{18.35}$$

Also, we may extend the domain of the functions $A(\omega)$, $B(\omega)$ from $(0, \pi]$ to $(-\pi, \pi]$ by regarding $A(\omega)$ as an even function such that $A(-\omega) = A(\omega)$ and by regarding $B(\omega)$ as an odd function such that $B(-\omega) = -B(\omega)$. Then we have

$$dZ^*(\omega) = dZ(-\omega). \tag{18.36}$$

From conditions under (18.33), it follows that

$$\begin{aligned} E\{dZ(\omega)dZ^*(\lambda)\} &= 0 \quad \text{if} \quad \omega \neq \lambda, \\ E\{dZ(\omega)dZ^*(\omega)\} &= dF(\omega), \end{aligned} \tag{18.37}$$

where the domain of $F(\omega)$ is now extended to the interval $(-\pi, \pi]$. These results may be used to re-express equation (18.31) as

$$\begin{aligned} y(t) &= \int_0^\pi \left\{ \frac{(e^{i\omega t} + e^{-i\omega t})}{2} dA(\omega) - i\frac{(e^{i\omega t} - e^{-i\omega t})}{2} dB(\omega) \right\} \\ &= \int_0^\pi \left\{ e^{i\omega t}\frac{\{dA(\omega) - idB(\omega)\}}{2} + e^{-i\omega t}\frac{\{dA(\omega) + idB(\omega)\}}{2} \right\} \\ &= \int_0^\pi \left\{ e^{i\omega t} dZ(\omega) + e^{-i\omega t} dZ^*(\omega) \right\}. \end{aligned} \tag{18.38}$$

When the integral is extended over the range $(-\pi, \pi]$, the equation becomes

$$y(t) = \int_{-\pi}^{\pi} e^{i\omega t} dZ(\omega). \tag{18.39}$$

This is commonly described as the spectral representation of the process $y(t)$.

The spectral representation is sufficiently general to accommodate the case of a discrete spectrum where the spectral distribution function $F(\omega)$ is constant except for jumps of magnitude $\sigma_1^2, \ldots, \sigma_k^2$ at the points $\omega_1, \ldots, \omega_k$. In that case, the spectral representation of the process reduces to

$$y(t) = \sum_{j=-k}^{k} e^{-i\omega_j t} \zeta_j, \tag{18.40}$$

where

$$\zeta_j = Z(\omega_j^+) - Z(\omega_j^-) \tag{18.41}$$

is the saltus in $Z(\omega)$ at the point ω_j, and where the conditions $\zeta_{-j} = \zeta_j^*$ and $\omega_{-j} = -\omega_j$ may be imposed to ensure that the process is real-valued. This is equation (18.17) again, which relates to a so-called cyclical process which can be regarded now as a prototype of a more general stationary stochastic process.

The Autocovariances and the Spectral Density Function

The sequence of the autocovariances of the process $y(t)$ may be expressed in terms of the spectrum of the process. From equation (18.39), it follows that the autocovariance of $y(t)$ at lag $\tau = t - s$ is given by

$$\begin{aligned}\gamma_\tau = C(y_t, y_s) &= E\left\{ \int_\omega e^{i\omega t} dZ(\omega) \int_\lambda e^{-i\lambda s} dZ(-\lambda) \right\} \\ &= \int_\omega \int_\lambda e^{i\omega t} e^{-i\lambda s} E\{dZ(\omega) dZ^*(\lambda)\} \\ &= \int_\omega e^{i\omega\tau} E\{dZ(\omega) dZ^*(\omega)\} \\ &= \int_{-\pi}^{\pi} e^{i\omega\tau} dF(\omega).\end{aligned} \tag{18.42}$$

Here the final equalities are derived by using the conditions under (18.36) and (18.37). The first of the conditions under (18.37) ensures that the autocovariance is a function only of the temporal separation $|t - s|$ of the elements of $y(t)$ and not of their absolute dates, as is required by the condition of stationarity. In the case of a continuous spectral distribution function, we may write $dF(\omega) = f(\omega)d\omega$ in the final expression, where $f(\omega)$ is the spectral density function.

The necessary and sufficient condition for the existence of a bounded spectral density function is the condition that the sequence of autocovariances is absolutely summable:

$$\sum_{\tau=-\infty}^{\infty} |\gamma_\tau| < \infty. \tag{18.43}$$

In that case, the autocovariances are the coefficients of an ordinary Fourier-series representation of the spectral density function which takes the form of

$$(18.44)\qquad \begin{aligned} f(\omega) &= \frac{1}{2\pi}\sum_{\tau=-\infty}^{\infty}\gamma_\tau e^{-i\omega\tau} \\ &= \frac{1}{2\pi}\Big\{\gamma_0 + 2\sum_{\tau=1}^{\infty}\gamma_\tau\cos(\omega\tau)\Big\}. \end{aligned}$$

The second expression, which depends upon the condition that $\gamma_{-\tau} = \gamma_\tau$, indicates that, for a real-valued sequence, the spectral density function is an even-valued function of ω such that $f(-\omega) = f(\omega)$. As a consequence, it is an almost invariable practice, in displaying the graph of $f(\omega)$, to plot the function only for positive values of ω and to ignore the remainder of the interval $(-\pi, \pi]$. Indeed, some authors define the spectrum of a real-valued process as $g(\omega) = 2f(\omega)$ with $\omega \in (0, \pi]$, which has the same effect. This definition is appropriate to a spectral representation of $y(t)$ which is in terms of trigonometrical functions rather than complex exponentials.

This function $f(\omega)$ is directly comparable to the periodogram of a data sequence which is represented by (14.35). However, the periodogram has T empirical autocovariances $c_0, \ldots, c_{T-1}$ in place of an indefinite number of theoretical autocovariances. Also, it differs from the spectrum by a scalar factor of 4π. In many texts, equation (18.44) serves as the primary definition of the spectrum.

To demonstrate the relationship which exists between equations (18.42) and (18.44) when the spectral density function exists, we may substitute the latter into the former to give

$$(18.45)\qquad \begin{aligned} \gamma_\tau &= \int_{-\pi}^{\pi} e^{i\omega\tau}\Big\{\frac{1}{2\pi}\sum_{\kappa=-\infty}^{\infty}\gamma_\kappa e^{-i\omega\kappa}\Big\}d\omega \\ &= \frac{1}{2\pi}\sum_{\kappa=-\infty}^{\infty}\gamma_\kappa\int_{-\pi}^{\pi} e^{i\omega(\tau-\kappa)}d\omega. \end{aligned}$$

From the fact that

$$(18.46)\qquad \int_{-\pi}^{\pi} e^{i\omega(\tau-\kappa)}d\omega = \begin{cases} 2\pi, & \text{if } \kappa = \tau; \\ 0, & \text{if } \kappa \neq \tau, \end{cases}$$

it can be seen that the RHS of the equation reduces to γ_τ. This serves to show that, when $dF(\omega) = f(\omega)d\omega$, equations (18.42) and (18.44) do indeed represent a Fourier transform and its inverse. The relationship between the sequence $\gamma(\tau)$ and the function $f(\omega)$ is liable to be described as the Wiener–Khintchine relationship.

The essential interpretation of the spectral distribution function is indicated by the equation

$$(18.47)\qquad \gamma_0 = \int_{-\pi}^{\pi} dF(\omega) = F(\pi),$$

which comes from setting $\tau = 0$ in equation (18.42). This equation shows how the variance or "power" of $y(t)$, which is γ_0, is attributed to the cyclical components of which the process is composed. The analogous equation for a cyclical process is to be found under (18.23).

It is easy to see that a flat spectral density function corresponds to the autocovariance function which characterises a white-noise process $\varepsilon(t)$. Let $f_\varepsilon = f_\varepsilon(\omega)$ be the flat spectrum. Then, from equation (18.44), it follows that

$$\begin{aligned}\gamma_0 &= \int_{-\pi}^{\pi} f_\varepsilon(\omega)d\omega \\ &= 2\pi f_\varepsilon,\end{aligned} \tag{18.48}$$

and, from equation (18.42), it follows that

$$\begin{aligned}\gamma_\tau &= \int_{-\pi}^{\pi} f_\varepsilon(\omega)e^{i\omega\tau}d\omega \\ &= f_\varepsilon\int_{-\pi}^{\pi} e^{i\omega\tau}d\omega \\ &= 0.\end{aligned} \tag{18.49}$$

These are the same as the conditions under (18.3) which have served to define a white-noise process. When the variance is denoted by σ_ε^2, the expression for the spectrum of the white-noise process becomes

$$f_\varepsilon(\omega) = \frac{\sigma_\varepsilon^2}{2\pi}. \tag{18.50}$$

The Theorem of Herglotz and the Decomposition of Wold

Some authorities base their account of the spectral theory of stationary processes primarily upon the relationship between the sequence of autocovariances and the spectral distribution function. This has the advantage of mathematical simplicity. If $y(t)$ is a purely nondeterministic process which possesses a continuous differentiable spectral distribution function, then, as we have shown already, there is an ordinary one-to-one Fourier relationship connecting the spectral density function to the sequence of autocovariances. Often this is the only aspect of spectral analysis which needs to exposed.

If a purely analytic approach to the spectral theory of stationary processes is taken, then it is necessary to establish the properties of the spectral distribution function by showing that it is a nondecreasing function of ω on the interval $(-\pi, \pi]$. This feature has already emerged naturally, via equation (18.34), in the constructive approach which we have been pursuing so far. Moreover, if the spectral density function $f(\omega)$ exists, then it is easy to show, via equation (18.44), that a nonnegative-definite autocovariance function implies that $f(\omega) \geq 0$, which is, once more, the required result.

When a discontinuous or partially discontinuous spectral distribution function is considered, it is more laborious to establish the requisite properties. In that case, the essential result which supports the analytic approach is a theorem of Herglotz:

(18.51) The sequence $\gamma(\tau)$ of the autocovariances is nonnegative-definite if and only if its elements can be expressed as

$$\gamma_\tau = \int_{-\pi}^{\pi} e^{i\omega\tau} dF(\omega),$$

where $F(\omega)$, is a nondecreasing function defined on the interval $(-\pi, \pi]$.

Proof. First assume that γ_τ can be expressed in terms of the formula above wherein $F(\omega)$ is a nondecreasing function of ω. Then an arbitrary quadratic function of the elements of a Tth-order dispersion matrix constructed from the autocovariances may be expressed as follows:

$$\begin{aligned}\sum_{t=0}^{T-1}\sum_{s=0}^{T-1} c_s\gamma_{|t-s|}c_t &= \int_{-\pi}^{\pi}\sum_{s=0}^{T-1}\sum_{t=0}^{T-1} c_sc_te^{i\omega(t-s)}dF(\omega)\\ &= \int_{-\pi}^{\pi}\left|\sum_{\tau=0}^{T-1} c_\tau e^{i\omega\tau}\right|^2 dF(\omega) \geq 0,\end{aligned} \tag{18.52}$$

where the squared term is a complex modulus. Since this holds for all T, it follows that $\gamma(\tau)$ is a nonnegative-definite function.

Now, in order to prove the converse, assume that $\gamma(\tau)$ is a nonnegative-definite function. Take $c_t = e^{-i\omega t}$ and $c_s = e^{i\omega s}$ in the LHS of equation (18.52) and divide by $2\pi T$. This gives

$$\begin{aligned}f_T(\omega) &= \frac{1}{2\pi T}\sum_{t=0}^{T-1}\sum_{s=0}^{T-1}\gamma_{|t-s|}e^{-i\omega(t-s)}\\ &= \frac{1}{2\pi}\sum_{\tau=1-T}^{T-1}\gamma_\tau\left(1-\frac{|\tau|}{T}\right)e^{-i\omega\tau} \geq 0.\end{aligned} \tag{18.53}$$

Here $f_T(\omega)$ is expressed as a finite Fourier sum wherein the coefficients are given by an inverse Fourier transformation:

$$\begin{aligned}\gamma_\tau\left(1-\frac{|\tau|}{T}\right) &= \int_{-\pi}^{\pi} e^{i\omega\tau}f_T(\omega)d\omega\\ &= \int_{-\pi}^{\pi} e^{i\omega\tau}dF_T(\omega).\end{aligned} \tag{18.54}$$

The final expression entails a nondecreasing cumulative distribution function

$$F_T(\omega) = \int_{-\pi}^{\omega} f_T(\omega)d\omega. \tag{18.55}$$

Now it is only necessary to show that $F_T(\omega) \to F(\omega)$ as $T \to \infty$. Since $F_T(\omega)$ is monotonically increasing and bounded by $F_T(\pi) = \gamma_0$, we can apply Helly's theorem to show that there is a subsequence $\{T_j\}$ of the values of $\{T\}$ such that $F_{T_j} \rightarrow F(\omega)$ as $T, T_j \rightarrow \infty$.

Observe that, in the case where the elements of the autocovariance function are absolutely summable in the manner of (18.43), and where, consequently, a spectral density function $f(\omega)$ exists, the condition that $f(\omega) \geq 0$ emerges directly from equation (18.53) via the convergence of $f_T(\omega) \rightarrow f(\omega)$ as $T \rightarrow \infty$.

Example 18.1. Consider the first-order moving average process $y(t) = \varepsilon(t) - \theta\varepsilon(t-1)$ to be found under (17.8). The autocovariances of the process are

$$(18.56) \qquad \gamma_\tau = \begin{cases} \sigma_\varepsilon^2(1+\theta^2) & \text{if } \tau = 0; \\ -\sigma_\varepsilon^2\theta & \text{if } \tau = 1; \\ 0 & \text{if } \tau > 1. \end{cases}$$

Therefore, according to (18.44), the spectral density function is

$$(18.57) \qquad \begin{aligned} f(\omega) &= \frac{1}{2\pi}\{\gamma_0 + 2\gamma_1\cos(\omega)\} \\ &= \frac{\sigma_\varepsilon^2}{2\pi}(1+\theta^2)\{1 + 2\rho\cos(\theta)\}, \end{aligned}$$

wherein $\rho = \gamma_1/\gamma_0$ is an autocorrelation coefficient. This function is nonnegative if and only if $|\rho| \leq \frac{1}{2}$. It has been shown already, in Example 17.2, that the latter is a necessary condition for the positive-definiteness of the autocovariance function. It now transpires that it is both a necessary and a sufficient condition.

The theorem of Herglotz may be coupled with a result which establishes the inverse mapping from the autocovariances to the spectral distribution function which subsumes equation (18.44) as a special case:

(18.58) Let $F(\omega)$ be the spectral distribution function corresponding to autocovariance function $\gamma(\tau)$ defined in (18.51), and let λ and $\mu > \lambda$ be any two points of continuity of $F(\omega)$. Then

$$F(\mu) - F(\lambda) = \lim_{n\rightarrow\infty} \frac{1}{2\pi} \sum_{\tau=-n}^{n} \gamma_\tau \frac{e^{-i\mu\tau} - e^{-i\lambda\tau}}{-i\tau}.$$

Proof. Consider the term

$$(18.59) \qquad b_\tau = \frac{e^{-i\mu\tau} - e^{-i\lambda\tau}}{-i\tau} = \int_\lambda^\mu e^{-i\omega\tau}d\omega = \int_{-\pi}^{\pi} e^{-i\omega\tau}\beta(\omega)d\omega,$$

where we are employing a rectangular function defined on the interval $(-\pi, \pi]$:

$$(18.60) \qquad \beta(\omega) = \begin{cases} 1, & \text{if } \omega \in (\lambda, \mu); \\ \frac{1}{2}, & \text{if } \omega = \lambda, \mu; \\ 0, & \text{if } \omega \notin [\lambda, \mu]. \end{cases}$$

We may observe in passing that $-i\tau$ in the denominator of b_τ cancels with terms in the series expansion of the numerator. Therefore the term is also defined at $\tau = 0$ where it takes the value of $b_0 = \mu - \lambda$. Now b_τ is nothing but the τth coefficient in the Fourier-series expansion of $\beta(\omega)$, from which it follows that $\sum |b_\tau| < \infty$. Hence $\sum b_\tau \gamma_\tau$ is bounded, and, taking the expression for γ_τ from (18.51), we have

(18.61)
$$\begin{aligned}\lim_{n\to\infty} \frac{1}{2\pi} \sum_{\tau=-n}^{n} b_\tau \gamma_\tau &= \lim_{n\to\infty} \int_{-\pi}^{\pi} \sum_{\tau=-n}^{n} b_\tau e^{i\omega\tau} dF(\omega) \\ &= \int_{-\pi}^{\pi} \beta(\omega) dF(\omega) = F(\mu) - F(\lambda),\end{aligned}$$

which is the required result.

The results of this section point to a decomposition of the spectral distribution function into three components:

(18.62)
$$F(\omega) = F_1(\omega) + F_2(\omega) + F_3(\omega).$$

The first of these is a continuous differentiable function. The second is a function which increases by jumps only at the points of discontinuity of $F(\omega)$. The third is a singular function defined on a set of measure zero where $F(\omega)$ is continuous but where the derivative $f(\omega)$ either does not exist or is infinite. Indeed such a decomposition is available for any monotonic nondecreasing function defined on a finite interval which has a finite or a denumerably infinite number of discontinuities.

In the so-called decomposition of Wold—which Wold [530, p. 68], in fact, attributes to Cramér [129]—$F_1(\omega)$ corresponds to the spectral distribution function of a purely nondeterministic process whilst $F_2(\omega)$ is the spectral distribution function of cyclical or "almost periodic" process of the sort depicted by equation (18.17). The singular component $F_3(\omega)$, which is essentially negligible, has no straightforward interpretation. Since their spectral distribution functions are confined to disjoint subsets which form a partition $(-\pi, \pi]$, the three process are mutually uncorrelated.

The Frequency-Domain Analysis of Filtering

It is a straightforward matter to derive the spectrum of a process $y(t) = \mu(L)x(t)$ which is formed by mapping the process $x(t)$ through a linear filter.

Taking the spectral representation of the process $x(t)$ to be

(18.63)
$$x(t) = \int_\omega e^{i\omega t} dZ_x(\omega),$$

we have

(18.64)
$$\begin{aligned}y(t) &= \sum_j \mu_j x(t-j) \\ &= \sum_j \mu_j \left\{ \int_\omega e^{i\omega(t-j)} dZ_x(\omega) \right\} \\ &= \int_\omega e^{i\omega t} \left(\sum_j \mu_j e^{-i\omega j} \right) dZ_x(\omega).\end{aligned}$$

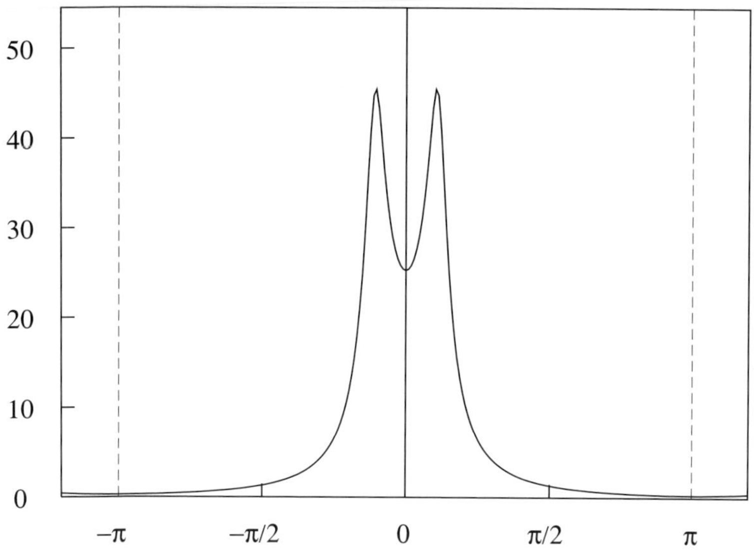

Figure 18.2. The gain of the transfer function $(1 + 2L^2)/(1 - 1.69L + 0.81L^2)$.

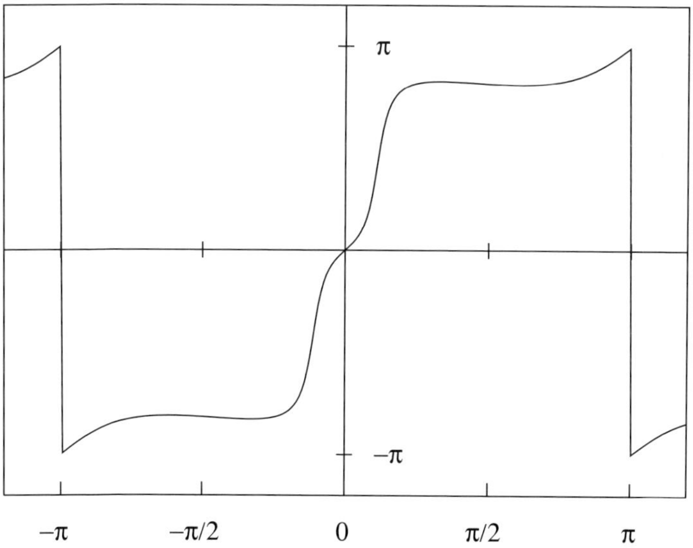

Figure 18.3. The phase diagram of the transfer function $(1+2L^2)/(1-1.69L+0.81L^2)$.

On writing $\sum \mu_j e^{-i\omega j} = \mu(\omega)$, this becomes

$$(18.65)\qquad \begin{aligned} y(t) &= \int_\omega e^{i\omega t}\mu(\omega)dZ_x(\omega) \\ &= \int_\omega e^{i\omega t}dZ_y(\omega). \end{aligned}$$

If the process $x(t)$ has a spectral density function $f_x(\omega)$, which allows one to write $dF(\omega) = f(\omega)d\omega$ in equation (18.37), then the spectral density function $f_y(\omega)$ of the filtered process $y(t)$ will be given by

$$(18.66)\qquad \begin{aligned} f_y(\omega)d\omega &= E\{dZ_y(\omega)dZ_y^*(\omega)\} \\ &= \mu(\omega)\mu^*(\omega)E\{dZ_x(\omega)dZ_x^*(\omega)\} \\ &= |\mu(\omega)|^2 f_x(\omega)d\omega. \end{aligned}$$

The complex-valued function $\mu(\omega)$, which is entailed in the process of linear filtering, can be written as

$$(18.67)\qquad \mu(\omega) = |\mu(\omega)|e^{-i\theta(\omega)},$$

where

$$(18.68)\qquad \begin{aligned} |\mu(\omega)|^2 &= \left\{\sum_{j=0}^{\infty}\mu_j\cos(\omega j)\right\}^2 + \left\{\sum_{j=0}^{\infty}\mu_j\sin(\omega j)\right\}^2, \\ \theta(\omega) &= \arctan\left\{\frac{\sum\mu_j\sin(\omega j)}{\sum\mu_j\cos(\omega j)}\right\}. \end{aligned}$$

The function $|\mu(\omega)|$, which is described as the gain of the filter, indicates the extent to which the amplitude of the cyclical components of which $x(t)$ is composed are altered in the process of filtering.

The function $\theta(\omega)$, which is described as the phase displacement and which gives a measure in radians, indicates the extent to which the cyclical components are displaced along the time axis.

The substitution of expression (18.67) in equation (18.65) gives

$$(18.69)\qquad y(t) = \int_{-\pi}^{\pi} e^{i\{\omega t-\theta(\omega)\}}|\mu(\omega)|dZ_x(\omega).$$

The virtue of this equation is that it summarises the two effects of the filter.

Figures 18.2 and 18.3 represent respectively the gain and the phase effect of a simple rational transfer function.

The Spectral Density Functions of ARMA Processes

Autoregressive moving-average or ARMA models are obtained by applying one-sided linear filters to white-noise processes. In the case of a pure MA model, denoted by

$$(18.70)\qquad y(t) = \mu(L)\varepsilon(t),$$

the filter $\mu(L) = 1+\mu_1 L+\cdots+\mu_q L^q$ is a polynomial of degree q in the lag operator L. In the case of the ARMA model, which is denoted by

$$\alpha(L)y(t) = \mu(L)\varepsilon(t) \qquad \text{or by} \qquad y(t) = \frac{\mu(L)}{\alpha(L)}\varepsilon(t), \tag{18.71}$$

there is an additional operator $\alpha(L) = 1+\alpha_1 L+\cdots+\alpha_p L^p$ which is a polynomial of degree p. The condition is imposed that $\alpha(z)$ and $\mu(z)$ should have no factors in common. Also, the condition of stationarity requires that the roots of $\alpha(z)$, which are the poles of the rational function $\mu(z)/\alpha(z)$, should lie outside the unit circle. This implies that the region of convergence for the series expansion of the rational function includes the unit circle. Such an expansion gives rise to a so-called infinite-order moving-average representation of the ARMA process.

These features have been discussed at length in the previous chapter, where the effects of the condition of invertibility, which is that the roots of $\mu(z)$ must lie outside the unit circle, are also described. In the present context, we shall make no such requirement; and, indeed, we shall admit the case where $\mu(z)$ has zeros on the unit circle.

In Chapter 17, it has been shown that the autocovariance generating function of an ARMA process is given by

$$\gamma(z) = \sigma_\varepsilon^2 \frac{\mu(z)\mu(z^{-1})}{\alpha(z)\alpha(z^{-1})}. \tag{18.72}$$

Given the conditions on the roots of $\alpha(z)$ and $\mu(z)$, this represents a function which is analytic in an annulus surrounding the unit circle; and it is usually understood the LHS stands for the Laurent expansion of the RHS.

When $z = e^{-i\omega}$, it is convenient to denote the two polynomials by $\alpha(\omega)$ and $\mu(\omega)$ respectively. In that case, it follow from (18.44) that the spectral density function of the ARMA process is given by

$$f(\omega) = \frac{\sigma_\varepsilon^2}{2\pi}\frac{\alpha(\omega)\alpha^*(\omega)}{\mu(\omega)\mu^*(\omega)} = \frac{\sigma_\varepsilon^2}{2\pi}\frac{|\alpha(\omega)|^2}{|\mu(\omega)|^2}. \tag{18.73}$$

Example 18.2. The second-order moving-average MA(2) process denoted by the equation

$$y(t) = \varepsilon(t) + \mu_1\varepsilon(t-1) + \mu_2\varepsilon(t-2) \tag{18.74}$$

has the following nonzero autocovariances:

$$\begin{aligned}\gamma_0 &= \sigma_\varepsilon^2(1+\mu_1^2+\mu_2^2),\\ \gamma_1 &= \sigma_\varepsilon^2(\mu_1+\mu_1\mu_2),\\ \gamma_2 &= \sigma_\varepsilon^2\mu_2.\end{aligned} \tag{18.75}$$

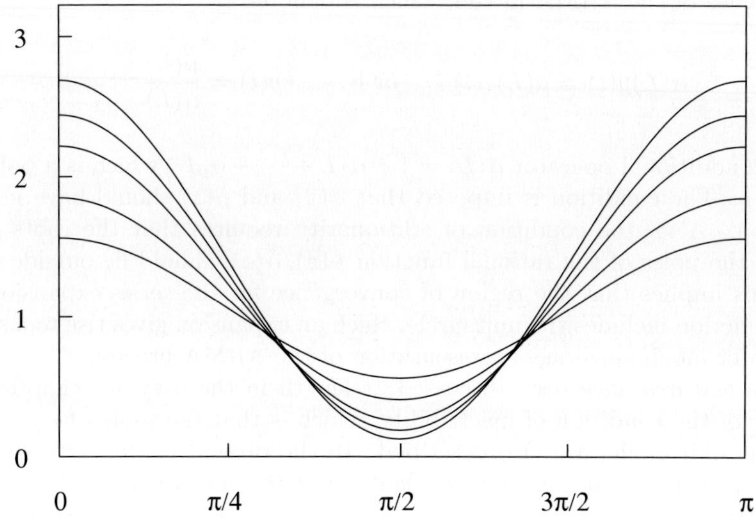

Figure 18.4. The spectral density functions of the MA(2) model $y(t) = (1 - \{2\rho\cos\omega_n\}L + \rho^2L^2)\varepsilon(t)$ when $\omega = 90°$ and $\rho = 0.8, 0.75, 0.7, 0.6, 0.5$.

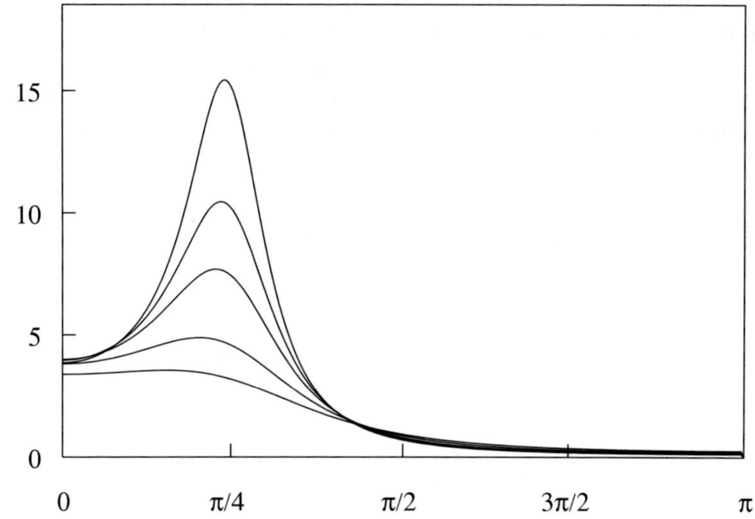

Figure 18.5. The spectral density functions of the AR(2) model when $\omega = 45°$ and $\rho = 0.8, 0.75, 0.7, 0.6, 0.5$ in the equation $(1 - \{2\rho\cos\omega_n\}L + \rho^2L^2)y(t) = \varepsilon(t)$.

The spectral density function is therefore

$$f(\omega) = \frac{1}{2\pi}\{\gamma_0 + 2\gamma_1\cos(\omega) + 2\gamma_2\cos(2\omega)\}. \tag{18.76}$$

To find the stationary points of the function, one may differentiate it with respect to ω and set the result to zero to obtain the equation

$$\begin{aligned} 0 &= -2\gamma_1\sin(\omega) - 4\gamma_2\sin(2\omega) \\ &= -2\gamma_1\sin(\omega) - 8\gamma_2\sin(\omega)\cos(\omega). \end{aligned} \tag{18.77}$$

Here the second equality comes from the trigonometrical identity $\sin(2A) = 2\cos(A)\sin(A)$ which is deduced from (13.125)(c). The solution to the equation is

$$\omega = \cos^{-1}\left\{\frac{-\gamma_1}{4\gamma_2}\right\} = \cos^{-1}\left\{\frac{-\mu_1(1+\mu_2)}{4\mu_2}\right\}. \tag{18.78}$$

Consider the case where the polynomial $1 + \mu_1 z + \mu_2 z^2$ has the conjugate complex roots $z = \rho\exp\{\pm i\lambda\}$. Then $\mu_1 = -2\rho\cos(\lambda)$ and $\mu_2 = \rho^2$, and hence

$$\omega = \cos^{-1}\left\{\frac{\rho(1+\rho^2)\cos\lambda}{2\rho^2}\right\}. \tag{18.79}$$

This is the frequency value for which the ordinate of the spectral density function is at a minimum. In the case where the roots are located on the unit circle, which is when $\rho = 1$, the equation delivers $\omega = \lambda$ which is, of course, a point where the spectral ordinate is zero-valued.

Figure 18.4 represents the spectral density functions of a range of MA(2) processes of which the MA operators have conjugate complex roots which differ only in respect of the value of the modulus ρ.

Example 18.3. The second-order autoregressive AR(2) process is denoted by the equation

$$y(t) + \alpha_1 y(t-1) + \alpha_2 y(t-2) = \varepsilon(t). \tag{18.80}$$

The corresponding autocovariance generating function is

$$\begin{aligned} \gamma(z) &= \frac{\sigma_\varepsilon^2}{(1+\alpha_1 z + \alpha_2 z^2)(1 + \alpha_1 z^{-1} + \alpha_2 z^{-2})} \\ &= \frac{\sigma_\varepsilon^2}{(1+\alpha_1^2+\alpha_2^2) + (\alpha_1 + \alpha_1\alpha_2)(z+z^{-1}) + \alpha_2(z^2+z^{-2})}. \end{aligned} \tag{18.81}$$

When $z = e^{-i\omega}$, the term $z^j + z^{-j}$ becomes

$$e^{i\omega j} + e^{-i\omega j} = 2\cos(j\omega). \tag{18.82}$$

Therefore, on setting $z = e^{-i\omega}$ in the autocovariance generating function and dividing by 2π, we obtain the spectral density function

$$(18.83) \qquad f(\omega) = \frac{\sigma_\varepsilon^2/2\pi}{(1+\alpha_1^2+\alpha_2^2) + 2(\alpha_1+\alpha_1\alpha_2)\cos(\omega) + 2\alpha_2\cos(2\omega)}.$$

Apart from a scalar factor, this is simply the inverse of the spectral density function of an MA(2) process. Moreover, it follows that the AR spectrum has a peak at the point where the corresponding MA function would have a trough.

Consider the AR(2) process with complex conjugate roots. These may be denoted by $z = \rho\exp\{\pm i\omega_n\}$, whence the equation for the process can be written as

$$(18.84) \qquad y(t) - (2\rho\cos\omega_n)y(t-2) + \rho^2 y(t-2) = \varepsilon(t).$$

Here $\rho \in [0, 1)$ is the modulus of the roots, which determines the damping of the system, whilst ω_n is their argument, which may be described as the natural frequency of an undamped system wherein $\rho = 1$. By drawing upon the results of the previous example, it is straightforward to show that the peak of the spectrum is located at the point

$$(18.85) \qquad \omega = \cos^{-1}\left\{\frac{\rho(1+\rho^2)\cos\omega_n}{2\rho^2}\right\}.$$

Moreover, as $\rho \to 1$, the value increases towards that of ω_n, which is the natural or resonant frequency of the undamped system.

Figure 18.5 depicts the spectral density functions of a range of AR(2) processes which differ only in respect of the modulus of the complex roots of the AR operator. This figure is reminiscent of Figure 5.2 which shows the frequency response of a second-order linear dynamic system with various damping ratios.

Canonical Factorisation of the Spectral Density Function

It is a basic result of time-series analysis that every stationary process $y(t)$ with a continuous spectrum may be represented as a moving-average process. In general, the moving-average operator entailed in such a representation is both two-sided and of infinite order. Nevertheless, under some fairly mild restrictions, it is possible to show that a one-sided operator is available which expresses the process is terms of the current and previous values of a white-noise sequence.

Often, the existence of a moving-average representation can be demonstrated, almost trivially, by showing that there exists an operator, say $\phi(L)$—not necessarily a linear filter—which reduces the process $y(t)$ to a white-noise sequence $\varepsilon(t) = \phi(L)y(t)$. Then, a linear filter $\theta(L)$ can be found can which reverses the operation, so that we have $y(t) = \theta(L)\varepsilon(t) = \{\theta(L)\phi(L)\}y(t)$, which is the desired representation.

The simplest case is where $y(t)$ is a stationary stochastic process with a spectrum $f_y(\omega) > 0$ which is everywhere nonzero. It is always possible to find a complex function $\mu(\omega)$ such that

$$(18.86) \qquad f_y(\omega) = \frac{1}{2\pi}\mu(\omega)\mu^*(\omega).$$

Given that $f_y(\omega) > 0$, it follows that $\mu(\omega)$ has none of its roots on the unit circle. Therefore

$$dZ_\varepsilon(\omega) = \frac{1}{\mu(\omega)} dZ_y(\omega) \tag{18.87}$$

exists for all values of ω, and the spectral representation of the process $y(t)$ given in equation (18.39), may be rewritten as

$$y(t) = \int_\omega e^{i\omega t}\mu(\omega)dZ_\varepsilon(\omega). \tag{18.88}$$

Expanding $\mu(\omega)$ as a Fourier series and interchanging the order of integration and summation gives

$$\begin{aligned} y(t) &= \int_\omega e^{i\omega t}\Big(\sum_j \mu_j e^{-i\omega j}\Big)dZ_\varepsilon(\omega) \\ &= \sum_j \mu_j\Big\{\int_\omega e^{i\omega(t-j)}dZ_\varepsilon(\omega)\Big\} \\ &= \sum_j \mu_j\varepsilon(t-j), \end{aligned} \tag{18.89}$$

where we have defined

$$\varepsilon(t) = \int_\omega e^{i\omega t}dZ_\varepsilon(\omega). \tag{18.90}$$

The spectrum of $\varepsilon(t)$ is determined by the equation

$$\begin{aligned} E\{dZ_\varepsilon(\omega)dZ_\varepsilon^*(\omega)\} &= E\left\{\frac{dZ_y(\omega)dZ_y^*(\omega)}{\mu(\omega)\mu^*(\omega)}\right\} \\ &= \frac{f_y(\omega)}{\mu(\omega)\mu^*(\omega)}d\omega \\ &= \frac{1}{2\pi}d\omega. \end{aligned} \tag{18.91}$$

Hence $\varepsilon(t)$ is identified as a white-noise process with unit variance. Therefore equation (18.89) represents a moving-average process.

In the case where $f(\omega) = 0$ in respect of a set of values of ω of a nonzero measure, it is no longer possible to transform $y(t)$, via a process of linear filtering, to a white-noise process $\varepsilon(t)$ with a flat spectrum. In this case, the conversion of $y(t)$ to white noise requires the supplementation of the process $Z_y(\omega)$ by a further process which adds power to the spectrum in the frequency ranges where there is none. Let $Z_s(\omega)$ be the relevant supplementary process which has uncorrelated increments $dZ_s(\omega)$ for which

$$E\{dZ_s(\omega)\} = 0 \quad \text{and} \quad V\{dZ_s(\omega)\} = E\{dZ_s(\omega)dZ_s^*(\omega)\} = d\omega. \tag{18.92}$$

Then the process entailed in spectral representation of the reconstructed white-noise process is specified by

(18.93) $$Z_\varepsilon(\omega) = \int_A dZ_s(\lambda) + \int_{A^c} \frac{1}{\mu(\lambda)} dZ_y(\lambda),$$

where A is the set of values of λ in the frequency interval $(-\pi, \omega]$ for which $f(\omega) = 0$, and where A^c is the complementary set.

In certain circumstances there exists a one-sided moving-average representation of the process $y(t)$:

(18.94) The stationary process $y(t)$ has a one-sided moving-average representation in the form of equation (18.70) if it has a spectrum $f(\omega) \geq 0$ which is zero-valued at most on a set of measure zero and if

$$\int_{-\pi}^{\pi} \ln\{f(\omega)\} d\omega > -\infty.$$

Proof. If the conditions of the theorem are true and $f(\omega)$ has only isolated zeros in the interval $(-\pi, \pi]$, then $\ln\{f(\omega)\}$ has a Fourier-series representation of the form

(18.95) $$\ln\{f(\omega)\} = \sum_{j=-\infty}^{\infty} \phi_j e^{-i\omega j},$$

with coefficients which are given by

(18.96) $$\phi_j = \frac{1}{2\pi} \int_{-\pi}^{\pi} e^{i\omega j} \ln\{f(\omega)\} d\omega = \phi_{-j},$$

where the condition that $\phi_j = \phi_{-j}$ reflects the fact that $f(\omega)$ is real-valued and symmetric. Since the sum in (18.95) converges, as do the separate sums over positive and negative integers, the following convergent Fourier series may also be defined:

(18.97) $$\mu(\omega) = \sum_{i=0}^{\infty} \gamma_i e^{-i\omega} = \exp\left\{ \sum_{j=1}^{\infty} \phi_j e^{-\omega j} \right\}.$$

Therefore the spectral density function may be factorised as

(18.98) $$f(\omega) = \exp\{\phi_0\} \mu(\omega) \mu^*(\omega).$$

If the factor $\exp\{\phi_0\}$ is identified with $\sigma_\varepsilon^2 / 2\pi$, then this factorisation is identical to the expression for the spectral density function for an MA process which is obtained from (18.72) when $\alpha(z) = 1$; and the condition emerges that

(18.99) $$\exp\{\phi_0\} = \frac{\sigma_\varepsilon^2}{2\pi} = \exp\left\{ \frac{1}{2\pi} \int_{-\pi}^{\pi} \ln\{f(\omega)\} d\omega \right\} > 0;$$

for which the condition stated under (18.94) is clearly necessary. If the condition is not met, then the process in question is deterministic.

It can be proved that the condition under (18.94) is also necessary for the existence of a one-sided moving-average representation of a stationary process—see, for example, Doob [162, pp. 106, 577]. The necessity of the condition is readily intelligible. In its absence, the case would be admitted where the spectrum of the process is zero-valued over a finite interval. Since the zeros of a polynomial $\mu(z) = 1 + \mu z + \cdots + \mu_q z^q$—or, indeed, those of a rational function—correspond to isolated points in the complex plane, it is clearly impossible for a finite-order ARMA model to generate such a spectrum. Equally it is impossible that such a spectrum could arise from a one-side moving average of infinite order. Nevertheless, it is one of the objectives of signal processing to construct filters which can virtually nullify the spectral density function over a designated frequency range. The foregoing results indicate that such an objective can never be achieved completely.

Bibliography

[129] Cramér, H., (1952), A Contribution to the Theory of Stochastic Processes, pp. 329–339 in *Proceedings of the 2nd Berkley Symposium on Mathematical Statistics and Probability*, University of California Press.

[162] Doob, J.L., (1953), *Stochastic Processes*, John Wiley and Sons, New York.

[229] Grenander, U., and M. Rosenblatt, (1957), *Statistical Analysis of Stationary Time Series*, John Wiley and Sons, New York.

[517] Whittle, P., (1951), *Hypothesis Testing in Time Series Analysis*, Almqvist and Wiksells Boktryckeri, Uppsala.

[518] Whittle, P., (1953), Estimation and Information in Stationary Time Series, *Arkiv för Matmatik*, **2**, 423–34.

[522] Wiener, N., (1930), Generalised Harmonic Analysis, *Acta Mathematica*, **55**, 117–258.

[530] Wold, H., (1938), *A Study in the Analysis of Stationary Time Series*, second edition 1954 with an Appendix by Peter Whittle, Almqvist and Wiksell, Stockholm.

[536] Yaglom, A.M., (1962), *An Introduction to the Theory of Stationary Random Processes*, revised English edition translated and edited by R.A. Silverman, Prentice-Hall, Englewood Cliffs New Jersey.

CHAPTER 19

Prediction and Signal Extraction

In classical time-series analysis, there are two branches of prediction theory. In the so-called problem of pure prediction, the object is to forecast the value of a time series for several steps ahead on the basis of the observed history of the series. In the problem of signal extraction, the object is to infer the value of a signal from a record which is affected by superimposed noise. In this case, the requirement may be for an estimate of the signal at any period: past, present or future. The signal-extraction problem is, therefore, more general than the problem of pure prediction. The fundamental problem is that of predicting one series from the record of another; and this is where we shall begin the account.

The dominant approach to the theory of prediction stems from the work of Wiener [523] and Kolmogorov [298], who reached their solutions independently at much the same time. Kolmogorov dealt with the problem in terms of the time domain, whereas Wiener used concepts from the frequency domain. Despite these differences of approach, the two solutions were essentially equivalent; and nowadays they are often expounded in a manner which transcends the distinctions. A good example is provided by Whittle's [519] account.

The statistical theory of Wiener and Kolmogorov is restricted to stationary stochastic processes. In the case of the pure prediction problem, such series are assumed to extend into the indefinite past. This is a convenient theoretical fiction; and, in practice, if the series is long enough and if its current values are not too strongly influenced by the past, then the falsity of the fiction does little harm to the quality of the forecasts. In the case of the signal-extraction problem, it is often assumed that the record of the process also extends into the indefinite future. This is also likely to be a harmless fiction if the record is long enough and if it can be played forwards and backwards at will, as in the case of a digital sound recording, for example.

Many of the problems of prediction in the social sciences, and in such scientific areas as climatology, concern time series which are manifestly nonstationary. The Wiener–Kolmogorov theory can be adapted to deal with such series as well, provided that they can be reduced to stationarity by use of a de-trending device, such as the difference operator, which may be accompanied by a prior logarithmic transformation. Nevertheless, there is a sleight of hand in this approach which can have deleterious consequences if insufficient care is taken.

The means which are best adapted to coping with problems of predicting nonstationary series from finite records are those which are provided by the recursive algorithms which we shall present in the second half of this chapter. The Kalman filter and the associated smoothing filter also provide appropriate methods in such cases. However, the algebra of the Kalman filter is extensive and the manner in which it relates to a specific problem may be obscure. This contrasts with the lucidity of the Wiener–Kolmogorov theory. Therefore, in this chapter, our recourse will be to develop the basic results within the context of the classical theory before translating them, where necessary, into the more elaborate versions which are associated with the finite-sample prediction algorithms and with the Kalman filter.

Before embarking on the main topics, we need to establish some basic results concerning minimum-mean-square-error prediction.

Mean-Square Error

The criterion which is commonly used in judging the performance of an estimator or predictor $\hat{y}$ of a random variable y is its mean-square error defined by $E\{(y-\hat{y})^2\}$. If all of the available information on y is summarised in its marginal distribution, then the minimum-mean-square-error prediction is simply the expected value $E(y)$. However, if y is statistically related to another random variable x whose value can be observed, and if the form of the joint distribution of x and y is known, then the minimum-mean-square-error prediction of y is the conditional expectation $E(y|x)$. This proposition may be stated formally:

(19.1) Let $\hat{y} = \hat{y}(x)$ be the conditional expectation of y given x, which is also expressed as $\hat{y} = E(y|x)$. Then $E\{(y-\hat{y})^2\} \leq E\{(y-\pi)^2\}$, where $\pi = \pi(x)$ is any other function of x.

Proof. Consider

$$\begin{aligned} E\{(y-\pi)^2\} &= E\Big[\{(y-\hat{y}) + (\hat{y}-\pi)\}^2\Big] \\ &= E\{(y-\hat{y})^2\} + 2E\{(y-\hat{y})(\hat{y}-\pi)\} + E\{(\hat{y}-\pi)^2\}. \end{aligned} \tag{19.2}$$

In the second term of the final expression, there is

$$\begin{aligned} E\{(y-\hat{y})(\hat{y}-\pi)\} &= \int_x\int_y (y-\hat{y})(\hat{y}-\pi)f(x,y)\partial y\partial x \\ &= \int_x\left\{\int_y (y-\hat{y})f(y|x)\partial y\right\}(\hat{y}-\pi)f(x)\partial x \\ &= 0. \end{aligned} \tag{19.3}$$

Here the second equality depends upon the factorisation $f(x,y) = f(y|x)f(x)$ which expresses the joint probability density function of x and y as the product of the conditional density function of y given x and the marginal density function of x. The final equality depends upon the fact that $\int(y-\hat{y})f(y|x)\partial y = E(y|x) - E(y|x) = 0$. Putting (19.3) into (19.2) shows that $E\{(y-\pi)^2\} = E\{(y-\hat{y})^2\} + E\{(\hat{y}-\pi)^2\} \geq E\{(y-\hat{y})^2\}$; and the assertion is proved.

The definition of the conditional expectation implies that

$$\begin{aligned} E(xy) &= \int_x \int_y xyf(x,y)\partial y \partial x \\ &= \int_x x\left\{ \int_y yf(y|x)\partial y \right\} f(x)\partial x \\ &= E(x\hat{y}). \end{aligned} \tag{19.4}$$

When the equation $E(xy) = E(x\hat{y})$ is rewritten as

$$E\{x(y - \hat{y})\} = 0, \tag{19.5}$$

it takes the form of an orthogonality condition. This condition indicates that the prediction error $y - \hat{y}$ is uncorrelated with x. The result is intuitively appealing; for, if the error were correlated with x, then some part of it would be predictable, which implies that the information of x could be used more efficiently in making the prediction of y.

The proposition of (19.1) is readily generalised to accommodate the case where, in place of the scalar x, there is a finite sequence $\{x_t, x_{t-1}, \ldots, x_{t-p}\}$ or even an indefinite sequence.

In practice, we are liable to form our predictions of y from linear combinations of a finite number of consecutive values of x and from lagged values of y. If the joint distribution of these various elements is a normal distribution, then, indeed, the conditional expectation of y will be a linear function of the remaining elements. However, even when the assumption of a normal distribution cannot be sustained realistically, we are still liable to predict y via a linear function.

The projection theorem, which we shall discuss more fully at a later stage, indicates that the linear combination $\hat{y}$ which fulfils the orthogonality condition of (19.5) is also the linear estimator with the minimum-mean-square prediction error. Therefore, we shall make extensive use of the orthogonality condition.

It is largely a matter of style whether, in our account of the theory of prediction, we choose to take the assumption of normality as a universal premise or whether we choose, instead, to remind ourselves that, in truth, we are dealing only with the linear theory of prediction. In fact, there is no other general theory of prediction. Having discussed these issues at the outset, we shall largely ignore them hereafter.

Predicting one Series from Another

Let $\{x_t, y_t\}$ be a sequence of jointly distributed random variables generated by a stationary stochastic process, and imagine that the intention is to use a linear function of the values in the information set $\mathcal{I}_t = \{x_{t-j}; j \in \mathcal{Q}\}$ to predict the value of y_t. Then the prediction may be denoted by

$$\hat{y}_t = \sum_{j \in \mathcal{Q}} \beta_j x_{t-j}. \tag{19.6}$$

For the moment, we may omit to specify the limits for the summation which would indicate the extent of the information set.

Let $e_t = y_t - \hat{y}_t$ be the error of prediction. Then the indefinite sequence $y(t) = \{y_t\}$ can be decomposed as

$$y(t) = \beta(L)x(t) + e(t), \tag{19.7}$$

where $\beta(L) = \sum_j \beta_j L^j$ is a polynomial or a power series of the lag operator, and where $x(t) = \{x_t\}$ and $e(t) = \{e_t\}$ are indefinite sequences.

The principle of orthogonality implies that the minimum-mean-square-error estimate of y_t will be obtained by finding the coefficients $\{\beta_j\}$ which satisfy the conditions

$$\begin{aligned} 0 &= E\{x_{t-k}(y_t - \hat{y}_t)\} \\ &= E(x_{t-k}y_t) - \sum_j \beta_j E(x_{t-k}x_{t-j}) \\ &= \gamma_k^{xy} - \sum_j \beta_j \gamma_{k-j}^{xx} \end{aligned} \tag{19.8}$$

for all $k \in \mathcal{Q}$.

The precise nature of the solution depends upon the extent of the information set which is determined by the index set $\mathcal{Q}$. In practice, the set will comprise only a finite number of elements. For theoretical purposes, however, it is also appropriate to consider sets with comprise an infinite number of elements from the past, as well as sets which comprise infinite numbers of elements stretching into the past and the future.

Let us begin by considering the case where $\mathcal{I}_t = \{x_t, x_{t-1}, \ldots, x_{t-p}\}$. Then, with $k = 0, 1, \ldots, p$, the orthogonality conditions of (19.8) become the normal equations of a linear regression. These equations can be compiled into a matrix format as follows:

$$\begin{bmatrix} \gamma_0^{xy} \\ \gamma_1^{xy} \\ \vdots \\ \gamma_p^{xy} \end{bmatrix} = \begin{bmatrix} \gamma_0^{xx} & \gamma_1^{xx} & \cdots & \gamma_p^{xx} \\ \gamma_1^{xx} & \gamma_0^{xx} & \cdots & \gamma_{p-1}^{xx} \\ \vdots & \vdots & \ddots & \vdots \\ \gamma_p^{xx} & \gamma_{p-1}^{xx} & \cdots & \gamma_0^{xx} \end{bmatrix} \begin{bmatrix} \beta_0 \\ \beta_1 \\ \vdots \\ \beta_p \end{bmatrix}. \tag{19.9}$$

The normal equations can also be expressed in terms of z-transform polynomials. Multiplying the jth equation by z^j gives

$$\gamma_j^{xy} z^j = \gamma_j^{xx} z^j \beta_0 + \gamma_{j-1}^{xx} z^{j-1} \beta_1 z + \cdots + \gamma_{p-j}^{xx} z^{j-p} \beta_p z^p. \tag{19.10}$$

The full set of equations for $j = 0, 1, \ldots, p$ can now be expressed as

$$\gamma^{xy}(z)_{(0,p)} = \left[\gamma^{xx}(z)\beta(z)\right]_{(0,p)}, \tag{19.11}$$

where $\beta(z) = \beta_0 + \beta_1 z + \cdots + \beta_p z^p$ stands for the z-transform of the sequence of coefficients and where the subscript $(0, p)$ indicates that only the terms associated with $z^0, z^1, \ldots, z^p$ have been taken from $\gamma^{xy}(z)$ and $\gamma^{xx}(z)\beta(z)$.

The virtue of this formulation is that it accommodates the case where the sequence $\{\beta_j\}$ comprises an indefinite number of nonzero coefficients, as is the case when $\beta(z) = \delta(z)/\alpha(z)$ represents the series expansion of a rational function. In that case, the sequence of equations obtained by letting p increase indefinitely is denoted by

$$\gamma^{xy}(z)_+ = \left[\gamma^{xx}(z)\beta(z)\right]_+, \tag{19.12}$$

where the subscripted + sign indicates that only nonnegative powers of z have been taken from $\gamma^{xy}(z)$ and $\gamma^{xx}(z)\beta(z)$. This notation is due to Whittle [519].

The Technique of Prewhitening

At first sight, there seems to be little hope of finding the coefficients of $\beta(z)$ from those of $\gamma^{xx}(z)$ and $\gamma^{xy}(z)$ unless there is a finite number of them. However, let us consider the canonical factorisation of the autocovariance function of the stationary process $x(t)$. This can be written as

$$\gamma^{xx}(z) = \sigma_\xi^2\psi(z)\psi(z^{-1}), \tag{19.13}$$

where $\psi(z) = \{\psi_0 + \psi_1 z + \cdots\}$. Now define the white-noise process $\xi(t) = \psi^{-1}(L)x(t)$. Then, on setting $x(t) = \psi(L)\xi(t)$ in equation (19.7), we get

$$\begin{aligned} y(t) &= \beta(L)\psi(L)\xi(t) + e(t) \\ &= \rho(L)\xi(t) + e(t), \end{aligned} \tag{19.14}$$

where $\rho(L) = \beta(L)\psi(L)$. Now observe that

$$E(\xi_{t-k}\xi_{t-j}) = \begin{cases} \sigma_\xi^2, & \text{if } j = k; \\ 0, & \text{if } j \neq k. \end{cases} \tag{19.15}$$

In effect, the autocovariance generating function of $\xi(t)$ takes the form of $\gamma^{\xi\xi}(z) = \sigma_\xi^2$. Thus, when $\xi(t) = \psi^{-1}(L)x(t)$ replaces $x(t)$, the matrix on the RHS of equation (19.9) is replaced by an identity matrix scaled by the value of σ_ξ^2. Also, equation (19.12) assumes the simplified form of

$$\gamma^{\xi y}(z)_+ = \left[\gamma^{\xi\xi}(z)\rho(z)\right]_+ = \sigma_\xi^2\rho(z). \tag{19.16}$$

Now $\gamma^{\xi y}(z) = \gamma^{xy}(z)/\psi(z^{-1})$ and $\beta(z) = \rho(z)/\psi(z)$, so it follows that

$$\beta(z) = \frac{1}{\sigma_\xi^2\psi(z)}\left[\frac{\gamma^{xy}(z)}{\psi(z^{-1})}\right]_+. \tag{19.17}$$

This is an equation which we shall depend upon later.

The virtue of replacing the signal $x(t)$ by the white-noise sequence $\xi(t)$ is that it enables the coefficients of $\rho(z)$ to be found one at a time from the elements of the covariance function $\gamma^{\xi y}(z)$. The coefficients of $\beta(z) = \rho(z)/\psi(z)$ can also be found one at a time by the simple algorithm for expanding a rational function described in Chapter 3.

The coefficients of $\beta(z)$ may also be found directly from the elements of the covariance function $\gamma^{\xi q}(z)$ of $\xi(t)$ and $q(t) = \psi^{-1}(L)y(t)$. To understand this, consider premultiplying equation (19.7) by $\psi^{-1}(L)$ to give

$$\begin{aligned} q(t) &= \psi^{-1}(L)y(t) \\ &= \beta(L)\psi^{-1}(L)x(t) + \psi^{-1}(L)\varepsilon(t) \\ &= \beta(L)\xi(t) + \eta(t). \end{aligned} \tag{19.18}$$

Then

$$\gamma^{\xi q}(z)_{+} = \left[\gamma^{\xi\xi}(z)\beta(z)\right]_{+} = \sigma_{\xi}^{2}\beta(z), \tag{19.19}$$

which is to say that the coefficients $\beta_j = C(\xi_{t-j}, q_t)/V(\xi_t)$ can be obtained from simple univariate regressions. This method of obtaining the coefficients is described as the prewhitening technique. In practice, it will usually require the estimation of the polynomial $\psi(z)$ which is unlikely to be known *a priori*. This is a matter of building an autoregressive moving-average (ARMA) model for $x(t)$.

If $\beta(z) = \delta(z)/\alpha(z)$ represents the series expansion of a rational function, then it should be straightforward to recover the coefficients of the polynomials $\delta(z)$ and $\alpha(z)$ from the leading coefficients of the series. Having recovered $\delta(z)$ and $\alpha(z)$, we are in a position to generate an indefinite number of the coefficients of $\beta(z)$ via a simple recursion. Thus the technique of prewhitening enables the parameters of an infinite-impulse-response (IIR) transfer function to be estimated without undue difficulty.

Extrapolation of Univariate Series

Now let us consider the problem of predicting the value of a stationary univariate series $y(t)$ for h steps ahead on the basis of the values which have been observed up to time t. Eventually, we shall show that this problem can be placed in the same framework as that of predicting one series from another. However, we shall begin by treating the problem in its own terms.

For a start, it can be assumed, without any loss of generality, that $y(t)$ is generated by an ARMA process. A stationary and invertible ARMA model can be represented in three different ways:

$$\alpha(L)y(t) = \mu(L)\varepsilon(t), \qquad \textit{Difference-equation form} \tag{19.20}$$

$$y(t) = \frac{\mu(L)}{\alpha(L)}\varepsilon(t) = \psi(L)\varepsilon(t), \qquad \textit{Moving-average form} \tag{19.21}$$

$$\frac{\alpha(L)}{\mu(L)}y(t) = \pi(L)y(t) = \varepsilon(t). \qquad \textit{Autoregressive form} \tag{19.22}$$

The moving-average representation provides the simplest context in which to establish the basic results concerning the minimum-mean-square-error predictors.

Consider making a prediction at time t for h steps ahead on the basis of an information set which stretches into the indefinite past. If $\psi(L)$ is known, then the sequence $\{\varepsilon_t, \varepsilon_{t-1}, \varepsilon_{t-2}, \ldots\}$ can be recovered from the sequence $\{y_t, y_{t-1}, y_{t-2}, \ldots\}$ and vice versa; so either of these constitute the information set. In terms of the moving-average representation, the value of the process at time $t+h$ is

$$\begin{aligned} y_{t+h} = &\{\psi_0\varepsilon_{t+h} + \psi_1\varepsilon_{t+h-1} + \cdots + \psi_{h-1}\varepsilon_{t+1}\} \\ &+ \{\psi_h\varepsilon_t + \psi_{h+1}\varepsilon_{t-1} + \cdots\}. \end{aligned} \tag{19.23}$$

The first term on the RHS embodies disturbances subsequent to the time t, and the second term embodies disturbances which are within the information set $\{\varepsilon_t, \varepsilon_{t-1}, \varepsilon_{t-2}, \ldots\}$. A linear forecasting function, based on the information set, takes the form of

$$\hat{y}_{t+h|t} = \{\rho_h\varepsilon_t + \rho_{h+1}\varepsilon_{t-1} + \cdots\}. \tag{19.24}$$

Then, given that $\varepsilon(t)$ is a white-noise process, it follows that the mean square of the error in the forecast h periods ahead is

$$E\{(y_{t+h} - \hat{y}_{t+h|t})^2\} = \sigma_\varepsilon^2 \sum_{i=0}^{h-1} \psi_i^2 + \sigma_\varepsilon^2 \sum_{i=h}^{\infty} (\psi_i - \rho_i)^2. \tag{19.25}$$

This is minimised by setting $\rho_i = \psi_i$; and so the optimal forecast is given by

$$\hat{y}_{t+h|t} = \{\psi_h\varepsilon_t + \psi_{h+1}\varepsilon_{t-1} + \cdots\}. \tag{19.26}$$

This forecasting formula might have been derived from the equation $y(t+h) = \psi(L)\varepsilon(t+h)$, which generates the true value of y_{t+h}, simply by putting zeros in place of the unobserved disturbances $\varepsilon_{t+1}, \varepsilon_{t+2}, \ldots, \varepsilon_{t+h}$ which lie in the future when the forecast is made.

On the assumption that the process is stationary, the mean-square error of the forecast, which is given by

$$E\{(y_{t+h} - \hat{y}_{t+h|t})^2\} = \sigma_\varepsilon^2 \sum_{i=0}^{h-1} \psi_i^2, \tag{19.27}$$

tends to the value of the variance of the process $y(t)$ as the lead time h of the forecast increases.

The optimal forecast may also be derived by specifying that the forecast error should be uncorrelated with the disturbances up to the time of making the forecast. For, if the forecast errors were correlated with some of the elements of the information set, then, as we have noted before, we would not be using the information efficiently, and we could generate better forecasts.

Now let us reconcile the results of this section with those of the previous section which relate to the problem of predicting the values of $y(t)$ from previous values of $x(t)$. Consider writing

$$\begin{aligned} y(t+h) &= \beta(L)y(t) + e(t) \\ &= \rho(L)\varepsilon(t) + e(t), \end{aligned} \tag{19.28}$$

where $\rho(L) = \beta(L)\psi(L)$. This is to be compared with equation (19.14). The only difference is the replacement of $x(t) = \psi(L)\xi(t)$ on the RHS of (19.14) by $y(t) = \psi(L)\varepsilon(t)$. The autocovariance generating function of $x(t)$, which was denoted by $\gamma^{xx}(z)$ and which is to be found under (19.13), is now replaced by the autocovariance generating function of $y(t)$:

$$\gamma(z) = \sigma_\varepsilon^2 \psi(z)\psi(z^{-1}). \tag{19.29}$$

By making the appropriate replacements in the formula of (19.17), it can be seen that the expression for $\beta(L)$ in (19.28) is

$$\beta(z) = \frac{1}{\sigma_\varepsilon^2 \psi(z)} \left[\frac{z^{-h}\gamma(z)}{\psi(z^{-1})} \right]_+ . \tag{19.30}$$

Here $z^{-h}\gamma(z)$ stands for the covariance of $y(t+h)$ and $y(t)$. However, in view of the expression for $\gamma(z)$, this can also be written as

$$\beta(z) = \frac{1}{\psi(z)} \left[z^{-h}\psi(z) \right]_+ , \tag{19.31}$$

from which it follows that

$$\rho(z) = \beta(z)\psi(z) = \left[z^{-h}\psi(z) \right]_+ . \tag{19.32}$$

This agrees with the formula under (19.26).

Example 19.1. Consider the ARMA(1,1) process $y(t)$ described by the equation

$$(1 - \phi L)y(t) = (1 - \theta L)\varepsilon(t). \tag{19.33}$$

Then $\psi(z) = (1 - \theta z)(1 - \phi z)^{-1}$; and it follows from equation (19.32) that

$$\rho(z) = \left[z^{-h} \frac{(1-\theta z)}{(1-\phi z)} \right]_+ = \left[z^{-h} \left\{ 1 + \frac{(\phi - \theta)z}{(1-\phi z)} \right\} \right]_+ = \phi^{h-1} \frac{(\phi - \theta)}{(1-\phi z)}, \tag{19.34}$$

Therefore, the filter which is applied to the sequence $y(t)$ for the purpose of generating the forecasts is

$$\beta(z) = \frac{\rho(z)}{\psi(z)} = \phi^{h-1} \frac{(\phi - \theta)}{(1 - \theta z)}. \tag{19.35}$$

It may be instructive to follow a more direct derivation of the one-step-ahead forecast. Consider taking expectations in the equation

$$y(t+1) = \phi y(t) + \varepsilon(t+1) - \theta\varepsilon(t) \tag{19.36}$$

conditional upon the information available at time t. Since $E(\varepsilon_{t+h}|\mathcal{I}_t) = 0$ for all $h > 0$, this gives

$$\begin{aligned}\hat{y}(t+1|t) &= \phi y(t) - \theta\varepsilon(t) \\ &= \phi y(t) - \theta\frac{(1-\phi L)}{(1-\theta L)}y(t) \\ &= \frac{(\phi-\theta)}{(1-\theta L)}y(t).\end{aligned} \tag{19.37}$$

Also, the equation

$$\hat{y}(t+h|t) = \phi\hat{y}(t+h-1|t) \tag{19.38}$$

holds for all $h > 1$, from which it follows that

$$\hat{y}(t+h|t) = \phi^{h-1}\frac{(\phi-\theta)}{(1-\theta L)}y(t). \tag{19.39}$$

Forecasting with ARIMA Models

A common way of modelling nonstationary time series is to incorporate a number of difference factors $\nabla = I - L$ in the autoregressive operator of an ARMA model. The resulting model is called an autoregressive integrated moving-average (ARIMA) model in reference to the fact that the inverse of the difference operator, which is $\nabla^{-1} = \{I + L + L^2 + \cdots\}$, is a species of integration operator.

An ARIMA(p, d, q) model takes the form of

$$\alpha(L)\nabla^d y(t) = \mu(L)\varepsilon(t), \tag{19.40}$$

where $\nabla^d y(t) = z(t)$, which is the dth difference of the observable nonstationary process $y(t)$, follows a stationary ARMA(p, q) process.

The series $y(t)$ may be forecast by forecasting its dth difference $z(t)$. The forecasts of the difference can then be aggregated—or integrated d times, in other words—so as to generate the forecasts of the level of the series. In order to generate $\hat{y}_{t+1|t}, \ldots, \hat{y}_{t+h|t}$ by integrating the values $\hat{z}_{t+1|t}, \ldots, \hat{z}_{t+h|t}$, one needs d initial values $z_t, \ldots, z_{t-d+1}$.

An alternative procedure for forecasting the values of $y(t)$, which we shall outline shortly, makes use of a recursion based directly on (19.40) which represents the difference-equation form of the model.

Notwithstanding of the presence of the unit roots in the autoregressive operator $\alpha(L)(I - L)^d$, it is possible to represent an ARIMA model in any of the three forms given under (19.20)–(19.22).

In the case of the autoregressive form $\pi(L)y(t) = \varepsilon(t)$, there is

$$\begin{aligned}\pi(z) &= \frac{\alpha(z)}{\mu(z)}(1-z)^d \\ &= \{\pi_0 + \pi_1 z + \pi_2 z + \cdots\},\end{aligned} \tag{19.41}$$

where $\pi_0 = 1$ on the assumption that the leading coefficients of $\alpha(z)$ and $\mu(z)$ are also units. Setting $z = 1$ shows that the sum of the coefficients of $\pi(z)$ is zero, which is to say that $\sum_{j=1}^{\infty} \pi_j = -1$. Also, provided that the moving-average polynomial $\mu(z)$ is invertible, the coefficients form a convergent sequence with $\pi_j \to 0$ as $j \to \infty$. Therefore, provided that the elements in the information set $\{y_t, y_{t-1}, y_{t-2}, \ldots\}$ are bounded, it follows that the value of y_t can be approximated, to any degree of accuracy, by summing a sufficient number of the leading terms of the autoregressive expansion

$$y_t = \varepsilon_t - \{\pi_1 y_{t-1} + \pi_2 y_{t-2} + \cdots\}. \tag{19.42}$$

The moving-average-representation of the ARIMA model is more problematic. Given that the roots of $(1-z)^d$ are units, it follows that the series expansion of the moving-average operator

$$\begin{aligned}\psi(z) &= \frac{\mu(z)}{(1-z)^d\alpha(z)} \\ &= \{\psi_0 + \psi_1 z + \psi_2 z^2 + \cdots\},\end{aligned} \tag{19.43}$$

does not converge when $z = 1$; which is to say that the sum of the moving-average coefficients is infinite. The consequence is that the value of y_t cannot be obtained directly by summing the leading terms of the expression

$$y_t = \{\psi_0\varepsilon_t + \psi_1\varepsilon_{t-1} + \psi_2\varepsilon_{t-2} + \cdots\}. \tag{19.44}$$

This appears to place a serious impediment in the way of extending a theory of the prediction of stationary series which is based upon the moving-average representation. Nevertheless, the crucial results of that theory do retain their validity in the case of ARIMA processes; and the difficulties to which we have alluded can be largely circumvented by adopting the autoregressive representation of the process or by adopting the difference equation representation as we shall do in the next section.

The difficulties can be circumvented completely by espousing a more sophisticated theory of prediction, such the theory which results from the application of the Kalman filter to finite data sequences.

Example 19.2. Consider the integrated moving-average IMA(1, 1) process which is described by the equation

$$(1-L)y(t) = (1-\theta L)\varepsilon(t). \tag{19.45}$$

This comes from setting $\phi = 1$ in the equation of the ARMA(1, 1) process which is found under (19.33).

By following the example of the ARMA(1, 1) process, it can be seen that the one-step-ahead forecast function can be represented by

$$\begin{aligned}\hat{y}(t+1|t) &= \frac{(1-\theta)}{(1-\theta L)}y(t) \\ &= (1-\theta)\{y(t) + \theta y(t-1) + \theta^2 y(t-1) + \cdots\}.\end{aligned} \tag{19.46}$$

This is the autoregressive formulation of the forecast function, and it corresponds to the method of forecasting known as exponential smoothing. Notice that the sum of the coefficients is unity. Since $\theta^j \to 0$ as $j \to \infty$, it is possible to truncate the sum in the confidence that, if the past values of the sequence fall within sufficiently narrow bounds, then the discarded terms will have insignificant values.

The moving-average form of the forecast function is

$$\begin{aligned}\hat{y}(t+1|t) &= \frac{(1-\theta)}{(1-L)}\varepsilon(t)\\ &= (1-\theta)\big\{\varepsilon(t)+\varepsilon(t-1)+\varepsilon(t-2)+\cdots\big\}.\end{aligned} \tag{19.47}$$

The boundedness of $\hat{y}(t+1|t)$ entails the boundedness of the sum of the white-noise processes on the RHS. However, this indefinite sum cannot be approximated by any partial sum.

Generating the ARMA Forecasts Recursively

The optimal (minimum-mean-square error) forecast of y_{t+h} is the conditional expectation of y_{t+h} given the values of $\{\varepsilon_t, \varepsilon_{t-1}, \varepsilon_{t-2}, \ldots\}$ or the values of $\{y_t, y_{t-1}, y_{t-2}, \ldots\}$ which constitute the information set $\mathcal{I}_t$ equally when the parameters of the process are known. This forecast is denoted by $\hat{y}_{t+h|t} = E(y_{t+h}|\mathcal{I}_t)$. Elements, such as y_{t-j} and ε_{t-j}, which lie within the information set are unaffected by the operator which generates the conditional expectations. Also, the information of $\mathcal{I}_t$ is of no assistance in predicting an element of the disturbance sequence which lies in the future, such as ε_{t+k}. The effects of the expectations operator can be summarised as follows:

$$\begin{aligned} E(y_{t+k}|\mathcal{I}_t) &= \hat{y}_{t+k|t} \quad \text{if} \quad k>0,\\ E(y_{t-j}|\mathcal{I}_t) &= y_{t-j} \quad \text{if} \quad j\geq 0,\\ E(\varepsilon_{t+k}|\mathcal{I}_t) &= 0 \quad \text{if} \quad k>0,\\ E(\varepsilon_{t-j}|\mathcal{I}_t) &= \varepsilon_{t-j} \quad \text{if} \quad j\geq 0. \end{aligned} \tag{19.48}$$

In this notation, the forecast h periods ahead is

$$\begin{aligned} E(y_{t+h}|\mathcal{I}_t) &= \sum_{k=1}^{h}\psi_{h-k}E(\varepsilon_{t+k}|\mathcal{I}_t)+\sum_{j=0}^{\infty}\psi_{h+j}E(\varepsilon_{t-j}|\mathcal{I}_t)\\ &= \sum_{j=0}^{\infty}\psi_{h+j}\varepsilon_{t-j}. \end{aligned} \tag{19.49}$$

This is equation (19.26) again.

In practice, we may generate the forecasts using a recursion based on the difference-equation form of the ARMA model.

$$\begin{aligned} y(t) = &-\big\{\alpha_1 y(t-1)+\alpha_2 y(t-2)+\cdots+\alpha_p y(t-p)\big\}\\ &+\mu_0\varepsilon(t)+\mu_1\varepsilon(t-1)+\cdots+\mu_q\varepsilon(t-q). \end{aligned} \tag{19.50}$$

This equation can also represent an ARIMA model if the difference factors are absorbed into the autoregressive operator. By taking the conditional expectation of the function, we get

$$(19.51)\quad \begin{aligned}\hat{y}_{t+h} = &-\{\alpha_1\hat{y}_{t+h-1}+\cdots+\alpha_p y_{t+h-p}\}\\ &+\mu_h\varepsilon_t+\cdots+\mu_q\varepsilon_{t+h-q} \quad \text{when} \quad 0<h\leq p,q,\end{aligned}$$

$$(19.52)\quad \hat{y}_{t+h} = -\{\alpha_1\hat{y}_{t+h-1}+\cdots+\alpha_p y_{t+h-p}\} \quad \text{if} \quad q<h\leq p,$$

$$(19.53)\quad \begin{aligned}\hat{y}_{t+h} = &-\{\alpha_1\hat{y}_{t+h-1}+\cdots+\alpha_p \hat{y}_{t+h-p}\}\\ &+\mu_h\varepsilon_t+\cdots+\mu_q\varepsilon_{t+h-q} \quad \text{if} \quad p<h\leq q,\end{aligned}$$

and

$$(19.54)\quad \hat{y}_{t+h} = -\{\alpha_1\hat{y}_{t+h-1}+\cdots+\alpha_p \hat{y}_{t+h-p}\} \quad \text{when} \quad p,q<h.$$

Equation (19.54) indicates that when $h>p,q$, the forecasting function becomes a pth-order homogeneous difference equation in y. The p values of $y(t)$ from $t=r=\max(p,q)$ to $t=r-p+1$ serve as the starting values for the equation.

The behaviour of the forecast function beyond the reach of the starting values can be characterised in terms of the roots of the autoregressive operator. We can assume that none of the roots of $\alpha(L)=0$ lie inside the unit circle. If all of the roots are greater than unity in modulus, which is the case of a stationary process, then $\hat{y}_{t+h}$ will converge to zero as h increases. If one of the roots of $\alpha(L)=0$ is unity, then we have an ARIMA$(p,1,q)$ model; and the general solution of the homogeneous equation of (19.54) will include a constant term which represents the product of the unit root with a coefficient which is determined by the starting values. Hence the forecast will tend to a nonzero constant. If two of the roots are unity, then the general solution will embody a linear time trend which is the asymptote to which the forecasts will tend. In general, if d of the roots are unity, then the general solution will comprise a polynomial in t of order $d-1$.

The forecasts can be updated easily once the coefficients in the expansion of $\psi(L)=\mu(L)/\alpha(L)$ have been obtained. Consider

$$(19.55)\quad \begin{aligned}\hat{y}_{t+h|t+1} &= \{\psi_{h-1}\varepsilon_{t+1}+\psi_h\varepsilon_t+\psi_{h+1}\varepsilon_{t-1}+\cdots\} \quad \text{and}\\ \hat{y}_{t+h|t} &= \{\psi_h\varepsilon_t+\psi_{h+1}\varepsilon_{t-1}+\psi_{h+2}\varepsilon_{t-2}+\cdots\}.\end{aligned}$$

The first of these is the forecast for $h-1$ periods ahead made at time $t+1$ whilst the second is the forecast for h periods ahead made at time t. It can be seen that

$$(19.56)\quad \hat{y}_{t+h|t+1} = \hat{y}_{t+h|t}+\psi_{h-1}\varepsilon_{t+1},$$

where $\varepsilon_{t+1}=\hat{y}_{t+1|t}-y_{t+1}$ is the current disturbance at time $t+1$. The latter is also the prediction error of the one-step-ahead forecast made at time t.

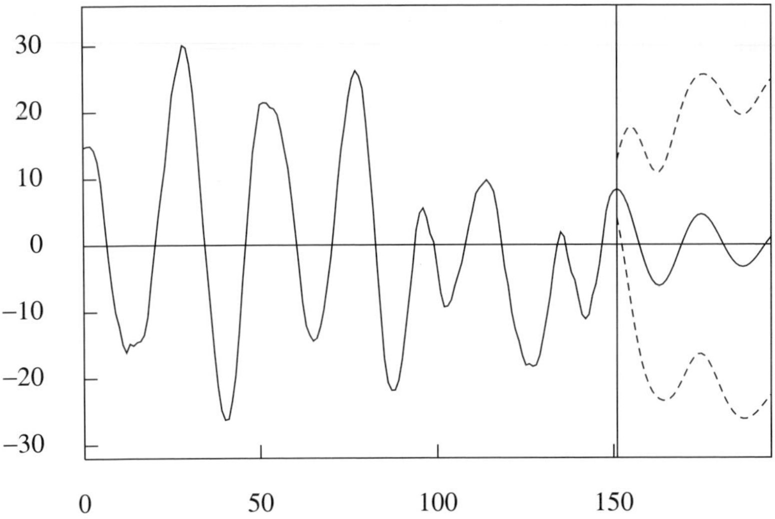

Figure 19.1. The graph of 150 observations on a simulated series generated by the AR(2) process $(1-1.884L+0.951L^2)y(t) = \varepsilon(t)$ followed by 45 forecast values. The broken lines represent a confidence interval.

Physical Analogies for the Forecast Function

To understand the nature of the forecasts, it is helpful to consider some physical analogies. One such analogy was provided by Yule [539], whose article of 1927 introduced the concept of a second-order autoregressive process and demonstrated that it is capable of generating a quasi-cyclical output. The equation of such a process can be written

$$y(t) = \{\rho\cos\omega\}y(t-1) - \rho^2 y(t-2) + \varepsilon(t), \tag{19.57}$$

where ω represents the natural frequency and $\rho \in [0, 1)$ represents the damping factor.

Yule likened the trajectory of the process to that of a pendulum bombarded by peas. The impacts of the peas were compared with the disturbances of the white-noise forcing function $\varepsilon(t)$ which drives the process. If these impacts were to cease, then the pendulum would swing with a regular oscillation of an amplitude which would be reduced gradually by frictional forces. The trajectory, from the moment that the bombardments cease, corresponds to the forecasts of the autoregressive process (see Figure 19.1).

Example 19.3. For an example of the analytic form of the forecast function, we may consider the integrated autoregressive IAR(1, 1) process defined by

$$\{1-(1+\phi)L+\phi L^2\}y(t) = \varepsilon(t), \tag{19.58}$$

wherein $\phi \in [0, 1)$. The roots of the auxiliary equation $z^2 - (1+\phi)z + \phi = 0$ are

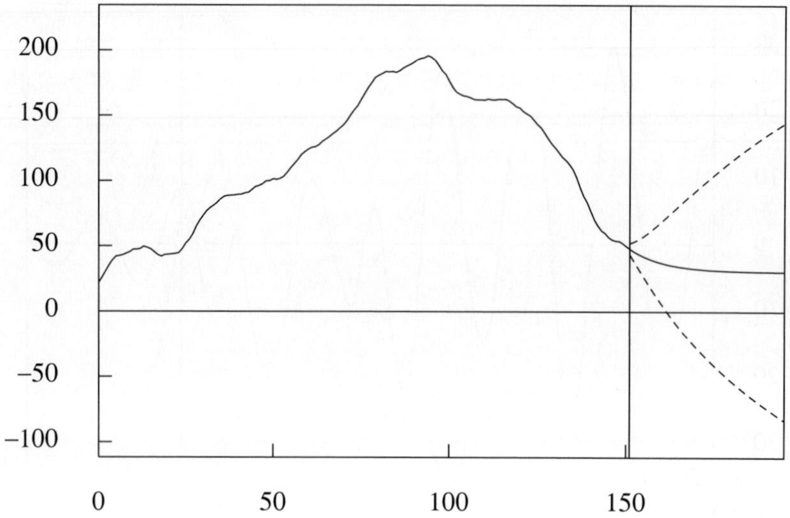

Figure 19.2. The graph of 150 observations on a simulated series generated by the IAR(1, 1) process $(1 - 1.9L + 0.9L^2)y(t) = \varepsilon(t)$ followed by 45 forecast values. The broken lines represent a confidence interval.

$z = 1$ and $z = \phi$. The solution of the homogeneous difference equation

$$\{1 - (1 + \phi)L + \phi L^2\}\hat{y}(t + h|t) = 0, \tag{19.59}$$

which defines the forecast function, is

$$\hat{y}(t + h|t) = c_1 + c_2\phi^h, \tag{19.60}$$

where c_1 and c_2 are constants which reflect the initial conditions. These constants are found by solving the equations

$$\begin{aligned} y_{t-1} &= c_1 + c_2\phi^{-1}, \\ y_t &= c_1 + c_2. \end{aligned} \tag{19.61}$$

The solutions are

$$c_1 = \frac{y_t - \phi y_{t-1}}{1 - \phi} \quad \text{and} \quad c_2 = \frac{\phi}{\phi - 1}(y_t - y_{t-1}). \tag{19.62}$$

The long-term forecast is $\bar{y} = c_1$ which is the asymptote to which the forecasts tend as the lead period h increases.

The model of this example has a straightforward physical analogy. One can imagine a particle moving in a viscous medium under the impacts of its molecules which are in constant motion. The particle is subject to a so-called Brownian motion. Its velocity $v(t)$ is governed by the equation $(1 - \phi L)v(t) = \varepsilon(t)$, where $\phi \in [0, 1)$ is a factor which reflects the viscosity of the medium and which governs the decay of the particle's velocity. The equation $(1-L)y(t)=v(t)$, which derives from equation (19.58), gives the position of the particle. The forecast function reflects the fact that, if the impacts which drive the particle through the medium were to cease, then it would come to rest at a point. Figure 19.2 represents these circumstances.

Interpolation and Signal Extraction

Now let us turn to the problem of interpolation. Imagine a sequence of observations $\{y_t\}$ on a stationary random signal $\{\xi_t\}$ which are afflicted by errors $\{\eta_t\}$ of zero mean which are independently and identically distributed and whose values are also independent of those of the signal. Then

$$y(t)=\xi(t)+\eta(t). \tag{19.63}$$

If the sequence $\xi(t)$ underlying the observations is serially correlated, then there will be some scope for deriving better estimates of its values than those which are provided by the sequence $y(t)$ of the observations. In that case, an estimate of $\xi(t)$ may be obtained by filtering $y(t)$ to give

$$x(t)=\beta(L)y(t). \tag{19.64}$$

For the sake of mathematical tractability, we may begin by assuming that, at any time t, an indefinite sequence of the values of $y(t)$ is available which stretches backwards and forwards in time. The use of data that lies ahead of t implies that the estimation of ξ_t cannot be conducted in real time and that it must be performed off-line.

A good example is provided by the processing of digital sound recordings where the object is to enhance the sound quality by removing various noise corruptions in the form of scratchings and hissings which overlie the music and which can be regarded in the same light as errors of observation. In such cases, the digital recording can be regarded as a sequence of indefinite length.

The coefficients of the filter $\beta(L)=\sum_j\beta_jL^j$ are estimated by invoking the minimum-mean-square-error criterion. The errors in question are the elements of the sequence $e(t)=\xi(t)-x(t)$. The principle of orthogonality, by which the criterion is fulfilled, indicates that the errors must be uncorrelated with the elements in the information set $\mathcal{I}_t=\{y_{t-k};k=0,\pm1,\pm2,\ldots\}$. Thus

$$\begin{aligned}0&=E\Big\{y_{t-k}(\xi_t-x_t)\Big\}\\&=E(y_{t-k}\xi_t)-\sum_j\beta_jE(y_{t-k}y_{t-j})\\&=\gamma_k^{y\xi}-\sum_j\beta_j\gamma_{k-j}^{yy}\end{aligned} \tag{19.65}$$

for all k. The equation may be expressed, in terms of the z transform, as

$$\gamma^{y\xi}(z)=\gamma^{yy}(z)\beta(z), \tag{19.66}$$

where $\beta(z)$ stands for an indefinite two-sided Laurent sequence comprising both positive and negative powers of z.

From the assumption that the elements of the noise sequence $\eta(t)$ are independent of those of the signal $\xi(t)$, it follows that

$$\text{(19.67)}\qquad \gamma^{yy}(z) = \gamma^{\xi\xi}(z) + \gamma^{\eta\eta}(z) \qquad \text{and} \qquad \gamma^{y\xi}(z) = \gamma^{\xi\xi}(z),$$

whence

$$\text{(19.68)}\qquad \beta(z) = \frac{\gamma^{y\xi}(z)}{\gamma^{yy}(z)} = \frac{\gamma^{\xi\xi}(z)}{\gamma^{\xi\xi}(z) + \gamma^{\eta\eta}(z)}.$$

Now, by setting $z = e^{i\omega}$, one can derive the frequency-response function of the filter which is used in estimating the signal $\xi(t)$. The effect of the filter is to multiply each of the frequency components of $y(t)$ by the fraction of its variance which is attributable to the signal. The same principle applies to the estimation of the noise component. The noise-estimation filter is just the complementary filter

$$\text{(19.69)}\qquad \rho(z) = 1 - \beta(z) = \frac{\gamma^{\eta\eta}(z)}{\gamma^{\xi\xi}(z) + \gamma^{\eta\eta}(z)}.$$

To provide some results of a more specific nature, let us assume that the signal $\xi(t)$ is generated by an autoregressive moving-average process such that $\phi(L)\xi(t) = \theta(L)\nu(t)$, where $\nu(t)$ is a white-noise sequence with $V\{\nu(t)\} = \sigma_\nu^2$. Also, let the variance of the white-noise error process be denoted by $V\{\eta(t)\} = \sigma_\eta^2$. Then

$$\text{(19.70)}\qquad y(t) = \frac{\theta(L)}{\phi(L)}\nu(t) + \eta(t),$$

whence

$$\text{(19.71)}\quad \gamma^{\xi\xi}(z) = \sigma_\nu^2 \frac{\theta(z)\theta(z^{-1})}{\phi(z)\phi(z^{-1})} \qquad \text{and} \qquad \gamma^{yy}(z) = \sigma_\nu^2 \frac{\theta(z)\theta(z^{-1})}{\phi(z)\phi(z^{-1})} + \sigma_\eta^2.$$

It follows from (19.68) that

$$\text{(19.72)}\qquad \beta(z) = \frac{\sigma_\nu^2\theta(z)\theta(z^{-1})}{\sigma_\nu^2\theta(z)\theta(z^{-1}) + \sigma_\eta^2\phi(z)\phi(z^{-1})} = \frac{\sigma_\nu^2}{\sigma_\varepsilon^2}\frac{\theta(z)\theta(z^{-1})}{\mu(z)\mu(z^{-1})}.$$

Here the denominator corresponds to the autocovariance generating function of a synthetic moving-average process

$$\text{(19.73)}\qquad \mu(L)\varepsilon(t) = \theta(L)\nu(t) + \phi(L)\eta(t).$$

The autocovariances of this process are obtained by adding the autocovariances of the constituent processes on the RHS. The coefficients of the Cramér–Wold factorisation of the generating function $\sigma_\varepsilon^2\mu(z)\mu(z^{-1})$ may be obtained via the procedure *Minit* of (17.39). The factors of the numerator of $\beta(z)$ are already known.

In order to realise the bidirectional filter $\beta(L)$, it is necessary to factorise it into two parts. The first part, which incorporates positive powers of the lag operator, runs forwards in time in the usual fashion. The second part of the filter, which incorporates negative powers of the lag operator, runs in reversed time.

Given this factorisation, the sequence $x(t)$, which estimates $\xi(t)$, can be found via two operations which are represented by

$$z(t) = \frac{\theta(L)}{\mu(L)}y(t) \quad \text{and} \quad x(t) = \frac{\theta(F)}{\mu(F)}z(t), \tag{19.74}$$

where $F = L^{-1}$ stands for the forward-shift operator whose effect is described by the equation $Fz(t) = z(t+1)$. The reversed-time filtering which converts $z(t)$ into $x(t)$ is analogous to the smoothing operation which is associated with the Kalman filter. Taken together, the two filtering operations will have no net phase effect.

It is notable that the filter which is appropriate for estimating the signal component $\xi(t) = \{\theta(L)/\phi(L)\}\nu(t)$ of the sequence $y(t)$ defined in (19.70) is equally appropriate for estimating the signal $\xi(t) = \theta(L)\nu(t)$ within

$$y(t) = \theta(L)\nu(t) + \phi(L)\eta(t). \tag{19.75}$$

Example 19.4. Consider the process specified by

$$\begin{aligned} y(t) &= \xi(t) + \eta(t) \\ &= (1+L)^n\nu(t) + (1-L)^n\varepsilon(t). \end{aligned} \tag{19.76}$$

Then the filter which estimates $\xi(t)$ takes the form of

$$\begin{aligned} \beta(z) &= \frac{\sigma_\nu^2(1+z)^n(1+z^{-1})^n}{\sigma_\nu^2(1+z)^n(1+z^{-1})^n + \sigma_\varepsilon^2(1-z)^n(1-z^{-1})^n} \\ &= \frac{1}{1+\lambda\left(i\dfrac{1-z}{1+z}\right)^{2n}}, \end{aligned} \tag{19.77}$$

where $\lambda = \sigma_\varepsilon^2/\sigma_\nu^2$. Reference to (16.113) will help to show that, when $z = e^{i\omega}$, this can be written as

$$\beta(e^{i\omega}) = \frac{1}{1-\lambda\{\tan(\omega/2)\}^{2n}}. \tag{19.78}$$

Here is the basis of the formula for the Butterworth lowpass digital filter which is given by (16.114). The latter has $\lambda = \{1/\tan(\omega_c)\}^{2n}$, where ω_c is the nominal cut-off point of the filter.

Extracting the Trend from a Nonstationary Sequence

The results of the previous section are not significantly altered in the case where the signal sequence $\xi(t)$ is generated by a process which owes its nonstationarity to the presence of unit roots in the autoregressive operator. In that case, $\xi(t)$ is liable to be described as the trend underlying the process $y(t) = \xi(t) + \eta(t)$.

Let us consider a leading example where

$$y(t) = \xi(t) + \eta(t) \tag{19.79}$$
$$= \frac{1}{(1-L)^2}\nu(t) + \eta(t).$$

This equation represents a second-order random walk which is obscured by errors of observation. The equation can be written alternatively as

$$(1-L)^2 y(t) = d(t) = \nu(t) + (1-L)^2\eta(t), \tag{19.80}$$

which can be construed as an instance of equation (19.75). Notice, in particular, that the elements in this expression are stationary processes. The filter which is designed to extract $\nu(t)$ from $d(t) = (1-L)^2 y(t)$ is

$$\beta(z) = \frac{\sigma_\nu^2}{\sigma_\eta^2(1-z)^2(1-z^{-1})^2 + \sigma_\nu^2} \tag{19.81}$$
$$= \frac{1}{\lambda(1-z)^2(1-z^{-1})^2 + 1},$$

where $\lambda = \sigma_\eta^2/\sigma_\nu^2$. This filter will also serve to extract $\xi(t)$ from $y(t)$.

Equation (19.81) defines the so-called Hodrick–Prescott filter (see Cogley and Nason [116] and Harvey and Jaeger [248]). The parameter λ^{-1} corresponds to the signal-to-noise ratio, which is the ratio of the variance of the white-noise process $\nu(t)$ which drives the random walk and the variance of the error process $\eta(t)$ which obscures its observations.

It is usual to describe $\lambda = \sigma_\eta^2/\sigma_\nu^2$ as the smoothing parameter. For any particular combination of the signal and noise variances, there is a unique value of the smoothing parameter which gives rise to the optimum (minimum-mean-square-error) trend-estimation filter. However, the filter can be used without any reference to the nature of the processes which generate the data. In such cases, the value of λ must be determined from heuristic considerations.

The Hodrick–Prescott filter is closely related to the Reinsch smoothing spline of which the algorithm has been presented in Chapter 11. The spline represents the optimal predictor of the trajectory of an integrated Wiener process which is obscured by white-noise errors of observation.

An integrated Wiener process is just the continuous-time analogue of a second-order random walk; and, as we shown in an appendix to Chapter 11, a sequence of equally spaced (exact) observations of the process follows a discrete-time integrated moving-average IMA(2, 1) process described by the equation

$$(I-L)^2\xi(t) = (1+\mu L)\nu(t). \tag{19.82}$$

Here $\mu = 2 - \sqrt{3}$ and $V\{\nu(t)\} = \sigma_\nu^2$ are solutions of the equations

$$4\kappa = \sigma_\nu^2(1+\mu^2) \qquad \text{and} \qquad \kappa = \sigma_\nu^2\mu, \tag{19.83}$$

wherein κ is a scale parameter which is affected only by σ_ν^2. These results have been presented previously under (11.116). The equations represent the autocovariances of the moving-average process on the RHS of equation (19.82). The formula of the signal-extraction filter for estimating $\xi(t)$ from $y(t) = \xi(t) + \eta(t)$ is

$$(19.84) \qquad \beta(z) = \frac{\sigma_\nu^2(1+\mu z)(1+\mu z^{-1})}{\sigma_\eta^2(1-z)^2(1-z^{-1})^2 + \sigma_\nu^2(1+\mu z)(1+\mu z^{-1})}.$$

Finite-Sample Predictions: Hilbert Space Terminology

The remaining sections of this chapter are devoted to the refinements which result from our taking proper account of the finite nature of the sample data from which the predictions are derived.

Our attention will be devoted to two closely related algorithms for generating optimal predictions. The first of these is the Durbin–Levinson algorithm which has been expounded already in Chapter 17. We shall re-examine the algorithm and we shall offer an alternative derivation. The second is the Gram–Schmidt prediction-error algorithm which is based upon the Cholesky factorisation of a positive-definite matrix described in Chapter 7.

The two algorithms are recursive in nature. That is to say, they generate sequences of predictions which keep step with the growing sample. As the size of the sample increases, the recursively generated finite-sample predictions should converge upon those generated by applying the practical versions of the infinite-sample methods which have been described in the previous sections. The simplest way of making such methods practical, in the case of a stationary process, is to replace the unobserved presample and postsample data values by zeros. Thus, in general, the methods of the ensuing sections can be seen as generalisations of the existing methods.

The exposition of the finite-sample methods is assisted by using some of the concepts and the terminology of infinite-dimensional Hilbert spaces. In this way, the intuitions which are associated with ordinary finite-dimensional Euclidean spaces can be used to elucidate relationships which might otherwise remain obscure.

We shall use the Hilbert-space concepts only to provide alternative proofs of propositions already established by other means; and we shall give the proofs a subsidiary status by confining them to examples. Brief accounts of Hilbert spaces are provided by Anderson [16] and Caines [95]. An exposition of the theory of linear regression in terms of finite-dimensional vector spaces has been provided by Pollock [397].

The essential idea is that the random variables $y_t; t \in \{0, \pm 1, \pm 2, \ldots\}$, which are elements of the sequence $y(t)$ defined over the set of positive and negative integers, may be construed as points in an infinite-dimensional Hilbert space. The dimension of the space corresponds not to the temporal dimension of the sequence but rather to the range of the individual random variables which are its elements. Finite linear combinations of the elements in the form of $\sum \phi_j y_{t-j}$ are also points in the space. The space is completed by including all random variables which are limits in the mean-square norm—as defined below—of the random variables or points already in the space.

For any two elements x, y in the space, an inner product is defined by

$$(19.85) \qquad \langle x, y\rangle = E(xy).$$

The norm or length of an element x is given by

$$\|x\| = \sqrt{\langle x, x\rangle}, \tag{19.86}$$

whilst the distance between two random variables x and y, conceived of as points in the space, is

$$\|x - y\| = \sqrt{\langle x - y, x - y\rangle}. \tag{19.87}$$

Since $E\{(y - \hat{y})^2\} = \|y - \hat{y}\|^2$, finding the minimum-mean-square-error linear predictor of y given the variables $x_1, \ldots, x_n$ which generate a subspace or manifold $\mathcal{M}$ is a matter of finding the point $\hat{y}$ in $\mathcal{M}$ which is closest to y. This result is expressed in the following theorem which is analogous to the theorem for a finite-dimensional linear space:

(19.88) *The Projection Theorem.* If $\mathcal{M}$ is a subspace of a Hilbert space $\mathcal{H}$ and y is an element of $\mathcal{H}$, then there exists a unique element $\hat{y} \in \mathcal{M}$ such that $\|y - \hat{y}\| \leq \|y - x\|$ and $\langle y - \hat{y}, x\rangle = 0$ for all $x \in \mathcal{M}$.

This theorem can be extended to the case where $\mathcal{M}$ is a linear manifold spanned by an infinite set of vectors. In particular, it follows that the best linear predictor of y_{t+h} given $\{y_t, y_{t-1}, \ldots\}$, which represents the entire history of the sequence $y(t)$, is given by the orthogonal projection of y_{t+h} on the space spanned by $\{y_t, y_{t-1}, \ldots\}$.

Let $\hat{y}_{t+h}$ be the best predictor based on an infinite history and let $\hat{y}^p_{t+h}$ be the predictor based on the elements $\{y_t, \ldots, y_{t-p}\}$. Then it may be shown that

$$\lim(p \to \infty) E\left\{(\hat{y}^p_{t+h} - \hat{y}_{t+h})^2\right\} = 0. \tag{19.89}$$

This implies that a prediction based on a finite history approximates the prediction based on an infinite history.

Recursive Prediction: The Durbin–Levinson Algorithm

Imagine that $y(t)$ is a stationary process with known autocovariances and with a zero expectation. The object is to form a sequence $\hat{y}_1, \hat{y}_2, \ldots, \hat{y}_{t+1}$ of one-step-ahead predictions on the basis of a growing sample of observations $y_0, y_1, \ldots, y_t$.

At the start, before any sample information is available, the predicted value of y_0 is its unconditional expectation $\hat{y}_0 = E(y_0) = 0$. The error of the prediction is the realised value $e_0 = y_0$. The minimum-mean-square-error predictions of the succeeding values are the conditional expectations $\hat{y}_1 = E(y_1|y_0), \hat{y}_2 = E(y_2|y_1, y_0), \ldots, \hat{y}_{t+1} = E(y_{t+1}|y_t, \ldots, y_1, y_0)$. If the process is normal, then these predictions are provided by a set of linear regression equations which take the form of

$$\begin{aligned} \hat{y}_1 &= -\alpha_{1(1)} y_0, \\ \hat{y}_2 &= -\alpha_{1(2)} y_1 - \alpha_{2(2)} y_0, \\ &\vdots \\ \hat{y}_{t+1} &= -\alpha_{1(t+1)} y_t - \cdots - \alpha_{2(t+1)} y_1 - \alpha_{t+1(t+1)} y_0. \end{aligned} \tag{19.90}$$

Even if the assumption of normality cannot be sustained, the same equations will continue, nevertheless, to represent the minimum-mean-square-error linear predictions.

The generic equation from (19.90) can be written as

$$\hat{y}_t = -\sum_{j=1}^{t} \alpha_{j(t)} y_{t-j}. \tag{19.91}$$

Defining $e_t = y_t - \hat{y}_t$ and setting $\alpha_{0(t)} = 1$ for all t gives rise to the following equation for the prediction error:

$$e_t = \sum_{j=0}^{t} \alpha_{j(t)} y_{t-j}; \quad \alpha_{0(t)} = 1. \tag{19.92}$$

The set of the first $t+1$ of such equations may be ordered in the following manner:

$$\begin{bmatrix} 1 & 0 & 0 & \dots & 0 \\ \alpha_{1(1)} & 1 & 0 & \dots & 0 \\ \alpha_{2(2)} & \alpha_{1(2)} & 1 & \dots & 0 \\ \vdots & \vdots & \vdots & \ddots & \vdots \\ \alpha_{t(t)} & \alpha_{t-1(t)} & \alpha_{t-2(t)} & \dots & 1 \end{bmatrix} \begin{bmatrix} y_0 \\ y_1 \\ y_2 \\ \vdots \\ y_t \end{bmatrix} = \begin{bmatrix} e_0 \\ e_1 \\ e_2 \\ \vdots \\ e_t \end{bmatrix}. \tag{19.93}$$

As the sample size t increases, the coefficients $\alpha_{t-j(t)}; j = 1, \dots, t$, together with the values $\nu^2_{(t)} = E(e^2_t)$ for the variance of the prediction errors, may be obtained by solving a succession of Yule–Walker systems of increasing order.

According to the projection theorem of (19.88), the sequence of prediction errors $\{e_0, e_1, \dots, e_t\}$ obey the conditions $E(e_t y_{t-j}) = 0; j = 1, \dots, t$. It follows from equation (19.92) that they also obey the conditions $E(e_t e_{t-j}) = 0; j = 1, \dots, t$. Therefore, the prediction errors form a sequence of mutually uncorrelated elements with a diagonal dispersion matrix. Thus the sequence of regressions entailed by the minimum-mean-square-error predictions are associated with the following diagonalisation of the dispersion matrix of the sample elements $\{y_0, y_1, \dots, y_t\}$:

$$\begin{aligned} &\begin{bmatrix} 1 & 0 & \dots & 0 \\ \alpha_{1(1)} & 1 & \dots & 0 \\ \vdots & \vdots & \ddots & \vdots \\ \alpha_{t(t)} & \alpha_{t-1(t)} & \dots & 1 \end{bmatrix} \begin{bmatrix} \gamma_0 & \gamma_1 & \dots & \gamma_t \\ \gamma_1 & \gamma_0 & \dots & \gamma_{t-1} \\ \vdots & \vdots & \ddots & \vdots \\ \gamma_t & \gamma_{t-1} & \dots & \gamma_0 \end{bmatrix} \begin{bmatrix} 1 & \alpha_{1(1)} & \dots & \alpha_{t(t)} \\ 0 & 1 & \dots & \alpha_{t-1(t)} \\ \vdots & \vdots & \ddots & \vdots \\ 0 & 0 & \dots & 1 \end{bmatrix} \\ &\qquad = \begin{bmatrix} \gamma_0 & 0 & \dots & 0 \\ 0 & \sigma^2_{(1)} & \dots & 0 \\ \vdots & \vdots & \ddots & \vdots \\ 0 & 0 & \dots & \sigma^2_{(t)} \end{bmatrix}. \end{aligned} \tag{19.94}$$

This process of diagonalisation is also associated with a Cholesky decomposition of the *inverse* of the dispersion matrix. Thus, if the equation of (19.94) is written is summary notation as $A\Gamma A' = D$, then it follows that

$$\Gamma = A^{-1}DA'^{-1} \quad \text{and} \quad \Gamma^{-1} = A'D^{-1}A. \tag{19.95}$$

The increasing burden of computation, imposed by the task of generating successive predictions, may be reduced by exploiting the previous calculations. A recursive scheme for calculating the coefficients is provided by the Durbin–Levinson algorithm which was presented in Chapter 16.

The algorithm, which is also displayed under (17.73), enables the coefficients of the $(t+1)$th iteration to be derived for those of the tth iteration:

$$\alpha_{t+1(t+1)} = \frac{-1}{\nu^2_{(t)}}\left\{\sum_{j=0}^{t}\alpha_{j(t)}\gamma_{t+1-j}\right\},$$

(19.96)
$$\begin{bmatrix}\alpha_{1(t+1)}\\ \vdots \\ \alpha_{t(t+1)}\end{bmatrix} = \begin{bmatrix}\alpha_{1(t)}\\ \vdots \\ \alpha_{t(t)}\end{bmatrix} + \alpha_{t+1(t+1)}\begin{bmatrix}\alpha_{t(t)}\\ \vdots \\ \alpha_{1(t)}\end{bmatrix},$$

$$\nu^2_{(t+1)} = \nu^2_{(t)}\left\{1-(\alpha_{t+1(t+1)})^2\right\}.$$

The starting values for the recursion are

(19.97)
$$\alpha_{1(1)} = -\gamma_1/\gamma_0 \quad\text{and}\quad \nu^2_{(1)} = \gamma_0\left\{1-(\alpha_{1(1)})^2\right\}.$$

The parameter $\alpha_{t+1(t+1)}$, which is the crucial element in the $(t+1)$th stage of the algorithm, is commonly described, in texts of signal processing, as a reflection coefficient. In a statistical context, it is described as a partial autocorrelation coefficient.

Example 19.5. The equations of the Durbin–Levinson algorithm can be derived in a way which makes use of the concepts of Hilbert space and which also alludes to the theory of linear regression. The coefficients of the best linear predictor of y_t given the previous p observations are obtained by projecting the random variable y_t onto the manifold $\mathcal{M}_p = \mathcal{M}\{y_{t-1}, \ldots, y_{t-p}\}$ spanned by the lagged values $y_{t-1}, \ldots, y_{t-p}$. Let the projection operator with respect to this manifold be denoted by P_p. Then, for the predicted value of y_t, we have

(19.98)
$$P_p y_t = -\sum_{j=1}^{p}\alpha_{j(p)}y_{t-j},$$

whilst the variance of the prediction error is

(19.99)
$$\nu^2_{(p)} = \|(I-P_p)y_t\|^2.$$

Now let the manifold be augmented by the addition of the lagged variable y_{t-p-1}. Then the augmented manifold $\mathcal{M}_{p+1}$ can be decomposed as the direct sum the original manifold $\mathcal{M}_p$ and a one-dimensional subspace which is obtained by projecting the augmenting variable y_{t-p-1} into the orthogonal complement of

$\mathcal{M}_p$. Thus $\mathcal{M}_{p+1} = \mathcal{M}_p \oplus (I - P_p)y_{t-p-1}$; and the appropriate projection for the prediction of y_t based on $p+1$ observations is

$$\begin{aligned} P_{p+1}y_t &= P_p y_t - \alpha_{p+1(p+1)}(I - P_p)y_{t-p-1} \\ &= -\alpha_{p+1(p+1)}y_{t-p-1} + \{P_p y_t + \alpha_{p+1(p+1)}P_p y_{t-p-1}\}, \end{aligned} \tag{19.100}$$

where

$$-\alpha_{p+1(p+1)} = \frac{\langle y_t, (I - P_p)y_{t-p-1}\rangle}{\|(I - P_p)y_{t-p-1}\|^2} \tag{19.101}$$

is the coefficient of the regression of y_t on $(I - P_p)y_{t-p-1}$.

Since the process $y(t)$ is stationary and reversible, the backwards prediction of y_{t-p-1} given $\mathcal{M}_p$ is formed using the same coefficients as the forward prediction of y_t provided by (19.98). The difference is that the coefficients which are associated with the variables are in reversed order. Thus

$$P_p y_{t-p-1} = -\sum_{j=1}^{p} \alpha_{p+1-j(p)} y_{t-j}. \tag{19.102}$$

Also, the denominator of (19.101) is

$$\|(I - P_p)y_{t-p-1}\|^2 = \|(I - P_p)y_t\|^2 = \nu^2_{(p)}. \tag{19.103}$$

On carrying these results of (19.102) and (19.103) into (19.101), we find that

$$\begin{aligned} \alpha_{p+1(p+1)} &= \frac{-1}{\nu^2_{(p)}}\left\{\langle y_t, y_{t-p-1}\rangle + \sum_{j=1}^{p} \alpha_{p+1-j(p)}\langle y_t, y_{t-j}\rangle\right\} \\ &= \frac{-1}{\nu^2_{(p)}}\left\{\sum_{j=0}^{p} \alpha_{j(p)}\gamma_{p+1-j}\right\}, \end{aligned} \tag{19.104}$$

since $\langle y_t, y_{t-j}\rangle = \gamma_j$ and $\alpha_{j(p)} = 1$. Also, using (19.98) and (19.102) in equation (19.100), it can be seen that the remaining coefficients of the projection on $\mathcal{M}_{p+1}$, which are associated with $y_t, \ldots, y_{t-p}$, are just

$$\alpha_{j(p+1)} = \alpha_{j(p)} + \alpha_{p+1(p+1)}\alpha_{p+1-j(p)}; \quad j = 1, \ldots, p. \tag{19.105}$$

To find a recursive formula for the variance of the prediction error, we refer to equation (19.100). Subtracting that equation from y_t and forming the modulus of the remainder gives

$$\begin{aligned} \big\|(I - P_{p+1})y_t\big\|^2 &= \big\|(I - P_p)y_t - \alpha_{p+1(p+1)}(I - P_p)y_{t-p-1}\big\|^2 \\ &= \big\|(I - P_p)y_t\big\|^2 + \alpha^2_{p+1(p+1)}\big\|(I - P_p)y_{t-p-1}\big\|^2 \\ &\quad -2\alpha_{p+1(p+1)}\big\langle (I - P_p)y_t, (I - P_p)y_{t-p-1}\big\rangle. \end{aligned} \tag{19.106}$$

On the RHS, there is

$$(19.107)\qquad \begin{aligned}\big\langle (I-P_p)y_t, (I-P_p)y_{t-p-1}\big\rangle &= \big\langle y_t, (I-P_p)y_{t-p-1}\big\rangle \\ &= -\alpha_{p+1(p+1)}\big\|(I-P_p)y_{t-p-1}\big\|^2.\end{aligned}$$

Here, the first equality follows from the symmetry and idempotency of the operator $I - P_p$, and the second equality comes directly from (19.101). On putting this back into (19.106) and and using (19.103), we get

$$(19.108)\qquad \big\|(I-P_{p+1})y_t\big\|^2 = \nu^2_{(p+1)} = \nu^2_{(p)} - \alpha^2_{p+1(p+1)}\nu^2_{(p)},$$

which is the result which we have been seeking. Equations (19.104), (19.105) and (19.108), which recapitulate the equations of (19.96), define the Durbin–Levinson algorithm.

The Durbin–Levinson algorithm provides a way of finding the minimum-mean-square-error predictions regardless of the nature of the stationary stochastic process to which the autocovariances correspond. There is no presumption that the process is autoregressive. However, if the process is described by an AR(p) equation of the form $(1 + \alpha_1 L + \cdots + \alpha_p L^p)y(t) = \varepsilon(t)$, then, for $t > p$, it will be found that $\alpha_{1(t)} = \alpha_1, \alpha_{2(t)} = \alpha_2, \ldots, \alpha_{p(t)} = \alpha_p$, whilst $\alpha_{p+1(t)} = \cdots = \alpha_{t(t)} = 0$. Also, for $t > p$, the prediction error will coincide with the white-noise disturbance, so that $e_t = \varepsilon_t$.

A facility for reducing a serially dependent sequence of random variables $y_0, \ldots, y_t$ to a sequence of uncorrelated prediction errors $e_0, \ldots, e_t$ greatly facilitates the computation of the exact likelihood function of the sample of data. We shall exploit this in a subsequent chapter which deals with the estimation of linear stochastic models.

Example 19.6. The transformation of the elements $y_0, \ldots, y_t$ into a sequence of independently distributed random variables $e_0, \ldots, e_t$ may be construed as a process of orthogonalisation. A set of random variables $y_0, \ldots, y_t$ with expected values of $E(y_j) = 0$ for all j and with covariances of $C(y_i, y_j) = \gamma_{ij}$ constitutes a basis of a vector space $\mathcal{S}$ of dimension t on the condition that their dispersion matrix $\Gamma = [\gamma_{ij}]$ is positive definite. The inner product of two elements $x, y \in \mathcal{S}$ may be defined as $\langle x, y\rangle = C(x, y)$ which is just their covariance, and the condition of orthogonality is the condition that $\langle x, y\rangle = C(x, y) = 0$.

The process of orthogonalisation begins by setting $e_0 = y_0$. Then e_1 is defined to be the component of y_1 which is orthogonal to y_0, and which is, therefore, uncorrelated with $e_0 = y_0$. This component is just the residual from the regression of y_0 on y_1. The next orthogonal element e_2 is the component of y_2 which is orthogonal to y_0 and y_1; and this is just the residual from the regression of y_2 on y_0 and y_1. Since e_2 is orthogonal to y_0 and y_1 and since $e_1 = y_0 + \alpha_1 y_1$ is a linear combination of these variables, it follows that e_2 is uncorrelated with e_1. The process of orthogonalisation continues in this fashion; and it is, of course, nothing but the process of solving a succession of Yule–Walker equations which we have detailed above.

A Lattice Structure for the Prediction Errors

There is an efficient method of generating the sequence of prediction errors which depends upon the Durbin–Levinson algorithm. Consider the prediction errors associated with a filter of order p which has the coefficients $1, \alpha_{1(p)}, \ldots, \alpha_{p(p)}$. Applying the filter to the the data sequence $y_t, y_{t-1}, \ldots, y_{t-p}$ generates a one-step-ahead prediction error:

$$\begin{aligned} e_{t(p)} &= y_t + \alpha_{1(p)} y_{t-1} + \cdots + \alpha_{p(p)} y_{t-p} \\ &= y_t - \hat{y}_t. \end{aligned} \tag{19.109}$$

Reversing the order of the coefficients and applying them again to the data generates the one-step-back prediction error:

$$\begin{aligned} b_{t(p)} &= \alpha_{p(p)} y_t + \cdots + \alpha_{1(p)} y_{t-p+1} + y_{t-p} \\ &= y_{t-p} - \hat{y}_{t-p}. \end{aligned} \tag{19.110}$$

Now suppose that the order of the filter is increased by one. Then the error of the one-step-ahead prediction becomes

$$\begin{aligned} e_{t(p+1)} &= \sum_{j=0}^{p+1} \alpha_{j(p+1)} y_{t-j} \\ &= y_t + \sum_{j=1}^{p} \alpha_{j(p+1)} y_{t-j} + c_{p+1} y_{t-p-1}, \end{aligned} \tag{19.111}$$

where $c_{p+1} = \alpha_{p+1(p+1)}$ is the reflection coefficient from the Durbin–Levinson algorithm. Equation (19.96) of the algorithm provides the expressions

$$\alpha_{j(p+1)} = \alpha_{j(p)} + c_{p+1}\alpha_{p+1-j(p)}, \quad j = 1, \ldots, p. \tag{19.112}$$

Substituting these into equation (19.111) gives

$$\begin{aligned} e_{t(p+1)} &= \left\{ y_t + \sum_{j=1}^{p} \alpha_{j(p)} y_{t-j} \right\} + c_{p+1} \left\{ \sum_{j=1}^{p} \alpha_{p+1-j(p)} y_{t-j} + y_{t-p-1} \right\} \\ &= e_{t(p)} + c_{p+1} b_{t-1(p)}. \end{aligned} \tag{19.113}$$

An analogous expression may be derived for the backwards prediction. Consider

$$\begin{aligned} b_{t(p+1)} &= \sum_{j=0}^{p+1} \alpha_{p+1-j(p+1)} y_{t-j} \\ &= c_{p+1} y_t + \sum_{j=1}^{p} \alpha_{p+1-j(p+1)} y_{t-j} + y_{t-p-1}. \end{aligned} \tag{19.114}$$

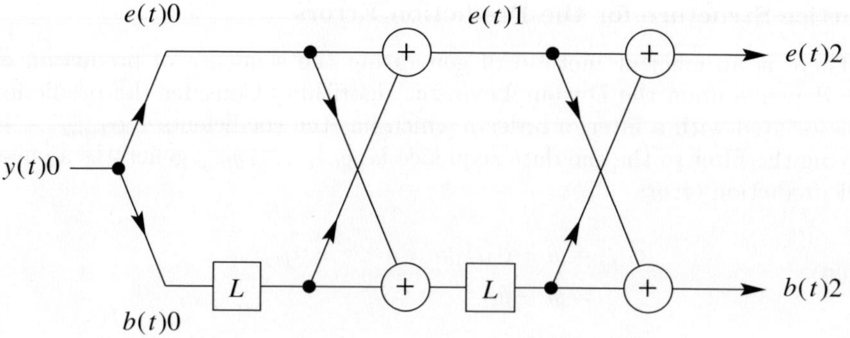

Figure 19.3. The initial segments of the signal diagram of the lattice filter used in generating forwards and backwards prediction errors.

Equation (19.96) provides the expressions

$$\alpha_{p+1-j(p+1)} = \alpha_{p+1-j(p)} + c_{p+1}\alpha_{j(p)}, \quad j = 1, \ldots, p. \tag{19.115}$$

Substituting these in (19.114) gives

$$\begin{aligned} b_{t(p+1)} &= c_{p+1}\left\{y_t + \sum_{j=1}^{p} \alpha_{j(p)} y_{t-j}\right\} + \left\{\sum_{j=1}^{p} \alpha_{p+1-j(p)} y_{t-j} + y_{t-p-1}\right\} \\ &= c_{p+1} e_{t(p)} + b_{t-1(p)}. \end{aligned} \tag{19.116}$$

On juxtaposing the results of (19.113) and (19.116), we find that

$$\begin{aligned} e_{t(p+1)} &= e_{t(p)} + c_{p+1} b_{t-1(p)}, \\ b_{t(p+1)} &= c_{p+1} e_{t(p)} + b_{t-1(p)}. \end{aligned} \tag{19.117}$$

These two equations describe an algorithm for constructing an autoregressive prediction filter in successive stages. Setting $p = 0$ gives the initial conditions

$$e_{t(0)} = b_{t(0)} = y_t. \tag{19.118}$$

Increasing the value of p by one at a time, adds successive stages to the filter, thereby increasing its order. The signal diagram of the filter (Figure 19.3) has a crisscross or lattice appearance in consequence of the cross-coupling of the backwards and forwards prediction errors which is evident in the equations.

It is interesting to discover that expressions for the reflection coefficient $c_{p+1} = \alpha_{p+1(p+1)}$ can be derived from the equations under (19.117). Consider the mean-square error of the forward prediction expressed as

$$\begin{aligned} E\left[\left\{e_{t(p+1)}\right\}^2\right] &= E\left[\left\{e_{t(p)} + c_{p+1} b_{t-1(p)}\right\}^2\right] \\ &= E\left[\left\{e_{t(p)}\right\}^2\right] + 2c_{p+1}E\left[e_{t(p)} b_{t-1(p)}\right] + c_{p+1}^2 E\left[\left\{b_{t-1(p)}\right\}^2\right]. \end{aligned} \tag{19.119}$$

Minimisation with respect to $c_{p+1} = \alpha_{p+1(p+1)}$ yields

$$(19.120) \qquad -\alpha_{p+1(p+1)} = \frac{E\left[e_{t(p)}b_{t-1(p)}\right]}{E\left[\left\{b_{t-1(p)}\right\}^2\right]}.$$

Since the forward and backwards mean-square errors are equal, the above equation can be written in the symmetrical form of

$$(19.121) \qquad -\alpha_{p+1(p+1)} = \frac{2E\left[e_{t(p)}b_{t-1(p)}\right]}{E\left[\left\{e_{t(p)}\right\}^2 + \left\{b_{t-1(p)}\right\}^2\right]}.$$

When the various expected values are replaced in this formula by sample analogues, we obtain Burg's [89] algorithm for estimating the prediction coefficients. This algorithm will be presented in a later chapter.

Recursive Prediction: The Gram–Schmidt Algorithm

Equation (19.93) indicates that the vector of one-step-ahead prediction errors can be expressed as a simple linear transformation of the vector of observations. The matrix of the transformation in this equation is manifestly nonsingular; and, therefore, an inverse relationship can be defined in the form of

$$(19.122) \qquad \begin{bmatrix} y_0 \\ y_1 \\ y_2 \\ \vdots \\ y_t \end{bmatrix} = \begin{bmatrix} 1 & 0 & 0 & \ldots & 0 \\ \mu_{1(1)} & 1 & 0 & \ldots & 0 \\ \mu_{2(2)} & \mu_{1(2)} & 1 & \ldots & 0 \\ \vdots & \vdots & \vdots & \ddots & \vdots \\ \mu_{t(t)} & \mu_{t-1(t)} & \mu_{t-2(t)} & \cdots & 1 \end{bmatrix} \begin{bmatrix} e_0 \\ e_1 \\ e_2 \\ \vdots \\ e_t \end{bmatrix}.$$

Thus the vector of observations can be expressed as a linear transform of the vector of prediction errors. If equation (19.93) is regarded as an autoregressive formulation, then, clearly, equation (19.122) is its moving-average counterpart. Observe that, if the matrix of the equation is denoted by M, if the dispersion matrix of the observations is denoted by Γ and if the diagonal dispersion matrix of the errors is denoted by D, then there is a Cholesky decomposition of Γ in the form of

$$(19.123) \qquad \Gamma = MDM'.$$

The counterpart to the Durbin–Levinson algorithm is an algorithm which generates recursively the values of the coefficients $\mu_{t-j(t)}; j = 0, \ldots, t-1$ for successive values of t. Consider the generic equation of (19.122) which takes the form of

$$(19.124) \qquad y_t = \sum_{j=0}^{t} \mu_{j(t)}e_{t-j}; \quad \mu_{0(t)} = 1.$$

Multiplying both sides of this equation by e_i, with $i \leq t$ and taking expectations, gives

$$E(y_t e_i) = \sum_{j=0}^{t} \mu_{j(t)} E(e_{t-j} e_i) = \mu_{t-i(t)} E(e_i^2), \tag{19.125}$$

since $E(e_{t-j} e_i) = 0$ if $(t-j) \neq i$. Using the notation $\nu^2_{(i)} = E(e_i^2)$, this can be written as

$$\mu_{t-i(t)} = \frac{1}{\nu^2_{(i)}} E(y_t e_i). \tag{19.126}$$

But equation (19.124) indicates that

$$e_i = y_i - \sum_{j=1}^{i} \mu_{j(i)} e_{i-j}, \tag{19.127}$$

so the expected value on the RHS of (19.126) becomes

$$\begin{aligned} E(y_t e_i) &= E(y_t y_i) - \sum_{j=1}^{i} \mu_{j(i)} E(y_t e_{i-j}) \\ &= E(y_t y_i) - \sum_{j=1}^{i} \mu_{j(i)} \mu_{t-i+j(t)} E(e_{i-j}^2), \end{aligned} \tag{19.128}$$

where the result that $E(y_t e_{i-j}) = \mu_{t-i+j(t)} E(e_{i-j}^2)$, which is indicated by (19.125), is used for the final equality. On defining $k = i - j$ and using the notation $\gamma_{t-i} = E(y_t y_i)$ and $\nu^2_{(i-j)} = E(e_{i-j}^2)$, the equation (19.126) can be written as

$$\mu_{t-i(t)} = \frac{1}{\nu^2_{(i)}} \left\{ \gamma_{t-i} - \sum_{k=0}^{i-1} \mu_{i-k(i)} \mu_{t-k(t)} \nu^2_{(k)} \right\}. \tag{19.129}$$

This is closely related to equation (17.33) which specifies the relationship between the parameters and the autocovariances of a moving-average process.

An expression must be found for $\nu^2_{(t)} = E\{(e_t)^2\} = E\{(y_t - \hat{y}_t)^2\}$ which is the variance of the prediction error. This may be derived by considering equation (19.124). Since the RHS of that equation is a sum of statistically independent terms, it follows that

$$\begin{aligned} \gamma_0 &= \sum_{j=0}^{t} \mu^2_{j(t)} \nu^2_{(t-j)} \\ &= \nu^2_{(t)} + \sum_{j=1}^{t} \mu^2_{j(t)} \nu^2_{(t-j)}, \end{aligned} \tag{19.130}$$

where $\gamma_0 = V(y_t)$ and $\nu^2_{(t-j)} = V(e^2_{t-j})$. This gives

$$\nu^2_{(t)} = \gamma_0 - \sum_{j=1}^{t} \mu^2_{j(t)} \nu^2_{(t-j)}. \tag{19.131}$$

Example 19.7. For an example of the recursive generation of the coefficients, consider the case where $t = 4$. Then equation (19.129), with $i = 0, \ldots, 3$, supplies the first four of the following equations:

$$\begin{aligned}
\mu_{4(4)} &= \frac{1}{\nu^2_{(0)}}\gamma_4, \\
\mu_{3(4)} &= \frac{1}{\nu^2_{(1)}}\left[\gamma_3 - \left\{\mu_{1(1)}\mu_{4(4)}\nu^2_{(0)}\right\}\right], \\
\mu_{2(4)} &= \frac{1}{\nu^2_{(2)}}\left[\gamma_2 - \left\{\mu_{2(2)}\mu_{4(4)}\nu^2_{(0)} + \mu_{1(2)}\mu_{3(4)}\nu^2_{(1)}\right\}\right], \\
\mu_{1(4)} &= \frac{1}{\nu^2_{(3)}}\left[\gamma_1 - \left\{\mu_{3(3)}\mu_{4(4)}\nu^2_{(0)} + \mu_{2(3)}\mu_{3(4)}\nu^2_{(1)} + \mu_{1(3)}\mu_{2(4)}\nu^2_{(2)}\right\}\right], \\
\nu^2_{(4)} &= \left[\gamma_0 - \left\{\mu^2_{4(4)}\nu^2_{(0)} + \mu^2_{3(4)}\nu^2_{(1)} + \mu^2_{2(4)}\nu^2_{(2)} + \mu^2_{1(4)}\nu^2_{(3)}\right\}\right].
\end{aligned} \tag{19.132}$$

The fifth equation of the scheme, which generates the variance $E(e_4^2) = \nu^2_{(4)}$ of the prediction error, is derived by setting $t = 4$ in equation (19.131). It can also be derived by setting $t = i = 4$ in (19.129) and rearranging the result in view of the identity $\mu_{0(4)} = 1$.

Example 19.8. The prediction-error algorithm may be construed as an implementation of the classical Gram–Schmidt orthogonalisation procedure. Let the random variables $y_0, \ldots, y_t$ be regarded as vectors which span a subspace of a Hilbert space $\mathcal{H}$. For any two elements $x, y \in \mathcal{H}$, an inner product is defined by $\langle x, y\rangle = E(x, y)$, and the condition that these elements are orthogonal is the condition that $\langle x, y\rangle = E(x, y) = 0$. By applying the Gram–Schmidt procedure to the set of random variables, the following sequence is derived:

$$\begin{aligned}
e_0 &= y_0, \\
e_1 &= y_1 - \frac{\langle y_1, e_0\rangle}{\langle e_0, e_0\rangle}e_0, \\
e_2 &= y_2 - \frac{\langle y_2, e_0\rangle}{\langle e_0, e_0\rangle}e_0 - \frac{\langle y_2, e_1\rangle}{\langle e_1, e_1\rangle}e_1, \\
&\vdots \\
e_t &= y_t - \sum_{i=0}^{t-1}\frac{\langle y_t, e_i\rangle}{\langle e_i, e_i\rangle}e_i.
\end{aligned} \tag{19.133}$$

The final equation can be written as

$$e_t = y_t - \sum_{i=0}^{t-1} \mu_{t-i(t)} e_i, \tag{19.134}$$

where

$$\mu_{t-i(t)} = \frac{\langle y_t, e_i \rangle}{\langle e_i, e_i \rangle}. \tag{19.135}$$

Substituting for $e_i = y_i - \sum_{k=0}^{i-1} \mu_{i-k(i)} e_k$ in the latter expression gives

$$\begin{aligned}
\mu_{t-i(t)} &= \frac{1}{\langle e_i, e_i \rangle}\left\{ \langle y_t, y_i \rangle - \sum_{k=0}^{i-1} \mu_{i-k(i)} \langle y_t, e_k \rangle \right\} \\
&= \frac{1}{\langle e_i, e_i \rangle}\left\{ \langle y_t, y_i \rangle - \sum_{k=0}^{i-1} \mu_{i-k(i)} \mu_{t-k(t)} \langle e_k, e_k \rangle \right\} \\
&= \frac{1}{\nu^2_{(i)}}\left\{ \gamma_{t-i} - \sum_{k=0}^{i-1} \mu_{i-k(i)} \mu_{t-k(t)} \nu^2_{(k)} \right\}.
\end{aligned} \tag{19.136}$$

This is a repetition of equation (19.129).

As in the case of the Durbin–Levinson algorithm, the present scheme for generating the minimum-mean-square-error predictions entails no presumption about the nature of the underlying process which corresponds to the sequence of autocovariances, other than that it is stationary and that it admits a moving-average representation. Nevertheless, it is clear, from a comparison of the equation (19.129) with the equation under (17.33), that the Gram–Schmidt prediction-error algorithm has a special affinity with finite-order moving-average processes. Thus if the underlying process is described by an MA(q) equation in the form of $y(t) = (1 + \mu_1 L + \cdots + \mu_q L^q)\varepsilon(t)$, then we can expect that, as t increases, the leading q coefficients in the sequence $\{\mu_{1(t)}, \mu_{2(t)}, \ldots, \mu_{q(t)}, \ldots\}$ will converge upon the values of $\mu_1, \mu_2, \ldots, \mu_q$ whilst the remaining coefficients will be identically zero.

In the case of an AR(p) process, the prediction equations of the Durbin–Levinson algorithm assume their asymptotic form after p steps when $t \geq p$. By contrast, when the underlying process is MA(q), the prediction equations of the prediction-error algorithm approach their asymptotic form gradually as $t \to \infty$. One can expect the convergence to be rapid nevertheless, so long as the roots of the MA process are reasonably remote from the boundary of the unit circle.

Example 19.9. Consider the MA(1) process defined by the equation $y(t) = \varepsilon(t) - \theta\varepsilon(t-1)$. The autocovariances are

$$\begin{aligned}
\gamma_0 &= \sigma^2_\varepsilon(1 + \theta^2), \\
\gamma_1 &= -\sigma^2_\varepsilon \theta, \\
\gamma_\tau &= 0 \quad \text{for} \quad \tau > 1.
\end{aligned} \tag{19.137}$$

It follows, from (19.129) and (19.131), that

$$\begin{aligned} \theta_{\tau(t)} &= 0 \quad \text{for} \quad \tau > 1, \\ \theta_{1(t)} &= -\frac{\sigma_\varepsilon^2}{\nu_{(t-1)}^2}\theta, \\ \nu_{(0)}^2 &= \sigma_\varepsilon^2(1+\theta^2) \quad \text{and} \\ \nu_{(t)}^2 &= \sigma_\varepsilon^2\left(1+\theta^2 - \frac{\sigma_\varepsilon^2}{\nu_{(t-1)}^2}\theta^2\right). \end{aligned} \tag{19.138}$$

As $t \to \infty$, it will be found that $\nu_{(t)}^2 \to \sigma_\varepsilon^2$.

In the following procedure for calculating the prediction errors, there is a provision from limiting the order q of the filter. This is appropriate to the case where $y(t)$ is indeed a qth-order moving-average process. The procedure takes as its input a sequence $\{y_0, \ldots, y_n\}$ of observations on $y(t)$ and returns, in its place, the corresponding vector of prediction errors.

(19.139)
```
procedure GSPrediction(gamma : vector;
        y : longVector;
        var mu : matrix;
        n, q : integer);

  var
    t : integer;

  procedure MuLine(t : integer);
    var
      i, k : integer;
  begin {t}
    for i := 0 to t do
      begin {i}
      mu[t − i, t] := gamma[t − i];
      for k := 0 to i − 1 do
      mu[t − i, t] := mu[t − i, t]
            −mu[i − k, i] * mu[t − k, t] * mu[0, k];
      if i < t then
      mu[t − i, t] := mu[t − i, t]/mu[0, i];
      end; {i}
  end; {MuLine}

  procedure Shiftmu;
    var
      t, j : integer;
  begin
    for t := 0 to q − 1 do
```

```
      for j := 0 to t do
        mu[j, t] := mu[j, t + 1];
  end; {Shiftmu}

  procedure PredictionError(q : integer);
    var
      k : integer;
  begin
    for k := 1 to q do
      y[t] := y[t] - y[t - k] * mu[q - k, q];
  end; {PredictionError}

begin {Main Procedure}

  for t := 0 to q do
    begin {t}
      MuLine(t);
      PredictionError(t);
    end; {t}

  for t := q + 1 to n do
    begin {t}
      Shiftmu;
      MuLine(q);
      PredictionError(q);
    end; {Main Procedure}

end; {GSPrediction}
```

It is interesting to compare the code of the sub-procedure *MuLine*, which implements equation (19.129), with that of the procedure *LDLprimeDecomposition* of (7.47) which factorises a symmetric matrix $A = LDL'$ in terms of a lower-triangular matrix L and a diagonal matrix $D = \text{diag}\{d_1, \ldots, d_n\}$. The inner loop of this procedure, which finds the ith row of L and the ith diagonal element of D, accomplishes the same task as *MuLine*; and a careful comparison, which takes account of different ways in which the variables are indexed in the two procedures, will reveal a one-to-one correspondence between the two sets of code.

The *GSPrediction* algorithm is sparing in its use of storage. It retains only as much of the matrix of moving-average coefficients as is needed for generating subsequent rows and for calculating the current prediction error.

The sub-procedure *Shiftmu* is responsible for moving the coefficients through the store, which is a square matrix of order $q+1$. It is activated as soon as $q+1$ rows of coefficients have been generated. Thereafter, at the start of each new iteration of the main procedure, it removes the oldest row from the store and it displaces each of the remaining coefficients. In terms of the matrix which is displayed in (19.122), the coefficients are moved one step to the left and one step upwards.

Signal Extraction from a Finite Sample

The task of adapting the Wiener–Kolmogorov theory of signal extraction to the circumstance of a limited sample often causes difficulties and perplexity. The problems arise from not knowing how to supply the initial conditions with which to start a recursive filtering process. By choosing inappropriate starting values for the forwards or the backwards pass, one can generate a so-called transient effect which is liable, in fact, to affect all of the processed values.

Of course, when the values of interest are remote from either end of a long sample, one can trust that they will be barely affected by the start-up conditions. However, in many applications, such as in the processing of economic data, the sample is short and the interest is concentrated at the upper end where the most recent observations are to be found.

One approach to the problem of the start-up conditions relies upon the ability to extend the sample by forecasting and backcasting. The additional extra-sample values can be used in a run-up to the filtering process wherein the filter is stabilised by providing it with a plausible history, if it is working in the direction of time, or with a plausible future, if it is working in reversed time. Sometimes, very lengthy extrapolations are called for (see Burman [92] for example).

The approach which we shall adopt in this chapter is to avoid the start-up problem altogether by deriving specialised finite-sample versions of the filters on the basis of the statistical theory of conditional expectations. This is the appropriate approach when the data is well matched by the statistical model which gives rise to the filter.

However, it must be recognised that filters are usually selected not for their conformity with a model of the processes generating the data, but, instead, with a view to their frequency-response characteristics. Therefore, there is no guarantee that our approach will be valid in general. We are pursuing it here because it has heuristic advantages and because it lends itself to a simpler exposition than do any of the alternatives such as, for example, the sophisticated methodology proposed by Chornoboy [105] or the methodology of the diffuse Kalman filter developed by de Jong [148].

Signal Extraction from a Finite Sample: the Stationary Case

Consider the case of a signal sequence $\xi(t)$ which is described by an ordinary ARMA model, and imagine that the observations $y(t)$ are contaminated by a white-noise error $\eta(t)$ which is assumed to be statistically independent of $\xi(t)$. Then we should have

$$y(t) = \xi(t) + \eta(t) \tag{19.140}$$

with

$$\alpha(L)\xi(t) = \mu(L)\nu(t), \tag{19.141}$$

where $\nu(t)$ is a white-noise sequence. A set of T observations running from $t = 0$ to $t = T - 1$ can be gathered in a vector

$$y = \xi + \eta. \tag{19.142}$$

Here, ξ is a vector which is assumed to have a normal distribution with

$$E(\xi) = 0 \quad \text{and} \quad D(\xi) = \sigma_\nu^2 Q, \tag{19.143}$$

whilst η is a normal vector with

$$E(\eta) = 0 \quad \text{and} \quad D(\eta) = \sigma_\eta^2 I. \tag{19.144}$$

The theory of conditional expectations indicates that an optimal estimate of the signal would be provided by

$$E(\xi|y) = E(\xi) + C(\xi, y)D^{-1}(y)\{y - E(y)\}. \tag{19.145}$$

From the assumptions under (19.143) and (19.144), and from the assumption that η and ξ are independent, it follows that

$$D(y) = \sigma_\nu^2 Q + \sigma_\eta^2 I \quad \text{and} \quad C(\xi, y) = \sigma_\nu^2 Q. \tag{19.146}$$

Therefore,

$$E(\xi|y) = x = \sigma_\nu^2 Q(\sigma_\nu^2 Q + \sigma_\eta^2 I)^{-1} y. \tag{19.147}$$

This equation must be solved in a way which does not make excessive demands on the memory of the computer—for the $T \times T$ matrix $\sigma_\nu^2 Q + \sigma_\eta^2 I$ may be very large, and it is clear that we cannot afford to find the inverse by a direct method which pays no attention to the structure of the matrix.

The practical methods of solving the equation (19.147) depend upon finding a factorisation of the matrix Q where the factors, or their inverses, are band-limited matrices. In Chapter 22, we provide a factorisation of the dispersion matrix of an ARMA(p, q) process which is in the form of

$$\sigma_\nu^2 Q = \sigma_\nu^2 A^{-1} V A'^{-1}, \tag{19.148}$$

where A is a banded lower-triangular Toeplitz matrix which has the autoregressive parameters $\alpha_0, \ldots, \alpha_p$ as its coefficients, and where V is a symmetric matrix with zeros above the rth supradiagonal band, where $r = \max(p, q)$, and with zeros below the rth subdiagonal band.

Consider rewriting equation (19.147) as

$$\begin{aligned} x &= A^{-1} V A'^{-1}(A^{-1} V A'^{-1} + \lambda I)^{-1} y \\ &= A^{-1} V (V + \lambda A A')^{-1} A y, \end{aligned} \tag{19.149}$$

where $\lambda = \sigma_\eta^2 / \sigma_\nu^2$. This can be recast as

$$h = (V + \lambda A A')g, \quad \text{where} \quad Vg = Ax \quad \text{and} \quad h = Ay. \tag{19.150}$$

The matrix $V + \lambda AA'$ is symmetric with zeros above the rth supradiagonal band and below the rth subdiagonal band. It is amenable to a Cholesky decomposition in the form of $V + \lambda AA' = LDL'$ where L is a lower-triangular matrix and D is a diagonal matrix. The system $LDL'g = h$ can be cast in the form of $Lk = h$ and solved for k. Then $L'g = D^{-1}k$ can be solved for g and, finally, x can be recovered from the equation $Vg = Ax$.

The equations for k, g and x entail triangular matrices with a limited number of subdiagonal or supradiagonal bands; and their solution involves only simple recursions. The vector x is our estimate of the trend.

Signal Extraction from a Finite Sample: the Nonstationary Case

Imagine that the observable sequence is described by

$$\begin{aligned} y(t) &= \xi(t) + \eta(t) \\ &= \frac{\mu(L)}{(1-L)^2}\nu(t) + \theta(L)\varepsilon(t), \end{aligned} \tag{19.151}$$

where $\nu(t)$ and $\varepsilon(t)$ are statistically independent sequences generated by normal white-noise processes. Here $\xi(t)$ follows an integrated moving-average process which can be thought of as a trend which underlies the data. The data, and the underlying trend, can be reduced to stationarity by taking second differences. Applying the difference operator to equation (19.151) gives

$$\begin{aligned} (1-L)^2 y(t) &= \mu(L)\nu(t) + (1-L)^2\theta(L)\varepsilon(t) \\ &= \zeta(t) + \kappa(t), \end{aligned} \tag{19.152}$$

where $\zeta(t) = (1-L)^2\xi(t) = \mu(L)\nu(t)$ and $\kappa(t) = (1-L)^2\eta(t) = (1-L)^2\theta(L)\varepsilon(t)$ both follow moving-average processes.

A set of T observations running from $t = 0$ to $t = T-1$ are represented, as before, by the equation (19.142). To find the finite-sample counterpart of equation (19.152), we need to represent the second-order difference operator $(1-L)^2$ in the form of a matrix. The matrix which finds the differences $d_2, \ldots, d_{T-1}$ of the data points $y_0, y_1, y_2, \ldots, y_{T-1}$ is in the form of

$$Q' = \begin{bmatrix} 1 & -2 & 1 & 0 & \ldots & 0 & 0 \\ 0 & 1 & -2 & 1 & \ldots & 0 & 0 \\ \vdots & \vdots & \vdots & \vdots & \ddots & \vdots & \vdots \\ 0 & 0 & 0 & 0 & \ldots & 1 & 0 \\ 0 & 0 & 0 & 0 & \ldots & -2 & 1 \end{bmatrix}. \tag{19.153}$$

Premultiplying equation (19.142) by this matrix gives

$$\begin{aligned} d = Q'y &= Q'\xi + Q'\eta \\ &= \zeta + \kappa, \end{aligned} \tag{19.154}$$

where $\zeta = Q'\xi$ and $\kappa = Q'\eta$. The first and second moments of the vector ζ may be denoted by

$$E(\zeta) = 0 \quad \text{and} \quad D(\zeta) = \sigma_\nu^2 M, \tag{19.155}$$

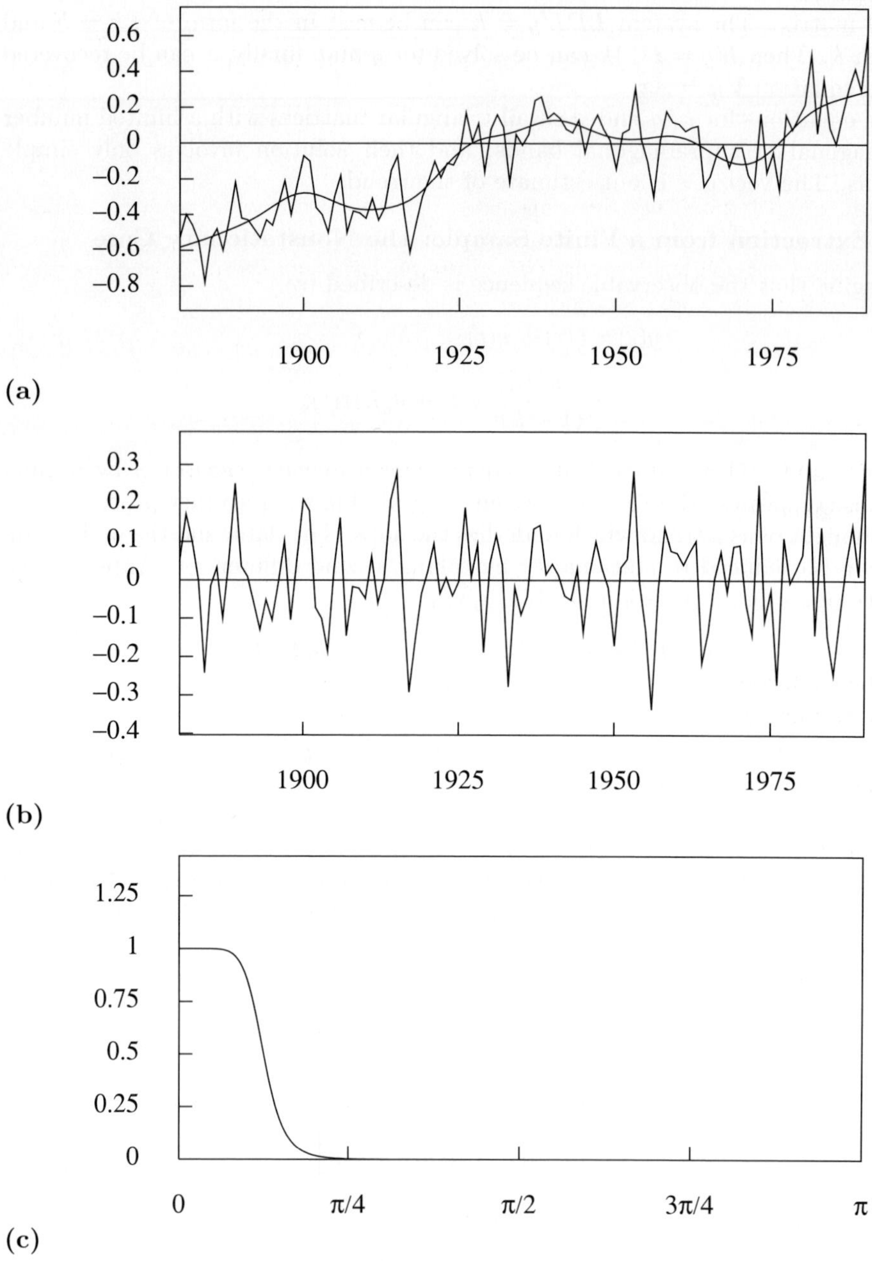

Figure 19.4. Northern hemisphere annual temperature anomalies 1880–1990: **(a)** the series with an interpolated trend, **(b)** the deviations from the trend and **(c)** the gain of the bidirectional fourth-order Butterworth lowpass filter, with a nominal cut-off frequency of $\pi/8$ radians, which is used in constructing the trend.

and those of κ by

$$E(\kappa)=0 \quad \text{and} \quad D(\kappa)=Q'D(\eta)Q, \qquad (19.156)$$
$$=\sigma_\varepsilon^2 Q'\Sigma Q,$$

where both M and Σ are symmetric Toeplitz matrices with a limited number of nonzero diagonal bands. The generating functions for the coefficients of these matrices are, respectively, $\mu(z)\mu(z^{-1})$ and $\theta(z)\theta(z^{-1})$.

The optimal predictor z of the twice-differenced signal vector $\zeta = Q'\xi$ is given by the following conditional expectation:

$$\begin{aligned} E(\zeta|d) &= E(\zeta)+C(\zeta,d)D^{-1}(d)\{d-E(d)\} \\ &= M(M+\lambda Q'\Sigma Q)^{-1}d = z, \end{aligned} \qquad (19.157)$$

where $\lambda = \sigma_\varepsilon^2/\sigma_\nu^2$. The optimal predictor k of the twice-differenced noise vector $\kappa = Q'\eta$ is given, likewise, by

$$\begin{aligned} E(\kappa|d) &= E(\kappa)+C(\kappa,d)D^{-1}(d)\{d-E(d)\} \\ &= \lambda Q'\Sigma Q(M+\lambda Q'\Sigma Q)^{-1}d = k. \end{aligned} \qquad (19.158)$$

It may be confirmed that $z+k=d$.

The estimates are calculated, first, by solving the equation

$$(M+\lambda Q'\Sigma Q)g = d \qquad (19.159)$$

for the value of g and, thereafter, by finding

$$z = Mg \quad \text{and} \quad k = \lambda Q'\Sigma Q g. \qquad (19.160)$$

The solution of equation (19.159) is found via a Cholesky factorisation which sets $M+\lambda Q'\Sigma Q = LDL'$, where L is a lower-triangular matrix and D is a diagonal matrix. The system $LDL'g = d$ may be cast in the form of $Lh = d$ and solved for h. Then $L'g = D^{-1}h$ can be solved for g.

Our object is to recover from z an estimate x of the trend vector ξ. This would be conceived, ordinarily, as a matter of integrating the vector z twice via a simple recursion which depends upon two initial conditions. The difficulty is in discovering the appropriate initial conditions with which to begin the recursion.

We can circumvent the problem of the initial conditions by seeking the solution to the following problem:

$$\text{Minimise} \quad (y-x)'\Sigma^{-1}(y-x) \quad \text{subject to} \quad Q'x = z. \qquad (19.161)$$

The problem is addressed by evaluating the Lagrangean function

$$L(x,\mu) = (y-x)'\Sigma^{-1}(y-x) + 2\mu'(Q'x-z). \qquad (19.162)$$

By differentiating the function with respect to x and setting the result to zero, we obtain the condition

$$\Sigma^{-1}(y-x) - Q\mu = 0. \tag{19.163}$$

Premultiplying by $Q'\Sigma$ gives

$$Q'(y-x) = Q'\Sigma Q\mu. \tag{19.164}$$

But, from (19.159) and (19.160), it follows that

$$\begin{aligned} Q'(y-x) &= d - z \\ &= \lambda Q'\Sigma Q g, \end{aligned} \tag{19.165}$$

whence we get

$$\begin{aligned} \mu &= (Q'\Sigma Q)^{-1}Q'(y-x) \\ &= \lambda g. \end{aligned} \tag{19.166}$$

Putting the final expression for μ into (19.163) gives

$$x = y - \lambda\Sigma Q g. \tag{19.167}$$

This is our solution to the problem of estimating the trend vector ξ. Notice that there is no need to find the value of z explicitly, since the value of x can be expressed more directly in terms of $g = M^{-1}z$.

It is notable that there is a criterion function which will enable us to derive the equation of the trend estimation filter in a single step. The function is

$$L(x) = (y-x)'\Sigma^{-1}(y-x) + \lambda x'QM^{-1}Q'x, \tag{19.168}$$

wherein $\lambda = \sigma_\varepsilon^2/\sigma_\nu^2$ as before.

After minimising this function with respect to x, we may use the identity $Q'x = z$ which comes from equation (19.165), and the identity $M^{-1}z = g$ which comes from equation (19.160). Then it will be found that criterion function is minimised by the value specified in (19.167).

The criterion becomes intelligible when we allude to the assumptions that $y \sim N(\xi, \sigma_\varepsilon^2\Sigma)$ and that $Q'\xi = \zeta \sim N(0, \sigma_\nu^2 M)$; for then it plainly resembles a combination of two independent chi-square variates.

An example of the use of the signal-extraction technique in estimating a trend is provided by Figure 19.4. This illustrates the effects of a bidirectional fourth-order Butterworth filter. Such a filter would be the appropriate device for estimating the sequence $\xi(t)$ in equation (19.151) in the case where $\mu(L) = (1+L)^4$ and $\theta(L) = (1-L)^2$.

Example 19.10. Consider the case of an integrated moving-average IMA(2, 1) process of which the observations are affected by independently and identically distributed errors. The sequence of the observations can be represented by the equation

$$(19.169)\qquad \begin{aligned} y(t) &= \xi(t) + \eta(t) \\ &= \frac{1+\mu L}{(1-L)^2}\nu(t) + \eta(t), \end{aligned}$$

and, when it is multiplied throughout by the operator $(1-L)^2$, this becomes

$$(19.170)\qquad \begin{aligned} (1-L)^2 y(t) &= (1+\mu L)\nu(t) + (1-L)^2\eta(t), \\ &= \zeta(t) + \kappa(t) = d(t). \end{aligned}$$

A set of T observations on $y(t)$ are represented, as before, by the equations $y = \xi+\eta$, and their second differences are represented by $Q'y = d = \zeta+\kappa$. Now the dispersion matrices are

$$(19.171)\qquad D(\zeta) = \sigma_\nu^2 M \qquad \text{and} \qquad D(\kappa) = \sigma_\eta^2 Q'Q.$$

A comparison of the latter with equation (19.156) shows that the present assumptions have resulted in the specialisation $\Sigma = I$. It follows that equation (19.157) now takes the form of

$$(19.172)\qquad z = M(M + \lambda Q'Q)^{-1}d.$$

This to be compared with an equation in the form of $z(t) = \beta(L)d(t)$ which represents the application of the Wiener–Kolmogorov filter $\beta(L)$ of (19.84) to the differenced sequence $d(t)$. The polynomial $(1+\mu z)(1+\mu z^{-1})$ of the numerator of equation (19.84) is the generating function of the coefficients of the matrix M, whilst the polynomial $(1-z)^2(1-z^{-1})^2$ from the denominator is the generating function of the coefficients of the matrix $Q'Q$.

It was asserted, in the context of equation (19.84), that the IMA(2, 1) process provides a model for a sequence of observations on an integrated Wiener process which are afflicted by white-noise errors. Subject to an appropriate choice of the smoothing parameter, the Reinsch smoothing spline of Chapter 11 provides an optimal predictor of the underlying Wiener process.

Now we are in a position to compare the equations of the spline with those of the signal-extraction algorithm. Consider, therefore, the specialisations of equations (19.159) and (19.163) which arise when $\Sigma = I$. These are, respectively, an equation in the form of

$$(19.173)\qquad (M + \lambda Q'Q)g = Q'y,$$

which is solved for g via a Cholesky factorisation, and an equation in the form of

$$(19.174)\qquad x = y - \lambda Qg,$$

which generates the estimate of the trend. After some minor changes of notation, these equations become identical, respectively, to the spline equations (11.80) and

(11.81), when $\Sigma = I$. We might also note that, subject to these specialisations, the concentrated criterion function of (19.168) is identical to the function under (11.77) which is entailed in the derivation of the smoothing spline.

Bibliography

[16] Anderson, T.W., (1971), *The Statistical Analysis of Time Series*, John Wiley and Sons, Chichester.

[45] Bell, W., (1984), Signal Extraction for Nonstationary Time Series, *The Annals of Statistics*, **12**, 646–664.

[66] Bloomfield, P., (1972), On the Error of Prediction of a Time Series, *Biometrika*, **59**, 501–507.

[89] Burg, J.P., (1967), Maximum Entropy Spectral Analysis, *Proceedings of the 37th Meeting of the Society of Exploration Geophysicists, Oklahoma City.* Reprinted in *Modern Spectral Analysis*, D.G. Childers (ed.), (1978), IEEE Press, New York.

[92] Burman, J.P., (1980), Seasonal Adjustment by Signal Extraction, *Journal of the Royal Statistical Society, Series A*, **143**, 321–337.

[93] Burridge, P., and K.F. Wallis, (1988), Prediction Theory for Autoregressive Moving Average Processes, *Econometric Reviews*, **7**, 65–95.

[95] Caines, P.E., (1988), *Linear Stochastic Systems*, John Wiley and Sons, New York.

[105] Chornoboy, E.S., (1992), Initialisation for Improved IIR Filter Performance, *IEEE Transactions on Signal Processing*, **40**, 543–550.

[112] Cleveland, W.P., and G.C. Tiao, (1976), Decomposition of Seasonal Time Series: A Model for the Census X-11 Program, *Journal of the American Statistical Association*, **71**, 581–587.

[116] Cogley, T., and J.M. Nason, (1995), Effects of the Hodrick–Prescott Filter on Trend and Difference Stationary Time Series, Implications for Business Cycle Research, *Journal of Economic Dynamics and Control*, **19**, 253–278.

[148] De Jong, P., (1991), The Diffuse Kalman Filter, *The Annals of Statistics*, **19**, 1073–1083.

[158] Diebold, F.X., (1986), The Exact Initial Covariance Matrix of the State Vector of a General MA(q) Process, *Economic Letters*, **22**, 27–31.

[191] Frances, P.H., (1991), Seasonality, Non-Stationarity and Forecasting of Monthly Time Series, *International Journal of Forecasting*, **7**, 199–208.

[248] Harvey, A.C., and A. Jaeger, (1993), Detrending, Stylised Facts and the Business Cycle, *Journal of Applied Econometrics*, **8**, 231–247.

[291] King, R.G., and S.G. Rebelo, (1993), Low Frequency Filtering and Real Business Cycles, *Journal of Economic Dynamics and Control*, **17**, 207–231.

[298] Kolmogorov, A.N., (1941), Interpolation and Extrapolation, *Bulletin de l'academie des sciences de U.S.S.R., Ser. Math.*, **5**, 3–14.

[305] Kydland, F.E., and C. Prescott, (1990), Business Cycles: Real Facts and a Monetary Myth, *Federal Reserve Bank of Minneapolis Quarterly Review*, **14**, 3–18.

[330] Maravall, A., and D.A. Pierce, (1987), A Prototypical Seasonal Adjustment Model, *Journal of Time Series Analysis*, **8**, 177–193.

[348] Mittnik, S., (1987), The Determination of the State Covariance Matrix of Moving-Average Processes without Computation, *Economic Letters*, **23**, 177–179.

[394] Pierce, D.A., (1979), Signal Extraction in Nonstationary Time Series, *The Annals of Statistics*, **6**, 1303–1320.

[397] Pollock, D.S.G., (1979), *The Algebra of Econometrics*, John Wiley and Sons, Chichester.

[519] Whittle, P., (1983), *Prediction and Regulation by Linear Least-Square Methods, Second Revised Edition*, Basil Blackwell, Oxford.

[523] Wiener, N., (1950), *Extrapolation, Interpolation and Smoothing of Stationary Time Series*, MIT Technology Press and John Wiley and Sons, New York.

[539] Yule, G.U., (1927), On a Method of Investigating the Periodicities in Disturbed Series, with Special Reference to Wolfer's Sunspot Numbers, *Philosophical Transactions of the Royal Society of London, Series A*, **226**, 267–298.

Time-Series Estimation

CHAPTER 20

Estimation of the Mean and the Autocovariances

A stationary stochastic process $x(t)$ can be characterised, for most practical purposes, by its first-order and second-order moments, which are its mean μ and its autocovariances which form the sequence $\gamma(\tau) = \{\gamma_\tau; \tau = 0, \pm 1, \pm 2, \ldots\}$. If the elements of the process are normally distributed, then these moments describe the statistical properties of the process completely.

The first-order and second-order moments are all that are needed in order to infer the parameters of a linear stochastic process of the ARMA variety. Therefore, a crucial role in the processes of inference is played by the corresponding estimates. Moreover, these estimates can be used in identifying the orders of the ARMA model which is supposed to have generated the data.

We shall begin this chapter by considering the estimation of the mean, for which we shall establish the requisite sampling properties. Then we shall consider the problems of estimating the autocovariance function and the autocorrelation function. The natural estimator of an autocovariance is the corresponding empirical product moment.

It transpires that, in the case of a finite-order moving-average model, the product-moment estimator of the autocovariances is statistically inefficient. This explains why the method of maximum likelihood is used, instead of the more direct method of moments, when efficient estimates are required of the parameters of a linear stochastic model with a moving-average component.

Estimating the Mean of a Stationary Process

Given a sample $x_0, \ldots, x_{T-1}$ of T observations from a stationary process $x(t)$, it is natural to estimate the mean $E(x_t) = \mu$ of the process by the sample mean $\bar{x} = T^{-1}\sum_{t=0}^{T-1} x_t$. This is an unbiased estimator, since its expected value is

$$E(\bar{x}) = \frac{1}{T}\sum_{t=0}^{T-1} E(x_t) = \mu. \tag{20.1}$$

The sample mean can also be expressed as $\bar{x} = (i'i)^{-1}i'x = i'x/T$, where $i' = [1, \ldots, 1]$ is a vector of T units and $x = [x_0, \ldots, x_{T-1}]'$ is the vector of sample elements. In terms of this notation, the variance of $\bar{x}$ is

$$V(\bar{x}) = T^{-2}V(i'x) = T^{-2}i'D(x)i, \tag{20.2}$$

where

$$(20.3) \qquad D(x) = \begin{bmatrix} \gamma_0 & \gamma_1 & \gamma_2 & \cdots & \gamma_{T-1} \\ \gamma_1 & \gamma_0 & \gamma_1 & \cdots & \gamma_{T-1} \\ \gamma_2 & \gamma_1 & \gamma_0 & \cdots & \gamma_{T-3} \\ \vdots & \vdots & \vdots & \ddots & \vdots \\ \gamma_{T-1} & \gamma_{T-2} & \gamma_{T-3} & \cdots & \gamma_0 \end{bmatrix}$$

is the variance–covariance matrix or dispersion matrix of x, of which the generic element is the autocovariance at lag τ defined by

$$(20.4) \qquad \begin{aligned} \gamma_\tau &= E\{(x_t - \mu)(x_{t-\tau} - \mu)\} \\ &= E(x_t x_{t-\tau}) - \mu^2. \end{aligned}$$

The expression $i'D(x)i$ stands for the sum of all the elements of the symmetric Toeplitz matrix $D(x)$; and, by examining the structure of this matrix, it can be seen that

$$(20.5) \qquad \begin{aligned} i'D(x)i &= \sum_{t=0}^{T-1}\sum_{s=0}^{T-1} \gamma_{|t-s|} \\ &= \sum_{\tau=1-T}^{T-1} \gamma_\tau (T - |\tau|). \end{aligned}$$

Thus the sum of the elements of a Toeplitz matrix becomes the so-called Cesàro sum of the sequence of autocovariances. There will be further instances in this chapter of the process by which the double summation over the indices t and s becomes a summation over the single index τ. The expression under (20.5) is the basis for the following expression for the variance of $\bar{x}$:

$$(20.6) \qquad \begin{aligned} V(\bar{x}) &= \frac{1}{T}\sum_{\tau=1-T}^{T-1} \gamma_\tau \left(1 - \frac{|\tau|}{T}\right) \\ &= \frac{2}{T}\sum_{\tau=0}^{T-1} \gamma_\tau \left(1 - \frac{\tau}{T}\right) - \frac{\gamma_0}{T}. \end{aligned}$$

A basic concern is to discover the conditions under which $\bar{x}$ represents a consistent estimator of μ. Since $\bar{x}$ is an unbiased estimator, it will be a consistent estimator as well if its variance tends to zero as the size of the sample increases. In fact, it can be proved that

(20.7) The sample mean $\bar{x}_T$ is a consistent estimator of μ, with $E(\bar{x}_T) = \mu$ for all T and $V(\bar{x}_T) \to 0$ as $T \to \infty$, if and only if the sequence of autocovariances $\{\gamma_\tau\}$ converges to zero in arithmetic mean, which is the condition that

$$\frac{1}{T}\sum_{\tau=0}^{T-1} \gamma_\tau \to 0 \qquad \text{as} \qquad T \to \infty.$$

Proof. Given that $T^{-1}\sum_{\tau=0}^{T-1}\gamma_\tau \geq T^{-1}\sum_{\tau=0}^{T-1}\gamma_\tau(1-\tau/T)$, it follows that the first term of the final expression of (20.6) will vanish if $\lim(T\to\infty)T^{-1}\sum_{\tau=0}^{T-1}\gamma_\tau = 0$. Also, the second term γ_0/T will vanish as $T\to\infty$. Therefore, the condition of (20.7) is sufficient for the consistency of the estimator. To show that it is also a necessary condition, consider writing

(20.8)
$$\begin{aligned}\frac{1}{T}\sum_{\tau=0}^{T-1}\gamma_\tau &= C(x_0, \{x_0 + x_1 + \cdots + x_{T-1}\}/T)\\ &= C(x_0, \bar{x}_T).\end{aligned}$$

The Cauchy–Schwarz inequality indicates that $\{C(x_0,\bar{x})\}^2 \leq V(x_0)V(\bar{x}) = \gamma_0 V(\bar{x})$; and thus it follows that $V(\bar{x}_T)\to 0$ implies $T^{-1}\sum_{\tau=0}^{T-1}\gamma_\tau \to 0$.

The condition for the consistency of the estimator $\bar{x}$ stated in (20.7) is a weak one. A sufficient condition which is stronger is that $\lim(T\to\infty)\gamma_\tau = 0$. Since it is usual to invoke the latter condition, it is useful to record the following result:

(20.9) The sample mean $\bar{x}_T$ is a consistent estimator of the expected value of the process $x(t)$ if the covariance $\gamma_{|t-s|}$ of the elements x_t and x_s tends to zero as their temporal separation $|t-s|$ increases.

Asymptotic Variance of the Sample Mean

The sequence of autocovariances generated by an ARMA process is absolutely summable such that $\sum_\tau |\gamma_\tau| < \infty$. This condition, which is stronger still than the condition that $\lim(\tau\to\infty)\gamma_\tau = 0$, enables an expression to be derived for the limiting form of the variance of $\bar{x}$. For, if $\sum_{\tau=-\infty}^{\infty}|\gamma_\tau| < \infty$, then $\sum_{\tau=-\infty}^{\infty}\gamma_\tau < \infty$; and, therefore,

(20.10)
$$\begin{aligned}\lim_{T\to\infty} TV(\bar{x}_T) &= \lim_{T\to\infty}\sum_{\tau=1-T}^{T-1}\gamma_\tau\left(1-\frac{|\tau|}{T}\right)\\ &= \sum_{\tau=-\infty}^{\infty}\gamma_\tau.\end{aligned}$$

Also, if $\sum_{\tau=-\infty}^{\infty}|\gamma_\tau| < \infty$, then the process $x(t)$ has a spectral density function of the form

(20.11)
$$f(\omega) = \frac{1}{2\pi}\left\{\gamma_0 + 2\sum_{\tau=1}^{\infty}\gamma_\tau\cos(\omega t)\right\};$$

and, therefore, it follows that

(20.12)
$$TV(\bar{x}) \to \sum_{\tau=-\infty}^{\infty}\gamma_\tau = 2\pi f(0).$$

Example 20.1. Consider an AR(1) process $x(t)$ such that $\{x(t) - \mu\} = \phi\{x(t-1) - \mu\} + \varepsilon(t)$, where $\varepsilon(t)$ is a white-noise process with $E(\varepsilon_t) = 0$ and $V(\varepsilon_t) = \sigma_\varepsilon^2$ for all t. We know, from (17.46), that

$$\gamma_\tau = \frac{\sigma_\varepsilon^2 \phi^{|\tau|}}{1-\phi^2} = \gamma_0 \phi^{|\tau|}. \tag{20.13}$$

Therefore,

$$\sum_{\tau-\infty}^{\infty} \gamma_\tau = \gamma_0 \sum_{\tau-\infty}^{\infty} \phi^{|\tau|} = \gamma_0 \frac{(1+\phi)}{(1-\phi)}, \tag{20.14}$$

where the final equality can be deduced from the fact that $\{1 + \phi + \phi^2 + \cdots\} = 1/(1-\phi)$ and $\{\cdots + \phi^3 + \phi^2 + \phi\} = \phi/(1-\phi)$. According to (20.10), there is the approximation

$$V(\bar{x}) \simeq \frac{\gamma_0}{T} \frac{(1+\phi)}{(1-\phi)}. \tag{20.15}$$

It can seen that, as $\phi \to 1$, the variance increases without bound. When $\phi = 0$, the process $x(t)$ generates a sequence of independent and identically distributed random variables; and the variance of the sample mean is given exactly by the familiar formula $V(\bar{x}) = T^{-1}\gamma_0$.

Estimating the Autocovariances of a Stationary Process

The autocovariance of lag τ is defined as

$$\begin{aligned} \gamma_\tau &= E\{(x_t - \mu)(x_{t-\tau} - \mu)\} \\ &= E(x_t x_{t-\tau}) - \mu^2. \end{aligned} \tag{20.16}$$

Given a sample $x_0, \ldots, x_{T-1}$ of T observations, it is usual to estimate γ_τ by

$$\begin{aligned} c_\tau &= \frac{1}{T} \sum_{t=\tau}^{T-1} (x_t - \bar{x})(x_{t-\tau} - \bar{x}) \\ &= \frac{1}{T} \sum_{t=0}^{T-1-\tau} (x_t - \bar{x})(x_{t+\tau} - \bar{x}). \end{aligned} \tag{20.17}$$

It is natural to question whether T is the appropriate divisor in the formula for the sample autocovariance c_τ which is the usual estimate of γ_τ. The choice of $T - \tau$ would seem more natural, since this is the number of elements comprised by the sum. The justification for the formula is that it guarantees that the matrix

$$C = \begin{bmatrix} c_0 & c_1 & c_2 & \ldots & c_{T-1} \\ c_1 & c_0 & c_1 & \ldots & c_{T-1} \\ c_2 & c_1 & c_0 & \ldots & c_{T-3} \\ \vdots & \vdots & \vdots & \ddots & \vdots \\ c_{T-1} & c_{T-2} & c_{T-3} & \ldots & c_0 \end{bmatrix} \tag{20.18}$$

will be positive-semidefinite; and this reflects an important property of the corresponding matrix of the true autocovariances. The fact the C is a positive-semidefinite matrix can be recognised when it is expressed as $C = T^{-1}\tilde{Y}'\tilde{Y}$, where

$$\tilde{Y} = \begin{bmatrix} \tilde{y}_0 & 0 & \dots & 0 \\ \tilde{y}_1 & \tilde{y}_0 & \dots & 0 \\ \vdots & \vdots & \ddots & \vdots \\ \tilde{y}_{T-1} & \tilde{y}_{T-2} & \dots & \tilde{y}_0 \\ 0 & \tilde{y}_{T-1} & \dots & \tilde{y}_1 \\ \vdots & \vdots & \ddots & \vdots \\ 0 & 0 & \dots & \tilde{y}_{T-1} \end{bmatrix} \tag{20.19}$$

is a matrix containing the deviations $\tilde{y}_t = x_t - \bar{x}$ of the sample elements about the sample mean.

For the purposes of analysing the properties of the estimator c_τ, it is often helpful to replace the formula for c_τ by the surrogate formula

$$\begin{aligned} c^*_\tau &= \frac{1}{T}\sum_{t=0}^{T-1}(x_t - \mu)(x_{t+\tau} - \mu) \\ &= \frac{1}{T}\sum_{t=0}^{T-1} y_t y_{t+\tau}, \end{aligned} \tag{20.20}$$

where $y_t = x_t - \mu$. Since $E(c^*_\tau) = \gamma_\tau$, this would provide an unbiased estimator of γ_τ. The formula depends upon the unknown value of the mean μ together with a set of elements $x_T, \ldots, x_{T+\tau-1}$ which lie outside the sample. Therefore, it cannot be used in practice. However, provided that $\bar{x}$ is a consistent estimator of μ and provided that τ is held constant, the practical estimator c_τ will tend to c^*_τ in probability as the sample size increases.

The consistency of the usual estimator c_τ can be investigated by seeking the conditions under which the surrogate estimator c^*_τ, which is unbiased, has a variance which tends asymptotically to zero as the sample size increases. Within the formula for c^*_τ under (20.20), the sum $\sum y_t y_{t-\tau}$ is analogous to the sum $\sum x_t$ entailed by the formula for the sample mean. According to (20.7), the consistency of the sample mean $\bar{x} = T^{-1}\sum x_t$ as an estimator of μ depends upon the convergence to zero in arithmetic mean of the sequence of the autocovariances

$$\gamma_{|t-s|} = C(x_t, x_s) = E(x_t x_s) - \mu^2 \tag{20.21}$$

when these are indexed on $\tau = |t - s|$. By the same reasoning, the consistency of the estimator c^*_τ of γ_τ depends upon the convergence to zero in arithmetic mean of the sequence of the fourth-order moments

$$\begin{aligned} \delta_{\tau,|t-s|} &= C\{(x_t - \mu)(x_{t-\tau} - \mu), (x_s - \mu)(x_{s-\tau} - \mu)\} \\ &= E\{(x_t - \mu)(x_{t-\tau} - \mu)(x_s - \mu)(x_{s-\tau} - \mu)\} - \gamma^2_\tau \end{aligned} \tag{20.22}$$

when these are indexed on $|t - s| = \kappa$.

The result is not affected if μ is replaced by $\bar{x}$, provided that $\bar{x}$ is a consistent estimator of μ by virtue of the convergence in arithmetic mean of the sequence of autocovariances. In that case, the surrogate estimator c_τ^* and the practical estimator c_τ are asymptotically equivalent. Therefore, it can be declared, in the manner of the statement under (20.7), that

(20.23) The covariance γ_τ is consistently estimated by the sample autocovariance c_τ if and only if both $T^{-1}\sum_\tau \gamma_\tau \rightarrow 0$ and $T^{-1}\sum_\kappa \delta_{\tau,\kappa} \rightarrow 0$ as $T \rightarrow \infty$.

These necessary and sufficient conditions are somewhat weaker than they need be, since we are usually prepared to assume that the moments $\delta_{\tau,\kappa}$ have a convergence which is more rapid than is demanded here. Thus, in the manner of (20.9), it can be asserted that that

(20.24) The sample autocovariance c_τ is a consistent estimator of the autocovariance γ_τ of the process $x(t)$ if the covariance $\gamma_{|t-s|}$ of x_t and x_s and the covariance $\delta_{\tau,|t-s|}$ of $(x_t-\mu)(x_{t-\tau}-\mu)$ and $(x_s-\mu)(x_{s-\tau}-\mu)$ both tend to zero as the temporal separation $|t-s|$ increases.

Asymptotic Moments of the Sample Autocovariances

If one is prepared to assume that $x(t)$ is generated by a linear stochastic process of the ARMA variety, then formulae can be derived which provide the asymptotic values of the dispersion parameters of the estimates of the autocovariances. Before establishing these formulae, we need to derive a preliminary result concerning the fourth-order moments of such processes.

Consider four sequences $a(t)=\{a_t\}$, $b(t)=\{b_t\}$, $c(t)=\{c_t\}$ and $d(t)=\{d_t\}$ generated by filtering the same white-noise process $\varepsilon(t)$ which has $E(\varepsilon_t)=0$, $E(\varepsilon_t^2)=\sigma^2$ and $E(\varepsilon_t^4)=\eta\sigma^4$. Here η is a scalar which is peculiar to the distribution of the elements of $\varepsilon(t)$ and which takes the value of $\eta=3$ when this is normal. The elements of the four sequences may be expressed as

$$\begin{aligned} a_t &= \sum_{i=-\infty}^{\infty} \alpha_i \varepsilon_{t-i}, \qquad & b_t &= \sum_{i=-\infty}^{\infty} \beta_i \varepsilon_{t-i}, \\ c_t &= \sum_{i=-\infty}^{\infty} \gamma_i \varepsilon_{t-i}, \qquad & d_t &= \sum_{i=-\infty}^{\infty} \delta_i \varepsilon_{t-i}. \end{aligned} \tag{20.25}$$

Then, using the result that

$$E(\varepsilon_q\varepsilon_r\varepsilon_s\varepsilon_t) = \begin{cases} \eta\sigma^4, & \text{if } q=r=s=t \\ \sigma^4, & \text{if } q=r\neq s=t \\ 0, & \text{if } q\neq r, q\neq s \text{ and } q\neq t, \end{cases} \tag{20.26}$$

it can be shown that

$$\begin{aligned} E(a_tb_tg_td_t) = {} & (\eta-3)\sigma^4\sum_i \alpha_i\beta_i\gamma_i\delta_i + \sigma^4\sum_i \alpha_i\beta_i\sum_j \gamma_i\delta_i \\ & + \sigma^4\sum_i \alpha_i\gamma_i\sum_j \beta_i\delta_i + \sigma^4\sum_i \alpha_i\delta_i\sum_j \beta_i\gamma_i. \end{aligned} \tag{20.27}$$

To be convinced of this formula, one needs to recognise that the subtraction of 3 within the factor $\eta - 3$ of the leading term is to prevent the same product $\alpha_i\beta_i\gamma_i\delta_i$ from being counted there and within the three remaining terms when $i = j$. It will also be recognised that $\sigma^2\sum_j \gamma_j\delta_j$, for example, is just the covariance of $g(t)$ and $d(t)$. Therefore, equation (20.27) can also be written as

(20.28)
$$\begin{aligned}E(a_tb_tg_td_t) = (\eta-3)\sigma^4\sum_i \alpha_i\beta_i\gamma_i\delta_i + C(a_t,b_t)C(g_t,d_t)\\ + C(a_t,g_t)C(b_t,d_t) + C(a_t,d_t)C(b_t,g_t).\end{aligned}$$

This result can be used in proving the following proposition:

(20.29) Let $y(t) = \psi(L)\varepsilon(t)$, where $\varepsilon(t)$ is a white-noise process such that $E(\varepsilon_t) = 0$, $V(\varepsilon_t) = \sigma^2$ and $E(\varepsilon_t^4) = \eta\sigma^4$ for all t, and where the coefficients of the operator $\psi(L)$ are absolutely summable such that $\sum_{i=-\infty}^{\infty}|\psi_i| < \infty$. Then the limiting value of the covariance of the estimates c_τ and c_κ of γ_τ and γ_κ is given by

$$\lim_{T\to\infty} TC(c_\tau, c_\kappa) = (\eta-3)\gamma_\tau\gamma_\kappa + \sum_{q=-\infty}^{\infty}\{\gamma_{q-\kappa}\gamma_{q-\tau} + \gamma_{q+\kappa}\gamma_{q-\tau}\}.$$

Proof. It simplifies the proof if c_τ and c_κ are replaced by the quantities c_τ^* and c_κ^*, defined by the equation (20.20), which are asymptotically equivalent on the supposition that $\bar{x} \to \mu$. Therefore, it is appropriate to consider the formula

(20.30)
$$\begin{aligned}C(c_\tau^*, c_\kappa^*) &= E(c_\tau^* c_\kappa^*) - E(c_\tau^*)E(c_\kappa^*)\\ &= \frac{1}{T^2}\sum_{t=\tau}^{T-1}\sum_{s=\kappa}^{T-1} E(y_t y_{t+\tau} y_s y_{s+\kappa}) - \gamma_\tau\gamma_\kappa.\end{aligned}$$

The elements under the expectation operator are

(20.31)
$$\begin{aligned} y_t &= \sum_{i=-\infty}^{\infty}\psi_i\varepsilon_{t-i}, & y_{t-\tau} &= \sum_{i=-\infty}^{\infty}\psi_{i+\tau}\varepsilon_{t-i},\\ y_s &= \sum_{i=-\infty}^{\infty}\psi_{i+s-t}\varepsilon_{t-i}, & y_{s-\kappa} &= \sum_{i=-\infty}^{\infty}\psi_{i+s-t+\kappa}\varepsilon_{t-i}.\end{aligned}$$

By applying the result under (20.28) to $E(y_t y_{t+\tau} y_s y_{s+\kappa})$, and by subtracting the term $\gamma_\tau\gamma_\kappa$, it is found that

(20.32)
$$C(c_\tau, c_\kappa) = A + B,$$

where

$$(20.33)\quad \begin{aligned} A &= \frac{1}{T^2}\sum_{t=0}^{T-1}\sum_{s=0}^{T-1}\{\gamma_{s-t}\gamma_{s-t+\kappa-\tau} + \gamma_{s-t+\kappa}\gamma_{s-t-\tau}\} \\ &= \frac{1}{T}\sum_{q=1-T}^{T-1}\left(1-\frac{|\tau|}{T}\right)\{\gamma_q\gamma_{q+\kappa-\tau} + \gamma_{q+\kappa}\gamma_{q-\tau}\}, \end{aligned}$$

and

$$(20.34)\quad \begin{aligned} B &= \frac{(\eta-3)\sigma^4}{T^2}\sum_{t=0}^{T-1}\sum_{s=0}^{T-1}\left\{\sum_i \psi_i\psi_{i+\tau}\psi_{i+s-t}\psi_{i+s-t+\kappa}\right\} \\ &= \frac{(\eta-3)\sigma^4}{T}\sum_{q=1-T}^{T-1}\left(1-\frac{|\tau|}{T}\right)\left\{\sum_i \psi_i\psi_{i+\tau}\psi_{i+q}\psi_{i+q+\kappa}\right\}. \end{aligned}$$

The second expression in either case represents a Cesàro sum which has been obtained by defining a new index $q = s - t$. By taking limits as $T \to \infty$, it will be found that

$$(20.35)\quad \lim_{T\to\infty} TA = \sum_{q=-\infty}^{\infty}\{\gamma_q\gamma_{q-\tau+\kappa} + \gamma_{q+\kappa}\gamma_{q-\tau}\}.$$

Here it will be observed that, since the sum is infinite, one can set $\sum \gamma_q\gamma_{q-\tau+\kappa} = \sum \gamma_{q-\kappa}\gamma_{q-\tau}$. Likewise, it will be found that

$$(20.36)\quad \begin{aligned} \lim_{T\to\infty} TB &= (\eta-3)\sigma^4\sum_{q=-\infty}^{\infty}\sum_{i=-\infty}^{\infty}\psi_i\psi_{i+\tau}\psi_q\psi_{q+\kappa} \\ &= (\eta-3)\gamma_\tau\gamma_\kappa. \end{aligned}$$

Putting the expressions together to form $\lim TC(c^*_\tau, c^*_\kappa) = \lim TC(c_\tau, c_\kappa)$ gives the result of (20.29).

When $y(t)$ is a normal process, there is $E(\varepsilon_t^4) = 3\sigma^4$. Therefore, the formula for the covariance of c_τ and c_κ simplifies to give

$$(20.37)\quad \lim_{T\to\infty} TC(c_\tau, c_\kappa) = \sum_{q=-\infty}^{\infty}\{\gamma_{q-\kappa}\gamma_{q-\tau} + \gamma_{q+\kappa}\gamma_{q-\tau}\}.$$

Asymptotic Moments of the Sample Autocorrelations

We are also concerned to discover the sampling properties of the estimates of the autocorrelation coefficients of a stationary process. The autocorrelation at lag τ is defined as

$$(20.38)\quad \rho_\tau = \frac{\gamma_\tau}{\gamma_0};$$

and it is reasonable to estimate this by

(20.39)
$$r_\tau = \frac{c_\tau}{c_0}.$$

The following theorem, due to Bartlett [35], provides an approximation for the covariance of a pair of empirical autocorrelation coefficients:

(20.40) *Bartlett's Formula.* Let $y(t) = \psi(L)\varepsilon(t)$, where $\varepsilon(t)$ is a white-noise process with $E(\varepsilon_t) = 0$, $V(\varepsilon_t) = \sigma^2$ and $E(\varepsilon_t^4) = \eta\sigma^4$ for all t. Let the sequence of coefficients of $\psi(L)$ be absolutely convergent such that $\sum_{i=-\infty}^{\infty} |\psi_i| < \infty$. Then the limiting value of the covariance of r_τ and r_κ is given by

$$\begin{aligned}\lim_{T\to\infty} TC(r_\tau, r_\kappa) = \sum_{s=-\infty}^{\infty} \{&\rho_{s+\tau}\rho_{s+\kappa} + \rho_{s+\kappa}\rho_{s-\tau} \\ &-2\rho_s\rho_\kappa\rho_{s-\tau} - 2\rho_s\rho_\tau\rho_{s-\kappa} + 2\rho_\tau\rho_\kappa\rho_s^2\}.\end{aligned}$$

Proof. The deviations of the estimated autocorrelation r_τ about its expected value can be approximated via the following formula:

(20.41)
$$d\left(\frac{u}{v}\right) \simeq \frac{vdu - udv}{v^2}.$$

We may set $u = E(c_\tau)$ and $v = E(c_0)$. Then the differentials are $du = c_\tau - E(c_\tau)$, $dv = c_0 - E(c_0)$ and $d(u/v) = r_\tau - E(r_\tau)$. Equation (20.41) indicates that

(20.42)
$$\begin{aligned}r_\tau - E(r_\tau) &\simeq \frac{c_\tau - E(c_\tau)}{E(c_0)} - \frac{E(c_\tau)\{c_0 - E(c_0)\}}{\{E(c_0)\}^2} \\ &\simeq \frac{c_\tau - \gamma_\tau}{\gamma_0} - \frac{\gamma_\tau\{c_0 - \gamma_0\}}{\gamma_0^2}.\end{aligned}$$

A similar approximation is available for $\{r_\kappa - E(r_\kappa)\}$. Therefore,

(20.43)
$$\begin{aligned}\{r_\tau - E(r_\tau)\}\{r_\kappa - E(r_\kappa)\} \simeq &\frac{(c_\tau - \gamma_\tau)(c_\kappa - \gamma_\kappa)}{\gamma_0^2} - \gamma_\tau\frac{(c_0 - \gamma_0)(c_\kappa - \gamma_\kappa)}{\gamma_0^3} \\ &-\gamma_\kappa\frac{(c_\tau - \gamma_\tau)(c_0 - \gamma_0)}{\gamma_0^3} + \gamma_\tau\gamma_\kappa\frac{(c_0 - \gamma_0)^2}{\gamma_0^4}.\end{aligned}$$

Taking expectations gives

(20.44)
$$\begin{aligned}C(r_\tau, r_\kappa) \simeq \frac{1}{\gamma_0^2}\{&C(c_\tau, c_\kappa) - \rho_\tau C(c_0, c_\kappa) \\ &-\rho_\kappa C(c_\tau, c_0) + \rho_\tau\rho_\kappa V(c_0)\},\end{aligned}$$

where the expressions for $C(c_\tau, c_\kappa)$ etc. are provided by the formula of (20.29). When these are drafted into the equation, the terms in $(\eta - 3)$ vanish. The resulting expression is

$$\begin{aligned} C(r_\tau, r_\kappa) \simeq \frac{1}{T}\sum_{s=-\infty}^{\infty} \{ & \rho_{s-\kappa}\rho_{s-\tau} + \rho_{s+\kappa}\rho_{s-\tau} \\ & -2\rho_s\rho_\kappa\rho_{s-\tau} - 2\rho_s\rho_\tau\rho_{s-\kappa} + 2\rho_\tau\rho_\kappa\rho_s^2\}, \end{aligned} \tag{20.45}$$

from which the result of (20.40) follows directly.

It is useful to recognise the fact that the expression

$$\rho_{s-\kappa}\rho_{s-\tau} + \rho_{s+\kappa}\rho_{s-\tau} - 2\rho_s\rho_\kappa\rho_{s-\tau} - 2\rho_s\rho_\tau\rho_{s-\kappa} + 2\rho_\tau\rho_\kappa\rho_s^2 \tag{20.46}$$

is reduced to zero by setting $s = 0$. This helps in verifying the identity

$$\begin{aligned} & \sum_{s=-\infty}^{\infty} \{\rho_s\rho_{s-\tau+\kappa} + \rho_{s+\kappa}\rho_{s-\tau} - 2\rho_s\rho_\kappa\rho_{s-\tau} - 2\rho_s\rho_\tau\rho_{s-\kappa} + 2\rho_\tau\rho_\kappa\rho_s^2\} \\ & = \sum_{s=1}^{\infty} \{\rho_{s+\tau} + \rho_{s-\tau} - 2\rho_s\rho_\tau\}\{\rho_{s+\kappa} + \rho_{s-\kappa} - 2\rho_s\rho_\kappa\}, \end{aligned} \tag{20.47}$$

which indicates a form of the expression for $C(r_\tau, r_\kappa)$ which may be more convenient for the purposes of computing.

The asymptotic expression for the variance of the estimated autocorrelation coefficients is obtained by specialising the formula of (20.40) to give

$$V(r_\tau) \simeq \frac{1}{T}\sum_{s=-\infty}^{\infty} \{\rho_s^2 + \rho_{s-\tau}\rho_{s+\tau} - 4\rho_s\rho_\tau\rho_{s-\tau} + 2\rho_\tau^2\rho_s^2\}. \tag{20.48}$$

In the case of a moving-average process of order q, there is $\rho_\tau = 0$ when $\tau > q$. Therefore, when $\tau > q$, all terms within the parentheses except the first term vanish leaving

$$V(r_\tau) \simeq \frac{1}{T}\{1 + 2(\rho_1^2 + \rho_2^2 + \cdots + \rho_q^2)\}. \tag{20.49}$$

The confidence intervals for the estimated autocovariances, which can be established with this result, may be used in assessing the hypothesis that a data sequence has been generated by an qth-order moving-average process. If any of the autocovariances after the qth exceeds 1.96 times the variance, then it is usually deemed to be significantly different from zero, and doubt is cast upon the hypothesis.

The corresponding expressions for the estimated autocorrelation coefficients of an autoregressive process are not so easily obtained. However, the difficulty is not as inconvenient as it might seem, since the order of an autoregressive process is usually assessed by examining the partial autocorrelation function rather than the autocorrelation function.

Example 20.2. Consider the AR(1) process defined by the equation $y(t) = \phi y(t-1)+\varepsilon(t)$. In order to determine the asymptotic form of the variance of the estimated autocorrelations, we can make use of the result that $\rho_\tau = \phi^{|\tau|}$. Using the formula under (20.48), it can be shown that

$$\begin{aligned} V(r_\tau) &\simeq \frac{1}{T}\sum_{s=1}^{\tau}\phi^{2\tau}\left(\phi^{-s}-\phi^{s}\right)^2 + \frac{1}{T}\sum_{s=\tau+1}^{\infty}\phi^{2s}\left(\phi^{-\tau}-\phi^{\tau}\right)^2 \\ &= \frac{1}{T}\left\{\frac{(1-\phi^{2\tau})(1+\phi^2)}{1-\phi^2} - 2\tau\phi^{2\tau}\right\}. \end{aligned} \tag{20.50}$$

Calculation of the Autocovariances

There are numerous ways of calculating the estimate

$$c_\tau = \frac{1}{T}\sum_{t=\tau}^{T-1}(x_t-\bar{x})(x_{t-\tau}-\bar{x}) \tag{20.51}$$

of the autocovariance of lag τ and of calculating approximations to it. One might consider expanding the formula to give

$$c_\tau = \frac{1}{T}\left\{\sum_{t=\tau}^{T-1}x_t x_{t-\tau} - \bar{x}\sum_{t=\tau}^{T-1}(x_t + x_{t-\tau}) + (T-\tau)\bar{x}^2\right\}. \tag{20.52}$$

Then, since the differences between

$$\sum_{t=\tau}^{T-1}x_t,\quad \sum_{t=\tau}^{T-1}x_{t-\tau}\quad \text{and}\quad (T-\tau)\bar{x} = \frac{T-\tau}{T}\sum_{t=0}^{T-1}x_t$$

are marginal, the approximation

$$c_\tau \simeq \frac{1}{T}\left\{\sum_{t=\tau}^{T-1}x_t x_{t-\tau} - (T-\tau)\bar{x}^2\right\}, \tag{20.53}$$

which entails considerably fewer numerical operations, recommends itself. However, only if the speed of calculation is a priority is there any virtue in this formula, which is prone to rounding errors and which may give rise to a matrix of autocovariances which violates the condition of positive-definiteness. The preferred way of calculating the autocovariances is to apply the formula

$$c_\tau = \frac{1}{T}\sum_{t=\tau}^{T-1}y_t y_{t-\tau} \tag{20.54}$$

to the deviations $y_t = x_t - \bar{x}$. The virtue of this procedure lies in the fact that there is less danger of rounding error in cumulating the sum of the adjusted elements $y_t y_{t-\tau}$ than there is in cumulating the unadjusted elements $x_t x_{t-\tau}$.

To ensure that the deviations are calculated accurately, it may be worth calculating the sample mean $\bar{x}$ in two passes. This is a useful precaution in cases where the coefficient of variation—which is the ratio of the standard deviation of the data to its mean—is small.

(20.55)
```
procedure Autocovariances(Tcap, lag : integer;
                    var y : longVector;
                    var acovar : vector);

  var
    ybar, mean : real;
    t, j : integer;

begin {Autocovariances}

{Calculate the mean}
  ybar := 0.0;
  mean := 0.0;
  for t := 0 to Tcap − 1 do {first pass}
    mean := mean + y[t];
  mean := mean/Tcap;
  for t := 0 to Tcap − 1 do {second pass}
    ybar := ybar + y[t] − mean;
  ybar := ybar/Tcap + mean;
  Writeln('ybar =  ', ybar : 4 : 5);

{Calculate the deviations}
  for t := 0 to Tcap − 1 do
    y[t] := y[t] − ybar;

{Calculate the autocovariances}
  for j := 0 to lag do
    begin {j}
      acovar[j] := 0.0;
      for t := j to Tcap − 1 do
        acovar[j] := acovar[j] + y[t − j] * y[t];
      acovar[j] := acovar[j]/Tcap;
    end; {j}

end; {Autocovariances}
```

An alternative way of calculating the sequence of autocovariances depends upon the Fourier transform. To explain the method, we should begin by considering the z-transform of the mean-adjusted data sequence sequence $\{y_0, \ldots, y_{T-1}\}$. This is the polynomial

$$y(z) = y_0 + y_1 z + \cdots + y_{T-1} z^{T-1}. \tag{20.56}$$

It is easy to see that the sequence of autocovariances is simply the sequence of coefficients $\{c_{1-T}, \ldots, c_0, \ldots, c_{T-1}\}$ associated with the positive and negative powers of z in the product

$$\begin{aligned}\frac{1}{T}y(z)y(z^{-1}) &= \frac{1}{T}\sum_{t=0}^{T-1}\sum_{s=0}^{T-1} y_t y_s z^{t-s}\\ &= \sum_{\tau=1-T}^{T-1}\left(\frac{1}{T}\sum_{t=\tau}^{T-1} y_t y_{t-\tau}\right) z^{\tau}\\ &= \sum_{\tau=1-T}^{T-1} c_{\tau} z^{\tau}.\end{aligned} \tag{20.57}$$

Here the final expression depends upon the fact that $c_{-\tau} = c_{\tau}$.

The z-transform of the sequence $\{y_0, \ldots, y_{T-1}\}$ becomes a Fourier transform when $z^{\tau} = e^{-i\omega\tau}$. In particular, the discrete Fourier transform results from setting $\omega = \omega_j = 2\pi j/T$ with $j = 0, \ldots, T-1$. In that case, z^t becomes T-periodic such that $z^{\tau-T} = z^{\tau}$. Also, there is the condition $c_{\tau-T} = c_{T-\tau}$. It follows that the product in (20.57) can be written as

$$\begin{aligned}\frac{1}{T}y(z)y(z^{-1}) &= \sum_{\tau=1-T}^{T-1} c_{\tau} z^{\tau}\\ &= c_0 + \sum_{\tau=1}^{T-1}(c_{\tau} + c_{T-\tau})z^{\tau}\\ &= \sum_{\tau=0}^{T-1} c_{\tau}^{\circ} z^{\tau},\end{aligned} \tag{20.58}$$

where $c_{\tau}^{\circ} = c_{\tau} + c_{T-\tau}$ is a so-called circular autocovariance.

The periodogram of $y(t)$ is just the function $I(z) = (2/T)y(z)y(z^{-1})$ evaluated over the set of points $z_j = \exp(-i2\pi j/T); j = 0, \ldots, T-1$ which lie at equal intervals around the circumference of the unit circle in the complex plane. Therefore, we might expect to be able to recover the sequence of autocovariances $\{c_{\tau}; \tau = 0, \ldots, c_{T-1}\}$ by applying an inverse Fourier transform to the ordinates of a periodogram evaluated over the set of so-called Fourier frequencies $\omega_j; j = 0, \ldots, T-1$. In fact, the inverse Fourier transform would deliver the sequence of coefficients $c_{\tau}^{\circ} = c_{\tau} + c_{T-\tau}; j = 0, \ldots, T-1$; from which the ordinary autocovariances cannot be recovered. Nevertheless, the low-order coefficients c_{τ}° will be close to the corresponding autocovariances c_{τ} if the higher-order autocovariances are close to zero.

To isolate the ordinary autocovariances, we should would have to turn the higher-order autocovariances into zeros. This is easily accomplished by padding the tail of the sequence $\{y_t\}$ with zero elements. Thus, when p zero elements are added to the tail of $\{y_t\}$, the same number of zero elements will appear in the tail of the sequence of autocovariances. Therefore, the first p coefficients which are delivered by the inverse Fourier transform will coincide with the ordinary autocovariances.

The following procedure calculates the autocovariances by the Fourier method described above. The value of p, which is the number of autocovariances we wish to calculate, is specified in the parameter *lag*. The sequence $y(t)$, which is assumed to be in deviation form, is padded by a number of zeros which is no less that p and which is sufficient to bring the total length to a number which is a power of 2. This is to allow the use of the base-2 FFT.

This algorithm is presented in order to show how the autocovariances may be calculated as a side-product of the calculation of the ordinates of the periodogram. If there is no interest in the periodogram and if only small handful of autocovariances are required, then it is not appropriate to use the algorithm.

(20.59)

```
procedure FourierACV(var y : longVector;
                        lag, Tcap : integer);

  var
    Ncap, Nover2, g, t, j : integer;

begin {FourierACV}

  Ncap := 1;
  g := 0;
  repeat
    begin
      Ncap := Ncap * 2;
      g := g + 1
    end;
  until Ncap >= (Tcap + lag);
  Nover2 := Ncap div 2;

  for t := Tcap to Ncap do
    y[t] := 0.0;

  CompactRealFFT(y, Ncap, g);

  y[0] := Sqr(y[0])/(Ncap * Tcap);
  y[Nover2] := Sqr(y[Nover2])/(Ncap * Tcap);

  for j := 1 to Nover2 - 1 do
    y[j] := (Sqr(y[j]) + Sqr(y[Nover2 + j]))/(Ncap * Tcap);
  for j := 1 to Nover2 - 1 do
    y[Ncap - j] := y[j];

  CompactRealFFT(y, Ncap, g);
end; {FourierACV};
```

Inefficient Estimation of the MA Autocovariances

The sample autocovariances do not provide statistically efficient estimates of the autocovariances of a moving-average process $y(t) = \mu(L)\varepsilon(t)$. This becomes apparent when we pursue a maximum-likelihood approach in deriving estimates of the parameters of the moving-average operator $\mu(L)$. The method of maximum likelihood is known to generate efficient estimates under very general conditions. Therefore, the mere fact that there is no direct connection between the maximum-likelihood estimates of the MA parameters and the product-moment estimates of the corresponding autocovariances is virtually a proof of the inefficiency of the latter.

The inefficiency is confirmed when the product-moment estimates are used as a basis for estimating the parameters of the moving-average operator. The asymptotic value of the sampling variance of the resulting estimates is usually far greater than the asymptotic sampling variance of the corresponding maximum-likelihood estimates.

Example 20.3. Consider the case of the MA(1) process defined by the equation $y(t) = \varepsilon(t) - \theta\varepsilon(t-1)$. The autocovariances are given by

$$\begin{aligned} \gamma_0 &= \sigma_\varepsilon^2(1+\theta^2), \\ \gamma_1 &= -\sigma_\varepsilon^2\theta \quad \text{and} \\ \gamma_\tau &= 0 \quad \text{for} \quad \tau > 1; \end{aligned} \tag{20.60}$$

and, therefore, the autocorrelation for a lag of one period is

$$\rho_1 = \frac{\gamma_1}{\gamma_0} = \frac{-\theta}{1+\theta^2}. \tag{20.61}$$

It follows that the parameter θ satisfies the equation $\theta^2\rho_1 + \theta + \rho_1 = 0$. By solving the equation subject to $|\theta| < 1$, which is the condition of invertibility, it is found that

$$\theta = \frac{-1+\sqrt{1-4\rho_1^2}}{2\rho_1}. \tag{20.62}$$

Since θ is real-valued, the condition that $0 < |\rho_1| \leq 0.5$ must prevail.

Replacing ρ_1 in equation (20.62) by the estimate $r_1 = c_1/c_0$ gives rise to following system for estimating θ:

$$\begin{aligned} \hat{\theta} &= \frac{-1+\sqrt{1-4r_1^2}}{2r_1} && \text{if} \quad 0 < |r_1| \leq 0.5, \\ &= 1 && \text{if} \quad r_1 < -0.5, \\ &= -1 && \text{if} \quad r_1 > 0.5. \end{aligned} \tag{20.63}$$

The variance of the estimate is approximated by

$$V(\hat{\theta}) \simeq \left\{\frac{\partial\hat{\theta}}{\partial r_1}\right\}^2 V(r_1). \tag{20.64}$$

This may be expressed in terms of θ by taking the inverse of

$$\frac{\partial\rho_1}{\partial\theta} = \frac{\theta^2-1}{(1+\theta^2)^2} \tag{20.65}$$

to be the asymptotic form of the derivative $\partial\hat{\theta}/\partial r_1$. The asymptotic form of the variance of r_1 is obtained from equation (20.41):

$$TV(r_1) \simeq 1 - 3\rho_1^2 + 4\rho_1^4 \tag{20.66}$$
$$= \frac{1 + \theta^2 + 4\theta^4 + \theta^6 + \theta^8}{(1 + \theta^2)^4}.$$

It follows that

$$TV(\hat{\theta}) \simeq \frac{1 + \theta^2 + 4\theta^4 + \theta^6 + \theta^8}{(1 - \theta^2)^2}. \tag{20.67}$$

This is to be compared with the asymptotic variance of the maximum-likelihood estimate of θ which has the value of $T^{-1}(1 - \theta^2)$ (see Figure 20.1).

The formula of (20.67) is valid only for large values of T and it is of little worth when θ is near the boundary values of -1 and 1, where it suggests that the limiting variance of $\sqrt{T}\hat{\theta}$ is unbounded. Clearly, the variance of the estimator is affected by the truncation of its distribution at the boundary points, and no account is taken of this in the asymptotic formula. Nevertheless it is worthwhile plotting the ratio of the asymptotic variance of the maximum likelihood and the asymptotic variance of $\hat{\theta}$, if only to emphasise the inefficiency of the latter.

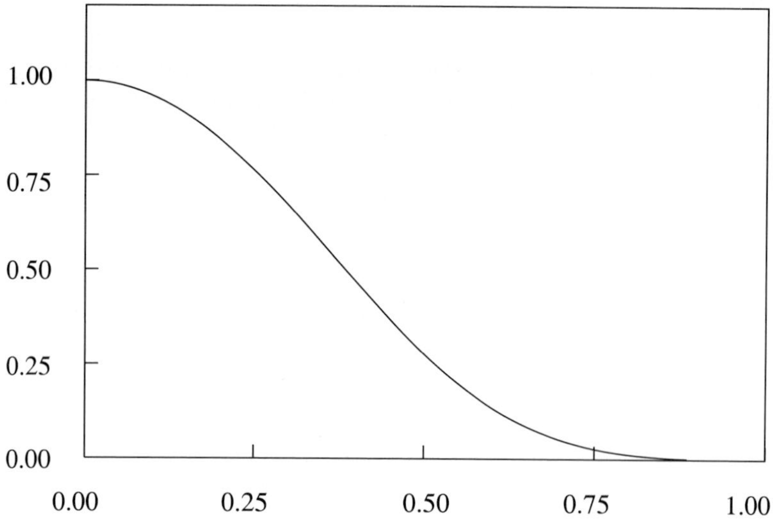

Figure 20.1. The asymptotic efficiency of the moments estimator of the parameter θ of the MA(1) process $y(t) = \varepsilon(t) - \theta\varepsilon(t)$ relative to that of the maximum-likelihood estimator for values of θ from 0 to 1.

Efficient Estimates of the MA Autocorrelations

One reason which can given for the inefficiency of the sample autocorrelations in estimating the corresponding population autocorrelations of a moving-average process is that they take no account of the fact that, for a process of order q, there is $\gamma_\tau = 0$ when $\tau > q$. Some more efficient estimates of the autocorrelations can be obtained by imposing this restriction.

One way of deriving such estimates, which is due to Walker [504], makes use of the information which is conveyed by a set of $M > q$ sample autocorrelations. It can be assumed, for the sake of the derivation, that the sample autocorrelations, which are contained in the vector $r = [r_1, \ldots, r_q, \ldots, r_M]'$, have a normal distribution which is given by

$$N(r; \rho, \Sigma) = (2\pi)^{-T/2}|\Sigma|^{-1/2} \exp\big\{ -\tfrac{1}{2}(r-\rho)'\Sigma^{-1}(r-\rho)\big\}, \tag{20.68}$$

wherein $\rho = [\rho_1, \ldots, \rho_q, 0, \ldots 0]'$ is a vector of order M representing the true autocorrelations. This expression also comprises the dispersion matrix $\Sigma = D(r)$ of the sample autocorrelations. The log of the likelihood function is

$$\ln L = -\frac{T}{2}\ln(2\pi) - \frac{1}{2}\ln|\Sigma| - \frac{1}{2}(r-\rho)'\Sigma^{-1}(r-\rho). \tag{20.69}$$

The dominant term within the function is the quadratic form $(r-\rho)'\Sigma^{-1}(r-\rho)$, in comparison with which the other terms become negligible as $T \to \infty$. Also, when T is large, the dispersion matrix Σ is adequately approximated by a matrix W whose elements are given by the formula under (20.40). Therefore, estimates of the autocorrelations may be obtained from the vector ρ which minimises the function $(r-\rho)'W^{-1}(r-\rho)$ subject to the restriction that $\rho_{q+1} = \cdots = \rho_M = 0$.

The restrictions on ρ can be expressed in the equations $H\rho = 0$, wherein $H = [0, I_{M-q}]$ is a matrix of order $(M-q) \times M$. The corresponding Lagrangean criterion function, for which the restricted estimates are derived, is

$$Q = (r-\rho)'W^{-1}(r-\rho) - 2\lambda' H\rho. \tag{20.70}$$

From the condition for minimisation, which is that $\partial Q/\partial\rho = 0$, we get

$$W^{-1}(r-\rho) - H'\lambda = 0. \tag{20.71}$$

Rearranging this gives

$$\rho = r - WH'\lambda. \tag{20.72}$$

When latter is premultiplied by $(HWH')^{-1}H$ and when the condition $H\rho = 0$ is invoked, it is found that

$$\lambda = (HWH')^{-1}Hr. \tag{20.73}$$

Putting λ back into equation (20.72) gives the vector of estimates in the form of

$$\hat{\rho} = \big\{I - WH(HWH')^{-1}H\big\}r. \tag{20.74}$$

The matrix $I - WH(HWH')^{-1}H$, which is idempotent, may be described as a minimum-distance projector of the M-dimensional space, in which the vector r

resides, onto the null space of the matrix H; and it will be observed that $H\hat{p} = 0$, which is to say only the leading q elements of $\hat{p}$ are nonzero. The matrix W may be partitioned, in a manner which is conformable with the partitioning of $H = [0, I_{N-q}]$, to give

$$W = \begin{bmatrix} W_{11} & W_{12} \\ W_{21} & W_{22} \end{bmatrix}. \tag{20.75}$$

The vector $\hat{\rho}_\diamond = [\hat{\rho}_1, \ldots, \hat{\rho}_q]'$, which contains all of the estimates, can then be expressed as

$$\hat{\rho}_\diamond = r_\diamond - W_{12}W_{22}^{-1}r_{\diamond\diamond}, \tag{20.76}$$

where $r_\diamond$ contains the first q elements of r and $r_{\diamond\diamond}$ contains the remainder. In this way, $\hat{\rho}_\diamond$ is depicted as a corrected version of $r_\diamond$.

Bibliography

[16] Anderson, T.W., (1971), *The Statistical Analysis of Time Series*, John Wiley and Sons, Chichester.

[35] Bartlett, M.S., (1946), On the Theoretical Specification and Sampling Properties of Autocorrelated Time Series, *Journal of the Royal Statistical Society, Series B*, **8**, 27–41.

[37] Bartlett, M.S., (1966), *An Introduction to Stochastic Processes, with Special Reference to Methods and Applications, Third Edition*, Cambridge University Press, Cambridge.

[55] Bhansali, R.J., (1980), Autoregressive Window Estimates of the Inverse Correlation Function, *Biometrika*, **67**, 551–66.

[79] Brockwell, P.J., and R.A. Davis, (1987), *Time Series: Theory and Methods*, Springer-Verlag, New York.

[113] Cleveland, W.S., (1972), The Inverse Autocorrelations of a Time Series and their Applications, *Technometrics*, **14**, 277–293.

[197] Fuller, W.A., (1976), *Introduction to Statistical Time Series*, John Wiley and Sons, New York.

[411] Priestley, M.B., (1981), *Spectral Analysis and Time Series*, Academic Press, London.

[504] Walker, A.M., (1961), Large-Sample Estimation of the Parameters for Moving-Average Models, *Biometrika*, **48**, 343–357.

CHAPTER 21

Least-Squares Methods of ARMA Estimation

In this chapter, we shall describe methods of estimating autoregressive moving-average (ARMA) models which fulfil the criterion of minimising the sum of squares of the one-step-ahead prediction errors within the compass of the sample period. We shall describe the resulting estimates as the least-squares estimates. In the following chapter, we shall consider alternative methods of ARMA estimation which are derived from the maximum-likelihood criterion.

The least-squares estimators, which are simpler to implement, are appropriate to circumstances where there is ample data. When they are correctly constructed, they are guaranteed to fulfil the conditions of stationarity and invertibility. However, when the sample is small, the moduli of the roots of the autoregressive operator tend to be underestimated; and the severity of this bias increases as the roots approach the unit circle. The maximum-likelihood methods pay more attention to the problems arising when the size of the sample is limited; and they tend, in these circumstances, to suffer less from bias.

The criteria of least-squares estimation and of maximum-likelihood estimation converge as the sample size increases; and it is reasonable to depict the least-squares estimates as approximate versions of their maximum-likelihood counterparts. However, the complexities of the exact maximum-likelihood estimators can obscure the features which are shared with the least-squares counterparts. Moreover, the calculations of the maximum-likelihood estimators have to be executed in a different manner from those of the least-squares estimators. These circumstances justify our dividing the topic of ARMA estimation between two chapters.

The present chapter pursues an additional theme. ARMA models may be defined in terms of a simple algebra of rational functions. However, some of the simplicity is lost when we work with approximations which are constrained to fit within finite-series representations. These approximations are conveniently expressed in terms of a matrix algebra; and throughout the chapter, we shall be seeking to elucidate the relationship between the polynomial algebra and the corresponding matrix algebra. Some of these relationships have been explored already in Chapter 2.

Representations of the ARMA Equations

Imagine that T observations, running from $t = 0$ to $t = T-1$, have been taken on a stationary and invertible ARMA(p, q) process $y(t)$ which is described by the equation

$$(1 + \alpha_1 L + \cdots + \alpha_p L^p)y(t) = (1 + \mu_1 L + \cdots + \mu_q L^q)\varepsilon(t), \tag{21.1}$$

wherein $\varepsilon(t)$ is a white-noise sequence of independently and identically distributed random variables of zero mean. Corresponding to the observations, there is a set of T equations which can be arrayed in a matrix format:

$$\begin{bmatrix} y_0 & y_{-1} & \cdots & y_{-p} \\ y_1 & y_0 & \cdots & y_{1-p} \\ \vdots & \vdots & \ddots & \vdots \\ y_p & y_{p-1} & \cdots & y_0 \\ \vdots & \vdots & \ddots & \vdots \\ y_{T-1} & y_{T-2} & \cdots & y_{T-p-1} \end{bmatrix} \begin{bmatrix} 1 \\ \alpha_1 \\ \vdots \\ \alpha_p \end{bmatrix} = \begin{bmatrix} \varepsilon_0 & \varepsilon_{-1} & \cdots & \varepsilon_{-q} \\ \varepsilon_1 & \varepsilon_0 & \cdots & \varepsilon_{1-q} \\ \vdots & \vdots & \ddots & \vdots \\ \varepsilon_q & \varepsilon_{q-1} & \cdots & \varepsilon_0 \\ \vdots & \vdots & \ddots & \vdots \\ \varepsilon_{T-1} & \varepsilon_{T-2} & \cdots & \varepsilon_{T-q-1} \end{bmatrix} \begin{bmatrix} 1 \\ \mu_1 \\ \vdots \\ \mu_q \end{bmatrix}. \tag{21.2}$$

Here the generic equation is

$$\sum_{i=0}^{p} \alpha_i y_{t-i} = \sum_{i=0}^{q} \mu_i \varepsilon_{t-i}, \qquad \text{where} \qquad \alpha_0 = \mu_0 = 1. \tag{21.3}$$

Apart from the elements $y_0, y_1, \ldots, y_{T-1}$ and $\varepsilon_0, \varepsilon_1, \ldots, \varepsilon_{T-1}$ which fall within the sample period, these equations comprise the presample values $y_{-p}, \ldots, y_{-1}$ and $\varepsilon_{-q}, \ldots, \varepsilon_{-1}$ which are to be found in the top-right corners of the matrices.

An alternative representation of the system of equations can be given which is in terms of polynomials. Thus, if

$$\begin{aligned} y(z) &= y_{-p}z^{-p} + \cdots + y_0 + y_1 z + \cdots + y_{T-1}z^{T-1}, \\ \varepsilon(z) &= \varepsilon_{-q}z^{-q} + \cdots + \varepsilon_0 + \varepsilon_1 z + \cdots + \varepsilon_{T-1}z^{T-1}, \\ \alpha(z) &= 1 + \alpha_1 z + \cdots + \alpha_p z^p \quad \text{and} \\ \mu(z) &= 1 + \mu_1 z + \cdots + \mu_q z^q, \end{aligned} \tag{21.4}$$

then

$$y(z)\alpha(z) = \varepsilon(z)\mu(z). \tag{21.5}$$

By performing the polynomial multiplication of both sides of (21.5) and by equating the coefficients of the same powers of z, it will be found that the equation associated with z^t is precisely the generic equation under (21.3).

In estimating the ARMA model from the data series $y_0, \ldots, y_{T-1}$, it is common to set the presample elements $y_{-p}, \ldots, y_{-1}$ and $\varepsilon_{-q}, \ldots, \varepsilon_{-1}$ to zeros. In general, when the presample elements have been set to zero and when the coefficients of $\alpha(z)$ and $\mu(z)$ assume arbitrary values, the equality of (21.5) can be maintained only by allowing the residual polynomial $\varepsilon(z)$ to be replaced by an indefinite series. There are exceptions.

First, if $\mu(z) = 1$, then the equality can be maintained by allowing the residual polynomial to take the form of $\varepsilon(z) = \varepsilon_0 + \varepsilon_1 z + \cdots + \varepsilon_{T-1+p}z^{T-1+p}$, which is a polynomial of degree $T - 1 + p$.

Secondly, if the polynomial argument z^j is nilpotent of degree T in the index j, such that $z^j = 0$ for all $j \geq T$, then all polynomial products are of degree $T - 1$

at most, and the equality may be maintained without the degree of $\varepsilon(z)$ exceeding $T-1$. Making z^j nilpotent of degree T will enable us to construct a correspondence between the algebra of the polynomials of degree $T-1$ and the algebra of the class of $T \times T$ lower-triangular Toeplitz matrices.

Thirdly, if the polynomial argument z^j is a T-periodic function of the index j, such that $z^{j+T} = z^j$ for all j, then, again, all polynomial products are of degree $T-1$ at most. Making z^j a T-periodic function, enables us to construct a correspondence between the algebra of the polynomials of degree $T-1$ and the algebra of the class of $T \times T$ circulant Toeplitz matrices.

A polynomial of degree $T-1$ is completely specified by the values which it assumes at T equally spaced points on the circumference of the unit circle in the complex plane which are $e^{i\omega_j}; j = 0, \ldots, T-1$, where $\omega_j = 2\pi j/T$. In particular, a product $\gamma(z) = \alpha(z)\beta(z)$ of two periodic polynomials is completely specified by $\gamma(e^{i\omega_j}) = \alpha(e^{i\omega_j})\beta(e^{i\omega_j}); j = 0, \ldots, T-1$. Therefore, when the polynomials have an T-periodic argument, the time-consuming business of polynomial multiplication can be circumvented by performing an equivalent but a simpler set of operations at the frequency points ω_j.

The Least-Squares Criterion Function

Using the polynomial algebra, we can define a criterion function for ARMA estimation which takes the form of

$$S = \frac{1}{2\pi i} \oint \varepsilon(z^{-1})\varepsilon(z)\frac{dz}{z}. \tag{21.6}$$

The value of S is nothing more than the coefficient associated with z^0 in the Laurent expansion of $\varepsilon(z^{-1})\varepsilon(z)$. Now consider the expression

$$\varepsilon(z) = \frac{\alpha(z)}{\mu(z)}y(z), \tag{21.7}$$

which comes from equation (21.5). Substituting this in (21.6) gives

$$\begin{aligned} S &= \frac{1}{2\pi i} \oint y(z^{-1})y(z)\frac{\alpha(z^{-1})\alpha(z)}{\mu(z^{-1})\mu(z)}\frac{dz}{z} \\ &= T\frac{\sigma_\varepsilon^2}{2\pi i} \oint \frac{c(z)}{\gamma(z)}\frac{dz}{z}. \end{aligned} \tag{21.8}$$

Here the denominator of the second expression is

$$\begin{aligned} \gamma(z) &= \sigma_\varepsilon^2 \frac{\mu(z^{-1})\mu(z)}{\alpha(z^{-1})\alpha(z)} \\ &= \{\gamma_0 + \gamma_1(z + z^{-1}) + \cdots + \gamma_{T-1}(z^{T-1} + z^{1-T}) + \cdots\}. \end{aligned} \tag{21.9}$$

When the coefficients in $\alpha(z)$ and $\mu(z)$ assume their true values, this becomes the autocovariance generating function of the ARMA process. On setting $z = e^{i\omega}$, it becomes the spectral density function of the process. The numerator of the expression is

$$(21.10)\qquad \begin{aligned} c(z) &= \frac{1}{T}y(z^{-1})y(z) \\ &= c_0 + c_1(z + z^{-1}) + \cdots + c_{T-1}(z^{T-1} + z^{1-T}). \end{aligned}$$

On the assumption that the presample elements within $y(z)$ have been set to zeros, it will be found that

$$(21.11)\qquad c_\tau = \frac{1}{T}\sum_{t=\tau}^{T-1} y_t y_{t-\tau}$$

is the ordinary empirical autocovariance of lag τ for the case where $E\{y(t)\} = 0$. Thus $c(z)$ becomes the empirical autocovariance generating function. When $z = \exp\{-i2\pi j/T\}$, it becomes the periodogram which is defined on the points $j = 0, \ldots, T/2$ when T is even and on the points $j = 0, \ldots, (T-1)/2$ if T is odd.

The criterion function of (21.8) is amenable to some further analysis. For a start, it may be observed that, if the coefficients of $\alpha(z)$ and $\mu(z)$ were to assume their true values and if the presample elements $y_{-p}, \ldots, y_{-1}$ were incorporated in $y(z)$, then S would be equal to the sum of squares of the disturbances $\varepsilon_{-q}, \ldots, \varepsilon_{T-1}$.

Next, we may consider how the criterion function evolves as T increases. According to the result under (20.23), the sample covariance c_τ is a consistent estimator of the true autocovariance $\bar{\gamma}_\tau$. It follows that, as T increases, the empirical autocovariance generating function $c(z)$ converges to its theoretical counterpart $\bar{\gamma}(z)$ which is a product of the true parameter values. On this basis, it is straightforward to prove that the functions $\alpha(z)$ and $\mu(z)$ obtained by minimising S will likewise converge to their true values $\bar{\alpha}(z)$ and $\bar{\mu}(z)$. Therefore, to prove the consistency of the estimates, it is enough to show that

(21.12) If $c(z) = \bar{\gamma}(z)$, then the function S achieves its minimum value of σ_ε^2 when $\gamma(z) = \bar{\gamma}(z)$, or equivalently when $\alpha(z) = \bar{\alpha}(z)$ and $\mu(z) = \bar{\mu}(z)$.

Proof. The true autocovariance generating function of the ARMA process may be written as $\bar{\gamma}(z) = \sigma_\varepsilon^2\bar{\omega}(z^{-1})\bar{\omega}(z)$ where $\bar{\mu}(z)/\bar{\alpha}(z) = \bar{\omega}(z) = \{1 + \bar{\omega}_1 z + \bar{\omega}_2 z^2 + \cdots\}$. Here the leading term of the expansion of $\bar{\omega}(z)$ is unity on account of the normalisation of the leading coefficients of $\bar{\mu}(z)$ and $\bar{\alpha}(z)$. In the same manner, the ratio of the estimates may be written as $\mu(z)/\alpha(z) = \omega(z) = \{1+\omega_1 z+\omega_2 z^2+\cdots\}$. Thus the criterion function S may be expressed as

$$(21.13)\qquad S = T\frac{\sigma_\varepsilon^2}{2\pi i}\oint \frac{\bar{\omega}(z^{-1})\bar{\omega}(z)}{\omega(z^{-1})\omega(z)}\frac{dz}{z} = T\frac{\sigma_\varepsilon^2}{2\pi i}\oint \theta(z^{-1})\theta(z)\frac{dz}{z},$$

where $\bar{\omega}(z)/\omega(z) = \theta(z) = \{1+\theta_1 z+\theta_2 z^2+\cdots\}$. Therefore the value of the function is given by the sum of squares $S = \sigma_\varepsilon^2\{1 + \theta_1^2 + \theta_2^2 + \cdots\}$. The latter assumes its minimum value of $S = \sigma_\varepsilon^2$ when $\omega(z) = \bar{\omega}(z)$. But, on the assumption that there are no factors common to $\bar{\alpha}(z)$ and $\bar{\mu}(z)$, this entails the equalities $\alpha(z) = \bar{\alpha}(z)$ and $\mu = \bar{\mu}(z)$.

The Yule–Walker Estimates

In the case of the pure autoregressive (AR) model, the criterion function of (21.8) assumes the simplified form of

$$
\begin{aligned}
S &= \frac{1}{2\pi i}\oint y(z^{-1})y(z)\alpha(z^{-1})\alpha(z)\frac{dz}{z}\\
&= T\sum_{j=0}^{p}\sum_{k=0}^{p}\alpha_k\alpha_j c_{|k-j|}\\
&= \sum_{t=0}^{T-1}\sum_{s=0}^{T-1} y_t y_s \lambda_{|t-s|}.
\end{aligned}
\tag{21.14}
$$

Here $\lambda_{|t-s|}$ is a coefficient of the self-reciprocal polynomial

$$
\begin{aligned}
\lambda(z) &= \alpha(z^{-1})\alpha(z)\\
&= \{\lambda_0 + \lambda_1(z+z^{-1}) + \cdots + \lambda_p(z^p + z^{-p})\}
\end{aligned}
\tag{21.15}
$$

which corresponds to the autocovariance generating function of a synthetic pth-order moving-average process $y(t) = \alpha(L)\varepsilon(t)$ based on a sequence $\varepsilon(t)$ of unit-variance white-noise disturbances.

The estimating equations for the autoregressive parameters are obtained by differentiating $S/T = \sum_j\sum_k \alpha_k\alpha_l c_{|k-j|}$ with respect to $\alpha_1,\ldots,\alpha_p$ and setting the results to zero. Taking account of the normalisation $\alpha_0 = 1$, the resulting first-order conditions may be assembled to produce the following matrix equation:

$$
\begin{bmatrix}
c_0 & c_1 & \dots & c_{p-1}\\
c_1 & c_0 & \dots & c_{p-2}\\
\vdots & \vdots & \ddots & \vdots\\
c_{p-1} & c_{p-2} & \dots & c_0
\end{bmatrix}
\begin{bmatrix}\alpha_1\\ \alpha_2\\ \vdots\\ \alpha_p\end{bmatrix}
= -\begin{bmatrix}c_1\\ c_2\\ \vdots\\ c_p\end{bmatrix}.
\tag{21.16}
$$

These are the empirical Yule–Walker equations which are derived from their theoretical counterparts found under (17.55) by replacing the autocovariances $\gamma_\tau; \tau = 0,\ldots,p$ by their estimates $c_\tau; \tau = 0,\ldots,p$ obtained from the formula of (21.11). The matrix $C = [c_{|j-k|}]$ of the leading $p+1$ empirical autocovariances is positive-definite. Therefore the following result applies:

(21.17) Let $S/T = \sum_{j=0}^{p}\sum_{k=0}^{p}\alpha_k\alpha_j c_{|j-k|} = \alpha' C\alpha$, where $C = [c_{|j-k|}]$ is a symmetric positive-definite Toeplitz matrix and where $\alpha = [1,\alpha_1,\ldots,\alpha_p]'$. If α is chosen so as to minimise the value of S, then all the roots of the equation $\alpha(z) = 1 + \alpha_1 z + \cdots + \alpha_p z^p = 0$ will lie outside the unit circle.

This theorem, which establishes that the Yule–Walker estimates obey the condition of stationarity, is simply a restatement of the assertion under (17.62) which has been proved already.

A procedure for estimating the parameters of an AR model may be derived by joining the *Autocovariances* procedure, which is to be found under (20.55), with the *YuleWalker* procedure from (17.67) which serves to solve the equations under (21.16) as well as their theoretical counterparts under (17.55). An alternative iterative procedure may be derived by using the *LevinsonDurbin* procedure of (17.75) which solves the equations recursively and which, in the process, generates estimates of a sequence of AR models of increasing orders up to the required order of p.

An alternative estimator the AR parameters, which might be preferred to the Yule–Walker method when the sample size is small, is the Burg estimator. This will be presented later in this chapter.

Estimation of MA Models

Now let us consider the form of the criterion function in the case of a pure moving-average (MA) model. Let the series expansion of the inverse of $\mu(z)$ be written as $\mu^{-1}(z) = \{1 + \psi_1 z + \psi_2 z^2 + \cdots\}$. Then the criterion function, which is obtained by simplifying the expression of (21.8), can be expressed as

$$\begin{aligned} S &= \frac{1}{2\pi i}\oint \frac{y(z^{-1})y(z)}{\mu(z^{-1})\mu(z)}\frac{dz}{z} \\ &= T\sum_{j=0}^{\infty}\sum_{k=0}^{\infty}\psi_k\psi_j c_{|k-j|} \\ &= \sum_{t=0}^{T-1}\sum_{s=0}^{T-1} y_t y_s \delta_{|t-s|}. \end{aligned} \tag{21.18}$$

Here $\delta_{|t-s|}$ is a coefficient of the self-reciprocal polynomial

$$\begin{aligned} \delta(z) &= \frac{1}{\mu(z^{-1})\mu(z)} \\ &= \{\delta_0 + \delta_1(z + z^{-1}) + \delta_2(z^2 + z^{-2}) + \cdots\} \end{aligned} \tag{21.19}$$

which corresponds to the autocovariance generating function of a synthetic qth-order AR process $y(t)$, defined by $\mu(L)y(t) = \varepsilon(t)$, which is based on a sequence $\varepsilon(t)$ of unit-variance white-noise disturbances.

The following theorem, which concerns the invertibility of the estimated MA model, is analogous to the theorem of (21.17) concerning the stationarity of the estimated AR model:

(21.20) Let $S = T\sum_{j=0}^{\infty}\sum_{k=0}^{\infty}\psi_k\psi_j c_{|k-j|}$ where ψ_k, ψ_j are coefficients in the expansion of the inverse polynomial $\mu^{-1}(z) = \{1 + \psi_1 z + \cdots\}$. If the coefficients of $\mu(z) = 1 + \mu_1 z + \cdots + \mu_q z^q$ are chosen so as to minimise the value of S, then all the roots of the equation $\mu(z) = 0$ will lie outside the unit circle.

Indeed, this theorem is almost self-evident. For S will have a finite value if and only if $\sum|\psi_i| < \infty$; and, for this to arise, it is necessary and sufficient that the roots $\mu(z) = 0$ lie outside the unit circle.

The feature which distinguishes the criterion function for MA estimation from its AR counterpart is the indefinite nature of the summations entailed in (21.18) and (21.19), which are due to the series expansions of $\mu^{-1}(z)$ and $\delta(z) = \{\mu(z^{-1})\mu(z)\}^{-1}$. The corresponding summations for the AR case, found under (21.14) and (21.15), are finite. In practice, the indefinite summations entailed by the MA criterion must be truncated. If the condition of invertibility is to be fulfilled by the estimates with any assurance, then it is important not truncate the summations too drastically. The closer the roots of $\mu(z)$ are to the perimeter of the unit circle, the slower is the rate of convergence of the coefficients of the expansion of $\mu^{-1}(z)$ and the greater must be the number of elements in the truncated summation.

The MA criterion function in the form of

$$\begin{aligned}\frac{1}{T}S &= \frac{1}{T}\sum_{\tau}\sum_{t} y_t y_{t-\tau}\delta_{\tau} \\ &= c_0 + 2\sum_{t=1}^{T-1} c_{\tau}\delta_{\tau}\end{aligned} \tag{21.21}$$

has been considered by Godolphin [216]. He has devised a specialised method for minimising the function which makes use of some approximations to the derivatives $d\delta_{\tau}/d\mu_j$ which are based on truncated power-series expansions. However, his iterative procedure has only a linear rate of convergence; and it is not clear how well the desirable properties of the criterion function survive the process of approximation.

Instead of deriving a specialised procedure for estimating a pure MA model, we shall concentrate of the producing a general algorithm of ARMA estimation which can also be used to estimate pure AR and pure MA models. It will be observed that the ARMA criterion function under (21.8) combines the features of the specialised AR and MA criteria functions of (21.14) and (21.18). The theorems under (21.17) and (21.20) together serve to show that the values of $\alpha_1, \ldots, \alpha_p$ and $\mu_1, \ldots, \mu_q$ which minimise the function correspond to a model which satisfies conditions both of stationarity and invertibility.

Representations via LT Toeplitz Matrices

Now let us look for the appropriate matrix representations of the ARMA model and of the criterion function. Consider setting to zero the presample elements $y_{-p}, \ldots, y_{-1}$ and $\varepsilon_{-q}, \ldots, \varepsilon_{-1}$ within the matrices of the equations under (21.2). If the matrices are also extended to a full set of T columns, then the resulting system will take the form of

$$\begin{bmatrix} y_0 & 0 & \ldots & 0 \\ y_1 & y_0 & \ldots & 0 \\ \vdots & \vdots & \ddots & \vdots \\ y_{T-1} & y_{T-2} & \ldots & y_0 \end{bmatrix}\begin{bmatrix} 1 \\ \alpha_1 \\ \vdots \\ 0 \end{bmatrix} = \begin{bmatrix} \varepsilon_0 & 0 & \ldots & 0 \\ \varepsilon_1 & \varepsilon_0 & \ldots & 0 \\ \vdots & \vdots & \ddots & \vdots \\ \varepsilon_{T-1} & \varepsilon_{T-2} & \ldots & \varepsilon_0 \end{bmatrix}\begin{bmatrix} 1 \\ \mu_1 \\ \vdots \\ 0 \end{bmatrix}. \tag{21.22}$$

This may be represented, in summary notation, by

$$Y\alpha = \mathcal{E}\mu. \tag{21.23}$$

Here Y and $\mathcal{E}$ are so-called lower-triangular (LT) Toeplitz matrices which are completely characterised by their leading vectors. These vectors are given by $\mathcal{E}e_0 = \varepsilon = [\varepsilon_0, \ldots, \varepsilon_{T-1}]'$ and $Ye_0 = y = [y_0, \ldots, y_{T-1}]'$, where e_0 is the leading vector of the identity matrix of order T. On the same principle, we can define lower-triangular Toeplitz matrices A and M which are characterised by their respective leading vectors $\alpha = [1, \alpha_1, \ldots, \alpha_p, 0, \ldots, 0]'$ and $\mu = [1, \mu_1, \ldots, \mu_q, 0, \ldots, 0]'$ which are found in equations (21.22) and (21.23).

Lower-triangular Toeplitz matrices can be represented as polynomial functions of a matrix $L = [e_1, \ldots, e_{T-1}, 0]$ which has units on the first subdiagonal and zeros elsewhere and which is formed from the identity matrix $I = [e_0, e_1, \ldots, e_{T-1}]$ by deleting the leading vector and appending a zero vector to the end of the array. Thus the matrix $A = A(\alpha)$ can be written as

$$\begin{aligned} A &= \alpha(L) \\ &= I + \alpha_1 L + \cdots + \alpha_p L^p. \end{aligned} \tag{21.24}$$

We may note that L^j is nilpotent of degree T in the index j such that $L^j = 0$ for $j \geq T$.

In some respects, the algebra of the matrices resembles the ordinary algebra of polynomials with a real or complex argument:

(i) The matrices commute such that $AY = YA$, whence $Ay = (AY)e_0 = (YA)e_0 = Y\alpha$,

(ii) If A, Y are LT Toeplitz matrices, then so is $AY = YA$,

(iii) If A is an LT Toeplitz matrix, then so is A^{-1}. In particular, $A^{-1}e_0 = [\omega_0, \omega_1, \ldots, \omega_{T-1}]'$ has the leading coefficients of the expansion of $\alpha^{-1}(z)$ as its elements.

(iv) $G = M'M$ is not a Toeplitz matrix.

Now consider the matrix representation of the criterion function. As an approximation of the function S of (21.8), we have

$$S^z = e_0'(Y'A'M'^{-1}M^{-1}AY)e_0 = y'A'M'^{-1}M^{-1}Ay. \tag{21.25}$$

This is just the coefficient associated with $I = L^0$ in the expansion of

$$\begin{aligned} \varepsilon(L')\varepsilon(L) &= y(L')\alpha(L')\mu^{-1}(L')\mu^{-1}(L)\alpha(L)y(L) \\ &= \frac{y(L')\alpha(L')\alpha(L)y(L)}{\mu(L')\mu(L)}, \end{aligned} \tag{21.26}$$

where $L' = [0, e_0, \ldots, e_{T-2}]$. We can afford to write this function in rational form on account of the commutativity in multiplication of the LT Toeplitz matrices.

The vectors α and μ which minimise the function S^z contain what Box and Jenkins [70] described as the conditional least-squares estimators. These estimators are not guaranteed to satisfy the conditions of stationarity and invertibility.

Consider the specialisation of the criterion function to the case of a pure AR model. Scaling by T^{-1} gives

$$\begin{aligned}\frac{1}{T}S^z &= \frac{1}{T}y'A'Ay \\ &= \frac{1}{T}\alpha'Y'Y\alpha.\end{aligned} \tag{21.27}$$

The first way of gauging the difference between this function and the model least-squares criterion function of (21.14) is to compare the matrix $A'A$ with the Toeplitz matrix $\Lambda = [\lambda_{|t-s|}]$ which, as we have already indicated, corresponds to the dispersion matrix of a synthetic pth-order MA process.

The second way is to compare the matrix

$$\frac{1}{T}Y'Y = \begin{bmatrix} \hat{c}_{00} & \hat{c}_{01} & \dots & \hat{c}_{0,T-1} \\ \hat{c}_{10} & \hat{c}_{11} & \dots & \hat{c}_{1,T-1} \\ \vdots & \vdots & & \vdots \\ \hat{c}_{T-1,0} & \hat{c}_{T-1,1} & \dots & \hat{c}_{T-1,T-1} \end{bmatrix} \tag{21.28}$$

with the matrix $C = [c_{|j-k|}]$ which contains the usual estimates of the autocovariances of the process. The matrix $Y'Y/T$ does not have the Toeplitz form which is required if the condition of stationarity is to be fulfilled in all cases.

Imagine that p zero elements are added to the tail of the vector $y = [y_0, \dots, y_{T-1}]'$ and that an LT Toeplitz matrix $\bar{Y}$ of order $T + p$ is formed from the "padded" vector in the manner in which $Y = Y(y)$ is formed from y. Then we should find that the principal minor of order $p + 1$ of the matrix $\bar{Y}\bar{Y}/T$ would coincide with that of the matrix C. Now let $\bar{\alpha}$ be formed from the vector α by the addition of p zeros. Then $\bar{\alpha}'\bar{Y}\bar{Y}\bar{\alpha}/T = \alpha'C\alpha$, and it follows that the criterion function has become equivalent to the function which delivers the stationary Yule–Walker estimates.

Now consider specialising the criterion function to the case of the pure MA model. Then

$$\begin{aligned}S^z &= y'M'^{-1}M^{-1}y \\ &= \psi'Y'Y\psi,\end{aligned} \tag{21.29}$$

where $\psi = M^{-1}e_0$ contains the first T coefficients of the expansion of $\mu^{-1}(z)$. This function may be compared with the function S of (21.18) which can be written as $S = y'\Delta y$, where $\Delta = [\delta_{|t-s|}]$ is the dispersion of a synthetic AR process.

In pursuit of estimates which fulfil the condition of invertibility, we can improve the approximation of $M'^{-1}M^{-1}$ to $\Delta = \Delta(\mu)$ by adding extra rows to the matrix M^{-1} so as to include additional coefficients of the series expansion of $\mu^{-1}(z)$. In practice, this object may be achieved, in the context of a computer procedure for estimating the parameters, by padding the tail of the vector y with zeros.

Representations via Circulant Matrices

The following is as example of a circulant matrix:

$$Y = \begin{bmatrix} y_0 & y_3 & y_2 & y_1 \\ y_1 & y_0 & y_3 & y_2 \\ y_2 & y_1 & y_0 & y_3 \\ y_3 & y_2 & y_1 & y_0 \end{bmatrix}. \tag{21.30}$$

Circulant matrices can be represented as polynomial functions of a matrix $K = [e_1, \ldots, e_{T-1}, e_0]$ which is formed from the identity matrix $I = [e_0, e_1, \ldots, e_{T-1}]$ by moving the leading vector to the back of the array. Thus the circulant matrix $A = A(\alpha)$ can be written as

$$\begin{aligned} A &= \alpha(K) \\ &= I + \alpha_1 K + \cdots + \alpha_p K^p. \end{aligned} \tag{21.31}$$

We may note that $K^{j+T} = K^j$ for all j and that $K^{-1} = K'$.

The algebra of circulant matrices closely resembles that of polynomials:

(i) The matrices commute in multiplication, $AY = YA$,

(ii) If A, Y are circulant, then so is $AY = YA$,

(iii) If A is circulant, then so are A' and A^{-1},

(iv) If M is circulant, then $M'M = MM'$ and $(M'M)^{-1}$ are circulant Toeplitz matrices.

Let $Y = y(K)$, $A = \alpha(K)$ and $M = \mu(K)$ be circulant matrices constructed from the same vectors y, α, μ as were the corresponding LT Toeplitz matrices. Then we can construct the following criterion function:

$$S^c = e_0'(Y'A'M'^{-1}M^{-1}AY)e_0 = y'A'M'^{-1}M^{-1}Ay. \tag{21.32}$$

This is just the coefficient associated with $I = K^0$ in the expansion of

$$\begin{aligned} \varepsilon(K)\varepsilon(K^{-1}) &= \frac{y(K^{-1})\alpha(K^{-1})\alpha(K)y(K)}{\mu(K^{-1})\mu(K)} \\ &= T\sigma_\varepsilon^2 \frac{c(K)}{\gamma(K)}, \end{aligned} \tag{21.33}$$

where $K^{-1} = [e_{T-1}, e_0, \ldots, e_{T-2}]$, and where $c(K) = y(K^{-1})y(K)/T$ and $\gamma(K) = \alpha(K^{-1})\alpha(K)/\mu(K^{-1})\mu(K)$.

The role of the matrix K in the above expression is essentially that of an indeterminate algebraic symbol, and it may be replaced by any other quantity which is a T-periodic function of the index j. In particular, we may replace K^j by $e^{i\omega_j} = \exp\{i2\pi j/T\}$. Then we have the result that

$$\begin{aligned} S^c &= \sum_{j=0}^{T-1} \frac{y(e^{-i\omega_j})\alpha(e^{-i\omega_j})\alpha(e^{i\omega_j})y(e^{i\omega_j})}{\mu(e^{-i\omega_j})\mu(e^{i\omega_j})} \\ &= T\sigma_\varepsilon^2 \sum_{j=0}^{T-1} \frac{c(e^{i\omega_j})}{\gamma(e^{i\omega_j})}. \end{aligned} \tag{21.34}$$

This follows from the fact that

$$\sum_{j=0}^{T-1} e^{i\omega_j} = \begin{cases} 0 & \text{if } j \neq 0, \\ T & \text{if } j = 0, \end{cases} \tag{21.35}$$

which indicates that the coefficient associated with $e^{i\omega_0} = 1$ can be isolated by summing over j.

The sum in (21.34) is manifestly an approximation to the function

$$S = T\frac{\sigma_\varepsilon^2}{2\pi}\int_{-\pi}^{\pi}\frac{c(e^{i\omega})}{\gamma(e^{i\omega})}d\omega \tag{21.36}$$

which is derived from (21.8) by taking the unit circle as the contour of integration. The approximation of S^c to S can be made arbitrarily close by increasing the number of frequency points ω_j at which the function is evaluated or, equivalently, by increasing the order of the matrix K. If the data series is of a fixed length T, then this is achieved by padding the vector y with zeros.

Consider specialising the criterion function to the case of a pure AR model. Then $T^{-1}S^c = T^{-1}y'A'Ay = T^{-1}\alpha'Y'Y\alpha$ has the form of the function of (21.27) apart from the fact that the matrices $A = \alpha(K)$ and $Y = y(K)$ are circulant matrices instead of LT Toeplitz matrices. The matrix $c(K) = Y'Y/T$ is given by

$$\begin{aligned} c(K) &= c_{T-1}K^{1-T} + \cdots + c_1K^{-1} + c_0I + c_1K + \cdots + c_{T-1}K^{T-1} \\ &= c_0I + (c_1 + c_{T-1})K + (c_2 + c_{T-2})K^2 + \cdots + (c_{T-1} + c_1)K^{T-1}, \end{aligned} \tag{21.37}$$

where the elements $c_0, \ldots, c_{T-1}$ come from (21.11). The equality follows from the fact that $K^{j-T} = K^j$.

Given that $c(K) = Y'Y/T$ is a positive-definite Toeplitz matrix, it follows from the theorem of (21.17) that the values which minimise the AR criterion function will correspond to a model which satisfies the condition of stationarity. The estimates will differ slightly from the Yule–Walker estimates because of the differences between $c(K) = Y'Y/T$ and $C = [c_{|j-k|}]$.

Consider the effect of adding p zeros to the tail of the vector y to create a vector $\tilde{y}$ and a corresponding matrix $\tilde{Y} = \tilde{y}(K)$ where K is now of order $T+p$ and $K^{T+p} = I$. Then, if $\tilde{c}(K) = \tilde{Y}'\tilde{Y}/T$, we have

$$\begin{aligned} \tilde{c}(K) &= c_0I + \cdots + c_pK^p + (c_{p+1} + c_{T-1})K^{p+1} + \cdots \\ &\quad + (c_{T-1} + c_{p+1})K^{T-1} + \cdots + c_1K^{T-1+p}. \end{aligned} \tag{21.38}$$

It can be seen that the principal minor of order $p+1$ of the matrix $\tilde{C} = \tilde{c}(K)$ is identical to that of the matrix $C = [c_{|k-j|}]$, and it follows that the criterion function has become equivalent to the function which delivers the Yule–Walker estimates.

Finally, let us comment on the specialisation of the criterion function to the case of the pure MA model. The criterion function has the form of the function under (21.29), albeit with circulant matrices $Y = y(K)$ and $M = \mu(K)$ in place of LT Toeplitz matrices. The conditions of the theorem of (21.20), which guarantee that the MA estimates will satisfy the condition of invertibility, are no longer fulfilled. Nevertheless, if there is any danger that the condition of invertibility may be violated, the simple expedient of padding the tail of the vector y with a sufficient number of zeros will avert the problem.

The representation of the least-squares criterion function which is in terms of circulant matrices is of some practical interest since, in the guise of equation (21.34), it is the criterion function which is entailed in the frequency-domain estimation of ARMA models. This method of estimation was expounded by Hannan [239] and has been investigated in some detail by Pukkila [413], [414]. Cameron and Turner [96] have show how to implement the method within the context of a flexible regression package.

The distinguishing feature of a frequency-domain ARMA estimation is the use of the fast Fourier transform (FFT) in performing the convolutions which are entailed in the multiplication of the polynomials or in the multiplication of the analogous matrices. There is little, if anything, to be gained from using the FFT when the data sample contains fewer than several hundred points.

Nowadays ARMA models are being used increasingly in signal-processing applications where there may be an abundance of data and where speed of computation is important. In such cases, a well-coded frequency-domain method may be far superior to a corresponding time-domain method.

The Gauss–Newton Estimation of the ARMA Parameters

In describing the Gauss–Newton (G–N) method for estimating the parameters of an ARMA model, we shall have in mind, principally, an LT Toeplitz representation of the ARMA equations. One can imagine that the data has been supplemented by extensive zero-padding. By this device, the criterion function can be made to approximate, to any degree of accuracy, the ideal criterion of estimation given under (21.8). It will be clear, however, that every aspect of the algorithm is equally applicable to a representation of the ARMA equations which is in terms of circulant matrices.

The object is to estimate the unknown elements of the parameter vector $\theta = [\alpha_1, \ldots \alpha_p, \mu_1, \ldots, \mu_q]'$ by finding the values which minimise the criterion function

$$S^z(\theta) = \varepsilon'\varepsilon = y'A'M'^{-1}M^{-1}Ay \tag{21.39}$$

which has been given previously under (21.25). The minimisation is achieved via the Gauss–Newton procedure of (12.52) which is described by the algorithm

$$\theta_{(r+1)} = \theta_{(r)} - \lambda_{(r)} \left\{ \frac{\partial\varepsilon'}{\partial\theta}\frac{\partial\varepsilon}{\partial\theta} \right\}^{-1}_{(r)} \left\{ \frac{\partial\varepsilon'}{\partial\theta}\varepsilon \right\}_{(r)}, \tag{21.40}$$

where $(\partial\varepsilon/\partial\theta)_{(r)}$ is the derivative of $\varepsilon(\theta) = M^{-1}Ay$ evaluated at $\theta = \theta_{(r)}$ and where $\lambda_{(r)}$ is a step-adjustment scalar which will be set to unity in the sequel.

The immediate problem is to find the requisite derivatives. Consider the expression $\varepsilon(z) = \mu^{-1}(z)\alpha(z)y(z)$. The ordinary rules of differentiation can be applied to show that

$$\begin{aligned} \frac{\partial\varepsilon(z)}{\partial\alpha_j} &= \mu^{-1}(z)y(z)z^j \\ &= \alpha^{-1}(z)\varepsilon(z)z^j, \end{aligned} \tag{21.41}$$

and that

$$\begin{aligned} \frac{\partial\varepsilon(z)}{\partial\mu_j} &= -\mu^{-2}(z)\alpha(z)y(z)z^j \\ &= -\mu^{-1}(z)\varepsilon(z)z^j. \end{aligned} \tag{21.42}$$

When the matrix L is substituted for the argument z, we obtain the derivatives in the form of Toeplitz matrices. The leading vectors are isolated by postmultiplying the matrices by e_0, which is the leading vector of the identity matrix of order T. Thus $\partial\varepsilon/\partial\alpha_j = \{\partial\varepsilon(L)/\partial\alpha_j\}e_0$ and $\partial\varepsilon/\partial\mu_j = \{\partial\varepsilon(L)/\partial\mu_j\}e_0$. Using the result that $L^j e_0 = e_j$, it is found that

$$\text{(21.43)} \qquad \frac{\partial\varepsilon}{\partial\alpha_j} = M^{-1}Ye_j \quad \text{and} \quad \frac{\partial\varepsilon}{\partial\mu_j} = -M^{-1}\mathcal{E}e_j,$$

where, apart from $\mathcal{E} = \varepsilon(L)$, there are the LT Toeplitz matrices $Y = y(L)$ and $M = \mu(L)$. The full set of derivatives are gathered together to form

$$\text{(21.44)} \qquad \frac{\partial\varepsilon}{\partial\theta} = \begin{bmatrix} M^{-1}Y_{1.p} & -M^{-1}\mathcal{E}_{1.q} \end{bmatrix},$$

where $Y_{1.p} = Y[e_1, \ldots, e_p]$ and $\mathcal{E}_{1.q} = \mathcal{E}[e_1, \ldots, e_q]$ represent the appropriate selections from the columns of Y and $\mathcal{E}$ respectively. When this expression is inserted into the equation (21.40), the expression for the Gauss–Newton algorithm becomes

$$\text{(21.45)} \qquad \begin{aligned} \begin{bmatrix} \alpha_{1.p} \\ \mu_{1.q} \end{bmatrix}_{(r+1)} &= \begin{bmatrix} \alpha_{1.p} \\ \mu_{1.q} \end{bmatrix}_{(r)} - \\ &\begin{bmatrix} Y'_{1.p}M'^{-1}M^{-1}Y_{1.p} & -Y'_{1.p}M'^{-1}M^{-1}\mathcal{E}_{1.q} \\ -\mathcal{E}'_{1.q}M'^{-1}M^{-1}Y_{1.p} & \mathcal{E}'_{1.q}M'^{-1}M^{-1}\mathcal{E}_{1.q} \end{bmatrix}^{-1}_{(r)} \begin{bmatrix} Y'_{1.p}M'^{-1}\varepsilon \\ -\mathcal{E}'_{1.q}M'^{-1}\varepsilon \end{bmatrix}_{(r)}. \end{aligned}$$

However, the identity

$$\text{(21.46)} \qquad \varepsilon = M^{-1}Y\alpha = M^{-1}y + M^{-1}Y_{1.p}\alpha_{1.p},$$

wherein $\alpha_{1.p} = [\alpha_1, \ldots, \alpha_p]'$, indicates that

$$\text{(21.47)} \qquad \begin{bmatrix} Y'_{1.p}M'^{-1}\varepsilon \\ -\mathcal{E}'_{1.q}M'^{-1}\varepsilon \end{bmatrix} = \begin{bmatrix} Y'_{1.p}M'^{-1}M^{-1}Y_{1.p}\alpha_{1.p} + Y'_{1.p}M'^{-1}M^{-1}y \\ -\mathcal{E}'_{1.q}M'^{-1}M^{-1}Y_{1.p}\alpha_{1.p} - \mathcal{E}'_{1.q}M'^{-1}M^{-1}y \end{bmatrix}.$$

The latter can be used in deriving alternative expressions for the RHS of equation (21.45).

An Implementation of the Gauss–Newton Procedure

In deriving the algorithm for the Gauss–Newton procedure, we have regarded $Y = y(L)$ and $A = \alpha(L)$ and $M = \mu(L)$ as LT Toeplitz matrices. In that case, $\mathcal{E} = M^{-1}AY$ is also an LT Toeplitz matrix. However, the symmetric products

$$\text{(21.48)} \qquad \begin{aligned} \frac{\partial\varepsilon'}{\partial\alpha_{1.p}}\frac{\partial\varepsilon}{\partial\alpha_{1.p}} &= Y'_{1.p}M'^{-1}M^{-1}Y_{1.p} \quad \text{and} \\ \frac{\partial\varepsilon'}{\partial\mu_{1.q}}\frac{\partial\varepsilon}{\partial\mu_{1.q}} &= \mathcal{E}'_{1.q}M'^{-1}M^{-1}\mathcal{E}_{1.q}, \end{aligned}$$

which are to be found within the equation for the G–N algorithm, are not Toeplitz matrices. Nor is the cross-product matrix

$$\frac{\partial \varepsilon'}{\partial \mu_{1.q}}\frac{\partial \varepsilon}{\partial \alpha_{1.p}} = -\mathcal{E}'_{1.q}M'^{-1}M^{-1}Y_{1.p} \tag{21.49}$$

a Toeplitz matrix. Nevertheless, as the sample size T increases, these various products come to approximate Toeplitz matrices with increasing accuracy. The same convergence ensues when the data is supplemented or "padded" with zeros.

Since a Toeplitz matrix of order T has only T distinct elements compared with the $(T^2+T)/2$ elements of an arbitrary symmetric matrix, there are considerable computational advantages in adopting Toeplitz forms for the products which enter the G–N algorithm.

A further advantage from using Toeplitz matrices in the G–N algorithm, which is indicated by the theorem under (21.17), is that the resulting estimates of the autoregressive parameters are guaranteed to fulfil the conditions of stationarity. Also, the theorem under (21.20) indicates that the conditions of invertibility will be fulfilled almost certainly by the estimates of the moving-average parameters if a large number of the coefficients of the expansion of $\mu^{-1}(z)$ are used instead of the limited number contained within of $\mu^{-1}(L) = M^{-1}$.

The differences between the Toeplitz and the non-Toeplitz representations may be illustrated in terms of the cross-product under (21.49). Consider the generic element

$$\frac{\partial \varepsilon'}{\partial \mu_i}\frac{\partial \varepsilon}{\partial \alpha_j} = -e_i'\mathcal{E}'M'^{-1}M^{-1}Ye_j. \tag{21.50}$$

In the Toeplitz representation, this is replaced by

$$\frac{1}{2\pi i}\oint \frac{\partial \varepsilon(z^{-1})}{\partial \mu_i}\frac{\partial \varepsilon(z)}{\partial \alpha_j}\frac{dz}{z} = \frac{-1}{2\pi i}\oint z^{j-i}\frac{\varepsilon(z^{-1})y(z)}{\mu(z^{-1})\mu(z)}\frac{dz}{z}, \tag{21.51}$$

which is the coefficient associated with z^{j-i} in the Laurent expansion of

$$\frac{\varepsilon(z^{-1})y(z)}{\mu(z^{-1})\mu(z)} = \left\{\cdots + \frac{q_{-2}}{z^2} + \frac{q_{-1}}{z} + q_0 + q_1 z + q_2 z^2 + \cdots\right\}. \tag{21.52}$$

Whereas the element of (21.50) is a function of both its indices i and j, the element of (21.51) which replaces it is a function only of the difference of the indices.

The formula of (21.50) suggests that the elements might be found via operations of matrix multiplication. However, the formula of (21.51) indicates that, when a Toeplitz representation is adopted, they should be found by the processes of convolution and of rational expansion which are entailed in finding the coefficients of polynomial products and of rational functions. For this purpose, we can use the procedures *Convolution* and *RationalExpansion* which are to be found under (2.14) and (3.43) respectively.

These procedures must be modified to accommodate data vectors which are of the *longVector* type. Thus, for example, a series of $n > T$ coefficients of the expansion of $y(z)/\mu(z)$, where $\mu(z) = \mu_0 + \mu_1 z + \cdots + \mu_q z^q$ and $y = y_0 + y_1 z + \cdots + y_T z^T$, may be found via the call

$$RationalExpansion(mu, q, T, n, y) \tag{21.53}$$

On the completion of the procedure, the coefficients of the expansion will be found in the *longVector* array y which has previously contained the coefficients of $y(z)$. A similar call may be used in finding the coefficients of $\varepsilon(z^{-1})/\mu(z^{-1})$.

The central coefficients of a Laurent expansion of (21.52) may be found by a procedure which finds the covariances of two mean-adjusted data series $x_0, x_1, \ldots, x_n$ and $y_0, y_1, \ldots, y_n$. When $n = T-1$, the covariances correspond to the coefficients of the product $c_{xy}(z) = T^{-1}x(z^{-1})y(z)$. In particular, the covariance at lag j is the product

$$c_{(xy)j} = \frac{1}{T}\sum_{t=j}^{T-1} x_{t-j}y_t = \frac{1}{2T\pi i}\oint z^j x(z)y(z^{-1})\frac{dz}{z}. \tag{21.54}$$

The procedure, which would also serve to find the coefficients of the autocovariance generating function $c(z) = T^{-1}y(z)y(z^{-1})$, is as follows:

(21.55)

```
procedure Covariances(x, y : longVector;
                      var covar : jVector;
                      n, p, q : integer);

  var
    t, j, s, f : integer;

begin {Covariances}
  for j := -p to q do
    begin {j}
      s := Max(0, j);
      f := Min(n, n + j);
      covar[j] := 0.0;
      for t := s to f do
        covar[j] := covar[j] + x[t - j] * y[t];
      covar[j] := covar[j]/(n + 1);
    end; {j}
end; {Covariances}
```

It is helpful to have a procedure which places the elements of the Laurent product within a matrix of the form

$$\begin{bmatrix} c_{(yy)0} & \cdots & c_{(yy)p-1} & c_{(yx)0} & \cdots & c_{(yx)q-1} \\ \vdots & \ddots & \vdots & \vdots & \ddots & \vdots \\ c_{(yy)p-1} & \cdots & c_{(yy)0} & c_{(yx)q-1} & \cdots & c_{(yx)0} \\ c_{(xy)0} & \cdots & c_{(xy)p-1} & c_{(xx)0} & \cdots & c_{(xx)q-1} \\ \vdots & \ddots & \vdots & \vdots & \ddots & \vdots \\ c_{(xy)p-1} & \cdots & c_{(xy)0} & c_{(xx)q-1} & \cdots & c_{(xx)0} \end{bmatrix}. \tag{21.56}$$

The following procedure serves to construct such a matrix from the arrays *covarYY*, *covarXX* and *covarXY* which contain the requisite elements:

(21.57)
```
procedure MomentMatrix(covarYY, covarXX, covarXY : jVector;
                       p, q : integer;
                       var moments : matrix);

  var
    i, j : integer;

  begin {MomentMatrix}
    for i := 1 to p + q do
      for j := 1 to p + q do
        if i >= j then
          begin {i, j : fill the lower triangle}
            if (i <= p) and (j <= p) then
              moments[i, j] := covarYY[i - j];
            if (i > p) and (j <= p) then
              moments[i, j] := covarXY[(i - p) - j];
            if (i > p) and (j > p) then
              moments[i, j] := covarXX[(i - p) - (j - p)];
            moments[j, i] := moments[i, j]
          end; {i, j}
  end; {MomentMatrix}
```

The above procedure serves the purpose of constructing a Toeplitz version of the matrix on the RHS of the equation under (21.45) which describes the G–N algorithm. The vector on the RHS is constructed with the help of the following procedure, which depends upon the products of the previous one.

(21.58)
```
procedure RHSVector(moments : matrix;
                    covarYY, covarXY : jVector;
                    alpha : vector;
                    p, q : integer;
                    var rhVec : vector);

  var
    i, j : integer;

begin {RHSVector}
  for i := 1 to p do
    rhVec[i] := covarYY[i];
  for i := p + 1 to p + q do
    rhVec[i] := covarXY[i - p];
  for i := 1 to p + q do
    for j := 1 to p do
      rhVec[i] := rhVec[i] + moments[i, j] * alpha[j];
end; {RHSVector}
```

The Gauss–Newton procedure itself, which makes use of the procedures listed above, can be constructed with various elaborations designed to speed its convergence and to avert failures. We shall present only the simplest implementation of the algorithm. Additional features may be added at will.

(21.59)

```
procedure GaussNewtonARMA(p, q, n : integer;
                          y : longVector;
                          var alpha, mu : vector);

  var
    rhVec, delta, theta, newtheta, newAlpha, newMu : vector;
    crossYY, crossEY, crossEE : jVector;
    storeY, storeE : longVector;
    moments : matrix;
    stepScalar, oldSS, newSS : real;
    i, iterations : integer;
    convergence : boolean;

begin {GaussNewtonARMA}
  convergence := false;
  iterations := 0;
  alpha[0] := 1;
  mu[0] := 1;
  newAlpha[0] := 1;
  newMu[0] := 1;

{Prepare vectors using initial parameters}
  for i := 0 to n do
    storeE[i] := -y[i];
  RationalExpansion(mu, q, n, n, storeE);
  Convolution(alpha, storeE, p, n);
  Covariances(storeE, storeE, crossEE, n, 0, 0);
  oldSS := crossEE[0];

  while (convergence = false) and (iterations < 20) do
    begin {while : beginning of the major loop}

{Form the moment matrix and the RHS vector}
      if q > 0 then
        RationalExpansion(mu, q, n, n, storeE);
      for i := 0 to n do
        storeY[i] := y[i];
      if q > 0 then
        RationalExpansion(mu, q, n, n, storeY);
      Covariances(storeY, storeY, crossYY, n, 0, p);
      if q > 0 then
        begin
```

```
          Covariances(storeE, storeY, crossEY, n, p, q);
          Covariances(storeE, storeE, crossEE, n, 0, q);
        end;
      MomentMatrix(crossYY, crossEE, crossEY, p, q, moments);
      RHSVector(moments, crossYY, crossEY, alpha, p, q, rhVec);

    {Find the updating vector}
      Cholesky(p + q, moments, delta, rhVec);

    {Update the value of theta}
      stepScalar := -Sqrt(2);
      repeat {until newSS < oldSS}
        stepScalar := -stepScalar/sqrt(2);
        for i := 1 to p + q do
          delta[i] := stepScalar * delta[i];
        for i := 1 to p do
          newAlpha[i] := alpha[i] - delta[i];
        for i := 1 to q do
          newMu[i] := mu[i] - delta[p + i];
        for i := 0 to n do
          storeE[i] := -y[i];
        if q > 0 then
          RationalExpansion(newMu, q, n, n, storeE);
        if p > 0 then
          Convolution(newAlpha, storeE, p, n);
        Covariances(storeE, storeE, crossEE, n, 0, 0);
        newSS := crossEE[0];
      until (newSS < oldSS * 1.0001);

      iterations := iterations + 1;
      oldSS := newSS;

      for i := 1 to p + q do
        begin {i}
          if i <= p then
            begin
              alpha[i] := newAlpha[i];
              theta[i] := alpha[i];
            end
          else if i > p then
            begin
              mu[i - p] := newMu[i - p];
              theta[i] := mu[i - p];
            end;
        end; {i}

    {Check for convergence}
```

```
        if q = 0 then
          convergence := true
        else
          convergence := CheckDelta(p + q, delta, theta);

      end; {while : end of the major loop}
    end; {GaussNewtonARMA}
```

This implementation incorporates a step-adjustment scalar which is called into play only if the parameter values computed by a straightforward iteration of the G–N procedure fail to secure a reduction in the value of the criterion function. In that case, the scalar, which is represented by $\lambda_{(r)}$ in equation (21.40), is given a succession of trial values

$$\lambda = \left(\frac{-1}{\sqrt{2}}\right)^j; \quad j = \{1, 2, 3, \ldots\}, \tag{21.60}$$

which tend towards zero. The trials cease when a reduction has been secured or when something close to the previous value of the criterion function has been recovered.

The estimation procedure terminates either when the sequence of G–N estimates is deemed to have converged or when the number of iterations exceeds a predetermined value. The test of convergence is performed by the function *CheckDelta* which is listed under (17.40).

The success of the G–N procedure depends largely upon the quality of the initial values which are supplied to it as parameters. These initial values may be obtained by use of the procedure *ARMAParameters* of (17.106) which finds the parameters of an ARMA process from the values of its autocovariances. In the present context, the autocovariances are, of course, determined empirically from the data.

Asymptotic Properties of the Least-Squares Estimates

If $y(t)$ is generated by a causal and invertible ARMA process, then the least-squares estimates of the parameters are equivalent, asymptotically, to the maximum-likelihood estimates which will be derived in the next chapter.

In the Chapter 25, we derive the ordinary theory of maximum-likelihood estimation under the assumption that the sample points are distributed independently and identically with well-defined moments up to the fourth order. The theory presented in the appendix is the prototype of a more general theory of maximum-likelihood estimation which extends to cases where the sample points are serially correlated. The results of the ordinary theory also prevail in the more general context.

The inferential theory of linear stochastic ARMA models has been developed to a high level of generality by Hannan [240] who has extended the results of Walker [505] and Whittle [518]. The article of Hannan makes use of a sophisticated central limit theorem for martingales. A more limited exposition of the theory which relies upon a central limit theorem for m-dependent sequences, which is due to Hoeffding and Robbins [257], has been provided by Brockwell and Davis [79]. The observations on an m-dependent sequence are independent provided that they are separated by more than m time periods. Therefore the theory is best suited to finite-order moving-average processes. However, the value of m can be increased indefinitely; and therefore the theory accommodates autoregressive and autoregressive moving-average processes which can be construed as infinite-order moving-average processes.

It follows from the theory of maximum-likelihood estimation, as well as from the general theory of least-squares estimation, that the estimate $\hat{\theta}$ of the ARMA parameters will have a limiting distribution with an expected value equal to the vector θ_0 of the true parameter values and with a vanishing dispersion. Moreover, the appropriate version of the central limit theorem will show that the limiting distribution is a normal distribution. Thus the asymptotic tendency can be expressed by

$$T^{-1/2}(\hat{\theta} - \theta) \xrightarrow{D} N(0, V), \tag{21.61}$$

where

$$V = \sigma_\varepsilon^2 \text{plim} \left\{ \frac{1}{T} \frac{\partial \varepsilon'(\theta_0)}{\partial \theta} \frac{\partial \varepsilon(\theta_0)}{\partial \theta} \right\}^{-1}. \tag{21.62}$$

The notation of equation (21.62) is intended to indicate that the derivatives are evaluated at the point of the true parameters values.

Our purpose now is to demonstrate how an expression may be found for the dispersion matrix V which is in terms of the parameters which are to be estimated. However, we should note that the matrix can be approximated usefully using the matrix of derivatives which is part of equation (21.45). The law of large number guarantees that the cross-products of the derivatives scaled by T^{-1} will converge to the corresponding expected values at $T \to \infty$.

To demonstrate these results, we need first to find the limiting value of $T^{-1}\varepsilon(z)\varepsilon(z^{-1})$. Consider the assumption that $\varepsilon(t)$ is a white-noise process. It follows that, for any two elements $\varepsilon_t, \varepsilon_s$, there is

$$E(\varepsilon_t \varepsilon_s) = \begin{cases} \sigma_\varepsilon^2, & \text{if } t = s; \\ 0, & \text{if } t \neq s. \end{cases} \tag{21.63}$$

Now, if the coefficients of $\alpha(z)$ and $\mu(z)$ were to assume their true values and if the presample elements $y_{-p}, \ldots, y_{-1}$ where incorporated in $y(z)$, then the coefficients of $\varepsilon(z)$ would be the true disturbances. In that case, we should have

$$E\{\varepsilon(z)\varepsilon(z^{-1})\} = \sum_{t=0}^{T-1} \sum_{s=0}^{T-1} E(\varepsilon_t \varepsilon_s) z^{t-s} = T\sigma_\varepsilon^2. \tag{21.64}$$

An analogous result prevails in the limit as $T \to \infty$. For, as the parameter estimates converge to the true values, the residuals will converge to the corresponding disturbances. Therefore, it follows from the law of large numbers that

$$\text{plim}(T \to \infty)\frac{1}{T}\varepsilon(z)\varepsilon(z^{-1}) = \text{plim}\frac{1}{T}\sum_{t=0}^{T-1}\sum_{s=0}^{T-1}\varepsilon_t\varepsilon_s z^{t-s} = \sigma_\varepsilon^2. \tag{21.65}$$

To find the analytic expressions for the elements of V, we take the forms of the derivatives found under (21.41) and (21.42) and we apply limits:

$$\begin{aligned}\text{plim}\left\{\frac{1}{T}\frac{\partial\varepsilon(z^{-1})}{\partial\alpha_i}\frac{\partial\varepsilon(z)}{\partial\alpha_j}\right\} &= \text{plim}\left\{\frac{1}{T}\frac{\varepsilon(z)\varepsilon(z^{-1})}{\alpha(z^{-1})\alpha(z)}z^{j-i}\right\}\\ &= \frac{\sigma_\varepsilon^2}{\alpha(z^{-1})\alpha(z)}z^{j-i},\end{aligned} \tag{21.66}$$

$$\begin{aligned}\text{plim}\left\{\frac{1}{T}\frac{\partial\varepsilon(z^{-1})}{\partial\mu_i}\frac{\partial\varepsilon(z)}{\partial\alpha_j}\right\} &= \text{plim}\left\{\frac{1}{T}\frac{\varepsilon(z)\varepsilon(z^{-1})}{\mu(z^{-1})\mu(z)}z^{j-i}\right\}\\ &= \frac{\sigma_\varepsilon^2}{\mu(z^{-1})\mu(z)}z^{j-i},\end{aligned} \tag{21.67}$$

$$\begin{aligned}\text{plim}\left\{\frac{1}{T}\frac{\partial\varepsilon(z^{-1})}{\partial\mu_i}\frac{\partial\varepsilon(z)}{\partial\alpha_j}\right\} &= \text{plim}\left\{\frac{1}{T}\frac{\varepsilon(z)\varepsilon(z^{-1})}{\mu(z^{-1})\alpha(z)}z^{j-i}\right\}\\ &= \frac{\sigma_\varepsilon^2}{\mu(z^{-1})\alpha(z)}z^{j-i}.\end{aligned} \tag{21.68}$$

The products which we are seeking, which are the elements of the dispersion matrix V indexed by i, j, are the coefficients associated with z^{j-i} in the Laurent expansion of the expressions on the RHS of these equations. Whenever the roots of the polynomials $\mu(z)$ and $\alpha(z)$ are available, it is straightforward to evaluate the Laurent expansion using the partial-fraction formula of (3).

Example 21.1. In the case of an ARMA(1, 1) process described by the equation $(1 + \alpha L)y(t) = (1 + \mu L)\varepsilon(t)$, we find that the dispersion matrix is

$$\begin{aligned}V &= \begin{bmatrix}(1-\alpha^2)^{-1} & (1-\alpha\mu)^{-1}\\ (1-\alpha\mu)^{-1} & (1-\mu^2)^{-1}\end{bmatrix}^{-1}\\ &= \frac{1-\alpha\mu}{(\alpha-\mu)^2}\begin{bmatrix}(1-\alpha^2)(1-\alpha\mu) & -(1-\alpha^2)(1-\mu^2)\\ -(1-\alpha^2)(1-\mu^2) & (1-\mu^2)(1-\alpha\mu)\end{bmatrix}.\end{aligned} \tag{21.69}$$

The Sampling Properties of the Estimators

Concern is sometimes expressed over the small-sample properties of the Yule–Walker estimator of a pure AR process. In particular, it appears that, in small samples, the moduli of the roots of the AR operator tend to be underestimated; and the severity of this bias increases as the roots approach the unit circle. The peaks of the estimated AR spectral density function which correspond to these roots assume less prominence that they should, and they may even disappear altogether. Evidence on these matters has been gathered by Lysne and Tjøstheim [326] and by Tjøstheim and Paulsen [485].

There is less evidence on the small-sample properties of the estimator of a pure MA model. However, it appears that there is a tendency to underestimate the moduli of the roots of the MA operator in small samples; and this is exacerbated when one resorts to the device of padding.

There are alternative ways of reaching an intuitive explanation of the small-sample bias of the Yule–Walker estimates which lead to various suggestions for improving their properties. These explanations make reference either to the sequence of empirical autocovariances or to the sequence of periodogram ordinates which represents the Fourier transform of the autocovariances.

To begin, consider the empirical autocovariance of lag τ which, on the assumption of that $E(y_t) = 0$, is given by

$$c_\tau = \frac{1}{T}\sum_{t=\tau}^{T-1} y_t y_{t-\tau}. \tag{21.70}$$

The expected value is

$$E(c_\tau) = \gamma_\tau\left(1 - \frac{|\tau|}{T}\right), \tag{21.71}$$

where γ_τ is the true value. If T is small, then the sequence of the estimated autocovariances is liable to decline more rapidly than it should as the lag value τ increases.

To understand the consequences of the over-rapid decline of the empirical autocovariances, we may consider the fact there is a one-to-one correspondence between the sequence $c_0, \ldots, c_p$ and the Yule–Walker estimates of the parameters $\sigma^2 = V(\epsilon_t), \alpha_1, \ldots, \alpha_p$. In particular, the estimated parameters satisfy the equation

$$-c_\tau = \alpha_1 c_{\tau-1} + \cdots + \alpha_p c_{\tau-p} \quad \text{for} \quad \tau = 1, \ldots, p. \tag{21.72}$$

This is a difference equation in $\{c_\tau\}$. If $\{c_0, \ldots, c_p\}$ is declining too rapidly, then the solution of the difference equation is liable to be over-damped, which means that the roots of the polynomial equation $\alpha(z) = 0$ will be too far from the unit circle.

One way of addressing the problem of bias is to replace the divisor T in the formula for c_τ by a divisor of $T - \tau$ so as to obtain unbiased estimates of the autocovariances. However, the resulting matrix of autocovariances is no longer guaranteed to be positive-definite; and this can lead to the violation of the condition of stationarity.

Another recourse is to adopt a two-stage estimation procedure. The initial Yule–Walker estimates can be used in forecasting sequences of postsample and presample values which are added to the sample. The forecast values in both directions will converge or "taper" to zero as the distance from the sample increases. At the points where the values are judged to be sufficiently close to zero, the sequences may be terminated. The Yule–Walker estimates which are obtained from the supplemented data have less bias than the initial estimates.

Recently, Pukkila [415] has proposed a modified Yule–Walker estimator which is calculated from autocorrelations which are obtained indirectly via the partial autocorrelation function. His sampling experiments suggest that the properties of the estimator are as good as, if not better than, those of the Burg estimator which is to be treated in the next section (see, for example, Burg [89], Ulrych and Bishop [493] and Giordano and Hsu [212]). In recent years, this estimator has provided a benchmark for small-sample performance.

The alternative way of explaining the bias in the Yule–Walker estimates is to consider the expectation of the periodogram which is the Fourier transform of sequence of the expected values of the empirical autocovariances:

$$E\{c(e^{i\omega})\} = \frac{1}{2\pi}\int_{-\pi}^{\pi}\gamma(e^{i\lambda})\kappa(e^{i\{\omega-\lambda\}})d\lambda. \tag{21.73}$$

The expected periodogram is also the convolution of the spectral density function $\gamma(e^{i\omega})$ with the Fejér kernel $\kappa(e^{i\omega})$. The former represents the Fourier transform of the sequence of true autocovariances whilst the latter represents the Fourier transform of the triangular sequence of the weights

$$d_\tau = \begin{cases} 1 - |\tau|/T & \text{if } |\tau| < T, \\ 0 & \text{if } |\tau| \geq T, \end{cases} \tag{21.74}$$

which appear in the expression for $E(c_\tau)$ under (21.71).

The convolution represents a smoothing operation, performed upon the spectral density function, which has the Fejér kernel as its weighting function. The effect of the operation is to diffuse the spectral power which spreads from the peaks of the spectrum, where it is concentrated, into the valleys. This is described as spectral leakage. The dispersion of the Fejér kernel diminishes as T increases, and, in the limit, it becomes a Dirac delta function. When the Dirac function replaces the Fejér kernel, the convolution delivers the spectral density function $\gamma(e^{i\omega})$ unimpaired.

An explanation of the presence of the Fejér kernel can be obtained from the notion that the sample values $y_t; t := 0, \ldots, T-1$ are obtained by applying the weights

$$w_t = \begin{cases} 1 & \text{if } 0 \leq t < T, \\ 0 & \text{otherwise}, \end{cases} \tag{21.75}$$

of a rectangular data window to the elements of an infinite sequence. The triangular weighting function $d_\tau = T^{-1}\sum_t w_t w_{\tau-t} = 1 - |\tau|/T$ of (21.74), which affects the sequence of autocovariances, and whose transform is the Fejér kernel, is formed from the convolution of two rectangular widows. By modifying the data window, we may alter the kernel function so as to reduce the leakage. In general, the leakage may be reduced by applying a taper to the ends of the rectangular window.

Investigations into the use of data-tapering in autoregressive estimation were pioneered by Pukkila [413] who modified the rectangular window by removing its corners to create a trapezoidal form. More recently, Dahlhaus [135] has investigated the effects upon the leakage of a tapered window obtained by splitting a cosine bell and inserting a sequence of units between the two halves. The sampling experiments of both Pukkila and Dahlhaus reveal dramatic improvements in the bias of the autoregressive estimates and in the resolution of the spectral density function which is inferred from these estimates.

Ideally the degree of tapering—which, in the case of Dahlhaus, is the ratio of the width of the cosine bell to the width of the data window—should be attuned to the values of the roots of $\alpha(z)$. A high degree of tapering is called for when the modulus of the dominant root is close to unity, which is usually the case when there is a prominent peak in the spectral density function.

The emphasis which has been placed, in the literature, upon the sampling properties of AR estimators should not detract from the importance of the MA component in time-series models. Its presence can greatly enhance the flexibility of the model in approximating transfer functions. An example is provided by the case of an AR process which has been corrupted by a white-noise error.

A white-noise corruption, which might arise simply from rounding error in the observations, increases the variance of the data, leaving its autocovariances unaffected. The inflated variance increases the damping of the autocovariances at the start of the sequence. This can lead to a severe underestimation of the moduli of the autoregressive roots. Formally, an AR(p) model with added white noise gives rise to an ARMA(p,p) process. Nevertheless, the noise corruption can often be accommodated by adding just a few moving-average parameters to the model.

The Burg Estimator

The Burg estimator is a close relative of the Yule–Walker estimator. It entails a recursive algorithm which is aimed at finding the sequence of values which constitute the (empirical) partial autocorrelation function and which are also described as reflection coefficients. Successive stages of the algorithm correspond to autoregressive models of increasing orders. At each stage, the autoregressive parameters may be obtained from the reflection coefficients and from the autoregressive parameters generated in the previous stage. The procedure by which this is accomplished is shared with the Durbin–Levinson algorithm which is the means of generating the Yule–Walker estimates recursively.

There is a crucial difference in the criterion functions which the two estimators are designed to minimise. The Yule–Walker estimator finds the values of $\alpha_1,\ldots,\alpha_p$ which minimise the sum of squares of one-step-ahead prediction errors which is defined as

$$S = \sum_{t=0}^{T-1} e_t^2 = T \sum_{i=0}^{p} \sum_{j=0}^{p} \alpha_i \alpha_j c_{|i-j|} \quad \text{with} \quad \alpha_0 = 1, \tag{21.76}$$

where $c_{|i-j|}$ is the empirical autocovariance defined according to the formula of (21.11). The Burg algorithm depends upon a sequence of criterion functions which pertain to the estimation of successive reflection coefficients. In the rth stage of the algorithm, a reflection coefficient $c_r = \alpha_{r(r)}$ is determined which minimises the sum of squares of forward and backwards prediction errors:

$$S_{(r)} = \sum_{t=r}^{T-1} \{e_{t(r)}^2 + b_{t(r)}^2\}. \tag{21.77}$$

These errors are defined as

$$\begin{aligned} e_{t(r)} &= y_t + \alpha_{1(r)} y_{t-1} + \cdots + \alpha_{r(r)} y_{t-r}, \\ b_{t(r)} &= \alpha_{r(r)} y_t + \cdots + \alpha_{1(r)} y_{t-r+1} + y_{t-r}. \end{aligned} \tag{21.78}$$

The index t of the sum of squares takes an initial value of $t = r$. The problem of attributing values to presample elements of the sequence $y(t)$ is thereby avoided. In the case of the Yule–Walker estimates, the presample elements are set to zero.

According to equation (19.117), the prediction errors under (21.78) may be formulated as

$$\begin{aligned} e_{t(r)} &= e_{t(r-1)} + c_r b_{t-1(r-1)}, \\ b_{t(r)} &= c_r e_{t(r-1)} + b_{t-1(r-1)}, \end{aligned} \tag{21.79}$$

where $c_r = \alpha_{r(r)}$ is the rth reflection coefficient and $e_{t(r-1)}$ and $b_{t-1(r-1)}$ are prediction errors generated by the autoregressive filter of order $r-1$. If the coefficients of the latter filter are given, then the reflection coefficient c_r is the only parameter which needs to be estimated directly in the rth stage. The remaining rth-order autoregressive parameters are obtained from the equations

$$\begin{bmatrix} \alpha_{1(r)} \\ \vdots \\ \alpha_{r-1(r)} \end{bmatrix} = \begin{bmatrix} \alpha_{1(r-1)} \\ \vdots \\ \alpha_{r-1(r-1)} \end{bmatrix} + \alpha_{r(r)} \begin{bmatrix} \alpha_{r-1(r-1)} \\ \vdots \\ \alpha_{1(r-1)} \end{bmatrix}. \tag{21.80}$$

The estimator of the rth reflection coefficient is derived from the function obtained by substituting the expressions for $e_{t(r)}$ and $b_{t(r)}$ under (21.79) into the sum of squares under (21.77). The result is

$$\begin{aligned} S_{(r)} &= \sum_{t=r}^{T-1} \Big[\big\{e_{t(r-1)} + c_r b_{t-1(r-1)}\big\}^2 + \big\{ c_r e_{t(r-1)} + b_{t-1(r-1)}\big\}^2 \Big] \\ &= (1 + c_r^2) \sum_{t=r}^{T-1} \{e_{t(r-1)}^2 + b_{t-1(r-1)}^2\} + 4c_r \sum_{t=r}^{T-1} e_{t(r-1)} b_{t-1(r-1)}. \end{aligned} \tag{21.81}$$

Differentiating the function with respect to c_r and setting the result to zero gives a first-order condition from which it is deduced that the optimum value of the reflection coefficient is

$$(21.82)\qquad c_r = -\frac{2\sum_{t=r}^{T-1} e_{t(r-1)}b_{t-1(r-1)}}{\sum_{t=r}^{T-1}\{e^2_{t(r-1)} + b^2_{t-1(r-1)}\}}.$$

The denominator in this expression is

$$(21.83)\qquad \sum_{t=r}^{T-1}\{e^2_{t(r-1)} + b^2_{t-1(r-1)}\} = S_{(r-1)} - e^2_{r-1(r-1)} - b^2_{T-1(r-1)},$$

which is easily calculated when the components on the RHS are available. Also, it may be confirmed that the minimised sum of squares of the prediction errors from the rth stage of the algorithm is

$$(21.84)\qquad S_{(r)} = (1 - c_r^2)\sum_{t=r}^{T-1}\{e^2_{t(r-1)} + b^2_{t-1(r-1)}\},$$

which is obtained immediately from the denominator. Therefore the labour in the rth stage is mainly in calculating the numerator of (21.82).

The starting values for the algorithm, which are obtained by setting $r = 0$ in (21.78), are

$$(21.85)\qquad e_{t(0)} = b_{t(0)} = y_t;\quad t = 0, \ldots, T-1.$$

It follows that

$$(21.86)\qquad S_{(0)} = 2\sum_{t=0}^{T-1} y_t^2.$$

In the following Pascal procedure, which implements the Burg algorithm, the segment which generates the autoregressive parameters via the scheme under (21.80) is shared with the *LevinsonDurbin* procedure which is listed under (17.75):

(21.87)
```
procedure BurgEstimation(var alpha, pacv : vector;
                         y : longVector;
                         p, Tcap : integer);

var
  t, r, n, j, jstop : integer;
  b, e : longVector;
  S, c, numer, denom, astore : real;

begin {BurgEstimation}
```

```
      n := Tcap - 1;
      S := 0;
      alpha[0] := 1;
      for t := 0 to n do
        begin {t}
          S := S + 2 * y[t] * y[t];
          e[t] := y[t];
          b[t] := y[t];
        end; {t}

      for r := 1 to p do
        begin {r}
          denom := S - e[r - 1] * e[r - 1] - b[n] * b[n];
          numer := 0.0;
          for t := r to n do
            numer := numer + 2 * e[t] * b[t - 1];

          c := -numer/denom;
          S := (1 - c * c) * denom;
          for t := n downto r do
            begin {t}
              b[t] := b[t - 1] + c * e[t];
              e[t] := e[t] + c * b[t - 1];
            end; {t}

        {Determine the autoregressive parameters}
          jstop := (r - 1) div 2;
          for j := 1 to jstop do
            begin {j}
              astore := alpha[j];
              alpha[j] := astore + c * alpha[r - j];
              alpha[r - j] := alpha[r - j] + c * astore;
            end; {j}
          j := jstop + 1;
          if odd(r - 1) then
            alpha[j] := alpha[j] * (1 + c);
          alpha[r] := c;
          pacv[r] := c

        end; {r}

    end; {BurgEstimation}
```

Bibliography

[17] Anderson, T.W., (1977), Estimation for Autoregressive Moving Average Models in Time and Frequency Domains, *The Annals of Statistics*, **5**, 842–86.

[30] Åström, K.J., and Söderström, (1974), Uniqueness of the Maximum Likelihood Estimates of the parameters of an ARMA Model, *IEEE Transactions on Automatic Control*, **AC-19**, 769–773.

[70] Box, G.E.P., and G.M. Jenkins, (1976), *Time Series Analysis: Forecasting and Control, Revised Edition*, Holden Day, San Francisco.

[79] Brockwell, P.J., and R.A. Davis, (1987), *Time Series: Theory and Methods*, Springer Verlag, New York.

[89] Burg, J.P., (1967), Maximum Entropy Spectral Analysis, *Proceedings of the 37th Meeting of the Society of Exploration Geophysicists, Oklahoma City.* Reprinted in *Modern Spectral Analysis*, D.G. Childers (ed.), (1978), IEEE Press, New York.

[90] Burg, J.P., (1968), A New Analysis Technique for Time Series Data, *NATO Advanced Study Institute on Signal Processing*, Reprinted in *Modern Spectral Analysis*, D.G. Childers (ed.), (1978), IEEE Press, New York.

[91] Burg, J.P., (1972), The Relationship between Maximum Likelihood and Maximum Entropy Spectra, *Geophysics*, **37**, 375–376.

[94] Cadzow, J.A., (1982), Spectral Estimation: An Overdetermined Rational Model Equation Approach, *Proceedings of the IEEE*, **70**, 907–937.

[96] Cameron, M.A., and T.R. Turner, (1987), Fitting Models to Spectra Using Regression Packages, *Applied Statistics*, **36**, 47–57.

[102] Childers, D.G., (1978), (ed.), *Modern Spectral Analysis*, IEEE Press, New York.

[103] Chiu, Shean-Tsong, (1988), Weighted Least Squares Estimators on the Frequency Domain for the Parameters of a Time Series, *The Annals of Statistics*, **16**, 1315–1326.

[104] Chiu, Shean-Tsong, (1991), A Linear Estimation Procedure for the Parameters of Autoregressive Moving-Average Processes, *Journal of Time Series Analysis*, **4**, 315–327.

[133] Cryer, J.D., J.C. Nankervis and N.E. Savin, (1989), Mirror-Image and Invariant Distributions in ARMA Models, *Econometric Theory*, **5**, 36–52.

[135] Dahlhaus, R., (1984), Parameter Estimation of Stationary Stochastic Processes with Spectra Containing Strong Peaks, in *Robust and Nonlinear Time Series Analysis: Lecture Notes in Statistics, vol. 26*, J. Franke, W. Hardle and D. Martin (eds.), Springer Verlag, New York.

[153] Denby, L., and D.R. Martin, (1979), Robust Estimation of the First-Order Autoregressive Parameter, *Journal of the American Statistical Association*, **74**, 140–146.

[165] Durbin, J., (1959), Efficient Estimation of Parameters in Moving Average Models, *Biometrika*, **46**, 306–316.

[212] Giordano, A.A., and F.M. Hsu, (1985), *Least Square Estimation with Applications to Digital Signal Processing*, John Wiley and Sons, New York.

[216] Godolphin, E.J., (1977), A Direct Representation for the Maximum Likelihood Estimator of a Gaussian Moving Average Process, *Biometrika*, **64**, 375–384.

[217] Godolphin, E.J., (1978), Modified Maximum Likelihood Estimation of a Gaussian Moving Average Using a Pseudoquadratic Convergence Criterion, *Biometrika*, **65**, 203–206.

[239] Hannan, E.J., (1969), The Estimation of Mixed Moving-Average Autoregressive Systems, *Biometrika*, **56**, 579–593.

[240] Hannan, E.J., (1973), The Asymptotic Theory of Linear Time-Series Models, *Journal of Applied Probability*, **10**, 130–145.

[257] Hoeffding W., and H. Robbins, (1948), The Central Limit Theorem for Dependent Variables, *Duke Mathematical Journal*, **15**, 773–780.

[268] Jaynes, E.T., (1982), On the Rationale of Maximum-Entropy Methods, *Proceedings of the IEEE*, **70**, 939–952.

[276] Kabaila, P., (1983), Parameter Values of ARMA Models Minimising the One-step-ahead Prediction Error when the True System is not in the Model Set, *Journal of Applied Probability*, **20**, 405–408.

[301] Koreisha, S., and T. Pukkila, (1990), A Generalised Least-Squares Approach for Estimation of Autoregressive Moving-Average Models, *Journal of Time Series Analysis*, **11**, 139–151.

[326] Lysne, D., and D. Tjøstheim, (1987), Loss of Spectral Peaks in Autoregressive Spectral Estimation, *Biometrika*, **74**, 200–206.

[344] Mentz, R.P., (1977), Estimation in the First-Order Moving Average Model through Finite Autoregressive Approximation: Some Asymptotic Results, *Journal of Econometrics*, **6**, 225–236.

[347] Milhøj, A., (1984), Bias Correction in the Frequency Domain Estimation of Time Series Models, *Biometrika*, **71**, 91–99.

[362] Nicholls, D.F., (1973), *Frequency Domain Estimation for Linear Models*, Biometrika.

[389] Phan-Dihn Tuan, (1979), The Estimation of Parameters for Autoregressive Moving Average Models with Sample Autocovariances, *Biometrika*, **66**, 555–560.

[413] Pukkila, T., (1977), *Fitting of Autoregressive Moving Average Models in the Frequency Domain*, Doctoral Thesis, Department of Mathematical Sciences, University of Tampere, Report A-6.

[414] Pukkila, T., (1979), The Bias in Periodogram Ordinates and the Estimation of ARMA Models in the Frequency Domain, *Australian Journal of Statistics*, **21**, 121–128.

[415] Pukkila, T., (1988), An Improved Estimation Method for Univariate Autoregressive Models, *Journal of Multivariate Analysis*, **27**, 422–433.

[485] Tjøstheim, D., and J. Paulsen, (1983), Bias of Some Commonly-Used Time Series Estimates, *Biometrika*, **70**, 389–399.

[493] Ulrych, T.J., and T.N. Bishop, (1975), Maximum Entropy Spectral Analysis and Autoregressive Decomposition, *Review of Geophysics and Space Physics*, **13**, 183–200.

[505] Walker, A.M., (1964), Asymptotic Properties of Least Squares Estimates of Parameters of the Spectrum of a Stationary Non-Deterministic Time-Series, *Journal of the Australian Mathematical Society*, **4**, 363–384.

[518] Whittle, P., (1953), Estimation and Information in Stationary Time Series, *Arkiv för Matematik*, **2**, 423–434.

CHAPTER 22

Maximum-Likelihood Methods of ARMA Estimation

The current value generated by a temporal stochastic process will depend upon the values generated in the past. Therefore, whenever a short sequence of consecutive values is recorded, much of the information which would help in explaining it is contained in unknown presample values. If the process is heavily dependent upon the past, then the decision of how to represent the presample values may have a significant effect upon the quality of the estimates of the parameters of the process. In particular, the more tractable methods of estimating the parameters are subject to the hazard that, if the roots of the autoregressive or moving-average polynomial operators of the underlying process are close to the boundary of the unit circle, then the stationarity and invertibility conditions are liable to be violated by the estimates.

There seems to be little doubt that the best method, in theory, of coping with the presample problem, and of fulfilling the conditions of stationarity and invertibility, is to use the estimating systems which are derived unequivocally from the principle of maximum likelihood. The resulting estimates are commonly described as the exact maximum-likelihood estimates. However, the exact criterion is difficult to fulfil, and the estimates are laborious to compute. Therefore, if the data is sufficiently abundant to make the starting-value problem negligible, or if there is a reason to believe that the estimates will be affected only slightly by the way the problem is handled, then more tractable estimation procedures, such as the ones given in the previous chapter, may be adopted.

In this chapter, we shall present a variety of maximum-likelihood methods. The primary purpose is to present the algorithms of exact maximum-likelihood estimation. A secondary purpose is to show how some of the methods of least-squares estimation may be construed as conditional maximum-likelihood methods which adopt differing approaches to the starting-value problem. In order to pursue these two purposes within the compass of a single chapter, we have to adopt a notation of sufficient generality which is also capable of showing the finer details. The burden of these details would be less if our only purpose were to present the algorithms of exact likelihood estimation.

Matrix Representations of Autoregressive Models

A stationary autoregressive process $y(t)$ of order p—known for short as an AR(p) process—is represented by the equation

$$\alpha(L)y(t) = \varepsilon(t), \tag{22.1}$$

where $\alpha(L) = 1+\alpha_1 L+\cdots+\alpha_p L^p$ and where $\varepsilon(t)$ is a sequence of independently and identically distributed random variables with expectations of zero. This equation can also be written as

$$y(t) = \alpha^{-1}(L)\varepsilon(t) = \phi(L)\varepsilon(t), \tag{22.2}$$

where $\phi(L) = \{1 + \phi_1 L + \phi_2 L^2 + \cdots\}$ is an infinite series; from which it appears that the current value of $y(t)$ depends upon the entire history of the disturbance sequence $\varepsilon(t)$. However, when we write $y(t) = -\alpha_1 y(t-1) - \cdots - \alpha_p y(t-p) + \varepsilon(t)$, we can see that the effects of this history are summarised by a few previous values of the sequence $y(t)$.

A sample of T elements from $y(t)$ may be represented in a corresponding vector equation:

$$A_* y_* + Ay = \varepsilon. \tag{22.3}$$

Here $y = [y_0, y_1, \ldots, y_{T-1}]'$ and $\varepsilon = [\varepsilon_0, \varepsilon_1, \ldots, \varepsilon_{T-1}]'$ are the vectors of order T containing, respectively, the elements of $y(t)$ and $\varepsilon(t)$ which fall within the sample period, whilst $y_* = [y_{-p}, \ldots, y_{-2}, y_{-1}]'$ contains the requisite presample elements of $y(t)$.

The banded lower-triangular matrix A, which is of order $T \times T$, is the analogue of the polynomial $\alpha(L)$, whilst the $T \times p$ matrix A_* is an extension of A which is due to the presample elements. The two matrices can be represented explicitly as follows:

$$A_* = \begin{bmatrix} \alpha_p & \ldots & \alpha_1 \\ \vdots & \ddots & \vdots \\ 0 & \ldots & \alpha_p \\ 0 & \ldots & 0 \\ \vdots & & \vdots \\ 0 & \ldots & 0 \\ 0 & \ldots & 0 \end{bmatrix}, \quad A = \begin{bmatrix} 1 & \ldots & 0 & 0 & \ldots & 0 & 0 \\ \vdots & \ddots & \vdots & \vdots & & \vdots & \vdots \\ \alpha_{p-1} & \ldots & 1 & 0 & \ldots & 0 & 0 \\ \alpha_p & \ldots & \alpha_1 & 1 & \ldots & 0 & 0 \\ \vdots & \ddots & \vdots & \vdots & \ddots & \vdots & \vdots \\ 0 & \ldots & \alpha_p & \alpha_{p-1} & \ldots & 1 & 0 \\ 0 & \ldots & 0 & \alpha_p & \ldots & \alpha_1 & 1 \end{bmatrix}. \tag{22.4}$$

It will prove helpful to express equation (22.2) in a more detailed notation which distinguishes the first p elements of the sample within y from the remainder:

$$\begin{bmatrix} A_{1*} & A_{11} & 0 \\ 0 & A_{21} & A_{22} \end{bmatrix} \begin{bmatrix} y_* \\ y_1 \\ y_2 \end{bmatrix} = \begin{bmatrix} \varepsilon_1 \\ \varepsilon_2 \end{bmatrix}. \tag{22.5}$$

The first p observations are contained in the vector $y_1 = [y_0, \ldots, y_{p-1}]'$ whilst the remainder are in $y_2 = [y_p, \ldots, y_{T-1}]'$. The vectors $\varepsilon_1 = [\varepsilon_0, \ldots, \varepsilon_{p-1}]'$ and $\varepsilon_2 = [\varepsilon_p, \ldots, \varepsilon_{T-1}]'$ contain the corresponding elements of $\varepsilon(t)$. The submatrices A_{1*} and A_{11}, which are both of order $p \times p$, are defined as

$$A_{1*} = \begin{bmatrix} \alpha_p & \alpha_{p-1} & \ldots & \alpha_1 \\ 0 & \alpha_p & \ldots & \alpha_2 \\ \vdots & \vdots & \ddots & \vdots \\ 0 & 0 & \ldots & \alpha_p \end{bmatrix}, \quad A_{11} = \begin{bmatrix} 1 & 0 & \ldots & 0 \\ \alpha_1 & 1 & \ldots & 0 \\ \vdots & \vdots & \ddots & \vdots \\ \alpha_{p-1} & \alpha_{p-2} & \ldots & 1 \end{bmatrix}. \tag{22.6}$$

The submatrix A_{21} may be formed by taking the first $T-p$ rows of A_*, whilst the submatrix A_{22} is a principal minor of order $(T-p)\times(T-p)$ taken from A. A comparison of equations (22.3) and (22.5) indicates the following identities:

$$A_* = \begin{bmatrix} A_{1*} \\ 0 \end{bmatrix}, \quad A = \begin{bmatrix} A_{11} & 0 \\ A_{21} & A_{22} \end{bmatrix}, \tag{22.7}$$

$$y = \begin{bmatrix} y_1 \\ y_2 \end{bmatrix}, \quad \varepsilon = \begin{bmatrix} \varepsilon_1 \\ \varepsilon_2 \end{bmatrix}. \tag{22.8}$$

Combining the T equations of (22.3) with the trivial identity $y_* = I_p y_*$ leads to the equations

$$\begin{bmatrix} y_* \\ \varepsilon \end{bmatrix} = \begin{bmatrix} I_p & 0 \\ A_* & A \end{bmatrix} \begin{bmatrix} y_* \\ y \end{bmatrix}, \tag{22.9}$$

$$\begin{bmatrix} y_* \\ y \end{bmatrix} = \begin{bmatrix} I_p & 0 \\ -A^{-1}A_* & A^{-1} \end{bmatrix} \begin{bmatrix} y_* \\ \varepsilon \end{bmatrix}. \tag{22.10}$$

The vectors y_* and ε in these equations are statistically independent with a zero covariance matrix $C(y_*, \varepsilon) = 0$. Therefore, with $D(y_*) = \sigma_\varepsilon^2 Q_p$ and $D(\varepsilon) = \sigma_\varepsilon^2 I_T$, the joint dispersion matrix is

$$D(y_*, \varepsilon) = \sigma_\varepsilon^2 \begin{bmatrix} Q_p & 0 \\ 0 & I_T \end{bmatrix}. \tag{22.11}$$

It follows from (22.10) that the joint dispersion matrix of y and y_* is

$$\begin{aligned} D(y_*, y) &= \sigma_\varepsilon^2 \begin{bmatrix} I & 0 \\ -A^{-1}A_* & A^{-1} \end{bmatrix} \begin{bmatrix} Q_p & 0 \\ 0 & I \end{bmatrix} \begin{bmatrix} I & -A_*'A'^{-1} \\ 0 & A'^{-1} \end{bmatrix} \\ &= \sigma_\varepsilon^2 \begin{bmatrix} Q_p & -Q_p A_*' A'^{-1} \\ -A^{-1}A_* Q_p & A^{-1}(A_* Q_p A_*' + I)A'^{-1} \end{bmatrix}. \end{aligned} \tag{22.12}$$

The inverse of this matrix is given by

$$D^{-1}(y_*, y) = \frac{1}{\sigma_\varepsilon^2} \begin{bmatrix} A_*'A_* + Q_p^{-1} & A_*'A \\ A'A_* & A'A \end{bmatrix}. \tag{22.13}$$

The AR Dispersion Matrix and its Inverse

The matrix of (22.10) has units on its principal diagonal and zeros above. Therefore its determinant is unity. It follows from the factorisation under (22.12) that $D(y_*, y)$ has a determinant which is equal to that of $D(y_*, \varepsilon)$. This, in turn, is equal to $\sigma_\varepsilon^{2(T+p)}|Q_p|$. Thus $\det D(y_*, y) = \sigma_\varepsilon^{2(T+p)}|Q_p|$. Similar reasoning leads to the result that

$$\det D(y) = \sigma_\varepsilon^{2T}|Q_T| = \sigma_\varepsilon^{2T}|Q_p|. \tag{22.14}$$

Our procedures for evaluating the exact likelihood function depend upon having tractable expressions for the quadratic form $y'Q_T^{-1}y$ and for the determinant $\det D(y)$. Consider, therefore, the dispersion matrix

$$D(y) = \sigma_\varepsilon^2 A^{-1}(A_* Q_p A_*' + I)A'^{-1}. \tag{22.15}$$

This is a submatrix of $D(y, y_*)$; and it is contained within the final expression under (22.12). Using the formula for the inverse of a sum of matrices found under (9.12), we get

$$D^{-1}(y) = \frac{1}{\sigma_\varepsilon^2}\{A'A - A'A_*(A_*'A_* + Q_p^{-1})^{-1}A_*'A\}. \tag{22.16}$$

However, the latter expression comprises the matrix Q_p^{-1} which is, as yet, of an unknown form. To find an expression for Q_p^{-1}, consider

$$D^{-1}(y_*, y_1) = \frac{1}{\sigma_\varepsilon^2}\begin{bmatrix} A_{1*}'A_{1*} + Q_p^{-1} & A_{1*}'A_{11} \\ A_{11}'A_{1*} & A_{11}'A_{11} \end{bmatrix}, \tag{22.17}$$

which is a matrix of order $2p$ derived by specialising the expression under (22.13) using the notation of equation (22.7). The symmetry of this matrix about its NE–SW axis implies that

$$A_{1*}'A_{1*} + Q_p^{-1} = A_{11}A_{11}'. \tag{22.18}$$

Here it should be recognised that, if $B^{\#}$ denotes the transpose of a matrix B about its NE–SW diagonal, then $(A_{11}'A_{11})^{\#} = A_{11}A_{11}'$. This is a consequence of the fact that A_{11} is a lower-triangular Toeplitz matrix. On writing $A_{11}A_{11}'$ in place of $A_*'A_* + Q_p^{-1} = A_{1*}'A_{1*} + Q_p^{-1}$ in equation (22.16), we get

$$\begin{aligned} D^{-1}(y) &= \frac{1}{\sigma_\varepsilon^2}\{A'A - A'A_*(A_{11}A_{11}')^{-1}A_*'A\} \\ &= \frac{1}{\sigma_\varepsilon^2}\left\{A'A - \begin{bmatrix} A_{11}'A_{1*}(A_{11}A_{11}')^{-1}A_{1*}'A_{11} & 0 \\ 0 & 0 \end{bmatrix}\right\}. \end{aligned} \tag{22.19}$$

Next, since they are banded upper-triangular matrices of the same order, A_{11}' and A_{1*} commute in multiplication to give $A_{11}'A_{1*} = A_{1*}A_{11}'$. Therefore, within (22.19), there is

$$\begin{aligned} A_{11}'A_{1*}(A_{11}A_{11}')^{-1}A_{1*}'A_{11} &= A_{1*}A_{11}'(A_{11}A_{11}')^{-1}A_{11}A_{1*}' \\ &= A_{1*}A_{1*}'; \end{aligned} \tag{22.20}$$

and it follows that

$$D^{-1}(y) = \frac{1}{\sigma_\varepsilon^2}\{A'A - A_*A_*'\}. \tag{22.21}$$

This is the result which we have been seeking. By using the expression for the inverse of a sum of matrices, it can be shown that

$$D(y) = \sigma_\varepsilon^2 A^{-1}\{I - A_*(A_*'A_* + I)^{-1}A_*'\}A'^{-1}, \tag{22.22}$$

and this expression replaces the one under (22.15) which is in terms of Q_p.

In forming the equation $L'L = Q_p^{-1}$ which is to be solved for L, we may use the expression

$$Q_p^{-1} = A_{11}'A_{11} - A_{1*}A_{1*}', \tag{22.23}$$

which is derived by specialising the expression $Q_T^{-1} = A'A + A_*A_*'$ from (22.21).

The matrix $A'A - A_*A_*'$ has been encountered already under (5.165) in connection with the conditions which are necessary and sufficient for the stability of a pth-order difference equation. These are known as the Schur–Cohn conditions. The difference equation is stable if and only if the matrix $A'A - A_*A_*'$ is positive-definite. This condition may be evaluated via the Cholesky decomposition of the matrix.

Example 22.1. In the case of an AR(2) process, the matrix $Q_p^{-1} = A_{11}'A_{11} - A_{1*}A_{1*}' = L'L$ is given by

$$\begin{bmatrix} \alpha_0^2 - \alpha_2^2 & \alpha_1\alpha_0 - \alpha_1\alpha_2 \\ \alpha_1\alpha_0 - \alpha_1\alpha_2 & \alpha_0^2 - \alpha_2^2 \end{bmatrix} = \begin{bmatrix} l_{11}^2 + l_{21}^2 & l_{21}l_{22} \\ l_{22}l_{21} & l_{22}^2 \end{bmatrix}, \tag{22.24}$$

where $\alpha_0 = 1$. The solutions of the equations

$$\begin{aligned} l_{22}^2 &= \alpha_0^2 - \alpha_2^2, \\ l_{21}l_{22} &= \alpha_1\alpha_0 - \alpha_1\alpha_2, \\ l_{11}^2 + l_{21}^2 &= \alpha_0^2 - \alpha_2^2, \end{aligned} \tag{22.25}$$

are given by

$$\begin{aligned} l_{22} &= \sqrt{(\alpha_0^2 - \alpha_2^2)}, \\ l_{21} &= \frac{\alpha_1(\alpha_0 - \alpha_2)}{\sqrt{(\alpha_0^2 - \alpha_2^2)}} = \alpha_1\frac{\sqrt{(\alpha_0 - \alpha_2)}}{\sqrt{(\alpha_0 + \alpha_2)}}, \\ l_{11} &= \left[\frac{(\alpha_0 - \alpha_2)}{(\alpha_0 + \alpha_2)}\left\{(\alpha_0 + \alpha_2)^2 - \alpha_1^2\right\}\right]^{1/2}. \end{aligned} \tag{22.26}$$

The solutions will be real-valued if and only if

$$\alpha_0^2 - \alpha_2^2 > 0 \quad \text{and} \quad (\alpha_0 + \alpha_2)^2 > \alpha_1^2; \tag{22.27}$$

and these are the conditions for the stability of a second-order difference equation which are to be found under (5.148).

Density Functions of the AR Model

Now let us assume that the elements of $\varepsilon(t)$ are distributed independently, identically and normally. Then $\varepsilon \sim N(0, \sigma_\varepsilon^2 I)$. It follows that the elements of $y(t)$ will be normally distributed also, so that $y_* \sim N(0, \sigma_\varepsilon^2 Q_p)$ and $y \sim N(0, \sigma_\varepsilon^2 Q_T)$. Given that they are statistically independent, vectors ε and y_* have a joint distribution which is just the product of the marginal distributions:

$$\begin{aligned} N(y_*, \varepsilon) &= N(y_*)N(\varepsilon) \\ &= (2\pi\sigma_\varepsilon^2)^{-p/2}|Q_p|^{-1/2} \exp\left\{\frac{-1}{2\sigma_\varepsilon^2} y_*' Q_p^{-1} y_*\right\} \\ &\quad \times (2\pi\sigma_\varepsilon^2)^{-T/2} \exp\left\{\frac{-1}{2\sigma_\varepsilon^2}\varepsilon'\varepsilon\right\}. \end{aligned} \tag{22.28}$$

The joint distribution of y_* and y is given by $N(y_*, y) = N\{y_*, \varepsilon(y_*, y)\}|J|$, where $\varepsilon(y_*, y)$ stands for the expression which gives ε in terms of y_* and y, and where $|J|$ is the Jacobian of the transformation of (22.9) from (y_*, y) to (y_*, ε). Since the Jacobian matrix is triangular with units on the principal diagonal, it follows that $|J| = 1$. Therefore,

$$\begin{aligned} N(y_*, y) &= N(y_*)N(y|y_*) \\ &= (2\pi\sigma_\varepsilon^2)^{-p/2}|Q_p|^{-1/2} \exp\left\{\frac{-1}{2\sigma_\varepsilon^2} y_*' Q_p^{-1} y_*\right\} \\ &\quad \times (2\pi\sigma_\varepsilon^2)^{-T/2} \exp\left\{\frac{-1}{2\sigma_\varepsilon^2}(A_* y_* + Ay)'(A_* y_* + Ay)\right\}. \end{aligned} \tag{22.29}$$

The marginal distribution of y may be expressed as

$$N(y) = (2\pi\sigma_\varepsilon^2)^{-T/2}|Q_T|^{-1/2} \exp\left\{\frac{-1}{2\sigma_\varepsilon^2} y' Q_T^{-1} y\right\}. \tag{22.30}$$

When y is partitioned into y_1 and y_2, the corresponding form for the inverse of Q_T is

$$Q_T^{-1} = \begin{bmatrix} A_{21}'A_{21} + Q_p^{-1} & A_{21}'A_{22} \\ A_{22}'A_{21} & A_{22}'A_{22} \end{bmatrix}. \tag{22.31}$$

Since $|Q_T| = |Q_p|$, it follows that $N(y)$ can be factorised as

$$\begin{aligned} N(y) &= N(y_1)N(y_2|y_1) \\ &= (2\pi\sigma_\varepsilon^2)^{-p/2}|Q_p|^{-1/2} \exp\left\{\frac{-1}{2\sigma_\varepsilon^2} y_1' Q_p^{-1} y_1\right\} \\ &\quad \times (2\pi\sigma_\varepsilon^2)^{(p-T)/2} \exp\left\{\frac{-1}{2\sigma_\varepsilon^2}(A_{21}y_1 + A_{22}y_2)'(A_{21}y_1 + A_{22}y_2)\right\}. \end{aligned} \tag{22.32}$$

This has the same form as the expression for $N(y_*, y)$ under (22.29). It serves to show that the starting-value problem is a transient one which affects the density function of y only via first p elements of the sample.

A further expression which can be derived in the context of the normal density function is the conditional expectation of y_* given y:

$$
\begin{aligned}
E(y_*|y) &= C(y_*, y)D^{-1}(y)y \\
&= -Q_p A_*'\big(A_* Q_p A_*' + I\big)^{-1} Ay \\
&= -\big(A_*' A_* + Q_p^{-1}\big)^{-1} A_*' Ay.
\end{aligned} \tag{22.33}
$$

Here the second equality depends upon the expression for $C(y_*, y)$ from (22.12) and the expression for $D(y)$ from (22.15). The final equality is by virtue of the matrix identity $AB'(BAB' + I)^{-1} = (A^{-1} + B'B)^{-1}B'$. The final expression can also be obtained directly from the elements of the matrix under (22.13).

The Exact M-L Estimator of an AR Model

The criterion function for the exact maximum-likelihood estimation of the parameter $\sigma_\varepsilon^2, \alpha_1, \ldots, \alpha_p$ of an AR(p) process is based on the marginal density function $N(y)$ of (22.30). The log of the likelihood function is

$$
L = -\frac{T}{2}\log(2\pi) - \frac{T}{2}\log\sigma_\varepsilon^2 - \frac{1}{2}\log|Q_T| - \frac{1}{2\sigma_\varepsilon^2}y'Q_T^{-1}y. \tag{22.34}
$$

By maximising L partially with respect to σ_ε^2, we obtain the estimating equation

$$
\sigma_\varepsilon^2 = \frac{1}{T}y'Q_T^{-1}y. \tag{22.35}
$$

By substituting this expression back into (22.34), and by taking $|Q_T| = |Q_p|$, we get the concentrated function

$$
L^c = -\frac{T}{2}\big\{\log(2\pi) + 1\big\} - \frac{T}{2}\log\left(\frac{y'Q_T^{-1}y}{T}\right) - \frac{1}{2}\log|Q_p|. \tag{22.36}
$$

It can be seen that maximising the likelihood function in respect of the autoregressive parameters is equivalent to minimising the function

$$
S^* = \frac{y'Q_T^{-1}y}{T}|Q_p|^{1/T}. \tag{22.37}
$$

Since $|Q_p|^{1/T}$ tends to unity as the sample size T increases, it is tempting to adopt the criterion of minimising simply the quadratic function

$$
y'Q_T^{-1}y = y_1'Q_p^{-1}y_1 + (A_{21}y_1 + A_{22}y_2)'(A_{21}y_1 + A_{22}y_2), \tag{22.38}
$$

of which the expression on the RHS is obtained from (22.31). However, it is the presence of the determinant $|Q_p|^{1/T}$ which ensures that the exact maximum-likelihood estimates satisfy the conditions of stationarity. This can be understood in view of an expression provided by Anderson and Mentz [19]:

$$
|Q_p|^{-1} = \prod_{i,j=1}^{p}(1 - \lambda_i\lambda_j), \tag{22.39}
$$

where $\lambda_1, \ldots, \lambda_p$ are the roots of the polynomial $1 + \alpha_1 z + \cdots + \alpha_p z^p = 0$. Whenever any of these roots approaches the boundary of the unit circle, the factor $|Q_p|^{1/T}$ will tend to infinity.

The exact maximum-likelihood estimates of an AR process may be found by an iterative procedure using one of the algorithms for nonlinear optimisation which are to found in Chapter 12. These algorithms depend upon a facility for evaluating the criterion function and its derivatives at an arbitrary point within an admissible parameter space. Since it is difficult to find analytic expressions for the derivatives of the present criterion function, these must be found by numerical means.

It is appropriate to take the Yule–Walker estimates or the Burg estimates as starting values; for these are guaranteed to satisfy the conditions of stationarity or stability. If the maximising values of the criterion function fall close to the boundary of the region of stationarity, then there is a danger that, in taking a small step away from a stationary value, one may cross the boundary. To guard against this, it is important to check the parameter values for stability prior to evaluating the criterion function.

The following Pascal procedure *ARLikelihood* is designed to provide the value of the expression S^* of (22.37). It may be used in conjunction with the optimisation routines of Chapter 12 by embedding it in a function which has the generic form of *Funct*(*lambda*, *theta*, *pvec*, *n*), wherein *lambda* is the step-adjustment scalar, *theta* is the value of the function's argument from the end of the previous iteration, *pvec* is the new direction vector and *n* is the order of *theta* and *pvec*. Such a function will serve to pass the value of S^* to the optimisation procedure.

(22.40)

```
procedure ARLikelihood(var S, varEpsilon : real;
                       var y : longVector;
                       alpha : vector;
                       Tcap, p : integer;
                       var stable : boolean);

  var
    i, j, k, t : integer;
    e, det : real;
    q : matrix;

begin {ARLikelihood}

{Find the Inverse Dispersion matrix}
  S := 0.0;
  for i := 0 to p - 1 do
    for j := 0 to i do
      begin {i, j}
        q[i, j] := 0.0;
        for k := 0 to p - 1 - i do
          begin {k}
            q[i, j] := q[i, j] + alpha[k] * alpha[k + i - j];
            q[i, j] := q[i, j] - alpha[p - k] * alpha[p - k - i + j];
```

```
            end; {k}
          if i <> j then
            S := S + 2 * y[i] * q[i, j] * y[j]
          else {if i = j}
            S := S + y[i] * q[i, i] * y[i];
        end; {i, j}

  {Evaluate the determinant}
    det := 1;
    stable := true;
    for i := 0 to p - 1 do
      begin {i}
        for j := 0 to i do
          begin {j}
            for k := 0 to j - 1 do
              q[i, j] := q[i, j] - q[k, k] * q[i, k] * q[j, k];
            if i > j then
              q[i, j] := q[i, j]/q[j, j];
          end; {j}
        det := det * q[i, i];
        if det <= 0 then
          stable := false;
      end; {i}

  {Evaluate the criterion function}
    if stable then
      begin {if}
        for t := p to Tcap - 1 do
          begin {t}
            e := y[t];
            for j := 1 to p do
              e := e + alpha[j] * y[t - j];
            S := S + e * e;
          end; {i}
        varEpsilon := S/Tcap;
        S := varEpsilon * Exp(Ln(1/det)/Tcap)
      end; {if}

  end; {ARLikelihood}
```

The first operation within the procedure *ARLikelihood* is to form the matrix Q_p^{-1}. Then the determinant of the matrix is evaluated by forming the product of the elements of the diagonal matrix $D = \text{diag}\{d_0, \ldots, d_{p-1}\}$ which is a factor of the Cholesky decomposition of $Q_p^{-1} = LDL'$ wherein L is a lower-triangular matrix with units for its diagonal elements. The code of the segment which performs these operations is borrowed from the procedure *LDLPrimeDecomposition* which is to be found under (7.48).

The condition of stability, which is that Q_p^{-1} must be positive-definite, is fulfilled if and only if the successive products $\prod_{i=0}^{r} d_i; r = 0, \ldots, p-1$ are all positive. If this condition is fulfilled, then the quadratic term $y'Q_T^{-1}y$ is calculated in the manner indicated by the RHS of (22.38). Otherwise the procedure is aborted; and this will be indicted by the Boolean variable *stability* which will take the value of *false*.

In cases where the condition of stability is violated, it should be possible to overcome the problem by reducing the length of the step which departs from a previous value of the parameter vector which must have fulfilled the condition. To accommodate such a facility, the function $Funct(lambda, theta, pvec, n)$ must be replaced by a procedure which is capable of redetermining step-length parameter *lambda*. The likelihood of a violation of the condition of stability is much reduced by using accurate estimates for the starting values of the optimisation procedure. The latter may be generated by the procedure *YuleWalker* of (17.67) or by the procedure *BurgEstimation* of (21.87).

The above procedure is presented in the belief that it represents the most efficient way of computing the value of the likelihood function of an AR model. However, an alternative procedure of a similar efficiency can be devised which is based upon the Levinson–Durbin algorithm which was first presented in Chapter 17 and which has been discussed further in Chapter 19.

Given the dispersion matrix $\Gamma = \sigma_\varepsilon^2 Q_p$, the Levinson–Durbin procedure will find the factors A and $D = \mathrm{diag}\{d_0, \ldots, d_{p-1}\}$ of $\Gamma^{-1} = A'D^{-1}A = Q_p^{-1}/\sigma_\varepsilon^2$ and of $\Gamma = A^{-1}DA'^{-1}$. Since A is a lower-triangular matrix with units on the diagonal, it follows that $\prod_i d_i = |\Gamma| = \sigma_\varepsilon^{2p}|Q_p|$. These are the essential elements which are needed for evaluating the likelihood function.

The circumstance which favours the method represented by the procedure *ARLikelihood* is the relative ease with which the inverse matrix Q_p^{-1} may be generated from the values of $\alpha_1, \ldots, \alpha_p$ compared with the labour of finding Q_p by generating the autocovariances.

Conditional M-L Estimates of an AR Model

The complexities of the exact maximum-likelihood estimator are due, in large measure, to the effect of the starting values. The estimation can be simplified considerably if these values are preassigned or if their estimation can be separated from that of the principal parameters of the model. This may be achieved via a conditional-likelihood approach,

The conditional distribution of y given y_*, which may be extracted from (22.29), is

$$(22.41) \qquad N(y|y_*) = (2\pi\sigma_\varepsilon^2)^{-T/2} \exp\left\{\frac{-1}{2\sigma_\varepsilon^2}(A_*y_* + Ay)'(A_*y_* + Ay)\right\}.$$

Once a value has been attributed to y_*, the conditional maximum-likelihood estimates of the parameters $\alpha_1, \ldots, \alpha_p$ may be found by locating the values which minimise the function

$$(22.42) \qquad S = (A_*y_* + Ay)'(A_*y_* + Ay).$$

The simplest way of representing the presample elements is to set them equal to their unconditional expectations, which are zeros, to give $S = y'A'Ay$.

Let Y be the banded lower-triangular matrix which has $y = Ye_0$ as its leading vector, where e_0 is the leading vector of an identity matrix of order T. This matrix has the same structure as the matrix A, and therefore the two matrices commute in multiplication such that $AY = YA$. It follows that the criterion function of (22.42) with $y_* = 0$ can also be written as

$$\begin{aligned} S &= y'A'Ay \\ &= e_0'(Y'A'AY)e_0 = e_0'(A'Y'YA)e_0 \\ &= \alpha'Y'Y\alpha, \end{aligned} \tag{22.43}$$

where $\alpha = Ae_0 = [1, \alpha_1, \ldots, \alpha_p, 0, \ldots, 0]'$ is the leading vector of the matrix A. Thus S is a quadratic function of both y and α. To express S in a more revealing form, let us define

$$Y_{1.p} = Y[e_1, \ldots, e_p] \quad \text{and} \quad \alpha_{1.p}' = \alpha'[e_1, \ldots, e_p], \tag{22.44}$$

where e_j stands for the $(j+1)$th column of the identity matrix of order T. Then the criterion may be written as

$$S = (y + Y_{1.p}\alpha_{1.p})'(y + Y_{1.p}\alpha_{1.p}). \tag{22.45}$$

This is the criterion function of an ordinary linear least-squares regression.

One is not bound to set the presample elements to zero. Instead, they might regarded as nuisance parameters which should be estimated at the same time as the parameters in $\alpha_1, \ldots, \alpha_p$. An estimating equation for the vector of presample values may be found by minimising the function $S = (A_*y_* + Ay)'(A_*y_* + Ay)$ partially in respect of y_* when the other quantities are fixed. The minimising vector is

$$y_* = -(A_*'A_*)^{-1}A_*'Ay; \tag{22.46}$$

and it is notable that this estimate differs from the conditional expectation given under (22.33). On substituting the estimate back into the S, we get the concentrated criterion function

$$\begin{aligned} S^c &= y'A'\{I - A_*(A_*'A_*)^{-1}A_*'\}Ay \\ &= \begin{bmatrix} y_1' & y_2' \end{bmatrix} \begin{bmatrix} A_{11}' & A_{21}' \\ 0 & A_{22}' \end{bmatrix} \begin{bmatrix} 0 & 0 \\ 0 & I_{T-p} \end{bmatrix} \begin{bmatrix} A_{11} & 0 \\ A_{21} & A_{22} \end{bmatrix} \begin{bmatrix} y_1 \\ y_2 \end{bmatrix} \\ &= (A_{21}y_1 + A_{22}y_2)'(A_{21}y_1 + A_{22}y_2). \end{aligned} \tag{22.47}$$

This criterion function has a structure which is analogous to that of the original function $S = (A_*y_* + Ay)'(A_*y_* + Ay)$. However, in the new function, the role of the vector $y_* = [y_{-p}, \ldots, y_{-1}]'$ of the p presample values is played by the vector $y_1 = [y_0, \ldots, y_{p-1}]'$ comprising the first p observations.

It is clear that the estimates which are obtained by minimising S^c are equivalent to those which would be obtained by minimising the conditional likelihood function

$$N(y_2|y_1) = (2\pi\sigma_\varepsilon^2)^{(p-T)/2} \exp\left\{\frac{-1}{2\sigma_\varepsilon^2}(A_{21}y_1 + A_{22}y_2)'(A_{21}y_1 + A_{22}y_2)\right\}. \tag{22.48}$$

The conditional-likelihood estimators do not impose the condition of stability upon the autoregressive parameters. This could be a desirable property in some circumstances.

Matrix Representations of Moving-Average Models

Consider a qth-order moving-average process—which is described as an MA(q) process for short. This may be represented by the equation

$$y(t) = (1 + \mu_1 L + \cdots + \mu_q L^q)\varepsilon(t), \tag{22.49}$$

where $\varepsilon(t)$ is a sequence of independently and identically distributed random variables. The current value of the process is explained in terms of a finite number of unobservable disturbances. If the roots of $\mu(z) = 1 + \mu_1 z + \cdots + \mu_q z^q$ lie outside the unit circle, then the process is invertible and it can be represented by

$$\mu^{-1}(L)y(t) = \psi(L)y(t) = \varepsilon(t), \tag{22.50}$$

where $\psi(L) = \{1+\psi_1 L+\psi_2 L^2+\cdots\}$. Then $y(t) = \varepsilon(t)-\{\psi_1 y(t-1)+\psi_2 y(t-2)+\cdots\}$. This shows that the current value of $y(t)$ and the current value of the prediction-error sequence $\varepsilon(t)$ depend upon the entire past history of the observable sequence. This is in contrast to the AR(p) process where the current values of $y(t)$ and $\varepsilon(t)$ depend only on p lagged values. The unlimited dependence of the prediction errors of the MA process on the past values of $y(t)$ greatly complicates the evaluation of the MA likelihood function.

A set of T realisations of $y(t)$ may be represented in matrix form by

$$y = M_*\varepsilon_* + M\varepsilon. \tag{22.51}$$

Here $y = [y_0, y_1, \ldots, y_{T-1}]'$ and $\varepsilon = [\varepsilon_0, \varepsilon_1, \ldots, \varepsilon_{T-1}]'$ are the vectors of order T containing the elements of $y(t)$ and $\varepsilon(t)$, respectively, which fall within the sample period, whilst $\varepsilon_* = [\varepsilon_q, \ldots, \varepsilon_{-2}, \varepsilon_{-1}]'$ contains the requisite presample elements of $\varepsilon(t)$. The matrices M_* and M can be represented explicitly as follows:

$$M_* = \begin{bmatrix} \mu_q & \ldots & \mu_1 \\ \vdots & \ddots & \vdots \\ 0 & \ldots & \mu_q \\ 0 & \ldots & 0 \\ \vdots & & \vdots \\ 0 & \ldots & 0 \\ 0 & \ldots & 0 \end{bmatrix}, \quad M = \begin{bmatrix} 1 & \ldots & 0 & 0 & \ldots & 0 & 0 \\ \vdots & \ddots & \vdots & \vdots & & \vdots & \vdots \\ \mu_{q-1} & \ldots & 1 & 0 & \ldots & 0 & 0 \\ \mu_q & \ldots & \mu_1 & 1 & \ldots & 0 & 0 \\ \vdots & \ddots & \vdots & \vdots & \ddots & \vdots & \vdots \\ 0 & \ldots & \mu_q & \mu_{q-1} & \ldots & 1 & 0 \\ 0 & \ldots & 0 & \mu_q & \ldots & \mu_1 & 1 \end{bmatrix}. \tag{22.52}$$

Combining the T realisations of the MA(q) model given in equation (22.51) with the trivial identity $\varepsilon_* = I_q\varepsilon_*$, leads to the equations

$$\begin{bmatrix} \varepsilon_* \\ y \end{bmatrix} = \begin{bmatrix} I_q & 0 \\ M_* & M \end{bmatrix} \begin{bmatrix} \varepsilon_* \\ \varepsilon \end{bmatrix}, \tag{22.53}$$

for which the inverse is

$$\begin{aligned} \begin{bmatrix} \varepsilon_* \\ \varepsilon \end{bmatrix} &= \begin{bmatrix} I_q & 0 \\ -M^{-1}M_* & M^{-1} \end{bmatrix} \begin{bmatrix} \varepsilon_* \\ y \end{bmatrix} \\ &= K\varepsilon_* + Ny, \end{aligned} \tag{22.54}$$

where

$$K = \begin{bmatrix} I_q \\ -M^{-1}M_* \end{bmatrix}, \quad N = \begin{bmatrix} 0 \\ M^{-1} \end{bmatrix}. \tag{22.55}$$

Given that the vectors ε_* and ε contain the elements of a white-noise process, it follows that $D(\varepsilon_*) = \sigma_\varepsilon^2 I_q$ and $D(\varepsilon) = \sigma_\varepsilon^2 I_T$. Therefore, the joint dispersion matrix of ε_* and y is

$$D(\varepsilon_*, y) = \sigma_\varepsilon^2 \begin{bmatrix} I_p & M_*' \\ M_* & MM' + M_*M_*' \end{bmatrix} = \sigma_\varepsilon^2 \begin{bmatrix} I_p & 0 \\ M_* & M \end{bmatrix} \begin{bmatrix} I_p & M_*' \\ 0 & M' \end{bmatrix}. \tag{22.56}$$

The inverse of this matrix is

$$\begin{aligned} D^{-1}(\varepsilon_*, y) &= \frac{1}{\sigma_\varepsilon^2} \begin{bmatrix} I + M_*'(MM')^{-1}M_* & -M_*'(MM')^{-1} \\ -(MM')^{-1}M_*' & (MM')^{-1} \end{bmatrix} \\ &= \frac{1}{\sigma_\varepsilon^2} \begin{bmatrix} K'K & K'N \\ N'K & N'N \end{bmatrix}. \end{aligned} \tag{22.57}$$

The MA Dispersion Matrix and its Determinant

Within (22.56), is found the dispersion matrix

$$D(y) = \sigma_\varepsilon^2 (MM' + M_*M_*'). \tag{22.58}$$

By using the formula for the inverse of a sum of matrices which is found under (9.12), we may obtain the inverse of this matrix:

$$\begin{aligned} D^{-1}(y) &= \frac{1}{\sigma_\varepsilon^2} M'^{-1} \Big[I_T - M^{-1}M_* \big\{ I_q + M_*'(MM')^{-1}M_* \big\}^{-1} M_*' M'^{-1} \Big] M^{-1} \\ &= \frac{1}{\sigma_\varepsilon^2} N' \{ I_{T+q} - K(K'K)^{-1}K' \} N. \end{aligned} \tag{22.59}$$

A tractable expression for $\det D(y)$ is also required. Therefore, let us reconsider the triangular factorisation of $D(\varepsilon_*, y)$ which appears under (22.56). Clearly, $\det D(\varepsilon_*, y) = \sigma_\varepsilon^{2(T+q)}$, since the diagonal elements of its triangular factors are all units. However, by applying the formula

$$\det\begin{bmatrix} E & F \\ G & H \end{bmatrix} = |E||H - GE^{-1}F| = |H||E - FH^{-1}G| \tag{22.60}$$

to (22.57), which stands for the inverse of the dispersion matrix, we find that

$$\begin{aligned}\sigma_\varepsilon^{2(T+q)}\det D^{-1}(\varepsilon_*, y) &= |K'K||N'N - N'K(K'K)^{-1}K'N| \\ &= \sigma_\varepsilon^{2T}|K'K||D^{-1}(y)|,\end{aligned} \tag{22.61}$$

where the final equality comes from (22.59). Since the value of this expression is unity, it follows immediately that

$$\det D(y) = \sigma_\varepsilon^{2T}|K'K|. \tag{22.62}$$

The implication of (22.62) is that we can evaluate the determinant of the dispersion matrix of order T by assessing the determinant of an equivalent matrix of order q. This represents an advantage from the point of view of computation; for, given that q is a relatively small order, it is possible to form and to store the matrix $K'K$ in its entirely. By contrast, unless the number T of the observations is strictly limited, there is no possibility of forming the matrix $D(y)$ in its entirety. However, given that there are only $q + 1$ distinct nonzero autocovariances in the dispersion matrix, we should not think of devoting more than $q + 1$ registers to its storage. In fact, as we shall see later, an efficient way of calculating the value of $|Q_T| = |K'K|$ is to form the product of the elements of the diagonal matrix D of the factorisation $Q_T = LDL'$.

Density Functions of the MA Model

Now let us assume that the elements of $\varepsilon(t)$ are distributed independently identically and normally. Then

$$\begin{aligned}N(\varepsilon_*, \varepsilon) &= N(\varepsilon_*)N(\varepsilon) \\ &= (2\pi\sigma_\varepsilon^2)^{-(q+T)/2}\exp\left\{\frac{-1}{2\sigma_\varepsilon^2}(\varepsilon_*'\varepsilon_* + \varepsilon'\varepsilon)\right\}.\end{aligned} \tag{22.63}$$

The joint distribution of ε_* and y is given by $N(\varepsilon_*, y) = N\{\varepsilon_*, \varepsilon(\varepsilon_*, y)\}|J|$, where $\varepsilon(\varepsilon_*, y)$ stands for the expression which gives ε in terms of ε_* and y, and where $|J|$ is the Jacobian of the transformation of (22.54) from (ε_*, y) to $(\varepsilon_*, \varepsilon)$. The value of the Jacobian is unity. Therefore,

$$\begin{aligned}N(\varepsilon_*, y) &= N(y|\varepsilon_*)N(\varepsilon_*) \\ &= (2\pi\sigma_\varepsilon^2)^{-T/2}\exp\left\{\frac{-1}{2\sigma_\varepsilon^2}(y - M_*\varepsilon_*)'(MM')^{-1}(y - M_*\varepsilon_*)\right\} \\ &\quad\times(2\pi\sigma_\varepsilon^2)^{-q/2}\exp\left\{\frac{-1}{2\sigma_\varepsilon^2}\varepsilon_*'\varepsilon_*\right\}.\end{aligned} \tag{22.64}$$

With the help of the expressions for $D^{-1}(y) = (1/\sigma_\varepsilon^2)Q_T^{-1}$ and $\det D(y)$, which may be found under (22.59) and (22.62), the marginal distribution of y may be expressed as

$$\begin{aligned} N(y) &= (2\pi\sigma_\varepsilon^2)^{-T/2}|Q_T|^{-1/2}\exp\left\{\frac{-1}{2\sigma_\varepsilon^2}y'Q_T^{-1}y\right\} \\ &= (2\pi\sigma_\varepsilon^2)^{-T/2}|K'K|^{-1/2}\exp\left\{\frac{-1}{2\sigma_\varepsilon^2}y'N'\big[I - K(K'K)^{-1}K'\big]Ny\right\}. \end{aligned} \tag{22.65}$$

It is also appropriate, at this point, to derive the conditional expectation of ε_* given y:

$$\begin{aligned} E(\varepsilon_*|y) &= M_*'(MM' + M_*M_*')^{-1}y \\ &= \big\{I + M_*'(MM')^{-1}M_*\big\}^{-1}M_*'(MM')^{-1}y \\ &= -(K'K)^{-1}K'Ny. \end{aligned} \tag{22.66}$$

The first expression on the RHS comes from putting the expressions for $C(\varepsilon_*, y)$ and $D(y)$ from (22.56) into the formula $E(\varepsilon_*|y) = C(\varepsilon_*, y)D^{-1}(y)y$. The second expression is by virtue of the identity $B'(A + BB')^{-1} = (I + B'A^{-1}B)^{-1}B'A^{-1}$. The final expression, which depends upon the definition of K under (22.55), can also be obtained directly from the elements of the matrix under (22.57).

The Exact M-L Estimator of an MA Model

The criterion function for the exact maximum-likelihood estimation of the parameters $\sigma_\varepsilon^2, \mu_1, \ldots, \mu_q$ of an MA(q) process is based on the marginal density function $N(y)$ of (22.65). The logarithm of the likelihood function is

$$L(y, \sigma_\varepsilon^2) = -\frac{T}{2}\log(2\pi) - \frac{T}{2}\log(\sigma_\varepsilon^2) - \frac{1}{2}\log|Q_T| - \frac{1}{2\sigma_\varepsilon^2}y'Q_T^{-1}y \tag{22.67}$$

which, as it stands, is indistinguishable from the corresponding expression for the likelihood equation for the autoregressive model.

The variance σ_ε^2 may be eliminated from the expression by replacing it by the relevant estimating equation. Differentiating $L(y, \sigma_\varepsilon^2)$ with respect to σ_ε^2 and setting the result to zero gives a first-order condition which leads to the estimating equation

$$\sigma_\varepsilon^2 = \frac{1}{T}y'Q_T^{-1}y. \tag{22.68}$$

When the latter is substituted into equation (22.67), a concentrated function is obtained in the form of

$$L^c = -\frac{T}{2}\big\{\log(2\pi) + 1\big\} - \frac{T}{2}\log(y'Q_T^{-1}y/T) - \frac{1}{2}\log|Q_T|. \tag{22.69}$$

Maximising L^c is equivalent to minimising the function

$$S^* = \frac{y'Q_T^{-1}y}{T}|Q_T|^{1/T}. \tag{22.70}$$

Substituting the expression of (22.59) for the inverse of the dispersion matrix $D(y) = \sigma_\varepsilon^2 Q_T$ gives

$$\begin{aligned} y'Q_T^{-1}y &= y'N'\{I_{T+q} - K(K'K)^{-1}K'\}Ny \\ &= \{Ny - K(K'K)^{-1}K'Ny\}'\{Ny - K(K'K)^{-1}K'Ny\} \\ &= \{Ny + KE(\varepsilon_*|y)\}'\{Ny + KE(\varepsilon_*|y)\}, \end{aligned} \tag{22.71}$$

wherein the expression for $E(\varepsilon_*|y)$ is from (22.66). In fact, this quadratic form may be derived by minimising the sum of squares

$$\begin{aligned} \varepsilon'\varepsilon + \varepsilon_*'\varepsilon_* &= (Ny + K\varepsilon_*)'(Ny + K\varepsilon_*) \\ &= (y - M_*\varepsilon_*)'(MM')^{-1}(y - M_*\varepsilon_*) + \varepsilon_*'\varepsilon_* \end{aligned} \tag{22.72}$$

in respect of ε_*.

There are two contrasting approaches which have been followed in constructing algorithms for evaluating the exact likelihood function of an MA model. The first of these approaches is in the spirit of Box and Jenkins [70] who originally proposed to ignore the determinant $|Q_T|^{1/T}$, which should be close to unity for large values of T, and to adopt the quadratic term $S = y'Q_T^{-1}y$ as a minimand. However, several authors, including Kang [283] and Osborn [374], observed that to omit the determinant was to run the risk that the estimates would violate the condition of invertibility; and, in an appendix to the second edition of their book, Box and Jenkins [70, p. 284] also drew attention to the danger.

Imagine that values are available for $\mu_1, \ldots, \mu_q$ and for σ_ε^2. Then the first task in evaluating the criterion function is to find the vector $E(\varepsilon_*|y)$ of the conditional expectations of presample disturbances. This can be obtained via the formula of (22.66). However, as Box and Jenkins [70, p. 213] have indicated, its values can be generated by a simple procedure of "back-forecasting".

Given a value for $\varepsilon_* = E(\varepsilon_*|y)$, one can proceed to find $\zeta = y - M_*\varepsilon_*$. This is a matter of adjusting the first q elements of the vector y. Next, the elements of $\varepsilon = M^{-1}(y - M_*\varepsilon_*)$ may be formed in turn, one at a time. The lower-triangular Toeplitz matrix M^{-1} is characterised by its leading vector $\psi = M^{-1}e_0$ of which all the elements are liable to be nonzero. Therefore, the tth element of ε might be formed from a weighted sum of the first t elements of ζ wherein the corresponding elements of ψ are the weights. This operation becomes more laborious as the value of t increases; and the method is not to be recommended.

A more efficient way of calculating the elements of ε is to employ the procedure *RationalExpansion*, found under (3.43), which generates the coefficients $\varepsilon_0, \ldots, \varepsilon_{T-1}$ of the series expansion of the rational function $\varepsilon(z) = \zeta(z)/\mu(z)$, where $\zeta(z) = \zeta_0 + \zeta_1 z + \cdots + \zeta_{T-1}z^{T-1}$ and $\mu(x) = 1 + \mu_1 z + \cdots + \mu_q z^q$. The remaining task of forming the sum of squares $S = \varepsilon'\varepsilon + \varepsilon_*'\varepsilon_*$ is straightforward.

To find the value of the criterion function for exact likelihood estimation, the determinant $|Q_T|^{1/T} = |K'K|^{1/T}$ must also be evaluated. To calculate this via the matrix $K'K$ requires the formation of the $T \times q$ matrix $M^{-1}M_*$. The labour involved in this operation and in forming the elements of $K'K$ is proportional to the size T of the sample. However, the matrix $K'K$ is of order q; and its determinant is easily evaluated.

The second method of evaluating the exact likelihood function of an MA model depends upon the use of the Gram–Schmidt prediction-error algorithm which has been presented in Chapter 19. Once the matrix Q_T has been determined from the values of the parameters $\mu_1, \ldots, \mu_q$, the algorithm may be used in finding the factors of the Cholesky decomposition $Q_T = LDL'$. Here L is a lower-triangular matrix with units along its diagonal and with q nonzero subdiagonal bands, whilst $D = \text{diag}\{d_0, \ldots, d_{T-1}\}$ is a diagonal matrix. From the matrix L and the vector y, a vector $\eta = L^{-1}y$ of one-step-ahead prediction errors may be found such that

$$\eta' D^{-1}\eta = y' L'^{-1} D^{-1} L^{-1} y = y' Q_T^{-1} y. \tag{22.73}$$

The determinant of the matrix Q_T is available as the product $|Q_T| = \prod_{t=0}^{T-1} d_t$ of the diagonal elements of D. Thus, all of the essential elements of the criterion function S^* of (22.70) are generated by the algorithm.

The prediction-error algorithm requires very little storage space. In the tth iteration, the q off-diagonal nonzero elements, $l_{1(t)}, \ldots, l_{q(t)}$ of a new row of L are computed from a $q \times q$ lower-triangular matrix whose elements are taken from the previous q rows. Also, the element d_t, which is used in evaluating the determinant and in rescaling the prediction errors, is calculated and stored as $l_{0(t)}$. The tth prediction error $\eta_t = y_t - l_{1(t)}\eta_{t-1} - \cdots - l_{r(t)}\eta_{t-r}$ is formed from the newly calculated coefficients and from a store of $r = \min(q, t)$ previous prediction errors. The matrix inversion which is indicated by the expression $\eta = L^{-1}y$ is notional rather than actual.

(22.74)
```
procedure MALikelihood(var S, varEpsilon : real;
                         var y : longVector;
                         mu : vector;
                         Tcap, q : integer);

  var
    i, j, k, t : integer;
    det : real;
    gamma, eta : vector;
    L : matrix;

  procedure MuLine(t : integer);
    var
      i, k : integer;
  begin
    for i := 0 to t do
      begin {i}
        L[t - i, t] := gamma[Abs(t - i)];
        for k := 0 to i - 1 do
          L[t - i, t] := L[t - i, t] - L[i - k, i] * L[t - k, t] * L[0, k];
        if i < t then
          L[t - i, t] := L[t - i, t]/L[0, i];
      end; {i}
  end; {MuLine}
```

```
procedure ShiftL;
  var
    t, j : integer;
  begin
    for t := 0 to q − 1 do
      begin {t}
        eta[t] := eta[t + 1];
        for j := 0 to t do
          L[j, t] := L[j, t + 1];
      end; {t}
  end; {ShiftL}

  procedure FormError(i : integer);
  var
    j : integer;
  begin
    eta[i] := y[t];
    for j := 1 to q do
      eta[i] := eta[i] − L[j, i] ∗ eta[i − j];
  end; {FormError}

begin {MALikelihood}

{Find the elements of the matrix Q}
  for j := 0 to q do
    begin {j}
      gamma[j] := 0.0;
      for k := p to q − j do
        gamma[j] := gamma[j] + mu[k] ∗ mu[k + j];
    end; {j}

  det := 1.0;
  S := 0.0;

  for t := 0 to q do
    begin {t}
      MuLine(t);
      FormError(t);
      det := det ∗ L[0, t];
      S := S + eta[t] ∗ eta[t]/L[0, t];
    end; {t}

  for t := q + 1 to Tcap − 1 do
    begin {t}
      ShiftL;
      MuLine(q);
      FormError(q);
```

```
        det := det * L[0, q];
        S := S + eta[q] * eta[q]/L[0, q];
      end; {t}

    varEpsilon := S/Tcap;
    S := varEpsilon * Exp(Ln(det)/Tcap);

  end; {MALikelihood}
```

This procedure for evaluating the maximum-likelihood criterion function must be used in association with a procedure for minimising a multivariate function. In common with the procedure *ARLikelihood*, of (22.40) it may be used in conjunction with the optimisation routines of Chapter 12 by embedding it in a function which has the form of *Funct(lambda, theta, pvec, n)*. This is to enable the value of S^* to be passed to the optimisation procedure. The starting values for the optimisation procedure, which are the initial estimates of the moving-average parameters, may be generated by passing the empirical autocovariances to the procedure *MAParameters* of (17.35) or to the procedure *Minit* of (17.39).

One might wish to ensure that the estimates of the moving-average parameters fulfil the condition of invertibility. However, one can be assured that the procedure *MALikelihood* will operate over both invertible and noninvertible regions of the parameter space with no danger of numerical overflow. If minimising values are found in the noninvertible region of the parameter space, then there will always exist a corresponding set of invertible values which can be recovered by inverting some of the roots of the polynomial $\mu(z) = 1 + \mu_1 z + \cdots + \mu_q z^q$.

An effective way of recovering the invertible parameters is to form the values of the autocovariances $\gamma_0, \ldots, \gamma_q$ from the available parameter values and to pass them to the procedure *MAParameters* or to the procedure *Minit*. These procedures will deliver the invertible values in return.

Conditional M-L Estimates of an MA Model

The conditional distribution of y given ε_*, which may be extracted from (22.64), is

$$(22.75)\qquad N(y|\varepsilon_*) = (2\pi\sigma_\varepsilon^2)^{-T/2}\exp\left\{\frac{-1}{2\sigma_\varepsilon^2}(y - M_*\varepsilon_*)'(MM')^{-1}(y - M_*\varepsilon_*)\right\}.$$

Once a value has been attributed to ε_*, we can find conditional maximum-likelihood estimates of $\mu_1, \ldots, \mu_q$ by determining the values which minimise the least-squares function

$$(22.76)\qquad S(y, \varepsilon_*) = (y - M_*\varepsilon_*)'(MM')^{-1}(y - M_*\varepsilon_*).$$

Setting ε_* to its unconditional expectation $E(\varepsilon_*) = 0$ gives the simple criterion of minimising the function

$$(22.77)\qquad S_*(y) = y'(MM')^{-1}y.$$

To find the minimising values, one may employ the a Gauss–Newton procedure of the sort which has already been described fully in the previous chapter, where strategies for avoiding the problem of non-invertibility are also discussed.

A more sophisticated way of dealing with the presample values was advocated by Phillips [390]. His suggestion was that ε_* might be regarded as a nuisance parameter to be estimated in common with the moving-average parameters. It might be supposed that, by taking more care of the presample values, the danger of violating the conditions of invertibility, which besets the simpler conditional least-squares estimator, will be avoided or at least mitigated.

To derive an expression for the estimator of the presample vector, the quadratic function $S(y,\varepsilon_*)$ of equation (22.76) is differentiated with respect to ε_* and the result is set to zero. The solution of the resulting first-order condition is

$$\varepsilon_* = \left\{M_*'(MM')^{-1}M_*\right\}^{-1}M_*'(MM')^{-1}y. \tag{22.78}$$

This differs from the expression for $E(\varepsilon|y)$ given under (22.66). Substituting this estimate into the function $S(y,\varepsilon_*)$ gives

$$S_p^*(y) = y'M'^{-1}\Big[I_T - M^{-1}M_*\left\{M_*'(MM')^{-1}M_*\right\}^{-1}M_*'M'^{-1}\Big]M^{-1}y. \tag{22.79}$$

The criterion function of Phillips differs only marginally from the so-called unconditional least-squares function of Box and Jenkins [70] which is, in fact, the quadratic function $y'Q_T^{-1}y$. This becomes apparent when it is recognised that unconditional least-squares function may be derived by minimising the sum of squares under (22.72) in respect of ε. The latter sum of squares differs from $S(y,\varepsilon_*)$ of (22.76) only in so far as it incorporates the extra term $\varepsilon_*'\varepsilon_*$.

Matrix Representations of ARMA models

Consider an ARMA(p,q) model which is represented by the equation

$$\alpha(L)y(t) = \mu(L)\varepsilon(t), \tag{22.80}$$

where $\alpha(L) = 1 + \alpha_1 L + \cdots + \alpha_p L^p$ and $\mu(L) = 1 + \mu_1 L + \cdots + \mu_q L^q$. A segment of T values from the ARMA(p,q) model from $t = 0$ to $t = T-1$ is comprised in the equation

$$A_* y_* + Ay = M_*\varepsilon_* + M\varepsilon. \tag{22.81}$$

The explicit forms of the various matrices in this expression have been given under (22.4) and (22.52) in the contexts of the AR and the MA models respectively.

Equation (22.81) may be written alternatively as

$$Ay = M\varepsilon + Vu_*, \tag{22.82}$$

where V and u_* are respectively a $T\times(p+q)$ matrix and a $(p+q)\times 1$ vector defined as

$$V = \begin{bmatrix} -A_* & M_* \end{bmatrix} \quad \text{and} \quad u_* = \begin{bmatrix} y_* \\ \varepsilon_* \end{bmatrix}. \tag{22.83}$$

The matrix V may be written alternatively as

$$V = \begin{bmatrix} V_1 \\ 0 \end{bmatrix}, \tag{22.84}$$

where V_1 is a matrix of order $r \times (p+q)$, with $r = \max(p, q)$.

Solving for ε and y in equation (22.82) and combining the results with the trivial identity $u_* = I_{p+q}u_*$ leads to the equation

$$\begin{bmatrix} u_* \\ y \end{bmatrix} = \begin{bmatrix} I_{p+q} & 0 \\ A^{-1}V & A^{-1}M \end{bmatrix} \begin{bmatrix} u_* \\ \varepsilon \end{bmatrix}, \tag{22.85}$$

and to its inverse

$$\begin{bmatrix} u_* \\ \varepsilon \end{bmatrix} = \begin{bmatrix} I_{p+q} & 0 \\ -M^{-1}V & M^{-1}A \end{bmatrix} \begin{bmatrix} u_* \\ y \end{bmatrix}. \tag{22.86}$$

Since the elements of $\varepsilon(t)$ are assumed to be independently and identically distributed, it follows that $\varepsilon \sim N(0, \sigma_\varepsilon^2 I_T)$. Also $u_* \sim N(0, \sigma_\varepsilon^2 \Omega)$, and, consequently,

$$D(u_*, \varepsilon) = \sigma_\varepsilon^2 \begin{bmatrix} \Omega & 0 \\ 0 & I_T \end{bmatrix}. \tag{22.87}$$

It follows that the joint dispersion matrix of u_* and y is

$$D(u_*, y) = \sigma_\varepsilon^2 \begin{bmatrix} \Omega & \Omega V' A'^{-1} \\ A^{-1}V\Omega & A^{-1}(V\Omega V' + MM')A'^{-1} \end{bmatrix}. \tag{22.88}$$

The dispersion matrix for y is

$$D(y) = \sigma_\varepsilon^2 A^{-1}(V\Omega V' + MM')A'^{-1}; \tag{22.89}$$

and, using the formula of (9.12), its inverse is found to be

(22.90)
$$D^{-1}(y) = \frac{1}{\sigma_\varepsilon^2} A'M'^{-1}[I_T - M^{-1}V\{\Omega^{-1} + V'(MM')^{-1}V\}^{-1}V'M'^{-1}]M^{-1}A.$$

It is easy to verify that the matrices under (22.89) and (22.90) can be specialised to give their AR counterparts under (22.15) and (22.16) and their MA counterparts under (22.58) and (22.59).

Density Functions of the ARMA Model

Now let us assume that the elements of $\varepsilon(t)$ are distributed independently identically and normally. Then the joint distribution of u^* and ε is

$$\begin{aligned} N(u_*, \varepsilon) &= N(u_*)N(\varepsilon) \\ &= (2\pi\sigma_\varepsilon^2)^{-(p+q)/2} |\Omega|^{-1/2} \exp\left\{ \frac{-1}{2\sigma_\varepsilon^2}(u_*'\Omega^{-1}u_*) \right\} \\ &\quad \times (2\pi\sigma_\varepsilon^2)^{-T/2} \exp\left\{ \frac{-1}{2\sigma_\varepsilon^2}(\varepsilon'\varepsilon) \right\}. \end{aligned} \tag{22.91}$$

The joint distribution of u_* and y is given by $N(u_*, y) = N\{u_*, \varepsilon(u_*, y)\}|J|$, where $\varepsilon(u_*, y)$ stands for the expression from (22.86) which gives ε in terms of u_* and y, and where $|J|$ is the Jacobian of the transformation from (ε_*, y) to (u_*, ε). The value of the Jacobian is unity. Therefore,

$$\begin{aligned} N(u_*, y) &= N(y|u_*)N(u_*) \\ &= (2\pi\sigma_\varepsilon^2)^{-T/2} \exp\left\{\frac{-1}{2\sigma_\varepsilon^2}(Ay_* - Vu_*)'(MM')^{-1}(Ay_* - Vu_*)\right\} \\ &\quad \times (2\pi\sigma_\varepsilon^2)^{-(p+q)/2}|\Omega|^{-1/2} \exp\left\{\frac{-1}{2\sigma_\varepsilon^2}u_*'\Omega^{-1}u_*\right\}. \end{aligned} \tag{22.92}$$

Using the expression for $D(y)$ from (22.89), the marginal distribution of y may be expressed as

$$\begin{aligned} N(y) &= (2\pi\sigma_\varepsilon^2)^{-T/2}|Q_T|^{-1/2} \exp\left\{\frac{-1}{2\sigma_\varepsilon^2}y'Q_T^{-1}y\right\} \\ &= (2\pi\sigma_\varepsilon^2)^{-T/2}|Q_T|^{-1/2} \exp\left\{\frac{-1}{2\sigma_\varepsilon^2}y'A'(V\Omega V' + MM')^{-1}Ay\right\}. \end{aligned} \tag{22.93}$$

Exact M-L Estimator of an ARMA Model

The exact maximum-likelihood of estimates of the ARMA model may be found by minimising the criterion function

$$|Q_T|^{1/T}y'Q_T^{-1}y = |V\Omega V' + MM'|^{1/T}y'A'(V\Omega V' + MM')^{-1}Ay, \tag{22.94}$$

or by minimising the logarithm of the function. Here the determinant on the RHS is explained by the identity

$$\begin{aligned} |Q_T| &= |A^{-1}(V\Omega V' + MM')A'^{-1}| \\ &= |A|^{-2}|V\Omega V' + MM| \\ &= |V\Omega V' + MM'|, \end{aligned} \tag{22.95}$$

where the final equality follows from the fact that $|A| = 1$.

The requirement is for a facility for evaluating the criterion function at an arbitrary point in the admissible parameter space which consists of the set of parameters fulfilling the conditions of stationarity and invertibility.

The central problem is that of evaluating the quadratic term

$$\begin{aligned} yQ_T^{-1}y &= y'A'(V\Omega V' + MM')^{-1}Ay \\ &= z'L'^{-1}D^{-1}L^{-1}z, \end{aligned} \tag{22.96}$$

where $z = Ay$, and where the lower-triangular matrix L and the diagonal matrix $D = \text{diag}\{d_0, \ldots, d_{T-1}\}$ are the Cholesky factors of

$$\begin{aligned} W &= V\Omega V' + MM' \\ &= LDL'. \end{aligned} \tag{22.97}$$

Here W is a symmetric matrix which has zeros below the rth subdiagonal and above the rth supradiagonal, where $r = \max(p, q)$. Also, beyond the leading submatrix of order r, it is identical to the matrix MM'. It follows that L also has zeros below the rth subdiagonal. Therefore, it is possible to compute the rows of L one at a time, with only a small set of elements from the previous r rows held in memory. Moreover, the contents of the matrix W, which is the subject of the decomposition, is represented completely by the leading submatrix of order $r + 1$, where the final row and column, which contain the elements of MM', serve to characterise the remainder of the matrix.

The elements of $\eta = z'L'^{-1}$ may be obtained by a simple recursion, based on a row of the equation $L\eta = z$, which keeps step with the process of Cholesky decomposition and which avoids inverting the matrix L. The determinant of (22.95) is calculated as a product of the diagonal elements of D.

The procedure for calculating the exact likelihood of an ARMA model has the same basic structure as the procedure *MALikelihood* of (22.74) above. However, In spite of the simplicity of its summary description, it is complex and extensive. Therefore, we shall break it into parts which will become the local procedures organised within a main procedure called *ARMALikelihood*. It will be necessary to insert the code of the local procedures into the body of the main procedure at places which are indicated by comment statements.

The greatest labour is in forming the leading submatrix of $W = V\Omega V' + MM'$. The calculation of this matrix is simplest when $r = p = q$. In fact, this becomes the general case if, whenever the orders p and q are unequal, we are prepared to supplement the autoregressive parameters $\alpha_0, \ldots, \alpha_p$ by $r - p \geq 0$ zeros and the moving-average parameters $\mu_0, \ldots, \mu_q$ by $r - q \geq 0$ zeros, where $r = \max(p, q)$.

When $r = p = q$, the matrices M_* and M may be partitioned in the same way as the matrices A_* and A are partitioned under (22.7):

$$\text{(22.98)} \qquad M_* = \begin{bmatrix} M_{1*} \\ 0 \end{bmatrix}, \quad M = \begin{bmatrix} M_{11} & 0 \\ M_{21} & M_{22} \end{bmatrix}.$$

Then, given that

$$\text{(22.99)} \qquad \sigma_\varepsilon^2 \Omega = \begin{bmatrix} D(y_*) & C(y_*, \varepsilon_*) \\ C(\varepsilon_*, y_*) & D(\varepsilon_*) \end{bmatrix} = \sigma_\varepsilon^2 \begin{bmatrix} Q_r & \Delta' \\ \Delta & I_r \end{bmatrix},$$

the leading submatrix of W of order r becomes

$$\text{(22.100)} \qquad \begin{aligned} W_{11} &= V_1 \Omega V_1' + M_{11} M_{11}' \\ &= \begin{bmatrix} -A_{1*} & M_{1*} \end{bmatrix} \begin{bmatrix} Q_r & \Delta' \\ \Delta & I_r \end{bmatrix} \begin{bmatrix} -A_{1*}' \\ M_{1*}' \end{bmatrix} + M_{11} M_{11}' \\ &= A_{1*} Q_r A_{1*}' - M_{1*} \Delta A_{1*}' - A_{1*} \Delta' M_{1*}' + M_{1*} M_{1*}' + M_{11} M_{11}'. \end{aligned}$$

It is notable that, in the case of a pure MA model, this reduces to the matrix $M_{1*} M_{1*}' + M_{11} M_{11}'$ which conforms with the expression under (22.58). In the case of a pure AR model, the reduction is to a matrix $A_{1*} Q_p A_{1*}' + I$ which is a factor in the expression under (22.15).

The elements of $D(y_*) = \Gamma = \sigma_\varepsilon^2 Q$ and of $C(\varepsilon_*, y_*) = \sigma_\varepsilon^2 \Delta$, which are needed in forming W_{11}, are delivered, in the arrays *gamma* and *delta* respectively, by the procedure *ARMACovariances*, which is listed under (17.98).

The following procedure generates the leading submatrix of W of order $r+1$ which comprises the matrix W_{11} together with a final row and column which also belong to the matrix MM' and which serve to characterise the remainder of W:

(22.101)
```
procedure FormW;
  var
    i, j, k, t, u, v : integer;
    temp, store : real;

  begin {FormW}
    for i := 0 to r - 1 do
      for j := 0 to r - 1 do
        begin {i, j}
          if j >= i then
            begin
              w[i, j] := 0.0;
              w[j, i] := 0.0;
            end;

        {Form M1*ΔA1* + A1*Δ'M'1* }
          store := 0.0;
          u := Max(0, j - i);
          v := Max(0, i - j);
          for k := 0 to (r - 1) - Max(i, j) do
            begin {k}
              temp := 0.0;
              for t := 0 to k + u do
                temp := temp + mu[r - t] * delta[k + u - t];
              store := store + temp * alpha[r - (k + v)];
            end; {k}
          w[i, j] := w[i, j] - store;
          w[j, i] := w[j, i] - store;

          if j >= i then
            begin {if j >= i}

          {Form A1*QrA'1* }
              store := 0.0;
              for k := 0 to r - 1 - j do
                begin {k}
                temp := 0.0;
                for t := 0 to r - 1 - i do
                begin
                v := Abs(t + i - j - k);
```

```
            temp := temp + alpha[r - t] * gamma[v];
            end;
            store := store + temp * alpha[r - k];
            end; {k}
          w[i, j] := w[i, j] + store;
          w[j, i] := w[i, j];

      {Form M1*M'1* }
          store := 0.0;
          v := Abs(i - j);
          for k := 0 to r - 1 - Max(i, j) do
            store := store + mu[r - k - v] * mu[r - k];
          w[i, j] := w[i, j] + store;
          w[j, i] := w[i, j];

      {Form M11M'11 }
          store := 0.0;
          for k := 0 to Min(i, j) do
            store := store + mu[i - k] * mu[j - k];
          w[i, j] := w[i, j] + store;
          w[j, i] := w[i, j];

        end{if j >= i}
      end; {i, j}

    {Calculate the final row and column}
      for j := 0 to r do
        begin {j}
          w[r, j] := 0.0;
          for i := 0 to j do
            begin
              w[r, j] := w[r, j] + mu[i] * mu[r - j + i];
              w[j, r] := w[r, j];
            end;
        end{j}

    end; {FormW}
```

Having generated the elements of the matrix $W = LDL'$, we are in a position to find the rows of the lower-triangular matrix L and the corresponding elements of the diagonal matrix D. These are generated one at a time by the procedure *CholeskyRow* which is equivalent to the procedure *MuLine* within *MALikelihood*.

(22.102)
```
procedure CholeskyRow(t : integer);
  var
    j, k : integer;

begin
```

```
    for j := 0 to t do
      begin {j}
        l[t, j] := w[t, j];
        for k := 0 to j - 1 do
          l[t, j] := l[t, j] - l[k, k] * l[t, k] * l[j, k];
        if t > j then
          l[t, j] := l[t, j]/l[j, j];
      end; {j}
  end; {CholeskyRow}
```

The process of generating rows continues uniformly until the factorisation of W_{11} is complete. Before generating the next row of L, which is the $(r+2)$th row indexed by $t = r + 1$, the first row and column, which are no longer needed in the subsequent computations, are discarded and the remaining rows and columns are shifted by moving every element one place up and one place to the left within the array which stores it. Likewise, the previous prediction errors are shifted upwards within the array *eta* to make space of a new value to be appended at the bottom. Similar shifts occur for all subsequent values of t.

(22.103)

```
procedure ShiftL;

  var
    t, j : integer;

  begin
    for t := 0 to r - 1 do
      begin {t}
        eta[t] := eta[t + 1];
        for j := 0 to t do
          l[t, j] := l[t + 1, j + 1]
      end; {t}
  end; {ShiftL}
```

The final subprocedure to be invoked by the main procedure is for finding the current value of the prediction error. When $t > r$, the prediction error η_t is given by the equations:

$$
\begin{aligned}
z_t &= y_t + \alpha_1 y_{t-1} + \cdots + \alpha y_{t-p},\\
\eta_t &= z_t - l_{t,t-1}\eta_{t-1} - \cdots - l_{t,t-q}\eta_{t-q}.
\end{aligned}
\tag{22.104}
$$

(22.105)

```
procedure FormError(i : integer);

  var
    j : integer;

  begin
    eta[i] := y[t];
    for j := 1 to Min(p, i) do
      eta[i] := eta[i] + alpha[j] * y[t - j];
    for j := 1 to Min(r, i) do
```

```
        eta[i] := eta[i] - l[i, i - j] * eta[i - j];
    end; {FormError}
```

The body of the procedure for evaluating the likelihood of an ARMA model is as follows:

(22.106)

```
procedure ARMALikelihood(var S, varEpsilon : real;
                         alpha, mu : vector;
                           y : longVector;
                           Tcap, p, q : integer);

  var
    i, t, r : integer;
    det : real;
    eta, gamma, delta : vector;
    W, L : matrix;

  {Insert FormW here}

  {Insert CholeskyRow here}

  {Insert ShiftL here}

  {Insert FormError here}

  begin {MainProcedure}

    r := Max(p, q);
    for i := p + 1 to r do
      alpha[i] := 0.0;
    for i := q + 1 to r do
      mu[i] := 0.0;

    ARMACovariances(alpha, mu, gamma, delta, 1, r, r);
    FormW;

    S := 0.0;
    det := 1.0;

    for t := 0 to r do
      begin {t}
        CholeskyRow(t);
        FormError(t);
        det := det * l[t, t];
        S := S + eta[t] * eta[t]/l[t, t];
      end; {t}

    for t := r + 1 to n do
```

```
begin {t}
  ShiftL;
  CholeskyRow(r);
  FormError(r);
  det := det * l[r, r];
  S := S + eta[r] * eta[r]/l[r, r];
end; {t}

VarEpsilon := S/Tcap;
S := varEpsilon * Exp(Ln(det)/Tcap);

end; {ARMALikelihood}
```

Bibliography

[19] Anderson, T.W., and R.P. Mentz, (1980), On the Structure of the Likelihood Function of Autoregressive and Moving Average Models, *Journal of Time Series Analysis*, **1**, 83–94.

[22] Ansley, C.F., (1979), An Algorithm for the Exact Likelihood of a Mixed Autoregressive Moving Average Process, *Biometrika*, **66**, 59–65.

[70] Box, G.E.P., and G.M. Jenkins, (1976), *Time Series Analysis: Forecasting and Control, Revised Edition*, Holden Day, San Francisco.

[132] Cryer, J.D., and J. Ledolter, (1981), Small-Sample Properties of the Maximum Likelihood Estimator in the First-Order Moving Average Model, *Biometrika*, **68**, 691–694.

[142] de Grooijer, J.G., (1978), On the Inverse of the Autocovariance Matrix for a General Mixed Autoregressive Moving Average Process, *Statistische Hefte*, **19**, 114–133.

[155] Dent, W., (1977), Computation of the Exact Likelihood Function of an ARIMA Process, *Journal of Statistical Computation and Simulation*, **5**, 193–206.

[158] Diebold, F.X., (1986), Exact Maximum-Likelihood Estimation of Autoregressive Models via the Kalman Filter, *Economic Letters*, **22**, 197–201.

[200] Galbraith, R.F., and J.I. Galbraith, (1974), On the Inverses of Some Partitioned Matrices Arising in the Theory of Stationary Time Series, *Journal of Applied Probability*, **11**, 63–71.

[202] Gardner, G., A.C. Harvey and G.D.A. Phillips, (1980), An Algorithm for Exact Maximum Likelihood Estimation of Autoregressive Moving Average Models by Means of Kalman Filtering., *Applied Statistics*, **29**, 311–322.

[249] Harvey, A.C, and G.D.A. Phillips, (1979), Maximum Likelihood Estimation of Regression Models with Autoregressive Moving Average Disturbances, *Biometrika*, **66**, 49–58.

[283] Kang, K.M., (1975), *A Comparison of the Least-Squares and Maximum-Likelihood Estimation for Moving-Average Processes*, Bureau of Census and Statistics, Canberra.

[313] Lempers, F.B., and T. Kloek, (1973), On a Simple Transformation for Second-Order Autocorrelated Disturbances in Regression Analysis, *Statistica Neerlandica*, **27**, 69–75.

[321] Ljung, Greta M., and G.E.P. Box, (1979), The Likelihood Function of Stationary Autoregressive Moving Average Models, *Biometrika*, **66**, 265–70.

[342] Mélard, G., (1983), Algorithm AS 197: A Fast Algorithm for the Exact Likelihood of Autoregressive Moving Average Time Series, *Applied Statistics*, **32**, 104–114.

[357] Murthy, D.N.P., (1974), On the Inverse of the Covariance Matrix of a First-Order Moving Average, *Sankhya: The Indian Journal of Statistics, Series A*, **36**, 223–225.

[360] Newbold, P., (1974), The Exact Likelihood Function for a Mixed Autoregressive Moving Average Process, *Biometrika*, **61**, 423–426.

[364] Nicholls, D.F., and A.D. Hall, (1979), The Exact Likelihood Function of Multivariate Autoregressive Moving Average Models, *Biometrika*, **66**, 259–264.

[374] Osborn, Densie R., (1976), Maximum Likelihood Estimation of Moving Average Processes, *Annals of Economic and Social Measurement*, **5**, 75–87.

[387] Phadke, M.S., and G. Kedem, (1978), Computation of the Exact Likelihood Function of Multivariate Moving Average Models, *Biometrika*, **65**, 511–9.

[390] Phillips, A.W., (1966), *Estimates of Systems of Difference Equations with Moving Average Disturbances*, paper presented at the San Francisco meeting of the Econometric Society. Reprinted in *Stability and Inflation*, A.R. Bergstrom et al. (eds.), (1978), pp. 181–199, John Wiley and Sons, Chichester.

[446] Schweppe, F.C., (1965), Evaluation of Likelihood Functions for Gaussian Signals, *IEEE Transactions on Information Theory*, **11**, 61–70.

[451] Sharman, P., (1969), On the Inverse of the Covariance Matrix of a First Order Moving Average, *Biometrika*, **56**, 595–599.

[454] Siddiqui, M.M., (1958), On the Inversion of the Sample Covariance Matrix in a Stationary Autoregressive Process, *Annals of Mathematical Statistics*, **29**, 585–588.

[494] Uppuluri, W.R.R., and J.A. Carpenter, (1969), The Inverse of a Matrix Occurring in First-Order Moving-Average Models, *Sankhya, Indian Journal of Statistics, Series A*, **31**, 79–82.

[496] Vallis Pereira, P.L., (1987), Exact Likelihood Function for a Regression Model with MA(1) Errors, *Economic Letters*, **24**, 145–149.

CHAPTER 23

Nonparametric Estimation of the Spectral Density Function

The topic of this chapter is the nonparametric estimation of the spectral density function of a stationary stochastic process. The basis of the nonparametric spectral estimates is the periodogram of a time series which has been described at length in Chapter 14 which is devoted to the discrete Fourier transform. The periodogram was viewed, in that context, as a device for uncovering periodic components hidden within data series, such as the harmonic components in a record of a mechanical vibration or the periodic components in a record of the luminosity of a pulsating star.

The applicability of the periodogram to other kinds of data, such as socioeconomic data, where the regularities are of a more tenuous nature, seems, at first sight, to be questionable; and the immediate results from applying the device often appear to be quite incoherent. This was the experience of such pioneers as Moore [352] in 1914, who used the periodogram to study the rainfall in the Ohio valley, and of Beveridge [52] in 1922, who studied a long series of wheat prices from Western Europe—see Figure 14.4. Such experiences tended to suggest that the practical usefulness of periodogram analysis was limited.

The subsequent development of the theory of stationary stochastic processes broadened the scope of the problems which could be treated to include stationary series exhibiting irregular cycles; and it reaffirmed the importance of the periodogram, which was recognised as the empirical counterpart of the spectral density function of a stationary process.

The profile of the spectral density function of a stationary stochastic process, such as an ARMA process, is expected to be smooth, whereas the profile of a periodogram calculated from the corresponding data has a distinctly rough or volatile appearance. Some means of smoothing the profile of the periodogram must be applied in order to obtain an estimate which has the appearance of a spectrum—see Figure 23.1. A substantial part of this chapter is devoted, therefore, to the problem of smoothing.

The question arises of why one is motivated to use a nonparametric estimate of the spectral density function when an estimated spectrum with the appropriate properties is readily available as a product of the parameters of an ARMA model which could be fitted to the data. However, in fitting a model, one is making a judgment about the nature of the underlying stochastic process; and to estimate a parametric spectrum without first examining a nonparametric version is to prejudge the issue of how the model should be specified.

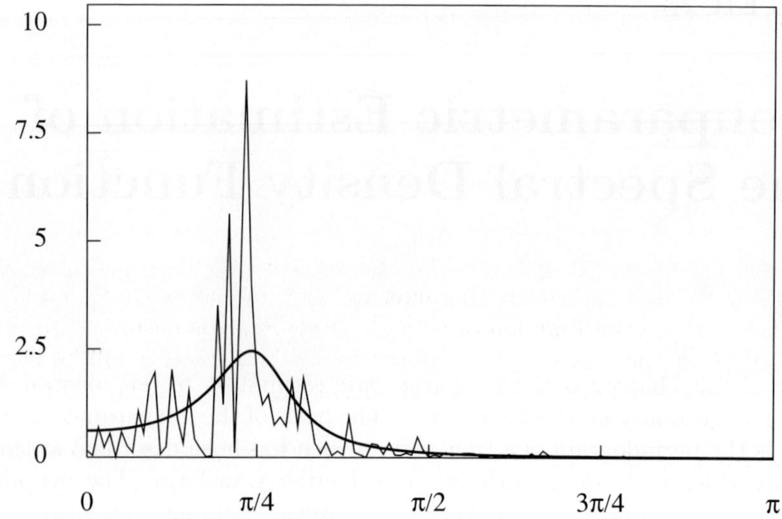

Figure 23.1. The graph of a periodogram calculated from 245 observations on a simulated series generated by an AR(2) process $(1 - 1.131L + 0.640L^2)y(t) = \varepsilon(t)$. The spectral density function belonging to the model which has generated the data has been superimposed upon the periodogram.

The form of the parametric spectrum of an ARMA model is crucially affected by the choice of the orders of the autoregressive and moving-average components; and it is in helping to discern the appropriate orders that the nonparametric spectral estimator is particularly effective. By applying varying degrees of smoothing to the periodogram, a judgment can be made as to which of its features represent systematic effects and which of them are purely accidental. The orders of an ARMA model can be chosen to reflect such judgments.

The traditional way of determining the orders of an ARMA model has been by inspecting the empirical autocovariances generated by the data series in question. When they are defined in the appropriate manner, the periodogram ordinates and the sequence of the empirical autocovariances bear a one-to-one relationship to each other; and so it should follow that any judgment which is reached by inspecting one of them can be reached just as well by inspecting the other. However, the possibility of applying varying degrees of smoothing to the periodogram makes the frequency-domain approach to the determination of the model orders more flexible than the time-domain methods which rely upon an unmodified sequence of empirical autocovariances.

The Spectrum and the Periodogram

The spectral density function or "spectrum" of a stationary stochastic process $y(t)$ is obtained via the discrete-time Fourier transform (DTFT) of the sequence $\gamma(\tau) = \{\gamma_\tau; \tau = 0, \pm 1, \pm 2, \ldots\}$ of the autocovariances of the process. The spectrum, which may be represented by the expressions

$$
\begin{aligned}
f(\omega) &= \frac{1}{2\pi}\sum_{\tau=-\infty}^{\infty}\gamma_\tau e^{-i\omega\tau} \\
&= \frac{1}{2\pi}\left\{\gamma_0 + 2\sum_{\tau=1}^{\infty}\gamma_\tau\cos(\omega\tau)\right\},
\end{aligned}
\tag{23.1}
$$

is, therefore, a continuous periodic function with a period of 2π. The symmetry condition $\gamma_\tau = \gamma_{-\tau}$, which characterises the autocovariances, implies that $f(\omega) = f(-\omega)$ is an even function of ω; and, therefore, it is customary to plot the function only over the interval $[0, \pi]$. However, in the sequel, it will be important to remember that the function is properly defined in terms of the values which it attains over the interval $(-\pi, \pi]$.

In fact, any interval of length 2π will serve in defining the function, and it is also convenient to use the interval $[0, 2\pi)$, as we shall do in most of the sequel. This proves to be the more convenient interval when we come to consider a set of T equally spaced frequency points; for we wish to avoid splitting the set in two at the point of zero frequency. The split is bound to be unequal if we are to avoid including both endpoints.

An natural way of attempting to estimate the spectrum is to replace the unknown autocovariances $\{\gamma_\tau\}$ by the corresponding empirical moments $\{c_\tau\}$ calculated from the data vector $[y_0, y_1, \ldots, y_{T-1}]$. These are defined by

$$
c_\tau = \frac{1}{T}\sum_{t=\tau}^{T-1}(y_{t-\tau} - \bar{y})(y_t - \bar{y}) \quad \text{if} \quad \tau \leq T-1. \tag{23.2}
$$

Notice that, beyond a lag of $\tau = T-1$, the autocovariances are not estimable, since

$$
c_{T-1} = \frac{1}{T}(y_0 - \bar{y})(y_{T-1} - \bar{y}) \tag{23.3}
$$

comprises the first and the last elements of the sample. Therefore, $c_\tau = 0$ when $\tau > T-1$. Thus a sample spectrum is obtained in the form of

$$
\begin{aligned}
f^r(\omega) &= \frac{1}{2\pi}\sum_{\tau=1-T}^{T-1} c_\tau e^{-i\omega\tau} \\
&= \frac{1}{2\pi}\left\{c_0 + 2\sum_{\tau=1}^{T-1} c_\tau\cos(\omega\tau)\right\}.
\end{aligned}
\tag{23.4}
$$

This may be associated with an inverse function in the form of

$$
c_\tau = \int_{-\pi}^{\pi} f^r(\omega)e^{i\omega\tau}d\omega. \tag{23.5}
$$

which maps from the continuous sample spectrum, which is defined over the real line, to the sequence of autocovariances, which may be defined formally over the entire set of all positive and negative integers without regard to the fact that $c_\tau = 0$ when $|\tau| > T-1$. We describe such a infinite sequence as the ordinary extension of the finite sequence $\{c_\tau; \tau = 0, \pm 1, \ldots, \pm(T-1)\}$, and we may denote it by $c(\tau)$.

In fact, the empirical autocovariances constitute a finite sequence, comprising T distinct values which must be specified by enumeration. The information in the sample spectrum can be summarised likewise by a set of T ordinates. In recovering the autocovariances, it is appropriate to apply a discrete Fourier transform (DFT) to a finite set of ordinates of the sample spectrum. This is instead of attempting the process of Fourier integration which is suggested by equation (23.5).

In practice, the DFT should be applied to a sequence of $2T-1$ ordinates corresponding to a set of equally spaced values in the frequency interval $(-\pi, \pi]$; and, in consequence of the symmetry of the sample spectrum, there are indeed only T distinct values amongst these ordinates. The product of the DFT will be the (symmetric) autocovariance sequence $\{c_{1-T}, \ldots, c_{-1}, c_0, c_1, \ldots, c_{T-1}\}$, wherein $c_\tau = c_{-\tau}$, which also comprises $2T-1$ elements that take T distinct values.

The sample spectrum is closely related to the periodogram which is defined under (14.31) by the expression

$$
\begin{aligned}
I(\omega_j) &= 2\left\{c_0 + 2\sum_{\tau=1}^{T-1} c_\tau \cos(\omega_j\tau)\right\} \\
&= 2\left\{c_0 + \sum_{\tau=1}^{T-1} (c_\tau + c_{T-\tau}) \cos(\omega_j\tau)\right\}.
\end{aligned} \tag{23.6}
$$

This stands for a periodic function of period 2π which is defined over the discrete the set of Fourier frequency values

$$
\omega_j = \frac{2\pi j}{T}; \quad j = \{0, \pm 1, \pm 2, \ldots\}. \tag{23.7}
$$

When $I(\omega_j)$ is scaled by a factor of $1/(4\pi)$, its ordinates coincide with those of the sample spectrum at the points ω_j. Thus we have the identity

$$
f^r(\omega_j) = \frac{1}{4\pi} I_j. \tag{23.8}
$$

In fact, various scalar factors are applied to the periodogram by different authors; and our own choice of scale has been made in view of certain physical analogies rather than for mathematical convenience.

An alternative expression for periodogram, which uses the notation $W_T = \exp(-i2\pi/T)$ which is conventionally associated with the discrete Fourier transform (DFT), has been given, together with and expression for its Fourier inverse, under (14.66) and (14.67):

$$
I_j = 2\sum_{\tau=0}^{T-1} c_\tau^\circ (W_T^j)^\tau, \tag{23.9}
$$

(23.10) $$c^{\circ}_{\tau} = \frac{1}{2T}\sum_{j=0}^{T-1} I_j \left(W_T^{-\tau}\right)^j.$$

Here

(23.11) $$c^{\circ}_0 = c_0 \quad \text{and} \quad c^{\circ}_{\tau} = c_{\tau} + c_{T-\tau}$$

are the so-called circular autocovariances; and these are to be regarded as elements of a periodic function. Equations (23.9) and (23.10) define a one-to-one relationship between the sequence of periodogram ordinates $\{I_j; j = 0, \ldots, T-1\}$, defined on the frequency values $\omega_j = 2\pi j/T \in [0, 2\pi)$, and the sequence of empirical circular autocovariances $\{c^{\circ}_{\tau}; \tau = 0, \ldots, T-1\}$; and it is notable that the ordinary autocovariances $\{c_{\tau}; \tau = 1, \ldots, T-1\}$ defined by (23.2) are not recoverable from this limited set of T periodogram ordinates.

As we have already asserted, in order to recover the ordinary autocovariances, the number of frequency points at which the periodogram or the sample-spectrum is sampled must be all but doubled. Let $N = 2T - 1$, and define a new set of frequency points

(23.12) $$\omega'_j = \frac{2\pi j}{N}; \quad j = \{0, \pm 1, \pm 2, \ldots\} \quad \text{with} \quad N = 2T - 1.$$

Also, let $W_N = \exp(-i2\pi/N)$ and let $\tilde{c}(\tau) = \{\tilde{c}_{\tau}\}$ stand for the periodic extension of the sequence of ordinary autocovariances defined by

(23.13) $$\tilde{c}_{\tau} = \begin{cases} c_{\tau} & \text{if } |\tau| \leq T; \\ c_{(\tau \bmod T)}, & \text{otherwise.} \end{cases}$$

Then the following one-to-one relationship may be defined between the periodogram ordinates $\{I'_j; j = 0, \ldots, N-1\}$ at the new frequency points ω'_j and the ordinary autocovariances $\{\tilde{c}_{\tau}; \tau = 0, \ldots, N-1\}$:

(23.14) $$I'_j = 2\sum_{\tau=0}^{N-1} \tilde{c}_{\tau} \left(W_N^j\right)^{\tau},$$

(23.15) $$\tilde{c}_{\tau} = \frac{1}{2N}\sum_{j=0}^{N-1} I'_j \left(W_N^{-\tau}\right)^j.$$

The inability to recover the ordinary autocovariances from a sequence of T periodogram ordinates may be regarded as an instance of the problem of aliasing. According to the result represented by equation (14.77), the Fourier transform of the sampled spectrum, when the sample points are separated by intervals of $2\pi/T$ radians, is given by

(23.16) $$c^{\circ}(\tau) = \sum_{j=-\infty}^{\infty} c(\tau - jT).$$

This represents a periodic function constructed from copies of the ordinary extension $c(\tau)$ of the original autocovariance function superimposed at equal intervals separated by T time periods. At the rate of sampling in question, there is a complete overlap between the elements of $c(\tau)$ and $c(T-\tau)$ for $\tau = 1,\ldots,T-1$; and this explains how the sequence of circular autocovariances $\{c_\tau^\circ = c_\tau + c_{T-\tau}; \tau = 1,\ldots,T-1\}$ arises. By sampling the spectrum at points separated by $2\pi/N$ radians, where $N = 2T-1$, which implies virtually doubling the sampling rate, one can ensure that there is no overlap between the nonzero elements of $c(\tau)$ and $c(N-\tau)$.

In practice, the problem of frequency-domain aliasing can be overcome by the crude but effective device of supplementing the data vector $[y_0,\ldots,y_{T-1}]$, from which the autocovariances and the periodogram are calculated, by a set of $T-1$ zero-valued elements. This technique, which is described as padding the vector, has been discussed in Chapter 20.

The Expected Value of the Sample Spectrum

At first sight, the sample spectrum defined under (23.4) appears to be the natural and appropriate estimator of the spectral density function defined under (23.1). Apart from the replacement of the unknown autocovariances γ_τ by their product-moment estimates c_τ, the only other modification to equation (23.1) is the truncation of the sum which is occasioned by the fact that $c_\tau = 0$ for $\tau > T-1$.

The sample autocovariances are asymptotically unbiased estimates of the population parameters such that $\lim(T\rightarrow\infty)E(c_\tau) = \gamma_\tau$; and it is easy to show that this convergence implies that the sample spectrum is, likewise, an asymptotically unbiased estimator of the spectral density function. The result may be expressed formally as follows:

(23.17) Let $y(t)$ be a stationary stochastic process with $E(y_t) = 0$ for all t and with an autocovariance function $\{\gamma_\tau; t = 0,\pm1,\pm2,\ldots\}$ which is absolutely summable such that $\sum_\tau |\gamma_\tau| < \infty$. Then the sample spectrum $f^r(\omega)$ is an asymptotically unbiased estimator of the spectral density function such that $\lim(T\rightarrow\infty)E\{f^r(\omega)\} = f(\omega)$.

Proof. The expected value of the autocovariance c_τ is

$$E(c_\tau) = \frac{1}{T}\sum_{t=\tau}^{T-1} E(y_t y_{t-\tau}) = \frac{T-\tau}{T}\gamma_\tau. \tag{23.18}$$

Putting this into the expression for $f^r(\omega)$ under (23.4) gives

$$E\{f^r(\omega)\} = \left\{\gamma_0 + 2\sum_{\tau=1}^{T-1}\gamma_\tau\cos(\omega\tau)\right\} - 2\sum_{\tau=1}^{T-1}\frac{\tau}{T}\gamma_\tau\cos(\omega\tau). \tag{23.19}$$

Here the term in braces on the RHS tends to the value of the spectral density function at ω as $T\rightarrow\infty$. The remaining term obeys the inequality

$$2\sum_{\tau=1}^{T-1}\frac{\tau}{T}\gamma_\tau\cos(\omega\tau) < 2\sum_{\tau=1}^{T-1}\frac{\tau}{T}|\gamma_\tau|. \tag{23.20}$$

Now, for any T and $N < T$, there is the further inequality

$$\sum_{\tau=1}^{T-1} \frac{\tau}{T}|\gamma_\tau| < \sum_{\tau=1}^{N} \frac{\tau}{T}|\gamma_\tau| + \sum_{\tau=N+1}^{\infty} |\gamma_\tau|. \tag{23.21}$$

But the first term on the RHS of this inequality vanishes as $T \to \infty$, and the second term may be made arbitrarily small by choosing N to be large enough. Thus the second term on the RHS of (23.19) vanishes as $T \to \infty$; and the result under (23.17) is proved.

Although the sample spectrum is an asymptotically unbiased estimator of the spectral density function, its dispersion does not decrease as the sample size T increases. It follows that it is not a consistent estimator. This will be demonstrated in the following section where the limiting values of the covariances of the periodogram ordinates are derived. However, the result can be apprehended easily for the special case where the $y(t)$ constitutes a sequence of random variables with independent and identical zero-mean normal distributions.

Consider representing the ordinates of the periodogram, in the manner of (14.29), as

$$I(\omega_j) = \frac{T}{2}\Big(\alpha_j^2 + \beta_j^2\Big), \tag{23.22}$$

where

$$\alpha_j = (c_j'c_j)^{-1}c_j'y = \frac{2}{T}\sum_{t=1}^{T-1} y_t \cos\omega_j t, \tag{23.23}$$

$$\beta_j = (s_j's_j)^{-1}s_j'y = \frac{2}{T}\sum_{t=1}^{T-1} y_t \sin\omega_j t. \tag{23.24}$$

Here $c_j = [c_{0j}, \ldots, c_{T-1,j}]'$ and $s_j = [s_{0j}, \ldots, s_{T-1,j}]'$ represent vectors of T values of the functions $\cos(\omega_j t)$ and $\sin(\omega_j t)$ respectively, whilst $y = [y_0, \ldots, y_{T-1}]'$ is the vector of the observations.

When $\omega_j \neq 0, \pi$, it is found that $c_j'c_j = s_j's_j = T/2$. These conditions have been recorded under (14.13). We may also recall the orthogonality conditions of (14.12). In particular, we note that $c_i'c_j = s_i's_j = 0$ if $i \neq j$ and that $c_i's_j = 0$ for all i, j.

Given a dispersion matrix for the data vector in the form of $D(y) = \sigma^2 I$ and assuming that $E(y) = 0$, it follows, in line with the familiar algebra of ordinary least-squares regression, that

$$\begin{aligned} V(\alpha_j) &= E\big\{(c_j'c_j)^{-1}c_j'yy'c_j(c_j'c_j)^{-1}\big\} \\ &= (c_j'c_j)^{-1}c_j'D(y)c_j(c_j'c_j)^{-1} \\ &= \frac{2\sigma^2}{T}, \end{aligned} \tag{23.25}$$

that

$$\begin{aligned}V(\beta_j) &= E\{(s_j's_j)^{-1}s_j'yy's_j(s_j's_j)^{-1}\}\\ &= (s_j's_j)^{-1}s_j'D(y)s_j(s_j's_j)^{-1}\\ &= \frac{2\sigma^2}{T},\end{aligned}\tag{23.26}$$

and that

$$\begin{aligned}C(\alpha_j,\beta_j) &= E\{(c_j'c_j)^{-1}c_j'yy's_j(s_j's_j)^{-1}\}\\ &= (c_j'c_j)^{-1}c_j'D(y)s_j(s_j's_j)^{-1}\\ &= 0.\end{aligned}\tag{23.27}$$

Therefore, on the assumption that $y \sim N(0, \sigma^2 I)$, we find that

$$\alpha_j \sim N\left(0, \frac{2\sigma^2}{T}\right) \quad \text{and} \quad \beta_j \sim N\left(0, \frac{2\sigma^2}{T}\right)\tag{23.28}$$

are independently distributed normal variates, and it follows that

$$I(\omega_j) = \frac{T}{2}\left(\alpha_j^2 + \beta_j^2\right) \sim \sigma^2\chi^2(2).\tag{23.29}$$

The expected value of a $\chi^2(2)$ variate is 2 and its variance is 4; and, from the identity of (23.8), it is deduced that, in this case, the sample spectrum is an unbiased estimator of the spectral density function with a variance of

$$V\{f^r(\omega_j)\} = \frac{\sigma^4}{4\pi^2} = f^2(\omega_j).\tag{23.30}$$

Here, the final equality corresponds to the fact that a white-noise process with a variance of σ^2 has a constant spectral density function of $f(\omega) = \sigma^2/2\pi$. Clearly, the variance does not diminish as T increases and, therefore, the sample spectrum $f^r(\omega)$ does not provide a consistent estimator of the true spectrum.

It also follows from the orthogonality conditions affecting the vectors c_i and s_j that, in the case of white-noise data which is normally distributed, the ordinates of the periodogram will be statistically independent.

In fact, the Fourier transform which generates the coefficients α_j, β_j constitutes an orthogonal transformation of the vector y which amounts to a rotation. If the elements of this data vector are both normally distributed and mutually independent, then the vector has a spherical distribution . The spherical property survives the transformation. Thus it follow that the periodogram ordinates I_j, which are compounded from the Fourier coefficients, will also be independently distributed.

The properties which we have discussed are evident in Figure 23.2 which represents the periodogram of a normal white-noise process. The volatility of the periodogram, which contrasts markedly with the smoothness of the white-noise spectrum, is easily accounted for at an intuitive level. For, if we recall that the

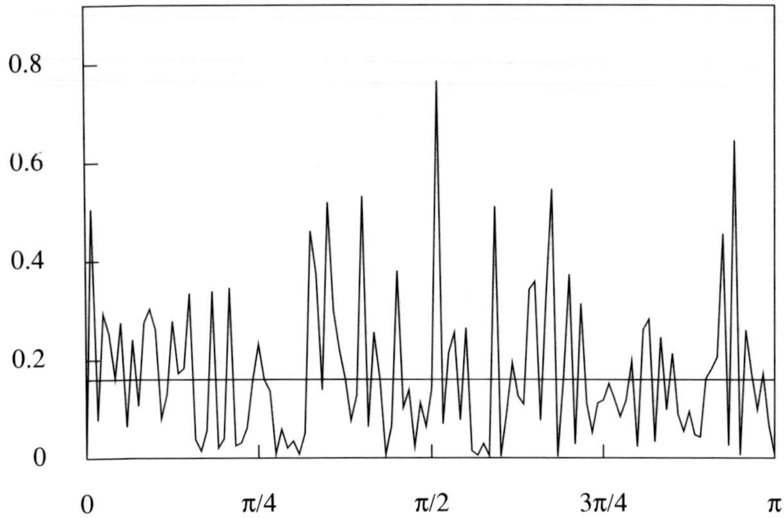

Figure 23.2. The graph of a periodogram calculated from 245 observations on a simulated series generated by a white-noise process. The uniform spectral density function of the process is superimposed upon the periodogram.

periodogram has half as many ordinates as the original white-noise data series, then we can envisage that it has inherited its volatile nature directly from the data.

A similar but a more sophisticated intuition will serve to explain the contrast between the periodogram of an AR(2) process represented in Figure 23.1 and the spectrum of the process which is superimposed upon it.

Asymptotic Distribution of The Periodogram

The result of (23.30) may be generalised so that it applies, in an asymptotic sense, to any linear stochastic process. It is somewhat laborious to demonstrate this proposition; and it is easiest to begin by finding the asymptotic form of the second-order moments for a white-noise process, which need not be a normal process.

(23.31) Let $y(t)$ be a white-noise process such that $E(y_t) = 0$, $V(y_t) = \sigma^2$ and $E(y_t^4) = \eta\sigma^4$ for all t, and let $I(\omega_j)$, which has $E\{I(\omega_j)\} = 2\sigma^2$, be a value of the periodogram calculated from T consecutive observations. Then

$$V\{I(\omega_j)\} = \begin{cases} 4T^{-1}(\eta - 3)\sigma^4 + 8\sigma^4, & \text{if } \omega_j = 0, \text{ or } \pi, \\ 4T^{-1}(\eta - 3)\sigma^4 + 4\sigma^4, & \text{if } 0 < \omega_j < \pi, \end{cases}$$

and

$$C\{I(\omega_j), I(\omega_k)\} = 4T^{-1}(\eta - 3)\sigma^4, \quad \text{if} \quad \omega_j \neq \omega_k.$$

Proof. The covariance of the spectral ordinates is given by

$$C\{I(\omega_j), I(\omega_k)\} = E\{I(\omega_j)I(\omega_k)\} - E\{I(\omega_j)\}E\{I(\omega_k)\}. \tag{23.32}$$

Here

$$\begin{aligned}\frac{T}{2}E\{I(\omega_j)\} &= \sum_{s=0}^{T-1}\sum_{t=0}^{T-1} E(y_s y_t)e^{i\omega_j(t-s)} \\ &= T\sigma^2,\end{aligned} \tag{23.33}$$

and

$$\frac{T^2}{4}E\{I(\omega_j)I(\omega_k)\} = \sum_{q=0}^{T-1}\sum_{r=0}^{T-1}\sum_{s=0}^{T-1}\sum_{t=0}^{T-1} E(y_q y_r y_s y_t)e^{i\omega_j(r-q)}e^{i\omega_k(t-s)}. \tag{23.34}$$

To evaluate the latter, we use the result that

$$E(y_q y_r y_s y_t) = \begin{cases} \eta\sigma^4, & \text{if } q = r = s = t, \\ \sigma^4, & \text{if } q = r \neq s = t, \\ \sigma^4, & \text{if } q = s \neq r = t, \\ \sigma^4, & \text{if } q = t \neq r = s, \\ 0, & \text{otherwise.} \end{cases} \tag{23.35}$$

Here η is a scalar which characterises the distribution of the data points. In the case of a normal distribution, the value is $\eta = 3$.

There are, therefore, four cases to consider which correspond to the nonzero terms in (23.34). First, there are T instances of the case where $q = r = s = t$, each of which gives rise to a term in (23.34) in the form of

$$\begin{aligned}E(y_q y_r y_s y_t)e^{i\omega_j(r-q)}e^{i\omega_k(t-s)} &= E(y_q^4) \\ &= \eta\sigma^4.\end{aligned} \tag{23.36}$$

Next, the condition that $(q = r) \neq (s = t)$ expropriates a set of terms in (23.34) which can be written as

$$\sum_{q\neq s}\sum_{s\neq q} E(y_q^2 y_s^2) = (T^2 - T)\sigma^4. \tag{23.37}$$

Third is the condition $(q = s) \neq (r = t)$. This gives rise to a set of terms which can be written as

$$\Big\{\sum_{q\neq r} E(y_q^2)e^{-i(\omega_j+\omega_k)q}\Big\}\Big\{\sum_{r\neq q} E(y_r^2)e^{i(\omega_j+\omega_k)r}\Big\}. \tag{23.38}$$

Finally, there is the condition $(q = t) \neq (r = s)$. This gives rise to a set of terms which can be written as

$$\Big\{\sum_{q\neq r} E(y_q^2)e^{-i(\omega_j-\omega_k)q}\Big\}\Big\{\sum_{r\neq q} E(y_r^2)e^{i(\omega_j-\omega_k)r}\Big\}. \tag{23.39}$$

By combining the terms, it is found that

$$\begin{aligned}\frac{T^2}{4}E\{I(\omega_j)I(\omega_k)\} &= T(\eta-3)\sigma^4 \\ &+\sigma^4\left(T^2 + \left|\sum_{t=0}^{T-1} e^{i(\omega_j+\omega_k)t}\right|^2 + \left|\sum_{s=0}^{T-1} e^{i(\omega_j-\omega_k)s}\right|^2\right),\end{aligned} \tag{23.40}$$

where the squared terms within the final parentheses stand for complex moduli. Notice that both of these terms include T elements which are due to the squares of the quadratics. To avoid counting these elements, which do not belong to (23.34), an adjustment is made to the leading term on the RHS.

The sums of the exponentials are evaluated in view of the result that

$$\sum_{t=0}^{T-1} e^{i\omega_j t} = \begin{cases} 0, & \text{if } \omega_j \neq 0 \text{ or } n\pi; \\ T, & \text{if } \omega_j = 0 \text{ or } 2n\pi. \end{cases} \tag{23.41}$$

When $\omega_j \neq \omega_k$, both sums in (23.40) vanish. When $\omega_j = \omega_k \neq 0$ or π, then one of the sums vanishes and the other assumes the value of T^2. When $\omega_j = \omega_k = 0$ or π, then both sums assume the value of T^2. By taking account of these results, and by subtracting $T^2E\{I(\omega_j)\}E\{I(\omega_k)\}/4 = T^2\sigma^4$ from (23.40), the various expressions under (23.31) may be derived.

At this stage, we can pause to confirm that the results which have just been proved are, indeed, generalisations of those which were derived on the assumption of a normally distributed white-noise process.

In the case of a normal distribution, the fourth moment is $E(y^4) = \eta\sigma^4 = 3\sigma^4$. Setting $\eta = 3$ in the expressions of (23.31) gives $V\{I(\omega_j)\} = 4\sigma^4$ for $0 < \omega_j < \pi$ and $C\{I(\omega_i, \omega_j)\} = 0$ for $\omega_i \neq \omega_j$. These are in accordance with the previous results. It is also true that these two results correspond to the limiting case as $T \to \infty$. That is to say, the periodogram ordinates retain a finite variance as the sample size increases and also, in the limit, they are mutually uncorrelated.

The next object is to find the values of the variances and the covariances of the periodogram ordinates in the case where $y(t) = \psi(L)\varepsilon(t)$ represents a linear stochastic process. For the sake generality, and for convenience, we shall assume that $\psi(L) = \sum_j \psi L^j$ is a two-sided moving-average operator which extends indefinitely in both directions. Then the essential result, which relates the spectrum $f_y(\omega)$ of the process to the spectrum $f_\varepsilon(\omega) = \sigma_\varepsilon^2/2\pi$ of the white noise, is that

$$f_y(\omega) = |\psi(\omega)|^2 f_\varepsilon(\omega) = \frac{\sigma_\varepsilon^2}{2\pi}|\psi(\omega)|^2. \tag{23.42}$$

In view of this relationship, we would expect the variance of the sample spectrum of $y(t)$ to be equal—asymptotically, at least—to the variance of the sample spectrum of $\varepsilon(t)$ times a scalar factor which is due to the gain of the transfer function $\psi(\omega)$. In order to show that this is the case, we need to show that the relationship which exists between the periodogram ordinates, which are $I_y(\omega_k)$ and $I_\varepsilon(\omega_k)$, is essentially analogous to the relationship which exists between the spectral ordinates $f_y(\omega)$ and $f_\varepsilon(\omega)$.

(23.43) Let $y(t) = \psi(L)\varepsilon(t)$, where $\varepsilon(t)$ is a white-noise process such that $E(\varepsilon_t) = 0$, $V(\varepsilon_t) = \sigma^2$ for all t, and where the coefficients of the operator $\psi(L)$ are absolutely summable such that $\sum_{j=-\infty}^{\infty} |\psi_j| < \infty$. Then the periodogram of $y(t)$, based on T observations, can be represented, at the Fourier frequency ω_k, by

$$I_y(\omega_k) = |\psi(\omega_k)|^2 I_\varepsilon(\omega_k) + R(\omega_k),$$

where $E\{R(\omega_k)\} \to 0$ as $T \to \infty$.

Proof. Using equation (14.65), we may express the ordinate of periodogram of $y(t)$ at the Fourier frequency ω as $I_y(\omega) = 2T\zeta_y(\omega)\zeta_y^*(\omega)$, where $\zeta_y(\omega) = T^{-1}\sum_t y_t e^{-i\omega t}$. (We omit the index of the Fourier frequency ω_k to simplify the notation.) Likewise the periodogram of $\varepsilon(t)$ is $I_\varepsilon(\omega) = 2T\zeta_\varepsilon(\omega)\zeta_\varepsilon^*(\omega)$, where $\zeta_\varepsilon(\omega) = T^{-1}\sum_t \varepsilon_t e^{-i\omega t}$. Since $y_t = \sum_j \psi_j \varepsilon_{t-j}$, it follows that

$$\begin{aligned} T\zeta_y(\omega) &= \sum_{t=0}^{T-1}\left\{\sum_{j=-\infty}^{\infty}\psi_j\varepsilon_{t-j}\right\}e^{-i\omega t} \\ &= \sum_{j=-\infty}^{\infty}\psi_j e^{-i\omega j}\left\{\sum_{t=0}^{T-1}\varepsilon_{t-j}e^{-i\omega(t-j)}\right\}. \end{aligned} \tag{23.44}$$

Here, within the braces, there is

$$\begin{aligned} S_j(\omega) &= \sum_{t=0}^{T-1}\varepsilon_{t-j}e^{-i\omega(t-j)} = \sum_{t=-j}^{T-j-1}\varepsilon_t e^{-i\omega t} \\ &= \sum_{t=0}^{T-1}\varepsilon_t e^{-i\omega t} + P_j(\omega) = T\zeta_\varepsilon(\omega) + P_j(\omega), \end{aligned} \tag{23.45}$$

where $P_j(\omega)$ is a term which arises out of the disparity between $S_j(\omega)$ and $T\zeta_\varepsilon(\omega)$ in respect of the range of summation. Using (23.45) in equation in (23.44), and defining $\psi(\omega) = \sum_j \psi_j e^{-i\omega j}$, gives

$$T\zeta_y(\omega) = T\psi(\omega)\zeta_\varepsilon(\omega) + Q_j(\omega). \tag{23.46}$$

where

$$Q_j(\omega) = \sum_{j=-\infty}^{\infty}\psi_j e^{-i\omega j}P_j(\omega). \tag{23.47}$$

Since $I_y(\omega) = 2T\zeta_y(\omega)\zeta_y^*(\omega)$ and $I_\varepsilon(\omega) = 2T\zeta_\varepsilon(\omega)\zeta_\varepsilon^*(\omega)$, it follows that

$$I_y(\omega) = |\psi(\omega)|^2 I_\varepsilon(\omega) + R_j(\omega), \tag{23.48}$$

where

$$(23.49)\quad \frac{1}{2}R_j(\omega) = Q_j(\omega)\zeta_\varepsilon^*(\omega)\psi^*(\omega) + \psi(\omega)\zeta_\varepsilon(\omega)Q_j^*(\omega) + T^{-1}|Q_j(\omega)|^2.$$

It is remains to show that the term $R_j(\omega)$ has an expected value which tends to zero asymptotically in consequence of $T^{-1} \to 0$.

Therefore, consider the term $P_j(\omega) = S_j(\omega) - T\zeta_\varepsilon(\omega)$ which is a factor of $Q_j(\omega)$. If $|j| < T$, which is to say that $S_j(\omega)$ and $T\zeta_\varepsilon(\omega)$ have overlapping summations, then $P_j(\omega)$ is a sum comprising $2|j|$ elements of the white-noise process $\varepsilon(t)$ each multiplied by a complex number of unit modulus. However, If $|j| \geq T$, which is to say that the ranges of the two summations do not overlap, then $P_j(\omega)$ is of a sum of $2T$ such elements. Thus

$$(23.50)\quad \begin{aligned} &E\{|P_j(\omega)|^2\} = 2\sigma_\varepsilon^2 \min\{|j|, T\} \quad \text{and} \\ &E\{|Q_j(\omega)|^2\} \leq 2\sigma_\varepsilon^2 \Big[\sum_j |\psi_j| \big(\min\{|j|, T\}\big)^{1/2}\Big]^2. \end{aligned}$$

Now take any fixed positive integer $N < T$ and consider

$$(23.51)\quad T^{-1/2}\sum_j |\psi_j|\big(\min\{|j|, T\}\big)^{1/2} \leq T^{-1/2}\sum_{|j|\leq N} |\psi_j||j|^{1/2} + \sum_{|j|>N} |\psi_j|.$$

As $T \to \infty$, the first term on the RHS vanishes. Also, the second term can be made arbitrarily small by making N large enough. It follows from (23.50) that

$$(23.52)\quad \lim_{T\to\infty} T^{-1}E\{|Q_j(\omega)|^2\} = 0,$$

which indicates that the expected value of the final term on the RHS of (23.49) tends to zero as $T \to \infty$.

Next we observe that the Cauchy–Schwarz inequality implies that

$$(23.53)\quad \big[E\{|Q_j(\omega)\zeta_\varepsilon^*(\omega)|^2\}\big]^2 \leq T^{-1}E\{|Q_j(\omega)|^2\}E\{|T^{1/2}\zeta_\varepsilon(\omega)|^2\}.$$

Also, we know from (23.30) that $2E\{|T^{1/2}\zeta_\varepsilon(\omega)|^2\} = E\{I_\varepsilon(\omega)\} = 2\sigma^2$. It follows that the expected values of the remaining terms of $R_j(\omega)$ tend to zero; and thus the proposition of (23.43) is established.

The relationship of (23.43) enables us to establish the asymptotic sampling properties of the periodogram ordinates $I_y(\omega_j)$ directly from the the results regarding the properties of $I_\varepsilon(\omega_j)$ which were given under (23.31).

(23.54) Let $y(t) = \psi(L)\varepsilon(t)$, where $\varepsilon(t)$ is a white-noise process such that $E(\varepsilon_t) = 0$, $V(\varepsilon_t) = \sigma^2$ and $E(\varepsilon_t^4) = \eta\sigma^4$ for all t. Let $f(\omega)$ be the spectral density function of $y(t)$. Then the second-order moments of the periodogram of $y(t)$, which is denoted by $I_y(\omega_j)$ have the following limiting values:

$$\lim_{T\to\infty} V\{I(\omega_j)\} = \begin{cases} 2(4\pi)^2 f^2(0), & \text{if } \omega_j = 0, \text{ or } \pi; \\ (4\pi)^2 f^2(\omega_j), & \text{if } 0 < \omega_j < \pi = k, \end{cases}$$

and

$$\lim_{T\to\infty} C\{I(\omega_j), I(\omega_k)\} = 0, \quad \text{if} \quad \omega_j \neq \omega_k.$$

Smoothing the Periodogram

One way of improving the properties of the estimate of the spectral ordinate $f(\omega_j)$ is to comprise within the estimator several adjacent values from the periodogram. Thus we may define a smoothing estimator in the form of

$$f^s(\omega_j) = \sum_{k=1-M}^{k=M-1} \mu_k f^r(\omega_{j-k}). \tag{23.55}$$

In addition to the value of the periodogram at the central point ω_j, this comprises a further $M-1$ adjacent values falling on either side. The set of smoothing coefficients

$$\{\mu_{1-M}, \ldots, \mu_{-1}, \mu_0, \mu_1, \ldots, \mu_{M-1}\} \tag{23.56}$$

should sum to unity as well being symmetric in the sense that $\mu_{-k} = \mu_k$. They define what is known as a spectral window.

As its stands, the formula of (23.55) appears to be defined only for the values $j = M-1, \ldots, T-M$. For $j = 0, \ldots, M-2$ and for $j = T-M+1, \ldots, T-1$ there are end-effects which seem to require special treatment. However, the problem disappears when it is recognised that $f^r(\omega_j)$ is a periodic function with a domain which extends indefinitely beyond the interval $[0, 2\pi]$.

In order to accommodate the end effects, it is appropriate to replace the ordinary convolution in equation (23.55) by a periodic convolution. For this purpose, it is necessary define a periodic smoothing sequence to replace the finite sequence of weights entailed in equation (23.55). Let $\mu(k)$ be the ordinary extension of the sequence $\{\mu_{1-M}, \ldots, \mu_{-1}, \mu_0, \mu_1, \ldots, \mu_{M-1}\}$ which is obtained by appending to that sequence an indefinite number of preceding and succeeding zeros. Then the periodic smoothing sequence is defined by

$$\tilde{\mu}(k) = \sum_{j=-\infty}^{\infty} \mu(k + jT). \tag{23.57}$$

The elements of $\tilde{\mu}(k)$, which can be specified by writing

(23.58)
$$\tilde{\mu}_k = \begin{cases} \mu_k, & \text{for } k = 0, \ldots, M-1, \\ 0, & \text{for } k = M, \ldots, T-M, \\ \mu_{T-k}, & \text{for } k = T-M+1, \ldots, T-1, \end{cases}$$

fulfil the condition of symmetry, which is $\tilde{\mu}_k = \tilde{\mu}_{-k}$ for all k, and the condition of periodicity, which is that $\tilde{\mu}_k = \tilde{\mu}_{T-k}$ for all k.

In these terms, the smoothed estimator of the spectral ordinate at the frequency ω_j may be specified as

(23.59)
$$f^s(\omega_j) = \frac{1}{4\pi} I(j) * \tilde{\mu}(j) = \frac{1}{4\pi} \sum_{k=0}^{T-1} \tilde{\mu}_k I(\omega_{j-k}),$$

where $*$ denotes the operation of circular convolution and where $I(j)$ denotes the periodogram defined over the set of the integers which are the indices of the frequency points.

The estimate $f^s(\omega_j)$ comprises a total of $2M - 1$ ordinates of the periodogram which span an interval of $Q = 4(M-1)\pi/T$ radians. This number of radians is the so-called bandwidth of the estimator. If M increases at the same rate as T, then Q will remain constant. This means that, in spite of the increasing sample size, we are denied the advantage of increasing the acuity or resolution of our estimation; so that narrow peaks in the spectrum, which have been smoothed over, may escape detection. Conversely, if we maintain the value of M, then the size of the bandwidth will decrease with T, and we may retain some of the disadvantages of the original periodogram. Ideally, we should allow M to increase at a slower rate than T so that, as $M \to \infty$, we will also have $Q \to 0$. This would be achieved, for example, by making M proportional to $\sqrt{T}$. In fact, the object can be achieved by making M proportional to T^λ for any λ in the open interval $(0, 1)$.

In choosing a smoothing function for a specified value of M, one is striking a compromise between the bias of the spectral estimator and its variance. A smoothing function which assigns equal weights to a broad band of frequencies will produce an estimate which achieves smoothness at the expense of bias. Conversely, a smoothing function which assigns most of the weight to a narrow band of central frequencies will produce an estimator with a small bias but with a relatively large variance.

There is no definitive set of smoothing coefficients which can be recommended above all others. The best approach is to experiment with a variety of smoothing schemes in pursuit of one which smooths the periodogram adequately while preserving the features which are considered to be systematic rather than accidental.

Figure 23.3 shows an estimate of the spectral density function plotted in Figure 23.1. The remarkable accuracy of the estimate is attributable to a well-chosen set of smoothing coefficients.

It is helpful if the choice of the smoothing coefficients can be systematised. An approach which seems to be reasonably flexible is to adopt a generating function in the form of

(23.60)
$$\mu(z) = \frac{1}{(p+1)^{2n}} (1 + z^{-1} + \cdots + z^{-p})^n (1 + z + \cdots + z^p)^n$$

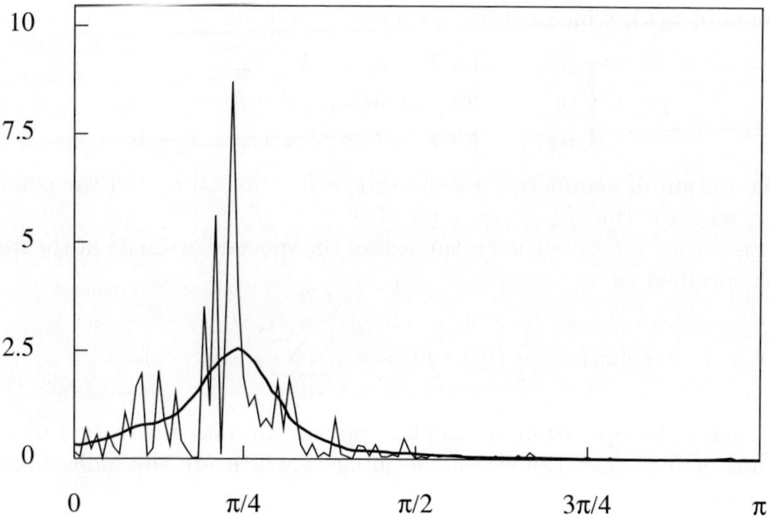

Figure 23.3. An estimated spectrum calculated from 245 data points. The spectrum is calculated by smoothing the periodogram. The smoothing function is a weighted average which spans a total of 21 frequency points. The estimated spectrum has been superimposed upon the periodogram.

which depends upon two parameters p and n. The smoothing coefficients are the coefficients of the resulting polynomial; and they span a set of $2np + 1$ frequency points. In general, the coefficients have a bell-shaped profile and, when $p = 1$, they correspond to the ordinates of a binomial distribution. Conversely, when $n = 1$, the coefficients have a triangular profile.

An alternative family of smoothing functions, which includes an ordinary average of adjacent ordinates of the periodogram, is generated by the polynomial

$$\mu(z) = \frac{1}{(2p+1)^n}(z^{-p} + \cdots + z^{-1} + 1 + z + \cdots z^p)^n. \tag{23.61}$$

The coefficients span a total of $2pn + 1$ frequency points, as in the previous case.

A feature of such smoothing schemes is that they can be implemented either in a single pass or in several successive applications. Thus, by setting $n = 1$, a basic smoothing sequence is derived which can be applied n times in succession. After each application, the results can be inspected so as to gain a useful impression of how the estimate of the spectrum is evolving.

An alternative approach to smoothing the periodogram involves splitting the sample of T points into K segments of M points each. For each segment, the periodogram is computed. Then the average of the periodograms becomes the spectral estimate; and the variance of the periodogram ordinates is thereby reduced by a factor of K.

This procedure can save time, since it is more efficient to compute K short FFTs that it is to compute one long one. One should recall that the number of operations entailed in an FFT of $T = 2^g$ points is proportional to $T \log T$.

Therefore, splitting the sample into K segments is liable to reduce the time spent in computing by a factor of $\log K$. Splitting the sample is the appropriate way of dealing with a long run of data which has been generated by a process which can be relied upon to remain unchanged over the sample period.

Weighting the Autocovariance Function

An alternative approach to spectral estimation is to give differential weighting to the estimated autocovariances comprised in the formula for the sample spectrum, so that diminishing weights are given to the values of c_τ as τ increases. This seems reasonable, since the precision of these estimates decreases as τ increases. If the series of weights associated with the autocovariances $c_0, c_1, \ldots, c_{T-1}$ are denoted by $m_0, m_1, \ldots, m_{T-1}$, then our revised estimator for the spectrum takes the form of

(23.62) $$f^w(\omega) = \frac{1}{2\pi}\left\{m_0 c_0 + 2\sum_{\tau=1}^{T-1} m_\tau c_\tau \cos(\omega_\tau)\right\}.$$

The series of weights define what is described as a lag window. If the weights are zero-valued beyond w_{M-1}, then we describe M as the truncation point.

From a theoretical perspective, weighting the autocovariance function is equivalent to smoothing the periodogram. Weighting the autocovariance function is tantamount to the modulation in the time domain of the sequence $c(\tau)$ by the sequence $m(\tau)$. On the other hand, smoothing the periodogram is tantamount to the frequency-domain convolution of the periodogram, which is the Fourier transform of the autocovariance sequence, and the smoothing sequence, which is the Fourier transform of the weighting sequence $m(\tau)$ which has been applied to the autocovariances. The equivalence of the two sets of operations may be asserted formally as follows:

(23.63) Let $c^\circ(\tau) = \{c^\circ_\tau\}$ be the empirically determined sequence of circular autocovariances and let $I(j)$ be the periodogram. Likewise, let $\tilde{m}(\tau) = \{m_\tau\}$ be a periodic weighting sequence and $\tilde{\mu}(j) = \{\tilde{\mu}_j\}$ be the corresponding sequence of Fourier coefficients from the DFT of $\tilde{m}(\tau)$. Then the following relationships prevail:

$$\tilde{m}_\tau = \frac{1}{T}\sum_k \tilde{\mu}_k W^{-k\tau} \longleftrightarrow \tilde{\mu}_k = \sum_\tau \tilde{m}_\tau W^{k\tau},$$
$$c^\circ_\tau = \frac{1}{T}\sum_k \frac{I_j}{2} W^{-j\tau} \longleftrightarrow \frac{I_j}{2} = \sum_\tau c^\circ_\tau W^{j\tau};$$

and, on defining the (circular) convolution of $\tilde{\mu}(\omega_j)$ and $I(\omega_j)$ to be the sequence $\tilde{\mu}(\omega_j) * I(\omega_j)$ whose generic elements is $I^s_j = \sum \tilde{\mu}_k I_{j-k}$, we find that

$$c^\circ(\tau)\tilde{m}(\tau) \longleftrightarrow \tilde{\mu}(\omega_j) * \frac{I(\omega_j)}{2}.$$

Proof. We have

$$
(23.64)\qquad
\begin{aligned}
\sum_k \mu_k \frac{I_{j-k}}{2} &= \frac{1}{T}\sum_k \mu_k \Big\{ \sum_\tau c_\tau^\circ W^{(j-k)\tau} \Big\} \\
&= \sum_\tau c_\tau^\circ W^{j\tau} \Big\{ \frac{1}{T}\sum_k \mu_k W^{-k\tau} \Big\} \\
&= \sum_k \tilde{m}_\tau c_\tau^\circ W^{j\tau},
\end{aligned}
$$

which represents the mapping from the weighted autocovariance function to the smoothed periodogram. The inverse mapping can be represented likewise.

The proposition of (23.63) seems to indicate that it is a matter of indifference whether the spectrum is estimated by smoothing the periodogram or by weighting the autocovariance function. The two approaches appear to be equivalent. In fact, when a number of practical issues are taken into account, some firm distinctions arise which serve to separate the approaches.

For a start, it should be recognised that, notwithstanding the features of our formal representation of the smoothing operation in equation (23.59), it is unlikely that the estimates of each of the spectral ordinates will be based on a weighted average of the full set of T ordinates of the periodogram. In practice, the smoothing function is liable to be a band-limited function comprising a set of $2M-1<T$ elements, such as those of the finite sequence of (23.56).

The limitation in the number of nonzero elements of the smoothing function $\mu(j)$ serves to reduce the amount of computation which is entailed in smoothing the periodogram. Moreover, one is also inclined to limit the bandwidth of the smoothing function in order to impose limits on the problem of leakage, which is the tendency of the smoothing operation to disperse the power of the periodogram over a range of neighbouring frequencies.

On the other hand, it is likely that we should wish to limit the number of the weighted autocovariances which are subject to the cosine Fourier transform of (23.62). In the first place, by limiting the number of nonzero elements in the weighting function $m(\tau)$, we can to reduce burden of computing the transform. In the second place, we are inclined to doubt the worth of the estimates of the autocovariances at high lag values; and this also leads us to imposes a lag limit (i.e. a time limit) on the weighting function $m(\tau)$.

Now let us recall that a time-limited function cannot possess a band-limited transform. Then we can see that the equivalence of the two methods of spectral estimation, which we have established in theory, breaks down in practice. That is to say, we recognise that a practical estimation which entails weighting and truncating the autocovariance function cannot be wholly equivalent to any of the alternative methods of estimation which involve the band-limited smoothing of the periodogram.

Weights and Kernel Functions

The original approach to nonparametric spectral estimation, which predated the advent of the fast Fourier transform (FFT), was based upon the method weighting the autocovariance function. The conventional presentation of the method is in terms of the continuous periodic function

$$f^w(\omega) = \frac{1}{2\pi}\sum_{\tau=1-T}^{T-1} m_\tau c_\tau e^{-i\omega\tau}, \tag{23.65}$$

which comes from applying the discrete-time Fourier transform (DTFT) to a truncated and weighted sequence of autocovariances. Here it is to be assumed that $m_\tau = 0$ for $|\tau| > M - 1$. The DTFT of the weighting sequence $\{m_\tau\}$ is itself a continuous periodic function $\mu(\omega)$ which is described as the smoothing kernel; and this replaces the smoothing sequence of the previous section which resulted from applying the discrete Fourier transform (DFT) to the weighting sequence. Thus

$$\mu(\omega) = \frac{1}{2\pi}\sum_{\tau=-\infty}^{\infty} m_\tau e^{-i\omega_\tau} \tag{23.66}$$

and

$$m_\tau = \int_{-\pi}^{\pi} \mu(\omega) e^{i\omega_\tau} d\omega \tag{23.67}$$

constitute a Fourier pair which bear the same relationship to each other as do the sample spectrum defined in (23.1) and the autocovariance function defined in (23.5).

It is desirable that the weighting function should integrate to unity over the relevant range, and this requires us to set $m_0 = 1$. The latter is exactly the value by which we would expect to weight the estimated variance c_0 within the formula in (23.65) which defines the spectral estimator $f^w(\omega)$.

On substituting the expression for c_τ from (23.5) into (23.65), we get

$$\begin{aligned} f^w(\omega) &= \frac{1}{2\pi}\sum_{\tau=1-T}^{T-1} m_\tau\left\{\int_{-\pi}^{\pi} f(\lambda)e^{i\lambda\tau}d\lambda\right\}e^{-i\omega\tau} \\ &= \int_{-\pi}^{\pi} f(\lambda)\left\{\frac{1}{2\pi}\sum_{\tau=1-T}^{T-1} m_\tau e^{-i(\omega-\lambda)\tau}\right\}d\lambda \\ &= \int_{-\pi}^{\pi} f(\lambda)\mu(\omega-\lambda)d\lambda. \end{aligned} \tag{23.68}$$

This shows, once more, that the technique of weighting the autocovariance function corresponds, in general, to a technique of smoothing the periodogram; albeit that, in the present instance, the operation of smoothing employs a convolution integral as compared with the circular convolution of the previous presentation. (See equation (23.59), for example.)

In the classical treatment of the problem of nonparametric spectral estimation, the interest centres on the nature of the smoothing kernel $\mu(\omega)$. The object is to find a form which minimises the spectral leakage while smoothing the profile of the sample spectrum; and this can be achieved only by a careful choice of the weighting function.

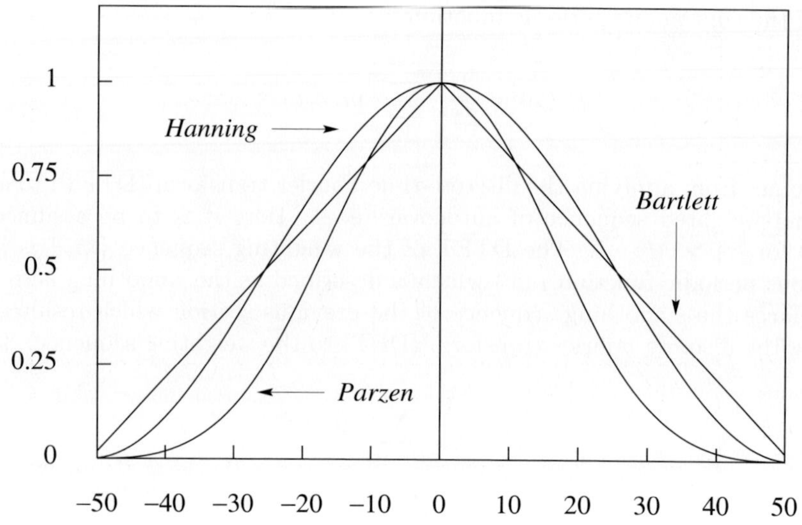

Figure 23.4. The profiles of three of the window functions which may be used to weight and to truncate a sequence of empirical autocovariances. An estimate of the spectral density function is obtained from the Fourier transform of the modified sequence.

The results which indicate how the choice should be made are provided by the theory of FIR filters which has been presented already in Chapter 16. In that context, the object was to discover how best to approximate the square-wave frequency response of an ideal lowpass filter using a finite sequence of filter coefficients. First a limited number of central coefficients were taken from the Fourier-series representation of the square wave. Then the leakage occasioned by the truncation of the Fourier series was minimised by applying the appropriate weights to the coefficients.

The general effect of a weighting function was demonstrated by plotting its Fourier transform, which is the corresponding frequency-domain kernel function. The specific effect of applying the weights to the filter coefficients was demonstrated by plotting the Fourier transform of the weighted coefficients, which is also the function which results from the convolution of the kernel with the frequency response of the unweighted filter. Examples are provided by Figures 16.14 to 16.18.

In the present context, our purpose is to reveal the effect of applying such weights to the coefficients of the autocovariance function. This can be demonstrated by plotting Fourier transform of the weighted autocovariances, which is the function which results from the convolution of the kernel function with the sample spectrum defined in (23.2).

Below, we record a variety of window functions together with some of the Fourier transforms which are the corresponding kernel functions. In Figure 23.4 we plot the profiles of the window functions, and in Figure 23.5 we plot the kernels.

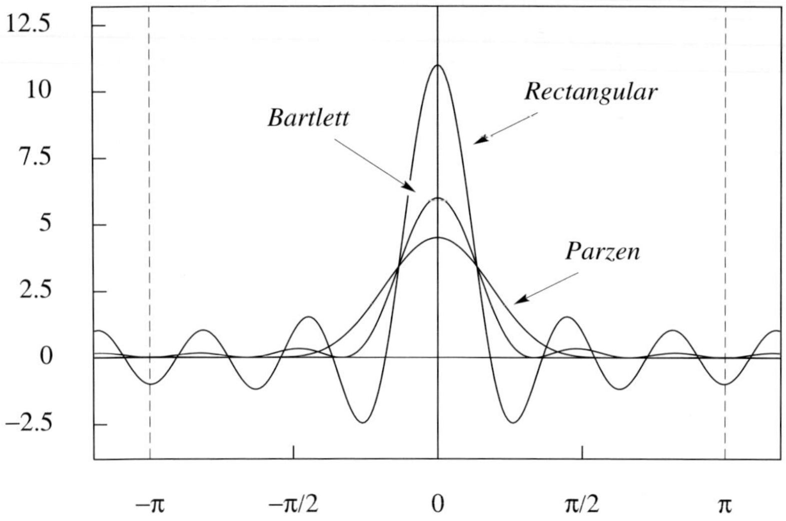

Figure 23.5. The kernel functions corresponding to three of the windows which may be applied to a sequence of empirical autocovariances. The Fourier transform of the modified sequence represents an estimate of the spectral density function.

Rectangular window $\longleftrightarrow$ *Dirichlet kernel:* Ref. (16.72)

$$(23.69) \qquad m_\tau = \begin{Bmatrix} 1, & \text{if } |\tau| < M; \\ 0, & \text{if } |\tau| \geq M. \end{Bmatrix} \longleftrightarrow \mu(\omega) = \frac{\sin\{\omega(2M-1)/2\}}{\sin(\omega/2)}.$$

Bartlett (triangular) window $\longleftrightarrow$ *Fejér kernel:* Ref. (16.78)

$$(23.70) \qquad m_\tau = \begin{Bmatrix} 1 - \dfrac{|\tau|}{M}, & \text{if } |\tau| \leq M; \\ 0, & \text{if } |\tau| \geq M. \end{Bmatrix} \longleftrightarrow \mu(\omega) = \frac{\sin^2\{\omega M/2\}}{M\sin^2(\omega/2)}.$$

Hanning (raised cosine) window: Ref. (16.79)

$$(23.71) \qquad m_\tau = \begin{Bmatrix} \frac{1}{2}\left\{1 + \cos\left(\dfrac{\pi\tau}{M}\right)\right\}, & \text{if } |\tau| \leq M; \\ 0, & \text{if } |\tau| \geq M. \end{Bmatrix}$$

Hamming window: Ref. (16.83)

$$(23.72) \qquad m_\tau = \begin{Bmatrix} 0.54 + 0.46\cos\left(\dfrac{\pi\tau}{M}\right), & \text{if } |\tau| \leq M; \\ 0, & \text{if } |\tau| \geq M. \end{Bmatrix}$$

Parzen window

$$(23.73)\qquad m_\tau = \begin{cases} 1-6\left(\dfrac{\tau}{M}\right)^2+6\left|\dfrac{\tau}{M}\right|^3, & \text{if } 0\leq|\tau|\leq\dfrac{M}{2};\\ 2\left(1-\dfrac{\tau}{M}\right)^3, & \text{if } \dfrac{M}{2}\leq|\tau|\leq M;\\ 0, & \text{if } |\tau|\geq M. \end{cases}$$

The rectangular window gives rise to the Dirichlet kernel, which is seriously affected by the problem of leakage. It also has the disadvantage that its ordinates become negative for certain values of ω. Therefore, the spectral estimates obtained from the truncated periodogram may also, on occasion, assume negative values, which is in conflict with the fact that the spectral density function is a nonnegative function—as is the sample spectrum also. The purpose of considering the rectangular window together with the truncated periodogram is to provide a benchmark against which any improvements which result from a more careful choice of the window can be assessed.

The second proposal which bears investigating is the use of the triangular window in generating the so-called Bartlett [36] spectral estimate. This window gives rise to a Fejér kernel function. The coefficients of the triangular window are formed from the convolution of two rectangular sequences of M units. It follows that the Fejér kernel takes the form of the square of the Dirichlet kernel. Since this kernel function is nonnegative and since the original periodogram ordinates as well as the ordinates of the sample spectrum are nonnegative, it also follows that the Bartlett estimate has the desirable property that its ordinates are guaranteed to be nonnegative.

The Bartlett [36] estimate has an interesting provenance. It originates in the idea that the spectrum may be estimated by taking the average of K sample spectra, each of which is derived from a separate segment of M successive elements belonging to a sample of $T = KM$ elements.

Recall that, if the autocovariance of lag τ is calculated from a sample of T points in the usual way, then, according to (23.18), an estimate $c_{T,\tau}$ of the underlying parameter γ_τ is produced which has an expected value of

$$(23.74)\qquad E(c_{T,\tau}) = \frac{T-|\tau|}{T}\gamma_\tau.$$

The expected value of an estimate obtained from a sample of M points is given likewise by

$$(23.75)\qquad \begin{aligned} E(c_{M,\tau}) &= \frac{M-\tau}{M}\gamma_{|\tau|}\\ &= \left(1-\frac{|\tau|}{M}\right)\left(1-\frac{|\tau|}{T}\right)^{-1}E(c_{T,\tau}). \end{aligned}$$

Now consider a sample spectrum calculated from a segment of M elements. The formula would be

$$(23.76)\qquad f^r_M(\omega) = \frac{1}{2\pi}\sum_{\tau=1-M}^{M-1} c_{M,\tau}e^{-i\omega\tau}.$$

But, in view of the formula of (23.75), we can see that averaging the sample spectra obtained from K such segments would produce much the same result as would the replacement of $c_{M,\tau}$ in (23.76) by the quantity $(1-|\tau|/M)(1-|\tau|/T)^{-1}c_{T,\tau}$. In this way, we would obtain the spectral estimator

$$f^b(\omega) = \frac{1}{2\pi}\sum_{\tau=1-M}^{M-1}\left(1-\frac{|\tau|}{M}\right)\left(1-\frac{|\tau|}{T}\right)^{-1} c_\tau e^{-i\omega\tau}, \tag{23.77}$$

where c_τ is the usual estimate based on a sample of T points.

The estimate of (23.77) is the one originally proposed by Bartlett [36]; and it provides a good illustration of how a technique of averaging (i.e. smoothing) the ordinates of the periodogram or of the sample spectrum can be commuted into a technique of weighting the autocovariance function. If the length M of the segments, which is the maximum value of τ, is much smaller than the length T of the entire sample, then the factor $(1-|\tau|/T)^{-1}$ can be approximated by unity, and the weighting scheme of Bartlett assumes the simpler form given in (23.70) above.

Reference to Figure 23.5—or to Figure 16.15, equally—shows that that the Bartlett window is still associated with a high degree of leakage. The Hanning (or Tukey–Hanning) window fares somewhat better. However, the Hamming [237] window of equation (23.72) is usually regarded as marginally superior for reasons which have been stated in Chapter 16. (See, also, Blackman and Tukey [65].) Nevertheless, it is doubtful whether the differences are perceptible in the context of the nonparametric estimation of the spectrum.

Of the window functions listed above, the one which is undoubtedly the most frequently employed in spectral estimation is the Parzen [384] window. This shows very little leakage beyond the range of its principal lobe. However, for a given value of M, the principal lobe has the widest dispersion. This is the reason why the Parzen window is not used in signal-processing applications for reducing the leakage of practical lowpass filters, for example. For the cost of reducing the leakage would be an unacceptable widening of the transition band, which obscures the cut-off point which is supposed to separate the pass band from the stop band.

The width of the transition band may be reduced by increasing the value of M, which means increasing the span of the averaging filter; but, in practical signal processing applications, this often signifies increasing the number of hardware components and incurring extra costs.

Bibliography

[36] Bartlett, M.S., (1950), Periodogram Analysis and Continuous Spectra, *Biometrika*, **37**, 1–16.

[40] Beamish, N., and M.B. Priestley, (1981), A Study of Autoregressive and Window Spectral Estimation, *Applied Statistics*, **30**, 41–58.

[51] Beveridge, W.H., (1921), Weather and Harvest Cycles, *The Economic Journal*, **31**, 429–452.

[52] Beveridge, W.H., (1922), Wheat Prices and Rainfall in Western Europe, *Journal of the Royal Statistical Society*, **85**, 412–478.

[65] Blackman, R.B., and T.W. Tukey, (1959), *The Measurement of the Power Spectrum from the Point of View of Communications Engineering*, Dover Publications: New York.

[102] Childers, D.G., (ed.), (1978), *Modern Spectral Analysis*, IEEE Press, New York.

[115] Cogburn, R., and H.T. Davis, (1974), Periodic Splines and Spectral Estimation, *The Annals of Statistics*, **2**, 1108–1126.

[169] Durbin, J., (1969), Tests for Serial Correlation in Regression Analysis Based on the Periodogram of Least-Squares Residuals, *Biometrika*, **56**, 1–15.

[171] Dzhaparidze, K., (1986), *Parameter Estimation and Hypothesis Testing in Spectral Analysis of Stationary Time Series*, Springer Verlag, Berlin.

[186] Fishman, G.S., (1969), *Spectral Methods in Econometrics*, Harvard University Press, Cambridge, Massachusetts.

[237] Hamming, R.W., (1989), *Digital Filters, Third Edition*, Prentice-Hall, Englewood Cliffs, New Jersey.

[244] Harris, F.J., (1978), On the Use of Windows for Harmonic Analysis with the Discrete Fourier Transform, *IEEE*, **66**, 51–84.

[269] Jenkins, G.M., and D.G. Watts, (1968), *Spectral Analysis and its Applications*, Holden-Day, San Francisco.

[285] Kay, S.M., (1988), *Modern Spectral Analysis*, Prentice-Hall, Englewood Cliffs, New Jersey.

[286] Kay, S.M., and S.L. Marple, (1981), Spectrum Analysis: A Modern Perspective, *Proceedings of the IEEE*, **69**, 1380–1419.

[352] Moore, H.L., (1914), *Economic Cycles: Their Law and Cause*, Macmillan Company, New York.

[383] Parzen, E., (1957), On Consistent Estimates of the Spectrum of a Stationary Time Series, *Annals of Mathematical Statistics*, **28**, 329–348.

[384] Parzen, E., (1961), Mathematical Considerations in the Estimation of Spectra, *Technometrics*, **3**, 167–190.

[411] Priestley, M.B., (1981), *Spectral Analysis and Time Series*, Academic Press, London.

[427] Robinson, E.A., (1982), A Historical Perspective of Spectrum Estimation, *Proceedings of the IEEE*, **70**, 885–906.

[499] Wahba, Grace, (1980), Automatic Smoothing of the Log Periodogram, *Journal of the American Statistical Association*, **75**, 122–132.

[511] Welch, P.D., (1967), The Use of Fast Fourier Transform for the Estimation of Power Spectra: A Method Based on Time-Averaging Over Short, Modified Periodograms, *IEEE Transactions on Audio and Electroacoustics*, **AU-15**, 70–73.

Index